Physical and Numerical Constants

Acceleration of gravity (on Earth), g = 9.8 m/s^2
Ångstrom unit, Å = 10^{-10} m
Gravitational constant, G = 6.67 × 10^{-11} N m^2/kg^2
Electron mass = 1/1,836 proton mass
Planck constant, h = 6.63 × 10^{-34} joule seconds
Proton mass = 1.67 × 10^{-27} kg
Radian = 57.3° = 206,265 seconds of arc
Speed of light in a vacuum, c = 299,792 km/s
Stefan-Boltzmann constant, σ = 5.67 × 10^{-8} joule/s/m^2/Kelvin^4

Astronomical Constants

Aberration of starlight (on Earth) = 20.5"
Astronomical Unit, AU = 1.496 × 10^8 km
Cosmic background temperature = 2.735 K
Earth year = 365.24219... days
Eclipse year = 346 days
Hubble constant, H$_0$ = 50 to 100 km/s/Mpc
Light-year = 9.46 × 10^{12} km = 0.306 pc = 63,300 AU
Obliquity of the (Earth's) ecliptic = 23°27'
Parsec, pc = 3.09 × 10^{13} km = 3.26 light-years = 206,265 AU
Seconds in a day = 86,400
Seconds in a year = 31,556,926
Solar luminosity = 3.83 × 10^{26} joules/s
Solar mass = 1.99 × 10^{30} kg
Solar radius = 6.96 × 10^5 km = 109 R_{Earth}

Astronomy!

Astronomy!

JAMES B. KALER

University of Illinois, Urbana-Champaign

HarperCollinsCollegePublishers

Executive Editor:	Doug Humphrey
Developmental Editors:	Don Gecewicz, Rebecca Strehlow
Project Editor:	Cathy Wacaser
Design Administrator:	Jess Schaal
Text and Cover Design:	Lesiak/Crampton Design Inc: Lucy Lesiak
Photo Researcher:	Lynn Mooney
Production Administrator:	Randee Wire
Composition/Art Production:	Precision Graphics
Printer and Binder:	R.R. Donnelley & Sons Company

Cover
Front   Anglo-Australian Telescope Board
Back   Anglo-Australian Telescope Board
Back Top Inset   David F. Malin, Anglo-Australian Telescope Board
Back Right Inset   Anglo-Australian Telescope Board
Back Center Inset   NASA/JPL
Back Bottom Inset   NASA/JPL

For permission to use copyrighted material, grateful acknowledgment is made to the copyright holders on pp. 589–596, which are hereby made part of this copyright page.

Astronomy!
Copyright © 1994 by James B. Kaler

Library of Congress Cataloging-in-Publication Data
Kaler, James B.
    Astronomy!/James B. Kaler.
      p.   cm.
    Includes index.
    ISBN 0-06-500004-8 (Student Edition)
    ISBN 0-06-502352-3 (Free Copy Edition)
    1. Astronomy. I. Title
QB43.2.K36 1994
520—dc20   93-28290
                                       CIP

93 94 95 96 9 8 7 6 5 4 3 2 1

To my stars: Lauren, Bruce, Lisa, and Jill

Contents in Brief

Part I CLASSICAL ASTRONOMY 1

  1 From Earth to Universe 2
  2 The Earth and the Sky 13
  3 The Earth and the Sun 29
  4 The Face of the Sky 49
  5 The Earth and the Moon 67
  6 The Planets 80

Part II PHYSICAL ASTRONOMY 95

  7 Newton, Einstein, and Gravity 96
  8 Atoms and Light 116
  9 The Tools of Astronomy 136

Part III PLANETARY ASTRONOMY 155

  10 The Earth 156
  11 The Moon 177
  12 Hot Worlds 197
  13 Intriguing Mars 215
  14 Magnificent Jupiter 233
  15 Beautiful Saturn 251
  16 Outer Worlds 267
  17 Planetary Creation and Its Debris 285

Part IV STELLAR ASTRONOMY 309

  18 The Sun 310
  19 The Stars 333
  20 Stellar Groupings: Doubles, Multiples, and Clusters 356
  21 Unstable Stars 373

Part V BIRTH AND DEATH IN THE GALAXY 391

  22 The Interstellar Medium 392
  23 Star Formation 407
  24 The Life and Death of Stars 422
  25 Catastrophic Evolution 436
  26 The Galaxy 457

Part VI GALAXIES AND THE UNIVERSE 473

  27 Galaxies 474
  28 The Expansion and Construction of the Universe 491
  29 Active Galaxies and Quasars 505
  30 The Universe 520

Contents

Tables xvii
Preface xix
Student Preface xxiv

Part I

CLASSICAL ASTRONOMY

1 From Earth to Universe 2

1.1 Numbers 3

1.2 A Quick Tour of the Universe 4

    1.2.1 Earth, Moon, and Sun 4

MATHHELP 1.1 *Exponential Numbers* 5

    1.2.2 The Solar System 7

    1.2.3 Stars 8

    1.2.4 Our Galaxy 9

    1.2.5 Other Galaxies 10

1.3 The Universe 11

2 The Earth and the Sky 13

2.1 Science and the Shape of the Earth 14

BACKGROUND 2.1 *Greek Learning and Astronomy* 15

2.2 The Size of the Earth 16

2.3 The Rotation of the Earth 17

2.4 Latitude and Longitude 18

2.5 The Celestial Sphere 18

MATHHELP 2.1 *Angles and Arcs* 19

    2.5.1 Points and Circles on the Sphere 20

    2.5.2 The Orientation of the Celestial Sphere 20

    2.5.3 Locating the Stars 21

2.6 Motions in the Sky 21

    2.6.1 Daily Paths 21

    2.6.2 Circumpolar Stars 22

    2.6.3 The Traveling Observer 22

2.7 Demonstrations That the Earth Spins 25

3 The Earth and the Sun 29

MATHHELP 3.1 *The Ellipse* 30

3.1 The Apparent Path of the Sun 30

3.2 The Seasons 32

3.3 Tropics, the Arctic, and the Antarctic 33

BACKGROUND 3.1 *Ancient Monuments to the Sun* 34

3.4 Time 37

    3.4.1 Solar Time 37

    3.4.2 Longitude and Standard Time 39

    3.4.3 The Calendar 39

BACKGROUND 3.2 *Our Astronomical Heritage* 40

3.5 Time and the Stars 40

    3.5.1 The Sidereal Day and the Motion of the Night Sky 41

    3.5.2 Sidereal Time 41

    3.5.3 Right Ascension 42

    3.5.4 The Determination of Time and Position 42

3.6 Demonstrations That the Earth Revolves 43

MATHHELP 3.2 *Thin Triangles in Astronomy* 44

3.7 Precession 44

4 The Face of the Sky 49
  4.1 The Constellations 50
  4.2 The Ancient Forty-eight 50
    4.2.1 The Zodiac 52
    4.2.2 The Bears 53
    4.2.3 Mighty Orion 54
    4.2.4 Perseus 54
    4.2.5 The Ship, the Centaur, and the Serpent 55
    4.2.6 Single Ancient Constellations 55
  4.3 The Modern Constellations 57
  4.4 Asterisms 59
  4.5 Stars and Their Names 59
  4.6 The Milky Way 61
  4.7 Star Maps 63

5 The Earth and the Moon 67
  5.1 The Lunar Orbit 68
  BACKGROUND 5.1 *Lunar Calendars* 69
  5.2 The Phases of the Moon 69
  5.3 Daily Paths, Moonrise, and Moonset 72
  5.4 Synodic and Sidereal Periods 72
  5.5 Eclipses of the Moon 73
  BACKGROUND 5.2 *Ancient Measures of Distance* 74
  5.6 Eclipses of the Sun 75

6 The Planets 80
  6.1 Basic Orbital Characteristics 81
  6.2 The Apparent Motions of the Planets 82
    6.2.1 The Superior Planets 82
    6.2.2 The Inferior Planets 85
  6.3 Early Theories of the Solar System 86
  6.4 The Revolution in Thought 88
    6.4.1 Copernicus 88
    6.4.2 Tycho 89
    6.4.3 Johannes Kepler 90
    6.4.4 Kepler's Laws of Planetary Motion 91
    6.4.5 Galileo 92

Part II

PHYSICAL ASTRONOMY

7 Newton, Einstein, and Gravity 96
  7.1 Newton's Laws of Motion 97
  MATHHELP 7.1 *Direct and Inverse Relationships* 98
  7.2 Energy 99
  7.3 The Law of Gravity 100
  7.4 Kepler's Generalized Laws of Planetary Motion 102
  7.5 The Measurement of Mass 104
  7.6 Orbits and Discovery 104
  7.7 Spaceflight 106
  BACKGROUND 7.1 *The Beginnings of Spaceflight* 108
  7.8 Einstein and Relativity 109
  7.9 Chaos 112

8 Atoms and Light 116
  8.1 Atoms 117
    8.1.1 Atomic Constituents and Electromagnetism 117
    8.1.2 Construction of the Chemical Elements 117
    8.1.3 Radioactivity 119
  BACKGROUND 8.1 *Science and the Discovery of Radioactivity* 120
    8.1.4 Electrons and Ions 121
    8.1.5 Molecules 121
    8.1.6 States of Matter 122
  8.2 Electromagnetic Radiation 122
    8.2.1 The Nature of Light 122
    8.2.2 The Electromagnetic Spectrum 123
    8.2.3 Photons and Energy 124

8.3   Radiation and Matter   125
    8.3.1   The Blackbody and Continuous Radiation   125
    8.3.2   Spectrum Lines   127
    8.3.3   The Formation of Spectrum Lines   129
BACKGROUND 8.2 *Lasers*   132
8.4   The Doppler Effect   132

9   *The Tools of Astronomy*   136

9.1   Optical Principles   137
9.2   Lenses and the Refracting Telescope   138
9.3   The Reflecting Telescope   140
9.4   Observatories   143
BACKGROUND 9.1 *Do It Yourself*   144
9.5   Analysis and Instrumentation   145
9.6   The Invisible Universe   147
    9.6.1   The Infrared   147
BACKGROUND 9.2 *Astronomical Research*   148
    9.6.2   Radio Telescopes   148
    9.6.3   Radio Resolution and Interferometers   150
9.7   Space Observatories   151

Part III

PLANETARY ASTRONOMY

10   *The Earth*   156

10.1   Age of the Earth   157
10.2   The Active Surface   158
    10.2.1   Land, Sea, and Mountains   158
    10.2.2   Surface Activity   159
10.3   The Earth's Structure   162
10.4   Continental Drift and Plate Tectonics   164
10.5   The Atmosphere   167
    10.5.1   Composition and Structure   167
BACKGROUND 10.1 *Changes in the Earth's Atmosphere*   168
BACKGROUND 10.2 *Optics and the Earth's Atmosphere*   170
    10.5.2   Weather and Climate   170
    10.5.3   The Origin of the Atmosphere   172
10.6   The Magnetosphere   173

11   *The Moon*   177

11.1   The Moon and the Earth   178
    11.1.1   Lunar Properties   178
    11.1.2   The Tides   178
11.2   Surface Features   181
    11.2.1   The Global View   181
    11.2.2   Exploration   181
    11.2.3   Craters   181
BACKGROUND 11.1 *Apollo to the Moon*   183
11.3   The Lunar Highlands   186
11.4   The Maria   187
BACKGROUND 11.2 *A Day on the Moon*   190
11.5   The Lunar Interior   191
11.6   Impacts and the Earth   192
11.7   The Origin of the Moon   193

12 Hot Worlds 197

12.1 The View from Earth 198
12.2 Modern Observation 200
12.3 Mercury 201
    12.3.1 Rotation 201
    12.3.2 The Atmosphere 202
    12.3.3 Thermal Poles 203
    12.3.4 The Surface 203
BACKGROUND 12.1 *A Day on Mercury* 204
    12.3.5 Interior and Origin 204
12.4 Venus 205
    12.4.1 Rotation 206
    12.4.2 The Atmosphere 206
    12.4.3 Surface Topography 207
    12.4.4 Surface Activity 208
BACKGROUND 12.2 *A Day on Venus* 212
    12.4.5 The Interior 213

13 Intriguing Mars 215

13.1 The View from Earth 216
BACKGROUND 13.1 *Martian Canals and the Scientific Method* 218
13.2 Flights to Mars 218
13.3 The Martian Atmosphere 219
13.4 The Beaten Volcanic Surface 220
    13.4.1 Topography 220
    13.4.2 Structure and Development 220
    13.4.3 Water 224
    13.4.4 Viking and the Surface 226

BACKGROUND 13.2 *A Day on Mars* 228
    13.4.5 The Poles 228
13.5 The Interior 229
13.6 Life 229
13.7 The Martian Moons 230

14 Magnificent Jupiter 233

14.1 The View from Earth 234
14.2 Spacecraft 236
14.3 Jupiter's Interior 237
14.4 The Atmosphere 238
    14.4.1 Cloud Temperatures and Winds 238
    14.4.2 Chemical Composition 239
    14.4.3 Ovals and the Great Red Spot 240
14.5 The Magnetosphere 241
14.6 Jupiter's Satellites 243
    14.6.1 The Galilean Satellites 243
BACKGROUND 14.1 *Jupiter and Science* 244
    14.6.2 The Inner and Outer Satellites 246
BACKGROUND 14.2 *A Visit to Jupiter* 247
    14.6.3 The Rings 248

15 Beautiful Saturn 251

15.1 The View from Earth 253
15.2 Planetary Structure 254
15.3 The Atmosphere 255
15.4 The Rings 256
15.5 Saturn's Satellites 260
BACKGROUND 15.1 *Saturn and the Joy of Science* 263

16 Outer Worlds 267

16.1 The View from Earth 269
16.2 Interiors of Uranus and Neptune 270
16.3 Atmospheres 271
    16.3.1 Uranus 271
    16.3.2 Neptune 273
16.4 Rings 275
16.5 Satellites 277
    16.5.1 The Moons of Uranus 277
    16.5.2 The Moons of Neptune 278
16.6 Pluto 280
BACKGROUND 16.1 *Life at the End of the Solar System* 283

17 **Planetary Creation and Its Debris** 285

17.1 Meteorites 286

17.2 Asteroids 288

    17.2.1 Discovery 288

    17.2.2 Orbits 290

    17.2.3 Physical Properties 291

    17.2.4 Origins of Asteroids and Meteorites 292

17.3 Comets 295

    17.3.1 Appearance and Orbits 295

Background 17.1 *Great Comets and Halley Madness* 297

    17.3.2 Physical Nature 298

    17.3.3 Meteors, Meteor Showers, and the Zodiacal Light 301

    17.3.4 The Sources of Comets 304

17.4 The Creation of the Solar System 304

Part IV

STELLAR ASTRONOMY

18 **The Sun** 310

18.1 Basic Properties 311

18.2 Special Tools 312

18.3 The Quiet Sun 313

    18.3.1 The Thin Grainy Photosphere 313

    18.3.2 The Solar Spectrum and Chemical Composition 315

    18.3.3 The Solar Chromosphere 317

    18.3.4 The Corona 318

18.4 The Active Sun 319

    18.4.1 Sunspots and the Magnetic Activity Cycle 319

BACKGROUND 18.1 *The Earth and the Solar Cycle* 322

    18.4.2 The Origin of the Solar Cycle 322

    18.4.3 Other Active Phenomena and the Solar Wind 323

18.5 The Solar Interior 326

    18.5.1 The Sun's Source of Energy 326

    18.5.2 The Solar Model 328

    18.5.3 Neutrino Trouble 329

19 **The Stars** 333

19.1 Distance 334

19.2 Stellar Motions 337

19.3 Magnitudes 339

19.4 Stellar Spectra and the Spectral Sequence 342

MATHHELP 19.1 *Logarithms* 343

19.5 The Hertzsprung-Russell Diagram 346

    19.5.1 Giants and Dwarfs 346

    19.5.2 Luminosity Classes 348

    19.5.3 Numerical Distribution 350

BACKGROUND 19.1 *The Discovery of the Stellar Populations* 351

    19.5.4 Galactic Distribution 351

19.6 Rotation and Stellar Activity 352

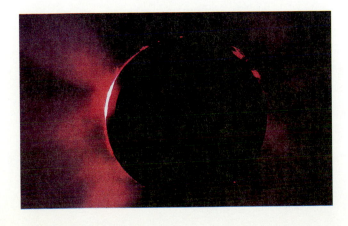

20 Stellar Groupings: Doubles, Multiples, and Clusters 356

20.1 Binary Stars 357
    20.1.1 Visual Binaries 357
    20.1.2 Spectroscopic Binaries 360
    20.1.3 Eclipsing Binaries 361
    20.1.4 Unseen Companions 363
20.2 Masses and the Mass-Luminosity Relation 363
20.3 Open Clusters 364
    20.3.1 Structure and Organization 364
    20.3.2 Distances 365
20.4 Associations 367

20.5 Globular Clusters 368
    20.5.1 Properties 368
    20.5.2 Distances and Distribution 369

21 Unstable Stars 373

21.1 Cepheid Variables 374
21.2 Other Regular Pulsators 378
BACKGROUND 21.1 *Progress in Science: The Battle for Distance* 379
21.3 Long-Period (Mira) Variables 379
21.4 Erratic Variables 381
21.5 Eruptive Variables and Interacting Binaries 383
21.6 Supernovae 386

Part V

BIRTH AND DEATH IN THE GALAXY

22 The Interstellar Medium 392

22.1 Diffuse Nebulae 393
22.2 The General Interstellar Medium 397
22.3 Dust 398
22.4 Interstellar Molecules 402

23 Star Formation 407

23.1 Infant Stars 408
23.2 The Formation of Solar-Type Stars 411
23.3 Higher-Mass Stars 414
23.4 Brown Dwarfs 414

23.5 Planets 415
Background 23.1 *Sunlike Stars* 417
23.6 Life 417
    23.6.1 Life on Earth 417
Background 23.2 *Unidentified Flying Objects* 418
    23.6.2 Life in Deep Space 418
    23.6.3 The Search for Life 419

24 The Life and Death of Stars 422

24.1 The Main Sequence 423
    24.1.1 Conditions Along the Main Sequence 423
    24.1.2 Stellar Lifetimes 424
    24.1.3 Evolution on the Main Sequence 425
24.2 Giants and Supergiants 425
    24.2.1 Realms of the Main Sequence 425
    24.2.2 Giants 426
    24.2.3 Supergiants 429
24.3 The Asymptotic Giant Branch and the Degenerate Core 429
BACKGROUND 24.1 B^2FH *and the Creation of the Elements* 430
24.4 Planetary Nebulae 431
24.5 White Dwarfs 432
24.6 Binary Evolution 433

25 Catastrophic Evolution 436

25.1 Supergiant Evolution 437

25.2 Advanced Evolution: The Road to Disaster 440

25.3 Supernovae 441

    25.3.1 Core Collapse 441

    25.3.2 Type Ia Supernovae and Supernova Significance 442

    25.3.3 Supernova 1987A 443

25.4 Supernova Remnants 444

25.5 Neutron Stars and Pulsars 445

    25.5.1 Discovery 445

    25.5.2 Physical Nature 446

    25.5.3 Binary Systems 447

25.6 Black Holes 450

    25.6.1 Theory 450

    25.6.2 Observation 452

BACKGROUND 25.1 *Mystery and Reality* 453

25.7 The Cycle of Evolution 454

26 The Galaxy 457

26.1 Basic Structure 458

26.2 The Galactic Disk 460

    26.2.1 Structure, Thickness, and Composition 460

    26.2.2 Galactic Rotation 460

    26.2.3 Spiral Structure 462

    26.2.4 The Magnetic Field and Cosmic Rays 464

26.3 The Halo and the Bulge 464

26.4 The Galactic Nucleus 465

26.5 The Mass of the Galaxy and Dark Matter 465

26.6 The Origin and Evolution of the Galaxy 468

    26.6.1 Age 468

    26.6.2 Evolution 469

Part VI

GALAXIES AND THE UNIVERSE

27 Galaxies 474

27.1 Revelation 475

27.2 Types of Galaxies 476

    27.2.1 Elliptical Galaxies 476

    27.2.2 Spiral Galaxies 477

    27.2.3 Irregular Galaxies 478

27.3 Distances 478

BACKGROUND 27.1 *Observe a Galaxy* 479

27.4 Clusters of Galaxies 481

27.5 Frequency of Galaxy Types 483

27.6 Dynamics and Masses 484

    27.6.1 Masses 484

    27.6.2 More Dark Matter 485

    27.6.3 More Distances 485

    27.6.4 Galactic Nuclei 485

27.7 Peculiar and Interacting Galaxies 486

27.8 The Origins of Galactic Forms 487

28 The Expansion and Construction of the Universe 491

28.1 The Velocity-Distance Relation 492

28.2 The Meaning of the Redshift 494

28.3 The Hubble Constant 496

28.4 The Big Bang and the Age of the Universe 497

28.5 The Distribution of Galaxies 498

    28.5.1 Redshift Distances 498

    28.5.2 The Local Supercluster 499

    28.5.3 Filaments, Sheets, and Voids 499

BACKGROUND 28.1 *Cosmology and Technology* 501

29 Active Galaxies and Quasars   505

29.1   Active Galaxies   506

29.1.1   Seyfert Galaxies   506

29.1.2   Radio Galaxies   507

29.1.3   The Origins of Active Galaxies   510

29.2   Quasars   511

29.2.1   Properties   511

29.2.2   Interpretation and Controversy   512

29.2.3   Number and Evolution   514

29.3   Gravitational Lenses   515

29.4   An Alternative View   517

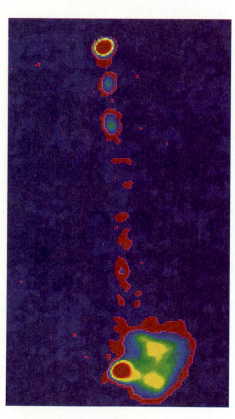

30 The Universe   520

30.1   A Fundamental Observation   521

30.2   Cosmological Principles   521

30.3   Cosmic Background Radiation   523

30.4   The Structure of the Universe   524

30.4.1   Possible Expansion Models   524

30.4.2   Curved Spacetime   524

30.4.3   The Age of the Universe   525

30.4.4   Distances, Look-Back Times, and the Cosmic Horizon   526

30.4.5   Observational Tests   528

30.5   The Origin and Evolution of the Universe   529

30.5.1   Back to the Beginning   529

30.5.2   Particles and the Unification of Forces   530

30.5.3   Challenges to Theory   531

30.5.4   An Evolutionary Model of the Universe   532

30.5.5   Transition   533

BACKGROUND 30.1   *Interactions Within Science*   534

30.5.6   The Formation of Galaxies   534

30.6   Alternative Cosmologies   536

30.7   The Future of the Universe   536

30.8   The Earth and the Universe   539

Appendix 1   *Star Maps*   541

Appendix 2   *The Messier Catalogue*   549

Bibliography   553

Glossary   563

Acknowledgments   589

Index   597

Tables

1.1 Exponential Notation and Powers of Ten   5

1.2 Distance Units   5

1.3 The Planets   7

2.1 The Greek Alphabet   19

2.2 Conversion Between Arc and Time Units   22

3.1 Solar and Vernal Equinox Passage Through the Constellations of the Zodiac   31

4.1 The Ancient Constellations   51

4.2 The Modern Constellations   58

4.3 Some Prominent Asterisms   60

4.4 Proper Names of Selected Stars   62

5.1 Eclipses of the Moon, 1994–2019   75

5.2 Eclipses of the Sun, 1994–2020   76

6.1 Planetary Orbital Data   81

8.1 The Chemical Elements   118

9.1 The Largest Telescopes   142

9.2 Some Astronomical Spacecraft   151

10.1 A Profile of the Earth   157

11.1 A Profile of the Moon   179

12.1 A Profile of Mercury   199

12.2 A Profile of Venus   199

13.1 A Profile of Mars   216

13.2 The Moons of Mars   231

14.1 A Profile of Jupiter   234

14.2 The Jovian Satellites   243

14.3 Jupiter's Rings   248

15.1 A Profile of Saturn   252

15.2 Saturn's Rings   257

15.3 The Saturnian Satellites   260

16.1 A Profile of Uranus   268

16.2 A Profile of Neptune   269

16.3 The Rings of Uranus   275

16.4 The Rings of Neptune   276

16.5 The Satellites of Uranus   277

16.6 The Satellites of Neptune   279

16.7 A Profile of Pluto   281

17.1 Types of Meteorites   287

17.2 The Fifteen Largest Asteroids   289

17.3 Selected Comets   296

17.4 Prominent Meteor Showers   302

18.1 A Profile of the Sun   311

19.1 The Closest Stars   335

19.2 Magnitudes and Brightness Ratios   339

19.3 The 40 Brightest Stars   340

19.4 Properties of the Spectral Classes   344

19.5 The Luminosity Classes   349

19.6 The Fractions of Stars in Different Spectral Classes per Unit Volume of Space   350

20.1 A Sampling of Visual Binaries   358

20.2 A Sampling of Bright Open Clusters   364

20.3 A Sampling of Bright Globular Clusters   370

21.1 Prominent Cepheids   375

21.2 Bright Mira Variables   380

21.3 Famous Twentieth Century Novae   384

21.4 The Characteristics of Variable Stars   388

22.1 Prominent Diffuse Nebulae   393

22.2 Some Famous Dark Nebulae   399

22.3 A Selection of Interstellar Molecules   403

23.1 Other Suns   417

26.1 Population Characteristics   458

27.1 The Local Group of Galaxies   480

27.2 Prominent Galaxies Not in the Local Group   481

27.3 Some Important Clusters of Galaxies   482

Preface

Welcome to *Astronomy!* Beginning with motions of the Earth, moving outward through the planets to the stars, and finishing with galaxies and the Universe, this text builds naturally on prior knowledge and leads the student into unfamiliar territory. Throughout, astronomy is associated with the reader's world. The sky is conceptually related to Earth by a substantial initial focus on classical astronomy; a unique chapter on stars and constellations allows the reader to relate the science to visible aspects of the sky; and a series of 33 Background essays ties astronomy to the familiar world, allows the student to experience days on the planets, and presents the practice and joy of science. Coverage of each subject area moves from observational results to explanation and theory. Controversy and uncertainty are highlighted to show the workings of an active science.

Features

Though mathematics is a constant presence, *Astronomy!* keeps equations to a minimum. Six separate discussions explain and refresh mathematical concepts. The history of astronomy, including the contributions of ancient astronomers, is made an integral part of the flow of astronomical discovery. Sixty detailed tables, set within the text, summarize planetary conditions, list eclipse predictions, classify a variety of astronomical objects (asteroids, comets, stars, clusters, and galaxies), and describe the different types of stars. A set of six star maps locates many of the stars and non-stellar objects discussed in the text. The brighter stars are colored according to their classes to help students visualize their distribution. A detailed map allows both naked-eye and telescopic exploration of the Moon.

Each chapter ends with a summary of key concepts and relationships. An average of 30 chapter exercises includes comparisons, numerical questions, thought and discussion questions, research suggestions, and activities. Each chapter concludes with a pair of scientific writing exercises.

The next two pages illustrate key features in *Astronomy!*

Organization

Astronomy! is divided into six parts. Part I, Classical Astronomy, demonstrates the motions of the Earth and how they appear as motions in the sky, and explains how the stars are used to measure time and position. Part I includes the chapter on constellations, explains the motion of the moon and eclipses, and concludes with the motions of the planets and the work of Copernicus, Kepler, and Galileo. Part II, Physical Astronomy, provides a foundation for the rest of the text by exploring the nature of gravity through both Newton and Einstein, the nature of light and the atom, and the tools of astronomy—telescopes and the equipment used for the analysis of radiation. Part III, Planetary Astronomy, begins with the Earth, then moves the reader through the characteristics of the planets with a strong emphasis on comparative planetology and how studying the planets allows us to understand our own world. It concludes with a chapter on the formation of the planets and the debris left behind.

The second half of the book concentrates on stars and their assemblies. Part IV, Stellar Astronomy, begins with the Sun as a turning point, treating it as both the central controlling body of the Solar System and as an example of a star. The natures of other stars are then explained, followed by chapters on stellar groupings and variability. Part V, Birth and Death in the Galaxy, takes the reader

MathHelp 3.1 The Ellipse

The **ellipse** (Figure 3.1), commonly encountered in astronomy, is a closed curve defined by two **foci** (plural of *focus*). Along the path of an ellipse the sum of the distances to the foci is a constant. You can easily draw an ellipse by placing two tacks on a board and tying a loose string between them. Pull the string taut with a pencil and trace out the curve. The line through the ellipse is called the *major axis*, and that perpendicular to it at the center is the *minor axis*. The size of the ellipse is characterized by the length of half the major axis, the **semimajor axis** (*a*). The shape, or the **eccentricity** (*e*), is the distance from the center to one of the foci divided by the length of the semimajor axis, or

$$e = CF_1/a = CF_2/a.$$

If the two foci are brought together at the center, $e = 0$, and the ellipse becomes a *circle*. If they are separated while the semiminor axis remains the same, the curve becomes more and more elongated and e approaches 1.

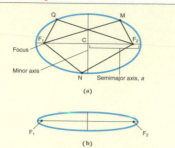

Figure 3.1 (a) *In any ellipse, the sum of the distances of any points along the curve to each of two focus points (F_1 and F_2) is a constant: MF_1 + MF_2 = QF_1 + QF_2 = NF_1 + NF_2, and so on. The semimajor axis (a) is half the major axis. The eccentricity e = CF_2/a = 0.84;* **(b)** *in this flatter ellipse, e = 0.95.*

The Earth exhibits a variety of motions, of which daily rotation is the most obvious. Of equal importance is the steady annual movement, the revolution, of the Earth around the Sun, which causes the changes of the seasons and gives great variety to human life.

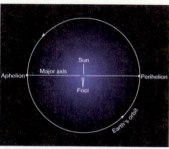

Figure 3.2 *The elliptical orbit of the Earth and the Sun are drawn to scale and the Sun placed at the right-hand focus. The center of the orbit is between the two foci. The Earth is closest to the Sun at perihelion and farthest at aphelion. (The dot representing the Sun is drawn to the correct scale.)*

3.1
The Apparent Path of the Sun

The Earth revolves around the Sun in the same direction as its rotation, that is, counterclockwise from the point of view of an observer looking down from above the north pole. (*Revolution* means one body going about another; *rotation* means *spin*.) Its period—the time it takes to return to its starting place—is a full year, during which an observer counts 365 [...] circular, bu [...] 3.1). The a[...] distance be[...] the orbital [...] lion km.

The Su[...] ellipse but [...] between th[...] ing. Howe[...] 0.017, the S[...] ter of the e[...] the Earth [...] orbit called[...] est," and h[...] million km [...] million.

Astronomy! features an elaborate program of artwork, photos, and tables tied closely to the text. Additional pedagogical aids include MathHelps, Background essays, and end-of-chapter materials.

Special boxes called **MathHelps** review relevant mathematical concepts for students without interrupting running text. Topics covered include exponential numbers, angles and arcs, the ellipse, thin triangles, direct and inverse relationships, and logarithms.

Background features help students put astronomical concepts into perspective by offering historical explanations, insights into the process of science, and fresh approaches that make abstractions more concrete.

Background 3.1 Ancient Monuments to the Sun

On the Salisbury Plain of southern England is an extraordinary ring of massive stones and other structures called *Stonehenge* (Figure 3.8a) that was built between 3000 and 1000 B.C. Its chief characteristic is a circular ring of vertical stones about 3 meters high placed in pairs that are topped by capstones. Outside the ring to the northeast is another, smaller, stone called the heelstone. From the center of the structure, the summer solstice Sun rises exactly above this rock, demonstrating that Stonehenge was built at least in part as some sort of calendrical device, marking the day summer begins. Other lines of sight involve the Moon.

Stonehenge is only one of many such constructions scattered about the world. The Incas of Peru established astronomical sightlines near Cuzco, and the Mayans of Mexico built an observatory a thousand years ago that allowed their priests to place the position of the rising equinox Sun and that contains sightlines for the rising of the planet Venus. Closer to home is the 400-year-old Big Horn Medicine Wheel in the Rocky Mountains of Wyoming (Figure 3.8b), built by Native Americans. This construction directs the observer to the rising and setting of the summer solstice Sun and to the risings of several stars.

Clearly, it was important to people in earlier cultures to know the sky. Positions of celestial objects were used to tell the dates on which planting and harvesting should take place. The sky served as a clock, as a calendar, and as you will see in the next chapter, as a storyboard. The knowledge and engineering capability of these and many other cultures were remarkable indeed. As impressive as our scientific feats are today, we must remember that they are rooted deeply in the past.

(a)

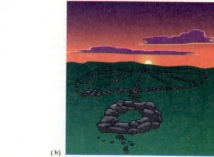

(b)

Figure 3.8 (a) *Stonehenge, in southern England, was used as a calendar; alignments of its stones mark the rising of the Sun on Midsummer Day, June 21. Other alignments may have been used to foretell lunar eclipses.* **(b)** *The Big Horn Medicine Wheel, constructed by Native Americans in Wyoming, is also astronomically aligned.*

An unusually varied and creative assortment of pedagogical aids concludes each chapter.

Key Concepts review the essential terminology introduced within the chapter. **Key Relationships** provide a summary of the chapter's main equations and measurements for astronomical phenomema.

he found that our Moon has mountains and valleys like the Earth, Jupiter has moons of its own, Venus exhibits phases just like those of the Moon, the Milky Way is made of faint stars, Saturn has "appendages" (later determined to be rings), and the Sun is spotted. All these revelations were announced in his famous tract of 1610, *The Sidereal Messenger*. Copernicus helped open our minds to the true nature of the Universe. Galileo now helped open our eyes, for the first time peering out into the darkness to see not just how things moved, but how they are *made*. He began our real study of the planets and the stars.

But he did not stop there. His greatness was that he continued to think for *years*, thinking about what he saw and, most importantly, drawing conclusions. It was evident to him that his view of the Solar System amply confirmed Copernican thought. The lunar surface looks remarkably like that of the Earth, with mountains and basins, so the Earth is not unique. The Jovian satellites go around Jupiter just as Copernicus said the planets go around the Sun; that is, a central body other than the Earth is surrounded by orbiting companions. Venus can pass through a full set of phases only if it has a path that takes it around the Sun, not the Earth. Galileo's astonishing labor helped put the Ptolemaic system to death. In his most famous work, the *Dialogue on Two Chief Systems of the World* (1632), he promoted Copernicanism by an imaginary argument between Simplicio, who believes Ptolemy, and Salviati, who follows Copernicus. A presumably impartial moderator consistently upholds Salviati. For so vigorously setting out to destroy the Aristotelian and Ptolemaic ideas, Galileo was called to the Inquisition at Rome. His fame protected him from harsh punishment, but he ... confined ... 1642. The ... cleared

... us how ... not *why*. ... responsi- ... manating ... ir orbits. ... ity—and ... nd Albert ... third law ... Solar Sys- ... he entire

Conjunction: The position in which a planet (as viewed from the Earth) is in the same direction as the Sun (or in which two planets are aligned with each other); in **opposition**, a planet is opposite the Sun as viewed from Earth (or two planets are opposite one another).

Copernican system: A system of circular heliocentric planetary orbits.

Elongation: The angle between the direction to a planet and the direction to the Sun as viewed from the Earth; **greatest elongation** is the maximum possible for an inferior planet.

Epicycle: In the Ptolemaic system, a secondary planet-carrying orbit centered on a **deferent,** which is centered on the Earth (or between the equant and the Earth).

Geocentric theory: A theory of the Solar System in which the planets orbit the Earth.

Heliocentric theory: A theory of the Solar System in which the planets orbit the Sun.

Inferior conjunction: A conjunction between Venus or Mercury and the Sun in which the planet lies between the Earth and the Sun; in a **superior conjunction** the planet is on the other side of the Sun.

Inferior planets: Mercury and Venus, the two inside Earth's orbit; the **superior planets** are Mars through Pluto, those outside the Earth's orbit.

Planets: The Sun's family of major orbiting bodies.

Ptolemaic system: A system of geocentric orbits that carry epicycles that carry planets.

Retrograde motion: The apparent westward motion of a planet relative to the stars as a result of the Earth's passing a superior planet in orbit or of the Earth being passed by an inferior planet.

Sidereal period (of a planet): The orbital period relative to the stars.

Synodic period (of a planet): The interval between successive conjunctions or oppositions of a planet with the Sun.

Kepler's laws of planetary motion:

1. **The law of ellipses:** The orbits of planets are ellipses with the Sun at one focus.
2. **The equal areas law:** The radius vector sweeps out equal areas in equal times.
3. **The harmonic law:** $P^2 = a^3$, where P is the sidereal period in years and a is the semimajor axis of the orbit in AU.

EXERCISES

Comparisons

1. What is the difference between the inferior and the superior planets?
2. What are the differences between opposition and conjunction, and between inferior and superior conjunctions?
3. Compare the visibility of Venus at greatest western elongation with that at greatest eastern elongation.
4. List the essential elements of the Ptolemaic and Copernican systems and describe the observational tests that can discriminate between them.
5. What is the difference between an epicycle and a deferent?

Numerical Problems

6. What is the maximum deflection of a star caused by the aberration of starlight as seen from Mercury?
7. NASA launches a spacecraft into orbit about the Sun with a period of six years. What is the craft's semimajor axis? If the orbit is circular, how fast in km/s does the craft travel in orbit?
8. NASA then launches another spacecraft into circular orbit about the Sun with a semimajor axis of 0.1 AU. What is its sidereal period in days? How fast does the craft travel? Draw a diagram of its orbit and with a protractor estimate the craft's greatest elongation as seen from Earth.

Thought and Discussion

9. Arrange the planets in order of maximum brightness. How can more-distant planets be brighter than closer ones?
10. Which three planets have the most eccentric orbits? Which has the most nearly circular orbit?
11. Pluto can come closer to the Sun than Neptune over a short portion of its orbit. Why do the two not crash into each other?
12. Why is the synodic period of Saturn shorter than the synodic orbit of Jupiter?
13. What phases can be seen for Venus and Mars in (a) the Ptolemaic system; (b) the Copernican system?
14. When do Mars and Venus move at their maximum angular speeds in the retrograde direction?
15. If Jupiter is in Capricornus at opposition, in what constellation will it appear at the next opposition?

16. Why does Mars have "favorable oppositions" whereas Jupiter does not?
17. Why is Mercury so hard to see even though it is quite bright?
18. Why is Venus at greatest brilliancy in a crescent phase rather than in its full phase?
19. What were Galileo's and Tycho's essential contributions toward proving the correctness of the Copernican system?
20. At what point in its orbit will a planet move most slowly in km/s?

Research Problem

21. Fit the works of Copernicus, Tycho, Kepler, and Galileo into their times. Using library materials, list the explorations of the world that were in progress and provide a brief summary on the state of Europe during that period.

Activities

22. Make a chart in which you organize the planets by their three categories: ancient and "modern" (discovered by telescope); inferior and superior; terrestrial and Jovian (see Chapter 1).
23. Using photocopies of the star maps in the Appendix, plot the positions of any observable planet over the course of the school term. Note any retrograde motion it may have. Locate the planet at opposition if appropriate.
24. If you have access to a telescope, use it to track the phases of Venus. Estimate the elongation of Venus from the Sun and, from a scale drawing of the orbits of the Earth and Venus, plot the position of Venus in orbit relative to the Earth, indicating the phase.
25. Use a popular astronomy magazine to find the time of the next maximum elongation of Mercury; then find the planet.

Scientific Writing

26. Assume that Venus is about to make its first appearance in the evening sky. Write a newspaper column explaining what the public can expect to see over the next seven or so months.
27. Some years ago, a television production called "Meeting of Minds" assembled actors portraying historical figures from different eras to discuss a variety of topics. You are on the show with Aristotle. Write a monologue for yourself in which you tell him of discoveries made in the sixteenth and seventeenth centuries regarding the construction of the Solar System.

Exercises are offered to reinforce learning in a variety of ways:

Comparisons draw on the student's abilities to distinguish among related concepts.

Numerical Problems test fundamental problem-solving skills, with answers provided in the *Instructor's Manual*.

Thought and Discussion questions provide topics for individual thought or classroom discussion.

Research Problems offer interesting puzzles requiring outside investigation and reporting.

Activities suggest ideas for individual or group projects.

Scientific Writing suggestions provide topics for essays that help students relate concepts to their daily lives.

along the entire course of stellar evolution, beginning with the interstellar medium and proceeding through star birth, the evolutionary cycles of sun-like stars, and the catastrophic evolution of high-mass stars. It ends with a discussion of the Galaxy that demonstrates the integral natures of stellar and galactic evolution. Life in the Universe is not treated separately, but as a possible part of star and planet formation.

Finally, Part VI, Galaxies and the Universe, opens with an empirical discussion of galaxies. Part VI then discusses the expanding Universe and its meaning, examines active galaxies and quasars, and concludes with views of the Universe, its origin, and evolution. By the end, we see the Earth's place in the Universe.

Supplements

The text is accompanied by a complete supplement program that includes a transparency set, a slide set, and a videotape featuring a variety of short selections that set the concepts of the book into motion. A test bank, available in print or on disk, offers 1,600 multiple-choice questions. An instructor's guide provides a detailed tour through the text, a route for a shorter course, and answers to end-of-chapter exercises.

Acknowledgments

I would like to thank the numerous individuals who provided great support and who helped draw this book together. Sandi Goldstein, a former student, made the initial contact and introduced me to HarperCollins. Don Gecewicz and Jack Pritchard brought me aboard, and Don led me through the first three drafts. Karen Bednarski, Jane Piro, Ed Moura, and Doug Humphrey provided administration and guidance. Becky Strehlow acquired the reviewers and masterfully edited the final draft. Nancy Brooks and Amy Johnson provided their usual superb and thoughtful copyediting; picture editor Lynn Mooney skillfully acquired the text's many images; and Cathy Wacaser provided superb coordination of editing and production. Thanks also go to Alison Ellis for assistance with the acquisitions of permissions and to Illinois' Jack Gladin for his fine photographic work. Jeff Mellander provided the excellent staff and facilities at Precision Graphics; Judy Taylor and Jim Gallagher, with the assistance of Kirsten Stigberg, accurately set a text with complex alterations; and Don Kesner patiently constructed wonderful line art from my sketches. I would like to express my appreciation to Lucy Lesiak for her superb design and layout.

The book would have been impossible without expert technical help. I would like to thank my Illinois colleagues Stefano Casertano, John Dickel, Icko Iben, Jim Truran, and Ken Yoss for particularly helpful discussions. I am also grateful to my friends Art Cox, George Jacoby, Karen Kwitter, and Tom Lutz for their wisdom and for answers to my many questions. I give my deep thanks to the many reviewers who kept me on straight and accurate pathways:

Lee Bonneau, *Foothill College*
Elizabeth P. Bozyan, *Providence College*
John C. Brandt, *University of Colorado, Boulder*
Daniel B. Caton, *Appalachian State University*
Margaret J. Clarke, *College of St. Scholastica*
Barney Conrath, *NASA Goddard Space Flight Center*
Bruce Dod, *Mercer University*
James L. Elliot, *Massachusetts Institute of Technology*
Paul B. Etzel, *San Diego State University*
John H. Evans, *University of Wisconsin-Oshkosh*
Donald Greenberg, *University of Alaska Southeast*
Donald E. Hall, *California State University, Sacramento*
David G. Iadevia, *Pima College-East*
Hollis R. Johnson, *Indiana University*
Steven D. Kawaler, *Iowa State University*
James Kirkpatrick, *University of Illinois*
Peter W. Knightes, *Central Texas College*
Karen B. Kwitter, *Williams College*
Paul D. Lee, *Louisiana State University*
Joel M. Levine, *Orange Coast College*
Thomas E. Lutz, *Washington State University*
Marles L. McCurdy, *Tarrant County Junior College Northeast*
A. R. Scott McRobbie, *Potsdam College-SUNY*
John Mathis, *University of Wisconsin-Madison*
James J. Merkel, *University of Wisconsin-Eau Claire*
J. Ward Moody, *Brigham Young University*
Leonard Muldawer, *Temple University*
Robert L. Mutel, *University of Iowa*
Gerald Newsom, *The Ohio State University*
Ronald A. Oriti, *Santa Rosa Junior College*
Michael O'Shea, *Kansas State University*
Tobias Owen, *University of Hawaii*
Robert S. Patterson, *Southwest Missouri State University*
Hans Plendl, *Florida State University*
B. E. Powell, *West Georgia College*
Stephen Ratcliff, *Middlebury College*
Allan M. Russell, *Hobart and William Smith Colleges*
Stephen J. Schulik, *Clarion University*

Richard L. Sears, *The University of Michigan*
John Simonetti, *Virginia Polytechnic Institute and State University*
Michael L. Sitko, *University of Cincinnati*
Paul Spudis, *Lunar and Planetary Institute*
John T. Stocke, *University of Colorado, Boulder*
Paul Tebbe, *Johnson County Community College*
David Theison, *University of Maryland*
Charles R. Tolbert, *University of Virginia*

Virginia Trimble, *University of California, Irvine*
Lois Veath, *Chadron State College*
Derek Wills, *University of Texas, Austin*
W. John Womarsley, *Florida State University*

Finally, I thank my wife Maxine, who through difficult times never wavered in her enthusiasm and her support.

Standing monumentally alone on the plains of Egypt are the legendary pyramids, the tombs of the great pharaohs who ruled the land. We can enjoy them in different ways. The casual tourist might stand at a distance and admire their beauty, symmetry, and serenity. The more curious might walk around or even attempt to climb them to try to fathom their construction, amazed at the ability of their builders. The explorers among them will go yet further and probe inside, delving into the chambers to see what is really there, making measurements to comprehend the design and engineering. All are valid means of appreciation but yield different things to the beholders. Howard Carter, in 1922, was one of the latter breed, entering for the first time a pyramid that had remained untouched over the ages. Upon reaching the burial vault of the ancient King Tutankhamen, he shone his flashlight into the dark and uttered words that still echo: "I see wonderful things." He had reached the great treasure, the gold, the art, the means to understand something of the minds of the ancient civilization.

So it is with the nighttime sky. Few of us have not stood outside in the dark simply to look at the flash and sparkle of the stars as they silently wheel across the vault above. It is, and always will be, a superb way of admiring nature's beauty. But some will inquire more deeply and learn the stars' names. They will delight at the ancient stories told about them and will wonder about their origins, their constructions, their lives, and their fates. The explorers will go deeper still and measure them to see how they work, how they fit into a master design, and perhaps most importantly, what they have to do with humankind. They will shine their flashlights into the dark, and they will see wonderful things.

This subject, astronomy, like any human endeavor, can be enjoyed simultaneously on many levels. Understanding the intricate details of the workings of the Universe does not erase the sense of magic and wonder that comes simply from letting the heavens surround you on a soft summer night, just as knowing how the blocks of stone are put together does not destroy the aesthetic quality of the pyramid. Indeed, greater comprehension of the design can only increase appreciation. So enter here the world of the explorer, the astronomer. Stand next to the one with the flashlight and perhaps hold it yourself, and in the pages to follow you too will see wonderful things.

Howard Carter peers into the tomb of King Tutankhamen.

1 *From Earth to Universe*

2 *The Earth and the Sky*

3 *The Earth and the Sun*

4 *The Face of the Sky*

5 *The Earth and the Moon*

6 *The Planets*

Classical Astronomy

PART
I

1

From Earth to Universe

A survey of the Universe to place the Earth in its surroundings

The sky is filled with thousands of stars in this view from a mountain observatory.

Figure 1.1 *The sky has long been populated with mystical and mythical figures.*

Come with me. We stand on a country hilltop, watching the late afternoon summer Sun glide slowly down a sapphire sky toward the west. It drops below the distant landscape, and the first stars begin to punctuate the onrushing darkness overhead. Before long, the sky is filled with sparkling lights. We have picked a night when the stars surround a slim bright crescent Moon, and as we watch, they too all creep westward. What are these bodies? Where are they, how far away? Why do they move? What else is out there? What is their role in our lives?

Thousands of years ago, people answered these questions by weaving rich mythologies into the celestial tapestry. The Sun, the Moon, and the patterns outlined by the stars—the **constellations**—personified gods and heroes and provided a medium to tell great stories (Figure 1.1). Over the centuries we have gradually dispelled the mystery to learn something of the true natures of the sky's inhabitants. But the **Universe**—the totality of everything in existence—is complex, and knowledge is not easily gained. Even now, some questions have only partial answers, and we are well aware that there are questions yet even to be asked.

Before painting a picture of the Universe as we now know it, we first need a frame, a survey of its contents and basic structure. The chapters that follow will then explore our Universe to reveal the depths of our knowledge, how it was obtained, and what it means to us as residents of Earth.

1.1
Numbers

The Sun is 93,000,000 miles away, the nearest star nearly 25,000,000,000,000. Such huge numbers give astronomy some of its intrigue. We speak easily of millions or billions, but what do these numbers really mean? Figure 1.2 shows 10,000 points, each new color demonstrating multiplication by 10. One lonely individual, singled out in yellow, represents the number 1, orange and yellow together make 10, and yellow, orange, and red constitute 100. These form one small block in the upper left. Multiply by 10 again. The total of 1,000 now stretches out in the row of yellow, orange, red, and green across the top. Another multiplication by 10 gives 10,000, represented by yellow, orange, red, green, and blue.

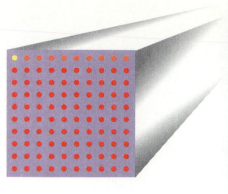

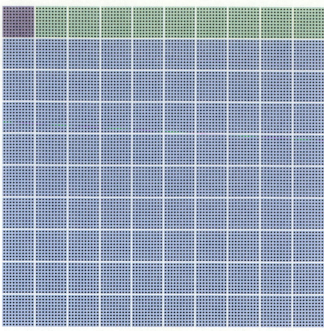

Figure 1.2   *Each color change repre-sents a multiplication by 10, progessing from 1 (yellow) to 10 (yellow plus orange) to 100 (yellow, orange, and red) to 1,000 (yellow, orange, red, and green) to 10,000 (all colors including blue).*

Ten grids like this one will give you 100,000 points. A *million* is ten times that. Imagine that the grid of 10,000 colored points fills the page; it would take 100 pages, about one-sixth the length of this book, to print 1,000,000 points.

Now you can have respect for a *billion,* which is a thousand million—1,000,000,000. To show a billion points would require a book 100,000 pages thick. One hundred or so copies of Tolstoy's *War and Peace* (each about 1,000 pages long) might serve. To print a *trillion* points (a thousand billion), we would need 100,000 copies and a library shelf nearly four miles long. A *quadrillion,* a thousand times that, would require a row of books stretching from New York through Los Angeles and a thousand miles out into the Pacific. *The number of stars in the observable Universe is over 10 million times more.*

1.2
A Quick Tour of the Universe

We now embark on a rapid tour of space in which we start with the Earth and look outward as far as our best instruments will allow.

1.2.1 *Earth, Moon, and Sun*

Begin with yourself (Figure 1.3), a person most likely somewhat short of 2 meters tall. The meter (m) is the fundamental unit of length in the **metric system** (Table 1.2), which will be used here. In conventional U.S. units it equals 39.4 inches or 1.09

yards. It is commonly multiplied and divided by powers of 10: a *kilometer* (km) is 1,000 meters (0.622 miles), a *centimeter* (cm) is a hundredth of a meter (there are 2.54 cm per inch), and a *millimeter* (mm) is a thousandth of a meter.

You stand on your world, the Earth (Figure 1.4), a sphere 12,700 km (7,900 miles) in diameter, some 7 million (7×10^6: see MathHelp 1.1) times your length. If the Earth were a basketball, you

Figure 1.3   *We stand two meters high and look out over the Earth, about to embark on a journey into deepest space.*

MathHelp 1.1 Exponential Numbers

To handle huge numbers reasonably, we must use powers of ten, or **exponential notation:** 100 is 10×10, which can be written as 10^2; a million—1,000,000—is six tens multiplied in succession, $10 \times 10 \times 10 \times 10 \times 10 \times 10$, or 10^6; 200 is 2×100 or $2 \times 10 \times 10$, or 2×10^2. Numbers less than 1 can be expressed the same way: $1/10 = 0.1 = 10^{-1}$; $1/1,000,000$ is 10^{-6}; and $2/1,000,000$ is two times $1/1,000,000$, or 2×10^{-6}. Table 1.1 provides additional examples. Numbers are commonly rounded, as there is no reason to carry decimal places beyond our needs. For example, the average distance to the Sun is 149,597,706.1 km, which for most purposes can be written as 1.496×10^8 km or even as 1.5×10^8 km.

Multiplication and division with exponential numbers is easier than with ordinary numbers. In multiplication, you multiply the prefixes and add the exponents. In division, you divide the prefixes and subtract the exponents. The rules are: $(a \times 10^x)$ times $(b \times 10^y) = (a \times b) \times 10^{x+y}$; $(a \times 10^x)$ divided by $(b \times 10^y)$ is $(a/b \times 10^{x-y})$. As an example, to find the length of a light-year (the distance light travels in a year) in kilometers, multiply the speed of light, 299,792 km per second (km/s), by the number of seconds in a year, 31,556,925. Without significant loss of accuracy, we can round 299,792 to 300,000 or 3.00×10^5, rewrite 31,556,925 as 3.16×10^7, and write the product as $3.00 \times 10^5 \times 3.16 \times 10^7$ to find 9.48×10^{12} (nearly 10 trillion) km.

TABLE 1.1
Exponential Notation and Powers of Ten

Number Name	Number	Exponential Form	Prefix[a]
trillion	1,000,000,000,000	10^{12}	(tera)
billion	1,000,000,000	10^9	giga
million	1,000,000	10^6	mega
thousand	1,000	10^3	kilo
hundred	100	10^2	(hecto)
ten	10	10^1	deca
zero	0	10^0	. . .
one-tenth	1/10	10^{-1}	(deci)
one-hundredth	1/100	10^{-2}	centi
one-thousandth	1/1,000	10^{-3}	milli
one-millionth	1/1,000,000	10^{-6}	micro

[a]Indicates a multiplying factor. For example, a thousand meters = 1 kilometer. Prefixes not used here are in parentheses.

TABLE 1.2
Distance Units

Units	Examples
1 centimeter (cm) ⊢———⊣ = 10 millimeters (mm) = 0.394 inches	Thickness of pencil: 7 mm Diameter of golf ball: 4 cm
1 meter (m) = 100 cm = 1,000 mm = 39.4 inches = 1.09 yards	Basketball player: 2 m Football field: 90 m
1 kilometer (km) = 1,000 m = 0.622 miles	United States: 4,900 km across Distance to the Moon: 384,000 km
1 astronomical unit (AU) = 1.50×10^8 km = 9.30×10^7 miles	Earth to Sun: 1 AU Sun to Pluto: 39 AU
1 light-year (ly) = 63,270 AU = 9.5×10^{12} km	Nearest star: 4 ly Galaxy diameter: 80,000 ly
1 parsec (pc) = 3.26 ly = 206,265 AU = 3.09×10^{13} km = 3.09×10^{18} cm	Nearest star: 1.3 pc

would be only 4 millionths of a millimeter high, and no ordinary microscope could see you. Even Mount Everest, at the limit of the breathable atmosphere, would project upward by less than two-tenths of a millimeter. No wonder it is so difficult to sense the Earth as round, and no wonder so many have believed it to be flat.

The nearest of all astronomical bodies is an old friend, figuring in the stories and songs of many cultures. The **Moon** (Figure 1.5) is our lone natural **satellite,** a body that circles around, or orbits, the Earth about once a month under the action of *gravity,* the attractive force that holds you to the ground.

Figure 1.4   *The Earth is 12,700 km across. South America is seen in the lower half of this Apollo 8 photograph. North America, toward the upper left, is swathed in clouds. The bulge of west Africa appears at the right.*

The Moon's diameter is a quarter that of the Earth, its distance is 384,000 km (238,000 miles or 30 Earth diameters), and it is the only other body in space to which human beings have traveled. With the Earth a basketball, the Moon is roughly the size of a baseball 7 m (6.5 yards) distant.

Now proceed to the **Sun** (Figure 1.6), which warms the day and keeps us all alive, and about which the Earth orbits once a year. *DO NOT EVER LOOK AT THE SUN DIRECTLY WITHOUT A PROPER SOLAR FILTER: YOU COULD DAMAGE YOUR EYES PERMANENTLY.* It is the nearest star, a vast gaseous ball 1.4×10^6 km (109 Earth diameters) across. It appears small in the sky because it is so far away, 1.5×10^8 km (93 million miles), a fundamental distance in astronomy called the **Astronomical Unit** (AU). The AU is roughly 100 times the Sun's diameter (which makes the Sun about 10^{-2} AU across) and about 10,000 times the diameter of Earth. On our Earth-as-a-basketball scale, the Sun would be nearly 30 m (98 feet) wide and 3 km (almost 2 miles) away.

The Moon and Sun have profoundly different physical characteristics. The Moon is made of rock like the Earth and is cold and solid. The Sun is hot, is composed mostly of hydrogen, the simplest of all chemical elements, and is gaseous throughout. The Moon shines by sunlight reflected from its surface, as does the Earth. The Sun, however, is self-luminous as a result of great internal energy, and shines as bright as 4×10^{24}—over a trillion trillion—standard 100-watt light bulbs. The Earth and the Sun, which were born together, are nearly 5 billion years old, and the Sun will stay alight for over 5 billion years more.

Figure 1.5   *The Earth and Moon were photographed together for the first time by Voyager 1 as it began its great flight to the outer reaches of the Solar System. We see only slivers of the daylight sides of the two bodies. Because of the position of the spacecraft, the Earth and Moon appear closer together than they really are. A true scale drawing appears below.*

384,000 km

1.2.2 The Solar System

Many celestial bodies (where *celestial* means "pertaining to the sky") orbit the Sun, among them a family of nine **planets** (Table 1.3 and Figure 1.7). The whole collection comprises the Solar System. The Earth is the third planet out. All the planets orbit in the same direction and in nearly the same plane, and six of the nine spin in the same direction. All shine by reflected sunlight. They are broadly divided into two groups. The inner four are within 1.5 AU of the Sun. Because they are constructed of rock like the Earth (which is the largest of them), they are called the *terrestrial planets*. (*Terrestrial*, from Terra, the Roman goddess of Earth, means "Earthlike.") The next four extend from 5 to 30 AU from the Sun, are much larger than Earth, are made partly or mostly of hydrogen and helium, and are called the *Jovian planets* after Jupiter, the biggest of them (Figure 1.8). Jupiter, however, is still only 0.001 AU across, and on the scale of Figure 1.7 would be invisible to the eye. The last planet is tiny Pluto. Only 2,400 km across, it fits into neither of the two categories, and averages 39 AU from the Sun. On the scale of a basketball Earth with the Sun 3 km distant, Pluto would be about 110 km (75 miles) away. Each planet has its own special characteristics; each is a different world to explore.

The terrestrial planets have few satellites: Mercury and Venus have none, the Earth one, and Mars only two tiny ones. In contrast, the Jovian planets have extensive satellite systems. Jupiter

Sun   Earth-Moon system

Figure 1.6   *The Sun, a typical star, is 1.4 million km across. The Earth-Moon system is shown projected onto its surface; 109 Earths and over 3 Earth-Moon systems could fit across its diameter. The dark spots on the Sun, some of which dwarf the Earth, are regions of intense magnetism. Across the bottom is a scale drawing of the Sun and Earth. The entire Earth-Moon system would fit inside the dot at the right, and the Earth itself would be a mere pinpoint.*

TABLE 1.3
The Planets

Planet	Average Distance from Sun (AU)	Characteristic
Terrestrial[a]		
Mercury	0.4	Second smallest
Venus	0.7	Brightest; just smaller than Earth
Earth	1.0	Carries life
Mars	1.5	Red color
Jovian[b]		
Jupiter	5.2	Largest; prominent cloud belts; 4 bright satellites
Saturn	9.5	Surrounded by bright rings
Uranus	19.2	Tipped on its side; nearly featureless
Neptune	30.1	Blue-green clouds
Neither Class		
Pluto	39.4	Smallest

[a]Like Earth
[b]Like Jupiter

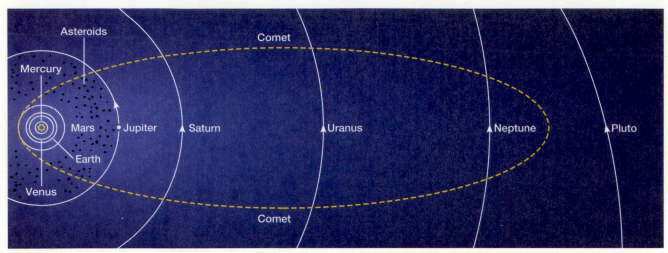

Figure 1.7   *The Solar System and portions of the orbital paths of the planets around the Sun are drawn to scale. The Sun is at the center of the orbits; on this scale it is only 0.04 mm in diameter. Jupiter is ten times smaller and the Earth ten times smaller yet. The dot that represents Jupiter encompasses the diameter of its whole satellite system. Most of the asteroids orbit between Mars and Jupiter. The dashed curve shows the path of a comet; most comets have orbits tilted out of the Solar System's plane.*

Figure 1.8   *Jupiter, the greatest of the planets, has a diameter 11 times that of Earth.*

Figure 1.9   *Comet Bennett passed close to the Sun in 1970, generating a long tail.*

has 16 known satellites, 4 of them easily visible in binoculars. Saturn has 18, Uranus 15, and Neptune 8. Pluto has a companion half as big as itself.

Minor bodies throng the Solar System. The **asteroids** are small chunks of rock or metal that orbit the Sun in paths that lie largely, but not exclusively, between Mars and Jupiter (see Figure 1.7). Thousands have been catalogued and there are countless more. The largest, Ceres, has a diameter only a quarter that of the Moon, and the smallest are mere pebbles. **Comets,** made of dusty ice, are typically a few kilometers in diameter and orbit the Sun on highly elongated paths. As a comet approaches the warmth of the Sun, the ice turns into gas that streams out in one or more *tails* that can be millions of kilometers long (Figure 1.9). We believe there are trillions of comets in clouds that extend perhaps as far as 100,000 AU away. Only a tiny fraction ever gets close enough to the Sun for us to see them.

Comets steadily disintegrate under the action of sunlight. As the Earth orbits, it continually collides with pieces of cometary debris and stray asteroids that heat up in our atmosphere and streak brilliantly across the sky as **meteors.** The cometary dust all burns up in the air, but asteroids more than about a centimeter or so across can survive to strike the ground to become **meteorites** (Figure 1.10).

1.2.3 Stars

In the nighttime sky, the **stars** and planets all seem to be at the same distance from Earth, but that is an illusion. The stars are distant suns, many far mightier than our own. The nearest of them (called Alpha Centauri) is an astonishing 271,000 AU away. At such distances, the stars appear as no more than points of light. In our scale model with a basketball Earth, Alpha Centauri would be 780,000 km (480,000 miles) distant, twice the actual distance to the Moon.

Such large distances require the use of a large unit. The **light-year** (ly) is the distance a ray of light will travel in a year at a speed of 299,792 km per second (km/s). It is 63,300 AU (almost 10^{13} km or 6×10^{12} miles) long. Therefore, Alpha Centauri is 4.3 ly away. The light we see actually left the star 4.3 years ago; as we look out into space, we are looking back into the past. Professional astronomers more commonly use an even larger unit, the parsec (pc). It has a length of 206,265 AU or 3.26 ly, making Alpha Centauri 1.3 pc away. As great as this distance is, most of the of the stars seen with the naked eye are vastly farther—hundreds, even thousands, of light-years or parsecs away.

We find great diversity among stars. Some are much hotter than the Sun, while others are considerably cooler. The smallest are no larger than a small city, and the biggest, if placed at the position of the Sun, would fill a good portion of the outer Solar System. A few stars are bright enough to be seen with the naked eye over distances of thousands of light-years, whereas the Sun would be invisible if only 70 ly away. Some others would barely be visible to the naked eye if they were a mere 0.01 ly distant. The varied properties are caused by differences in the amounts of matter the stars contain—which range from about a tenth that of the Sun to a hundred times as much—and by the aging process. Stellar aging, however, is very slow—life spans are typically measured in millions and billions of years—and as a result the stars appear immutable.

1.2.4 Our Galaxy

In the grand design, all stars are arranged in individual **galaxies,** vast collections of matter tied together by gravity (Figure 1.11). Individual galaxies are separated from one another by wide spaces. Some galaxies are huge, others very small. The one we live in, called simply **the Galaxy,** is quite large, most of its 200 billion stars contained within a volume roughly 80,000 ly or 25,000 pc across. If the Sun were a marble, our Galaxy would be 5 million km in diameter, 13 times the actual distance to the Moon!

We can obtain a good idea of the structure of our Galaxy by looking at others that are thought to be similar. Its most important component is a flat *disk* that bulges in the middle, like the galaxy in Figure 1.11a. The disk contains the Sun, which is locat-

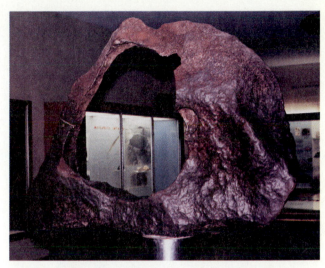

Figure 1.10 *This iron meteorite is a piece of an asteroid that collided with the Earth.*

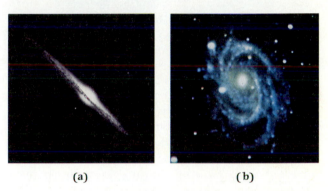

(a)   (b)

Figure 1.11 **(a)** *A galaxy called NGC 4565 is a flat collection of billions of stars about 100,000 ly across.* **(b)** *A similar system called NGC 4603 is seen face-on and displays spiral arms.*

ed about two-thirds of the way out from the center. A face-on view of our Galaxy would reveal a set of spiral arms something like those in Figure 1.11b. The spiral arms are where new stars are created, stars that replace others that are constantly dying.

As we look out into the sky from the Earth we see the disk of the Galaxy surrounding our heads in a great diffuse band of light that we call the **Milky Way** (Figure 1.12), which is made of the collected light of billions of stars that individually are too faint to be seen with the naked eye. To understand the nature of the Milky Way, sit in crowded roomful of people. The heads of the people in the room, including yours, collectively form a disk. As you survey the room, you see a band of heads surrounding you, just as the Milky Way surrounds the Earth. The Milky Way is not well known to town dwellers because its pale glow is easily swamped by artificial lighting, but seen from the dark countryside it can be spectacular. Because it is the most prominent

Figure 1.12 *The Milky Way, the combined light of the stars of the galactic disk, stretches upward from the horizon.*

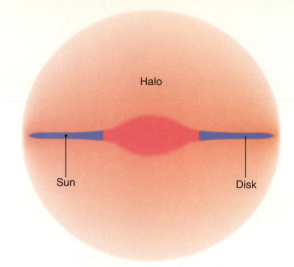

Figure 1.13 *The disk of our Galaxy, depicted edge-on, is surrounded by a dim halo of faint stars.*

manifestation of our Galaxy as a unit, our system is commonly called the **Milky Way Galaxy.**

A second component of our Galaxy consists of a vast spherical *halo* of stars that entirely surrounds and encloses the disk (Figure 1.13). Think of slicing a basketball in two and running a flat dinner plate through the middle. The plate represents the disk and the ball the halo. The disk is thickly populated with stars of all brightnesses, but those in the halo are faint and relatively few and far between. As a consequence, this component is hard to see in galaxies like those in Figure 1.11.

Because of the way in which stars are created, they tend strongly to aggregate into doubles, triples, multiples, and **clusters** that are held together by the stars' gravities. In the disk we find thousands of loose *open clusters* (Figure 1.14a) that typically contain a few hundred stars. The halo contains a very different kind, about 150 known huge *globular clusters* (Figure 1.14b), each populated by tens or hundreds of thousands of stars.

1.2.5 *Other Galaxies*

Our nearest neighboring galaxy of a size comparable to our own, called the *Andromeda Galaxy* or M 31 (Figure 1.15), is 2.3 million ly (700,000 pc) away, a distance of about 30 times the diameter of our Galaxy. With the Sun a marble, M 31 would be 140 million km away, or about as far as the real Sun. The range of properties exhibited by galaxies is as great as that displayed by stars. Our Galaxy

(a)

(b)

Figure 1.14 **(a)** *The Jewel Box, a typical open cluster, is a loose collection of a few hundred stars.* **(b)** *The globular cluster M 3 contains more than 100,000.*

and M 31 are spirals; *elliptical galaxies,* like the ones in Figure 1.15 that accompany M 31, have no spiral arms; and *irregular galaxies* are small systems that have little form or structure. Some elliptical galaxies are 10 times the size of our Milky Way Galaxy.

Like stars, galaxies tend to group in multiples (as in Figure 1.15) and clusters (Figure 1.16). Our Galaxy and the Andromeda Galaxy both belong to a sparse cluster with about two dozen members called the *Local Group.* Other clusters contain thousands of galaxies. There are even higher levels of organization, with clusters teamed together in complex ways that we are only beginning to understand.

1.3
The Universe

We can see galaxies and their clusters as far away as our technology will allow, out to distances of billions of light-years. Had we the time, we could count over 10 billion galaxies, each containing billions of stars. With improved scientific instruments, a trillion or more may be within our grasp. The radius of the potentially visible Universe extends outward to over 10 billion ly.

But size, immense as it is, is not the Universe's most remarkable property. With the exception of a few galaxies within the Local Group, every galaxy in the Universe is receding from us with a speed directly proportional to its distance; that is, if the distance is doubled, so is the speed. We in fact infer that all clusters of galaxies are moving away from all other clusters of galaxies with speeds that depend directly on the distances separating them. Our conclusion is that the Universe as a whole is expanding.

Because the speed is increasing in direct proportion to distance, all the matter in the Universe must once have been compressed into a very small volume, which began its expansion suddenly in an event called the **Big Bang.** From the speed of recession at a given distance, we can estimate how long it has been since that moment; that is, we can find the age of the Universe. The age is not known very well because of various difficulties of measurement and interpretation, but it seems clearly to fall between 10 and 20 billion years. The Sun, only 5 billion years old, is therefore a relative newcomer.

What were the discoveries and what was the progression of human thought that led us to our understanding of the structures of stars and the constitution of the Universe? What do we need yet to learn? To find out, we return to Earth and begin the story.

M 31   The Galaxy

Figure 1.15   *The Andromeda Galaxy, also called M 31, is the closest spiral galaxy comparable in size to our own. It is accompanied by a pair of small elliptical galaxies. Our Galaxy and M 31 are drawn to scale below. On that scale the separation between the Sun and Alpha Centauri is only 1/100,000 cm.*

Figure 1.16   *The central part of a large cluster of galaxies contains over a dozen bright members and many more fainter ones.*

KEY CONCEPTS

Asteroids: Rocky or metallic bodies, smaller than planets, that orbit between Mars and Jupiter.

Astronomical Unit (AU): The average distance between the Earth and the Sun.

Big Bang: The event that appears to have created the Universe.

Celestial: Pertaining to the sky.

Clusters: Groups of stars or galaxies held together by gravity.

Comets: Small icy bodies that move around the Sun on highly elongated orbits.

Constellations: Named patterns of naked-eye stars.

Earth: Our world, the third planet from the Sun.

Exponential notation: A means of expressing numbers by using powers of ten.

Galaxies: Basic units of the Universe in which stars are collected.

Galaxy, The: The collection of stars in which we live; also called the Milky Way Galaxy.

Light-year: The distance a ray of light travels in a year at 299,792 km/s.

Meteorites: Pieces of rock or metal from space that hit the surface of the Earth.

Meteors: Streaks of light in the sky caused by rocks and dust from space burning up in the Earth's atmosphere.

Metric system: A system of measures in which the units differ by multiples of ten.

Milky Way: The band of light around the sky caused by stars on the disk of our Galaxy.

Moon: The Earth's satellite, which orbits the Earth under the force of gravity.

Planets: The larger bodies of the Solar System orbiting the Sun.

Satellite: A body that orbits a planet.

Solar System: The Sun and its collection of orbiting bodies.

Stars: Gaseous, self-luminous bodies similar in nature to the Sun but with diverse properties.

Sun: The gaseous, self-luminous body that dominates the Solar System; the nearest star.

Universe: The all-encompassing structure that contains everything.

EXERCISES

Comparisons

1. What are the differences between terrestrial and Jovian planets?
2. What are the differences between the disk and the halo of our Galaxy?
3. What are the differences between comets and asteroids?
4. What is the difference between meteors and meteorites?
5. What is the fundamental difference between the Sun and a planet?
6. Compare the size of the Sun with the sizes of other stars.

Numerical Problems

7. Write the numbers 100, 1,000,000, 126, 0.335, and 0.000036 in exponential notation.
8. Write 22×10^2, 3.62×10^6, 2.34×10^{-3}, and 6.8456×10^{-5} in normal numerical form.
9. What are: **(a)** $(3 \times 10^2) \times (2 \times 10^5)$; **(b)** $(6 \times 10^{-2}) \times (3 \times 10^{-7})$; **(c)** $(1.5 \times 10^5) \times (1.4 \times 10^{-17})$?
10. If the Earth and Sun were separated by 100 meters, **(a)** what would be the diameters of the Earth and Sun in centimeters; **(b)** how far would the Moon be from the Earth in centimeters; **(c)** how big would the Solar System be in meters (out to the orbit of the most distant planet); **(d)** how far away would the nearest star be in kilometers?
11. The star Vega is 24 light-years away. How far is that in parsecs, astronomical units, and meters?
12. Roughly how far is the Sun from the center of the Milky Way Galaxy in parsecs and light-years?
13. How many of our Milky Way Galaxies could you line up from the Earth to the nearest large spiral galaxy, M 31?

Thought and Discussion

14. Describe the relations of planets, comets, and asteroids to the Sun.
15. How does the structure of the Galaxy give rise to the Milky Way?

Research Problem

16. What countries still use distance units like the inch, foot, and mile in daily life? Look in old astronomy books to see when the metric system began to be used as the astronomical standard.

Activity

17. Using ordinary household items, make scale models of the Solar System and the Galaxy.

Scientific Writing

18. Write a letter to a grade school student in which you explain and demonstrate how far even the nearest star is from the Earth.
19. Describe in a single typewritten page how the Earth fits into the Solar System, how the Solar System fits into the Galaxy, how the Galaxy fits into the Local Group, and how the Local Group fits into the Universe as a whole.

2

The Earth
and
the Sky

*The rotation
of the
Earth and
apparent motions
in the sky*

*The armillary sphere
and its circles
represent the sky,
its apparent motions,
and its measurement.*

As you stand outdoors on a sunny day, the sky appears as a blue hemispherical bowl inverted over your head. The countryside stretches outward to the **horizon,** where land seems to meet the sky. You watch the Sun move slowly across the bowl above and eventually drop below the horizon, or *set,* in the west. When the sky darkens and the stars come out, they too are seen to be in steady motion, some *rising* over the eastern horizon, others setting below the western. Gradually the sky becomes light, the stars fade, and the Sun reappears in the east. The performance repeats itself day after day. The Earth looks flat, and it appears that the Sun, the stars, and the whole sky are circling around you. Appearances, however, are deceiving. The truth is very different.

2.1
Science and the Shape of the Earth

We learn as children that the Earth is a ball. This concept can be traced back 2,500 years, to the Greek mathematician Pythagoras. But how do you really *know* the Earth is round, other than that someone told you? The subject makes a good illustration of how science works. The scientist does not accept authority but looks at evidence, endeavoring to see what **theory** best fits the observations. This word "theory" is often misused to suggest an idea without support. To the contrary, a theory is a logical or mathematical description, or model, of

Figure 2.2   *A cosmology, or world model, from old India shows a partially curved Earth that ultimately floats on an eternal sea. On such a world, an observer would still see a horizon and hull-down ships, and the stars would change positions as one traveled north or south.*

nature that incorporates and explains the observations. A theory allows the scientist to make predictions about how nature should behave and suggests new observations. The experimentalist or observer examines these predictions to see if they are true. If they are, the theory stands; if not, it falls and must be modified, or a new theory must be developed to accommodate the fresh data. Although science does not always work in such a methodical fashion—many discoveries are quite accidental—this *scientific method* must be followed if we are to find the truth.

Look at the evidence for the Earth's shape. The sharp horizon at first suggests a flat Earth with an edge. But as you walk toward the horizon, you never reach an edge. The horizon recedes ahead of you, implying that the Earth's surface is at least curved. You need not even travel. Look at a ship sailing out to sea (Figure 2.1a). It does not stay permanently visible, nor does it drop instantly away. First the hull disappears, then the superstructure, then the masts. When you see only the top of the ship, you know that the bottom must be still resting on an ocean that is below your line of sight (Figure 2.1b).

Additional evidence is found in the sky. As long ago as the fourth century B.C., Aristotle discussed how the stars and the Sun change their positions. As you walk north, the stars in front of you climb higher above the horizon, and those to your back drop toward it, an effect difficult to explain unless

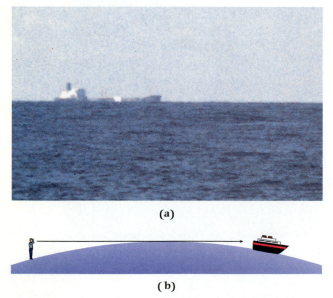

(a)

(b)

Figure 2.1   **(a)** *A photograph of a ship at sea shows it to be "hull down," its lower parts below the horizon.* **(b)** *The drawing shows how the curved Earth hides the lower hull.*

Background 2.1 Greek Learning and Astronomy

The years 600 B.C. to about A.D. 170 represent a remarkable interval in the history of human thought. This was the time of the flowering of Greek science and philosophy, upon which we have built so much of the knowledge of our world: its impact cannot be overestimated. The Greek arena of this period, with its cultural center in Athens, extended beyond the borders of modern Greece across an area that ultimately included Asia Minor (the Aegean coast of modern Turkey), what is now northern Egypt (particularly Alexandria), and parts of Italy (Figure 2.3).

Ancient Greece was a hodgepodge of separate often-warring states. Though slavery was common, the upper class had the political freedom, as well as sufficient wealth and free time, to be able to wonder about the nature of humanity and the Universe. Pythagoras, who lived in the sixth century B.C., sought to use mathematics to bring order to the Universe, and he founded a school of thought that was followed for over a hundred years. He also believed in things that could not be demonstrated, such as music creat-

ed by celestial bodies, the "music of the spheres." Other philosophers believed that pure thought alone, without contamination by the real world, could yield truth. The greatest of these were Socrates (469–399 B.C.) and Plato (427–347 B.C.), who expounded Socrates' teachings and philosophy. The belief in the sphericity of the Earth came largely not from observation but from the notion that the sphere was the perfect figure, and therefore the bodies of the known universe must be spheres. Plato's student Aristotle (384–322 B.C.), however, took a somewhat different point of view. He used arguments constructed not from logic alone, but also from what he could observe.

Shortly thereafter, Greek astronomers such as Eratosthenes, Aristarchus, and Hipparchus began actively measuring some of the basic properties of the Moon, Earth, and Sun. Our intellectual predecessors were an extraordinary group of people, open to new thoughts, always arguing different points of view. Out of the contention and the willingness to search for the truth came much of our modern view of life and of the Universe.

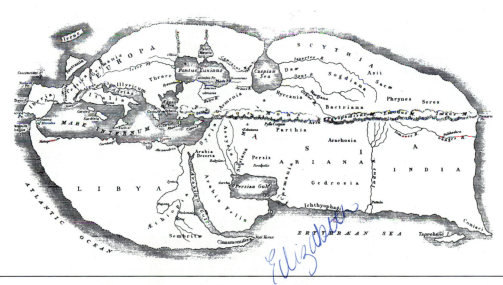

Figure 2.3 *The ancient Greek world, centered on modern Greece, extended great distances into Asia, Africa, and Italy. The first major cultural center was Athens, but after about 300 B.C. the focus of learning shifted to Alexandria in northern Egypt, a city founded by the conqueror from Macedonia (northern Greece), Alexander the Great.*

either the Earth is curved or the stars are very close. Still, these phenomena do not demonstrate that our home is a complete sphere, closing back upon itself. It might appear as in Figure 2.2.

Aristotle pointed out that eclipses of the Moon provide telling evidence. These events occur when the orbiting Moon passes through the shadow the

Earth casts into space. The shadow always has a circular outline (Figure 2.4). The model of the Earth in Figure 2.2 would throw a circular shadow only if the Sun were directly below and the Moon overhead. The model predicts that if the Sun were to the side and the eclipsed Moon near the horizon, which is not uncommon, the shadow would not be

Figure 2.4 *The shadow of the Earth cast upon the Moon during the entering phase of an eclipse is always circular.*

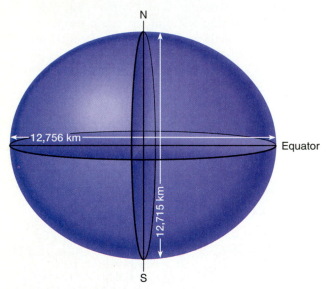

Figure 2.5 *The Earth is an oblate spheroid (greatly exaggerated here) rather than a perfect sphere. The diameter from the north to the south pole is 41 km less than it is across the equator.*

circular and would show the world's edge (or even the elephants). The model of a spherical Earth, however, predicts only circular shadows. We therefore conclude that the spherical model is the better.

Confirming evidence is furnished by circumnavigation. In 1522, one of Ferdinand Magellan's ships kept a steady course and returned to its starting point. Spaceflight, which allows us to see our planet from a distance, yields more compelling evidence. The first impression from a space observation is that the Earth is a flat disk (see Figure 1.4). However, as the spacecraft continues around it, a disk would present its edge. Instead, different conti-

nents come into view, and from whatever point of view the Earth always has a circular outline, again demonstrating sphericity.

No scientist, therefore, doubts the accuracy of the theory that the Earth is at least some approximation to a sphere, as the combination of evidence is overwhelming. We will encounter other theories, however, that rest on far less evidence, and then we must remember that no theory can actually be proved: it can only not be *dis*proved. Only one contradicting observation shows a theory to be incorrect and in need of revision or even replacement.

Such development is illustrated by further investigations of the Earth. Data obtained in the eighteenth century demonstrated that the Earth is actually about 40 km larger along one diameter than it is along another. The theory then had to be modified from a sphere to a flattened form called an **oblate spheroid** (Figure 2.5). More refined measurements made in our own century revealed an even more complicated shape that cannot be described by any simple figure.

2.2
The Size of the Earth

The next step is to measure the Earth's size, knowledge needed for travel and commerce. In the third century B.C., Eratosthenes of Cyrene knew that in the town of Syene in southern Egypt, the Sun would shine directly down a well at noon on the first day of summer, and therefore had to be directly overhead. However, to the north, in Alexandria, the Sun was 1/50 of a whole circle (7°) south of the overhead point at noon (Figure 2.6). If the Earth is a sphere (and the Sun very far away), the distance between Syene and Alexandria must then be 1/50 of the globe's circumference.

Such a distance was not easy to measure. From the time it took the king's messenger to run the distance, Eratosthenes estimated it at 5,000 *stadia*. The Earth's circumference must then be 50 × 5,000, or 250,000 stadia. Our best estimate of the length of the Greek stadium (the singular of stadia) is 157 m, so Eratosthenes' result was a circumference of 39,300 km. Modern measurement with the same technique gives 40,008 km around the polar axis of the oblate spheroid and 40,075 km around the equatorial axis. Eratosthenes was off by only 2%. The circumference of a circle equals π times its diameter, where π is close to 3.1416. The equatorial diameter of the Earth is 12,756 km and the polar (which

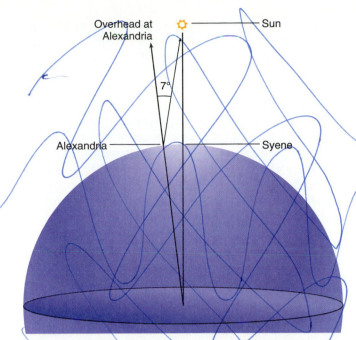

Figure 2.6   *The overhead point for an observer is always on a line directed outward from the Earth's center. Alexandria is 1/50 of the way around the Earth (7°) from Syene. If the Sun is overhead at Syene, it will be 1/50 of a full circle away from overhead at Alexandria.*

needs a different formula) is 12,715 km; the average is 12,735 km.

2.3
The Rotation of the Earth

Is it the Earth or the sky that moves? Aristotle, whose opinions swayed human thought until the seventeenth century, argued forcefully that the Earth was simply too heavy to move. His was not the only view, however. It was clear to his contemporary, Heraclides Ponticus (and suspected even by some of the Pythagoreans 200 years before), that the daily movement of the Sun, Moon, and stars is more easily explained by having the Earth rotate daily from west to east, in the direction opposite the motions seen in the sky. Stand in the middle of a room and begin to turn counterclockwise. *You* are spinning, but the room and its furnishings seem to go in the opposite direction, clockwise. The chairs and tables first "rise" into your view from the left and then disappear or "set" to the right when you turn away. The Earth is so large and the motion so smooth that you just do not feel it. You have to look outside the Earth to see that the movement is there at all.

The Earth (Figure 2.7) rotates about an imaginary **axis** that runs through its center and exits

through its **poles.** If you place your fingers on the opposite sides of a basketball and spin it, they become the poles of rotation. We discriminate between the two poles by calling the upper one north and the other south, corresponding to directions used by people living on the planet. The **equator** of the Earth is a circle that is everywhere equidistant between the poles. The equator is a special kind of circle called a **great circle,** one whose center is the same as the center of the sphere and that divides the sphere into two equal hemispheres. (The shortest distance between two points on Earth is always given by the great circle that connects them.) The Earth spins counterclockwise from the point of view of an observer located above the north pole. If you were standing there you would spin to your left, a motion that makes the stars appear to move to your right.

On the surface of the Earth, rotational motion must be parallel to the equator. Since this circle has the largest possible radius, a person there must move the fastest, at a speed of 1,670 km/hr, to make the full circuit in a day. No one is aware of this great velocity because everything, including

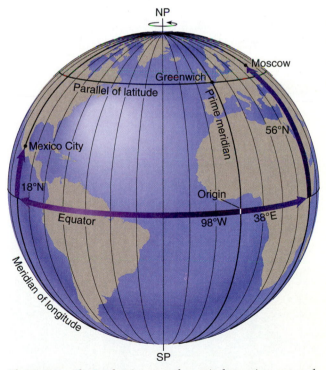

Figure 2.7   *The Earth spins around an axis that projects outward from the north and south poles (NP and SP). The equator is a great circle equidistant from both poles. Meridians of longitude run perpendicular to the equator through the poles. The prime meridian passes through Greenwich, England, and strikes the equator at the origin of longitude. Longitude is measured east and west of Greenwich along the equator, and latitude is measured along a meridian north and south of the equator.*

the air, moves at the same speed. The rotation of the Earth is responsible for its oblate shape because the greater speed at the equator slightly forces the Earth outward to a larger diameter.

2.4
Latitude and Longitude

Where are you? If you want to go somewhere else, how do you find your way? Simple, you say; "I use a map." But how did the mapmaker know where to place cities and countries? And what would you do if you were at sea? You need a way to locate position—otherwise, travel would be difficult and dangerous.

Position on Earth is given by a **coordinate system,** a grid that overlays the Earth's surface (Figure 2.7). The equator is the fundamental great circle of the system. Perpendicular to that we draw a series of secondary great circles (or semicircles) called **meridians** that run through the poles perpendicular to the equator. North and south are always directed along meridians, east and west perpendicular to them. One is selected as a **prime meridian.** Its intersection with the equator provides the *origin* or zero point of the coordinate system. Because England was the greatest sea power on Earth when this system was finalized, the prime meridian was

passed through the Royal Observatory at Greenwich in the outskirts of London (Figure 2.8). The origin then falls in the Atlantic Ocean off the coast of Africa.

Terrestrial position is measured in terms of angles: degrees, minutes, and seconds of arc (Math-Help 2.1). Select a place in Figure 2.7, perhaps Mexico City. Draw a meridian through it and the poles, and note where it crosses the equator. The arc between that intersection and the zero point is called **longitude,** denoted by the lower-case Greek letter lambda, λ (the Greek alphabet is given in Table 2.1). Longitude is measured from 0° to 180° east and west of the prime meridian. For Mexico City, the arc on the equator cut off by the two meridians is somewhat greater than a right angle (90°), and is measured at 98° W. Then measure the arc from the equator to Mexico City along its meridian to find the **latitude,** ϕ. Latitude is always measured north and south of the equator (respectively considered positive and negative) and, in this example, is 18° N. Circles around the globe of constant latitude are parallel to one another and are therefore called *parallels of latitude.* East and west are always measured along such parallels. All points on Earth can be specified by giving a longitude east or west of Greenwich and a latitude north or south of the equator. Meridians of longitude all converge to a point at the poles, where longitude disappears. The locations of the poles are specified merely by their latitudes of 90° N and 90° S.

If the Earth were a perfect sphere, the length of a degree of latitude would be constant from the equator to the poles. However, because of terrestrial oblateness, a degree of latitude is shorter at lower latitudes than at higher ones. Measurements of the lengths of degrees of latitude at various points on the globe made about 1735 established that the Earth is a good approximation to an oblate spheroid.

2.5
The Celestial Sphere

Now turn your eyes from the Earth to the sky. Ignore their different distances, and imagine the Moon, Sun, planets, and stars to be affixed to a great **celestial sphere** with the Earth at the center (Figure 2.10). During the day the celestial sphere is simply the apparent inverted bowl of the blue sky. Coordinates defined on it allow astronomers to locate the stars, and the navigator and the mapmaker to determine latitude and longitude.

Figure 2.8   *A brass strip in front of the Old Royal Observatory at Greenwich locates the prime meridian, which is defined by a telescope inside the building.*

MATHHELP **2.1** **Angles and Arcs**

Distances or separations are commonly measured in terms of some metric unit or, in astronomy, light-years or parsecs. However, astronomers also specify separations between two bodies in terms of *angles*. An angle is measured at the point of intersection of two lines in degrees (°) from 0 to 360 (Figure 2.9a). Each degree is subdivided into 60 parts called minutes (1° = 60') and each minute into 60 seconds (1' = 60"). The terms must not be confused with minutes and seconds of time measurement, which are specified by m and s.

The angle can also be specified by the *arc* of the surrounding circle isolated by the two lines. The length of the arc in degrees is the fraction of the circle isolated by the lines times 360°. If the physical length of the arc is L meters and the circumference of the circle is C meters, then the arc in degrees is 360° L/C. Arc and angular measurement are numerically identical. Although the physical length of the arc depends on the radius of the circle, the length of the arc in degrees does not.

The **angular diameter** of an astronomical body is the angle formed by lines that project from the eye to either side of the body (Figure 2.9b). The body is then said to *subtend* that angle or arc.

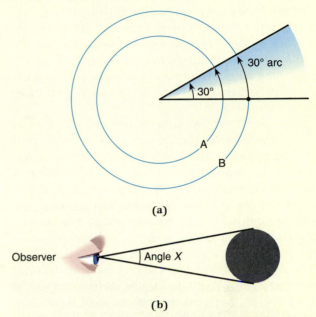

(a)

(b)

Figure 2.9 **(a)** *Angles are formed at the intersection of two lines and are measured in degrees, the whole circle being 360°. In the example shown, the lines come together at 30°. The lines intercept arcs on circles that are drawn around the point of intersection. The arcs have the same length in degrees as the angle at the center independent of the circle's size.* **(b)** *A body is seen to subtend an angle x of 20°.*

TABLE 2.1
The Greek Alphabet[a]

Letter	Upper case	Lower case	Letter	Upper case	Lower case
alpha	A	α	nu	N	ν
beta	B	β	xi	Ξ	ξ
gamma	Γ	γ	omicron	O	o
delta	Δ	δ	pi	Π	π
epsilon	E	ε	rho	P	ρ
zeta	Z	ζ	sigma	Σ	σ
eta	H	η	tau	T	τ
theta	Θ	θ	upsilon	Υ	υ
iota	I	ι	phi	Φ	ϕ
kappa	K	κ	chi	X	χ
lambda	Λ	λ	psi	ψ	Ψ
mu	M	μ	omega	Ω	ω

[a]Greek letters are commonly employed to denote scientific quantities and are used in the naming of stars.

2.5.1 Points and Circles on the Sphere

Assume you are observing from the northern United States, at a latitude of 45°N, maybe from Minneapolis, Minnesota, or Portland, Oregon. Because gravity pulls you downward toward the center of the Earth, wherever you are you feel that you are standing on top of our planet. Since you are actually standing halfway between the north pole and the equator, the Earth's axis must then be drawn tilted by the same 45°.

The point on the sphere directly above your head is the **zenith** and that directly beneath your feet is the **nadir.** The line between the two passes through the center of the Earth. Extend a plane at your feet outward in all directions perpendicular to the line to the zenith. This plane will intersect the celestial sphere at the **astronomical horizon,** a great circle that divides the sky into visible and invisible hemispheres and is approximated by the visible horizon at sea where there are no local obstructions.

The rotating Earth defines its own references in the sky. In Figure 2.10, extend the Earth's rotational axis outward through the north and south poles. It will pierce the celestial sphere at the **north celestial pole (NCP)** and the **south celestial pole**

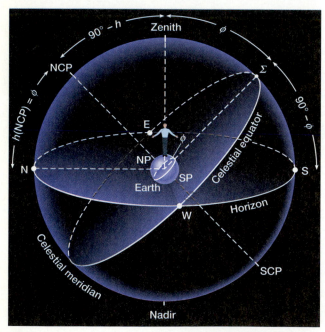

Figure 2.10 *The celestial sphere surrounds the Earth. NP and SP are the north and south poles of the Earth. The zenith is over the observer's head, the nadir opposite; the horizon is 90° from these two points. The axis of the Earth intersects the sphere at the north and south celestial poles (NCP and SCP). The celestial meridian runs through the poles, zenith, and nadir, and the celestial equator is parallel to the Earth's equator. North (N), south (S), east (E), west (W), and the equator point (Σ) are defined by the intersections of the three circles. The observer is at 45° N latitude (φ).*

(SCP). Next, extend the plane of the Earth's equator outward into space. It will slice through the celestial sphere at the **celestial equator,** another great circle that is everywhere 90° from the NCP and SCP. Just as the Earth's equator splits the planet into two hemispheres around the north and south poles, the celestial equator divides the heavens into equal northern and southern hemispheres centered on the NCP and SCP. Since the NCP is above the horizon (at your latitude of 45°N), it is visible, but the SCP, opposite the NCP and below the horizon, is not. However, you can still see a considerable portion of the southern celestial hemisphere. Finally, your **celestial meridian** is defined as the great circle that passes through the NCP, SCP, the zenith, and the nadir. It splits the sky into its eastern and western hemispheres. The intersections between the horizon and the celestial meridian define the directions north and south for the observer and those between the horizon and the celestial equator east and west.

2.5.2 The Orientation of the Celestial Sphere

In Figure 2.10, the celestial equator climbs the sky and, at its highest arch, intersects the celestial meridian at the equator point, Σ. Two lines emerge from the center of the Earth. One goes through the observer to the zenith and the other passes through the equator and strikes the sky at the equator point. They intersect at an angle ϕ equal to the observer's latitude. These lines define an arc on the celestial meridian between the zenith and the equator point that is also equal to ϕ. Arcs measured straight downward from the zenith to a point in the sky are called **zenith distances,** z. The observer's latitude is then the zenith distance of the equator point, or

$$\phi = z(\text{equator point}).$$

Since the celestial equator is 90° from the NCP, the zenith distance of the NCP must be $(90° - \phi)$. However, the zenith distance of the horizon is also 90°. As a result, the arc from the horizon to the NCP is equal to $90° - (90° - \phi)$ or just the latitude, ϕ. Arcs in the sky measured upward perpendicular to the horizon are called **altitudes,** h, where $h = 90° - z$. Therefore, the altitude of the pole is equal to the latitude of the observer, or

$$h(\text{NCP}) = \phi.$$

We thus identify the first rule of *celestial navigation,* the finding of our way by the stars. Locate the celestial pole and you know your latitude.

Observers in the northern hemisphere are fortunate to have a modestly bright star, called Polaris

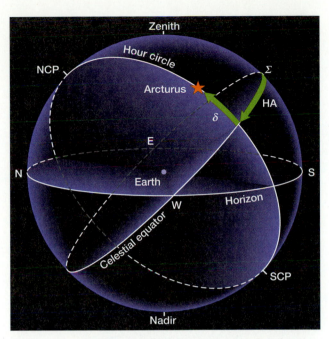

Figure 2.11 *Equatorial coordinates are defined by the celestial equator. An hour circle runs from the NCP through the star Arcturus to the SCP. The hour angle (HA) is the arc measured westward along the equator from the equator point (Σ) to the intersection of the hour circle and the equator, here 45° or 3^h. Declination (δ) is measured along the hour circle from the equator to the star. For Arcturus, it is 19°N.*

or the North Star, that is very close to the NCP. To locate it for the first time, find your latitude from an atlas, then face north and look up by that many degrees. Once you have the star, point one arm toward it, place the other 90° away, and trace out the celestial equator.

2.5.3 Locating the Stars

To find stars, astronomers set a variety of coordinate systems into the sky. They can then assign each star a position and make celestial maps. For example, a star can be located by its altitude, h (as we did the NCP), and by its *azimuth,* the direction along the horizon you have to face to see it (measured in degrees from north through east). For most purposes, however, it is better to use an **equatorial system** that is parallel to latitude and longitude, one based on the celestial equator. A bright star called Arcturus is placed in Figure 2.11. Run a great semicircle from the NCP through Arcturus to the SCP and note where it crosses the celestial equator. Such a circle is similar in concept to a meridian of longitude, but in the sky is called an **hour circle.** We now define two arcs. One, the **hour angle,** HA, is measured from the equator point westward to the intersection of the hour circle and the celestial equator. The other arc, the **declination,** δ, is mea-

sured from the celestial equator along the hour circle to the star. Like latitude, it is expressed in degrees north or south of the equator, respectively considered positive and negative.

The hour angle of the celestial meridian south of the observer in Figure 2.11 is 0°, and the hour angles of the west and east points of the horizon are 90° and 270°, respectively. The declination of the celestial equator is 0°, and the declinations of the NCP and SCP are respectively 90°N and 90°S. From Figure 2.10, the zenith distance of the equator point equals the latitude. This same arc, measured in the other direction, is also the declination of the zenith, so

$$\phi = z(\text{equator point}) = h(\text{NCP}) = \delta(\text{zenith}).$$

The hour circle through Arcturus in Figure 2.11 hits the equator halfway between the equator point and the west point of the horizon, so the hour angle must be 45°. The star's declination, δ, is 19°N.

Hour angle is not normally measured in degrees but in time units. Instead of dividing the circle into 360 equal parts, we split it into 24 equal units called *hours,* h (Figure 2.12). This "hour" is actually an angle, the equivalent of 360°/24, or 15°. As examples, the hour angle of the west point of the horizon is 90°/15 or 6^h and that of the east point is 18^h. The hour angle of the equator point is 0^h or 24^h and that of Arcturus as placed in Figure 2.10 is 3^h. The hour is subdivided into 60 *minutes* (just like a clock), symbolized by m, and each of these into 60 *seconds,* s. Unfortunately, the words "minutes" and "seconds" also refer to the subdivisions of the degree denoted by ' and " (see MathHelp 2.1). They are discriminated by context and by their symbols. One hour of time, 60^m, equals 15°, so 1° corresponds to 4^m. But 1° = 60' = 4^m, so 1^m = 15' (Figure 2.12 and Table 2.2).

2.6
Motions in the Sky

The Sun and stars appear to rise in the east and set in the west in response to the Earth's rotation. However, the apparent motions depend upon both declination and latitude. Some stars are visible all the time and others are never seen.

2.6.1 Daily Paths

As the Earth turns counterclockwise, or eastward (as viewed from above the north pole), the stars seem to move oppositely, clockwise, to the west. Your motion on Earth is parallel to the terrestrial

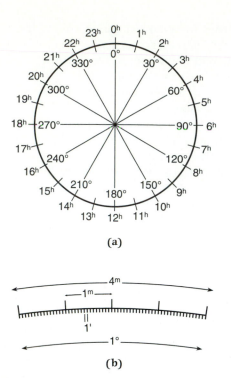

(a)

(b)

Figure 2.12 (a) *The circle is divided into 360 degrees or 24 hours; one hour equals 15°.* (b) *An arc with a length of 1° is divided into 60 minutes of arc, or 60'. Time units are shown along the top of the arc. If 15° = 1^h, then 1° = 4^m, and 1^m = 15'.*

TABLE 2.2
Conversion Between Arc and Time Units

$$1^h = 15° = 60^m$$
$$1° = 4^m = 60'$$
$$1^m = 15' = 60^s$$
$$1' = 4^s = 60''$$
$$1^s = 15''$$
$$1'' = 1/15^s = 0.0667^s$$

equator and, as a result, the stars seem to trace out **daily paths** (Figure 2.13) that are parallel to the celestial equator and have positions dependent on declination. Because the celestial equator is tilted relative to the horizon, so are the daily paths.

The daily paths show the convenience of the equatorial system. Stellar altitudes and azimuths change with terrestrial rotation in complex ways, but declinations remain constant. Moreover, hour angles change smoothly with time. For example, the Sun will take a full day, 24 hours, to make a complete circuit of its daily path, during which it will move through 360° or 24 hours of hour angle. It therefore moves at a rate of 15° or 1^h of angle per

hour of time, or 1° for every 4 minutes of time. If the Sun has an hour angle of 3^h, then one hour later the hour angle will be 4^h. We can therefore use a clock as a way of telling where to find a celestial object.

A star exactly on the celestial equator (Figure 2.14) will rise precisely at the east point and set exactly at the west. It will then cross or **transit** the meridian at a zenith distance equal to the latitude. However, if a star is north of the equator, like Arcturus, it will rise in the northeast (that part of the horizon between north and east), transit higher, and set in the northwest. The greater a star's declination (for example, Vega), the more toward the north are its rising and setting points. Stars progressively south of the celestial equator (Spica and Antares) transit ever lower and rise and set more and more in the southeast and southwest. Go outside and watch.

2.6.2 Circumpolar Stars

As we look toward more northerly declinations, the stars finally become so close to the NCP that they cannot reach the horizon and cannot set, like the star Mizar in Figure 2.14. Such stars are called **circumpolar,** and in typical northern latitudes they include many famous sky figures such as the Big and Little Dippers. If not for daylight, these stars would be perpetually visible as they go around the pole (Figure 2.15). Since the celestial pole's altitude is equal to the observer's latitude, any star that is closer to the pole than the latitude, or that has a declination greater than 90° minus the latitude (90° − ϕ), must be circumpolar.

Conversely, for a northern observer there is also a zone around the SCP in which the stars never rise. For example, Achernar in Figure 2.14 is too close to the southern pole to make it to the horizon. As a result, Chicagoans, New Yorkers, and San Franciscans can never see the famous Southern Cross from their hometowns. From Figure 2.10, the declination of the south point of the horizon is (90° − ϕ); therefore, any star with a declination more southerly than (90° − ϕ) cannot ever rise and must remain perpetually unseen.

2.6.3 The Traveling Observer

So far, the sky has been viewed from a fixed latitude of 45° N (Figure 2.16a). Now go toward the north pole, following the sequence of Figures 2.16a through 2.16c. Since the altitude of the NCP equals the observer's latitude, the north celestial pole climbs the sky. Daily paths also begin to flatten out,

Figure 2.13   *The daily paths of stars are seen in a time exposure. The photographer set the camera and allowed the stars to trail across the film as the Earth rotated.*

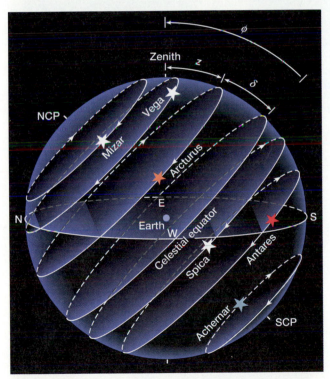

Figure 2.14   *Daily paths of stars are parallel to the celestial equator, and the motions are always to the west. A star on the celestial equator will rise exactly in the east and set exactly in the west. Arcturus and Vega, north of the equator, rise and set closer to the north point, whereas Spica and Antares, which are south of the equator, rise and set more toward the south point. For the observer at 45° N latitude, Mizar is so far north that it cannot reach the horizon and is circumpolar; Achernar is so far south that it cannot come up above the horizon and remains invisible.*

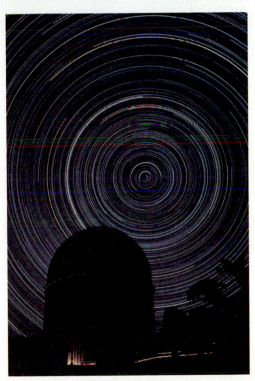

Figure 2.15   *Circumpolar stars tread their daily paths around the south celestial pole. The true pole is at the point of zero rotation.*

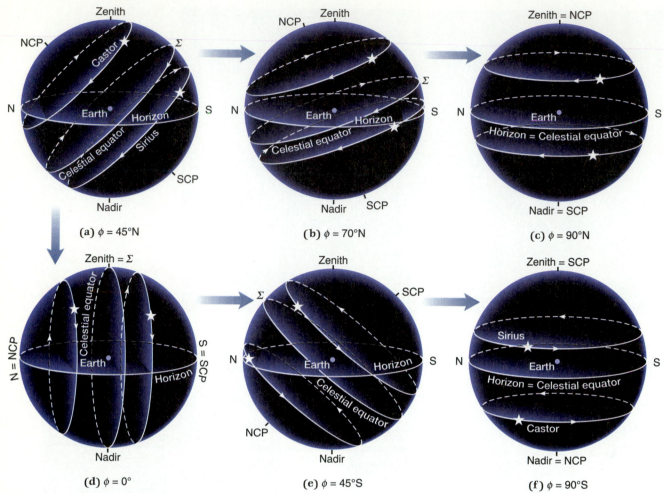

(a) $\phi = 45°$N   **(b)** $\phi = 70°$N   **(c)** $\phi = 90°$N

(d) $\phi = 0°$ **(e)** $\phi = 45°$S   **(f)** $\phi = 90°$S

Figure 2.16   *The sky seen from a series of different latitudes. In the sequence (a), (b), (c) you walk from 45° N to the north pole and watch the daily paths of Castor and Sirius flatten out and their rising and setting points move more toward the north and south points. At the pole, the NCP is in the zenith, the equator is on the horizon, the daily paths are parallel to the ground, and everything in the northern hemisphere is circumpolar. In the sequence (a), (d), (e), (f) you travel past the equator to the south pole and watch the daily paths first become perpendicular to the horizon and then tilt the other way. At the south pole, the SCP is in your zenith, all the southern stars are circumpolar, and they now move to your left rather than to your right.*

as exemplified by those of Castor (δ = 32°N) and Sirius (δ = 17°S). At 70°N latitude (Figure 2.16b), all stars north of declination +20° are circumpolar and we have lost all those south of –20°. When we finally stand at the north pole, the NCP is in the zenith and the celestial equator falls exactly on the horizon. The entire northern hemisphere of the sky is circumpolar and none of the southern is visible. Since daily paths are parallel to the equator, they are also parallel to the horizon. All stars move perpetually to the right, none ever rising or setting.

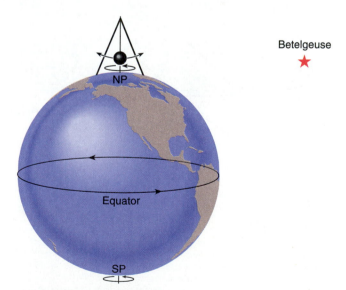

Figure 2.17   *A Foucault pendulum is erected at the north pole and swung toward the star Betelgeuse. As the Earth turns under it, the pendulum's plane stays fixed in space, always directed at the star moving on its daily path. To the observer on Earth, it is the plane of the pendulum that seems to turn.*

Next, travel south from 45°N by following the sequence of Figures 2.16a, d, e, and f. When you arrive at the Earth's equator (Figure 2.16d), latitude 0°, the NCP must lie on the horizon and have an altitude of 0°. If the NCP is on the northern horizon, the SCP must be on the southern. Here, none of the heavens is circumpolar. Everything rises and sets perpendicular to the horizon and, over the course of the day, everything is visible—or would be, if not for daylight.

As you step across the equator, the NCP disappears and the SCP rises in the sky (Figure 2.16e). Now some southern stars are circumpolar and some northern ones become perpetually invisible. When you finally arrive at the Earth's south pole (Figure 2.16f), it is the SCP that is in your zenith. The celestial equator again lies on the horizon, all the stars in the southern celestial hemisphere are circumpolar, and they are now moving to your left.

2.7
Demonstrations That the Earth Spins

What evidence do we have that it is really the Earth that is spinning and not the celestial sphere? At the north pole erect a pendulum, a heavy weight mounted on a long wire fastened to a support with a nearly frictionless bearing (Figure 2.17). Swing the pendulum toward a star, perhaps bright Betelgeuse, which you will find near the horizon. As Betelgeuse marches around the sky, the pendulum moves to follow it; that is, the plane of the swing rotates to the right with a **period** (the time it takes to return to its starting position) of a day. It looks at first as if there is some magical force that is making the pendulum's swing-plane move. The simple explanation, however, is that the weight is swinging freely in space. The Earth, with you on it, is turning beneath the pendulum, a demonstration that it is we who are moving, not the sky.

The first such pendulum was constructed in 1851 by Léon Foucault at the Pantheon in Paris (Figure 2.18). At locations other than the poles, however, the pendulum's rotation period is greater than 24 hours because the swing cannot be made perpendicular to the stars' daily paths. The period depends strongly on latitude, and at 45°N it is about 34 hours. Many museums around the world have these **Foucault pendulums** where you can easily watch the rotation of the Earth.

An older demonstration involves the flight of a projectile. Stand at some northern latitude and fire a cannonball to the north (Figure 2.19). The ball will veer to the right, to the east, and will miss its target. This **Coriolis effect** (named for a nineteenth-century French engineer) is very obvious. To make a hit, you must direct your cannon a bit to

Figure 2.18 *Léon Foucault demonstrates his pendulum and the rotation of the Earth.*

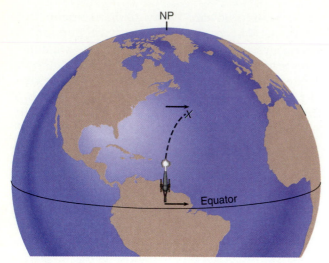

Figure 2.19 *Because of the rotation of the Earth, the ball leaving the muzzle of the cannon has a greater speed eastward than does the target to the north (marked by the X), so the cannonball seems to veer to the right.*

the left. The ability to make such a correction is one of the skills of warfare. The Coriolis effect is another consequence of the Earth's rotation. In Figure 2.19, the cannon, being farther away from the Earth's axis than the target, must be moving faster to the east. The ball then has an easterly motion relative to the point it must strike and falls to the east of it.

The Foucault pendulum and the Coriolis effect do not actually *prove* that the Earth rotates. However, rotation is the simplest and most logical theory that explains all the observations, including the daily paths of stars in the sky. There are no experiments or observations that demonstrate otherwise, and therefore we say that the Earth does indeed spin.

KEY CONCEPTS

Angular diameter: The angle formed by lines projecting to opposite sides of a body.

Axis: A line about which a body rotates or moves.

Altitude (*h*): The angular elevation of a body above the horizon.

Celestial equator: The great circle on the celestial sphere above the Earth's equator equidistant from the celestial poles.

Celestial meridian: The great circle through the celestial poles, the zenith, and the nadir.

Celestial sphere: The apparent sphere of the sky.

Circumpolar stars: Stars that do not set on their daily paths.

Coordinate system: A grid for measurement of position.

Coriolis effect: Motion of a body caused by the Earth's spin.

Daily paths: The apparent paths taken by celestial bodies as the Earth rotates.

Declination (δ): Arc measurement along an hour circle north or south of the celestial equator.

Equator (terrestrial): The great circle on the Earth that is equidistant from the poles.

Equator point: The intersection of the celestial equator and the celestial meridian.

Foucault pendulum: A swinging weight that demonstrates the Earth's rotation.

Great circle: A circle on a sphere whose center is coincident with the center of the sphere.

Horizon: The line where the land seems to meet the sky; the **astronomical horizon** is the great circle defined by the intersection between the celestial sphere and a plane at the observer's feet perpendicular to the line to the zenith.

Hour angle (HA): The arc along the celestial equator measured westward from the equator point to an hour circle.

Hour circle: A great semicircle that runs between the celestial poles through a celestial body.

Latitude: Arc measurement north and south of the equator.

Longitude: Arc measurement on the equator east and west of Greenwich.

Meridians (of longitude): Great semicircles that run between the poles of the Earth perpendicular to the equator.

Nadir: The point on the celestial sphere beneath your feet.

North and **South Celestial Poles** (NCP and SCP): The points of rotation of the celestial sphere that lie above the north and south poles of the Earth.

Oblate spheroid: A sphere flattened at its poles.

Period: The time required for a body to go from its starting point or position and return to that point or position.

Poles (terrestrial): The points where the Earth's rotation axis emerges.

Prime meridian: The meridian that runs through Greenwich and defines the origin for longitude.

Theory: A model that embraces and explains observational or experimental data.

Transit: Passage of a body across the celestial meridian.

Zenith: The point on the celestial sphere over your head.

Zenith distance (*z*): The vertical arc between the zenith and a point in the sky.

KEY RELATIONSHIPS

Circumpolar stars:

Stars north of declination $(90° - \phi)$ in the northern hemisphere and south of $-(90° + \phi)$ in the southern.

Latitude:

$$\phi = z(\text{equator point}) = h(\text{NCP}) = \delta(\text{zenith}).$$

Zenith distance and altitude:

$$z = 90° - h; \, h = 90° - z.$$

EXERCISES

Comparisons

1. What is the difference in concept and measurement between latitude and longitude?
2. What is the difference between the zenith and the nadir? Which can you see?
3. How do the celestial poles relate to the terrestrial poles?
4. What is the difference between minutes of arc and minutes of time?

Numerical Problems

5. How many degrees are there in a 1/60 of a circle?
6. If Eratosthenes had observed a 10° change in the position of the Sun in the sky between Syene and Alexandria, what would the circumference of the Earth have to be?
7. What are the latitude and longitude of a city that is one-third the way from the Earth's equator to the north pole and one-quarter of the way around the globe to the west of the prime meridian?
8. How many kilometers are there in a degree of latitude and how many meters are there in a second of arc of latitude? (Assume a spherical Earth.)
9. What are the altitudes of stars when they are seen **(a)** halfway up the sky; **(b)** setting; **(c)** two-thirds the way from the horizon to the zenith?
10. What is **(a)** 3 hours 16 minutes in degrees; **(b)** 37° in hours and minutes of time?
11. The zenith distance of a star is 37°. What is its altitude?
12. What is the hour angle of a star that rises exactly east?
13. What is the observer's latitude when the north celestial pole is **(a)** 27° above the horizon; **(b)** 33° below the horizon?
14. If you live at a latitude of 33°N, what is the declination of the most southerly star that is circumpolar?
15. If you live at a latitude of 42°N, what is the declination of the most southerly star you can see?

Thought and Discussion

16. What observations might demonstrate the shape of a flat Earth?
17. How does an oblate spheroid differ from a sphere?
18. If you shoot a cannonball exactly south from Kansas City, in which direction would it appear to travel?
19. What is the evidence for the rotation of the Earth?
20. What is the period of a Foucault pendulum at the Earth's equator?

Research Problems

21. What are the latitudes and longitudes of **(a)** Quito, Ecuador; **(b)** Melbourne, Australia; **(c)** Sacramento, California; **(d)** Tokyo, Japan; **(e)** your hometown?
22. What major cities are close to **(a)** 17°E, 51°N; **(b)** 88°E, 22.5°N; **(c)** 58°W, 34.5°S; **(d)** 112°W, 33.5°N.

Activities

23. Draw a sphere that represents the Earth and include the equator and the north and south poles. On it place Greenwich (latitude 52°N) and label it G. Draw in the prime meridian. Then with an S estimate the location of Santiago, Chile, 70°W, 33°S, and with a B that of Beijing, China, 116°E, 40°N. Indicate any points that may be on the back of the sphere.
24. Draw and properly label celestial spheres for observers with latitudes of 60°N, 10°N, 30°S, and 90°S. On each place a star with an hour angle of 0^h and a declination of 10°N. Then show the daily paths.
25. Draw a celestial sphere with the NCP 45° above the north point of the horizon. Draw and label the horizon, celestial poles, equator, and meridian. On it place a star that is exactly in the northwest at an altitude of 45°. What are the approximate hour angle and declination of the star? (Estimate from your drawing.)
26. Draw another celestial sphere like the one in Question 25 and on it place a star with an hour angle of 2^h and a declination of 30°S. What is the approximate altitude of the star? (Estimate from your drawing.)
27. Fasten two straight sticks together at their ends with a single nail. Find Polaris in the sky and turn the sticks until one points at Polaris and the other points at the horizon below Polaris. Measure the angle with a protractor. You now have a measurement of your latitude. How close did you come to the latitude taken from an atlas?
28. Draw a sketch of your western horizon. Then over a period of two or three hours watch a setting star move on its daily path. Sketch the path and note the angle it makes with the horizon.
29. On a night when the Moon is out, estimate its altitude, azimuth, hour angle, and declination. An hour

after your first observations estimate the same quantities again and compare the results.

30. Try making your own star trails. Load a 35 mm camera with fast film and mount it on a tripod. Go to a dark site and point the camera at the north celestial pole. Take a series of exposures from a few minutes to an hour or more and process at any convenient photo shop. Look at the negatives before you have any printed. When you have correct exposure times worked out, make star-trail photographs of the celestial equator and of intermediate declinations.

Scientific Writing

31. You have been invited to give a ten-minute speech to a high school science club. Write a script in which you explain the need for scientific theory and how it relates to scientific observations.

32. A magazine for a travel club describes a trip from New York to Buenos Aires. Write a sidebar (a short one-column essay of about 200 words that relates to the main article) in which you describe the changes the voyagers will see in the sky. One of your goals is to make your readers interested in sky-watching.

3

The Earth and the Sun

The orbit of the Earth and the apparent motion of the Sun

The setting Sun
floods the sky with light.
Our lives depend on the way
it appears to move across the heavens
in response to terrestrial motion.

MATHHELP **3.1** The Ellipse

The **ellipse** (Figure 3.1), commonly encountered in astronomy, is a closed curve defined by two **foci** (plural of *focus*). Along the path of an ellipse the sum of the distances to the foci is a constant. You can easily draw an ellipse by placing two tacks on a board and tying a loose string between them. Pull the string taut with a pencil and trace out the curve. The line through the foci is called the *major axis,* and that perpendicular to it at the center is the *minor axis.* The size of the ellipse is characterized by the length of half the major axis, the **semimajor axis** (*a*). The shape, or the **eccentricity** (*e*), is the distance from the center to one of the foci divided by the length of the semimajor axis, or

$$e = CF_1/a = CF_2/a.$$

If the two foci are brought together at the center, $e = 0$, and the ellipse becomes a *circle*. If they are separated while the semiminor axis remains the same, the curve becomes more and more elongated and e approaches 1.

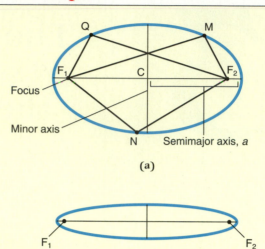

(a)

(b)

Figure 3.1   **(a)** *In any ellipse, the sum of the distances of any points along the curve to each of two focus points (F_1 and F_2) is a constant:* $MF_1 + MF_2 = QF_1 + QF_2 = NF_1 + NF_2$ *and so on. The semimajor axis (a) is half the major axis. The eccentricity* $e = CF_2/a = 0.84$; **(b)** *in this flatter ellipse, e = 0.95.*

The Earth exhibits a variety of motions, of which daily rotation is the most obvious. Of equal importance is the steady annual movement, the revolution, of the Earth around the Sun, which causes the changes of the seasons and gives great variety to human life.

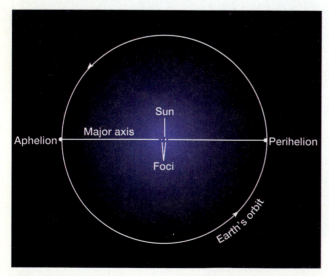

Figure 3.2   *The elliptical orbit of the Earth and the Sun are drawn to scale and the Sun placed at the right-hand focus. The center of the orbit is between the two foci. The Earth is closest to the Sun at perihelion and farthest at aphelion. (The dot representing the Sun is drawn to the correct scale.)*

3.1
The Apparent Path of the Sun

The Earth revolves around the Sun in the same direction as its rotation, that is, counterclockwise from the point of view of an observer looking down from above the north pole. (*Revolution* means one body going about another; *rotation* means *spin.*) Its period—the time it takes to return to its starting place—is a full year, during which an observer counts $365\frac{1}{4}$ (365.2422 . . .) days. The orbit is not circular, but has the shape of an *ellipse* (MathHelp 3.1). The astronomical unit, earlier described as the distance between the Earth and the Sun, is actually the orbital semimajor axis (Figure 3.2) of 149.6 million km.

The Sun is not located at the center of the ellipse but at a focus, and as a result, the distance between the Earth and Sun is continuously changing. However, because the eccentricity is only 0.017, the Sun is just 2.5 million km from the center of the ellipse and the variation is small. When the Earth and Sun are closest, at a point in the orbit called **perihelion** (from the Greek *peri,* "closest," and *helios,* "Sun"), the solar distance is 147.1 million km, and at its farthest, at **aphelion,** 152.1 million.

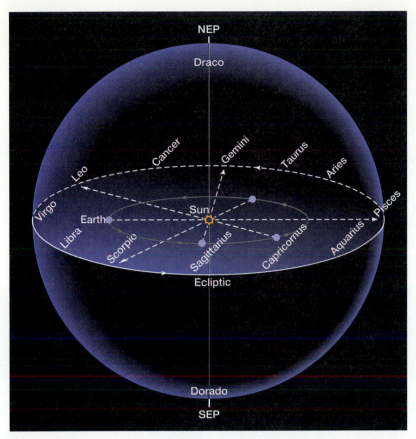

Figure 3.3   *The Earth's orbit is set into the celestial sphere. As the Earth moves counterclockwise around the Sun (symbolized by ✿), the Sun appears to move counterclockwise around the Earth, following the ecliptic, a great circle through the stars that passes through the constellations of the zodiac. The perpendicular to the ecliptic defines the north and south ecliptic poles, NEP and SEP, which are respectively in the constellations Draco and Dorado. At night, when you are facing away from the Sun, you see the stars in the opposite direction.*

TABLE 3.1
Solar and Vernal Equinox Passage Through the Constellations of the Zodiac

Constellation	Dates of Current Solar Passage	Years of Vernal Equinox Containment[a]
Aries	April 19–May 14	2000 B.C.–100 B.C.
Taurus	May 15–June 20	4500 B.C.–2000 B.C.
Gemini	June 21–July 20	6600 B.C.–4500 B.C.
Cancer	July 21–August 10	8100 B.C.–6600 B.C.
Leo	August 11–September 16	10800 B.C.–8100 B.C.
Virgo	September 17–October 30	A.D. 12000–A.D. 15300
Libra	November 1–November 23	A.D. 10300–A.D. 12000
Scorpius[b]	November 24–December 17	A.D. 8600–A.D. 10300
Sagittarius	December 18–January 19	A.D. 6300–A.D. 8600
Capricornus	January 20–February 15	A.D. 4400–A.D. 6300
Aquarius	February 16–March 11	A.D. 2700–A.D. 4400
Pisces	March 12–April 18	100 B.C.–A.D. 2700

[a]Discussed in Section 3.7.

[b]Scorpius is combined here with the nonzodiacal constellation Ophiuchus; the Sun passes through Ophiuchus's modern boundaries between November 30 and December 17.

The Earth's average orbital speed (the orbital circumference divided by the period in seconds) is 29.8 km/s, but we are unaware of the motion because everything is moving together. Instead, it looks as if the Sun is orbiting the Earth in a plane that annually takes the Sun counterclockwise—to the east—across a background of stars (invisible in daylight) at an average rate of just under 1° per day (Figure 3.3). If you walk around a chair placed in the middle of the room, it will appear to change its position in relation to the walls and will seem to be revolving around *you.* The apparent annual path of the Sun is a great circle on the celestial sphere called the **ecliptic** (from *eclipse*, because that is where they take place). The ecliptic passes through 12 ancient star patterns or constellations collectively called the **zodiac** (Table 3.1). Near this path we also find the Moon and planets.

The Earth has an orbital axis (see Figure 3.3) that is perpendicular to the orbital plane and points to the **north** and **south ecliptic poles** (NEP and SEP, respectively). The rotational axis does not align with it but is tipped by 23.5° (23° 27'), represented by ε (Figure 3.4). As a result, the ecliptic poles are displaced from the celestial poles by 23.5°. The ecliptic is inclined to the celestial equator by the same angle (Figure 3.5), which is therefore called the **obliquity of the ecliptic.** If ε were zero, and the rotational axis in Figure 3.4 were straight up, the Sun would always ride along the celestial equator as the year progresses. But because of the tilt the Sun can be seen as far as 23.5° below the celestial equator (as on the left in Figure 3.4) or 23.5° above it (as on the right). As the Sun glides along its apparent path it must move alternately north of the celestial equator and then south of it, between declinations 23.5°N and 23.5°S, crossing the equator twice a year at declination 0°.

3.2
The Seasons

Four specific points on the ecliptic (Figure 3.5) define our *seasons*, periods of the year related to temperature, weather, and climate. The Sun crosses the celestial equator on its way north on March 20 or 21 at the **vernal equinox** (symbolized by ♈, the zodiacal sign for Aries) to define the first day of northern-hemisphere *spring*. The vernal equinox is located in the direction of the constellation Pisces. When the Sun is at the vernal equinox, its daily path is along the celestial equator, and it must rise and set exactly east and west (Figure 3.6). Days and nights are then approximately equal, both about 12 hours long. (We ignore the half-degree

angular diameter of the Sun, which slightly extends daylight beyond what it would be if the Sun were a point, and *twilight,* the period of time before sunrise or after sunset when the Earth's atmosphere, bright from sunlight, illuminates the ground.)

As the Sun moves farther north along the ecliptic, it transits the meridian higher in the sky and rises and sets progressively more toward the north. It is therefore above the horizon for a longer period of time. About June 21, after a quarter of an orbit, the Sun reaches its most northerly declination (23.5°N) at a point called the **summer solstice** (in Gemini), which marks the beginning of northern-hemisphere *summer* (the right-hand Earth in Figure 3.4). The Sun now transits as high and rises and sets as far north as possible, as seen from the daily

Figure 3.4 *The Earth's rotational axis is tilted by 23.5° relative to the perpendicular to its orbit. The celestial poles are then directed 23.5° away from the ecliptic poles. On the usual first day of northern winter on December 22 (shown on the left), the Sun (☼) is 23.5° S of the celestial equator, and appears low in the sky from the Earth's northern hemisphere and high from the southern. Six months later (June 21, the usual first day of northern summer, shown on the right), the Earth has moved through half its orbit. The axis maintains the same orientation in space, the Sun is 23.5° N of the celestial equator, and appears high in the sky from the Earth's northern hemisphere and low from the southern.*

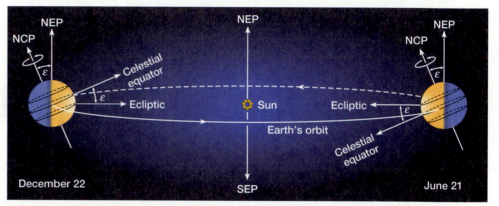

Figure 3.5 *The apparent path of the Sun, the ecliptic, is inclined by ε = 23.5° to the celestial equator. From the northern hemisphere, the Sun appears to move counterclockwise, crossing the ecliptic at the equinoxes, the vernal where it goes north and the autumnal where it passes south. Its farthest point north of the equator is at the summer solstice, declination 23.5° N, and the farthest point south is at the winter solstice, δ = 23.5° S.*

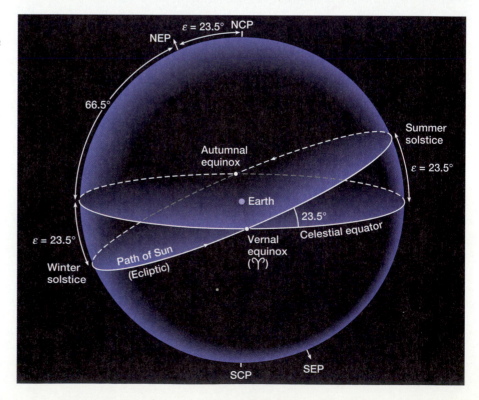

path of the summer solstice in Figure 3.6. Daylight hours are now at a maximum.

From the summer solstice, the Sun must move to the south. It crosses the celestial equator again at the **autumnal equinox,** in the constellation Virgo, about September 23, marking the beginning of northern-hemisphere *autumn.* Once again it rides the celestial equator and rises and sets due east and west. It then plunges into the southern hemisphere, transiting lower, rising and setting progressively more toward the south point, and coming up later and going down earlier. Finally, the Sun bottoms out at the **winter solstice** in Sagittarius about December 22 (the left-hand Earth in Figure 3.4), 23.5° south declination, and on this first day of northern-hemisphere *winter* begins to move back north. The continuous change of the solar declination with time is shown in Figure 3.7.

As you move closer to a fire you get warmer, and if farther away, colder. It is therefore commonly assumed that the change of seasons is caused by the varying distance between the Earth and Sun as the Earth moves on its elliptical orbit. However, the Earth passes perihelion about January 2, in the dead of northern hemisphere winter, and aphelion about July 4 during summer's heat. Moreover, if the varying solar distance were the cause of seasonal change, both the northern and southern hemispheres of the Earth would suffer summer or

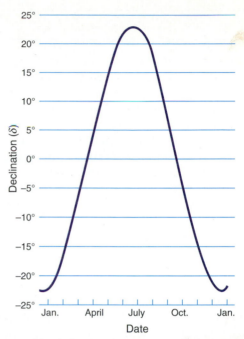

Figure 3.7   *Solar declination changes continuously with time. It is 0° when the Sun crosses the equinoxes on March 20 and September 23, 23.5° N at the summer solstice on June 21, and 23.5° S when the Sun arrives at the winter solstice on December 22.*

winter at the same time, but in fact they alternate: when it is summer in the north, it is winter in the south. Distance variation has little to do with seasonal temperature and weather changes.

The real cause is the tilt of the terrestrial axis. In summer, a ray of sunlight beams down from nearly overhead (Figures 3.6 and 3.9a). In winter, however, when the Sun is low in the sky, that same beam must spread itself over more ground (Figures 3.6 and 3.9b), with the result that the temperature cannot get as high. Therefore, the weather turns cold. The effect is intensified by the shortening of the day. However, when the Sun is low in the northern hemisphere, it is high in the southern (as we saw in Figure 2.16), so it is summer in Argentina when it is winter in Nebraska. The actual times of maximum high and low temperature at our northern latitudes do not coincide with the technical beginning of summer or winter but occur about a month later because of the time it takes the Earth's land and waters to heat and cool.

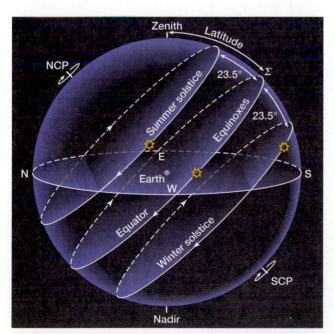

Figure 3.6   *The daily paths of the two equinoxes must be the celestial equator, so on March 20 and September 23 the Sun (☼) rises and sets exactly east and west. The daily paths of the solstices are parallel to and 23.5° north and south of the equator. On June 21, the Sun must rise and set well north of east and west. On December 21, the rising and setting points are to the south.*

3.3
Tropics, the Arctic, and the Antarctic

Because your latitude equals the declination of the zenith, and the Sun's maximum declination is 23.5° N, you can never see the Sun overhead at 45° N latitude (see Figure 3.6). On that parallel, on

BACKGROUND 3.1 Ancient Monuments to the Sun

On the Salisbury Plain of southern England is an extraordinary ring of massive stones and other structures called *Stonehenge* (Figure 3.8a) that was built between 3000 and 1000 B.C. Its chief characteristic is a circular ring of vertical stones about 3 meters high placed in pairs that are topped by capstones. Outside the ring to the northeast is another, smaller, stone called the heelstone. From the center of the structure, the summer solstice Sun rises exactly above this rock, demonstrating that Stonehenge was built at least in part as some sort of calendrical device, marking the day summer begins. Other lines of sight involve the Moon.

Stonehenge is only one of many such constructions scattered about the world. The Incas of Peru established astronomical sightlines near Cuzco, and the Mayans of Mexico built an observatory a thousand years ago that allowed their priests to place the position of the rising equinox Sun and that contains sightlines for the rising of the planet Venus. Closer to home is the 400-year-old Big Horn Medicine Wheel in the Rocky Mountains of Wyoming (Figure 3.8b), built by Native Americans. This construction directs the observer to the rising and setting of the summer solstice Sun and to the risings of several stars.

Clearly, it was important to people in earlier cultures to know the sky. Positions of celestial objects were used to tell the dates on which planting and harvesting should take place. The sky served as a clock, as a calendar, and as you will see in the next chapter, as a storyboard. The knowledge and engineering capability of these and many other cultures were remarkable indeed. As impressive as our scientific feats are today, we must remember that they are rooted deeply in the past.

(a)

Figure 3.8 (a) *Stonehenge, in southern England, was used as a calendar; alignments of its stones mark the rising of the Sun on Midsummer Day, June 21. Other alignments may have been used to foretell lunar eclipses.* (b) *The Big Horn Medicine Wheel, constructed by Native Americans in Wyoming, is also astronomically aligned.*

(b)

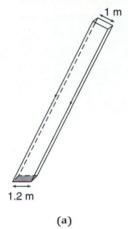
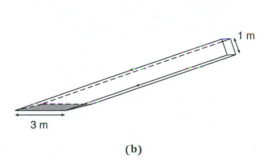

Figure 3.9 **(a)** *Shadows are short as a square shaft of sunlight 1 m across shines nearly straight down and illuminates a 1 × 1.2 m rectangle and an area of 1.2 square meters.* **(b)** *When it strikes at a higher angle, shadows are long, the shaft must cover a larger area (now 3 square meters), and the ground cannot get as warm.*

(a)   **(b)**

the first day of summer, the minimum solar zenith distance equals 45° – 23.5° = 21.5°. As you walk south, the summer solstice rises in the sky and the summer Sun transits ever higher. Eventually, when you stand at latitude 23.5°N (Figure 3.10a), where $\phi = \delta$ (summer solstice), perhaps near the southern tip of Baja California in Mexico, it transits your zenith. This special parallel of latitude is called the **tropic of Cancer** because, when the name was assigned, the vernal equinox was in that constellation. Farther south, the Sun will pass overhead twice during the year, once on its way north and then on its way south. When you reach the equator, $\phi = 0°$ in Figure 3.10b (maybe near Quito, Ecuador), the Sun transits the zenith at the time of its equinox passages. Here, solar daily paths are perpendicular to the horizon, and the days and nights are always 12 hours long no matter what the time of year. At 23.5°S latitude, the Sun will transit the zenith only when it reaches the winter solstice, and once below it, can no longer be overhead. This parallel is the **tropic of Capricorn.** The lands between the two limits, loosely called the tropics,

are the hottest places on Earth because sunlight always shines nearly straight down and seasonal changes are minimized.

Now return to 45°N latitude and walk north. The Sun transits ever lower and the weather gets cooler. Eventually, you arrive at latitude 90° – 23.5° = 66.5°N (Figure 3.10c), possibly Fort Yukon, Alaska, north of Fairbanks. On the first day of summer the Sun is 23.5° from the equator and therefore 66.5° from the NCP; but the NCP is 66.5° from the horizon, so that on June 21 the Sun is circumpolar (again assuming it is a point and not a disk); that is, the Sun cannot set, as vividly shown in Figure 3.11a. Alternatively, on the first day of winter, the Sun will not rise (Figure 3.11b) but will just graze the southern horizon. Latitude 66.5° is called the **arctic circle** (from the Greek *arktos*, "bear"; the constellation known as the Great Bear passes overhead in these regions). The analogue to the arctic circle in the south is the **antarctic circle** (which encompasses most of Antarctica) at $\phi = 66.5°$S. As you proceed above the arctic circle you will encounter more days of the circumpolar Sun dur-

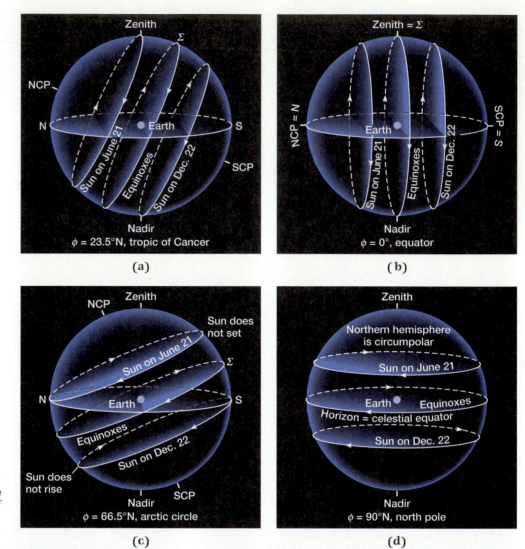

Figure 3.10 **(a)** *At the tropic of Cancer, 23.5° N, the Sun is overhead only on the first day of northern-hemisphere summer.* **(b)** *At the equator, 0°, the Sun sets vertically and passes through the zenith twice per year.* **(c)** *At the arctic circle, 66.5° N, the Sun is circumpolar on the first day of summer and technically does not rise on the first day of winter.* **(d)** *At the north pole, 90° N, the Sun is up between March 20 and September 23.*

Figure 3.11 **(a)** *The midnight Sun is real, as seen in this multiple exposure taken on June 21 north of Fairbanks, Alaska. At midnight the Sun simply passes below the NCP without grazing the horizon.* **(b)** *On the day of the winter solstice, however, the Sun just barely clears the southern horizon from just below the arctic circle. A bit farther north and the Sun would not rise at all.*

ing summer (that is, the Sun can be circumpolar at a lower declination) and a longer period of darkness during winter. Since the Sun never gets very high, it is always cold.

Finally, you arrive at the north pole in the Arctic Ocean (Figure 3.10d): the NCP is overhead, daily paths are parallel to the ground, and only the northern celestial hemisphere is seen. The Sun is now visible for six months straight, from the time of its vernal equinox passage on March 20 to its crossing of the autumnal equinox on September 23. It then sets and cannot be seen for the next six months. When it is dark at the north pole, it is daylight at the south. At these extreme latitudes the Sun never attains an altitude greater than 23.5°, the temperature rarely rises above freezing, and the poles are covered with thick sheets of ice.

3.4
Time

Our lives are largely run by the time of day. To synchronize the activities of our society we use clocks, which are set according to the positions of the Sun and the stars.

3.4.1 *Solar Time*

Time is told by the hour angles of celestial bodies. The most primitive kind of time is based on the visible Sun, which is called by astronomers the **apparent sun. Local apparent solar time** (LAST), is defined as the hour angle of the apparent sun plus 12 hours (Figure 3.12). When the hour angle is zero it is therefore 12^h or *noon*. Astronomical times are given by a 24-hour clock. On the average, the Sun sets with an hour angle of 6^h, and so the time is 18^h. It is common in the United States, however, to keep 12-hour clocks. Morning times are designated A.M., the abbreviation of the Latin *ante meridiem,* "before noon"; afternoon times are called P.M., *post meridiem.* A sunset time of 18^h then becomes 6 P.M. When the solar hour angle is 12^h, the time is 24^h or 0^h (since we start over) or *midnight.* A.M. and P.M. are never applied to noon or midnight.

Local apparent solar time can be read directly from a *sundial,* a device that consists of a stick that casts a shadow on a plate marked off in hour angle. If you have ever read time from a sundial you possibly noticed that the time did not coincide with the time on your watch. There are two reasons for this discrepancy. First, the apparent sun is not a very good timekeeper. Because the distance of the Sun from the Earth continuously changes, so does the Earth's orbital velocity: the Earth moves slightly

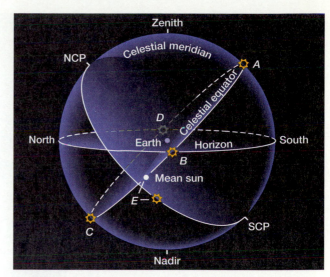

Figure 3.12 *Local apparent solar time is defined as the hour angle of the apparent sun (☼) plus 12 hours. In the five examples A, B, C, D, and E, the apparent sun has hour angles of 0^h, 6^h, 12^h, 18^h, and 8.5^h respectively, for which LAST is 12^h (noon), 18^h (6 P.M.), $24^h = 0^h$ (midnight), 6^h (6 A.M.) and 20^h30^m (8:30 P.M.). The apparent sun is positioned in E on December 7. On that date, its hour circle is 15 minutes ahead of the mean sun (symbolized by the white dot), and the apparent sun will set as early as possible.*

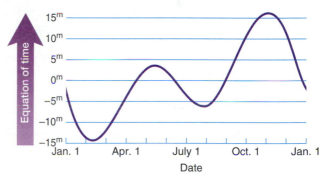

Figure 3.13 *The equation of time, E = LAST − LMST, is plotted against the date.*

slower than average at aphelion in July and slightly faster at perihelion in January. Consequently, the Sun moves at a variable rate along the ecliptic, and sundial time will alternately run fast, then slow. The effect is compounded by the tilt of the Earth's axis, since the apparent sun moves on the ecliptic but hour angles are read on the celestial equator.

To overcome this difficulty, astronomers invented a fictitious **mean** (that is, average) **sun** that moves at a constant rate along the celestial equator and that keeps pace with the average position of the hour circle through the apparent sun. We can now define **local mean solar time** (LMST), as the hour angle of the mean sun plus 12^h. The difference LAST − LMST, called the **equation of time**, E (Figure 3.13), can amount to as much as 17 minutes. The effect of the equation of time can be quite noticeable. The shortest day of

Figure 3.14 *The Sun was photographed in multiple exposures about every eight days at a specific mean solar time, allowing us to see the relation between the equation of time and declination. The figure traced out by the apparent sun, called an analemma, is often seen on terrestrial globes. The solar photographs, of very short duration, were combined with one standard exposure to record the foreground. On three days the Sun was allowed to trail on the picture to show its daily path.*

Figure 3.15 *The Sun (off the page far to the right) is on observer A's meridian (indicated by m_A) and the time is noon. Observer B, who sees meridian m_B, is 15° or one hour east of A where the solar hour angle (HA_B) is 15°, or 1 hour, west. It must then be 13^h (1 P.M.) for B.*

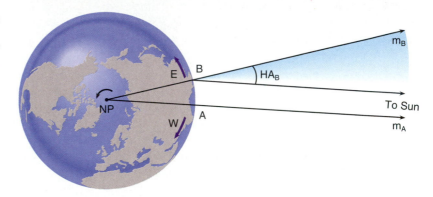

Figure 3.16 *The time zones of the world are centered on meridians of longitude that are 15° apart, though some zone boundaries deviate wildly from the basic scheme. The international date line closely follows the 180° meridian.*

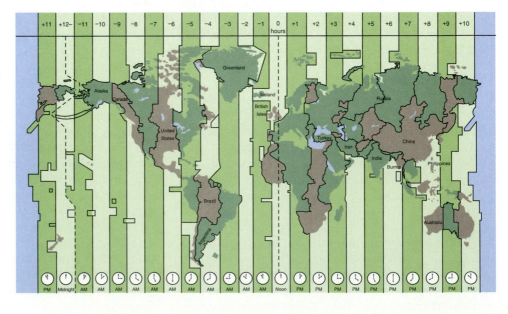

the year in the northern hemisphere is December 22, when the apparent sun passes the winter solstice. However, in early December, the apparent sun is well ahead of the mean sun (see Figure 3.12), which makes the date of earliest sunset fall around December 7. The change of the equation of time with solar declination is vividly seen photographically in Figure 3.14.

3.4.2 Longitude and Standard Time

The second reason for the difference between a sundial and your watch involves terrestrial position. If (ignoring the effect of the rotation of the Earth) you move east by a specific amount of longitude, the Sun and the stars will move to the west and increase their hour angles by exactly the same amount (Figure 3.15). Since solar time is given by the solar hour angle, changes in longitude are exactly equivalent to changes in time, explaining the use of the word "local" in defining time. The local clock of someone standing 1° to the west of you will read 4^m earlier than yours, and that of another 2° to the east will read 8^m later.

If you travel, local mean solar time is highly inconvenient. In 1883, railroads in the United States adopted Charles Dowd's invention of **standard time.** He divided the world into 24 *time zones* centered on *standard meridians* 15° apart starting at Greenwich. In principle, every location within 7.5° of a standard meridian will keep the local mean solar time of that meridian. A longitude change of 15° corresponds to 1 hour of time. Therefore, when you pass from one time zone to the next, the hour changes by 1 but the minute and the second remain the same. In the United States, the local times at 75°W, 90°W, 105°W, 120°W, 135°W, and 150°W are respectively called Eastern, Central, Mountain, Pacific, Alaskan, and Hawaiian Standard Times. In practice, the boundaries of time zones are highly distorted for social and administrative reasons (Figure 3.16), but the principle does not change.

To use a sundial to ascertain standard time, first read the local apparent solar time, then subtract the equation of time (obtained from Figure 3.13). If you are west of your standard meridian, add 4 minutes per degree (and fractions thereof) of difference; and if you are east of it, subtract. To find your local apparent solar time (and the hour angle of the apparent sun) from standard time, reverse the procedure.

The time at Greenwich, 0° longitude, is an international standard called **Universal Time** (UT). To find UT, divide your standard longitude by 15 and, in the western hemisphere, add the

result to your standard time. (If you are in the eastern hemisphere, you subtract it). Precise UT is announced continuously by radio station WWV (located in Ft. Collins, Colorado) at frequencies of 2.5, 5, 10, 15, and 20 megaHertz in the shortwave radio band and by telephone at (900) 410-TIME (there is a fifty-cent charge per call). It is also available from the Canadian shortwave station CHU at 3.3, 7.335, and 14.670 MHz.

Now we can finally navigate the Earth. Latitude is easy to find from the altitude of the pole or the declination of stars in the zenith. To find longitude, determine your local mean solar time and compare it to Universal Time:

$$\lambda(\text{west}) = \text{UT} - \text{LMST}; \lambda(\text{east}) = \text{LMST} - \text{UT}.$$

Today, all we need do to find UT is listen to the radio or make a telephone call. Before radio, however, you had to set your clock at Greenwich (or some other place of known longitude), then travel to your new location, measure the LMST, and compare the two. In the early days of world exploration, voyages could take years and the clocks were poor. As a result, longitude could not be accurately determined, maps were severely distorted, and sea travel exceedingly dangerous. Precise longitude was not readily available until the invention of the first accurate seagoing clock by the Englishman John Harrison in the middle 1700s.

3.4.3 The Calendar

We must keep track not only of hours, but also of days. Our common *calendar* is based on the number of days in the year and is therefore called a *solar calendar*. The much older Jewish calendar adds a unit based on the Moon, and the Moslem calendar is strictly lunar. Our year is 365.2422 . . . days long. If we adopt a calendar year of 365 days, the date of the vernal equinox would advance at the rate of about one day every four years. To remove this difficulty, Julius Caesar's astronomer, the Greek Sosigenes, established a calendar two millennia ago with a four-year cycle in which three years have 365 days and the fourth, a *leap year,* has 366 (added in February). Sosigenes also placed the first day of spring on March 25. However, the average number of days in this *Julian calendar* is 365.25, slightly longer than the duration of the true year, so the date of the vernal equinox creeps slowly backward. By late in the sixteenth century it had shifted to March 12, and Pope Gregory XIII ordered a calendar reform.

Gregory's astronomers slightly modified the Julian calendar by declaring that leap years were to

Our modern culture has removed us from familiarity with the night sky, making us unaware of the origins of so many familiar things. Look, for example, at your clock, which is simply a model of the apparently rotating sky. The Sun moves along its daily path from east to west, and in upper-northern latitudes from left to right, or clockwise if the celestial sphere is viewed from above the north celestial pole. Our clocks are northern-hemisphere inventions and their hands move in the same direction as the Sun. The Earth, therefore, is said by northerners to spin in the opposite direction, or counterclockwise. No matter what planet you might visit, its rotation would likely be called counterclockwise by its astronomers.

Astronomical motions are also responsible for many of the simple units used in daily life. The Earth orbits very nearly 1° per day around the Sun, which is no accident. There are 360° in the circle because the year contains $365\frac{1}{4}$ days, and 360 is the closest easily divisible number. There are 12 months in the year because it is the nearest whole number to the number of times (12.4) that the Moon orbits our planet annually. And there are probably seven days in the week to commemorate the seven moving bodies of the sky, the Sun, Moon, and the five planets known in antiquity. Even some of the names that we use are derived from the heavens: "month" comes from Moon, "Monday" is the Moon's day, and "Sunday" belongs to the Sun.

Our holidays are closely related to the passage of the Sun past the solstices and equinoxes. Christmas falls close to the winter solstice passage, its date placed there by the early church to supplant an ancient non-Christian holiday. Easter and Passover are near the time of the vernal equinox; the date for Easter is always the first Sunday past the first full Moon after the first day of spring. The Jewish high holidays, Rosh Hashanah andYom Kippur, are tied to the autumnal equinox. And Midsummer Night, the evening of the summer solstice, was a magical time in Britain (as seen from the work of Shakespeare).

In between the four major ecliptic points are the *cross-quarter days.* February 2 is Groundhog Day, which represents the approach of spring. May 1, May Day, celebrates the heart of spring. The cross-quarter is the night before, May Eve; in Scotland and Ireland it was known as Beltane, an ancient feast of fertility. The most familiar is Halloween, or All Saints' Day Eve, on October 31. August 1, Lammas Day, was a major harvest festival in what is now Great Britain, a time when loaves of bread were made from the first ripe grain. Our astronomical roots are deep indeed.

---

be skipped in century years not evenly divisible by 400. Thus in the **Gregorian calendar,** which we use today, the years 1700, 1800, and 1900 were *not* leap years despite their position in the four-year cycle, but the year 2000 will be. This scheme slightly shortens the average length of the calendar year, making it 365.2425 days long over a 400-year period—nearly perfect. At the same time, ten days were dropped from the calendar to make Spring begin on March 21, the date on which it fell in the year 325 at the time of the Council of Nicaea, a church assembly at which the rules for calculating the date of Easter were established. England and its colonies finally adopted the system in 1752.

If you could rapidly travel in an easterly direction around the world, your time would get steadily later. Eventually you would encounter midnight and enter a new day. Upon arriving back home, your calendar would read a day later than that of those you left behind. Somewhere on the trip a full day must be returned. By common agreement the change is done at the **international date line,** which closely follows 180° longitude. As you cross the date line from west to east, you go back one day (for example, Monday, August 2 becomes Sunday, August 1); if you go in the other direction, you advance one (Sunday the first becomes Monday the second). The change can be fun for eastbound travelers, who may get to celebrate a birthday or the New Year twice.

3.5
Time and the Stars

Our common day is naturally based on the life-giving Sun, and our time is geared to its position in the sky, but we could just as easily use the stars to establish a

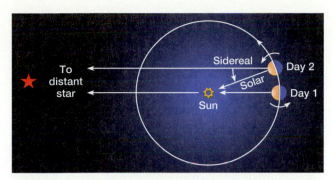

Figure 3.17 *As the Earth orbits, a given star will return to the celestial meridian more quickly than will the Sun. On Day 1 the two are lined up on the celestial meridian at noon. By Day 2 the Earth has moved in its orbit and the star returns to the celestial meridian before the Sun: the sidereal day is therefore shorter than the solar day.*

sidereal time, from the Latin *sidus,* meaning "constellation" or "star." Such a system is in fact needed both to locate the stars and to tell accurate solar time.

3.5.1 The Sidereal Day and the Motion of the Night Sky

The stars we see depend on the direction we face at night, roughly opposite the direction of the Sun (see Figure 3.3). In late June, the Sun is in Gemini, and that constellation is invisible because of daylight. When we rotate away from the Sun, we then see Sagittarius at night. As the Earth moves in its orbit and the Sun progresses along the ecliptic, we continuously face in a different direction. As a result, the appearance of the nighttime sky changes with the seasons. In December, the Sun is in Sagittarius, and Gemini appears high at midnight.

Place a star on the meridian at midnight. By tomorrow night the Earth will have moved 1/365 of its orbit, or a little less than 1°. At midnight you will be facing in a slightly different direction, and you will see the same star about 1° to the west. The following night at midnight the star will be another degree farther yet. In a month it will have moved some 30°, and in three months 90°. One degree corresponds to 4 minutes of time, and you must therefore look not quite 4 minutes earlier each night to see the same star at transit. The stars therefore appear to go around the Earth, not in 24 hours as does the Sun, but in just over 23^h56^m ($23^h56^m4.09^s$). This interval, the **sidereal day,** is the period of the Earth's rotation relative to the stars (Figure 3.17).

If the sidereal day is shorter than the solar day, there must be more sidereal days in a year. The exact nightly shift of a star to the west is $3^m56.56^s$ of hour angle, which when multiplied by the number

of days in the year (365.2422) equals one day. There are therefore 366.2422 sidereal days in a year. Place a chair in the middle of a room and consider the walls to represent the stars. Walk around a chair while facing it. Since the chair is always on your meridian there are *zero* chair-days in your year, but the walls will go about you once, giving you *one* sidereal day: the act of revolving in the same direction as rotation adds an extra day relative to the stars.

3.5.2 Sidereal Time

Just as we can divide the solar day into 24 hours, we can divide the sidereal day into 24 *sidereal hours* that have to be shorter than their solar counterparts. A *sidereal clock* then runs about 4^m (actually, $3^m56.56^s$) fast per day. **Local sidereal time** (LST) is not in fact defined by the stars, but by the *hour angle of the vernal equinox* (Figure 3.18).

Local mean solar time and local sidereal time are identical at the moment the mean sun crosses the autumnal equinox about September 21 (two days before the apparent sun crosses it to mark the beginning of northern-hemisphere fall). To demonstrate, say the vernal equinox is on the meridian on September 21, so that LST is 0^h. The mean sun, however, is opposite in the sky and then must have an hour angle of 12^h, so the LMST is 24^h, or also 0^h. For each full solar day past the moment the mean

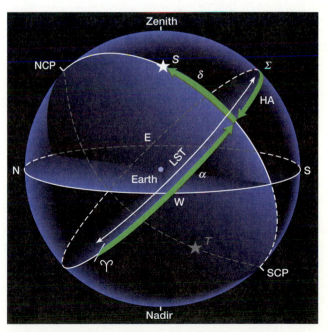

Figure 3.18 *Right ascension is measured counterclockwise (opposite from hour angle) from the vernal equinox to the intersection of the hour circle and the celestial equator. The hour angle of star S is 3^h, α is 6^h, and $\delta = +35°$ N. The local sidereal time is 8^h. (Exercise 13 asks you to estimate α and δ for star T.)*

Figure 3.19 *The meridian transit telescope of the U.S. Naval Observatory is fixed to rotate on one axis that always directs the instrument at the meridian. The right ascension of a star in the center of the field of view then defines LST. This particular instrument is used to measure right ascensions of stars on the meridian from the LST.*

sun goes across the autumnal equinox, the sidereal clock gains $3^m56.56^s$ over the solar clock; it gains a full day over the year. To calculate approximate sidereal time, count the number of days since September 21, multiply by 4 minutes, and add the result to LMST.

3.5.3 Right Ascension

The equatorial coordinate system introduced in Section 2.5.3 is not satisfactory for the permanent locations of the stars because hour angle changes continuously with time, increasing by 1° every 4 sidereal minutes and 15° every sidereal hour. The problem is solved by adopting the vernal equinox as the origin of a new coordinate. In Figure 3.18 draw an hour circle through star S. The star's **right ascension** (α) is defined as the arc measured along the celestial equator from the vernal equinox (Υ) *eastward* (counterclockwise as viewed from the north), opposite the direction from hour angle, to the intersection of the equator and the star's hour circle. Since the vernal equinox has a daily path around the sky just like the stars, α stays the same. Right ascension is also measured in hours, minutes, and seconds of time: for star S it is about 6^h, and the declination is about 35° N. The equatorial system of right ascension and declination is, like latitude and longitude, stable with time (slow changes are dis-

cussed in Section 3.7), allowing us to compile great catalogues of the positions of millions of stars.

In Figure 3.18 we see that the sum of star S's hour angle and right ascension equals the hour angle of the vernal equinox, which is the sidereal time. For any object in the sky,

$$LST = HA + \alpha \text{ and } HA = LST - \alpha.$$

To find a star, look up its right ascension and declination in any of a variety of catalogues, look at your sidereal clock (or calculate sidereal time from standard time), and subtract as shown to find the hour angle. The hour angle and the declination allow you to locate the star relative to the celestial meridian and to yourself. The sidereal time in Figure 3.18 is 8^h; the hour angle of star S is then 2^h. You now have the ability to wander through the sky.

3.5.4 The Determination of Time and Position

The Sun is a bright extended disk, and it is difficult to determine its exact hour angle. It is much more accurate to use the stars to measure the flow of time. From Figure 3.18 we see that the hour angle of the vernal equinox is the same as the right ascension of the equator point and of the celestial meridian. Therefore, local sidereal time is also the right ascension of the meridian. To find the sidereal time, we can point a *meridian transit telescope* (Figure 3.19) at the meridian and observe a star as it transits. The local sidereal time then equals the star's right ascension. From the exact date (which

Figure 3.20 *The sextant has mirrors that allow the navigator to measure a star's altitude and fix latitude and longitude.*

gives LST – LMST), we can calculate the mean solar time and, from the longitude of the observatory, the Universal Time and other standard times. We can then set or correct our clocks. The subject will be developed further as we examine other areas of astronomy.

To find the right ascension of a star, reverse the procedure. The moment the star transits the meridian, record the sidereal time, which at that moment equals the star's right ascension. Declination is found by measuring the arc between the star and the equator point with precisely calibrated circles attached to the meridian transit telescope, where the position of the equator point is known from the latitude.

The stars allow the navigator and the surveyor to travel the Earth with complete confidence. Right ascension and declination fix stellar positions relative to the celestial equator and the vernal equinox. Solar and sidereal times tell the rotational position of the Earth and fix the Sun, vernal equinox, and stars relative to the Greenwich meridian. For a particular time, the positions of stars relative to the navigator's horizon and meridian then depend only on latitude and longitude. In practice, the measurement of the altitudes of only three stars at a known Universal Time allows the calculation of position on Earth. At sea, measurement with the *sextant* (Figure 3.20) can give the location of a ship to within a few kilometers, and more stable instruments on land can fix coordinates to within a meter. The legal location of your house lot is ultimately fixed on maps relative to local meridians determined by the stars.

3.6
Demonstrations That the Earth Revolves

How do we really know the Earth goes about the Sun and not the Sun about the Earth? If you look at a nearby object from two points in space it will appear to be in different directions. Hold your finger in front of you and then look alternately at it with one eye and then the other (Figure 3.21a). It will appear to jump back and forth against the background of the room, an effect called **parallax,** which is the basis of three-dimensional vision. Now in your imagination place your eyes at opposite sides of the Earth's orbit, 2 AU apart, and do the same thing. A nearby star will now appear to jump back and forth (Figure 3.21b). A revolving Earth would allow you to make just such an observation as it moves from one side of its orbit to the other. The immense distances of the stars, render the

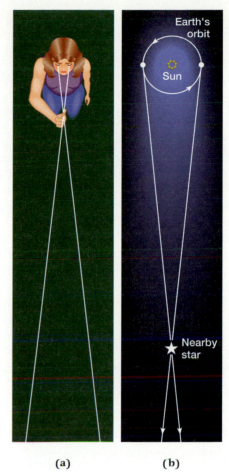

(a)   **(b)**

Figure 3.21 **(a)** *Each eye sees an outstretched finger from a different direction.* **(b)** *Similarly, you see a nearby star from different directions as the Earth orbits the Sun.*

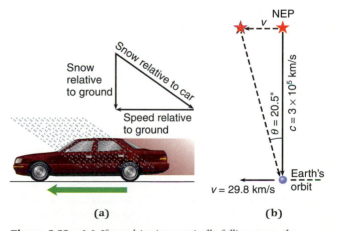

(a)   **(b)**

Figure 3.22 **(a)** *If you drive into vertically falling snow, the flakes appear to come from the forward direction. The faster you move, the more the direction of their fall seems to come from a point in front of you.* **(b)** *A ray of light and the apparent position of a star are affected the same way. The observed shift of 20.5 seconds of arc is defined by the relative speeds of the Earth (v) and light (c).*

MathHelp 3.2 Thin Triangles in Astronomy

Astronomers often deal with objects or motions at great distances, creating long **thin triangles** like that in Figure 3.22b, and it is necessary to understand them. The ratio of the circumference of a circle to its diameter, known as π (pi), is 3.1415 . . .; it is an irrational number, one that can never be divided evenly by any other number and that has an infinite number of decimal places. The circumference of a circle is then just $\pi d = 2\pi r$, where d and r are respectively the circle's diameter and radius. If the radius of the circle is one unit (Figure 3.23a), the circle is then called a *unit circle* and the circumference is simply 2π.

In arc measurement, a circle contains 2π angular units called *radians* that must be equivalent to 360°. One radian is an angle of $(360/2\pi)°$ = 57.30° (Figure 3.23a). Since a circle also contains $360 \times 60 \times 60 = 1,296,000$ seconds of arc, a radian is also the equivalent of $(1,296,000/2\pi) = 206,265$ seconds of arc. In a unit circle, the length of an arc of the circle then equals the angle in radians. You can take any circle at all (Figure 3.23b) and reduce it to a unit circle by dividing by the radius, so that the radius is one unit. Any angle measured in radians on a real circle is just the physical length of the arc divided by the circle's radius. In Figure 3.23b, the angle X (radians) = a/r, and the angle in seconds of arc is $206,265X$.

The triangle in Figure 3.22b is actually an ordinary triangle drawn on a plane, but it is so thin that to an excellent approximation the side labeled v can be thought of as the arc of a circle of radius c. Consequently, the angle θ at the triangle's apex in radians is just the length of the arc v divided by the radius c, rendering it a unit circle, and θ (radians) = v/c. Therefore, $\theta = v/c \times 206,265$, or 29.8 km/s (the Earth's orbital velocity)/300,000 km/s (the speed of light) $\times$ 206,265 = 20.5 seconds of arc.

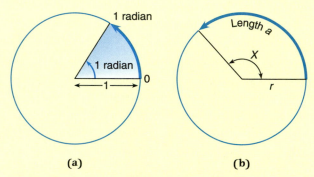

Figure 3.23 **(a)** *The unit circle has a radius of 1. The circumference of the circle is $2\pi r$, and since $r = 1$, the arc of the whole circle is 2π radians long. One radian is then $360/2\pi$, which is 57.3° or 206,265 seconds of arc. From the diagram you can see that the lengths of the radian and the radius of the circle are identical.* **(b)** *Another circle of radius r is marked off with an arc of length a, where a is 2.4 times r. By expressing the arc in radians as $a/r = 2.4$, the circle is reduced to a unit circle. The arc in degrees is then just the arc in radians times 57.3°, or 137.5°, and the arc in seconds of arc is that in radians times 206,265 or 495,036".*

annual shifts—which are at most only 1.5 seconds of arc—far below the ability of the unaided eye to see them. But their eventual discovery in 1837 is powerful evidence that the Earth revolves.

A more obvious effect, discovered in 1727 by the English astronomer James Bradley, is the **aberration of starlight.** Stand in a snowstorm on a windless day. The snow falls from overhead. But as you walk or drive (Figure 3.22a), it falls into your face or onto the windshield, appearing to come from a point in front of you. The faster you move, the greater the apparent shift from the overhead point. The same effect is produced when you move relative to a beam of light. In Figure 3.22b, the Earth is moving to the left in the ecliptic plane at speed v of 29.8 km/s, and is intercepted by a ray of light moving at speed c (300,000 km/s) from a star that is in the perpendicular direction, toward the north ecliptic pole. From a triangle constructed with sides v and c (greatly out of scale in the diagram), we can calculate that the star is shifted by 20.5 seconds of arc (MathHelp 3.2). And that is exactly what we see. As the Earth goes around the Sun, a star near the ecliptic pole will move in a little loop 20.5" in radius, providing a wonderful demonstration that it is the Earth that moves, not the Sun.

3.7
Precession

You can think of our planet as a sphere of matter with a "spare tire" at the equator, the so-called *equatorial bulge* (Section 2.1). However, both the Sun and (on the average) our nearby Moon are found in the plane of the ecliptic rather than in the plane of the celestial equator. In the process, they exert a gravitational pull on the equatorial bulge (Figure

3.24). In a sense they are trying to make the Earth's rotational axis perpendicular to the ecliptic plane, that is, to align the celestial and ecliptic poles. However, rotating bodies resist movement, as you can tell if you try to turn a spinning wheel away from the direction of its motion. As a result, the rotational axis wobbles or *precesses* about the orbital axis while maintaining the same angle of obliquity. **Precession** makes the north celestial pole describe a circle about the north ecliptic pole, and makes the south celestial pole move similarly about the south ecliptic pole, as shown in Figure 3.24. Measurement (which agrees with the calculation from the known forces) shows a period of oscillation, the time it takes the rotational axis to make a full turn, of 25,725 years.

Though this period seems long, the effects are quite noticeable and of great importance. Looking upward from the Earth in Figure 3.24, you would see the NCP describe a small circle of radius 23.5°

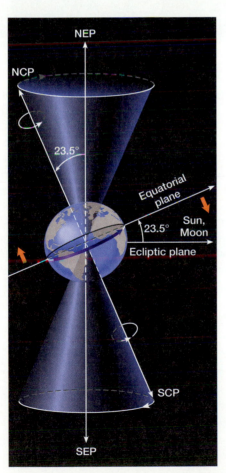

Figure 3.24   *The Sun and, on the average, the Moon are in the ecliptic plane and exert gravitational forces on the Earth's equatorial bulge (highly exaggerated here) in the direction of the orange arrows. Because the Earth is rotating, the spin axis wobbles, causing the NCP to trace out a circle of radius 23.5° about the NEP with a period of nearly 26,000 years.*

(outlined in the maps in the Appendix) around the NEP in the counterclockwise direction. You, however, are observing from the moving body. The NCP will always have an altitude that is equal to your latitude. It will therefore seem that the stars are moving past the pole rather than that the pole is moving through the stars.

The Earth's axis, which defines the NCP, is currently pointed nearly in the direction of the star Polaris, at the end of the handle of the Little Dipper (Figure 3.25). Polaris's great distinction, however, is only temporary. Around 1100 B.C., near the dawn of ancient Greek times, the star Kochab in the bowl of the Little Dipper was near the pole, and for the ancient Egyptians of 2700 B.C. the pole star was Thuban, in the constellation Draco. In 13,000 years, the NCP will be on the other side of the precessional circle near the star Vega, and our Polaris will be 23.5° × 2, or 47°, away, giving it a declination of 90° − 47° = 43°. The star that is now at our pole will one day pass nearly overhead and rise and set for all of the contiguous United States.

If the celestial poles move through space, so must the celestial equator, which is constrained between them. If the equator shifts, so must the points of intersection between it and the ecliptic. The effect is to move the equinoxes and the solstices clockwise (as viewed looking down from the NEP), or to the west, at the rate of 50" per year. Over a human lifetime, the total shift amounts to about a degree, easily noticeable even with crude instruments. By comparing two star maps made by naked eye observation only 150 years apart, Hipparchus of Nicaea discovered this remarkable motion over 2,000 years ago.

Since the vernal equinox continuously moves westward, it will traverse all the zodiacal constellations during its 26,000-year journey, so that on the average it will pass into a new one about every 2,000 years. Because we are on the moving body, however, it will appear that the constellations are sliding in the opposite direction past the vernal equinox as if on a moving belt. The progress of the vernal equinox is evident from its symbol, ♈, which represents the horns of Aries, the Ram, even though it is one constellation of the zodiac to the west, in Pisces. The vernal equinox *was* in Aries when the symbol was assigned. About the year 2700 it will move into Aquarius (see Table 3.1). Similarly, the tropics of Cancer and Capricorn bear these names even though the summer and winter solstices are respectively in Gemini and Sagittarius, each one constellation to the west. Even the newspaper astrology columns bear silent witness to this

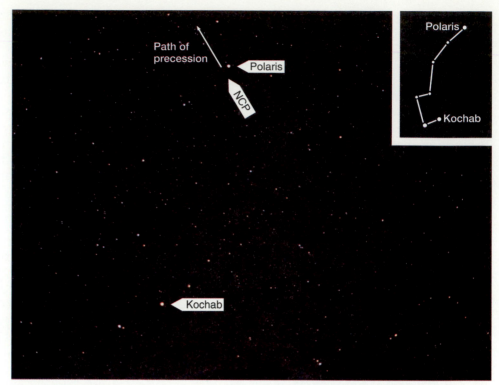

Figure 3.25 *The north celestial pole is found near the star Polaris in the constellation Ursa Minor, the Smaller Bear. The most notable part of the constellation is the figure of the Little Dipper, which is outlined here. The NCP is pursuing a path in the direction of the arrow. It will get a little closer to Polaris, come within 27 minutes of arc in a few decades, and then pull away.*

majestic motion. Aries, which announced spring when astrology was invented, still begins the list, not Pisces.

If the equator and the equinoxes are moving, then stellar coordinates must continuously change as well. The changes are complex, but over the course of the precessional period, all the right ascensions go through 360° (or 24^h) and the declinations change through a range of 47°, double the obliquity of the ecliptic. With precise instrumentation, the effect is actually visible almost night-to-night. The only way we can keep track of the stars is to express right ascensions and declinations for a particular time, the **epoch.** Star catalogues are always referred to specific epochs, usually at 50-year intervals. We are now leaving 1950 coordinates behind and adopting 2000 positions. Since we know exactly how precession changes the coordinates, we can then calculate what α and δ should be on any given night.

If the declinations of stars change, so must the stars' visibilities from specific latitudes. Some of the stars we see today will decrease their declinations sufficiently to drop for thousands of years below the southern horizon. New ones that we cannot now admire will then take their places. The Southern Cross, now only barely visible from the far

southern United States, was a common sight for Native Americans of the Great Plains only 6,000 years ago.

Precession is remarkably complex. Variations in the pull on the Earth's bulge caused by the orbital eccentricities of the Earth and Moon and by a slight tilt to the lunar orbit make the precessing poles oscillate back and forth by up to 22 seconds of arc. Moreover, the planets' gravitational effects on the Earth slowly cause the ecliptic plane to shift, which changes the obliquity, ε. The result is that the precessional path is not a true circle and does not close back on itself: in 26,000 years Polaris will not be as good a pole star as it is today. Nothing is stable, nothing is still.

KEY CONCEPTS

Aberration of starlight: The shift in the position of a star caused by the motion of the Earth relative to the velocity of light.

Apparent sun: The real, observed Sun.

Arctic and **antarctic circles:** The parallels of latitude (respectively 66.5°N and 66.5°S) above and below which it is possible to have a midnight sun.

Eccentricity: The degree of flattening of an ellipse, ranging from 0 for a circle up to 1.

Ecliptic: The apparent path of the Sun through the stars.

Ecliptic poles (north and south): The directions of the perpendicular to the Earth's orbit and to the ecliptic.

Ellipse: A curve defined by two **foci** such that sum of the distance from any point on the curve to each of the foci is constant.

Epoch: The moment for which right ascensions and declinations are valid.

Equation of time: The difference between local apparent solar time and local mean solar time.

Gregorian calendar: Our modern calendar of 365 days that has an extra day every four years except in century years not evenly divisible by 400; it replaced the Julian calendar.

International date line: A line near 180° longitude where you drop a day when going from west to east.

Mean sun: A point that travels the celestial equator keeping pace with the average position of the hour circle through the apparent sun.

Obliquity of the ecliptic: The tilt of the Earth's axis relative to the direction to the ecliptic poles, or the angle between the celestial equator and the ecliptic, equal to 23° 27'.

Parallax: The apparent shift in the position of a body when viewed from different directions.

Perihelion: The point of closest approach of a planet to the Sun; **aphelion** is the point of greatest distance between them.

Precession: The motion of the celestial poles and equinoxes; caused by a 26,000-year wobble of the Earth's axis.

Right ascension (α): The arc on the celestial equator between the vernal equinox and the hour circle through a celestial body, measured eastward.

Semimajor axis: Half the major axis, which characterizes the size of an ellipse.

Sidereal day: The day defined according to the stars, equal to $23^{h}56^{m}$ of solar time.

Standard time: Local mean solar time at standard meridians spaced 15° apart starting at Greenwich.

Summer and winter solstices: The points on the ecliptic farthest north and south (23.5° N and 23.5° S, respectively) of the celestial equator, which the Sun crosses on June 21 and December 22.

Tropics of Cancer and Capricorn: The parallels of latitude (respectively 23.5° N and 23.5° S) where the Sun passes overhead when it crosses the summer solstice (June 21) and winter solstice (December 22).

Universal Time: Local mean solar time at Greenwich, longitude 0°.

Vernal and autumnal equinoxes: The points where the Sun and the ecliptic cross the celestial equator, with the Sun moving respectively north (March 20) and south (September 23).

Zodiac: The band of constellations that contains the ecliptic.

KEY RELATIONSHIPS

Eccentricity of an ellipse:

e = (center-to-focus distance)/semimajor axis (a)

Local apparent solar time:

Time told by the apparent (real) sun;

$$LAST = HA(\text{apparent sun}) + 12^{h}$$

Local mean solar time:

Time told by the mean (average) sun;

$$LMST = HA(\text{mean sun}) + 12^{h}$$

Local sidereal time:

Time told according to the stars or vernal equinox (Υ);

$$LST = HA(\Upsilon) = \alpha(\text{meridian})$$

Longitude:

$$\lambda(\text{west}) = UT - LMST; \lambda(\text{east}) = LMST - UT$$

Right ascension and hour angle:

$$LST = HA + \alpha$$

Thin triangle:

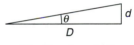

$$\theta(\text{radians}) = d/D;$$

$$\theta(\text{seconds of arc}) = 206{,}265d/D \text{ for small angles.}$$

EXERCISES

Comparisons

1. What is the difference between perihelion and aphelion?
2. Compare the visibility of the Sun at the south and north poles of the Earth.
3. Differentiate local mean solar time, local apparent solar time, Eastern Standard Time, and Universal Time.

Numerical Problems

4. Draw an ellipse with a semimajor axis of 10 cm and an eccentricity of 0.5. Note that the ends of the semimajor axis lie on the ellipse, and therefore you can find the sum of the distances from these points to the foci.
5. Assume that the Earth's orbit is a circle and show that the orbital speed is 29.8 km/s.

6. Tonight at 8 P.M. you see a certain star transiting (crossing) the celestial meridian. What will be the hour angle of that star at 8 P.M. in a day, a week, and a month?

7. A sundial reads 08^h43^m on October 1 for a person at longitude 75° W. What is the Eastern Standard Time?

8. What distance in kilometers corresponds to 1 minute of time at the Earth's equator?

9. Your local mean solar time is 14^h33^m at longitude 117° W. What is the Pacific Standard Time?

10. Central Standard Time is 2^h19^m. What is the Universal Time?

11. Your local mean solar time is 16^h32^m at 18^h36^m UT. What is your longitude in degrees and minutes of arc?

12. The hour angle of the vernal equinox is 14^h. What is the sidereal time?

13. What are the approximate α and δ for star T in Figure 3.18?

14. If the Earth were moving twice as fast in orbit as it does now, what would be the maximum shift caused by the aberration of starlight?

15. A sphere 200 meters away has a diameter of 1 meter. What is its angular diameter in radians and in seconds of arc? (See MathHelps 2.1 and 3.2.)

16. A body 100,000 km away subtends an angular diameter of 3 seconds of arc. What is its diameter in meters?

17. In what constellations of the zodiac will the vernal and autumnal equinoxes be in the year 7000?

18. After 12,900 years of precession, what will be the right ascensions of: (a) the autumnal equinox; (b) the summer solstice?

19. What is the largest possible angle between the north celestial pole and Polaris? What is the minimum declination that Polaris can have?

Thought and Discussion

20. If the axis of the Earth were tilted by an angle of 33° relative to the perpendicular to the ecliptic plane, what would be the declinations of (a) the winter solstice; (b) the vernal equinox?

21. What is the significance of (a) the tropic of Cancer; (b) the arctic circle?

22. Which latitudes on Earth receive the greatest solar heating, and which receive the least?

23. How can the coldest days of the northern hemisphere coincide with the Earth's perihelion?

24. At what latitude will the Sun be seen overhead on (a) March 20; (b) December 22; (c) August 1; (d) November 1?

25. Over what range of latitudes will the Sun be circumpolar on (a) June 21; (b) March 20; (c) December 22; (d) November 1; (e) February 1?

26. What century years in the next millennium will not be leap years?

27. What is the evidence that the Earth revolves about the Sun?

28. What is meant by the epoch of a star catalogue, and why is it necessary?

Research Problem

29. Early explorers were severely hampered by their difficulty in navigating. Examine the work of the Englishman Thomas Harrison and the voyages of Captain James Cook to find how they helped seafarers know their longitudes.

Activities

30. Make a sundial. Use a paper plate and onto it glue a straw at an angle equal to your latitude so that it can point at the NCP when the plate is horizontal (you will probably have to support the straw at the high end). Calculate the local mean solar time at which the local apparent solar time will be 1^h, 2^h, and so on. Take your construction outside at noon, point the straw north, and at each apparent solar hour draw a line along the shadow. Draw symmetrical lines on the other side of the plate for eastern hour angle. You can now read the LAST on any day.

31. Draw the sky as seen from latitude 35° N with a local sidereal time of 12^h. Locate stars with coordinates (a) $\alpha = 8^h$, $\delta = 30°$ N; (b) $\alpha = 13^h$, $\delta = 20°$ S. What are their hour angles?

32. Draw celestial spheres, like those in Figure 3.10, showing the daily paths of the Sun during the year at the tropic of Capricorn, the antarctic circle, and the south pole.

33. Locate a shortwave radio and set your watch with radio station WWV.

34. Go outside on a clear night. Find a star you recognize, look up its right ascension (you can estimate it from the star maps in the Appendix), and estimate your standard time from the star's hour angle, the date, and your longitude. Compare the result with the time given by your watch.

Scientific Writing

35. Write an article for a gardening magazine in which you explain the origin of the change of the seasons and the reason that growing conditions at lower latitudes are different from those at higher latitudes.

36. The same gardening magazine features an issue on garden decorations that includes sundials. Write an article on why sundials do not read the "correct" time, the one on the gardener's watch.

4

The Face of the Sky

Stars and constellations and the organization of the heavens

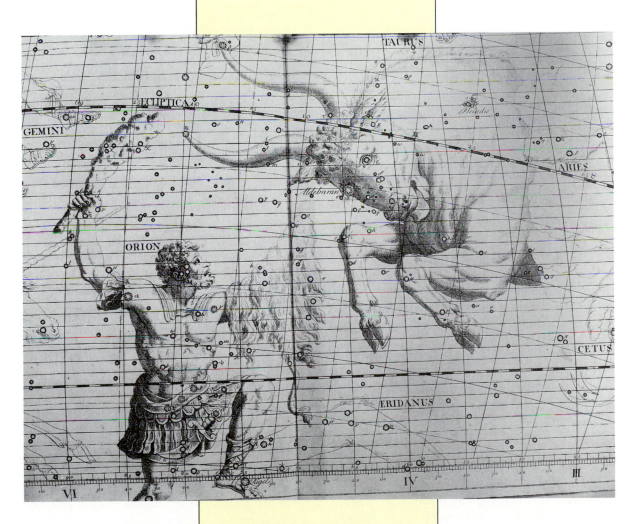

The sky is populated by creatures of our stories, dreams, and inventions. Here the hunter Orion prepares to strike the zodiacal bull Taurus; they are surrounded by a river, a whale, a ram, a pair of twins, and a unicorn.

A clear, dark night sky with no Moon or artificial lighting is a spectacular sight. Stars of different brightnesses and colors shimmer everywhere, so many that they shed a pale embracing light. They scatter into patterns, much like a handful of thrown sand. Here you see a circle of them, there a square, somewhere else figures of a man and a meandering river. Night after night they repeat their display, friendly, familiar, reigning supremely and untouchably over the Earth. As you learn to recognize stars, you begin to name them and the patterns they seem to form, translating to the sky images from your experience. Such is the deep meaning of the **constellations** as they carry forward an ancient lore that allows you to reach back and touch your ancestors of unrecorded time.

4.1
The Constellations

There is not one set of constellations but many. Everywhere, in every land, people have named the stars. In China, India, Arabia, North and South America, and elsewhere, people used the nightly patterns in different ways. Our Western constellations began to be developed some 4,000 years ago in the cradle of our civilization in the Middle East, which was centered around ancient Mesopotamia. The stars were used to tell stories, to honor gods and heroes, and to serve as reminders of days gone by. As the tales passed down from one age to another the legends grew and changed, and the sky became suffused with new ideas. The Greeks adopted the early constellations and eventually identified 48 of them with their own fables and religious beliefs. The Romans accepted these celestial figures whole, and after the gradual collapse of their great empire, the stellar patterns passed on to the Arabian lands, where they were further embellished and the stars graced with new names. Finally they returned to Europe in much the form we see them today, 48 **ancient constellations** engraved into the sky.

The old star patterns do not fill the heavens, especially since their northern inventors could not see below about 50° south declination. So beginning with the scientific renaissance of the late sixteenth century, and continuing for some 200 years, astronomers created fresh ones to fill in the blanks. Like their ancient forebears, they designed constellations around their own civilization, placing a hundred or more familiar objects, animals, kings, and heroes into the sky. Most did not survive the passage of time, and the astronomers of our own century finally officially adopted 38 **modern constellations.** With the breakup of one ancient constellation into

Figure 4.1 *Ptolemy was the last of the great Greek astronomers. Building on the work of many others, he gave us the final form of the ancient constellations and developed a theory of the Solar System that prevailed for 1,500 years.*

three parts, we now recognize a grand total of 88 spread out over the entire celestial sphere.

The constellations have a profound place in astronomy even today. The sky is vast, and these fanciful figures partition it into manageable segments that allow us easily to name and locate stars and other objects. They also frequently provide the spark that creates interest in our subject, that brings the children out to look, and that starts new scientists—whether amateur or professional—on the road to discovery and a lifetime of enjoyment.

It is often disconcerting to the beginner to find that only a few constellations resemble what they are supposed to be. But why should they? After all, on the small scale the stars are distributed largely at random. The constellations are meant to represent, not portray. So relax and use your imagination, and if you proceed with good humor you may see the sky come alive as it did for your predecessors so many years ago.

4.2
The Ancient Forty-eight

These oldest constellations were formally codified from far more ancient tales by the Greek mathematician Eudoxus of Cnidos about 375 B.C. and were lavishly embraced by Aratos (about 270 B.C.) in a famous poem called On the Phaenomena. The astronomer Hipparchus repeated the list a hundred or so years later, and the constellations were finally cast in their present form by Ptolemy (Figure 4.1) about A.D. 150 in his *Syntaxis* (meaning "compositions"), in which he set down the astronomical knowledge of the time. They are listed in Table 4.1, which also gives their central right ascensions (α)

TABLE 4.1
The Ancient Constellations

	Name	Meaning	α^a	δ^a	Genitive	Abbreviation
The Zodiac	Aries	Ram	3	+20	Arietis	Ari
	Taurus	Bull	5	+20	Tauri	Tau
	Gemini	Twins	7	+20	Geminorum	Gem
	Cancer	Crab	8.5	+15	Cancri	Cnc
	Leo	Lion	11	+15	Leonis	Leo
	Virgo	Virgin	13	0	Virginis	Vir
	Libra[b]	Scales	15	−15	Librae	Lib
	Scorpius	Scorpion	17	−30	Scorpii	Sco
	Sagittarius[c]	Archer	19	−25	Sagittarii	Sgr
	Capricornus[d]	Water-goat	21	−20	Capricorni	Cap
	Aquarius[d]	Water-carrier	22	−10	Aquarii	Aqr
	Pisces[d]	Fishes	1	+10	Piscium	Psc
Ursa Major	Ursa Major	Greater Bear	11	+60	Ursae Majoris	UMa
	Ursa Minor[e]	Smaller Bear	16	+80	Ursae Minoris	UMi
	Boötes	Herdsman	15	+30	Boötis	Boo
Orion	Orion	proper name: hunter, giant	6	0	Orionis	Ori
	Canis Major	Larger Dog	7	−20	Canis Majoris	CMa
	Canis Minor	Smaller Dog	8	+5	Canis Minoris	CMi
	Lepus	Hare	6	−20	Leporis	Lep
Perseus	Perseus	proper name: hero	3	+45	Persei	Per
	Andromeda	proper name: princess	1	+40	Andromedae	And
	Cassiopeia	proper name: queen	1	+60	Cassiopeiae	Cas
	Cepheus	proper name: king	22	+65	Cephei	Cep
	Cetus	Whale	2	−10	Ceti	Cet
	Pegasus	proper name: winged horse	23	+20	Pegasi	Peg
The Ship Argo: Jason and the Argonauts	Carina[f]	Keel	9	−60	Carinae	Car
	Puppis[f]	Stern	8	−30	Puppis	Pup
	Vela[f]	Sails	10	−45	Velorum	Vel
	Hercules[g]	proper name: hero	17	+30	Herculis	Her
	Hydra	Water Serpent	12	−25	Hydrae	Hya
	Aries[h]	Ram; here, the golden fleece	3	+20	Arietis	
Centaurus	Centaurus[i]	Centaur	13	−45	Centauri	Cen
	Lupus	Wolf	15	−45	Lupi	Lup
	Ara	Altar	17	−55	Arae	Ara
Ophiuchus	Ophiuchus[j]	Serpent-bearer	17	0	Ophiuchi	Oph
	Serpens[j]	Serpent	17	0	Serpentis	Ser
Single Constellations	Aquila	Eagle	20	+15	Aquilae	Aql
	Auriga	Charioteer	6	+40	Aurigae	Aur
	Corona Australis[k]	Southern Crown	19	+40	Coronae Australis	CrA
	Corona Borealis[l]	Northern Crown	16	+30	Coronae Borealis	CrB
	Corvus[m]	Crow, Raven	12	−20	Corvi	Crv
	Crater	Cup	11	−15	Crateris	Crt
	Cygnus	Swan	21	+40	Cygni	Cyg
	Delphinus[d]	Dolphin	21	+10	Delphini	Del
	Draco[n]	Dragon	15	+60	Draconis	Dra
	Equuleus	Little Horse	21	+10	Equulei	Equ
	Eridanus	proper name: river	4	−30	Eridani	Eri
	Lyra[m]	Lyre	19	+35	Lyrae	Lyr
	Piscis Austrinus[d]	Southern Fish	22	−30	Piscis Austrini	PsA
	Sagitta	Arrow	20	+20	Sagittae	Sge
	Triangulum	Triangle	2	+30	Trianguli	Tri

[a]α and δ are the coordinates of the approximate centers of the constellations.
[b]Originally the claws of Scorpius.
[c]Contains the galactic center.
[d]Constellations of the wet quarter.
[e]Contains the north celestial pole.
[f]Carina, Puppis, and Vela are modern subdivisions of the old constellation Argo and together make one of the ancient 48.
[g]One of the oldest constellations known, also called the Kneeler.

[h]Also included in the zodiac.
[i]Sometimes included in the Argo group.
[j]Ophiucus is identified with the physician Asclepius, and Serpens with the caduceus.
[k]Sometimes considered as Sagittarius' crown.
[l]Ariadne's crown.
[m]Lyra was Orpheus' harp, Corvus his companion.
[n]Contains the north ecliptic pole.

and declinations (δ) so you can find them easily on the star maps in the Appendix. Most ancient constellations do not stand alone, but are part of groups of patterns that tell the stories of one or more myths. The myths are interwoven, and the stories sometimes conflict, so at different times these groupings have shifted and recombined. As we proceed, refer to the star maps, and if the night is clear, try to find the real figures in the sky.

4.2.1 The Zodiac

The famed constellations of the zodiac served in the last chapter to illustrate the ecliptic and precession. They are among the oldest constellations known, and to the ancients were certainly the most important as they contain the Sun, Moon, and (as we will see in Chapter 6) the planets, all the moving bodies of the Solar System that were closely identified with the gods. Rare is the person who is not already familiar with them from their appearances in popular astrology columns.

There is an important distinction to be made between the astrological *signs* of the zodiac and the astronomical *constellations* of the zodiac. The constellations are real star-groups that can be seen on any clear night. The signs are symbolic concepts that are locked to the vernal equinox. They were once aligned with the constellations, but no longer are because precession has moved them westward. To the superstitious the signs possess magical powers that derive from the names of the parent constellations and that provide the means to foretell the future.

There are 12 of these constellations because the Moon runs through its phases (see Chapter 5)

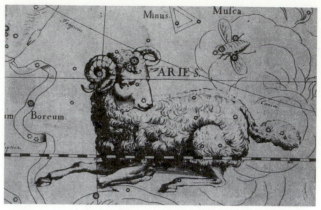

Figure 4.3 *Aries, the Ram, is generally represented in the sky by the flat triangle of stars in his head and horns. The carefully drawn figures were in part meant to locate and outline the boundaries of the constellations; now we just use dotted lines. Taurus is to the right, Pisces is to the left, and Cetus is below. The stars are presented backward, as they would appear on the outside of a celestial globe or sphere. The small triangle labeled "Minus" is no longer recognized. The defunct figure Musca, or Musca Borealis, the Northern Fly, is above his rump*

somewhat over 12 times a year. As a consequence, the Sun is roughly in a different one every time the Moon is "new" (when it cannot be seen) or "full," or every month. With one exception, the images are all of living creatures, hence the term *zodiac,* which is derived from the Greek and means "circle of animals." The exception is Libra, the Scales, a constellation only about 2,000 years old. At one time its only two prominent stars represented the claws of its eastern neighbor Scorpius (or Scorpio, Figure 4.2), rendering it a "double sign."

Some, like Taurus (see the figure that opens this chapter), Scorpius, Gemini, and Leo are among the most prominent patterns of the sky and actually bear some likeness to their names. Others—Virgo and Aquarius, for example—are just sprawls of stars with little in the way of prominent form, and faint Cancer would probably not have been distinguished at all were it not on the solar path. But the Sun and the planets were considered so important in human affairs that constellations *had* to exist along it even if imagination had to be stretched to invent them.

The constellations of the zodiac have deep meanings from our agrarian roots. Aries (Figure 4.3), the rutting ram that held the Sun on the first day of spring 2,500 years ago, is clearly a fertility symbol associated with planting. Taurus, the Bull, likely had similar significance even earlier in our history. And Virgo, the Virgin, assumed much the same kind of symbolism when she embraced the Sun at the time of the late summer harvest. The meaning of the balance, Libra, is obvious since the Sun once figuratively lay in its pans at the time of

Figure 4.2 *The deadly scorpion, Scorpius, with red Antares at its heart, presides over the summer southern landscape.*

the autumnal equinox. The last three constellations, Capricornus, Aquarius, and Pisces (collectively known as the wet quarter) represent the solar passage during a wet season prior to planting.

At night, our northern-hemisphere seasons are marked by the zodiacal constellations that are opposite the Sun. Leo rising in early evening is a classic sign of the coming of spring, when the Sun is in its opposite, Aquarius. In summer, while the Sun is far north of the ecliptic in Gemini, the southern zodiac, punctuated by prominent Scorpius and Sagittarius, graces the sky. When Aries rides high at midnight the leaves will soon turn, and the winter snows sometimes seem to reflect the sparkle of brilliant Gemini as the twins make their midnight transit. It is unfortunate that in our modern times we have removed ourselves from direct seasonal contact with the sky. It need not be so. The stars still delightfully mark the passage of our lives just as they did in days long past.

4.2.2 The Bears

The next six groups are listed in Table 4.1 more or less in order of renown, beginning with the most famous constellations in the sky, Ursa Major and Ursa Minor, the Greater and Smaller Bears, which respectively contain the Big and Little Dippers (Figure 4.4). Almost everyone can find the Big Dipper, its high northern declination giving it almost year-round visibility. Once you have that pattern you can extend it outward to find the bear, and then use it to locate other constellations, such as the Little Dipper and the third member of the group, Boötes (also in Figure 4.4). One legend has it that the bear was hurled into the sky by his tail, which stretched out into the string of stars that makes the Dipper's handle.

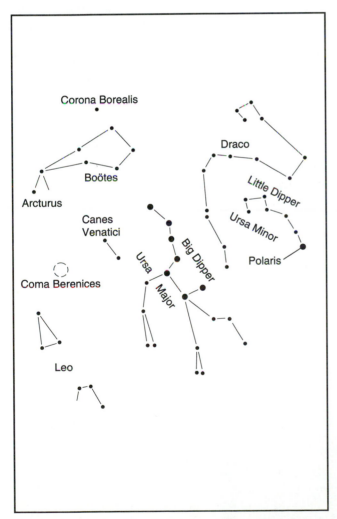

Figure 4.4 *The most famous star pattern of all, the Big Dipper, a part of the constellation Ursa Major, the Great Bear, falls silently in the northwest toward the Black Hills of South Dakota. The paws of the bear are below the Dipper, whose front bowl stars point directly at Polaris, at the end of the handle of the Little Dipper, which stands upright. Draco (which contains the NEP) coils between the two. The arc of the Big Dipper's handle points to orange Arcturus in Boötes. Below the curve is the pair that make the modern constellation Canes Venatici. Setting at lower left are the hindquarters and part of the head of Leo.*

The Smaller Bear is more difficult to find. Most of the stars of the Little Dipper, which make up the tail and body of the bear, are faint and cannot easily be seen from town. Boötes, the celestial herdsman, is less well known to the public, but it is very recognizable by its kitelike shape terminating in the great orange star Arcturus, which watches over the bear as it plods around the pole.

4.2.3 *Mighty Orion*

Next in fame is Orion (see the chapter-opening figure and Figure 4.5) and his companions. The great mythical hunter faces you, absolutely dominating the northern winter sky. Three bright stars practically on the celestial equator stud his belt, bright reddish Betelgeuse and bluish Rigel respectively mark his right shoulder and left foot, and another triplet of stars makes the sword hanging from his belt. He is usually depicted with upraised club, ready to strike Taurus, who is charging him. This constellation is such an imposing figure that the ancient Arabs called it Al Jauza, "the central one"— a mysterious woman.

All you have to do to find him (or her!) is to look along the celestial equator on a winter night and the figure will jump out at you. The constellation is a veritable signpost in the sky. If you follow a line upward through the two bright stars, Betelgeuse and Rigel, they point to Gemini. Follow his belt up and to your right to find the "V" that makes up the head of Taurus, lit brightly by orange Aldebaran; then follow it down to the left to the brightest star in the whole sky, to Sirius in Canis Major (see Figure 4.5), Orion's larger dog, who stands on his hind legs ready to do his master's bidding. The top two stars of the main seven-star figure of Orion point leftward to bright Procyon, the luminary, or brightest star, of Canis Minor, one of a pair that make Orion's smaller dog. Under the hunter's feet is some of his prey, little Lepus, the Hare.

In one story Orion met his end when his love, the goddess Diana, personification of the Moon, was tricked into shooting him with her bow. In another, he was stung by Scorpius, and when the gods honored the hero by placing him forever in the sky they put him opposite the celestial scorpion so that he need never look upon his killer: as one rises, the other sets.

4.2.4 *Perseus*

The Perseus myth, which ties together a number of major autumn constellations, may be the best loved of the celestial tales. The queen *Cassiopeia*, wife of *Cepheus*, boasted that her daughter *Andromeda* was

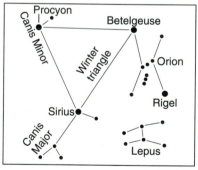

Figure 4.5 *Great Orion draws attention in the winter sky with his prominent belt and the brilliant stars Betelgeuse and Rigel. His hunting dogs, Canis Major and Canis Minor, follow him across the sky, and beneath his feet is Lepus, the Hare. Betelgeuse, Procyon, and Sirius form the Winter Triangle. The winter Milky Way runs to the left of the figure.*

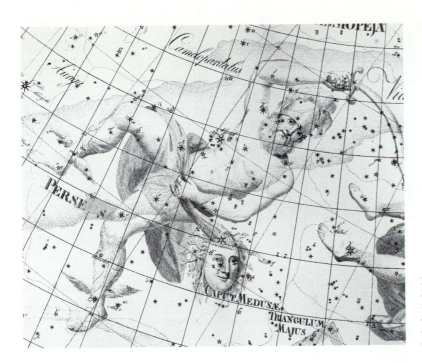

Figure 4.6 *Bold Perseus, the rescuer of Andromeda, holds the Medusa as he turns Cetus to stone. Constellation boundaries are drawn as curved dotted lines. The solid lines show the coordinate grid of right ascension and declination. The straight dashed lines represent a coordinate grid based on the ecliptic.*

more beautiful even than the sea nymphs. In anger, Neptune had Andromeda chained to a rock by the sea as a sacrifice to *Cetus* (barely seen in Figure 4.3), the sea monster or whale. But as Cetus bears down upon her, *Perseus* (Figure 4.6), riding his flying horse *Pegasus,* comes dramatically to the rescue. He has just slain the dreaded Medusa, a woman with hair of snakes who is so horrifying that one look will turn the viewer to stone. He pulls out her severed head, petrifies the monster, and rescues Andromeda.

In the sky the group is led by Cassiopeia, her conspicuous six-star chair climbing the northeast in autumn opposite the pole from the Big Dipper. She is followed by the graceful sweeps of stars that make Perseus and Andromeda and the solid square that represents powerful Pegasus.

4.2.5 The Ship, the Centaur, and the Serpent

Argo, or Argo Navis, the ship in which Jason sailed to find the golden fleece, is by far the largest constellation in the sky, so big that in more recent times it was broken into three more-manageable parts—Carina, the Keel; Puppis, the Stern (or poop); and Vela, the Sails—thereby raising the number of ancient constellations from 48 to 50. From most of the United States only the top parts of the ship can be seen to the east and south of Canis Major, which is a pity because these three still-large segments contain some of the loveliest parts of the sky, including Canopus, the second brightest star.

Jason did not foolishly go alone on his grand quest, which was plagued by disasters epitomized by Hydra, the Water Serpent. Hydra is the longest constellation, wrapping itself about one-third the way around the sky. Among Jason's crew was Hercules, the strongest man of ancient, or any other, times. His northern constellation stands quite well on its own. It was originally called the Kneeler, the origin of which is lost to time. The Argo group overlaps that associated with Centaurus, the Centaur, (Jason's foster father and tutor). This half man—half horse holds the third brightest and closest star, Alpha Centauri (Figure 4.7), and is sometimes shown sacrificing Lupus, the Wolf, upon Ara, the Altar.

Ophiuchus is sometimes called the thirteenth constellation of the zodiac. The modern boundaries of this large and sprawling pattern dip past the ecliptic to the west of Scorpius (see Table 3.1). Ophiuchus is a man depicted as holding and fighting a massive serpent that, although considered as one constellation, is separated in two: Serpens Caput, the head, seen to the west of Ophiuchus, and Serpens Cauda, the tail, to the east. Ophiuchus is identified with Asclepius, the mythic physician of the ancient Greeks. His serpent-wrapped body is the model for the caduceus, the staff entwined with snakes that is the modern physician's symbol—the sky brought back to Earth.

4.2.6 Single Ancient Constellations

Several prominent constellations have no or minimal relations with others. During the northern

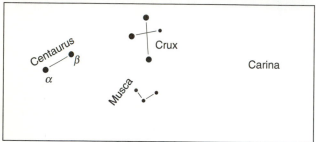

Figure 4.7 *Alpha and Beta Centauri lie in the Milky Way to the east of Crux, the famous Southern Cross. A dark cloud called the Coalsack is seen down and to the left of the Cross. Musca, the Fly, a modern constellation, is just below the Cross.*

summer months, two great birds, Cygnus, the Swan (Figure 4.8), and *Aquila*, the Eagle (Figures 4.8 and 4.9), fly through the Milky Way. Cygnus, its tail lit by the bright star Deneb, culminates near the zenith in typical northern latitudes. Aquila is formed by bright Altair and two flanking stars that give the impression of outstretched wings. West of Cygnus we find Lyra, the Lyre, played by Orpheus, the musician who sang so beautifully he could make the rocks applaud. It is marked by the brilliant white star Vega, from which hangs a small near-perfect parallelogram of faint stars.

In the northern winter, Auriga, the Charioteer, adds to the luster created by Orion, Canis Major, Canis Minor, and Gemini. You can find him directly above Orion's head, a bit farther north than Gemini. To the west of Orion springs Eridanus, the celestial depiction of Ocean Stream that in ancient times was thought to girdle the world. For most of the population of the United States it flows below the southern horizon before ending in bright Achernar. And there are more: fish, fowl, royal crowns, and an array of ancient artifacts all waiting in the stars for you to find.

Figure 4.8 *The northern summer Milky Way displays a rich variety of constellations. Cygnus, the Swan, which also forms the Northern Cross, stretches to the right of the star Deneb at lower left. Aquila, the Eagle, marked by its bright star Altair, is near right center, and Lyra, with Vega (which completes the Summer Triangle), is toward the top. The parallelogram that forms Lyra is down and to the right of Vega.*

4.3
The Modern Constellations

The great prominent ancient constellations take up only a part of the sky. The Greeks and their predecessors felt no desire to fill up the spaces, which they called the *amorphotoi,* "unformed." Presumably, the gods could use these voids, which contain only faint stars, to create new constellations honoring great heroes or deeds. It was not until science began to bloom near the end of the sixteenth century with the revelations of Copernicus, Galileo, and others that the new astronomers wanted all the sky to be filled with star figures, at least in part because of a need for heavenly organization.

Table 4.2 gives the official list of the 38 that now supplement the ancient constellations. The large majority of these were created by a handful of people. In the seventeenth century, the German astronomer Johann Hewel (his name commonly Latinized to Hevelius) concentrated on filling the spaces of the northern hemisphere, generally with animals. To him are attributed Canes Venatici, the Hunting Dogs (see Figure 4.4); Lacerta, the Lizard, seen northeast of Cygnus; the obviously named Lynx between Ursa Major and Auriga; and obscure Leo Minor, the Smaller Lion, who rides the back of Leo himself.

It was the southern hemisphere below 50°S that provided the great treasury of unnamed stars and patterns. Around the year 1600 the Dutch travelers Pieter Keyser and Frederik de Houtman organized the southern stars into nearly a dozen fanciful animals. These were immortalized in Johannes Bayer's great atlas, the *Uranometria* of 1603, its style exemplified by Figure 4.9. Over a century later, the Abbé Nicolas de Lacaille honored the artifacts of a burgeoning industrial revolution (Figure 4.10). Thus we see a furnace (Fornax), an air pump (Antlia), and a microscope (Microscopium) mixed in with a bird of paradise (Apus), a chameleon (Chamaeleon), and a peacock (Pavo).

Two modern constellations deserve special mention. The first is *Coma Berenices,* Berenice's hair, a lovely sprinkling of physically related faint stars south of the Big Dipper's handle (see Figure 4.4). The figure is quite old, perhaps going back to Eratosthenes, and rightly ought to take its place among the ancient constellations, but it was not included in Ptolemy's definitive list. Berenice was an Egyptian queen of about 250 B.C. who gave a lock of her hair in thanks to the gods for bringing her husband safely home from war. And the gods gave it to the sky.

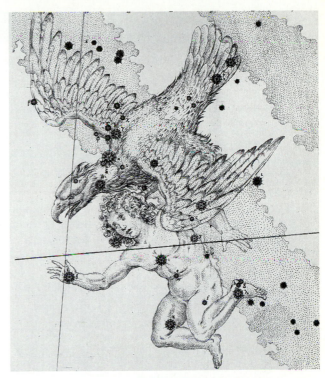

Figure 4.9   *Aquila soars the Milky Way south of Cygnus carrying Antinoüs, companion of the Roman emperor Hadrian. Aquila can be seen in the context of the Milky Way to the right of center in Figure 4.8.*

The other is that absolute gem of the southern hemisphere, *Crux,* the Southern Cross (see Figure 4.7). This striking constellation with its pair of brilliant luminaries was carved from the feet of Centaurus. From the southern hemisphere it rises just ahead of Centaurus' two bright stars, and as they all proceed across the celestial sphere they present a never-to-be-forgotten sight.

The sky is also filled with failed constellations, ones that never made it onto the modern official list. Many were highly nationalistic, the best known being Edmund Halley's *Robur Carolinium,* Charles's Oak. This figure was taken from the stars of Argo and commemorated Charles II of England, who was saved in battle by hiding in an oak tree. But no nation was about to accept the political inventions of another, so these creations eventually died. Only Scutum, set in the Milky Way just south of Aquila, passed the test of time. This constellation represents the shield of the seventeenth century Polish king John Sobieski, who won the affection of all the nations of Europe for his defense of Vienna against the Turks.

One failure involves a poignant and true story. Figure 4.9 shows Antinoüs clutched tightly in Aquila's talons, perhaps on his way to live with the gods. The Roman emperor Hadrian created this

TABLE 4.2
The Modern Constellations

Name	Meaning	α	δ	Genitive	Abbreviation
Antlia	air pump	10	−35	Antliae	Ant
Apus	bee	16	−75	Apodis	Aps
Caelum	graving tool	5	−40	Caeli	Cae
Camelopardalis	giraffe	6	+70	Camelopardalis	Cam
Canes Venatici	hunting dogs	13	+40	Canum Venaticorum	CVn
Chamaeleon	chameleon	10	−80	Chamaeleontis	Cha
Circinus	compasses	15	−65	Circini	Cir
Columba	dove	6	−35	Columbae	Col
Coma Berenices[a]	Berenice's hair	13	+20	Comae Berenices	Com
Crux[b]	southern cross	12	−60	Crucis	Cru
Dorado[c]	swordfish	6	−55	Doradus	Dor
Fornax	furnace	3	−30	Fornacis	For
Grus	crane	22	−45	Gruis	Gru
Horologium	clock	3	−55	Horologii	Hor
Hydrus	water snake	2	−70	Hydri	Hyi
Indus	Indian	22	−70	Indi	Ind
Lacerta	lizard	22	+45	Lacertae	Lac
Leo Minor	smaller lion	10	+35	Leonis Minoris	LMi
Lynx	lynx	8	+45	Lyncis	Lyn
Microscopium	microscope	21	−40	Microscopii	Mic
Monoceros[d]	unicorn	7	0	Monocerotis	Mon
Mensa	table	6	−75	Mensae	Men
Musca (Australis)[e]	(southern) fly	12	−70	Muscae (Australis)	Mus
Norma	carpenter's square	16	−50	Normae	Nor
Octans[f]	octant	—	−90	Octantis	Oct
Pavo	peacock	20	−70	Pavonis	Pav
Phoenix	phoenix	1	−50	Phoenicis	Phe
Pictor	painter's easel	6	−55	Pictoris	Pic
Pyxis[g]	compass	9	−30	Pyxidis	Pyx
Reticulum	net	4	−60	Reticuli	Ret
Sculptor[h]	sculptor's workshop	1	−30	Sculptoris	Scl
Scutum[i]	shield	19	−10	Scuti	Sct
Sextans	sextant	10	0	Sextantis	Sex
Telescopium	telescope	19	−50	Telescopii	Tel
Triangulum Australe	southern triangle	16	−65	Trianguli Australis	TrA
Tucana[j]	toucan	0	−65	Tucanae	Tuc
Volans	flying fish	8	−70	Volantis	Vol
Vulpecula	fox	20	+25	Vulpeculae	Vul

[a]Star cluster with many old references, but still not considered one of the ancient 48. Referred to as Ariadne's Hair by Eratosthenes. Contains the north galactic pole.

[b]Originally a part of Centaurus.

[c]Contains the Large Magellanic Cloud and the south ecliptic pole.

[d]Of older origin.

[e]Musca Australis was so called to distinguish from Musca Borealis, the Northern Fly, which is now defunct; "Australis" is now dropped.

[f]Contains the south celestial pole.

[g]Grouped with Argo.

[h]Contains the south galactic pole.

[i]Shield of the Polish King and hero John Sobieski.

[j]Contains the Small Magellanic Cloud.

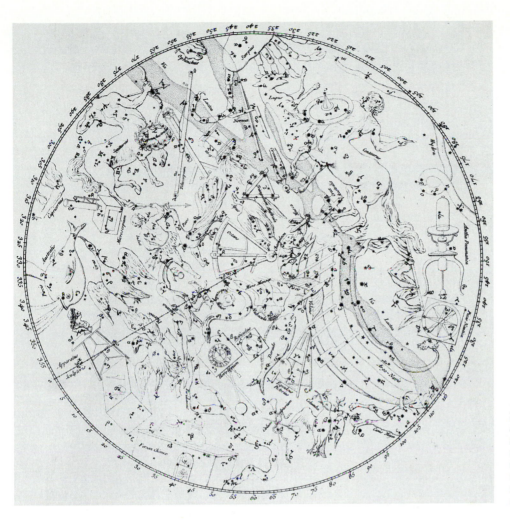

Figure 4.10 *Lacaille's southern sky, in which he includes his southern constellations together with the Greek letters that name the stars. Argo is down and to the right of center, and ancient Centaurus, with newly formed Crux, are to the upper right.*

constellation in the year 132 to honor his youthful companion, who had committed suicide in the belief that his years would go to his master.

4.4
Asterisms

It comes as a surprise to the beginner in astronomy that the two Dippers (see Figure 4.4) are not official constellations, but are only segments of Ursa Major and Ursa Minor. Such separately named groupings are called **asterisms.** Several are listed in Table 4.3. Look for the Sickle of Leo, which makes the lion's foreparts, and for the Northern Cross, which lies within the Swan (see Figure 4.8). Some asterisms extend over two or three constellations. The star in the northeastern corner of the Great Square of Pegasus belongs to Andromeda. The Summer Triangle (see Figure 4.8) is made of Deneb, Altair, and Vega, and the Winter Triangle (see Figure 4.5) of Betelgeuse, Procyon, and Sirius.

Among the most prominent asterisms are two star clusters, the Pleiades and the Hyades, both in Taurus (see the chapter-opening figure). The Pleiades, the Seven Sisters, is a lovely, very obvious compact group of stars northwest of Orion that rises in northern autumn evenings. In myth, the Pleiades were daughters of Atlas. Pursued by Orion, they are chased by him even today across the heavens. The Hyades, more of Atlas' daughters and the Pleiades' half-sisters, is a sprawl of stars around Aldebaran that make the head of the celestial bull as he bears down upon the hunter.

4.5
Stars and Their Names

There are roughly 6,000 to 8,000 stars that can be seen with the naked eye; the exact number is critically dependent upon the condition of the atmosphere and the sensitivity of the particular observer's vision. Their brightnesses are important

TABLE 4.3
Some Prominent Asterisms

Asterism	Constellation	Description
Big Dipper	Ursa Major	Seven bright stars (α–η UMa)
Circlet	Pisces	Head of the western fish
Hyades	Taurus	Star cluster around Aldebaran
Keystone	Hercules	Four stars in north Her (η, π, ε, ζ Her)
Kids	Auriga	Triangle south of Capella (ε, ζ, η Aur)
Little Dipper	Ursa Minor	Seven stars beginning at Polaris
Little Milk Dipper	Sagittarius	Five-star dipper (λ, τ, σ, ϕ, ξ Sgr)
Northern Cross	Cygnus	Deneb at top of cross
Pleiades	Taurus	Seven Sisters star cluster
Sickle	Leo	Terminates in Regulus
Great Square	Pegasus	Central figure of Peg (α And; α, β, γ Peg)
Teapot	Sagittarius	Little Milk Dipper plus stars to west
Urn (water jar)	Aquarius	Four stars in Y in northern Aqr
Summer Triangle	. . .	Deneb, Vega, Altair
Winter Triangle	. . .	Betelgeuse, Procyon, Sirius

Figure 4.11 *Arabian astronomers, depicted here with a variety of astronomical instruments, carried forward the work of the Greeks and made their own original contributions.*

qualifiers. Some are so luminous that they can be seen even from the centers of big cities. Others are only barely visible on the blackest of nights. About 130 B.C., Hipparchus, the discoverer of precession, devised a simple scheme to order stellar brightness. He arranged the naked eye stars into six categories called **magnitudes,** with first magnitude the brightest and sixth the faintest. The modern version has been decimalized (second magnitude ranges from 1.50 to 2.49, and so on) and extended with higher numbers to fainter stars that can be observed only with telescopes. In addition, the stars of Hiparchus' first magnitude have such a big range that the brightest stretch into the modern magnitude classes 0 and even –1. However, by tradition the brightest 21 stars are still informally lumped together as "first magnitude." (The technical basis of the modern magnitude system will be explained in Chapter 19.)

The luster of the sky is enhanced by stellar colors, which range across reddish, orange, yellow, white, and bluish-white. Only the brighter stars will show their colors to the naked eye, as the eye sees only in black and white at low light levels, but binoculars or a telescope will bring them out for the fainter stars. Color, another stellar qualifier, is related to the stars' temperatures, and is also discussed in Chapter 19.

The constellation patterns provide a framework within which to display and organize the stars. About a thousand of the brighter ones carry **proper names.** These come from a variety of languages and usually relate to the stars' properties or locations. Brilliant Sirius in Canis Major, for example, derives from a word that means "scorching" in Greek. Sirius is also known as the dog star, and Procyon means "before the dog," as this star in Canis Minor rises before Sirius and Canis Major.

The majority of proper names, however, are derived from Arabic, and are a heritage of the time after the fall of the Roman Empire when the great Arabian civilizations cultivated the knowledge of Greece and Rome (Figure 4.11). The Arabians had their own constellation lore, but they adopted the Greek figures and little of it survived. Their principal contribution was to apply quite a large number of names that generally depended on the positions of the stars within the Greek constellations. Deneb, for example, simply means "tail." That name is given to the first-magnitude star in the tail of Cygnus, the Swan, and appears as Denebola in Leo, the Lion, as Deneb Kaitos in Cetus, and as Deneb Algedi in Capricornus.

When the Arabic names were translated into Latin in the Middle Ages, and later, however, many were seriously shortened, distorted, or corrupted, so that their present form and meaning might be totally unintelligible to a modern Arab. For example, the name of the great reddish star in Orion, Betelgeuse, comes from the original phrase "yad al-jauza," meaning "the hand of the central one," but it has been mangled and mistranslated into the "armpit" of the hunter. A list of proper names and their meanings is given in Table 4.4. It includes most of the stars visible from the north brighter than magnitude 2.1 plus several others of interest.

Proper names, however, can be hard to remember and may be ambiguous, as one name may have been applied to several stars. In the early 1600s, Johannes Bayer developed a more methodical system. He applied Greek letters to stars within a constellation more or less in order of brightness (there are several exceptions), and then added the Latin genitive (or possessive) case to the constellation name. Vega, the brightest star in Lyra, then becomes "α of Lyra," or α Lyrae (abbreviated α Lyr), the second brightest β Lyrae, and so on. Tables 4.1 and 4.2 list the genitives and official abbreviations for all the constellations. After Bayer ran out of Greek letters he continued with lower- and upper-case Roman letters, but with a few exceptions these have been discontinued. The **Bayer Greek letters** are all prominent in Figures 4.6, 4.9, and 4.10.

A more extensive naming system was initiated in the eighteenth century by the Englishman John Flamsteed, who organized the stars roughly to fifth magnitude by right ascension within individual constellations. Shortly thereafter the French astronomer Joseph LaLande assigned numbers. By this system, Vega, α Lyrae, is 3 Lyrae as well. The general convention is to use proper or Greek-letter names for the first-magnitude stars, to use Greek-letter names for all the others when available, and finally to go to the **Flamsteed numbers.**

To name fainter stars, astronomers dispense with the constellations and turn to standard catalogues based on position. The most extensive such catalogue is the nineteenth-century *Bonner Durchmusterung* (the "Bonn Survey"), extended as the *Cordoba Durchmusterung* for stars below 23°S declination. The sky (at epoch 1855) was divided into 1° declination strips, and all the stars numbered consecutively by right ascension. These catalogues assign **BD** or **CD numbers** respectively for hundreds of thousands of stars. (Vega carries the BD number BD+38°3238.) The Yale *Bright Star Catalogue* assigns **HR numbers** to all the naked-eye stars in order of increasing right ascension. Other catalogues and naming systems will be introduced as needed.

4.6
The Milky Way

The Milky Way, the visible manifestation of the disk of our Galaxy (see Section 1.2.4), has a rich mythology equal to that of the constellations. In a dark sky it is an awesome sight, the billions of stars that compose it blending into a near-continuous stream with islands of darkness caused by the patchy distribution of light-absorbing interstellar dust (Figure 4.12). The Greeks named it the Milky Circle, and our word *galaxy* comes from the Greek word for "milk." To some cultures it is the street of souls on their way to heaven, to others a great river that brings water from the ocean to the irrigating streams. Its sheer beauty needs no mythology: in a clear dark sky it is a stunning sight.

Because of the absorbing interstellar dust and the Sun's off-center position in the Galaxy, the Milky Way varies considerably in brightness around its circle (follow it in the Appendix star maps). During the northern winter months, when we are facing away from the center, the Milky Way is weak and hard to see through Auriga, Taurus, and Orion. It brightens as the white band passes south to the horizon from Canis Major (see Figure 4.5) to Carina and Puppis, only to be lost from sight. Its beauty is revealed in the northern summer when the Earth is

TABLE 4.4 Proper Names of Selected Stars

Proper Name	Meaning[a]	Greek-Letter Name	Magnitude[b]
Achernar	end of the river	α Eri	0.46
Adhara	virgins	ε CMa	1.50
Albireo	(corr)	β Cyg	2.92
Alcyone[c]	proper name: one of the Pleiades (Gk)	η Tau	2.87
Aldebaran	follower	α Tau	0.85
Algenib	side	γ Peg	2.83
Algieba	forehead	γ Leo	2.30
Alioth	bull (corr)	ε Tau	1.77
Alnitak	girdle	ζ Ori	1.91
Alpheratz	horse's shoulder	α And	2.06
Alkaid	chief mourner	η UMa	1.86
Al Nair	bright one in the fish's tail	α Gru	1.74
Alnasl	point	γ^2 Sgr	2.99
Alphard	the solitary one	α Hya	1.98
Arcturus	bear-watcher (Gk)	α Boo	−0.04
Algol	the demon's head	β Per	2.12v
Alnilam	string of pearls (refers to Orion's belt)	ε Ori	1.70
Altair	flying eagle	α Aql	0.77
Antares	like Mars (Gk)	α Sco	0.96
Bellatrix	female warrior (Lat)	γ Ori	1.64
Betelgeuse	armpit of the central one (corr)	α Ori	0.50v
Canopus	proper name: pilot (Gk)	α Car	−0.72
Capella	she-goat (Lat)	α Aur	0.08
Castor	proper name: twin (Gk)	α Gem	1.58
Cor Caroli	Charles's heart (Lat)	α CVn	2.90
Deneb	tail	α Cyg	1.25
Denebola	tail	β Leo	2.14
Deneb Kaitos	sea monster's tail	β Cet	2.04
Dubhe	bear	α UMa	1.79
Elnath	the butting one	β Tau	1.65
Fomalhaut	fish's mouth	α PsA	1.16
Gemma	jewel (Lat)	β CrB	2.23
Hamal	lamb	α Ari	2.00
Kaus Borealis	north part of bow	λ Sgr	2.81
Kaus Media	middle of bow	δ Sgr	2.70
Kochab	star (?)	β UMi	2.08
Markab	the shoulder	α Peg	2.49
Megrez	root of the tail	δ UMa	3.31
Menkalinen	shoulder of the rein holder	β Aur	1.90
Merak	loin of the bear	β UMa	2.37
Miaplacidus	meaning uncertain; "placidus" = "calm" (Lat)	β Car	1.68
Mintaka	belt	δ Ori	2.21
Mirach	girdle	β And	2.06
Mirfak	elbow	α Per	1.79
Mirzam	announcer (of Sirius)	β CMa	1.98
Mizar	groin	ζ UMa	2.06
Nunki	celestial city (Bab)	σ Sgr	2.02
Phecda	thigh	γ UMa	2.44
Pherkad	calf	γ UMi	3.05
Polaris	pole star (Lat)	α UMi	2.02v
Pollux	proper name: twin (Lat)	β Gem	1.14
Procyon	before the dog (Gk)	β CMi	0.38
Rasalhague	head of serpent-bearer	α Oph	2.08
Regulus	little king (Lat)	α Leo	1.35
Rigel	foot of the central one	β Ori	0.12
Saiph	giant's sword	κ Ori	2.06
Scheat	the shin	β Peg	2.42
Sirius	scorching (Gk)	α CMa	−1.46
Spica	ear of wheat (Lat)	α Vir	0.98
Suhail	(?)	λ Vel	2.21
Thuban	the serpent (corr)	α Dra	3.65
Unukalhay	snake's head	α Ser	2.65
Vindemiatrix	female grape-gatherer (Lat)	ε Vir	2.83
Vega	swooping eagle	α Lyr	0.03
Wezen	weight	δ CMa	1.84
Zubenelgenubi	southern claw of the scorpion	α^2 Lib	2.75
Zubeneschamali	northern claw of the scorpion	β Lib	2.61

[a] "Lat," "Gk," and "Bab" refer respectively to Latin, Greek, and Babylonian names. Names are of Arabic origin unless otherwise noted. Serious corruption is indicated by "corr."

[b] "v" Indicates that the magnitude is variable.

[c] Brightest star in the Pleiades.

Figure 4.12 *The Milky Way, the combined light of the stars of the disk of our Galaxy, stretches from horizon to horizon with the Galaxy's center overhead.*

positioned so that we look in the other direction. It brightens again as it comes out of Cassiopeia into Cygnus (see Figure 4.8), then forms massive star clouds as it plunges through Scutum and down to Sagittarius, where our view is directed through the thickest part of the system toward the very center of the Galaxy. Then, sadly, it is lost to northerners as it passes below the horizon. It is in the southern hemisphere that the Milky Way shines in its full glory. In the middle southern latitudes, the galactic center passes overhead, the great silver stream brilliant as it extends through Centaurus and Crux and back north to float the mighty ship Argo.

4.7
Star Maps

To find your way on Earth you need a map. You also need one to roam the sky. Many early star maps were placed on globes, showing the positions of the stars and constellations on a three-dimensional representations of the celestial sphere. The first accurately plotted flat map came from the hands of a great artist, Albrecht Dürer, in 1515 (Figure 4.13). In addition to placing the stars, he created beautiful wood-block engravings to illustrate the celestial population and to outline the boundaries of the constellations. This tradition of fine ornamentation continued for another 200 years, as seen in the wonderful drawings that illustrate the con-

stellations in the chapter-opening figure and in Figures 4.3, 4.6, 4.9, and 4.10. After Dürer's work, the most famous of these appear in Bayer's *Uranometria,* completed in 1603, which has engravings by Jacobo de Gheyn, one of which is seen in Figure 4.9. These star maps are not merely fanciful depictions of the sky. They are scientifically accurate, with carefully drawn coordinate grids based either on the celestial equator or on the ecliptic.

The engraved figures do not partition the sky very well, however, and the maps hardly include every naked-eye star. Somewhat later, the constellations were outlined by dotted lines on the maps that matched the drawn figures reasonably well (see Figure 4.6), but there was frequently disagreement about where the lines should be, and even about which constellations to include. The confusion began to be resolved in 1922, when the International Astronomical Union, a worldwide organization of professional research astronomers, established a committee to decree the final constellations that were to populate the sky and to draw proper rectangular boundaries so that every star belongs uniquely to a particular constellation.

Today we dispense with the drawings for other than historic or aesthetic purposes as they produce too much clutter and obscure the fainter stars. Some astronomical atlases even do away with the constellation boundaries, providing only a grid of right ascension and declination. One of the most famous atlases is derived from the *Bonner Durchmusterung,* in which the catalogue of nearly a million stars as faint as the 10th magnitude (far below naked-eye vision) was hand-plotted on folio-sized charts. It survives today in a popular amateur atlas. A more modern fundamental atlas is the now out-of-print *Smithsonian Astrophysical Observatory* (SAO) atlas, which plots (and names in its accompanying catalogue) more than half a million stars (Figure 4.14 inset). Copies can usually be found in astronomy libraries. Numerous amateur atlases are available as well, the best of them superior to the older professional versions (see the Bibliography).

For most work, the professional astronomer turns not to plotted maps but to photographic star atlases that are generally too expensive to be in individual hands. The archetype is the *Palomar Observatory Sky Survey* (Figure 4.14), which shows stars as faint as magnitude 20 or so above about 40° south declination. (An even more extensive version is now in preparation.) It is coupled with the European Southern Observatory/Scientific Research Council *Southern Sky Survey,* which goes even fainter. The ultimate atlas-catalogue combination

Figure 4.13 *The planispheres of Albrecht Dürer were the first real flat star maps and are superb combinations of art and science. The northern and southern hemispheres, centered on the ecliptic poles, are shown at top and bottom respectively. The ecliptic runs around the periphery. The constellation figures are reversed, as you would see them on a globe. The gap in the southern hemisphere, centered roughly on the south celestial pole, had yet to be filled with constellations. The astronomers Aratos and Ptolemy appear in the upper corners of the map of the northern hemisphere.*

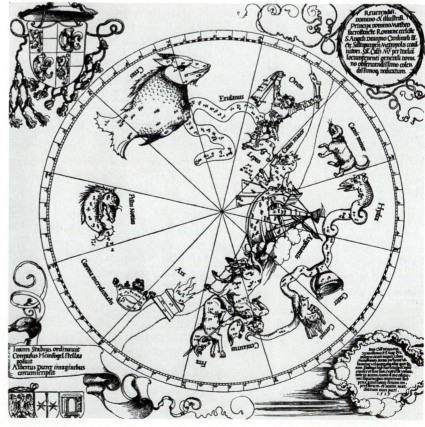

belongs to the Hubble Space Telescope (to be examined in Chapter 9). This *Guide Star Catalogue* lists over *15 million stars* to about the 15th magnitude. It is so large that it is available only through computers on compact disks similar to the familiar audio CDs. The whole sky, to very faint magnitude limits, is now readily available at a library or computer not too far from you.

But these deep atlases are neither necessary nor suitable as you begin to learn about the sky. Instead, examine the star maps in the Appendix, which show the whole sky to about the fourth magnitude and provide some star colors. Try to find the constellations and the stars we have discussed, and on a dark night watch the heavens come alive.

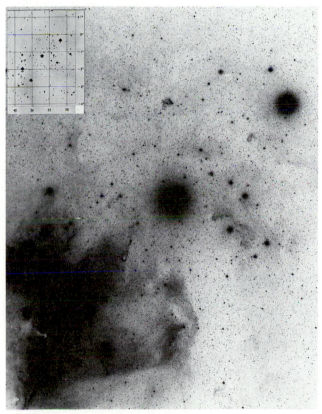

Figure 4.14 *The photograph shows a 2.3° × 3.5° segment of the blue print (that is, a black-and-white photograph taken through a blue filter) of the* Palomar Observatory Sky Survey *centered on Orion's belt. Astronomers commonly use such negative prints. The two bright stars at center and to the right are ε and δ Ori respectively. A cloud of bright gas and dark dust (invisible to the naked eye) nearly obscures ζ Orionis at left. The faintest stars are near magnitude 18. The inset at upper left shows that section of the sky from the Smithsonian Astrophysical Observatory atlas. The declination and the minutes of right ascension (5ʰ) are indicated. Try to match the stars.*

KEY CONCEPTS

Ancient constellations: The 48 constellations handed down by the ancient Greeks.
Asterisms: Small named portions of constellations, or stellar groupings that extend over constellation boundaries.
Bayer Greek letters: Greek letters assigned to stars, usually in order of brightness within a constellation.
BD, CD, or HR numbers: Numbers that catalogue stars by right ascension, ignoring constellations.
Constellations: Named star patterns.
Flamsteed numbers: Numbers assigned to stars in order of increasing right ascension within a constellation.
Magnitudes: Classes of star brightnesses.
Modern constellations: Constellations generally invented since about 1600.
Proper names: Individual names assigned to the brighter stars, usually reflecting the stars' properties or positions; most are of Arabic origin.

EXERCISES

Use the star maps in the Appendix where necessary.

Comparisons

1. What are the differences in time and concept between the ancient and the modern constellations?
2. What is the difference between a formal constellation and an asterism?

Numerical Problems

3. What is the highest northern latitude at which you can see all of **(a)** Crux; **(b)** Centaurus; **(c)** Grus?
4. What is the most southerly latitude at which you can see all of **(a)** Ursa Major; **(b)** Aquila; **(c)** Pegasus?

Thought and Discussion

5. What two constellations comprise the zodiac's "double sign"?
6. To what constellations do the Big and Little Dippers belong?
7. What are the constellations of the "wet quarter"?
8. What are the proper names of **(a)** the two bears; **(b)** the two constellations involving fish; **(c)** the three constellations involving dogs; **(d)** the two crowns?
9. What first-magnitude stars are found in the zodiac?
10. What constellation is divided into two noncontiguous parts?
11. What constellation was broken into three subdivisions, and what are their names?
12. In what principal northern seasons do you find the following constellations in the evening sky: **(a)** Orion; **(b)** Leo; **(c)** Sagittarius; **(d)** Ursa Minor; **(e)** Octans?

13. What constellations surround **(a)** Aquila; **(b)** Scorpius; **(c)** Virgo; **(d)** Monoceros; **(e)** Hydra; **(f)** Dorado?

14. In what hemispheres do you find the modern constellations **(a)** Lynx; **(b)** Fornax; **(c)** Lacerta; **(d)** Microscopium?

15. What constellations represent **(a)** a clock; **(b)** a table; **(c)** an air pump; **(d)** an arrow; **(e)** a cup; **(f)** a dove?

16. To what constellations do the following stars belong: **(a)** Arcturus; **(b)** Achernar; **(c)** Rigel; **(d)** Antares; **(e)** Deneb; **(f)** Adhara?

17. What are the prominent asterisms in **(a)** Taurus; **(b)** Auriga; **(c)** Sagittarius (name two); **(d)** Pisces; **(e)** Aquarius?

18. What are two asterisms that cross over constellation boundaries?

19. What are three ways in which the naked-eye stars are named?

20. What are five constellations that fall within the Milky Way?

Research Problems

21. Try to find an old star atlas or globe in a library or in a museum. Note the date and any constellations that are not in the accepted lists in this chapter. What constellations now contain these groupings?

22. Use a library to research Native American or Chinese constellations. Compare the constellations of the Chinese zodiac with those of the ancient Greek zodiac.

Activities

23. Some clear night go outside and make your own list of constellations. Find a few prominent groupings, outline them, and give them names. Compare your results with those of the ancients.

24. Select a first-magnitude or otherwise well known star. If your college library has the materials, find its name in as many different star catalogues as you can.

Scientific Writing

25. Write a short commentary for beginning readers of a popular astronomy magazine that explains the significance and importance of the constellations in the past and at present, and why astronomers should know them.

26. Write a three-page article for a nature magazine in which you explain how a teacher or a parent could begin to teach the stars and constellations to children.

5 The Earth and the Moon

The motion of the Moon and shadows in space

*A bright rising Moon
sheds its light
over a French shepherd and flock
in a painting
by Jean-François Millet.*

The Moon is a source of endless fascination as it continuously changes both its apparent shape and its location in the sky. We might see it as a lovely slim crescent in the deep blue of evening twilight, or as a full circular disk nearly overhead at midnight, lighting up the Earth so brightly that we can easily find our way home. Lunar lore is nearly endless. People have long and falsely believed that the orientation of the crescent can foretell rain, that crops must be planted at certain phases, and even that the full Moon can bring on insanity, hence the term *lunacy* after Luna, the Roman goddess of the Moon. Folklore aside, the Moon truly has a profound effect on our planet. It largely produces the precession of the equinoxes, causes the tides, and even slows the Earth's rotation, thus lengthening the day.

5.1
The Lunar Orbit

The Moon, our nearest neighbor, takes just under a month to orbit the Earth, moving counterclockwise, in the same direction as the Earth's rotation and revolution (Figure 5.1). The **inclination** of its orbit, the angle an orbit makes with the ecliptic plane, is only 5°; as a result, the Moon is always found within 5° of the ecliptic and within the constellations of the zodiac.

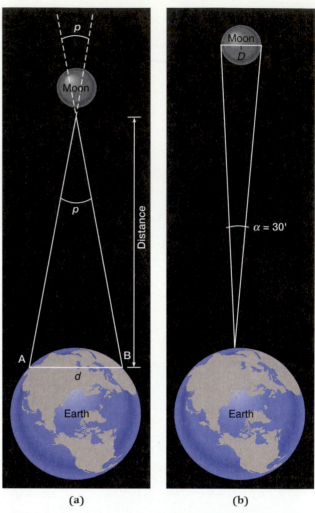

(a)   **(b)**

Figure 5.2   **(a)** *Observers A, B, and the Moon make a long thin triangle. Each observer sees the Moon in a different direction, which establishes a parallax, p. If the separation of the observers, d, is known, they can calculate the lunar distance.* **(b)** *The Moon subtends an angle α equal to 30 minutes of arc. With the distance of the Moon known, we can calculate the lunar diameter, D, in kilometers.*

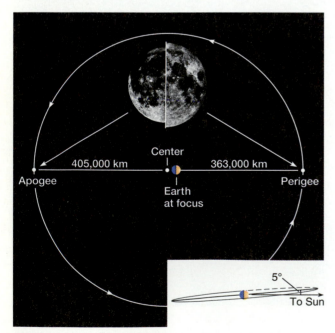

Figure 5.1   *The orbit of the Moon is an ellipse of eccentricity 0.055 with the Earth at one focus. At perigee, the Moon is 363,000 km away; at apogee, 405,000 km. The 11% difference in the Moon's angular diameter at apogee and perigee is readily noticeable from the photographs. The inset at lower right shows the orbit nearly on edge and the 5° tilt relative to the ecliptic plane.*

The Moon's average distance of 384,000 km (1/400 AU) is easily found from its parallax (Section 3.6), which is the apparent shift in the Moon's position when viewed from different locations on the Earth's surface (Figure 5.2a). The orbit is a reasonable approximation to an ellipse with the Earth at one focus (see Figure 5.1). The eccentricity is 0.055, more than three times that of the terrestrial orbit. The lunar distance therefore changes considerably over the month. The minimum distance of 363,000 km takes place at **perigee** (Greek *peri-*, "near," plus *ge*, for Earth). The maximum of 405,000 km is at **apogee** (Greek *apo*, "away from"). With the distance and the measured angular lunar diameter of 30 minutes an arc (Figure 5.2b), the physical

---

BACKGROUND 5.1 Lunar Calendars

The lunar phases have long been an important means of timekeeping. Native American tradition, for example, names the moons that fall in different months: "Snow Moon" for February, "Grass Moon" for April, "Thunder Moon" for July.

Formal calendars, such as that used in Islamic societies, incorporate the 29.5-day synodic period. A simple lunar calendar might use 12 months that alternate between 29 and 30 days. This pattern produces a calendar year of 354 days. However, twelve full phase cycles actually sum to 354.367 days, leaving the simple calendar short by 0.367 days per year. The Muslims reconcile the difference by adding 11 days over a 30-year period to make their average lunar year 354.355 days long, resulting in an error of only one day in 2,500 years. Each month begins with the first sighting of the crescent Moon after the new phase (in ancient times, this thin crescent was called the "new Moon"). Years are counted from A.D. 622 of the civil calendar, the date of Mohammed's escape from his persecutors in Mecca. Muslim years are designated Anno Hegirae, abbreviated A.H., "in the year of the flight." Because the Muslim calendar is nearly 11 days short of a solar year, the months cycle perpetual-ly through the seasons, taking about 36 years to go all the way around.

The Jewish calendar is reconciled with both the Moon and the Sun and is therefore more complicated. Like the Muslim calendar, its months have either 29 or 30 days (but do not alternate strictly for religious reasons) to match the lunar phases, adding the extra days as before. Again, each new month starts with the new Moon. However, the Jewish calendar also makes use of the *Metonic cycle,* discovered by the Greek astronomer Meton (born about 460 B.C.). In 19 years there are almost exactly 235 phase cycles. To keep the lunar calendar in synchronism with the seasons, an extra month—called Adar II, following the month of Adar—has to be added into 7 leap years during the 19-year period. The Jewish holidays then move around considerably from one year to the next relative to the civil calendar, but on the average, over the Metonic cycle, they are reconciled with the solar year. The new year, Rosh Hashana, falls near the time of the autumnal equinox, but can be found anywhere between September 6 and October 5. The Jewish calendar counts years from 3760 B.C.: 1995 of the civil calendar will start the Jewish year 5756.

---

diameter is found to be 3,500 km, roughly a quarter that of the Earth (see MathHelp 3.2).

5.2
The Phases of the Moon

Over the course of its orbit, the apparent shape of the Moon changes radically. These shapes, or **phases** (Figure 5.3), are strictly correlated with lunar visibility, that is, with the Moon's rising and setting times. The Moon produces no light of its own but shines purely by sunlight reflected from its surface. Like the Earth, one full hemisphere is bright with daylight, the other dark with night. The apparent shape of the Moon depends on how much of its daylight side is visible to us on Earth at different orbital positions.

In Figure 5.3, the rotating Earth is in the center. It is surrounded by the lunar orbit, which, for simplicity, has been drawn as a circle. The illuminating Sun is far off the page to the right. If the orbiting Moon is between the Earth and the Sun (position 1), the Moon and Sun must rise and set at the same time. The Moon's sunlit portion faces away from us, we look at the unilluminated nighttime side, and the Moon is not visible. In this position the Moon is said to be *new*.

Now let the Moon proceed a quarter of the way along its orbit to position 3, where it is 90° to the east of the Sun. From Earth, we now see half of each of the daylight and nighttime sides. This *first-quarter* phase (its name reflecting the amount of the orbit the Moon has traversed, not its appearance) looks like a half moon in the sky. When the observer (who is being carried around the Earth) passes from day into night, the Sun is setting on the horizon and the time (on the average) must be 6 P.M. or 18 hours. The first-quarter Moon will therefore be seen on the meridian at sunset and can also be viewed easily in the afternoon. Since the Moon continues to lag behind the Sun in its daily path, it will set at midnight and rise at noon. In between

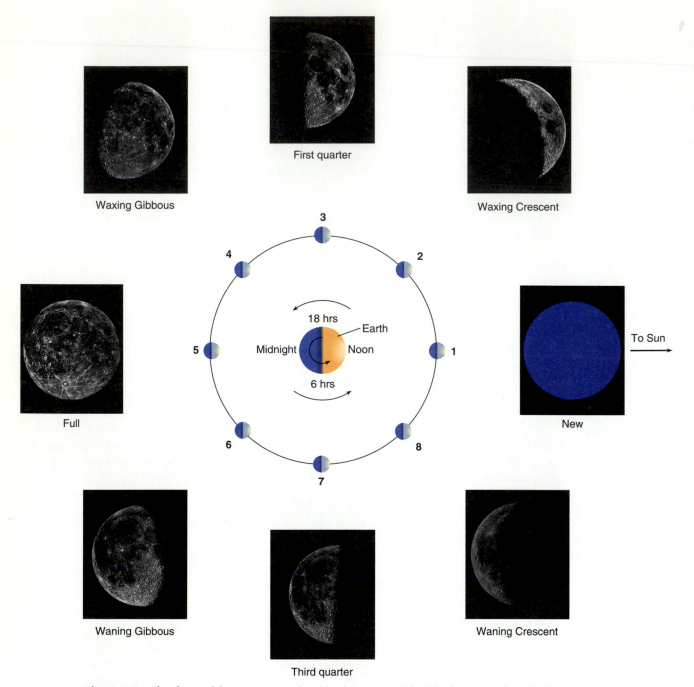

Figure 5.3   *The phases of the Moon are produced by the amount of daylight that is seen from Earth as our satellite revolves about us. At position 1, the new phase, we look only at the nighttime side, and the Moon is invisible as it rises and sets with the Sun. Opposite, at position 5, the full phase, we see the entire sunlit face, which will rise at sunset and set at sunrise. In between are the quarters (positions 3 and 7) where only half the daytime side can be viewed. First quarter will be on the celestial meridian at sunset, third at sunrise. Between the quarters and new we see only a crescent, and between the quarters and full the Moon appears gibbous. As seen from the Moon, the Earth would go through similar, but opposite, phases.*

new and first quarter, for example at position 2, we see less than half of the daylight side, and consequently the Moon takes on the appearance of a *crescent*. Since the Moon makes less than a 90° angle with the Sun, it must set in the evening, before midnight.

At position 5, the Moon is directly opposite the Sun, and we see the full daytime side, or the *full Moon*. The nighttime side now faces away from us. Since the Moon and Sun are opposite in the sky, the full Moon must rise at sunset, transit the meridian at midnight, and set at sunrise. In between first quarter and full, typified by position 4, we see more than half of the sunlit side and it is the *dark* portion that is the crescent. The visible part now takes on what is called a *gibbous* appearance, and the Moon will rise in the afternoon and set after midnight. From new to full the Moon grows in apparent size, and is in its **waxing** phases (from the Anglo-Saxon *weaxen,* "to grow").

Following full, the sequence is reversed. The Moon, now in its **waning** phases, seems to shrink. After full (see position 6) the Moon becomes gibbous again, now rising *after* sunset, and at position 7 it reaches its *third quarter,* where we again see only half the daylight face (the half that was dark at first quarter). Now the Moon *rises* at midnight and *sets* at noon, just the opposite of its behavior at first quarter. The Moon then passes into its waning crescent phase (as in position 8) and is seen in the early morning hours before sunrise. And finally, 29.5 days after the sequence began, the Moon is again new and disappears from the sky.

As you can see from the photographs in Figure 5.3, only one side of the Moon is ever visible from the Earth, the result of our satellite slowly rotating on its axis so that it faces us at all times. To someone on the Moon, the Earth would not appear to move. It would neither rise nor set, but would hang in space at a constant position. The invisible lunar **farside** is sometimes colloquially referred to as the "dark side," a misleading phrase that is meant to refer to mystery, not illumination. Figure 5.3 shows that the far side must experience just as much daylight as its opposite, the **nearside.**

During the crescent phases, you can see the whole lunar outline, the supposed nighttime side bathed in a quiet light (Figure 5.4). The full or nearly full Moon is remarkably bright, yielding enough light that you can almost read large print. Now consider what the night would look like if you were standing on the Moon instead of the Earth. First, the Earth would also go through a set of phases, like the ones in Figure 5.3 but in reverse. If the

Figure 5.4   *At the time of the crescents, you can see the entire outline of the Moon, the nighttime side gently glowing against the brilliant daylight side.*

Figure 5.5   *An astronaut on the Moon stands beneath the gibbous Earth. Terrestrial observers saw the Moon as a crescent.*

Moon is new, the Earth would appear full, and vice versa. If the Moon is seen in a crescent phase from Earth, the Earth would be gibbous as viewed from the Moon (Figure 5.5). Second, the Moon does not reflect much sunlight. Its **albedo,** or percentage reflectivity, is only about 7% because its surface is made of dark rock. In contrast, the Earth's oceans and white clouds reflect nearly 40% of the light that falls on them. Third, the Earth's diameter is four times that of the Moon. Since the area of a circle depends on the square of its radius, the reflecting area of the apparent terrestrial disk as seen from space is 16 times that of the apparent lunar disk. Full **earthlight,** the light from the Earth, on the Moon is then about 80 times brighter than full moonlight on the Earth! The Earth is bright enough

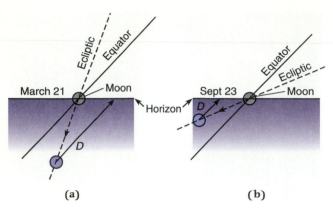

Figure 5.6 *(a) On or about March 21, the full Moon rises about 6 P.M. along with the autumnal equinox as the Sun sets with the vernal equinox. The ecliptic now makes a steep angle with the eastern horizon. At sunset the following night, the Moon has dropped well below the horizon, and the delay in moonrise (shown by the segment of the daily path, D) is large. (b) The vernal equinox is now on the eastern horizon and the ecliptic's pitch angle is small. The rising full Moon will be here in September, and the delay in moonrise from night to night will be smaller.*

to light up the Moon's nighttime side so well that it can easily be seen, producing a wonderful sight poetically referred to as "the old Moon in the new Moon's arms."

5.3
Daily Paths, Moonrise, and Moonset

Since the Moon is confined to within 5° of the ecliptic plane, its daily paths throughout the month will behave much like the solar daily paths throughout the year. When the orbiting Moon passes near the equinoxes in Pisces and Virgo, it will rise and set in the east and west. When it is near the summer solstice in Gemini, it will rise well in the northeast, set in the northwest, and transit high in the sky. When it is near the winter solstice in Sagittarius, it will rise and set in the southeast and southwest and transit low.

On the average, the Moon rises about 45 minutes later each night, but the delay is variable throughout the year and depends on the tilt of the ecliptic relative to the horizon. If you look to the east at sunset in northern spring you see the autumnal equinox rising (Figure 5.6a), and if the Moon is full, it will rise with the equinox at 6 P.M. The ecliptic is descending from the summer solstice and makes a steep angle with the horizon. The following night at 6 P.M., one day past full, the Moon will have dropped far below the horizon, and moonrise will occur over an hour later; the next night it will rise over two hours later. The waning gibbous Moon will then disappear rapidly from the evening sky.

In September, the configuration is reversed. The autumnal equinox is now at the west point and the full Moon will rise with the vernal equinox at 6 P.M. (Figure 5.6b). Now the ecliptic is descending from the winter solstice and the angle it makes with the horizon is quite flat. The delay in moonrise from one night to the next is only about half an hour, and for a few days the early evening is flooded with nearly full moonlight. For centuries this **Harvest Moon** was important to farmers because it provided additional light by which to bring in crops. Moonrise in October behaves similarly, and is called the *Hunter's Moon.*

5.4
Synodic and Sidereal Periods

Two periods, or repeating intervals of time, are needed to describe the orbit of the Moon. The 29.5-day period of the phases is called the **synodic period** (from the Greek *synodos,* "coming together" or "meeting"), as it is the interval between successive "meetings" between the Moon and the Sun.

The **sidereal period,** which relates to the stars, is shorter, 27.3 days. The difference is the result of the Earth's going about the Sun while the Moon goes about the Earth. Begin with the Moon, the Sun, and a star all aligned at position A in Figure 5.7. As the Moon circles the Earth, our own motion will cause the Sun to move steadily to the east of the star. The Moon will then first return to the star at position B as it completes a sidereal period, by which time the Sun has shifted (at the rate of 1° per

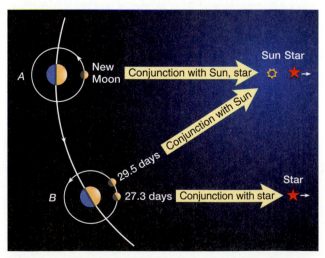

Figure 5.7 *The lunar sidereal period of 27.3 days is the time it takes the Moon to travel from a star in the zodiac back to that same star. The cycle of the phases, however, the synodic period, must be geared to the Sun, and takes longer, 29.5 days, because the Sun is continuously moving to the east of the star at the rate of about 1° per day.*

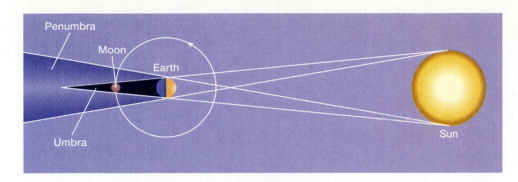

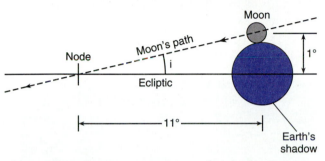

Figure 5.9 *As the Moon approaches a node, its path makes an angle to the ecliptic equal to the orbital inclination, i, of 5° (exaggerated here). For the Moon just barely to graze the terrestrial shadow, it must be within about a degree of the ecliptic and (on the average) within 11° of a node.*

day) about 27° to the east of where it started. To go about the Earth in 27.3 days requires that the Moon slide through the zodiac at a rate of about 13° per day. It will then take 2.2 more days (taking into account the additional motion of the Sun over that time) to catch up with the Sun and return to its new phase. Watch for yourself. Note the position of the Moon against the constellations at any particular phase. It will return to that constellation about two days before it completes its cycle of phases.

5.5
Eclipses of the Moon

The ancients were fascinated by eclipses of the Sun and Moon, and we are no less drawn to them today. Eclipses are produced in sunlight by terrestrial and lunar shadows. When the Moon passes into the shadow thrown by the Earth, it dims and we see an **eclipse of the Moon.** When the Moon places itself between the Earth and Sun and cuts off some sunlight for those of us within its shadow, we witness an **eclipse of the Sun.**

Figure 5.8 shows the Sun shining on the Earth, and the Earth casting a shadow in space (the diagram is very much out of scale: compare it with Figure 1.5). All shadows have two parts: an **umbra,**

in which no direct light falls, and a **penumbra,** a region of partial shade. If you were to stand in the umbra of the terrestrial shadow you could see none of the Sun; if you were in the penumbra you would see that the Earth cuts off only a part of the solar disk. The closer to the umbra you get, the darker the penumbra. The Earth's umbra is very long, projecting 1.38 million km into space. At the Moon's average distance of only 384,000 km, the umbral shadow is 9,700 km in diameter, 2.8 times the size of the lunar globe. Consequently the Moon can be entirely engulfed by the umbra. When it is, we see an eclipse.

From Figure 5.8 you see that the Moon can be eclipsed only when it is opposite the Sun, that is, when it is full. But experience tells you that most full Moons are *not* eclipsed. The reason is the 5° tilt of the lunar orbit, which carries the Moon as much as 5° above or below the ecliptic. The Moon therefore usually misses the umbral shadow, which at the lunar distance is only 1.4° across. The lunar orbit crosses the ecliptic at two points called the **nodes** of the orbit. Figure 5.9 demonstrates that any sort of umbral eclipse is possible only when the center of the full Moon is within about 1° of the ecliptic, half of the angular diameter of the shadow (0.7°) plus half the lunar angular diameter (0.25°). For that to happen the full Moon must be within 11° of one of the nodes (the number varies somewhat, depending on whether the Moon is near perigee or apogee). Even then, if it is *just* within this limit, only a portion of the Moon will cross the shadow for a **partial eclipse of the Moon.**

Just as the Earth's equatorial plane precesses in space, so does the orbital plane of the Moon. This wobble causes the nodes to *regress* or to move westward along the ecliptic at a rate of 19° per year, going all the way around in 18.6 years. Since the Sun is traveling in the opposite direction at the rate of a degree per day, it must encounter a specific node every 365 – 19 = 346 days, an interval called the **eclipse year.** Because the full Moon shifts east-

BACKGROUND 5.2 Ancient Measures of Distance

In the third century B.C., 2,300 years ago, Aristarchus of Samos actually made an estimate of the ratio of the distance of the Sun to that of the Moon by observing the angle between the Moon and the Sun at the time of the quarters. If the Sun is infinitely far away, the angle should be exactly 90°. The closer the Sun, the smaller the angle (Figure 5.10). Aristarchus found an angle of 87° and announced that the Sun is 20 times farther from the Earth than is the Moon. Unfortunately, his method is impossible to apply with any accuracy because the Sun is so far away (20 times more distant than he thought), and because the true angle is so close to 90° that it cannot be discriminated from a right angle. His result was produced by simple (and understandable) observational error. Nevertheless, the idea is ingenious, and even if his measurement was wrong, his conclusion of enormous distance was correct.

A century later, the great Hipparchus measured the distance of the Moon relative to the diameter of the Earth. By watching a total lunar eclipse, he could easily determine that the angular size of the Earth's shadow at the Moon's distance is 1.4°, from which he deduced that the

Moon is at a distance of 59 times the terrestrial radius: this figure is very close to the actual value of 60.3. In the second century A.D., Ptolemy found essentially the same ratio as had Hipparchus from the more direct measurement of lunar parallax. Eratosthenes' measure of the physical size of the Earth gives the true distance of the Moon, and Aristarchus' determination of the ratio of the distances of the Moon and Sun gives at least a measure of the true solar distance! Clearly, it is a mistake to underestimate the abilities of the ancient thinkers.

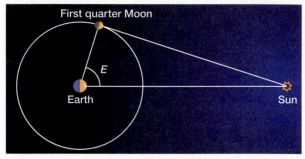

Figure 5.10 *If the Sun were infinitely far away, the first-quarter Moon would appear perpendicular to the solar direction; but if the Sun is nearby, the angle E between the Moon and the Sun at that phase must be significantly less than 90°.*

---

ward by about a zodiacal constellation each month, it will come close to a node only twice during the 346-day cycle, during periods called *eclipse seasons.* As a consequence, only two lunar eclipses are possible during an eclipse year. If one occurs in early January, however, we could get a third in late December during the next eclipse year, for a total of three during an ordinary solar year. It is also possible that during an eclipse season the full Moon may be just too far away from a node and we will see no eclipse at all: the odds are about two out of three that one will occur. Because the nodes keep shifting westward, the Sun encounters a specific one earlier each successive year and the dates of eclipses slowly retreat within the calendar.

For the full lunar disk to be completely immersed in shadow, its center must lie within about half a degree of the ecliptic, half the angular shadow diameter minus half the lunar, a circumstance that occurs only if the Moon is within about 5.5° of the node. Such a **total eclipse of the Moon** will take place at about half the events. The Moon

moves at a rate of about half a degree—its own angular diameter—in an hour, so under the most favorable circumstances (full Moon right at the node) it will completely transit the shadow in only about $3\frac{1}{2}$ hours. Therefore only a little over one hemisphere of the Earth can be witness to the sight, cutting the number of total eclipses seen in any one place nearly in half again. We might then see a total eclipse from any given location about once every three years.

Total lunar eclipses are great fun to watch and always attract large numbers of people. The sequence of an eclipse is diagrammed in Figure 5.11 and shown photographically in Figure 5.12. In the drawing, the Moon is moving to the left. Its passage through the penumbral shadow, which starts at P_1, is only barely noticeable. The visible event really starts at position 1, when the Moon encounters the umbra and we see a distinct dark bite taken out of the lunar disk. The eclipse will be partial until position 2, when the whole Moon is in shadow. Now until position 3 the eclipse will be total;

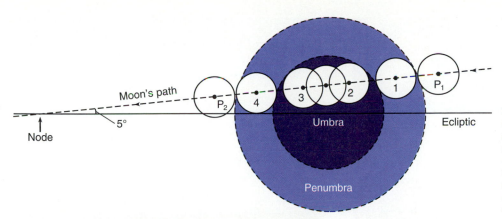

Figure 5.11   *The Moon first strikes the penumbra at P_1 and contacts the umbra, where the eclipse is partial, at position 1. Between positions 2 and 3 it is totally immersed in the umbra. A glimpse of sunlight is seen at position 3; the Moon leaves the umbra at position 4 and the penumbra at P_2.*

Figure 5.12   *The Moon proceeds through a total eclipse in this multiple exposure. Because the camera is fixed and the Moon is rapidly moving on a daily path, the eclipse proceeds from left to right, backward from the diagram in Figure 5.11.*

then the partial phase resumes and continues until position 4. The event is finally over when the Moon leaves the penumbra at P_2.

Remarkably, the Moon does not usually disappear when in full shadow, as the Earth's atmosphere scatters and refracts, or bends, considerable sunlight into the umbra. However, absorption in the air also causes the sunlight to be considerably reddened (the same reason that the Sun often looks red on the horizon), and consequently the eclipsed Moon takes on a dull, dark brick-red color (seen in Figure 5.12). The brightness of the eclipsed Moon depends heavily on terrestrial volcanic activity, which can raise obscuring dust into the atmosphere. Consequently, no two eclipses are quite alike. From the known orbital characteristics of the Moon, it is possible to predict eclipses with high precision far into the future. Table 5.1 gives a list of total eclipses through the year 2020.

5.6
Eclipses of the Sun

If new Moon occurs when the Moon is passing a node, the Moon can block out, or eclipse, the Sun. The Moon is only a quarter the diameter of the Earth and its shadow is correspondingly shorter than Earth's (Figure 5.13). As seen from Earth, the Moon and the Sun have very nearly the same angu-

TABLE 5.1
Eclipses of the Moon, 1994–2019[a]

Greenwich Date and Hour		Type	Moon Overhead at Mid-Eclipse
1994	May 25	3	P	53°W, 21°S; Brazil
1995	Apr. 15	12	P	176°E, 10°S; Micronesia
1996	Apr. 4	0	T	1°W, 6°S; Gulf of Guinea
1996	Sept. 27	3	T	46°W, 1°N; Brazil
1997	Mar. 24	5	P	69°W, 1°S; Colombia
1997	Sept. 16	19	T	77°E, 3°S; Indian Ocean
1999	July 28	12	P	172°W, 19°S; Samoa
2000	Jan. 21	5	T	68°W, 20°N; Haiti
2000	July 16	14	T	153°E, 21°S; Coral Sea
2001	Jan. 9	20	T	57°E, 22°N; Oman
2001	July 5	15	P	136°E, 23°S; Australia
2003	May 16	4	T	56°W, 19°S; Brazil
2003	Nov. 9	1	T	23°W, 16°N; Cape Verde Islands
2004	May 4	21	T	52°E, 16°S; Madagascar
2004	Oct. 28	3	T	50°W, 13°N; West Indies
2005	Oct. 17	12	P	176°E, 10°N; Central Pacific
2006	Sept. 7	19	P	76°E, 6°S; Indian Ocean
2007	Mar. 3	23	T	13°E, 7°N; E. Africa Coast
2007	Aug. 28	11	T	158°W, 10°S; Central Pacific
2008	Feb. 21	3	T	48°W, 11°N; W. Central Atlantic
2008	Aug. 16	21	P	44°E, 14°S; Madagascar
2009	Dec. 31	19	P	69°E, 23°N; W. India
2010	Jun. 26	12	P	174°W, 23°S; Samoa
2010	Dec. 21	8	T	125°W, 23°N; W. Mexican Coast
2011	Jun. 15	20	T	57°E, 23°S; Madagascar
2011	Dec. 10	15	T	140°E, 23°N; Marianas
2012	Jun. 4	11	P	166°W, 22°S; Cook Islands
2013	Apr. 25	20	P	57°E, 13°S; Madagascar
2014	Apr. 15	8	T	117°W, 9°S; Central Pacific
2014	Oct. 8	11	T	166°W, 6°N; Central Pacific
2015	Apr. 4	12	T	180°, 6°S; Central Pacific
2015	Sept. 28	3	T	44°W, 2°N; N. Brazil Coast
2017	Aug. 7	18	P	87°E, 16°S; Indian Ocean
2018	Jan. 31	14	T	161°E, 17°N; Central Pacific
2018	Jul. 27	20	T	56°E, 19°S; Mauritania
2019	Jan. 21	5	T	75°W, 20°N; Cuba
2019	Jul. 16	22	P	39°E, 21°S; Madagascar

[a]Twenty-three total (T) and 14 partial (P) lunar eclipses are scheduled to occur in the 27-year interval 1994 through 2020. Roughly, each is visible over the terrestrial hemisphere whose pole is located at the indicated latitude and longitude.

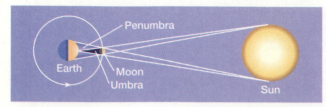

Figure 5.13 *When the Moon is new and near a node, it can cast its short umbral shadow upon the Earth.*

Figure 5.14 *The great "ring of fire" annular eclipse was seen in a brightly colored sunset from the West Coast of the United States in January 1992.*

TABLE 5.2
Eclipses of the Sun, 1994–2020[a]

Greenwich Date	Type	Path
1994 May 10	A	N. Pacific, U.S., N. Atlantic
1994 Nov. 3	T	S. Pacific, Cent. S. America, S. Atlantic
1995 Apr. 29	A	S. Pacific, N. South America
1995 Oct. 24	T	Iran, India, Thailand, W. Pacific Islands
1997 Mar. 9	T	Siberia, Arctic Ocean
1998 Feb. 26	T	Cent. Pacific, N. South America, N. Atlantic
1998 Aug. 22	A	Indonesia, S. Pacific
1999 Feb. 16	A	S. Africa, Indian Ocean, Australia
1999 Aug. 11	T	North Atlantic, Europe, India, Thailand
2001 June 21	T	S. Atlantic, S. Africa, Madagascar
2001 Dec. 14	A	Pacific Ocean, Central America
2002 June 10	A	Northern Pacific Ocean
2002 Dec. 4	T	S. Africa, Indian Ocean, Australia
2003 May 31	A	Greenland, North Atlantic
2003 Nov. 23	T	S. Indian Ocean, Antarctica
2005 Apr. 8	A-T	S. Pacific, N. South America
2005 Oct. 3	A	N. Atlantic, N. Africa, Indian Ocean
2006 Mar. 29	T	Atlantic, N. Africa, Central Asia
2006 Sept. 22	A	Indian Ocean, S. Atlantic, N. South America
2008 Feb. 7	A	Antarctica, S. Pacific
2008 Aug. 1	T	N. Canada, Arctic, Russia
2009 Jan. 26	A	Atlantic, Indian Ocean, Indonesia
2009 July 22	T	India, China, Central Pacific
2010 Jan. 15	A	Africa, Indian Ocean, China
2010 July 11	T	Central Pacific
2012 May 20	A	China, N. Pacific, United States
2012 Nov. 13	T	N. Australia, S. Pacific
2013 May 10	A	N. Australia, New Guinea, Cent. Pacific
2013 Nov. 3	A	Atlantic, Central Africa
2014 Apr. 29	A	Antarctica
2015 Mar. 20	T	N. Atlantic, Arctic
2016 Mar. 9	T	Indian Ocean, Borneo, Central Pacific
2016 Sept. 1	A	Atlantic, Africa, Indian Ocean
2017 Feb. 26	A	S. South America, Atlantic, Cent. Africa
2017 Aug. 21	T	Pacific, United States, Atlantic
2019 Jul. 2	T	Pacific, S. South America
2019 Dec. 26	A	Arabia, Indian Ocean, Borneo
2020 Jun. 21	A	Africa, China, Central Pacific
2020 Dec. 14	T	Pacific, S. America, Atlantic

[a]Twenty annular (A), 18 total (T), and 1 annular-total (A-T) eclipses are scheduled for the 27-year interval from 1994 through 2020. The path of each eclipse begins in the first-listed geographic area, sweeps sequentially across other regions indicated, and ends in the last-listed area. In addition, there are partial-only eclipses that are seen only at higher latitudes, generally near the poles. Saros pairs can be seen starting with 1994 May 10–2012 May 20, continuing to the end of the table with 2002 Dec. 4–2020 Dec. 14.

lar diameters, about half a degree. Even at the average lunar distance, however, the Moon's angular diameter is slightly the smaller of the two, and as a result, the umbral shadow cannot quite reach us. We then see an **annular eclipse** (Figure 5.14), one in which the darkened Moon is surrounded by a thin ring (or annulus) of sunlight.

For the new Moon to cover the Sun fully so that the umbra strikes the Earth to produce a **total eclipse of the Sun,** the Moon must also be near perigee. At best, the shadow on the ground directly below the Moon is then a spot no more than 269 km across (Figure 5.15). Since this shadow flies along at nearly 2,000 km/hour at the equator—the speed of the orbiting Moon minus that of the spinning Earth—an individual can be immersed in the lunar umbra for at most only about seven minutes! The shadow spot is so small that a given location will have an eclipse only on the average of once every 300 or so years. However, whether the cen-

tral eclipse is annular or total, the lunar penumbra is thousands of kilometers across, so that at any one time a large portion of the population can see a **partial eclipse of the Sun,** in which only a portion of the Sun is covered.

Just where on Earth the shadow strikes, or where we will see the annular or partial phases, depends upon how far the new Moon is from the node when it covers the Sun. If it is at the node, totality will be close to the equator; but if it is near

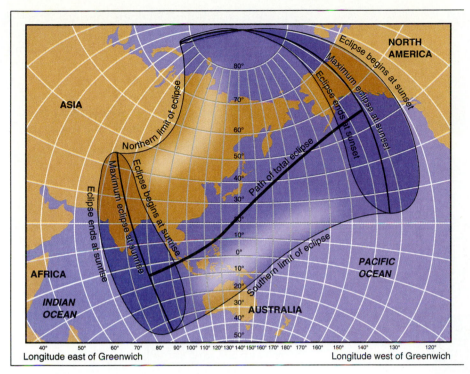

Figure 5.15 *The umbra of the Moon's shadow is so small that as it flies across the Earth (black path) only a few people see totality. The penumbra (shaded in darker blue), in which only a portion of the solar disk is covered, is much larger. This eclipse, which took place on March 18, 1988, crossed Sumatra and the Philippines and just missed the Aleutian Islands of Alaska. The apparent curvature of the eclipse path is caused by the distorted flat map.*

the edge of the allowed limit, well away from the node, the event will take place closer to one of the poles, and the eclipse may be partial only, with no total or annular phase. Predictions of total or annular solar eclipses through the year 2020 are given in Table 5.2.

Although solar eclipses are uncommon at any given location, worldwide they are actually more common than the lunar variety. It is possible for the Moon to avoid an eclipse at full, but the new Moon that takes place closest to nodal passage *must* produce an eclipse somewhere on Earth, and consequently there will be at least two every eclipse year. If the node is positioned about halfway between the locations of two successive new Moons, there will actually be *two* eclipses during a nodal passage, separated by a synodic month. In between, there has to be a total eclipse of the Moon. This triplet can occur twice during the eclipse year, for a total of four solar and two lunar eclipses. Because the eclipse year is shorter than the solar year, it is again possible to add a solar eclipse in early January or late December for a total of seven eclipses, five solar and two lunar. This rare occurrence last took place in 1935 and will not happen again until 2160.

As it happens, 223 synodic months, an interval of 18 years $11\frac{1}{3}$ days called a **saros,** is only 0.46 days shy of 19 eclipse years. (The word is of Chaldean origin: we inherited it from a people who lived and

discovered the interval more than 3,000 years ago!) As a result, an eclipse we see today will be followed by another one that is almost identical to it 18 years $11\frac{1}{3}$ days from now, but that will take place 120° of longitude to the west because of the extra one-third rotation of the Earth. After three such periods, another eclipse will occur at about the original longitude. A **saros cycle** begins with an eclipse near either the north or south pole. Each successive eclipse within the cycle will be separated by the saros interval and will be shifted slightly in latitude. It takes about 600 years and 30 to 35 eclipses for the cycle to work its way to the other pole and disappear. Since there must be at least two eclipses every year, there will be a good part of a hundred different saros cycles going on at any one time.

A total solar eclipse presents a number of remarkable phenomena (Figure 5.16). *Note, however, that only during totality is it safe to look directly at the Sun without an adequate solar filter. Any direct sunlight is bright enough to cause partial blindness.* During the partial phases use pinhole projection. Make a small hole in a piece of cardboard and project the image against paper or even onto the ground. At first you see only a tiny piece of the Moon over the Sun, but then the dark section gets larger, the Sun becomes a crescent (Figure 5.16a), and the sky begins to take on a cool deep-blue hue. Look beneath the trees shortly before totality and you will see hundreds of little crescents lying on

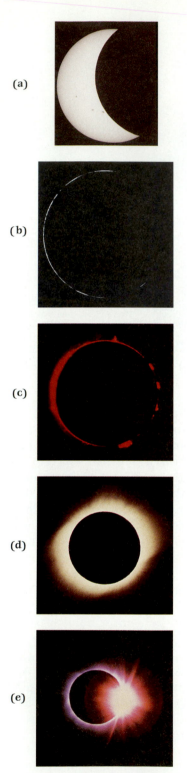

(a)

(b)

(c)

(d)

(e)

Figure 5.16   **(a)** *The Moon takes a small bite out of the Sun.* **(b)** *Just before totality, the last bit of sunlight shines through the lunar valleys to produce Baily's beads. Suddenly, the Sun is covered and the red chromosphere* **(c)** *and the corona* **(d)** *pop into view. At the end of the total eclipse, the first bit of sunlight coupled with the fading corona creates the beautiful "diamond ring"* **(e).**

the ground, the holes between the leaves acting like little pinhole projectors. Just before totality, when only a sliver of the Sun is showing, you may notice bands of shadows racing across the ground, and in the distance you can see the onrushing shadow of the Moon high in the Earth's atmosphere. The last bit of sunlight then comes shining through the valleys at the lunar edge, producing an effect called **Baily's beads** (Figure 5.16b), and finally the Sun is covered during totality. At first you see a fiery red ring surrounding the darkened lunar disk (Figure 5.16c)—the *chromosphere*—and then the glorious *corona* (Figure 5.16d), a pearly-white envelope that extends many solar diameters outward; the natures of these layers will be explained in Chapter 18. The corona can be seen by eye only during totality; the blue sky hides it during the day. (Solar eclipses are among the few chances we have to examine the corona, a hot, turbulent extended solar atmosphere that tells us a great deal about the workings of the Sun.) After only a few minutes, the first rays of brilliant sunlight break out, the corona disappears, and for a fraction of a second you see the stunning "diamond ring" (Figure 5.16e) surrounding the blackened disk of the new Moon. The partial phases now repeat in reverse. The event is now nearly over, with just the partial phases left to go. Once seen, it will never be forgotten.

KEY CONCEPTS

Albedo: Percentage reflectivity of a surface.

Earthlight: Light from the Earth seen illuminating the Moon.

Eclipse of the Moon: The passage of the Moon through the Earth's shadow; in a **partial eclipse of the Moon,** only part of the Moon passes through shadow; in a **total eclipse of the Moon,** the entire Moon is immersed.

Eclipse of the Sun: The passage of the Moon across the Sun and the lunar shadow across the Earth; in a **partial eclipse of the Sun,** the Moon blocks only part of the Sun; in a **total eclipse of the Sun,** the Moon covers the entire Sun; in an **annular eclipse,** the Moon is too far from the Earth to cover the Sun completely, leaving a ring.

Eclipse year: The interval of 346 days between successive passages of the Moon across a specific node.

Farside: The invisible side of the Moon away from Earth; the side facing us is the **nearside.**

Harvest Moon: The full Moon nearest the time of the autumnal equinox, following which there is a great deal of early evening moonlight.

Inclination: The angle that an orbital plane makes with the ecliptic plane.

Nodes: The intersections of an orbital path and the ecliptic as viewed from Earth.

Perigee: The point in the lunar orbit (or in any other orbit around Earth) closest to the Earth; **apogee** is the most distant point.

Phases (of the Moon): The different apparent shapes of the Moon caused by viewing different segments of the lighted side as the Moon orbits the Earth; the **waxing phases** are the growing phases between new and full Moon, and the **waning phases** are the diminishing phases between full and new Moon.

Saros: an interval of 18 years $11\frac{1}{3}$ days, after which the circumstances of an eclipse repeat themselves; a **saros cycle** is an interval of about 600 years in which similar eclipses separated by a saros work their way from one terrestrial pole to the other.

Sidereal period: The orbital period relative to the stars; for the Moon, 27.3 days.

Synodic period: The orbital period relative to the Sun; for the Moon, the 29.5-day period of the phases.

Umbra: A region of full shadow; the **penumbra** is a region of partial shadow.

EXERCISES

Comparisons

1. What is the difference between waning and waxing crescent Moons?
2. What is the difference (numerically and conceptually) between the Moon's sidereal and synodic periods?
3. What is the difference between the umbra and the penumbra of a shadow?
4. What is the conceptual difference between a partial eclipse of the Sun and a partial eclipse of the Moon?
5. What is the difference between total and annular eclipses of the Sun? How do they relate to the angular diameters of the Moon and Sun?

Numerical Problems

6. At approximately what time would you expect the Moon to set if it were a crescent halfway between new and first quarter? About what time would you expect it to rise if it were gibbous and halfway between full and third quarter?
7. How far in angle does the Moon move with respect to the background stars between two successive new Moons?
8. What are the maximum and minimum lunar declinations?
9. From the diameters and distances of the Moon and Sun given in this chapter and in Chapter 1, show that the lunar and solar angular diameters are both about half a degree.

10. From Table 5.2, give two saros pairs for total solar eclipses.

Thought and Discussion

11. If the Moon is full on March 12, what would be the phase on **(a)** March 19; **(b)** March 31?
12. If the Moon is in its third quarter, what is the phase of the Earth as viewed from the Moon?
13. Why do we not see the whole outline of the Moon when it is in its gibbous phase as we do when it is a crescent?
14. How would the crescent Moon look if the Earth had an albedo of zero?
15. How would the Moon appear to move over the month if viewed from the Earth's north pole?
16. Why can we see the full Moon during a total lunar eclipse?
17. What are the conditions under which you will see **(a)** a total lunar eclipse; **(b)** a total solar eclipse; **(c)** an annular solar eclipse?
18. A viewer on Earth sees **(a)** a total eclipse of the Moon; **(b)** a total eclipse of the Sun. What would these events look like from the Moon?
19. What is meant by the term *eclipse year*? Why is it shorter than the solar year?

Research Problems

20. Use popular astronomy magazines on file in the library to find information on the brightness of the Moon during the past six lunar eclipses. List the reported reasons for the eclipse illumination.
21. Use the same magazines as in Question 20 to compare the appearances of the past three total solar eclipses. Why did the eclipses appear different to the observer, and where did they occur?

Activities

22. Photocopy the equatorial star maps in the Appendix. Plot the apparent path of the Moon over the course of a month, noting the phases and the visibility of earthlight. Locate the nodes.
23. If the timing is right, observe a lunar eclipse. Note **(a)** the times of contact; **(b)** the appearance of the Moon during the total phase, especially the variation in brightness across the lunar surface. Why should the brightness vary?

Scientific Writing

24. It is commonly believed that the phases of the Moon are somehow caused by the shadow of the Earth. Write an illustrated article for a children's magazine that shows the difference between lunar phases and lunar eclipses.
25. Write a newspaper article describing the next total solar eclipse and what the observers might expect to see. Include a safety message.

6

The Planets

Planetary orbits and how we learned of their natures

Galileo is seen with his telescope in this fresco by the Florentine L. Sabatelli.

If you watch the constellations of the zodiac for any length of time you will see that five of their brightest "stars" do not stay fixed, but *move* (Figure 6.1). These bodies long ago received the name **planets,** from the Greek *planetai,* "wanderers." Because of their brilliance and their motions through the seemingly mystic constellations of the zodiac, the *ancient planets* (those known since antiquity) were long ago assigned the names and personalities of the gods: Mercury, the swift messenger, is the fastest; Venus, the goddess of beauty, is the loveliest; Mars, the god of war, is the red planet; Jupiter, the king of the gods, moves through the zodiac at a stately pace; and Saturn, Jupiter's father, is fittingly the faintest and slowest. The search for the laws that govern the apparent wanderings of these bodies lasted more than 2,000 years.

6.1
Basic Orbital Characteristics

The nine planets, including the Earth and the three discovered in modern times (Uranus, Neptune, and Pluto), are listed in Table 6.1 along with a set of fundamental characteristics to be explored. Like our Moon, the planets shine by reflected sunlight, their apparent brilliance depending on their physical

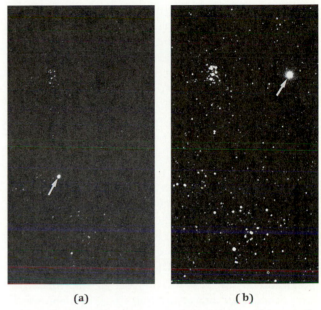

(a)   (b)

Figure 6.1 *Jupiter moves westerly against the stars of the constellation Taurus between October 1988* **(a)** *and January 1989* **(b)**. *In both photographs the Pleiades cluster is seen at upper left and the Hyades at lower center.*

TABLE 6.1
Planetary Orbital Data

Planet	m	a (AU)	e	i°	P Sidereal	P Synodic	n°/day	$v_{km/s}$	R
(1)	(2)	(3)	(4)	(5)	(6)	(7) (days)	(8)	(9)	(10)
Mercury	−1.8	0.387	0.206	7.00	88d	116	4.09	48	23
Venus	−4.7	0.723	0.007	3.39	225d	584	1.60	35	42
Earth	—	1.000	0.017	0.00	1.000y	—	0.99	30	—
Mars	−2.8[a]	1.524	0.093	1.85	1.88y	780	0.52	24	73
Ceres[b]	+7.9	2.77	0.097	10.6	4.61y	466	0.21	18	85
Jupiter	−2.9	5.20	0.048	1.31	11.9y	399	0.083	13	121
Saturn	+0.0	9.54	0.056	2.49	29.5y	378	0.033	10	138
Uranus	+5.5	19.2	0.047	0.77	84.0y	370	0.012	7	152
Neptune	+7.7	30.1	0.010	1.77	165y	367	0.006	5	158
Pluto	+14.0	39.5	0.248	17.1	249y	367	0.004	5	162

The columns are 1: names; 2: maximum brightnesses in magnitudes (m); 3: semi-major axes in AU (a); 4: orbital eccentricities (e); 5: inclinations of the orbits to the plane of the ecliptic (i); 6 and 7: sidereal and synodic periods; 8: average rates of motion along the orbit in degrees per day, n; 9: mean orbital speeds in km/s, v; 10: average number of days each planet spends in retrograde motion each year as viewed from Earth, R.

[a]At favorable opposition.
[b]Largest asteroid (Chapter 17).

sizes, albedos, and distances. The maximum brightnesses (minimum magnitude numbers, m) in the second column show that the ancient planets are bright and easily visible, all but the last outshining any star. Of the remaining three, only Uranus can be seen with the naked eye, and then just barely.

All planetary orbits (shown to scale in Figure 6.2) are ellipses with the Sun at one focus. The semimajor axes, which are the average distances, are given in the third column of Table 6.1 and range from only 0.4 AU to 39 AU. The fourth column of the table gives the eccentricities: only the paths of Mercury, Mars, and Pluto deviate much from circles. Pluto's orbit is so eccentric that near perihelion it can come closer to the Sun than Neptune (where it will be until 1999).

Most planets are always seen near the ecliptic, and therefore their orbital planes have low tilts, or inclinations, to the ecliptic plane (Figure 6.3). Values of inclination, i, are given for all the planets in the fifth column of Table 6.1 (recall that the Moon's orbital inclination is 5°). Pluto is the only real exception to the general flatness of the Solar System, as its value of i is so large that Pluto stays well away from Neptune near its perihelion and can leave the zodiac altogether.

The **sidereal period** of a planet is the time required for it to orbit the Sun, that is, to go from one point in its orbit and return. The farther a planet is from the Sun, the greater distance it must travel and the slower it moves. Consequently, more distant planets have longer sidereal periods. The sidereal periods of all the planets are given in the sixth column of Table 6.1, and the orbital speeds in the ninth.

6.2
The Apparent Motions of the Planets

As the planets orbit the Sun, they continuously appear to change their apparent positions relative to the Sun and to one another. Certain points in a planet's journey are of particular interest as they relate strongly to its visibility, which is quite different for the **inferior planets** (those closer to the Sun than the Earth: Mercury and Venus) and the **superior planets** (those farther away).

6.2.1 The Superior Planets

The superior planets are typified by Mars, whose orbital relation to the Sun and the Earth is shown in Figure 6.4. Planetary positions are defined by the angle that the line from the Earth to the body makes with that from the Earth to the Sun, called the **elongation,** E. If a planet is directly lined up with the Sun, $E = 0°$, it is said to be in **conjunction** with the Sun. At this point, like the new Moon, it rises and sets with the Sun and cannot be seen. If the Sun and a planet are in opposite directions in the sky, elongation 180°, they are in **opposition** to one another. Now, like the full Moon, the planet rises at sunset, sets at sunrise, and transits the celestial meridian at midnight. We then see it all night.

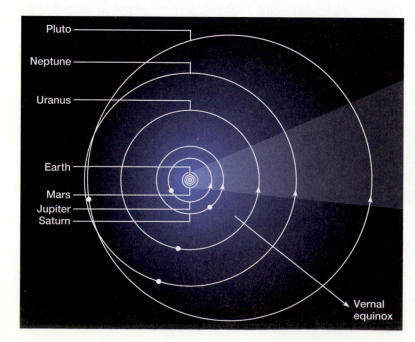

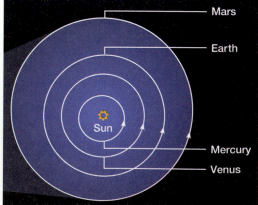

Figure 6.2   *The planets all orbit the Sun and, except for Mercury, Mars, and Pluto, on paths that are almost circular. The positions of Jupiter, Saturn, Uranus, Neptune, and Pluto are shown for the year 1995 (the others move too quickly to place). On this scale the actual sizes of the planets are microscopic; even huge Jupiter would be only 0.001 mm across.*

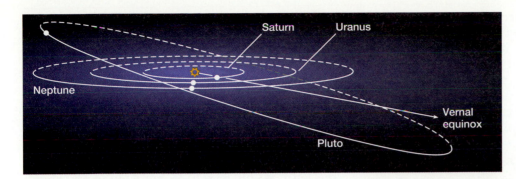

Figure 6.3 *A side view of the outer four planets clearly indicates the plane of the Solar System and Pluto's 17° inclination. The positions of the planets are shown for 1995.*

Like the Moon, the planets have synodic as well as sidereal periods. The **synodic period** of a superior planet is the interval between successive oppositions or conjunctions. It is the lapping time of the speedier Earth relative to the more slowly moving outer planet, and expresses the interval between successive times of best observability. Figure 6.5 demonstrates the synodic period of Jupiter. Start with position 1, where Jupiter is in opposition to the Sun. After half a year has elapsed, by position 2, the Earth has moved through half its orbit and Jupiter has moved through only a small arc. After a full terrestrial orbit, position 3, the Earth has returned to its starting point and Jupiter, with a sidereal period of 12 years, has traced out a twelfth of its path. It will take the Earth about another month to catch up with it, during which time Jupiter will have shifted a bit more, to position 4. Finally, after 399 days, or 1.09 years, Jupiter is back to opposition with the Sun. The farther the superior planet is from the Sun, the less it moves in orbit in a year, and the shorter the synodic period, as seen from the seventh column of Table 6.1.

The motions of the planets puzzled the ancient astronomers and ultimately provided one of the clues that allowed us to understand how the Solar System is constructed. Watch any of the superior planets. Because (like Mars in the example shown in Figure 6.6) it is moving counterclockwise around the Sun (as viewed from above the north eliptic pole) it will generally appear to be traveling eastward through the zodiac, in the same direction the Sun travels along the ecliptic, at the average angular rate given in the eighth column of Table 6.1. But as opposition approaches, the planet slowly grinds to a halt and then begins its **retrograde motion,** in which it goes *backward,* to the west (eastward motion is called *prograde*). The planet reaches maximum angular westward velocity at opposition, and then after a time it again slows, stops, and resumes traveling east.

The phenomenon occurs because we are observing from a moving body (Figure 6.7). As the

Figure 6.4 *When a superior planet like Mars is lined up with the Sun, elongation E = 0°, it is in conjunction with the Sun; when it is opposite the Sun in the sky, elongation E = 180°, it is in opposition to the Sun.*

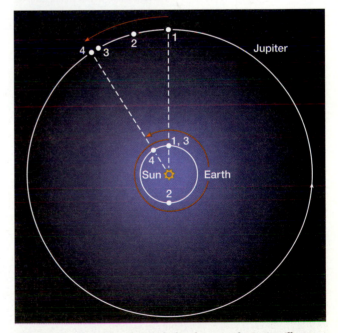

Figure 6.5 *The sidereal period of a planet (as for Jupiter illustrated here) is the time it takes to revolve entirely about its orbit, from position 1 back to position 1 again. The synodic period is the interval between successive oppositions or conjunctions, here the time between position 1, at which Jupiter is in opposition to the Sun, through 2 and 3, and then to 4.*

Earth overtakes a superior planet and passes between it and the Sun, the outer one simply *appears* to go in the reverse direction. As you pass a car on the highway while you both are driving north, the other seems to be going south. Since the planet has a small but significant orbital inclination relative to the orbit of the Earth, it will usually be moving away from or toward the ecliptic. The planet in its retrograde motion will then not just move back and forth but will trace out a loop or irregular path (see Figure 6.6).

The brightness of a superior planet depends on its distance from Earth. It will be greatest at closest approach, or at opposition to the Sun, the position

for which magnitudes are given in the second column of Table 6.1. Mars exhibits the greatest variation. Near conjunction, just before it disappears into or comes out of the solar glare, Mars averages a distance nearly five times farther from the Earth than at conjunction and is 3.4 magnitudes fainter. Of more importance is the relatively large eccentricity of its orbit (Figure 6.8). If opposition takes place when Mars is near perihelion, a circumstance known as a *favorable opposition,* the distance from the Earth is only 56 million km and the planet outshines Jupiter. But if Mars is at aphelion, the distance is almost twice as great, and the planet is 1.3 magnitudes fainter than when closest to us.

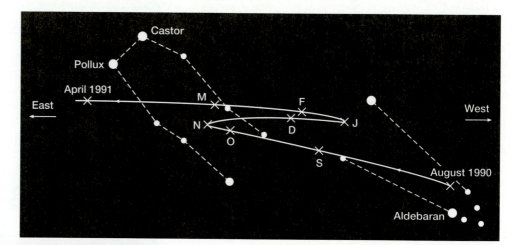

Figure 6.6   *The path traced out by Mars in 1990–91 is shown against the background of Taurus and Gemini. The letters indicate the Martian position at the beginnings of the months. Opposition occurred on November 27, 1990.*

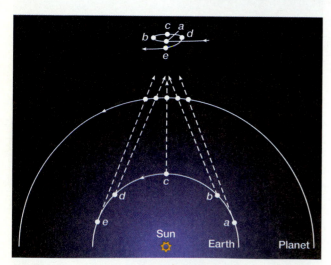

Figure 6.7   *Retrograde motion occurs when the Earth overtakes a superior planet (or is overtaken by one of the inferiors). The planet is orbiting counterclockwise and, if the Earth were stationary, would appear to be moving easterly through the stars. The dashed arrows show the direction of the planet as seen from Earth from points a to e; the loop above shows the appearance of the motion in the sky. From a to b the planet still moves easterly, but from b through c (opposition) to d it moves westerly. Between d and e it resumes easterly movement.*

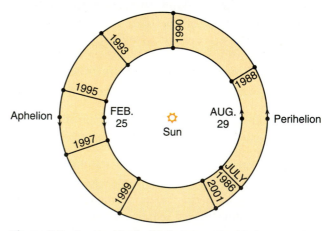

Figure 6.8   *Because Mars' orbit deviates noticeably from a circle, successive oppositions occur at quite different distances. Favorable oppositions take place every 17 years. The best possible time for an opposition to occur is August 29; the worst is February 25.*

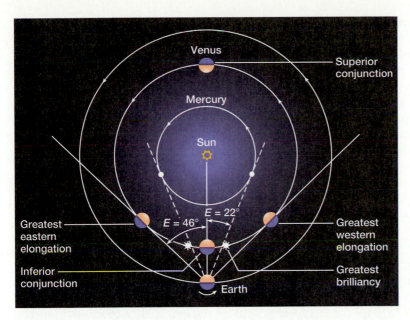

Figure 6.9 *Mercury and Venus go through two kinds of conjunctions, one with the planet between the Earth and Sun (inferior) and again with the planet on the far side of its orbit (superior). Mercury's orbit is shown as a circle for simplicity. Venus's average greatest elongation from the Sun (solid lines) is 46° and Mercury's (dashed lines) is only 22°. (It can actually reach 28° as a result of orbital eccentricity.) The positions of Venus's greatest brilliancy are indicated by asterisks.*

Because the Martian synodic period is 1.14 times the sidereal, each successive opposition will take place a little farther around the orbit, leading to a 17-year opposition cycle. At a favorable opposition, which last occurred in 1988 and will not happen again until 2005, the planet also has its largest angular diameter, allowing us to see a maximum of surface detail through the telescope.

6.2.2 The Inferior Planets

Since Venus and Mercury are closer to the Sun than is the Earth, they can never be in solar opposition; instead, they pass through two different kinds of conjunctions (Figure 6.9). **Inferior conjunction** occurs when an inferior planet is between Earth and Sun, and **superior conjunction** when the *Sun* is in the middle. The synodic period is the interval between successive inferior or superior conjunctions and is now the lapping time of the inferior planet relative to the slower-moving Earth.

The greatest angular separation from the Sun, or **greatest elongation,** will occur at the points where our line of sight just grazes the orbit. There are two of these, one where the planet appears in the evening to the east of the Sun, called *greatest eastern elongation,* and again where it appears west of the Sun in the morning, at *greatest western elongation.* At greatest elongation, Venus averages 46° from the Sun, so it can be seen in a dark sky well after sunset or before sunrise. Poor Mercury, however, is so close to the Sun that its greatest elongation averages only 22°. It therefore rises or sets close upon the Sun and is never seen in a dark sky, only in twilight, making it quite difficult to locate.

Like the Moon, the inferior planets go through phases. If you could turn a telescope onto Venus or Mercury at superior conjunction (see Figure 6.9), you would see the full sunlit face, or full phase. But at greatest elongation you see half the daylight and half the nighttime sides, and consequently the planet will look like the Moon at one of its quarters. At inferior conjunction, when you look at the nighttime side, it will be new, and invisible. Between greatest elongation and inferior conjunction it will take on the appearance of a crescent. Since the planet's distance from Earth is continuously decreasing between superior and inferior conjunctions, its angular size will steadily grow and correlate with the phases (Figure 6.10).

The brightness of an inferior planet depends on the amount of the visible illuminated angular area (the number of visible square seconds of arc that are in daylight), which in turn depends on a combination of distance and phase. Venus's *greatest brilliancy* will actually occur in the crescent phase between elongation and inferior conjunction. It is then a glorious sight at night, casts shadows in a dark location, and can be seen in full daylight.

The inferior planets show retrograde motions too as they swing past the Earth near inferior conjunction, but the motion is a bit more complicated. Follow Venus from superior conjunction (see Figure 6.10). The planet slowly climbs the western sky after sundown, brightening as it goes. It takes about a month to come into easy view and about seven months to reach greatest eastern elongation. Since the Sun is already moving easterly on the ecliptic, Venus must appear to move through the zodiac in the same direction, but even faster. After greatest elongation, the still-brightening planet begins slowly to fall back toward the horizon. The Sun starts to overtake it, but for a time the planet still moves to the east relative to the stars. Five weeks after greatest elongation, Venus reaches greatest brilliancy, and two weeks after that it finally enters retrograde motion as it prepares to swing between Earth and Sun. It then rapidly disappears from the western evening sky, arriving at inferior conjunction only five weeks after greatest brilliancy. But practically before you know it, it pops back up in the eastern sky before dawn, then repeats its cycle in reverse, successively quitting retrograde motion and reaching greatest brilliancy and greatest western elongation. Then follows a long seven-month fall to superior conjunction. Mercury behaves the same way, except that the apparent movement is enormously speeded up and further complicated by the eccentricity of its orbit.

6.3
Early Theories of the Solar System

To understand the development of the rules that govern planetary motion, as well as that of scientific thought, we return to ancient Greece. Although the great astronomer Aristarchus of Samos, who lived between 310 and 250 B.C., had introduced a **heliocentric theory** of the Solar System, one that had the planets going about the Sun, the common view nearly to modern times was **geocentric,** centered on the Earth. The Earth, after all, was the special creation of the gods and should hold a special place. But if the Earth is immobile, how do you explain the retrograde motions of the planets?

Constructions of the Solar System were predicated on the ingrained idea of the circle and the sphere as the perfect figures and the concept of uniform circular motion. The Athenian Eudoxus (409–365 B.C.) then used an ingenious mechanical system of 27 nested rotating crystalline spheres to match the observed planetary movements. The Earth was at the center. The Moon was mounted on

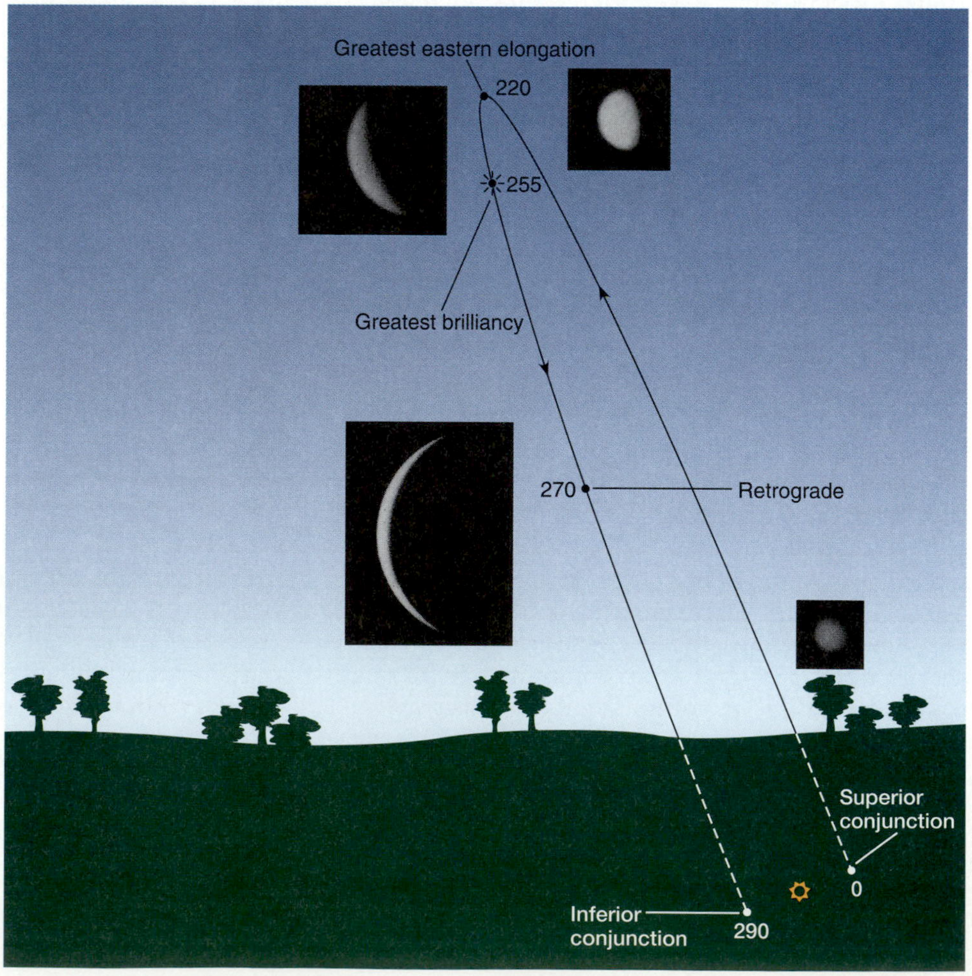

Figure 6.10 *Venus is marvelous to watch as it moves relative to the Sun. The numbers indicate the number of days since superior conjunction, and the photographs show the change of phase (changes in azimuth, that is, direction along the horizon, are ignored). Following elongation, the thinning crescent quickly reaches greatest brilliancy as it descends to the horizon, enters retrograde, and approaches inferior conjunction. The planet will then appear to go in the reverse direction in the morning sky.*

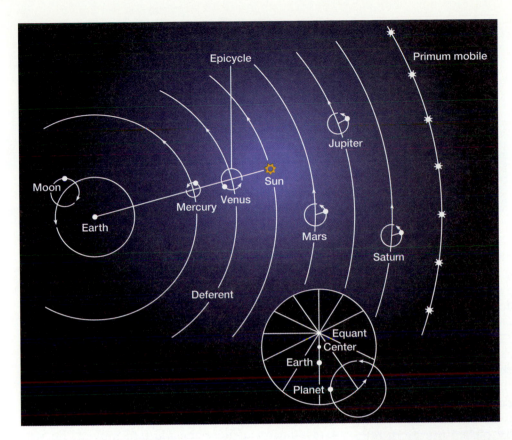

Figure 6.11 *Ptolemy's view of a geocentric Solar System had the planets moving on circular epicycles that in turn moved about the Earth (or about the point between the Earth and the equant in the inset) on circular deferents. The planets moved uniformly relative to the equant to account for variable planetary motion. The Moon moved backward on its epicycle to account for observed irregularities (The diagram is not to scale.)*

the first of three spheres, each one connected to the next at different angles to account for the lunar motion. We look through these spheres to see the Sun, which was carried by three more spheres. Around these were the planets, connected to four spheres each, and then finally we saw through the whole system to the sphere that transported the stars.

Others added to this great and ponderous scheme, which culminated in the work of Claudius Ptolemaeus, or Ptolemy, an Alexandrian Greek who lived in the second century A.D. In the year 140, he produced his great *Syntaxis* (meaning "compositions") or, as it became known to the eighth-century Arabs who translated and preserved it, the *Almagest* (from *al Magisti,* "the greatest"). Ptolemy did not use the cumbersome spheres but allowed the planets to travel independently about the Earth (Figure 6.11).

In the **Ptolemaic system,** each planet moves counterclockwise on a small circular orbit called an **epicycle,** taking a synodic period to go around. Each epicycle is centered on a large circular orbit called a **deferent,** and moves counterclockwise about the Sun with the sidereal period. As the planet swings between the Earth and its epicycle's center, it appears to go backward, or retrograde (Figure 6.12). The whole system, including the outer starry

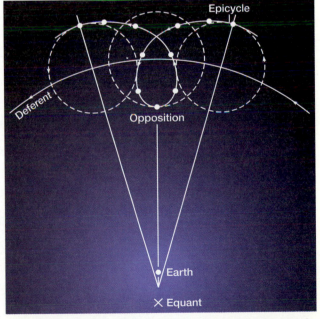

Figure 6.12 *A distant planet moves on its epicycle as the epicycle moves along its deferent. When the planet comes between the Earth and the epicycle's center, the planet goes into retrograde motion, moving westerly among the stars rather than easterly.*

Figure 6.13 *Nicolaus Copernicus revolutionized astronomical thought through an exact mathematical treatment of a heliocentric Solar System.*

sphere, rotates about the Earth once a day, causing the stars to rise and set.

For Mars, Jupiter, and Saturn, the line that connects the planet to the deferent is parallel to the one that connects the Earth to the Sun so that retrograde occurs at opposition. The epicyclic centers of Venus and Mercury are on that same line to account for the motions of the inferior planets back and forth across the Sun. The Moon orbits on an epicycle in the reverse direction to account for its highly variable angular movement, the result of a complex real orbit. Ptolemy placed the Earth off center and put a point called the *equant* (Figure 6.11 inset) on the other side at the same distance. The planetary orbits were offset so as to revolve at constant angular speed about the equant, helping to account for their variable motions through the zodiac. Finally, the whole external sphere rotates

against the equinoxes with a period then thought to be 36,000 years to emulate precession. Except for some minor fine-tuning, this system was regarded as the absolute standard for the next 1,400 years.

6.4
The Revolution in Thought

The story that leads to modern times involves six great scientists who revolutionized our concept of the Universe. The first was Nicolaus Copernicus, a Polish astronomer born in 1473, the last Albert Einstein, a German who immigrated to the United States and who died in 1955.

6.4.1 *Copernicus*

Nicolaus Copernicus (Figure 6.13) attended the University of Cracow, studied medicine as well as law in Italy, and worked as a church administrator in Prussia (now part of modern Germany and Poland). There he pursued his real interest, astronomy. Although he published very little until near the end of his life, he became known quite early as a considerable expert in the subject. Finding the Ptolemaic system unsatisfactory as a physical explanation of how the planets moved, he was struck by the old Greek idea of a heliocentric Solar System, and decided to examine it from a strict mathematical point of view. It was a labor of love that occupied his entire working life.

The results (summarized by Figure 6.14) appeared in a massive and complex work called *De*

Figure 6.14 *The Copernican view of the Solar System, taken from* De revolutionibus, *is drawn deliberately out of scale to enable placement on a page. Names of the bodies are given in Latin. Telluris refers to the Earth, about which the Moon is seen in orbit; Jovis refers to Jupiter. The largest circle is the celestial sphere of the "fixed stars."*

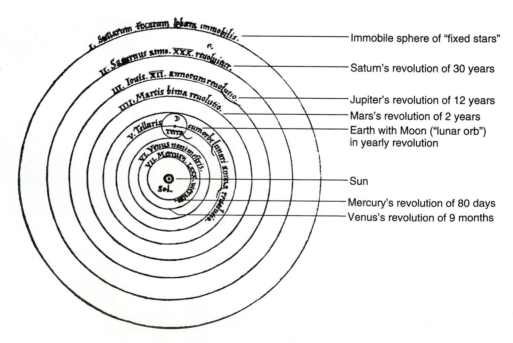

- Immobile sphere of "fixed stars"
- Saturn's revolution of 30 years
- Jupiter's revolution of 12 years
- Mars's revolution of 2 years
- Earth with Moon ("lunar orb") in yearly revolution
- Sun
- Mercury's revolution of 80 days
- Venus's revolution of 9 months

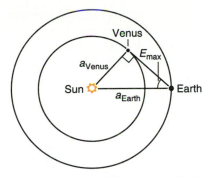

Figure 6.15 *Copernicus' method for obtaining the distance to Venus (and to Mercury as well) involves simply determining the angle at which the planet is at maximum elongation, E_{max}. Since Earth-Venus-Sun is at that time a right triangle, and a_{Earth} = 1 AU, the distance of Venus a_{Venus} is easily found to be 0.72 AU.*

revolutionibus orbium coelestium ("On the Revolutions of the Celestial Spheres"). The finished volume is said to have been handed to him on the day of his death in 1543. Much of the book involves the development of the mathematics needed to explain the subject. Before Copernicus' time, the planets were placed at distances from the Earth according to their sidereal periods. Copernicus, however, determined accurate relative distances for the first time by *calculation*. Figure 6.15 shows the planet Venus at greatest western elongation. Since Earth-Venus-Sun form a right triangle (one with an angle of 90°), Copernicus could easily calculate the relative lengths of the sides, and the distance Sun-Venus in terms of the distance Sun-Earth (or, in modern terms, the distance of Venus in AU). The calculation of the relative distances of the superior planets is similar but more complicated and involves observing the planet when the *Earth* would appear to be at maximum elongation as seen from the other body. The results are essentially the modern values, all determined from naked-eye observations.

There was as yet no way of knowing which theory, Copernican or Ptolemaic, was correct. However, the **Copernican system** had the advantage of simplicity, even though Copernicus still clung to the idea of circular orbits and consequently had to keep the concept of epicycles to explain the irregularities of planetary movements. The heliocentric theory was shunned for religious reasons, because it displaced the Earth from the center of the Universe. The Catholic Church eventually banned the volume and it remained on the *Index of Prohibited Books* until 1837, long after the heliocentric theory was firmly shown to be correct.

Copernicus represents an extraordinary turning point in the history of science and of human

thought. He displaced us forever from our centric view of ourselves, from the notion that the Universe was created with our planet as the focal point. Much of the subsequent history of astronomy displaces us ever further, until we now know ourselves to be at the edge of a galaxy of billions of stars in a Universe of billions of galaxies, with no center at all.

6.4.2 Tycho

We next encounter the greatest of all the pretelescopic naked-eye observers, Tycho Brahe. This brilliant, contentious Danish nobleman (Figure 6.16), born only three years after the death of Copernicus, played a pivotal role in the discovery of the true orbital properties of the planets. From his youth he had a strong inclination toward astronomy. As his prominence rose, he caught the interest of King Frederik II of Denmark, who was fascinated with science and who provided funds for the construction and support of an observatory on the island of Hveen, which Tycho named Uraniborg, or Castle of the Heavens (Figure 6.17). It contained large graduated circles for precise measurements of the positions of the stars and the planets. Tycho's brilliant contributions were to establish a catalogue of star positions that was far more accurate than anything done before and to compile a continuous record of the motions of the objects of the Solar System relative to the locations of the fixed stars, working to an astonishing precision of one or two minutes of arc.

Figure 6.16 *Tycho Brahe's dedication to precision would have made him great in any age*

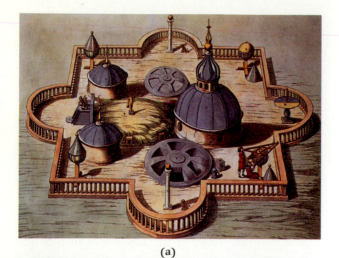

(a)

(b)

Figure 6.17 **(a)** Uraniborg, *Tycho's observatory, is seen from overhead.* **(b)** *Tycho works within the curve of one of his great graduated quadrants, instruments used to measure angles in the sky to an accuracy of one or two minutes of arc.*

This sort of accuracy eventually demanded a new and better theory for the motions of the planets.

After the king's death, Tycho's increasingly quarrelsome nature disenchanted the guardians of the young heir to the throne and eventually the new king himself, Christian IV. The result was the removal of Tycho's funding and the departure of the astronomer from Denmark. After reestablishing himself near Prague and publishing his catalogue of the positions of 777 stars, he died in 1601.

Tycho dramatically advanced the state of astronomical observations, but he did not have the foresight to extend the Copernican ideas. He clung to the geocentric theory, going only so far as to allow Mercury and Venus to orbit the Sun, which in turn still circled about the Earth along with the other planets.

6.4.3 Johannes Kepler

Unlike Tycho, Johannes Kepler (Figure 6.18) was more an interpreter of nature than an observer. He was born in Württemberg, which eventually became part of modern Germany, in 1571, nearly 30 years after the publication of *De revolutionibus.* He studied in Tübingen and eventually found an instructor's position in mathematics at Graz, in what is now Austria. Kepler was a Protestant, and in 1598 he was forced from his home during a period of persecution initiated by the Catholic Archduke Ferdinand. He was invited to Prague to be Tycho's assistant in 1600, and following the Danish astronomer's death, he succeeded to Tycho's academic position; most importantly, he acquired the great observer's data. Kepler was absorbed by the geometry of planetary orbits. His mathematical abilities, combined with his awesome dedication to hard work and insistence that every observation of planetary position be accounted for, led to his great discovery of the laws of planetary movement.

The Copernican theory, which had circular orbits and epicycles, predicted planetary positions that could differ from those observed by several degrees. It was obvious that something was wrong. Kepler set out to find how the planets truly move about the Sun by a thought process that was radically new: he would directly use Tycho's observations, particularly those of Mars, to create a new theory, letting the data lead the way.

He first tried circular orbits and adopted the equant of Ptolemy (see Figure 6.11) but with the Sun replacing the Earth. By setting up pairs of observations at intervals of a Martian sidereal year, at the beginning and end of which the Earth would be at two different places (Figure 6.19), he could

Figure 6.18 *Johannes Kepler, using Tycho's observations, discovered the laws that rule the planets.*

ground of the constellations. In Figure 6.20, the line that connects the planet with the Sun is called the *radius vector*. **Kepler's second law of planetary motion** states:

2. *The radius vector sweeps out equal areas in equal times.*

According to this **equal-areas law,** a planet must move farther at perihelion in a given amount of time (and therefore it must move faster) than it

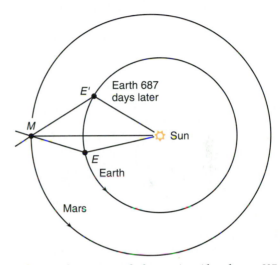

Figure 6.19 *After an interval of a Martian sidereal year, 687 days, Mars returns to the same place, (M), but the position of the Earth has shifted from E to E'. All the angles are known and distances can be computed.*

use the principles of geometry to determine the offsets of the Sun and the equant from the center. Then, taking similar pairs of observations at intervals of an *Earth* year, at the beginning and end of which *Mars* would be at two different places, he could find the analogous offsets to the Martian orbit. Computation of position still showed differences with observation larger than he believed Tycho could have made. Kepler then used pairs of observations like those shown in Figure 6.19 to measure the distances of Mars from the Sun at various positions and, rejecting the wisdom of the ages, found that (as he wrote in his *New Astronomy*) "the planetary orbit is no circle." He continued, "It is as if I awoke from sleep and saw a new light." The orbit, he saw, is an *ellipse*.

6.4.4. Kepler's Laws of Planetary Motion

After three years of crushing calculation, about 1604, Kepler had at last found the true shape of the Martian orbit and, by generalization, the shapes of the orbits of the other planets. After considerable publication delay, in 1609 he could finally announce **Kepler's first law of planetary motion:**

1. *The orbits of planets are ellipses with the Sun at one focus.*

More profoundly, with this **law of ellipses** he had discovered the first real indication that the Sun *controls* the movements of its tiny companions.

At the same time, Kepler explained the variation of Mars' angular motion against the back-

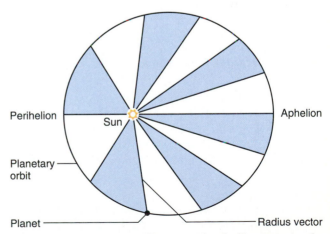

Figure 6.20 *Kepler's second law states that the line connecting the planet to the Sun (the radius vector) must sweep out equal areas in equal times. All the sectors have the same area. The planet takes the same time to move from one edge of each sector to the other edge. As it gets closer to the Sun, it therefore moves faster.*

would at aphelion in order that the two swept areas be the same. The movement is such that the average distance of a planet from the Sun over the course of its year equals the length of its semimajor axis.

The effects of Kepler's second law can easily be seen in the *inequality of the seasons*. Northern hemisphere spring and summer each have 93 days, but autumn has 90 and winter 89. The Earth is near its perihelion point only two weeks past the winter solstice and near aphelion two weeks past the beginning of summer. The Earth must then move faster during autumn and winter, making those pass quickly, and slower during spring and summer, stretching them out by a few days (and allowing those of us in the northern hemisphere to linger longer in the Sun's warmth).

Kepler finally set out to examine the orbits of the other visible planets and to find the relations among them. Tycho's data were not good enough to establish their elliptical natures, but logic and symmetry strongly suggested that they must be similar to that of Mars. Ten years after his publication of the first two laws, Kepler established his **third law of planetary motion:**

> 3. *The squares of the orbital periods of the planets about the Sun are proportional to the cubes of the orbital semimajor axes.*

This **harmonic law**—one of the central ideas in astronomy—is usually expressed relative to the Earth's orbit by the formula

$$P^2 = a^3,$$

where P is the planet's sidereal period expressed in years, and a is its orbital semimajor axis in AU.

The rule is gloriously simple. Jupiter is 5.2 AU from the Sun, which when multiplied by itself twice is 140.6, and which must then be equal to P^2. We then have to find the number that when squared is 140.6. A hand calculator, or a bit of trial and error, yields 11.9, Jupiter's sidereal period in years.

Look at the power of the law. Say you take your telescope outdoors and discover a new body in motion about the Sun. Continued observation shows it to have a sidereal period of 2.6 years. Just square 2.6 to get 6.76. Then find the number which when multiplied by itself twice (the cube root) gives 6.76 to find a semimajor axis of 1.89 AU. Together the three laws allow calculations of accu-

rate planetary positions both into the future and back into the past. We now know how the planetary system works, and we can account for the things we see.

6.4.5 Galileo

At about the same time Kepler produced his epochal first two laws, the Italian Galileo Galilei (depicted in the painting that opens this chapter) was observationally confirming the Copernican system. He was born in Pisa in 1564. Absorbed by science and astronomy, he eventually became a professor at the universities of Pisa and Padua. Although he lectured on the Ptolemaic system in his younger years, he was increasingly drawn toward the logic of the Copernican. He had an immense natural curiosity about the world around him. He did not invent the telescope (Figure 6.21), and may not even have been the first to turn it on the sky, but he was unquestionably the first to use it in a systematic way to learn something of the nature of the heavens. He first examined the sky during 1609 and 1610, and in those marvelous years

Figure 6.21 *A pair of Galileo's simple astronomical telescopes are mounted on a decorated stand. The first one he made gave a magnifying power of only 28, low by today's standards, but awesomely powerful compared to the unaided human eye.*

he found that our Moon has mountains and valleys like the Earth, Jupiter has moons of its own, Venus exhibits phases just like those of the Moon, the Milky Way is made of faint stars, Saturn has "appendages" (later determined to be rings), and the Sun is spotted. All these revelations were announced in his famous tract of 1610, *The Sidereal Messenger*. Copernicus helped open our minds to the true nature of the Universe. Galileo now helped open our eyes, for the first time peering out into the darkness to see not just how things moved, but how they are *made*. He began our real study of the planets and the stars.

But he did not stop there. His greatness was that he continued to observe for years, thinking about what he saw and, most importantly, drawing conclusions. It was evident to him that his view of the Solar System amply confirmed Copernican thought. The lunar surface looks remarkably like that of the Earth, with mountains and basins, so the Earth is not unique. The Jovian satellites go around Jupiter just as Copernicus said the planets go around the Sun; that is, a central body other than the Earth is surrounded by orbiting companions. Venus can pass through a full set of phases only if it has a path that takes it around the Sun, not the Earth. Galileo's astonishing labor helped put the Ptolemaic system to death. In his most famous work, the *Dialogue on Two Chief Systems of the World* (1632), he promoted Copernicanism by an imaginary argument between Simplicio, who believes Ptolemy, and Salviati, who follows Copernicus. A presumably impartial moderator consistently upholds Salviati. For so vigorously setting out to destroy the Aristotelian and Ptolemaic ideas, Galileo was called to the Inquisition at Rome. His fame protected him from harsh punishment, but he was forced to recant his positions and was confined to his home from 1633 until his death in 1642. The church ultimately admitted its error and cleared Galileo in 1992.

The combined work of these men told us *how* the planets move about the Sun, but not *why*. Kepler knew that the Sun was somehow responsible, and he believed that something emanating from it carried the planets along in their orbits. The discovery of the actual force—gravity—and how it works remained for Isaac Newton and Albert Einstein. In Newton's hands, Kepler's third law will become a way not just of probing the Solar System but of examining the structure of the entire Universe.

KEY CONCEPTS

Conjunction: The position in which a planet (as viewed from the Earth) is in the same direction as the Sun (or in which two planets are aligned with each other); in **opposition,** a planet is opposite the Sun as viewed from Earth (or two planets are opposite one another).

Copernican system: A system of circular heliocentric planetary orbits.

Elongation: The angle between the direction to a planet and the direction to the Sun as viewed from the Earth; **greatest elongation** is the maximum possible for an inferior planet.

Epicycle: In the Ptolemaic system, a secondary planet-carrying orbit centered on a **deferent,** which is centered on the Earth (or between the equant and the Earth).

Geocentric theory: A theory of the Solar System in which the planets orbit the Earth.

Heliocentric theory: A theory of the Solar System in which the planets orbit the Sun.

Inferior conjunction: A conjunction between Venus or Mercury and the Sun in which the planet lies between the Earth and the Sun; in a **superior conjunction** the planet is on the other side of the Sun.

Inferior planets: Mercury and Venus, the two inside Earth's orbit; the **superior planets** are Mars through Pluto, those outside the Earth's orbit.

Planets: The Sun's family of major orbiting bodies.

Ptolemaic system: A system of geocentric orbits that carry epicycles that carry planets.

Retrograde motion: The apparent westward motion of a planet relative to the stars as a result of the Earth's passing a superior planet in orbit or of the Earth being passed by an inferior planet.

Sidereal period (of a planet): The orbital period relative to the stars.

Synodic period (of a planet): The interval between successive conjunctions or oppositions of a planet with the Sun.

KEY RELATIONSHIPS

Kepler's laws of planetary motion:

1. **The law of ellipses:** The orbits of planets are ellipses with the Sun at one focus.
2. **The equal areas law:** The radius vector sweeps out equal areas in equal times.
3. **The harmonic law:** $P^2 = a^3$, where P is the sidereal period in years and a is the semimajor axis of the orbit in AU.

EXERCISES

Comparisons

1. What is the difference between the inferior and the superior planets?
2. What are the differences between opposition and conjunction, and between inferior and superior conjunctions?
3. Compare the visibility of Venus at greatest western elongation with that at greatest eastern elongation.
4. List the essential elements of the Ptolemaic and Copernican systems and describe the observational tests that can discriminate between them.
5. What is the difference between an epicycle and a deferent?

Numerical Problems

6. What is the maximum deflection of a star caused by the aberration of starlight as seen from Mercury?
7. NASA launches a spacecraft into orbit about the Sun with a period of six years. What is the craft's semimajor axis? If the orbit is circular, how fast in km/s does the craft travel in orbit?
8. NASA then launches another spacecraft into circular orbit about the Sun with a semimajor axis of 0.1 AU. What is its sidereal period in days? How fast does the craft travel? Draw a diagram of its orbit and with a protractor estimate the craft's greatest elongation as seen from Earth.

Thought and Discussion

9. Arrange the planets in order of maximum brightness. How can more-distant planets be brighter than closer ones?
10. Which three planets have the most eccentric orbits? Which has the most nearly circular orbit?
11. Pluto can come closer to the Sun than Neptune over a short portion of its orbit. Why do the two not crash into each other?
12. Why is the synodic period of Saturn shorter than the synodic orbit of Jupiter?
13. What phases can be seen for Venus and Mars in **(a)** the Ptolemaic system; **(b)** the Copernican system?
14. When do Mars and Venus move at their maximum angular speeds in the retrograde direction?
15. If Jupiter is in Capricornus at opposition, in what constellation will it appear at the next opposition?

16. Why does Mars have "favorable oppositions" whereas Jupiter does not?
17. Why is Mercury so hard to see even though it is quite bright?
18. Why is Venus at greatest brilliancy in a crescent phase rather than in its full phase?
19. What were Galileo's and Tycho's essential contributions toward proving the correctness of the Copernican system?
20. At what point in its orbit will a planet move most slowly in km/s?

Research Problem

21. Fit the works of Copernicus, Tycho, Kepler, and Galileo into their times. Using library materials, list the explorations of the world that were in progress and provide a brief summary on the state of Europe during that period.

Activities

22. Make a chart in which you organize the planets by their three categories: ancient and "modern" (discovered by telescope); inferior and superior; terrestrial and Jovian (see Chapter 1).
23. Using photocopies of the star maps in the Appendix, plot the positions of any observable planet over the course of the school term. Note any retrograde motion it may have. Locate the planet at opposition if appropriate.
24. If you have access to a telescope, use it to track the phases of Venus. Estimate the elongation of Venus from the Sun and, from a scale drawing of the orbits of the Earth and Venus, plot the position of Venus in orbit relative to the Earth, indicating the phase.
25. Use a popular astronomy magazine to find the time of the next maximum elongation of Mercury; then find the planet.

Scientific Writing

26. Assume that Venus is about to make its first appearance in the evening sky. Write a newspaper column explaining what the public can expect to see over the next seven or so months.
27. Some years ago, a television production called "Meeting of Minds" assembled actors portraying historical figures from different eras to discuss a variety of topics. You are on the show with Aristotle. Write a monologue for yourself in which you tell him of discoveries made in the sixteenth and seventeenth centuries regarding the construction of the Solar System.

7 *Newton, Einstein, and Gravity*
8 *Atoms and Light*
9 *The Tools of Astronomy*

Physical Astronomy

7

Newton, Einstein, and Gravity

The glue of the Universe and how it relates to celestial motions

A skydiver falls toward Earth, accelerating in the terrestrial gravitational field.

Throw a ball in the air and it returns to Earth. Why? Nothing seems to pull on it, nothing you can see. The ball is subject to **gravity,** a force of nature that acts over a distance to draw things together. Gravity pervades the Universe and holds you to the ground, the Moon to the Earth, the Earth and planets to the Sun, and the Sun to the Galaxy. Our understanding of it begins with the work of Kepler and Galileo, continues through the brilliant deductions of Isaac Newton (Figure 7.1), who founded the science of *mechanics,* the study of the way things move, and culminates in remarkable theories by Albert Einstein that allow us to see how the Universe actually works.

7.1
Newton's Laws of Motion

Isaac Newton was born in England in 1642, the year of Galileo's death. A founder of modern science, Newton searched for the physical principles that underlie the Copernican and Keplerian discoveries and applied rigorous mathematical treatments to understanding them. He studied at Cambridge University, where he became a professor of mathematics in 1669. For some two decades he roamed through science, making fundamental discoveries in mechanics, optics, and pure mathematics. His great work was the *Principia* (1687)—formally *Philosophiae Naturalis Principia Mathematica,* or "Mathematical Principles of Natural Philosophy." Here Newton laid down three basic rules of how things move, known to us since as **Newton's Laws of Motion,** and derived a law that describes gravity. To do so, he (simultaneously with Leibniz in Germany) invented a whole branch of mathematics, the calculus. His labors changed our view of the world forever.

Drawing from the writings of Galileo in his *Dialogue* and those of the French mathematician René Descartes, Newton stated his rigorous **first law of motion:**

> 1. *Left undisturbed, a body will continue in a state of rest or uniform straight-line motion.*

This concept is contrary to everyday experience; we are used to seeing moving things come to a halt. To reconcile fact and intuition we must look carefully at how things move and how they can be made to change their states of motion.

A moving body has a **speed,** the rate of change of distance with time, measured (for example) in meters per second (m/s) or kilometers per hour

Figure 7.1   *Isaac Newton discovered the natures of motion and gravity and is seen here investigating properties of light. His reflecting telescope, which he invented, is to the right of center.*

(km/hr). A moving body also has a direction. Taken together, speed and direction constitute the body's **velocity** (*v*). If you and a friend are each driving at 60 km/hr in opposite directions (Figure 7.2a), your speeds are the same but your velocities are different. An **acceleration** is any change in a body's velocity, meaning a change in speed, direction, or both. If you step on the gas pedal of a car—or even step on the brake—you accelerate. If you drive around a curve at constant speed (Figure 7.2b), you still accelerate because you are changing your direction. Since acceleration is the rate of change of velocity with time, it is measured in meters per second, per second, or m/s^2.

Acceleration is produced by the application of a **force.** If you kick a stationary football, you apply force and accelerate the ball to some velocity. A ball rolling on a level floor will always come to a stop because of the retarding force of friction caused by microscopic interactions between it and the floor. If friction were not present (and if there were no walls or air resistance), the ball would roll forever. Acceleration and force also have directions. If a force is applied along the direction of motion of a body, only the speed will change. If the force is applied at an angle to the velocity, the direction of motion will also change. As you turn a car through a curve, a frictional force is applied at

MATHHELP 7.1 Direct and Inverse Relationships

Many rules in science, such as Newton's second law of motion, make use of *inverse relationships.* If two numbers, *D* and *N,* are in direct proportion to each other, one varies directly with the other. If *N* doubles, so does *D.* The inverse or *reciprocal* (*R*) of a number *N* is the number 1 *divided* by *N,* or *R* = 1/*N.* We then say that *R* is proportional to 1/*N.* For example, if *N* = 100, *R* = 1/100 or 0.01. *N* is also then the reciprocal of *R.* Reciprocals behave oppositely to their corresponding numbers. As *N* increases, *R* = 1/*N* decreases and vice versa. If *N* = 200, *R* = 1/200 = 0.005. When *N* is large, as in the above examples, *R* is small; but when *N* is small, *R* is large: if *N* = 0.001, 1/*N* = 1,000.

Inverse relationships also commonly involve squares, cubes, and other powers, as they do in Newton's law of gravity. A number *S* may be the inverse square of *N,* or *S* = 1/N^2. In that case, we multiply *N* by itself before dividing into 1. For example, if *N* = 100, *S* = 1/100^2 = 1/10,000 = 0.0001. The rate of change in an inverse-square relationship or *inverse-square law* is much greater than it is for a simple reciprocal. As *N* goes from 2 to 4, *R* = 1/*N* goes from 1/2 to 1/4, whereas *S* = 1/N^2 goes from 1/4 to 1/16. Inverse cubes, *C* = 1/N^3, will be more extreme.

Direct and inverse relationships can be combined. *Z* may be directly proportional to *Y* and inversely proportional to *N* squared, so that *Z* = *Y*/N^2. If *Y* is doubled, so is *Z,* but if *N* is doubled, *Z* is quartered. If both are doubled, *Z* is increased by a factor of 2 but decreased by a factor of 4, or is multiplied by a factor of 2/4 = 1/2.

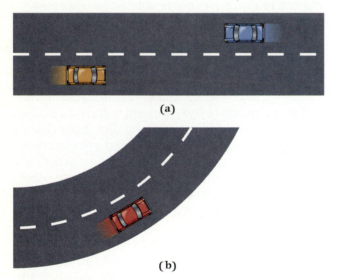

(a)

(b)

Figure 7.2 *Velocity combines speed and direction.* **(a)** *Although the two cars approach each other at identical constant speeds, their velocities are different because the directions of motion are not the same.* **(b)** *A car passes through a curve at constant speed. It is accelerating because it is changing its direction.*

an angle by the road to the tires that makes the vehicle accelerate and change its direction.

Gravity is a force, and as a consequence it can produce an acceleration. The **acceleration of gravity,** g, is measured to be 9.8 m/s^2 downward. Drop a ball. After one second it will be moving at a speed of 9.8 m/s, after 2 seconds double that at 19.6 m/s, after 3 seconds, triple, and so on. All falling bodies, as Galileo knew, accelerate at this same rate independently of their natures (neglecting air resistance).

The interaction of the force and the body is described by Newton's **second law of motion:**

> 2. *The size of an acceleration is directly proportional to the force applied, is inversely proportional to the mass of the body being accelerated, and the direction of the acceleration is the same as the direction of the force.*

(See MathHelp 7.1.) That is, acceleration (*A*) equals force (*F*) divided by mass (*M*):

$$A = F/M,$$

or, as more commonly expressed,

$$F = MA.$$

Mass is commonly thought of as the amount of matter in a body. Mass is actually defined as the degree to which the motion of a body resists a force. Its unit of measure is the **kilogram** (kg).

Place two things with different masses, M_1 and M_2, in front of you, perhaps a toy block and a brick. Put a hand on each and push the two objects with the same force. They will accelerate in the direction in which you push. The block will achieve the greater acceleration and velocity. Acceleration is inversely proportional to mass, so that with equal forces,

$$A_1/A_2 = M_2/M_1.$$

If one body has half the mass of the other, its acceleration will be twice as great. The mass of one body relative to the other can therefore be found by measuring the relative accelerations.

To express force, a unit is needed that is the product of mass times acceleration, or kg m/s^2, called the newton (N). A body on the Earth is subject to a force induced by gravity that is equal to the body's mass times the acceleration of gravity, or

$$F = Mg,$$

which is called **weight.** A 100-kg mass is pressed downward (that is, in the direction of the force) against the Earth with a weight equal to 100 kg × 9.8 m/s^2 = 980 N.

You know, however, that the weight of a box of cereal in a grocery store is not given in newtons but in both pounds and kilograms (or grams, g, where 1 kg = 1,000 g). Weight in kilograms is a social and political convention by which the weight is numerically set equal to the mass. In scientific usage, the kilogram applies strictly to mass, never weight. Masses can be measured by comparing their weights on the surface of the Earth with the weight of an agreed-upon standard mass.

Newton's **third law of motion** builds further on these concepts, and deals with the problem of *two* bodies when one applies a force on the other:

3. *For every force applied to a body, there is an equal force exerted in the opposite direction (action equals reaction).*

A pair of skaters face one another on the ice. She has a mass of 50 kg, he 100 kg. She gives him a push. He will move backward, *and so will she.* Moreover, even though she applied the force, she will accelerate twice as much because her mass is half his. This principle lies behind the simple act of walking. You apply a force on the sidewalk to your rear through friction. The Earth is so massive, its acceleration in that direction is negligible, but you accelerate considerably in the forward direction. It is also the principle behind a rocket or jet aircraft (Figure 7.3).

Figure 7.3   *A jet aircraft takes off from a runway. The engines are combustion chambers with open ends. Burning fuel causes hot gases to expand. They apply a force against the forward part of the chamber and the chamber pushes back. The gases rapidly exit through the nozzle, and the plane moves off in the other direction.*

7.2
Energy

If you apply a force to a body while it is moving over a certain distance, you accelerate it, and are said to be doing **work** on it. A moving body has **energy,** which is the body's capacity for doing work. The unit of work and energy is the *joule,* which measures the work done when a force of 1 newton is applied to a body over a distance of 1 meter.

There are many kinds of energy. **Kinetic energy** is energy of motion. A moving body with a kinetic energy of x joules can strike a second body and do up to x joules of work on it (some energy is wasted), thus accelerating it to some velocity. The kinetic energy (KE) possessed by a moving body is one-half its mass times the square of its velocity, or

$$KE = \tfrac{1}{2}Mv^2.$$

This formula is not often appreciated by automobile drivers. At 100 km/hr your car possesses four times the energy that it does at 50 km/hr, and can do four times the damage in a collision.

Potential energy involves the energy a body has as a result of its position or configuration. If you raise a bowling ball above the floor you are doing work on it against the Earth's gravity. When you hold it high, you have given it the potential to have kinetic energy. When you drop it, gravity exchanges the ball's potential energy for kinetic energy.

Energy of motion and its potential counterpart are *mechanical energy.* A beam of light represents *radiant energy* (to be explored in Chapter 8). All matter contains *heat energy,* a microscopic form of

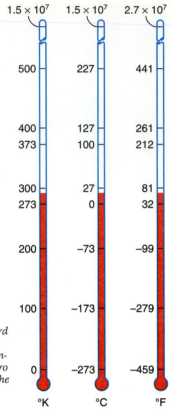

Figure 7.4   *The three standard temperature scales, Kelvin, centigrade (Celsius), and Fahrenheit, are lined up at absolute zero and show room temperature. The Kelvin and centigrade scales always differ by 273°.*

Within a closed system, one free of outside influence, energy can change its form but not its amount: this concept is called the **conservation of energy.** A ski jumper acquires internal heat energy from breakfast cereal, which in turn acquired its energy from sunlight when the grain was grown. The skier expends this energy in keeping warm and in moving. In climbing the jump, the skier converts internal energy into potential energy that will be converted again into kinetic energy during descent and into heat energy created by friction with the snow. There are a great number of other energy exchanges, but in the end, all the energy is accounted for.

7.3
The Law of Gravity

The work of Galileo and others demonstrated that the Sun had to play an active role in holding the Earth in orbit and preventing it from flying away in a straight line (Figure 7.5). Newton's friend Edmund Halley had even discovered (from Kepler's laws) that the force required to accelerate the planets toward the Sun, as well as the accelerations themselves (the more distant planets have less curvatures to their paths), vary according to the inverse squares of their distances from the Sun (see MathHelp 7.1). But although several scientists found and fit together pieces of the puzzle of the nature of the Solar System, only the great Newton saw the picture whole.

Newton independently derived the inverse-square law of planetary accelerations and forces, and furthermore pondered its origin. The Moon orbits and is accelerated toward the Earth just as the planets orbit and are accelerated toward the Sun. If gravity, the downward force felt at the surface of the Earth, extends as far as the Moon, per-

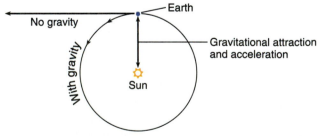

Figure 7.5   *The Earth is accelerated in the direction of the Sun by a gravitational attraction that keeps the Earth moving in a curved path. If the gravity could be switched off (which it cannot), then according to Newton's first law of motion, the Earth would fly away in a straight line.*

mechanical energy related to the speeds at which particles—atoms and molecules—move in a gas or vibrate in a solid. Their kinetic energies, hence their speeds, and the amount of heat energy per unit mass, are indicated by **temperature.**

The common scale of temperature measurement in the United States is degrees Fahrenheit (°F), invented by Gabriel Fahrenheit in 1709. It was originally established with the freezing point of salt water at 0° and body temperature at 100°. The worldwide standard is degrees centigrade, or Celsius (°C), developed by Anders Celsius in 1742. It is defined by the freezing and boiling points of water, which (at sea level) are placed at 0°C and 100°C respectively. As you remove heat from a body, its particles slow, and its temperature goes down. The body reaches minimum energy (there is always some present) at absolute zero and can get no colder. That happens at −459°F = −273°C. To avoid negative temperatures, the Kelvin (K) scale starts at 0°K at absolute zero and then counts centigrade degrees upward (Figure 7.4). The freezing point of water is then +273°K and the boiling point +373°K. The degree symbol is now usually dropped, and "degrees Kelvin" are referred to as "Kelvins," or K.

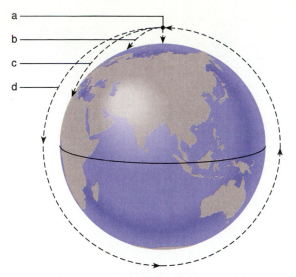

Figure 7.6   *If you drop a rock* **(a)** *it will fall toward the center of the Earth and land at your feet. If the rock is thrown* **(b),** *it still falls toward the Earth's center, but now drops a good distance away because of its horizontal motion. The faster it is thrown* **(c),** *the farther it goes, in part because the Earth drops away beneath it. If it could be thrown fast enough, the Earth would curve away at the same rate at which the rock drops* **(d).** *The rock cannot ever reach the Earth and is now in orbit.*

haps it is the force causing the lunar acceleration. It should then be the same kind of force that causes the planetary accelerations toward the Sun. If so, then the Earth's gravity should also behave according to an inverse-square law. Does it?

Newton tested his idea on the Moon. He showed first that if you add together the attracting power of all the matter in the Earth, it collectively behaves as if it were all concentrated to a point at the Earth's center. Consequently, for gravitational calculations, your "distance from the Earth" is not zero, but the terrestrial radius of 6,400 km. The ratio of the acceleration of the Moon to the acceleration of gravity at the Earth's surface should then just be $(R/D)^2$, where R and D are respectively the Earth's radius and the lunar distance. The value of g was reasonably well known in Newton's time. The acceleration of the Moon could be calculated from the degree of curvature of its orbit, or from its velocity and distance from the Earth's center. The accelerations are in the expected ratio! Newton showed that gravity controls orbiting bodies. He then assumed that everything in the Universe attracts every other thing in the Universe with a force proportional to the inverses of the squares of the distances between them. Every mass is surrounded by a gravitational *field* or field of force (through which it mutually attracts other masses) that weakens with distance, but is infinite in extent.

Newton showed that orbiting bodies are falling bodies. Although the Moon is accelerating and falling toward the Earth, it does not reach us because it is also moving in a direction perpendicular to the line between the two bodies. Drop a rock and see how long it takes to strike the ground (Figure 7.6). Then throw the rock horizontally. As it moves away, it drops at the same rate as before and strikes the ground after the same amount of time. Now imagine throwing it very hard, at such a speed that as it falls, the Earth curves below it. The rock will now take longer before it strikes. If you hurl it quickly enough, at 28,400 km/hr (and ignore the resistance of the Earth's atmosphere), the rock drops at the same rate at which the Earth curves and is in an orbit just like the Moon.

From Section 7.1, the force F with which a body of mass M_{body} pushes against the surface of the Earth (that is, its weight) is equal to M_{body}g. Since g is proportional to $1/R^2$, F is proportional to M_{body}/R^2. (If you double M_{body}, you double the weight.) However, from Newton's third law, the Earth must push back with equal force on the body: otherwise, the body would move. Think next of the *Earth* resting on the *body*. The force now has to be proportional to M_{Earth}/R^2 (if you double M_{Earth} you would also double the weight). Since the force is proportional to *both* the mass of the Earth and the mass of the body, it has to be proportional to their product, $M_{Earth} \times M_{body}$, or

$$F = G \frac{M_{Earth} \times M_{body}}{R^2}.$$

G is a constant of proportionality called the *gravitational constant,* which from laboratory measurements is found to be 6.67×10^{-11} N m^2/kg^2. Since $F = M_{body}$g, the downward acceleration of the body, or the acceleration of gravity, must (at the Earth's surface) be

$$g = G \frac{M_{Earth}}{R_{Earth}^2}.$$

The acceleration of a falling body is independent of its mass, and depends only on the mass and radius of the Earth.

Gravity was assumed by Newton to be universal. The acceleration of gravity toward any mass M is GM/R^2, where R is the distance to the body's center. The gravitational force F that any two bodies exert on one another is

$$F = G \frac{M_1 M_2}{R^2},$$

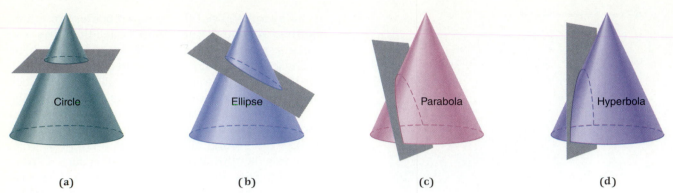

Figure 7.7   *The conic sections are created by intersecting a cone with a plane.* **(a)** *If the plane is parallel to the cone's base, the intersection is a circle.* **(b)** *If you tip the plane but leave the pitch angle less than that of the cone's side, the intersection is an ellipse.* **(c)** *A parabola is generated by a plane that cuts the cone parallel to the side.* **(d)** *A hyperbola is made by the plane with a pitch angle greater than that of the side.*

where M_1 and M_2 are the two masses and R is the distance between their centers. Here, in its final form, is Newton's famous **law of gravity,** with which we can follow the motions of the planets and stars in exquisite detail.

7.4
Kepler's Generalized Laws of Planetary Motion

Edmund Halley asked Newton what the shape of an orbit should be if a planet were accelerating according to an inverse-square law. Newton instantly responded that he had already calculated the curve: it would be an ellipse. He had reproduced Kepler's first law! Upon application of his laws of motion and the law of gravity, Newton eventually discovered a more general version of Kepler's first law:

> 1. *The paths of orbiting bodies are conic sections with the other body at one focus of the curve.*

The **conic sections** (Figure 7.7) are the curves created when a cone is intersected at different angles by a plane: the circle, ellipse, parabola, and hyperbola. When the plane is parallel to the cone's base, the intersection is a circle. As the plane is tilted, the circle turns into an ellipse that becomes more and more elongated as its eccentricity e goes from 0 to 1. The parabola is the limiting case of the ellipse in which the plane is parallel to the cone's side, one end does not close, and $e = 1$. If the cone is cut at a steeper angle, e becomes greater than 1 and open-ended hyperbolas are created.

Imagine a body in uniform circular motion around the Sun, curve A in Figure 7.8. If you apply

a force at point X to slow or speed the body, you can place it into an elliptical orbit (curves B or C), wherein X respectively becomes the aphelion or perihelion point. The body will follow Kepler's second law of motion, accelerating as it falls toward the Sun, slowing down as it recedes. Next, apply so much force at X that solar gravity cannot bring the outbound speed to zero. The body is now in a one-way, open-ended hyperbolic orbit (E) *and can never return.* There is a special case in which the velocity will decrease to zero, but only after an infinite time has passed. In that case, the path is that of a parabola (D).

Throw a ball upward. It has a certain kinetic energy that depends on its mass and velocity. It also has a certain potential energy within the Earth's gravitational field as measured from the Earth's center. The potential energy is considered negative because gravity is directed downward. The total energy is the sum of the potential and the kinetic. If the ball's potential energy is the greater, its total energy is negative, and it falls back to the ground. If you throw the ball upward with higher velocity, the total energy increases and the ball goes higher. At a critical velocity at which the total energy is zero, the **escape velocity** (v_{esc}), the ball never returns. The escape velocity is that needed to launch the ball into a parabolic orbit. From the equations for kinetic and gravitational potential energy,

$$v_{esc}^2 = 2GM/R.$$

Given the mass and radius of the Earth and the constant of gravity, G, the escape velocity at the Earth's surface is 11.1 km/s or 39,960 km/hr. This is the minimum speed that must be attained relative to the Earth to send a spacecraft on an interplanetary mission.

Kepler's second law in its generalized form involves a concept known as **angular momentum.** Tie a weight to a string and whirl it in a circle. The angular momentum (L) is the rock's mass (M) times its velocity (v) times the string's length (the orbital radius, R), or

$$L = MvR.$$

The general restatement of Kepler's second law is:

2. *In any closed system, angular momentum is conserved,*

that is, angular momentum does not change unless there is an action by an outside force.

The effect of the **conservation of angular momentum** can be quite impressive. A skater (Figure 7.9) will usually set a spin with outstretched arms and establish the original angular momentum of the body by applying an outside force with the skates. Once the spin is going, the arms are brought inward to reduce the body's average radius, and the speed dramatically increases. In an elliptical orbit, the distance of the planet from the Sun is continuously varying. As the distance goes down, the speed goes up to satisfy the law of the conservation of angular momentum and to produce the equal areas law.

In Kepler's third law, $P^2 = a^3$, where P and a (the sidereal period and semimajor axis respectively) are expressed in years and astronomical units. But that does little good if we wish to analyze orbits

Figure 7.9 *A skater in a spin employs the principle of the conservation of angular momentum. The smaller she can make her body's average radius by bringing in her arms, the faster she will rotate.*

in terms of physical units, in seconds and meters. To use these units, we must write $P^2 = ka^3$. The problem is then to find the value of k. The gravitational force depends on the separation between the two bodies and the product of the their masses. The parameter k must then contain the masses. The law of gravity leads to a generalization of Kepler's third law:

3.
$$P^2 = \frac{4\pi^2}{G} \times \frac{a^3}{\left(M_1 + M_2\right)},$$

where the masses of the two bodies are given by M_1 and M_2. (Note that there is no dependence on eccentricity, so orbits with different e but the same semimajor axes have the same periods.)

Kepler's third law works in its original form for the Solar System because the mass of the Sun (M_1) is so much greater than the masses of any of the planets, M_2. Consequently, $M_1 + M_2$ is practically the same for all the planets and the sum is very nearly a constant, where k = $4\pi^2/G(M_1 + M_2)$. So if we write $P^2_{\text{planet}} = ka^3_{\text{planet}}$ and $P^2_{\text{Earth}} = ka^3_{\text{Earth}}$ in seconds and meters, and divide one by the other, we just get $(P_{\text{planet}}/P_{\text{Earth}})^2 = (a_{\text{planet}}/a_{\text{Earth}})^3$, which is $P^2 = a^3$ in years and AU. The generalized law, however, is applicable to any orbit that involves two bodies.

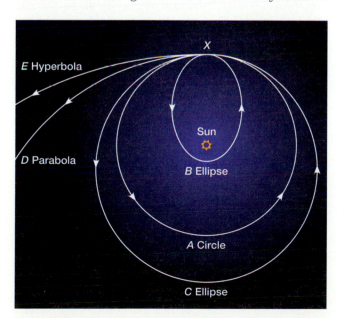

Figure 7.8 *Orbit A about the Sun is a circle. An elliptical orbit can be created by either slowing the body at X (curve B) or by speeding it up (curve C). If the body is given enough speed, the orbit can be parabolic (D) or hyperbolic (E).*

7.5
The Measurement of Mass

Gravitational theory allows us to determine the masses of celestial bodies. All you need do to find the mass of the Earth is drop a weight and measure the acceleration of gravity (9.8 m/s^2). Since g = GM_{Earth}/R^2, and G and R are known, the mass can be calculated. The result is 5.97×10^{24} kg. The mass of the Sun is found from Kepler's generalized third law. The Earth's orbital period is 3.16×10^7 seconds and its semimajor axis (the AU) is 1.50×10^8 km. The unknown quantity in the equation is the sum of the masses, which is readily found to be 1.99×10^{30} kg. The mass of the Earth is negligible by comparison, and when subtracted away, the result is still 1.99×10^{30} kg.

We can apply Kepler's generalized third law to any orbit to find the sum of the masses of the two bodies involved. However, gravity is really a mutual affair, each body pulling on and accelerating the other. As a consequence, one body never actually orbits the other. The bodies are *both* orbiting around a common point called the *center of mass* that lies on a line between them (Figure 7.10). The distance of each body from the center of mass is inversely proportional to that body's mass. Figure 7.10 shows two bodies, M_1 and M_2, in simple circular orbits. Their orbital semimajor axes about the center of mass are respectively a_1 and a_2, and therefore

$$M_1/M_2 = a_2/a_1,$$

or

$$M_1 a_1 = M_2 a_2.$$

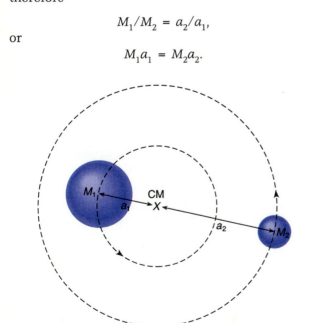

Figure 7.10 *Masses M_1 and M_2, where M_1 is twice M_2, have circular orbits about a common center of mass (CM) that lies exactly between them, and a_1 is half of a_2.*

The concept can be applied with equal validity to elliptical orbits, where the center of mass is at a focus of each of the orbits and always lies between the two bodies.

In reality, the Earth and Moon both revolve about a common center of mass that is only 4,700 km from the Earth's center and is actually *inside* our planet. Still, it is common to simplify orbits and to think of the less massive body as going about the more massive, with the latter at an orbital focus. In that case, the semimajor axis, a, is $a_1 + a_2$. This is the value that enters into Kepler's third law to derive the sum of the masses. The location of the center of mass then gives the *ratio* of the masses. Combination of the sum of the masses and the ratio of the masses provides the *individual* masses. As an example, say the sum of the masses $M_1 + M_2 = 2$ units and the ratio is $M_1/M_2 = 2$. Then $M_1 = 2M_2$. If we substitute this value of M_1 into the equation for the sum, $2M_2 + M_2 = 3M_2 = 2$ units, $M_2 = 2/3$ unit and $M_1 = 4/3$ unit.

Here is a key for understanding the Universe. In later chapters we will find the masses of the planets from the orbits of their satellites, we will measure the masses of other stars by observing double-star systems, and we will even find the masses of whole galaxies.

7.6
Orbits and Discovery

Newton's laws and the law of gravity allow the calculation of the size, shape, and orientation of the orbit of a body about the Sun from observations of its changing right ascension and declination. In the late 1600s, Edmund Halley worked out the paths of several comets and noted that the orbits of bright comets appearing in the years 1531, 1607, and 1682 were almost identical. He concluded that they were actually the same comet returning on an elongated elliptical orbit. Halley then calculated that the comet should return in 1758–59. He had been born in 1656 and had little chance of seeing it himself, but it returned on schedule 16 years after his death, providing solid support for Newton's laws. The comet, which has a 76-year period, and has since made reappearances in 1846, 1910, and 1985, was named in his honor. We now know that Halley's Comet was seen as long ago as 240 B.C. and has made at least 29 encores.

Nature, however, is not quite so simple. There are actually no pure two-body systems in the Solar System, a fact that complicates predictions of position. The Earth is indeed dominated by the Sun, but

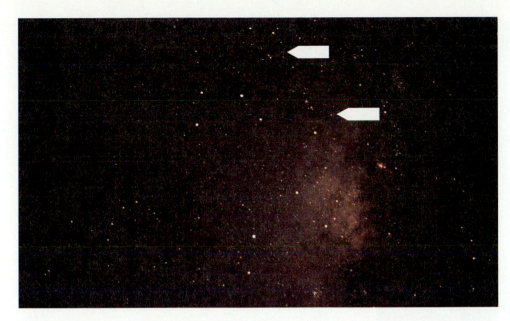

Figure 7.11   *The Little Milk Dipper of Sagittarius is just above center. Uranus (lower arrow), magnitude 5.7, is prominent and visible to the naked eye. Neptune, indicated by the upper arrow, is at the limit of the picture at magnitude 8.0.*

it is also pulled upon by Jupiter as well as by all the other planets. Furthermore, all affect the *solar* location, and pull on one another as well. With nine planets and the Sun there is a total of 45 interactions! As a result, the Earth cannot maintain a perfectly elliptical orbit, nor can any of the other bodies of the Solar System. If we wish to calculate the correct orbit of a planet, it is necessary to take these small forces, or gravitational **perturbations,** into account. Newton's laws still work, however. We first find the elliptical orbit a planet would have under the influence of the Sun alone. Then we calculate all the accelerations imposed by the other planets, enabling us to compute the way in which the semimajor axis, the eccentricity, inclination to the ecliptic, and the orbital orientations change with time, allowing us to predict planetary positions well into the future.

Theory could now be an important part in the exploration of the Solar System. At the time of the American Revolution, only six planets were known. However, a modern era was being ushered in by one of the great figures of astronomical history, William Herschel. Herschel was born in Hanover (now a part of Germany) in 1738. For reasons of health (and avoidance of service in the Hanoverian army) he was sent to England by his family to earn his living composing and performing music. His interests, however, also ranged over mathematics and astronomy. By 1774 he had built his first telescope and had begun to scan the skies in earnest, discovering, cataloguing, and studying new celestial sights. In 1781 he came across an unusual object in the constellation Gemini that he suspected was a comet. This new object, however, did not have a typical elongated cometary path but remained at a constant and great distance. Kepler's laws showed it to be orbiting at 19 AU, and consequently it had to be a planet well outside the orbit of Saturn. Herschel rushed to name it Georgius Sidus ("George's star") after his king, George III. But the only way to get international agreement on a name was to continue the use of mythology, and Uranus, Saturn's father and lord of the heavens, was chosen as appropriate.

Uranus is surprisingly bright. At opposition it has a magnitude between 5 and 6 and is visible to the naked eye (Figure 7.11). Astronomers quickly found that it had in fact been seen and recorded as a faint star as early as 1690. Oddly, neither the old observations nor the new ones that were rapidly being acquired quite agreed with the positions expected from orbital calculations made with Newton's laws. In the early 1840s, John Couch Adams in England and Urbain Leverrier in France decided that the culprit must be the gravitational pull of yet another planet even farther from the Sun than Uranus. From the deviation between the observed and expected positions of Uranus—which never amounted to more than two minutes of arc—they independently calculated (in 1845 and 1846 respectively) the trans-Uranian planet's orbit and position.

Adams had little success in persuading the English astronomers to look for the body. Leverrier transmitted his results to the Berlin Observatory, where the new planet—to be named Neptune, after the god of the sea—was found almost immediately

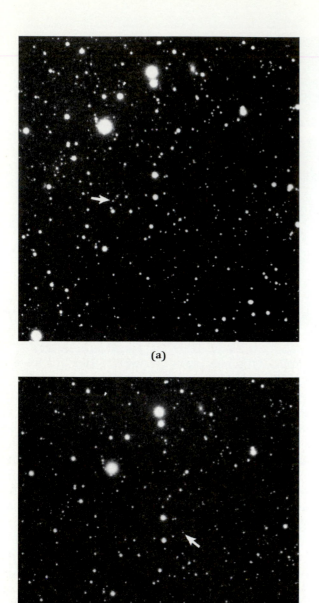

(a)

(b)

Figure 7.12 *Portions of Clyde Tombaugh's original discovery photographs of Pluto, taken January 23* **(a)** *and January 29* **(b)**, *1930, reveal the planet's motion.*

by Johann Galle just where expected. This discovery was rightly considered one of the triumphs of Newtonian theory.

With Neptune's perturbations factored into the calculations, Uranus now was closer to its predicted position, but the fit was still not perfect. The American philanthropist and astronomer Percival Lowell suggested the existence of yet another planet, this one outside the orbit of Neptune. Clyde

Tombaugh was hired in 1929 at the observatory founded by Lowell to look for the mysterious planet, and a year later he came across dim Pluto (Figure 7.12), named after the god of Hades, or the underworld. However, although Pluto was rather near its calculated location, we know it to have such a low mass that it could not cause any significant changes in Uranus's orbit. Pluto's discovery had been accidental.

The ecliptic plane has been thoroughly searched and nothing but small bodies have so far been found. Moreover, modern calculations fit Uranus's observed position very well. There is no evidence for any more planets like Uranus and Neptune.

7.7
Spaceflight

The Newtonian rules have allowed us to send our spacecraft around the Earth, to the Moon and planets, and even out of the Solar System. With a great blast of hot gas—Newton's third law in action—the space shuttle rises straight up from the launching pad at Cape Canaveral. As it accelerates, it slowly begins to tip horizontally (Figure 7.13a). By the time it reaches an altitude of 300 km and a speed of 27,700 km/hr, it is moving parallel to the Earth's surface and is in orbit. The engines then shut down and the shuttle coasts around the planet, taking 91 minutes to make the journey.

The farther away a spacecraft is from Earth, the less the acceleration of gravity, the slower the craft moves, the greater the orbital circumference, and the longer the period in accordance with Kepler's third law, $P^2 = a^3$. At an altitude of 1,000 km the period is up to 105 minutes. At a distance of 42,200 km (26,200 miles), there is a special place where the period is 24 hours, equal to that of the Earth's rotation. If the satellite is above the Earth's equator, it will appear stationary from the ground and we say that it is in *geosynchronous orbit*. That is where communications satellites are placed and why a communications receiver—a TV satellite dish, for example—always points to the same place in the sky.

The astronauts riding the shuttle are accelerated toward the Earth along with the shuttle cabin, and all orbit at the same speed (Figure 7.13b). There is nothing to press the space voyagers to the shuttle walls or floor, and consequently they are weightless. The passengers will just float around the cabin, bouncing off the walls. It is erroneous to say—as you will hear on many news broadcasts—that the astronauts are "out of the Earth's gravity."

One is *never* out of the Earth's gravity. It is the gravity that makes the craft and its inhabitants orbit. The astronauts simply feel no weight.

To get to the Moon with minimum energy, a spacecraft must accelerate into an elliptical orbit around the Earth that has an apogee at the lunar distance (Figure 7.14a). Ignoring the lunar gravity (which would perturb the elliptical path), the semi-

major axis of such an orbit is about 190,000 km (half the sum of the lunar distance and the Earth's radius). Since a near-Earth orbit (6,500-km radius) has a period of about 90 minutes, Kepler's third law shows that such a round trip would require 10 days and the one-way journey 5 days. The Moon's sidereal period is 27.3 days, so that in 5 days it will move 5/27.3 = 0.18 of the way around its orbit, or

(a)

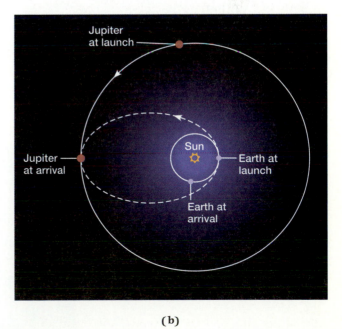

(b)

Figure 7.13   (**a**) *The space shuttle is accelerating and is tilting parallel to the Earth's surface.* (**b**) *The astronauts are orbiting around the world with the same acceleration as the cabin and are therefore weightless.*

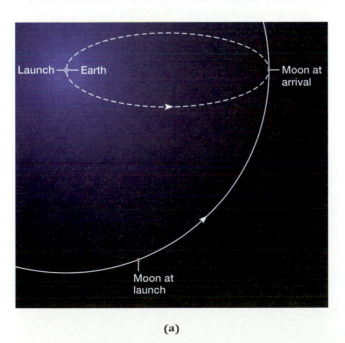

(a)

(b)

Figure 7.14   (**a**) *A spacecraft is launched into an elliptical orbit so that it reaches apogee at the lunar distance and at the same time as the Moon.* (**b**) *A spacecraft to Jupiter is launched from Earth into an elliptical solar orbit with the Sun at perihelion and Jupiter at aphelion.*

BACKGROUND 7.1 The Beginnings of Spaceflight

From the time of ancient Greece, we have dreamed of going to the stars or, at the very least, to the Moon. The master exponent of a lunar trip was the writer Jules Verne, who in the late nineteenth century piqued public interest in a variety of science-fiction travels, including one to the bottom of the sea. His method of space travel was to pack the voyagers' ship into a great cannon that shot them Moonward.

Real spaceflight depends on rocket power and Newton's third law of motion. Unlike a jet aircraft, a rocket does not need oxygen from the surrounding air. It carries its combustion supplies on board and can therefore go into space. From the time of their invention in the thirteenth century by the Chinese, rockets have been developed primarily for warfare. Inspired by Verne, however, the Russian scientist Konstantin Tsiolkovsky looked skyward, and between the years 1883 and 1898 developed the theory of rocket-powered spaceflight. The theory included the multistage rocket, without which spaceflight is impossible. Most of the weight of a rocket is fuel. To keep the craft as light as possible, one rocket is set atop another. When the first one uses its fuel in giving the machine an initial speed, it is cast away, thus decreasing the weight. The second one is then fired, increasing the speed further.

By the time Verne died, Robert Goddard (Figure 7.15), the father of modern rocketry, was 23 years old. Working quietly and almost alone in the years 1914 to 1941, he pursued the engineering technology needed to propel spacebound rockets, experimenting with fuels and multistaging devices. Goddard was the first to create solid rockets (those in which the fuel is in solid form, like the big boosters of the space shuttle), and from 1920 onward, he experimented with liquid-fueled engines.

The modern era of rocketry was actually ushered in by German scientists and engineers in World War II. Under Werner von Braun, the program culminated in the powerful 46-foot-long V2, which the Nazi government used principally against Britain in 1944 and 1945. The V2 could fly from its launch site in occupied Holland to London in a mere five minutes, falling on its target at nearly 6,000 km/hr. After the war, several V2s, as well as von Braun, were brought back to the United States. The rockets were further developed and used for scientific purposes, such as the examination of the upper atmosphere.

The cold war with the Soviet Union accelerated rocket research to enable long-range delivery of both conventional and nuclear bombs. Von Braun was convinced that these powerful machines could place satellites in orbit. The U.S. military first attempted orbital flights in 1956, but before they were successful, the Soviet Union changed the world with the launch of *Sputnik 1* on October 4, 1957. Over the last 35 years, enormous numbers of satellites have been launched, so many that we have even developed a severe "space garbage" problem. Flights to the Moon and the planets were then a natural, though difficult, extension of orbital missions. We can now routinely send such flights anywhere we like within the Solar System, and have even launched four packages free of the Sun, ships that will someday roam silently among the stars.

Figure 7.15 *Robert Goddard (1882–1945), second from the right, developed modern rocketry.*

Figure 7.16 *Albert Einstein formulated the theories of relativity and gave us our modern view of space, time, and gravity.*

65°. The craft must be shot at a point in the lunar orbit that at the time of launch is 65° ahead of the Moon. In 5 days, the craft and the Moon will arrive at the same place. Rockets can then be fired to slow the ship and enable it to land on the lunar surface. The time can be (and in the actual event was) cut by using a higher speed and a hyperbolic orbit, but at greater cost.

A minimum-energy trip to Jupiter involves the same principles (Figure 7.14b), except that the craft is fired into an elliptical orbit around the Sun, with the Earth's orbit at perihelion and Jupiter's at aphelion. Such an orbit has a major axis of 6.2 AU, the sum of the semimajor axes of the Earth and Jupiter. The craft's orbital semimajor axis is then 3.1 AU. Again from Kepler's third law, the period is 5.46 years and the travel time to Jupiter half that, or 2.73 years. Now the launch must lead Jupiter by 83°. In reality, the gravitational effects of the Earth and Jupiter, as well as launching errors, must be taken into account. As the spaceship cruises on its almost-elliptical path, small rockets must be fired to effect midcourse corrections to arrive at the rendezvous point at the right time.

Celestial engineers have progressed far beyond these simple examples. They now regularly use the planets themselves to accelerate and direct spacecraft. For example, as the interplanetary probe *Voyager 2* neared Jupiter in 1979 it was directed behind the giant planet. Jupiter's gravity changed the direction of the craft and caused it to speed up enough to be hurled to Saturn. The same trick was used there to throw it to Uranus. Newton probably would have enjoyed it all immensely.

7.8
Einstein and Relativity

About 1870, Urbain Leverrier, the codiscoverer of Neptune, found a small discrepancy in Newtonian theory. The perihelion point of Mercury's orbit should move clockwise around the Sun at the rate of 527 seconds of arc per century as a result of the perturbations of the other planets. The observations, however, showed the increase to be 38 seconds of arc larger, later set at a more precise value of 43 seconds. The difference between theory and observation defied explanation. The problem was finally resolved between 1908 and 1914 by a young man named Albert Einstein (Figure 7.16) in a pair of publications that extended Newton's science of mechanics and that again revolutionized scientific thought.

Einstein was born in 1879 in Germany and educated there and in Switzerland. He was more or less undistinguished in school and made his living in Switzerland between 1901 and 1909 as a patent examiner, a position that gave him considerable free time to explore mathematical physics. As a result of his theoretical discoveries he succeeded to a number of academic positions in Prague, Zurich, and Berlin, where he continued his extraordinary work. In 1933 he emigrated to the United States to avoid Nazi persecution and took a post at Princeton, where he remained until his death in 1955. His theory, called **relativity,** comes in two forms: the **special theory of relativity** involves bodies moving in constant relative motion (hence the name), and the **general theory of relativity** includes accelerated motion and gravity.

Look first at the special theory. Run an experiment in your mind. A man on a flatbed truck moving at constant speed passes by a woman standing on the side of the road (Figure 7.17). Ignore the effects of wind and vibration and pretend that any observation or measurement is possible: the mech-

Figure 7.17 *A truck moves past a woman standing by the roadside. A man on the truck shines a light in the direction of the truck's motion and measures a speed c. In spite of the truck's motion, the woman also sees the light move at c. The woman also sees lightning bolts strike the ground simultaneously. Because the man is moving, he sees them strike at different times.*

anism of observation is unimportant, only the results. To her, the truck is in motion. To him, the woman appears to be moving in the opposite direction. There is no such thing as absolute motion, only *relative* motion. He throws a ball in the direction of his motion at 20 m/s. If the truck is moving at 50 m/s, she will see the ball move at 70 m/s. She just adds the velocities, a basic Newtonian concept. Now say the truck is moving at half the speed of light and that he shines a light in the direction of his motion. He sees the beam move away from him at a speed (called c) of 299,792 km/s. You might intuitively say that the stationary woman should measure the beam moving forward at the sum of the two speeds, or 1.5 c. Remarkably, however, she *too* measures the speed of the light at c. The speed of light is the upper limit to speeds in the Universe and is independent of the speeds of the source or the observer, facts well documented in real laboratory experiments.

Suddenly the woman sees two lightning bolts strike the road simultaneously, one in front of the truck, the other behind it. Because the man is moving, however, the light from the forward stroke will reach him before the light from the rearward one arrives. To him, the strokes are *not* simultaneous. His sense of time and space must then be different from hers. If the woman and the man are carrying identical clocks, each will see the other's clock appear to run slow relative to the one in hand, an effect called *time dilation.* Each person also holds a 100-g meter stick in the direction of motion. As they move past each other, each makes a measurement of the length of the meter stick that the other is carrying. To each person the other's stick looks to be *shorter* than a meter, a phenomenon sometimes called the *Lorentz contraction* (named for the Dutch physicist, Hendrik Antoon Lorentz). Moreover, each can apply a force to measure the mass of the other's meter stick. Her meter stick will appear to have a mass of *more* than 100 g, and so will hers as determined by him.

The degrees to which the quantities of time, space, and mass are changed are controlled by a factor

$$\gamma = \sqrt{\left(1 - v^2/c^2\right)},$$

where v is the relative velocity between a body and an observer. The symbols L, T, and M represent a length, a time interval, and a mass for a body at *rest* ($v = 0$). The values of these quantities when the body is in motion relative to the observer are $L\gamma$, T/γ, and M/γ. The Lorentz contraction, time dilation, and mass increase are apparent only when v

comes close to c. At the speed of a real truck, say 100 km/hr (0.03 km/s), v^2/c^2 is only 10^{-14}, and γ is effectively 1. Even at a tenth of c, γ is still 0.99. Consequently, the effects of relativity—*relativistic effects*—are unnoticed at low speeds, and Newtonian mechanics works very well. But as soon as the speeds go high enough, close to that of light, Einsteinian rules must be used. If v = c, v/c becomes 1, and $(1 - v^2/c^2) = 0$. *The length L then becomes zero, and the time interval T and the mass M become infinite.* Since M cannot be greater than infinity, v cannot exceed c, which is consistent with the speed of light as the maximum allowed speed. Since acceleration requires energy, it would take an infinite amount of energy to accelerate a mass even *to* the speed of light. Consequently, no material particle can attain that speed, but can only approach it.

The relation between measured mass and relative velocity demonstrates that mass (M) and energy (E) are two manifestations of a more general concept called mass-energy. Either can readily be converted into the other through the famous relation

$$E = Mc^2.$$

Because of the large value of c, a tiny amount of mass can produce an enormous quantity of energy. If fully converted, a kilogram of matter could produce nearly 10^{17} joules, enough to serve your household energy needs for a million years! The

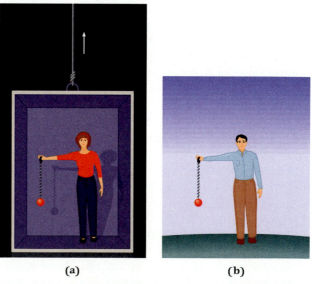

(a)   (b)

Figure 7.18   (**a**) *A woman rides an elevator in interstellar space and feels the upward acceleration as a downward weight. She holds a mass on a spring, which stretches to the floor under the acceleration.* (**b**) *A man stands on the Earth, holding an identical mass. The Earth's gravitational field pulls him downward and also stretches his spring. If the elevator accelerates at 9.8 m/s^{-2} each person feels exactly the same.*

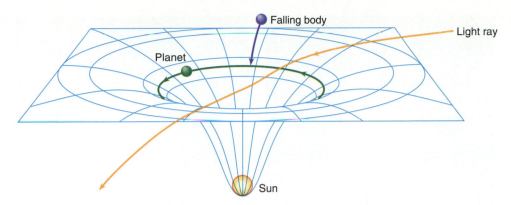

Figure 7.19 *It is impossible to draw four-dimensional spacetime on a sheet of paper. Instead, imagine that space has only two dimensions, so that spacetime has a total of three. In a perspective drawing you can get a sense that a mass causes a distortion in spacetime's fabric that results in a gravitational field. The central mass will make bodies fall toward it, will accelerate planets to keep them in orbit, and will bend light rays (greatly exaggerated here).*

relation forms the basis for understanding solar and atomic energy, subjects to be explored in later chapters. It also changes the law of the conservation of energy (see Section 7.2) into a more general law of the **conservation of mass-energy.** There is only a certain amount of mass-energy in a closed system, and although mass can change into energy and energy into mass (through Einstein's equation), the total amount remains constant.

Proofs of special relativity are ample. Time dilation has been measured with precise clocks in relative motion. Modern atomic accelerators can speed particles up nearly to c, and observation agrees perfectly with prediction. But special relativity still does not solve Leverrier's problem of the advance of Mercury's perihelion. For that, Einstein had to expand the theory from uniform motion at constant speed to include accelerations, which led to general relativity and to his greatest discovery, the modern theory of gravity.

We are used to thinking of "space" as consisting of three dimensions, whose directions are easily seen where the walls and ceiling meet in the corner of a room. We naturally expect space to go on forever and that the rules learned in geometry—that the angles of a triangle add up to 180° and parallel lines never meet—are true over every distance and on every scale. This space is called "Euclidean," after the great Greek mathematician of the fourth century B.C. who invented the formality of geometry; for short, it is just called "flat." We are also used to thinking of "time" as independent of space. To include the effect of accelerations in the theory of relativity, however, it is necessary to combine space and time into one *four*-dimensional system called **spacetime.** We move through the Universe in a continuum of four dimensions, our location given by three coordinates in space, *x, y,* and *z,* and by one in time, *t.* The four cannot be separated, and the distortions of time and space that involved the truck experiment above cannot be separated in

time and space either. We have to deal with distortions in spacetime as a unit.

Run another imaginary experiment. A woman floats in a weightless state into an elevator placed deep in interstellar space away from any significant gravitational fields. The doors close and she accelerates at a steady rate (Figure 7.18). She is then forced to the floor and acquires weight. To her, it feels exactly as if a gravitational field has been switched on. If she drops a ball, it will accelerate to the floor, and if the elevator is accelerating at 9.8 m/s², she feels as if she is back on Earth. There is no way for her to tell the difference. The identity of the experience of the acceleration of the elevator and the acceleration of gravity is called the *principle of equivalence.*

A ray of light now passes through the accelerating elevator. To the passenger the light will appear to move in a curved path. Because of the principle of equivalence, the same effect must be seen in a gravitational field. The light beam, however, defines a straight line. Therefore, it is *spacetime* that is curved. Einstein's great discovery is that you fall to Earth not because of a mysterious "force" of gravity, but because of the curvature of spacetime caused by the very existence of mass.

Within a gravitational field, the simple rules of three-dimensional Euclidean or flat geometry do not work. The effect is something like a well in spacetime in which bodies will be accelerated (Figure 7.19). The planets orbit around the Sun within this gravitational well. The greater the mass, the more spacetime curves, the greater the acceleration, and the greater the apparent force that is felt. As you move away from the mass and the curvature decreases, so do the force and acceleration, and the planets move more slowly.

Mercury is the fastest-moving planet, and although its average speed of 48 km/s is far below that of light, its proximity to the Sun is enough to affect its motion in the deep solar gravitational

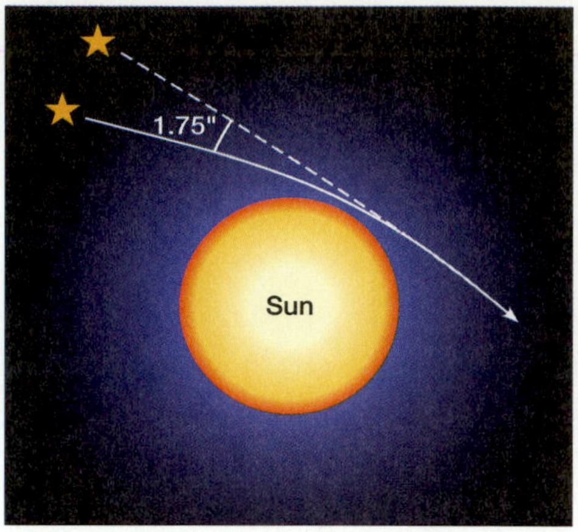

Figure 7.20 **(a)** *A ray of light from a distant star brushes past the Sun and is deflected by 1.75 seconds of arc (considerably exaggerated here).* **(b)** *We look toward the Sun during a total solar eclipse and see stars apparently displaced from the solar center by angles proportional to their angular separations from the solar edge.*

(a)   (b)

well. The elliptical orbit keeps the planet moving through different curvatures of spacetime, making Mercury's perihelion advance by the missing 43 seconds of arc per century, something that does not happen under Newtonian gravitational theory.

Einstein suggested another proof of the theory. When a ray of light passes the Sun, it must curve. As a result, the stars that surround the Sun in the daytime ought to be slightly displaced from the solar center (Figure 7.20). The effect should be visible during a solar eclipse, when the stars can be seen, but even at the edge of the Sun's apparent disk, the shift is only 1.75 seconds of arc, and the observation is extremely difficult. The first attempt was made at the total eclipse of 1919 under the leadership of the British astrophysicist Sir Arthur Eddington. The results, although crude, accommodated the theory. A much more definitive test run in 1973 revealed the stars placed within about 10% of their expected positions. Similar observations of celestial radio sources near the Sun show the agreement with theory to be essentially perfect.

General relativity also requires that the rate of a clock depends on the strength of the gravitational field in which it runs. Precise clocks on the ground are observed to run slower than those placed on top of tall towers, where the strength of the gravitational field is diminished. The effect must be factored into keeping time. A clock in space runs faster than one on the ground, and clocks in the orbits of any of the planets depend on their distances from the Sun. The time system used for plotting the motions of the planets actually corrects for these effects. Relativity is not just an abstract theory. It is real, it is true, and it helps describes our world.

7.9
Chaos

In the eighteenth and nineteenth centuries, partly as a result of the great successes of Newtonian mechanics, scientists tended toward a deterministic attitude. Given the positions of all the atoms and their velocities, it should—so they thought—be possible to predict the state of the Universe for all times in all places. The inclusion of relativity does not change that idea.

Two other discoveries, however, destroyed that belief. The first was that of the natural uncertainty associated with *quantum mechanics,* the realization that atoms and subatomic particles play by different rules (we will look at that subject in Chapter 8). The second involves **chaos.** Many mechanical systems are inherently unstable, so that a small disturbance in position or velocity can be greatly and quickly magnified. Assume that we have exact equations whose solutions predict an eventual state of a mechanical system given specified initial conditions. As an example, start with the position of a body in *A*, calculate the forces on it, and predict its course, path *a.* Now shift the location even a small amount, to *B*, and the body follows a new path *b.* If you place *B* close to *A*, paths *b* and *a* will follow one another for a while, but after sufficient time will diverge. No matter how close you place *B* to *A*, you cannot get the same path—the two will *always* diverge. It is not possible to specify the precise positions of all things and the strengths and directions of all forces. Therefore, it is inherently impossible to predict the path for an unspecified time into the future: the system is chaotic.

Not all systems are chaotic; many are indeed quite predictable. Some important ones, however, like the weather (Figure 7.21), are chaotic. Minute instabilities quickly grow into big ones: it has been colorfully said that "a butterfly flaps its wings in Africa and a hurricane follows in Florida." Long-term detailed weather prediction is then impossible. Pluto's orbit is chaotic over only short time scales. The smallest displacement of the planet, or changes in the influencing gravitational forces, grow into major shifts in position. The calculations are so dependent on initial location that it is not possible to calculate Pluto's position more than about 100,000 years into the future or back in time. The orbits of the other planets (including that of Earth) are probably chaotic as well, though—fortunately for us—on time scales longer than the lifetime of the Sun.

Should we feel distressed by this development? Certainly not. As scientists, we set out to learn how the world works and cannot change it by our beliefs. There is always a challenge to be met. In the nineteenth century it was the perihelion of Mercury, solved by relativity theory. Now it is a different problem, chaos, which has its own mathematical solutions that will help lead us to the comprehension of the Universe.

Figure 7.21 *Weather is an example of a chaotic system in which small instabilities grow rapidly into large ones.*

KEY CONCEPTS

Acceleration: A change in velocity (speed or direction) with time.

Chaos: A branch of mechanics that deals with unstable systems.

Conic sections: The curves (circle, ellipse, parabola, and hyperbola) defined by the intersections of a cone and a plane.

Conservation of angular momentum: The concept that total angular momentum is always constant in a closed system.

Conservation of energy: The concept that energy can neither be created nor destroyed, only changed in form; in relativity, the law becomes the **conservation of mass-energy.**

Energy: The capacity of a body to do work on (or accelerate) or heat another body.

Force (*F*): That which produces an acceleration of a mass.

Gravity: An attractive force; the curvature of spacetime caused by the presence of mass.

Kilogram (kg): The basic unit of mass, equal to 1,000 grams.

Mass (*M*): The amount of matter in a body or the degree to which a force is resisted.

Perturbations: Orbital changes induced by outside gravitational forces.

Potential energy: Energy held by virtue of position or configuration.

Relativity: The branch of mechanics developed by Albert Einstein that lets the speed of light be independent of the speeds of the source or observer; the **special theory of relativity** involves constant speed, the **general theory of relativity** involves accelerations.

Spacetime: A four-dimensional construction that consists of the three dimensions of space and the one of time.

Speed: The rate at which a body changes its distance with time.

Temperature: A measure of the average velocities of the particles in a gas.

Velocity (*v*): The combination of speed and direction.

Work: Force applied over a distance.

KEY RELATIONSHIPS

Acceleration of gravity (g):

The acceleration of a falling body at the Earth's surface is

$$g = G \frac{M_{\text{Earth}}}{R^2}.$$

Angular momentum:

$$L = MvR$$

Center of mass:

The point at the mutual focus of two orbiting bodies such that $a_1/a_2 = M_2/M_1$.

Escape velocity (v_{esc}):

The velocity needed to achieve a parabolic orbit so that a body will not return; $v_{esc}^2 = 2GM/R$ (where R is the distance to the body's center).

Kepler's generalized laws of planetary motion:

1. The path of an orbiting body is a conic section with the other body at one focus of the curve.
2. In any closed system, angular momentum is conserved.
3. $P^2 = \dfrac{4\pi^2}{G} \times \dfrac{a^3}{(M_1 + M_2)}$.

Kinetic energy:

Energy of motion; $KE = \frac{1}{2} Mv^2$

Law of gravity:

$$F = G\frac{M_1 M_2}{R^2}.$$

Newton's laws of motion:

1. Left undisturbed, a body will continue in a state of rest or uniform straight-line motion.
2. The size of an acceleration is directly proportional to the force applied, is inversely proportional to the mass of the body being accelerated, and the direction of the acceleration is the same as the direction of the force.
3. For every force applied to a body, there is an equal force exerted in the opposite direction (action equals reaction).

Weight (W):

The force with which a body is pressed to the surface of another body as a result of gravity; $W = Mg$ (mass times the acceleration of gravity).

EXERCISES

Comparisons

1. Distinguish between velocity and speed.
2. Distinguish between mass and weight.
3. What is the difference between potential and kinetic energy?
4. Distinguish a body in an elliptical orbit from one in a hyperbolic orbit.
5. What distinguishes special from general relativity?

Numerical Problems

6. A body has a mass of 2 kg. You apply a force of two newtons to give it a certain acceleration. How many newtons would you have to apply to a 4-kg mass to give it the same acceleration?
7. You and a friend who is one-third your mass stand on ice. You give her a push to the north and she glides away at 1 m/s. How fast are you going and in what direction?
8. What is the kinetic energy of a 100-kg man running at 4 m/s, and how does his energy compare with that of a second man of the same mass running at 2 m/s and with that of a 50-kg mass woman also running at 4 m/s?
9. The temperature in your room is 20°C. What is it in Kelvins?
10. What are the periods of Earth satellites in circular orbits 2,000 km and 10,000 km above the ground?
11. How many times more (or less) would you weigh if you could travel to a planet that has **(a)** the same radius as the Earth but one-third the mass; **(b)** the same mass as the Earth but one-third the radius; **(c)** one-third the mass of the Earth and one-third the radius?
12. How fast must a body be launched from the Moon to escape?
13. Tie a string 2 m long to a 1-kg weight and whirl it at 10 m/s. What is its angular momentum? Draw in the string to a radius of 1 m. How fast is the weight moving now?
14. Use the orbit of Mars to determine the mass of the Sun.
15. Bodies with masses of 20 and 2 times that of the Earth are in circular orbit about each other 11 AU apart. How far is the center of mass from each body?
16. What would your travel time to the planet Saturn be if you were launched into a minimum-energy orbit?
17. A friend of yours has a mass of 50 kg when sitting next to you. When you next see him he is shooting past you at 0.97c. What would you measure his mass to be then?

Thought and Discussion

18. What is meant by the term *acceleration*?
19. If you shut off the engine of a car and coast in neutral on a straight, level road, the car will come to a stop. Is this behavior a violation of Newton's first law of motion?
20. Why can a body not be cooled below absolute zero?
21. How might you use the acceleration of gravity to measure the shape of the Earth?
22. Write a formula for the acceleration of gravity at the surface of Mars.
23. If you could increase the mass of the Sun but keep the semimajor axis of the Earth a constant, what would happen to the Earth's sidereal period?

24. If you could decrease the mass of the Sun but hold the sidereal period of the Earth constant, what would happen to the semimajor axis?

25. What is meant by a chaotic orbit?

Research Problem

26. Construct a list of the different important functions of Earth satellites.

Activities

27. Try Galileo's proposition regarding the acceleration of gravity for yourself. Use two different weights that are both heavy enough to be unaffected by air resistance. Drop them at the same time and have someone with a stopwatch try to detect any difference in the times at which they strike. What might cause any such difference?

28. Find Uranus in the nighttime sky with or without binoculars using an annual map from a popular astronomy magazine like *Sky and Telescope* or *Astronomy*.

Scientific Writing

29. You read in the newspaper that "the shuttle astronauts are floating around in their cabin because they have been launched out of the gravitational field of the Earth." Write a letter to the editor in which you set the matter straight.

30. Write a one- or two-page article on the importance of Kepler's laws to modern science, as he derived them and as Newton derived them.

8

Atoms and Light

*To know
the very large
we must first
understand
the very small*

*Laser beams,
high-intensity
light rays created by atoms,
rocket from mirror to mirror
around the laboratory.*

Our view so far has been directed outward, into the sky. Now we reverse direction and focus inward. To understand astronomical bodies, we need to examine their constituents, the atoms and molecules from which they are built, and the means by which we can analyze their natures over great distances.

8.1
Atoms

As early as 400 B.C., the great Greek scholar Democritus suggested that all matter is ultimately composed of tiny indivisible particles called **atoms,** from the Greek word meaning "uncuttable." It took over 2,000 years to find and understand them, but we now have a good idea of how they are constructed and how they work.

8.1.1 *Atomic Constituents and Electromagnetism*

All matter—the stuff around us, the Sun, stars, grass, the air we breathe—is made of chemical **elements,** each composed of a different kind of atom. Of the 109 known elements, 90 are present in the Earth. The names and symbols of 105 are given in Table 8.1 (on page 118). Many are known to us in our daily lives, including the oxygen and nitrogen we breathe, the carbon out of which we are largely made, the metals—iron, aluminum, copper—with which we build. Others—iridium, cesium, europium—are uncommon but still play an important role in life and astronomy. Still others, like radium and uranium, are familiar by name, but for reasons of safety are kept at a distance.

Atoms are made of three *subatomic* or *elementary particles* that can be arranged in a variety of ways: **protons, neutrons,** and **electrons** (Figure 8.1). Chapter 7 presented gravity as a basic force of nature. Protons and electrons are carriers of the second of four such basic forces, the **electromagnetic force.** Although gravity is the more familiar of the two and appears to be the more prominent, on an atom-to-atom basis electromagnetism is over 10^{30} times more powerful. Gravity just seems stronger because all the atoms in the Earth are pulling on us, whereas electromagnetism, as we will see, is largely neutralized.

Electromagnetism has two manifestations, the **electric charge,** which produces a surrounding **electric field,** and the **magnetic field.** Like gravity, the electric field behaves according to an inverse-square law, its strength dropping off as $1/\text{distance}^2$. Unlike gravity, however, electric charge has two forms, positive and negative. Oppo-

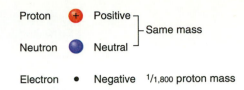

Figure 8.1 *Protons, neutrons, and electrons are the building blocks of atoms. The proton and electron have opposite electric charges of the same strength, whereas the neutron is neutral. The proton and neutron are about 10^{-13} cm across and have about the same mass. The electron has 1/1,800 the mass of the proton.*

site charges will attract each other, similar charges will repel. Though magnetic fields are more complex, they too have positive and negative poles or directions. Opposite poles of two common magnets will stick together, and similar ones will repel.

The proton has a positive charge. Its mass is a mere 1.67×10^{-24} g and its radius only about 10^{-13} cm. There are more protons in a gram—one cubic centimeter—of water than there are stars in the observable Universe. The electron has an equal and opposite negative charge. Its mass, however, is 1,800 times smaller. The neutron has very nearly the same mass as a proton, but is electrically neutral, carrying no charge at all.

Electricity and magnetism are tightly related. A flow of electrons in one direction constitutes an electric current, which is responsible for the glow of a light bulb, a bolt of lightning, and a huge variety of other natural phenomena. An electric current generates a surrounding magnetic field, and a wire moving through a magnetic field produces an electric current.

The hierarchy of atomic particles descends even further. Protons and neutrons are constructed from smaller particles called **quarks.** These particles have fractional electric charges and other properties that can be combined in different ways. The quarks and the electron are thought to be truly elementary and not made of anything else.

8.1.2. *Construction of the Chemical Elements*

The proton, electron, and neutron are combined into the more familiar forms of matter. Since they have opposite charges, a proton and an electron will attract one another. An electron can then attach itself to a proton to create the simplest kind of atom, which is called hydrogen (Figure 8.2a). The central proton is then also called the **nucleus** of the atom. The atom as a whole is electrically neutral, the negative and positive charges effectively balancing one another.

Each kind of atom has a different number of protons—called the **atomic number**—in its nucle-

TABLE 8.1
The Chemical Elements

Name	Symbol	A	M	Name	Symbol	A	M
Hydrogen	H	1	1	Xenon	Xe	54	132
Helium	He	2	4	Cesium	Cs	55	133
Lithium	Li	3	7	Barium	Ba	56	138
Beryllium	Be	4	9	Lanthanum	La	57	139
Boron	B	5	11	Cerium	Cs	58	140
Carbon	C	6	12	Praseodymium	Pr	59	141
Nitrogen	N	7	14	Neodymium	Nd	60	142
Oxygen	O	8	16	Promethium[a]	Pr	61	147
Fluorine	F	9	19	Samarium	Sm	62	152
Neon	Ne	10	20	Europium	Eu	63	153
Sodium	Na	11	23	Gadolinium	Gd	64	158
Magnesium	Mg	12	24	Terbium	Tb	65	159
Aluminum	Al	13	27	Dysprosium	Dy	66	164
Silicon	Si	14	28	Holmium	Ho	67	165
Phosphorus	P	15	31	Erbium	Er	68	166
Sulfur	S	16	32	Thulium	Tm	69	169
Chlorine	Cl	17	35	Ytterbium	Yb	70	174
Argon	Ar	18	40	Lutecium	Lu	71	175
Potassium	K	19	39	Hafnium	Hf	72	180
Calcium	Ca	20	40	Tantalum	Ta	73	181
Scandium	Sc	21	45	Tungsten	W	74	184
Titanium	Ti	22	48	Rhenium	Re	75	187
Vanadium	V	23	51	Osmium	Os	76	192
Chromium	Cr	24	52	Iridium	Ir	77	193
Manganese	Mn	25	55	Platinum	Pt	78	195
Iron	Fe	26	56	Gold	Au	79	197
Cobalt	Co	27	59	Mercury	Hg	80	202
Nickel	Ni	28	58	Thallium	Tl	81	205
Copper	Cu	29	63	Lead	Pb	82	208
Zinc	Zn	30	64	Bismuth	Bi	83	209
Gallium	Ga	31	69	Polonium[b]	Po	84	210
Germanium	Ge	32	74	Astatine	At	85	210
Arsenic	As	33	75	Radon	Rn	86	222
Selenium	Se	34	80	Francium	Fr	87	223
Bromine	Br	35	79	Radium	Ra	88	226
Krypton	Kr	36	84	Actinium	Ac	89	227
Rubidium	Rb	37	85	Thorium	Th	90	232
Strontium	Sr	38	88	Protactinium	Pa	91	231
Yttrium	Y	39	89	Uranium	U	92	238
Zirconium	Zr	40	90	Neptunium[a]	Np	93	237
Niobium	Nb	41	93	Plutonium[a]	Pu	94	242
Molybdenum	Mo	42	98	Americium[a]	Am	95	243
Technetium[a]	Tc	43	99	Curium[a]	Cm	96	247
Ruthenium	Ru	44	102	Berkelium[a]	Bk	97	249
Rhodium	Rh	45	103	Californium[a]	Cf	98	251
Palladium	Pd	46	106	Einsteinium[a]	Es	99	254
Silver	Ag	47	107	Fermium[a]	Fm	100	253
Cadmium	Cd	48	114	Mendelevium[a]	Md	101	256
Indium	In	49	115	Nobelium[a]	No	102	254
Tin	Sn	50	120	Lawrencium[a]	Lw	103	257
Antimony	Sb	51	121	Kurchatovium[a]	Ku	104	—
Tellurium	Te	52	130	Hahnium[a]	Ha	105	—
Iodine	I	53	127				

The column headed A gives the atomic number. The column headed M gives the mass number or atomic weight (number of nucleons, or protons plus neutrons) of the most abundant isotope of each element, or in the case of radioactive elements, the weight of the most stable isotope.

[a]Radioactive elements that do not exist in the Earth and are made in the laboratory.
[b]All elements heavier than bismuth are radioactive; M gives mass of most stable isotope.

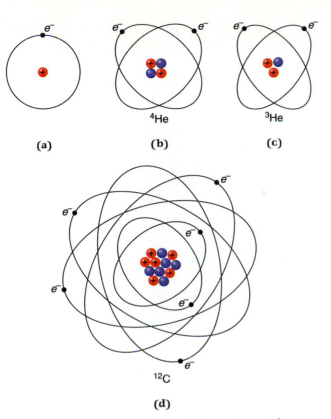

(a)   **(b)**   **(c)**

(d)

Figure 8.2 **(a)** *The simplest atom, hydrogen, has a single proton (red, +) surrounded by an electron (e^-). The size of the proton is about 1/100,000 the size of the atom.* **(b, c)** *The helium (He) atom consists of coupled pairs of protons and electrons. Neutrons (blue) are attached to the nuclei to help bind them together. Two neutrons make the ^4He isotope* **(b)** *and one makes ^3He* **(c)**. *The carbon atom in* **(d)** *has 6 protons and 6 neutrons, making it ^{12}C. Carbon's electrons are arranged in two shells.*

us. The atomic number of hydrogen is 1. The next atom up the scale is helium, which has a nucleus that contains *two* protons tightly bound together (Figure 8.2b). Since similar electric charges repel one another, such a construction seems like a contradiction. The combination is affected by the **strong** or **nuclear** force, the third and most powerful of all known forces. The strong force, however, does not behave according to an inverse-square law. It extends over only a very short distance, a mere 10^{-13} cm, the size of the nucleus itself. Within that realm, the strong force, which in its net action is attractive, overpowers the electric repulsion, and keeps the protons tied within its grip.

Even the strong force, however, is not sufficient to join two lone protons in an atomic nucleus. The repulsive force must be further moderated by the addition of neutrons. Different numbers of neutrons are possible and the different varieties of resulting atoms are called **isotopes** of one another. The most common form of helium has two protons and two neutrons (collectively called *nucleons*), so it is four times as massive as the hydrogen atom.

The total number of nucleons is consequently called the **atomic weight,** which is commonly expressed as a superscript placed to the upper left of the chemical symbol. This kind of helium atom is then called ^{4}He, or helium-4. The other allowed isotope of helium has only one neutron and is called ^{3}He because it has three nucleons (Figure 8.2c). Even hydrogen has isotopes. A neutron can attach itself to the proton in Figure 8.2a to make ^{2}H, a variety of hydrogen called deuterium. There are, however, limits to the number of possible isotopes. If there are too few or too many neutrons, the nucleus will fall apart.

The progression of elements continues with additional protons. Lithium has three protons and two allowed isotopes, ^{6}Li and ^{7}Li, beryllium (^{7}Be) has four protons, and common carbon, ^{12}C (Figure 8.2d), has six protons and six neutrons. We also find carbon atoms with seven neutrons in the form of ^{13}C. Table 8.1 gives the atomic weight of the most common isotope of each element.

8.1.3 *Radioactivity*

There is a middle ground between the isotopes of an element that are stable and permanent (like ^{3}He and ^{4}He) and those that do not exist, such as ^{4}H or ^{2}He. Most atoms have one or more unstable isotopes that can survive for a period of time but will eventually disintegrate. Examples are tritium (^{3}H) and ^{14}C. Tritium can be manufactured in the laboratory and ^{14}C occurs in nature. Moreover, several of the chemical elements, including *all* those heavier than bismuth (atomic number 83) have no stable isotopes at all.

When an unstable isotope decays it can change into a different isotope or into a different atom altogether (Figure 8.3). Some of the decay processes involve the fourth, or *weak* force. Though weaker

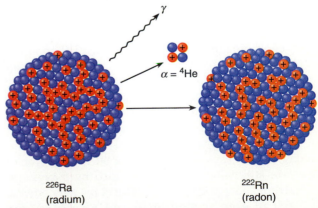

^{226}Ra
(radium)

^{222}Rn
(radon)

Figure 8.3 *An atom of ^{226}Ra (radium) will decay into ^{222}Rn (radon) with the ejection of an alpha particle (^{4}He) and a gamma ray. The radium was created in a chain that started with ^{238}U.*

BACKGROUND 8.1 Science and the Discovery of Radioactivity

The scientific method is a procedural concept that involves the interplay between theory and experiment or observation. A theory predicts an outcome of an experiment, and the scientist runs the experiment to see if the theory is correct. If the experiment contradicts the theory, the theory must be replaced or modified. However, in the real world, science is frequently not that neat. Many discoveries are made by accident. There is so much that we do not know that the act of investigating one property of nature often brings new properties—usually quite undreamed of—to light. You look for one thing and find something else, and the world changes as a result.

Radioactivity provides a wonderful example. In 1896, a French scientist named Henri Becquerel was experimenting with phosphorescent chemicals, those that after exposure to light will glow in the dark, and wanted to see how phosphorescent radiation would affect a photographic plate (it is common to spread photographic emulsions on glass plates rather than on film). He had completed an experiment with a compound containing uranium and then stored both the uranium and a wrapped plate in a dark cabinet. He subsequently discovered that the plate was fogged even though the uranium had not been exposed to light. The uranium was capable was emitting radiation that could penetrate opaque materials *all by itself.*

The greatest early exponents of the discovery were Marie and Pierre Curie (Figure 8.4). Together they began finding new radioactive elements. By 1898 they had isolated polonium (named after Marie's native Poland) and, most

Figure 8.4 *Marie Curie, with her husband Pierre, discovered radium.*

important, radium, one of the daughter products of the radioactive decay of uranium. They and Becquerel were awarded the 1903 Nobel prize in physics for their monumental discoveries. Pierre was killed in an accident in 1906, and Marie went on to greater fame for her continuing scientific investigations. She was awarded her second Nobel prize, this time in chemistry, also in 1906.

At the time of its discovery, and for many decades thereafter, radium was thought to be harmless. People used it casually, even for such things as watch dials that would glow in the dark, with little thought to its dangerous effects. Marie Curie, who more than anyone else laid the groundwork for the enormous number of modern uses of radioactivity, herself fell victim to it, dying of cancer in 1934.

than electromagnetism, it is far stronger than gravity, and also acts over only a short range. The decay is accompanied by the release of energy in the form of radiation and the isotope is called **radioactive.** Three kinds of radiation, which in the early days of study were called alpha (α), beta (β), and gamma (γ) rays, are commonly emitted in the decay. The α rays were found to be helium nuclei and are still called α particles, and the β rays are electrons (now β particles). The γ rays, however, consist of **electromagnetic radiation:** energy transported by electric and magnetic fields, an example of which is ordinary light.

Degrees of stability among radioactive isotopes vary. Some take billions of years to disintegrate, others only fractions of a second. The decay time is indicated by the isotope's **half-life,** the time it takes half of a given amount of an element to change into something else. For example, ^{238}U (uranium) is found in varying small degrees in most of the rocks and soils of the Earth. It decays into a variety of subproducts, eventually ending as stable ^{206}Pb (lead), emitting an abundance of α and β particles and γ rays along the way. It has a long half-life, 4.5 billion years. If you start with a kilogram of ^{238}U, after 4.5 billion years you would have only 0.5

kg left, and after 9 billion only 0.25 kg. Other isotopes have much shorter half-lives and decay so quickly that their radiation can be exceedingly dangerous. One of the by-products of uranium decay, ^{226}R (radium, seen in Figure 8.3), has a half-life of only 1,100 years. It is so energetic that you can see pure radium glow in the dark.

Radioactive elements have an enormous number of uses. Some are employed in violent weaponry and are used to generate heat and electric power. Others are used to trace the flow of medicines through the body, and still others to date old artifacts and even the Earth itself.

8.1.4 Electrons and Ions

Electrons are sometimes said to orbit their nuclei, although the analogy with the planets and the Sun is at best very loose. A major difference is that only specific values of orbital radii are allowed and there is a minimum orbital radius below which the electron cannot go. Atomic electrons are at great relative distances from their nuclei. For hydrogen, the minimum distance is about 100,000 times the size of the proton. Atoms are therefore mostly empty space. What we feel as the solid surface of an object is not the atoms themselves, but the electrical forces of their constituents.

For an atom to be neutral it must have just as many negative charges as positive and therefore as many electrons as protons. The electrons are not bound very tightly to their nuclei, however, and they can be removed with relative ease. Atoms with missing electrons take on positive electric charges and are called **ions.** The removal of one electron creates an ion with an excess positive charge of +1 called the singly ionized state. The removal of two electrons yields a net charge of +2 or the doubly ionized state, and so on. A helium atom with a missing electron is called He^+ and, if both electrons are gone, revealing a bare nucleus, the atom is He^{+2}. A lone proton is then also H^+. The carbon atom in Figure 8.2d can have up to six electrons taken away and so has six possible ionization states, C^+ through C^{+6}. It is even possible to have negative ions: in Chapter 18, we will encounter H^-.

8.1.5 Molecules

Much of the stuff of life does not consist of the elements in their pure form but of combinations of elements called **molecules** or **compounds.** As atomic number increases from hydrogen, the additional electrons are arranged in a series of shells, like those seen in the carbon atom in Figure 8.2d.

Molecules are formed when the outer electrons are shared by two or more atoms.

The simplest possible example is the joining of a pair of hydrogen atoms into *diatomic* (that is, two-atom) molecular hydrogen, which is called H_2 (Figure 8.5a). We count the number of individual atoms in the molecule and indicate the number by a subscript. The nuclei are not so close that their positive charges can repel each other, and the two electrons surround the protons, keeping them locked together.

Atoms can combine in a limitless number of ways. The air you breathe is mostly nitrogen and oxygen in the molecular forms of N_2 and O_2: atomic oxygen is poisonous. Oxygen can also form a *triatomic* (three-atom) molecule called ozone (O_3). You are made mostly of water, H_2O, and you exhale carbon dioxide, CO_2. More complex molecules, like those that make soap (Figure 8.5b) and gasoline, can consist of dozens of atoms. Molecules

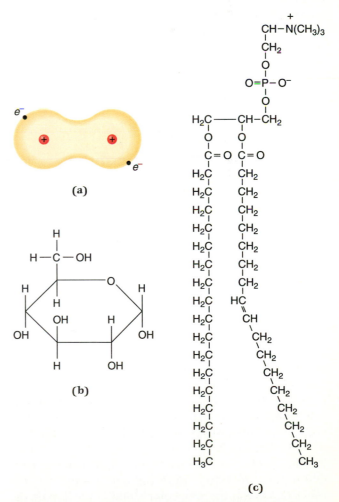

Figure 8.5 *A hydrogen molecule, H_2 **(a),** is held together because the protons share their electrons. Molecules can be quite complicated, as indicated by one of soap **(b)** and a complex organic (carbon-containing) one called phosphatidylcholine **(c).***

that contain carbon are called *organic* because they constitute and control the processes of life. They can be incredibly complicated and contain hundreds of atoms (Figure 8.5c).

Like atoms, molecules can be ionized. Moreover, the electron bonds that tie atoms together are not generally very strong. As a result, molecules can be broken apart by collisions that take place among them or with atoms. Molecules can therefore exist only at temperatures that are low, where the speeds and kinetic energies of the particles are relatively small.

8.1.6 States of Matter

Together, atoms and molecules make the three basic **states of matter.** In a **gas** (for example, the air you breathe), these particles are free to move past one another, interacting only electrically or by collisions. A gas has neither shape nor definite volume but can readily expand and contract. In a **solid** (like a rock), the particles are locked together, which gives the substance both definite shape and specific volume. A **liquid,** like water from a tap, is intermediate. A liquid has specific volume, but the particles can still move past one another, which allows it to change its shape. Astronomy deals mostly with gases and solids. Liquids are rare.

8.2
Electromagnetic Radiation

Radiation, of which ordinary light is an example, is the principle means by which energy and information are transported in the Universe. It is produced and affected by atoms, and almost all astronomical knowledge comes from its analysis.

8.2.1 The Nature of Light

Light is the fastest phenomenon known: its velocity in a vacuum (c) is an astounding 299,792 km/s. Our basis for understanding its properties goes back once again to Newton, who passed a beam of sunlight through a glass prism and discovered that light could be split into a **spectrum** of myriad colors (Figures 7.1 and 8.6), those seen in the common rainbow. Light therefore can be decomposed. But what constitutes the pure colors?

A famous experiment performed in 1803 by the English physicist Thomas Young began to clarify the nature of light. Shine light of a single color (*monochromatic* or "one-color" light) onto a pair of narrow slits (Figure 8.7). You might expect to see simple images of the two slits projected onto a dis-

Figure 8.6   *Sunlight is resolved by a glass prism into a spectrum of all the visible colors from red to violet.*

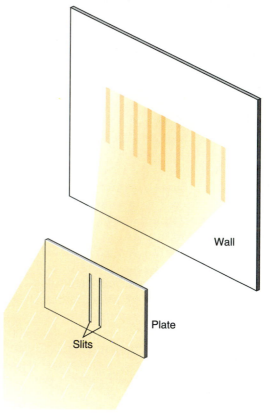

Figure 8.7   *A light shining on a pair of slits will produce a pattern of fringes on a wall or screen.*

tant wall. Instead, you see a pattern of light and dark *fringes,* a phenomenon that can be understood only if light is made of waves.

A wave is characterized by two numbers (Figure 8.8). The distance between two consecutive wave crests or wave troughs is called the **wavelength,** λ (Greek letter *lambda*), and is measured in centimeters or meters. The number of wave crests that pass a specific point per second is the **frequency,** ν (Greek letter *nu*), and is measured in cycles per second, more commonly called Hertz (Hz), after the nineteenth-century German physicist Heinrich Hertz. Wavelength, frequency, and velocity are related. Look at the crest marked by the vertical arrow in Figure 8.8. In one second, additional v waves will pass that point, each λ apart. The marked crest will then have moved a distance to the left equal to v times λ. Distance per unit time is velocity, so

$$\lambda v = c.$$

Wavelength and frequency are therefore expressible in terms of each other:

$$\lambda = c/v \quad \text{and} \quad v = c/\lambda.$$

In Young's experiment, each slit acts as a new source of waves, spreading the light into what would otherwise be dark areas. This phenomenon is called **diffraction** (Figure 8.9). The waves that emanate from each slit can interfere with one another. At the point on the wall centered between the slits (C), the distance to each slit is the same and the waves from each slit fall on top of each other. They add together in **constructive interfer-**

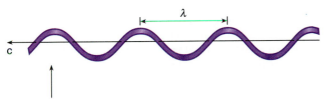

Figure 8.8   *A wave moves to the left at velocity c. The wavelength, λ, is the distance between crests or troughs. The frequency of the wave, v, is the number of wave crests passing a fixed point (vertical arrow) per second.*

ence to make a central bright fringe. At the points labeled C_1, the distances to each slit differ by exactly one wavelength, and the waves again overlap to produce bright fringes. However, in between, the difference in path length is one-half a wavelength. The crests of one wave fall into the troughs of the other, and the waves cancel to produce a dark fringe in **destructive interference.** A difference of two wavelengths makes more bright fringes, $2\frac{1}{2}$ dark fringes, and so on.

The nature of the wave was revealed by the English physicist James Clerk Maxwell, who in the 1860s formulated four basic equations that describe electric and magnetic fields. If an electric charge is accelerated, it produces waves in the electric field and in the resulting magnetic field. A solution to his complex equations showed that the waves must move in a vacuum at 3×10^8 m/s, the known speed of light. Maxwell then immediately identified light as an electromagnetic wave, or electromagnetic radiation. Unlike a water wave, an *electromagnetic wave* does not travel through anything but is an alternating oscillation of electric and magnetic fields that move at c.

8.2.2 The Electromagnetic Spectrum

There are no restrictions on the lengths or frequencies of electromagnetic waves. The full array of wavelengths or frequencies is called the **electromagnetic spectrum** (Figure 8.10). Different names are assigned to different domains, but all the waves move at the same speed, c. The most famil-

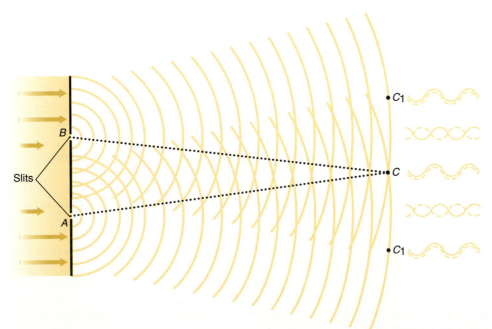

Figure 8.9   *At C, the paths from the two slits (A and B) to the wall have identical lengths. The waves from each fall on top of one another and constructively interfere to produce a central fringe. At C_1, the path lengths differ by 1 wavelength. The waves again constructively interfere, producing a pair of bright fringes to the side of C. In between, the difference is $\frac{1}{2}\lambda$. The waves then destructively interfere and cancel.*

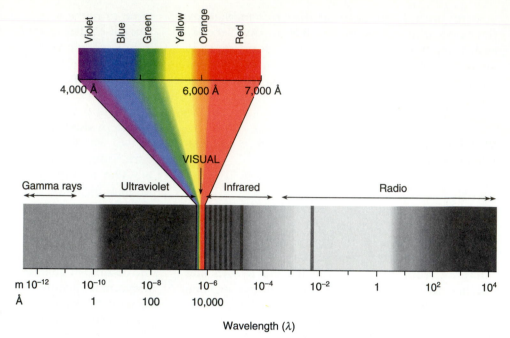

Figure 8.10 *The electromagnetic spectrum is composed of different wavelength domains. The boundaries of the descriptive terms are not physically real, nor are they sharp: the X-ray section gradually overlaps the ultraviolet, and so on. The shading roughly indicates the degree to which light is blocked by the Earth's atmosphere. None of the darkly shaded parts of the spectrum can penetrate to the ground.*

iar region, ordinary visible light, or *optical radiation,* is made of relatively short waves. The human eye is sensitive to electromagnetic radiation with wavelengths between about 4 and 7×10^{-7} m. These numbers are so small, it is convenient to use a unit called the *Ångstrom* (Å), which is 10^{-10} m (10^{-8} cm). The range of our vision is then expressed as 4,000–7,000 Å. The frequencies of light waves are high. Near the middle of the range, at 6,000 Å, ν is 3×10^8 m/s divided by $6,000 \times 10^{-10}$ m, or 5×10^{14} Hz.

Within the optical domain, we see different-length waves as the different colors of Figures 8.6 and 8.10. At the long-wave end, light appears red, and at the short-wave end violet, with the other colors—orange, yellow, green, blue, and hundreds of recognizable intermediate shades—in between. Waves with lengths greater than about 7,000 Å, called infrared (IR), are not able to excite the retina of the eye. Those just beyond red—the *near infrared*—are felt by the skin as heat. The warmth you feel under a sunny sky comes from solar infrared radiation.

Proceeding to longer wavelengths through the *far infrared,* near a tenth of a millimeter, we begin to enter the *radio* region which is divided into sub-regions more or less according to use. The shorter wavelengths, between perhaps a millimeter and a few centimeters, with frequencies of billions of Hertz (gigaHertz, GHz), are called *microwaves,* and are used for radar and for heating food. Television and FM radio are broadcast at frequencies of mil-

lions of Hertz (megaHertz, MHz), or at wavelengths of several centimeters to a few meters. Commercial AM radio has much lower frequencies, in the thousands of Hertz (kiloHertz, kHz), and wavelengths that approach a kilometer.

At wavelengths below 4,000 Å, the waves again cannot excite the eye. This *ultraviolet (UV) radiation* extends from the *near ultraviolet* of a few thousand Å downward to the *far ultraviolet* at about 100 Å, at which point it becomes known as *X rays.* Discovered in 1895 by the German physicist Wilhelm Röntgen, X rays can penetrate the tissue of a human body and are indispensable in medicine. Finally, below about an Ångstrom are the gamma rays produced by atomic decay.

8.2.3 *Photons and Energy*

The wave theory of light successfully explains Young's experiment as well as many others. However, in still other experiments, the wave theory fails completely. An example is the *photoelectric effect,* in which a beam of light knocks electrons loose from a substance to produce a flow of electricity. The phenomenon, exploited in "electric-eye" door openers and motion-picture sound systems, was first explained by Albert Einstein (who won a Nobel prize for his work). To produce the photoelectric effect, light (and electromagnetic radiation in general) must behave not as waves but as particles. But the particle theory cannot explain Young's double slit experiment (see Figure 8.7). Light then acts like both a wave and a particle *at the same time,* a concept outside our everyday experience. In fact, light—which is completely describable mathematically—is something greater than either a particle or a wave, and its apparent behavior depends on how it is viewed, that is, on the experiment.

The particle of light, the *photon,* is a packet or *quantum* of energy with an underlying wave nature. Each photon carries an amount of energy, *E,* that depends only upon its frequency, *v*, where

$$E = h v,$$

and h (Planck's constant, named after the early-twentieth-century German physicist Max Planck) is 6.63×10^{-34} joule seconds. The amount of energy carried by a single optical photon is quite small. The frequency of yellow light, at 5,500 Å, is 5.45×10^{14} Hz and the energy of a yellow photon (the *photon* itself is not colored: the term simply describes the wavelength) is then $5.45 \times 10^{14} \times 6.63 \times 10^{-34} = 3.6 \times 10^{-19}$ joule. A 100-watt bulb uses 100 joules of energy per second. To a crude approximation, the bulb radiates predominantly yellow light. It must then generate $10^2 / 3.6 \times 10^{-19}$ or very roughly 10^{20} photons per second.

Optical radiation and most radio waves are benign and can cause little damage (the exceptions are the microwaves that are absorbed by water and used in cooking). Ultraviolet photons, however, have enough energy to produce severe burns. Your tan is caused by your body's protective response to ultraviolet sunlight between 3,000 and 4,000 Å. The Earth's atmosphere becomes opaque below about

Figure 8.12   *The optical spectrum shown here is continuous because it has no breaks, gaps, or sudden jumps in brightness.*

3,000 Å, largely as a result of absorption by ozone (O_3). Were it not for our air, a human being would receive a lethal dose of short-wave ultraviolet radiation in a few seconds. Both X rays and gamma rays can pass into the body, where they have the potential to ionize atoms and cause considerable cellular damage. The result can be burns or even cancer. Gamma rays, produced by radioactive materials, are created in copious amounts by nuclear explosions and were a principal cause of death in the atomic bombings of Japan in 1945.

Since the electromagnetic force behaves according to an inverse-square law, so does the apparent brightness of a light. The amount of energy produced per second (joules/s) by a source of radiation like that in Figure 8.11 is called its **luminosity,** *L*. At any distance *r*, the photons and the energy radiated by the source must spread over a sphere of that radius. The **flux** of radiation is the amount of energy that passes each second through a surface with an area of one square meter, measured in joules/m²/s. The surface area of a sphere is $4\pi r^2$. The apparent brightness of the light at distance *r* is its flux at that distance, or the luminosity, *L*, divided by $4\pi r^2$. As the size of the sphere increases, the flux and apparent brightness decrease according to $1/r^2$. If you double your distance from a light, it will appear only a quarter as bright.

8.3
Radiation and Matter

Electromagnetic radiation is both emitted and absorbed by matter, specifically by charged particles—electrons, protons, and the atoms, ions, and molecules that contain them—that are accelerated or otherwise change their energies. Analysis of the radiation allows us to study and understand the astronomical bodies that we cannot reach or touch.

8.3.1. The Blackbody and Continuous Radiation

Radiation is said to be *continuous* if its spectrum has no breaks or gaps nor any sudden changes in brightness. In the continuous spectrum in Figure 8.12, one color blends smoothly into another. Many

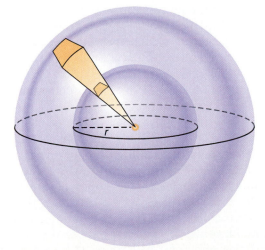

Figure 8.11   *The energy radiated by a source of electromagnetic radiation per second (its luminosity, L) spreads out over a sphere of radius r. The energy passing through one square meter of surface of the sphere (the shaded area) per second (the flux) is the luminosity divided by the sphere's surface area. The outer sphere has twice the diameter of the inner one but has four times the surface area, so the energy must spread itself out four times as thinly.*

physical processes produce such continuous radiation. One kind of radiator, however, the **blackbody,** is particularly important because the Sun, stars, and planets can bear a close similarity to it.

A blackbody is defined as any kind of body or surface that absorbs all the radiation that falls upon it. A blackbody reflects no radiation, hence its name. Solids, liquids, and gases can all be blackbodies. The absorption of radiative energy will raise a blackbody to some particular temperature. To maintain that temperature even though the body is constantly receiving energy, the blackbody must *emit* just as much energy as it receives. Consequently, a blackbody can be bright to the eye.

The flux of radiation from the blackbody, *F,* the amount of energy that leaves its surface per square meter per second, joules/m²/s, is very sensitive to its temperature. The greater the temperature, the more energetic the atomic particles within the body, and the more radiation they can emit. Observational and theoretical studies in the 1870s and 1880s by Josef Stefan and Ludwig Boltzmann showed that the radiated flux is given by

$$F = \sigma T^4,$$

where σ, the Stefan-Boltzmann constant, is 5.67×10^{-8} joule/s/m²/Kelvin⁴. This **Stefan-Boltzmann law** shows that if you double a blackbody's temperature, it becomes 2^4, or 16, times brighter!

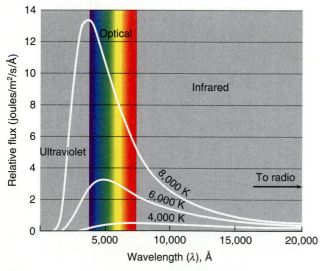

Figure 8.13 *The graphs plot the energy produced per unit wavelength interval per square meter per second per Å (flux per Ångstrom) against wavelength for three different temperatures. All the curves have the same shape, rising slowly to a peak, then dropping rapidly. As temperature (indicated on the curves) rises, the amount of radiation increases at all wavelengths and the peak of the flux shifts to shorter wavelengths. The 8,000-K blackbody produces 16 times as much energy as the one at 4,000 K*

Where in the spectrum the emitted radiation falls also depends on temperature. As temperature rises, so does the average atomic speed and energy and the average energy of the emitted photons. At very low temperatures, from a few to a few tens of Kelvins, the blackbody produces only low-energy, long-wavelength photons in the radio spectrum. As temperature climbs into the hundreds of Kelvins, it begins to radiate infrared photons *in addition to* those in the radio. When the temperature reaches into the thousands of Kelvins, the energies are high enough to produce optical photons (in addition to infrared and radio), and we can see the body glow. If the temperature is elevated into the millions of Kelvins, even X and gamma rays will be produced.

Figure 8.13 shows graphs of the flux F_λ of a blackbody plotted against wavelength for three temperatures, where F_λ represents the energy radiated per square meter per second per Ångstrom (joules/s/m²/Å). F_λ climbs slowly from the radio region of the spectrum (far off the graph to the right) to a sharp peak, then it suddenly drops. In the 1890s, Wilhelm Wien showed that the wavelength of the peak of the curve, where the emitted flux is at a maximum, λ_{max}, depends inversely on the temperature. The **Wien law** states that

$$\lambda_{max} = 2.898 \times 10^{-3}/T,$$

where λ_{max} is in meters (multiply by 10^{10} to get it in Ångstroms). As the temperature of a blackbody climbs, the flux per Ångstrom, F_λ, increases at all wavelengths, and λ_{max} shifts to shorter wavelengths, or to the left. At room temperature (295 K), $\lambda_{max} = 9.82 \times 10^{-6}$ m, or 98,200 Å, which lies in the far infrared. Such a blackbody would be invisible to the human eye. The walls of your room therefore do not visibly glow in the dark.

In spite of the work of Stefan, Boltzmann, and Wien, the shapes of the *blackbody curves* in Figure 8.13 defied explanation. Finally, in 1900, Max Planck derived the law that governs blackbody radiation by assuming that radiation is emitted in discrete units—the quanta or photons described above—each one having $E = h\nu$. Planck's work gave the first indication of the quantum nature of light and the atom and was the beginning of **quantum mechanics,** the study of the behavior of atomic particles and their radiation. The formula that relates F_λ to λ is complex, but from it Planck could derive both the Stefan-Boltzmann and the Wien laws. Blackbody curves are therefore commonly called *Planck curves.*

Figure 8.14 *Redder stars (seen here in Gemini) are cooler than bluer ones, in accordance with the Wien law.*

A hot blackbody's color depends on its temperature. At 2,000 K, the body emits only a small amount of radiation in the red part of the spectrum and very little at shorter wavelengths. It therefore looks red to the eye. At 4,000 K you see the body as orange, at 6,000 K yellowish (like sunlight), at 10,000 K white, and above that there is so much blue and violet light that the body takes on a bluish cast. Star colors (Figure 8.14) are the result of different stellar surface temperatures.

If a blackbody's λ_{max} can be measured from observation, we can find the temperature from the Wien law, where

$$T(\text{K}) = 2.898 \times 10^{-3}/\lambda_{max},$$

(λ_{max} is in meters) or from the Stefan-Boltzmann law. Assume a spherical blackbody to have a radius of r meters and temperature T. The surface area of the sphere is $4\pi r^2$ square meters. Each square meter radiates a total flux of σT^4 joules/s/m². The luminosity of a spherical blackbody, L, in joules/s is then the product of the area and the total flux, or

$$L \text{ (joules/s)} = 4\pi r^2 \sigma T^4,$$

one of the most important formulae in astronomy. From it, we find that the temperature is just the fourth root of $L/4\pi r^2\sigma$.

Even though stars and planets are not perfect blackbodies, we can still apply this formula. The result is the object's **effective temperature,** the temperature it *would* have if it *were* a perfect blackbody of the same radius and luminosity. Effective temperatures provide a superb way of comparing astronomical bodies with one another.

8.3.2 Spectrum Lines

Newton and his followers saw only the continuous spectra of Figures 8.6 and 8.12. Imagine then the surprise of William Wollaston when, in 1802, he discovered dark lines in the solar spectrum that ran perpendicular to the flow of colors. He mistakenly thought they were lines that separated the colors.

Within a dozen years, Joseph von Fraunhofer had catalogued over 300 such gaps (Figure 8.15) where light and color seemed to be missing. These features have become known as **absorption lines.** Fraunhofer gave the most prominent of them letter names, starting with A at the red end and continuing into the violet. A few of these designations (such as D for neutral sodium at 5,893 Å and H and K for ionized calcium in the violet) are still in common use: do not confuse them with chemical symbols. It was soon found that if light was made to pass through a substance, such as a rarefied or low-density gas, absorption lines could be made in the laboratory. (**Density** is mass per unit volume, measured in kilograms per cubic meter, kg/m³, or more commonly in astronomy, in g/cm³). Furthermore, if a substance were made to burn so as to *emit* light from a hot but rarefied gas, the absorption lines (Figure 8.16a) were replaced by **emission lines** (Figure 8.16b), bright lines of color, at exactly the same wavelengths.

It quickly became obvious that individual kinds of atoms could be identified by their **spectrum lines,** which could be either absorption or emission lines. Different materials always produce spectrum lines at different wavelengths. Burning sodium, for example, always causes a pair of bright orange-yellow lines (Figure 8.16c) at 5,890 and 5,896 Å (Fraunhofer D); other metals and hot gases produce quite different features. Some, like iron (Figure 8.16d), are extremely complex, with thousands—even millions—of lines. *Every ion of every atom was seen to produce a unique pattern;* there is no possibility of confusing one with another.

By 1859, the chemist Gustav Kirchhoff had determined basic rules, now known as **Kirchhoff's**

Figure 8.15 *Fraunhofer's spectrum of the Sun showed an enormous number of dark lines. He labeled the strongest ones with letters. Each line is produced by a particular atom, ion, or molecule in the solar atmosphere.*

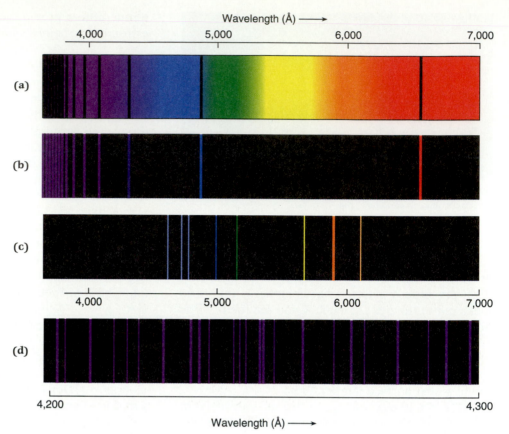

Figure 8.16 *Hydrogen displays* **(a)** *absorption and* **(b)** *emission lines at exactly the same wavelengths. Neutral sodium* **(c)** *produces a pair of bright orange-yellow emissions (seen here as one line) as well as several others. The spectrum of iron* **(d)** *is so complex that only a 100 Å segment is shown.*

laws of spectral analysis, under which the different types of spectra are produced. Spectra can be observed by a device called a *spectroscope* (to be examined in Chapter 9). In Figure 8.17, spectroscope A looks only at a hot blackbody that produces a continuous spectrum. But spectroscope B looks at the blackbody though a transparent box filled with a rarefied hydrogen gas that is cooler than the blackbody. The observer now sees dark lines superimposed on the continuous spectrum as the hydrogen atoms absorb light at their characteristic wavelengths. Spectroscope C points directly at the box and avoids the blackbody radiation. Now the observer sees emission lines that create the exact reversal of the spectrum seen in spectroscope B.

Kirchhoff's laws are thence summarized as:

1. *An incandescent solid or hot high-density gas produces a continuous spectrum (Figure 8.17a).*

2. *A hot low-density gas will produce an emission-line spectrum (Figure 8.17c).*

3. *A source of continuous radiation viewed through a cooler low-density gas will produce an absorption-line spectrum (Figure 8.17b).*

If we replace the laboratory apparatus with nature and look at a star, we find an absorption-line spectrum like that seen in spectroscope B, showing that a light from a source of continuous radiation is passing through a lower-density cooler gas, in this case, the star's low-density atmosphere.

Experimental physicists gradually uncovered patterns in the array of absorption and emission lines that helped lead to the explanations of how the lines were formed. Hydrogen presents the simplest case. Its optically observed spectrum (see Figures 8.16a and b) consists of a series of lines (called the *Balmer series* after the Swiss physicist J. J. Balmer, who studied it in 1885) that starts in the red at 6,563 Å at a line called Hα. It continues into the blue and violet with Hβ at 4,861 Å, Hγ at 4,340 Å, Hδ at 4,101 Å. The lines become progressively closer together and continue into the ultraviolet, where they pile up against the *Balmer limit* at 3,646 Å. In the early 1900s, similar series were found in the deep ultraviolet (the *Lyman series,* whose lines lie between Lyman α at 1,216 Å and the *Lyman limit* at 912 Å) and in the infrared (the Paschen series). Working independently, Balmer and the Swedish physicist J. R. Rydberg found laws that give the wavelengths of the lines. In a modern form, the *Rydberg formula* becomes

$$\lambda = 911.5 \text{ Å} \frac{1}{1/m^2 - 1/n^2}.$$

The letter m represents a series number that starts in the ultraviolet and increases toward longer wavelengths: $m = 1$, 2, and 3 for the Lyman, Balmer, and Paschen series respectively. The letter n, which must be greater than m, represents the number of the line within the series. For example, the Balmer lines Hα, Hβ, Hγ, Hδ . . . have n = 3, 4, 5, 6, ... The formula predicts a vast number of additional series that extend progressively farther into the infrared and even into the radio spectrum.

8.3.3 The Formation of Spectrum Lines

The mechanism by which the absorption and emission lines are produced still presented a mystery to the physicists of the first decade of the century. They knew from experiment that negative electrons surrounded positive nuclei, and they surmised that the electrons moved in orbits, held there by electrical attraction. But an accelerating electric charge should radiate its energy away. As a result, the electrons should spiral into the nuclei, destroying the integrity of the atoms. Since atoms and matter continue to exist, the electrons must keep their places. How?

The internal nature of the atom began to be revealed by the work of the Danish scientist Niels Bohr in 1913. He adopted the idea of electron orbits. However, he made the assumption that the orbits are *quantized,* that is, that the angular momenta of the orbiting electrons can have only specific values (or quantities) that are multiples of

Planck's constant divided by 2π. The radii of the orbits and the electrons' energies must then also be quantized (Figure 8.18), and there must be an orbit with a minimum radius and energy, called the *ground state,* in which the angular momentum is h/2π. The second orbit has twice that value, the third three times as much, and so on. From this information, it is possible to construct an **energy-level diagram** (inset, Figure 8.18) that gives the potential energies (see Section 7.2) of the orbiting electrons above the ground state. The energy levels climb like ascending steps that get closer together, converging on a limit that represents the energy needed to tear the electron from the proton, that is, to ionize the atom.

The electrons are free to jump between any two orbits or energy levels. If an electron jumps downward between two levels, it will give up its energy in the form of a photon that has an energy equal to the energy difference between the levels and that will contibute to an emission line. Assume an upper level, m, and a lower one, n (Figure 8.19a). The energies of these levels are E_m and E_n. The radiated photon then has energy $E_m - E_n$. The energy of a photon corresponds to a specific frequency through $E = h\nu$, so that $\nu_{mn} = (E_m - E_n)/h$. Since the wavelength of the photon $\lambda = c/\nu$,

$$\lambda_{mn} = hc/(E_m - E_n).$$

Conversely, an electron in a lower level n in Figure 8.19b can *absorb* a photon of wavelength λ_{mn} and be raised upward into level m. The essence of the process, however, is that *all* the energy of the incoming photon must be absorbed, which means

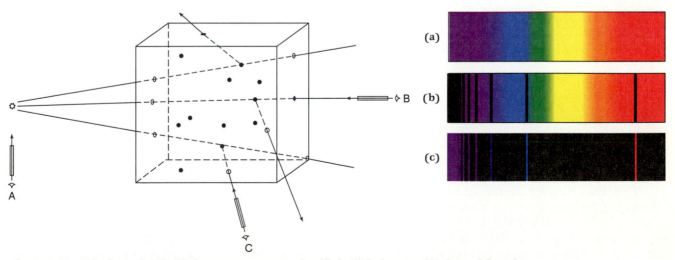

Figure 8.17 *Light from a hot blackbody enters a transparent box filled with hydrogen and is observed through spectroscopes A, B, and C. The blackbody seen through A has a continuous spectrum, as shown in* **(a)**. *If the light is looked at through the box, through B, the continuous spectrum is seen to be crossed by absorption lines, as in* **(b)**. *If the box alone is looked at, through C, the observer sees only emission lines, as in* **(c)**.

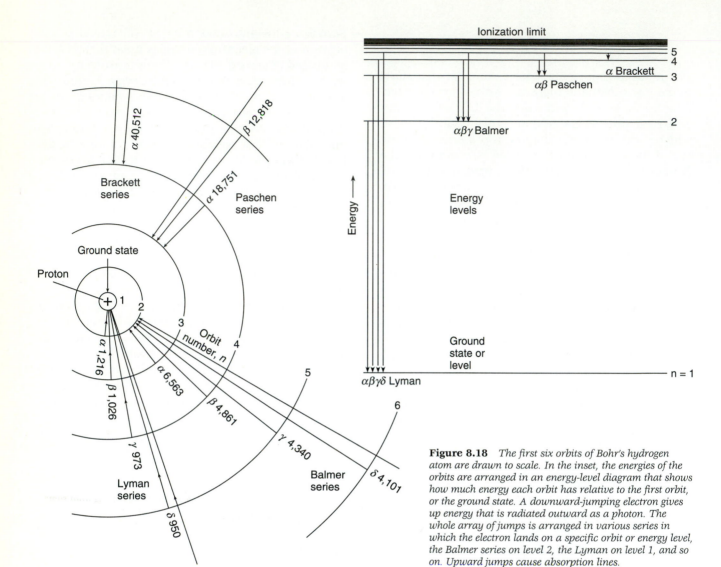

Figure 8.18 *The first six orbits of Bohr's hydrogen atom are drawn to scale. In the inset, the energies of the orbits are arranged in an energy-level diagram that shows how much energy each orbit has relative to the first orbit, or the ground state. A downward-jumping electron gives up energy that is radiated outward as a photon. The whole array of jumps is arranged in various series in which the electron lands on a specific orbit or energy level, the Balmer series on level 2, the Lyman on level 1, and so on. Upward jumps cause absorption lines.*

that the photon must have energy $E_m - E_n$ (and wavelength λ_{mn}) exactly. With this theory, Bohr could derive the Rydberg formula above and the wavelengths of all the hydrogen lines. The Hα emission line is created by downward electron jumps between levels 3 and 2, the Hβ emission line by jumps between 4 and 2, and so forth (see Figure 8.18), and the respective absorption lines by upward jumps between 2 and 3, 2 and 4, and so on.

The Bohr theory of the atom was only the beginning. Electromagnetic radiation has properties of particles and waves at the same time. The theory of relativity, however, shows that mass and energy are simply different manifestations of the larger concept of mass-energy. As a result, the French physicist Prince Louis de Broglie suggested in 1924 that electrons and other so-called particles must

also have a dual nature and remarkably can behave like *waves*. The greater the particle's energy, the shorter its wavelength. The truth of this contention was demonstrated only three years later when physicists at Bell Laboratories succeeded in diffracting speeding electrons through a crystal to produce interference fringes! An atoms's electrons therefore do not orbit their nucleus like tiny planets: instead, they can be thought of as energy-carrying waves that surround the nucleus. The concept of the energy-level diagram, however, remains the same.

Hydrogen's energy-level diagram is simple because there is only one electron. Two or more electrons will interact with one another to create more complex energy-level structures and diagrams. With more levels there are many more possible transitions. The complexity can be awesome,

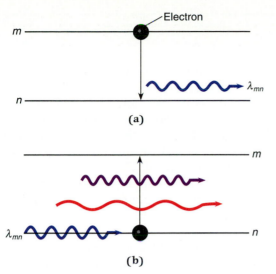

(a)

(b)

Figure 8.19 **(a)** *An electron in a high energy level m jumps downward to a lower level n and gives up its energy as a photon of energy $E_m - E_n$ and wavelength λ_{mn}.* **(b)** *Photons of various wavelengths pass an electron in lower level n. The electron can absorb only the photon with wavelength λ_{mn} (energy $E_m - E_n$), which elevates the electron to the upper level m.*

as seen in the partial energy-level diagram of singly ionized oxygen in Figure 8.20. This diagram does not even include many of the levels that are responsible for ultraviolet and infrared transitions. Metals like iron are far more complicated.

The champions of complexity, however, are not atoms or ions, but molecules. A molecule can rotate around various axes (depending on how it is constructed) and can vibrate along the lines connecting the individual atoms. The energies of rotation and vibration are also quantized. As a result, an electronic energy level is broken into many different sublevels that in turn are broken into sub-sublevels. What then would be a single spectrum line for an atom becomes a series of bands, each band consisting of hundreds or even thousands of individual lines (Figure 8.21).

This theory of the atom readily explains Kirchhoff's laws. Whatever the atoms (or molecules) in the box in Figure 8.17 may be, they are constantly colliding with one another. The collisions elevate

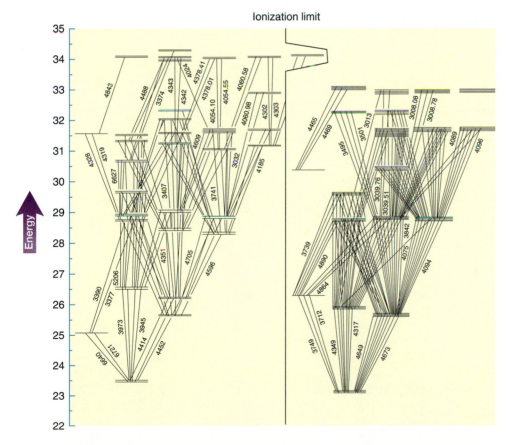

Ionization limit

Energy

Violet   Red

Figure 8.20 *The energy-level diagram of singly ionized oxygen gives an example of the complexity of atoms with multiple electrons. The scale on the left gives energies, which actually go all the way to zero in a region of the diagram also filled with levels. The numbers give the wavelengths (in Å) of spectrum lines arising from some of the jumps between levels. The top of the diagram is the ionization limit.*

Figure 8.21 *Molecular spectra, like that of C_2 shown here, consist of many series of bands, each band containing hundreds or thousands of individual lines.*

BACKGROUND **8.2** Lasers

The laser (see the chapter-opening photograph), an acronym for "*l*ight *a*mplification by the stimu*late*d *e*mission of *r*adiation," has become a familiar part of our technology. Used for everything from eye surgery to supermarket scanners, it is a remarkable device that has its origin in atomic structure and quantum mechanics.

Figure 8.22 shows two energy levels separated by an energy E_0 that corresponds to a photon with wavelength λ_0. As already explained, an electron in the lower level can absorb a photon and move upward; one in the upper level can spontaneously emit a photon and jump downward. There is yet another possibility. If the electron is in the upper level and is struck by the photon of wavelength λ_0 it can be *stimulated* into making a downward jump. The first photon is unaffected by the process, and the pair go flying off in exactly the same direction and in phase with each other (that is, the waves go up and down at the same time), as on the left-hand side of Figure 8.22.

In a normal gas, more atoms have their electrons in the lower levels than in the upper as a result of atomic collisions. With an external input of energy, however, it is possible to place more electrons in the upper levels than in the lower: a single photon entering the gas (from the left, in Figure 8.22) can then start a chain reaction that will cause the excited electrons suddenly to cascade downward. The result is a powerful beam of radiation in which all the photons move in phase and in the same direction. Laser radiation can be so concentrated that it can drill a hole through a steel plate, be made fine enough to repair the retina of the eye, and be sufficiently strengthened to make a round-trip journey to the Moon.

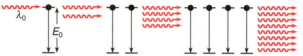

Figure 8.22   *Several atoms have their electrons in the top of a pair of energy levels. The photon entering from the left has a wavelength that corresponds to the energy difference between the levels, and stimulates the first electron to jump down, creating a second, identical photon. These create two more, and so on, resulting in a powerful beam of laser radiation that goes off to the right.*

electrons into upper levels. These energetic or *excited* atoms (as well as those with electrons in the ground state) absorb the photons from the incoming beam of continuous radiation, but only those photons whose energies correspond to energy differences between the various levels. Absorption lines are then seen in spectroscope B. In addition, if a photon's energy is greater than the difference between an energy level and the top of the energy-level diagram, the electron can be ripped away from the nucleus, and the atom or molecule becomes ionized. Electrons in upper states are also free to jump downward. When they do, they radiate the emission lines seen in spectroscope C.

The absorption lines seen in spectroscope B will not be completely dark, as some photons at the critical wavelengths will get through. By measuring the *strengths* of absorption lines (the amounts of energy extracted from the continuum), the temperature and density of the gas and the amount of the chemical element or molecule forming the lines can be determined. If we can detect lines of more than one element or molecule, we can also derive the relative numbers of the different kinds of atoms. Similar procedures can be applied to the brightnesses of emission lines seen in spectroscope C. When these concepts are applied to real astronomical objects, we can analyze their conditions and compositions and begin to understand the nature of the Universe.

8.4
The Doppler Effect

The wavelengths of spectrum lines are fixed by the energy differences between orbits or energy levels. However, the apparent (measured) wavelengths can be changed by relative motion between the observer and the source. In Figure 8.23, person A stands next to the wave and counts the number of wave crests that go by every second to determine the wave's frequency. Person B, however, does not stand still, but moves to the right, into the waves. The waves hit this observer at a faster rate, the frequency will appear higher, and the wavelength—the distance between crests—will appear to be less.

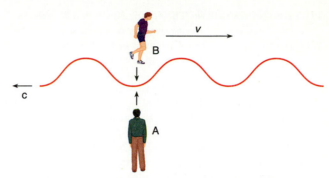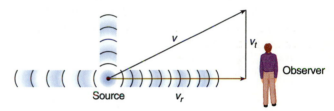

The faster the motion, the shorter the observed wavelength. Conversely, if B reverses direction and moves to the left, the frequency will appear lower and the wavelength greater. The same effect will be observed if the source is moving instead of the observer.

This change in measured wavelength with motion is called the **Doppler effect,** after Christian Doppler, who explained it in 1842. It is most commonly observed with sound waves. Stand by the side of a road and listen to the noise made by a car. The pitch will be higher as the car approaches and will drop as it recedes. Any velocity with respect to an observer can be broken into two components, one along the line of sight, the **radial velocity,** v_r, and another across the line of sight, the **tangential velocity,** v_t (Figure 8.24). All that matters in the Doppler effect is the radial component. If the car is moving across the line of sight, there is no radial velocity and no Doppler effect.

Electromagnetic radiation is also subject to the Doppler effect. A body coming at you at sufficiently high speed would appear somewhat bluer than normal, and one going away somewhat redder. As long as the relative velocity is much less than the velocity of light (so that relativity is not important), the magnitude of the Doppler effect, the **Doppler shift,** is given by the Doppler formula,

$$\lambda_{obs} - \lambda_{rest} = \lambda_{rest}(v_r/c),$$

where λ_{rest} is the wavelength that would be observed if the source and observer were at rest with respect to one another, λ_{obs} is that observed when the two are moving at the relative radial velocity v_r, and c is the speed of light. The radial velocity is considered negative for approach and positive for recession.

The speed of light is so great that the optical Doppler effect is usually subtle. For a yellow light (5,500 Å) to appear green (5,000 Å), the source of light would have to be moving at 9% c, or 27,000 km/s, and (except for distant galaxies) most astronomical sources are moving far more slowly. The velocities of stars, for example, are typically a few tens of kilometers per second relative to the Sun. The effect is accurately measurable by observations of spectrum lines. The wavelengths at rest are known from laboratory measurements. We need then only measure the wavelengths of absorption or emission lines in stars to determine their radial velocities.

The ramifications of the Doppler effect are profound. It is used to assemble critical informa-

Figure 8.23 *A wave—light, sound, or even water—moves to the left at velocity c. The wave has a well-defined wavelength or frequency for observer A, who is watching it go by. Observer B, however, is moving to the right into the wave at some velocity v. This person will encounter the wave crests more frequently and will measure a higher frequency and shorter wavelength. If B moves with the wave to the left, the frequency will appear lower and the wavelength longer.*

Figure 8.24 *A source of radiation moves at velocity v relative to the observer. This motion can be broken down into a motion in the line of sight, the radial velocity, v_r, and one across the line of sight, the tangential velocity, v_t. Only the radial motion produces a Doppler effect.*

tion about the construction of the Universe, the dynamics of the Galaxy, and even to search for planets orbiting other stars. We will find out how to measure it in the next chapter, where we assemble the tools astronomers use to learn about the Universe.

KEY CONCEPTS

Atom: The basic unit that forms the chemical elements.
Atomic number: The number of protons in an atomic nucleus; this number defines the chemical element.
Atomic weight: The number of protons and neutrons in an atomic nucleus.
Blackbody: A body that absorbs all the radiation that falls upon it; if temperature is constant, a blackbody emits as much radiation as it absorbs.
Density: Mass per unit volume, measured in kg/m³ or g/cm³.
Diffraction: A phenomenon that spreads light around barriers into areas that otherwise would be dark.
Doppler effect: The observed shift in wavelength or frequency caused by relative radial motion.

Effective temperature: The temperature of a blackbody that has the size and luminosity of the radiant source.

Electromagnetic force: The force of nature that combines electricity and magnetism; manifested by the **electric charge,** which produces **electric** and **magnetic fields** that act over a distance.

Electromagnetic radiation: Energy transported by electric and magnetic fields (that is, electromagnetic waves or photons).

Electromagnetic spectrum: The array of the kinds (wavelengths) of electromagnetic waves, including gamma rays, X rays, ultraviolet, optical, infrared, and radio waves.

Electrons: Negatively charged atomic particles.

Elements: The basic constituents of common matter; each element is defined by a different number of nuclear protons.

Energy-level diagram: A graph of energies of atomic electrons.

Flux of radiation (F): The amount of radiation flowing per second through a surface with an area of one square meter (joules/m^2/s).

Frequency (ν): The number of waves passing a location per second..

Half-life: The time it takes for a specific amount of a radioactive element to decay to half that amount.

Interference (constructive and destructive): Reinforcement or destruction of waves that fall on top of one another.

Ions: Atoms or molecules that are electrically charged because electrons are missing or added.

Isotopes: Variations of a chemical element caused by differences in neutron number.

Luminosity (L): The amount of energy radiated by a body per second (joules/s).

Molecules or **compounds:** Combinations of atoms.

Neutrons: Neutral atomic particles.

Nucleus: The combined protons and neutrons at the atomic center.

Photons: Particles of electromagnetic radiation that also incorporate wave motion.

Protons: Positively charged atomic particles.

Quantum mechanics: The science that deals with the atom and its constituents, with photons, and with the quantization of their energies.

Quarks: Particles that make protons and neutrons.

Radial velocity: The relative speed of a body along the line of sight.

Radioactive isotopes: Isotopes whose nuclei decay into other nuclei with the release of particles and radiation.

Spectrum: An array of properties; here, the array of electromagnetic waves.

Spectrum lines: Radiation at specific wavelengths **(emission lines)** or gaps in a continuous spectrum **(absorption lines).**

States of matter: The forms matter may take that include **gases** (that do not have fixed shapes or volumes); **solids** (fixed shapes and volumes), and **liquids** (fixed volumes but not shapes).

Strong (nuclear) force: The force that binds atomic nuclei together.

Tangential velocity: The relative speed of a body perpendicular to the line of sight.

Wavelength (λ): The distance between crests of a wave .

KEY RELATIONSHIPS

Doppler shift (for velocities much less than c):
$$\lambda_{obs} - \lambda_{rest} = \lambda_{rest}(v_r/c).$$

Effective temperature:
$$T^4 = 4\pi r^2 \sigma/L, \text{ where } r = \text{radius of a sphere.}$$

Frequency and energy:
$$E = h\nu = hc/\lambda.$$

Kirchhoff's laws:

1. An incandescent solid or hot high-density gas produces a continuous spectrum.
2. A hot low-density gas will produce an emission line spectrum.
3. A source of continuous radiation viewed through a cooler low-density gas will produce an absorption-line spectrum.

Luminosity of a spherical blackbody:
$$L = 4\pi r^2 \sigma T^4.$$

Stefan-Boltzmann law:
$$F = \sigma T^4.$$

Wavelength and frequency:
$$\lambda \nu = c.$$

Wien law:
$$\lambda_{max} (\text{Å}) = 2.898 \times 10^7/T.$$

EXERCISES

Comparisons

1. Compare the four known forces of nature with regard to their strengths, what they do, and their ranges of effectiveness.
2. Compare the electric charges and masses of protons, neutrons, and electrons.
3. What is the difference between atomic number and atomic weight?
4. How do ions differ from isotopes?

5. Compare alpha, beta, and gamma rays.
6. In what ways do gamma rays differ from infrared rays?
7. What is the difference between absorption and emission lines?
8. How does the Balmer series of hydrogen differ from the Paschen series?
9. Why is the spectrum of neutral helium different from the spectrum of hydrogen?

Numerical Problems

10. How many neutrons are there in the nuclei of ^{52}Cr, ^{120}Sn, and ^{202}Hg?
11. A kilogram of a radioactive isotope has a half-life of 25 years. How much of the element will be left in 100 years?
12. The wavelength of a photon is 1 mm. What is its frequency?
13. What are the energies (in joules) of photons with wavelengths of 1 kilometer, 5,000 Ångstroms, and 1 Ångstrom?
14. At what wavelengths will blackbodies with temperatures of 125 K, 3,300 K, 50,000 K, and 10^6 K produce their maximum radiation? In what wavelength domain does each maximum fall?
15. How much more radiant energy per square meter of surface will a blackbody with a temperature of 3,000 K produce than a blackbody with a temperature of 1,000 K?
16. Spherical blackbodies A and B have the same luminosity. Blackbody A is twice as hot as blackbody B. What is the radius of B compared with that of A?
17. What are the wavelengths of the hydrogen emission lines that are created when an electron jumps from orbit 7 to 6 and from orbit 4 to 1? In what wavelength domains do they fall?
18. By how many Ångstroms is the Hα line of hydrogen shifted in the spectrum of a star that is approaching at a speed of 300 km/s? In what direction is the line shifted?

Thought and Discussion

19. What kind of particle defines a specific element?
20. Following the example of Figure 8.2, what do atoms of ^{7}Li, ^{11}B^{+2}, and ^{14}C$^+$ look like?
21. What is meant by a "radioactive element"? Name three.
22. What part of the electromagnetic spectrum lies between **(a)** the optical and the X ray; **(b)** the optical and the radio?

23. How do we know that light has the natures of both waves and particles?
24. How can a blackbody be bright in the optical spectrum?
25. What is the idea behind the concept of effective temperature and why are effective temperatures needed?
26. What is meant by "continuous radiation"?
27. What does the spectrum of ionized hydrogen look like?
28. Why are the wavelengths of photons apparently shortened when a source of light is approaching you?

Activities

29. Observe the effect of temperature on the color of a blackbody. Turn on a toaster or electric stove and watch as it heats. Record the order of colors. From the color and the Wien law, estimate the temperature of the element at various stages. What other blackbodies are you likely to have in your home? What might be their temperatures?
30. On a moonless night, record the different star colors you can see with the naked eye. Estimate the temperatures of the stars.
31. Your radio is actually a form of spectroscope. As you tune across the dial, you find radio stations broadcasting at specific frequencies. These are actually emission lines in the radio spectrum. Use your AM radio to draw the spectrum as it might appear if you could photograph it. Estimate the strengths of these spectrum lines.

Research Problems

32. When were nuclear forces first used for destruction and when were they first used for peaceful purposes? List the ways in which nuclear forces are used beneficially.
33. Find ways in which each domain of the electromagnetic spectrum is used for beneficial purposes.

Scientific Writing

34. Your company is in the business of providing radioactive isotopes for industry and medicine. Write a two-page section for a stockholder's report that explains the background of radioactivity, specifically the structure of the atom and the nature of radioactive elements.
35. A children's encyclopedia has asked you to write an article on light and radio. Explain the similarities and differences between them in the simplest possible terms.

9

The Tools of Astronomy

How we look outward from the Earth

The ridge of the Kitt Peak National Observatory, seen through the shutter of the 4-m telescope, holds one of the greatest collections of telescopes in the world.

Light falls from the distant stars upon the Earth, carrying a coded message that tells us what they are. How are we to decipher it? Until 1609, the only tools we had were our own eyes, and all we could do was measure celestial positions and estimate brightnesses. We had no way of knowing what the stars were made of, and little indication that we could ever know. The great invention that made our modern learning possible was the **telescope,** whose purpose is to make distant objects appear nearer, larger, and brighter. The flow of telescopic discoveries initiated by the great Galileo has continued unabated to this day because of rapid and dramatic improvements in the instrument that have allowed us to detect and record weak signals from the sky at all wavelengths. In our own era we have even loosed the strings that long tied the telescopes to Earth to achieve brilliantly clear views of the Universe from the cold clear depths of space.

9.1
Optical Principles

The construction of telescopes and the accompanying instruments that detect radiation depend on a variety of optical effects. The most obvious is **reflection,** in which a ray of light is returned from a surface. In Figure 9.1, the *angle of incidence* (i) between the incoming ray and the perpendicular is always equal to the *angle of reflection, r*. As a result, you always see a faithful (though backward) image of yourself in a bathroom mirror.

Refraction involves the bending of light rays as they travel from one substance *into* another. Figure 9.2 shows rays 1 and 2 of Figure 9.1 falling from the air onto a smooth transparent surface, perhaps glass or water. Some of the light will be reflected. The rest will enter the substance. Light moves at c only in a vacuum. When light travels in a transmitting medium, it interacts with that substance's atoms and slows down, the speed in general corresponding to the medium's density. In glass, light travels only at 0.65 to 0.7c and in diamond at 0.45c. The slowing causes the light ray to change direction and to bend toward the perpendicular, so that the *angle of refraction, r* (the angle between the bent ray and the perpendicular) is *less* than i. The degree to which the light is bent ($i - r$) increases with i. It also increases with the ratio of c to the velocity (v) in the substance (the *index of refraction* is c/v). Glass will cause more bending than water, diamond more than glass. The paths of the light rays in Figure 9.2 are independent of direction. If a ray starts

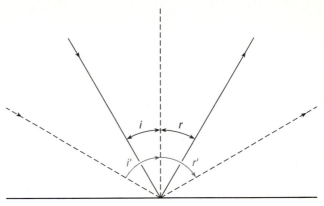

Figure 9.1 *Rays fall downward from the left onto a smooth surface at angles i and i' (the angles of incidence) to the perpendicular to the surface. They are reflected up and to the right at the angles of reflection, r and r'. The angle of reflection equals the angle of incidence for any ray, so i = r and i' = r'.*

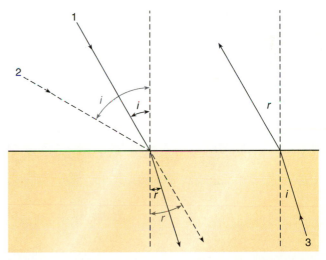

Figure 9.2 *Light rays 1 and 2 enter a dense substance from air and are refracted toward the perpendicular. The angle of refraction, r, is less than the angle of incidence, i. If the light source is in the denser substance (ray 3) and the ray travels in the opposite direction, it is bent away from the perpendicular.*

in the substance (ray 3 in Figure 9.2) and goes into the air or a vacuum, it bends *away* from the perpendicular, i and r are reversed, and the angle of refraction is *larger* than the angle of incidence.

Refraction produces numerous distortions in nature. Drop a penny into a bowl of water, and if you look at it from across the surface, it will appear to lie well above its real position. Or go to a swimming pool and look at the lane lines. Although they are really parallel, they seem to curve.

The index of refraction depends on both the light-transmitting material and on wavelength. In Figure 9.3, a beam of white light, which is a mixture of colors, enters a glass prism from the air.

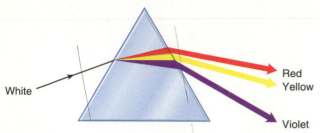

Figure 9.3   *White light is dispersed as it enters the face of a triangular prism and is dispersed even more as it exits. The colors can now fall on a screen and be seen.*

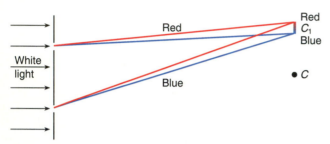

Figure 9.4   *White light lands on a double slit. Since red waves have longer wavelengths than blue, they interfere constructively at a greater distance away from the center, C. The result is a fringe at C_1 colored red at one edge and blue or even violet at the other.*

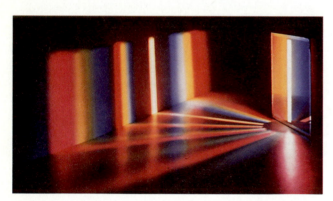

Figure 9.5   *A grating (a plate with closely spaced slits) diffracts light to produce multiple spectra, one for each fringe.*

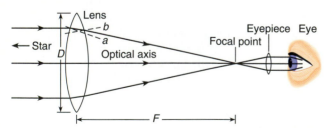

Figure 9.6   *A simple refracting lens of diameter D bends starlight. Light that strikes or leaves the lens's surface at an angle to the perpendicular (lines a and b, respectively) are refracted toward the optical axis and come to a focus at the focal point a distance F (the focal length) from the lens. The result is an image of the distant star. To view the image, we place an eyepiece behind it.*

The blue photons in the light are refracted more than the yellow, the yellow more than the red, and so the white light is *dispersed* or spread into its component colors. It is dispersed even more when it leaves the prism. We can now let it fall on a screen and, like Newton (see Figure 7.1), see a beautiful spectrum. Refractive **dispersion** is also part of the everyday world. Look at a bright light through an aquarium and you will see it wreathed in colors, or watch colored sunlight sparkle from icy prisms of new-fallen snow.

Diffraction (see Section 8.2.1 and Figure 8.7) also disperses light. The longer the wavelength, the farther from the center the first fringe (C_1) will fall (Figure 9.4). If white light falls upon the double slit, the red waves constructively interfere best at the outer edge of C_1, the blue waves best at the inner. Each diffraction fringe therefore becomes a crude spectrum. If four slits are used, the positions at which the waves of different length will constructively interfere are more tightly defined, and the colors become clearer. Thousands of slits like those in the *grating* in Figure 9.5 (a grating is a plate ruled with parallel grooves) will produce beautiful spectra. Diffractive dispersion is produced by reflection from the surface of an audio CD, which has a microscopic spiral groove that has the effect of a grating.

9.2
Lenses and the Refracting Telescope

The convex **lens** in Figure 9.6 is a circular piece of glass of diameter, or **aperture** (*D*), whose surfaces are shaped into sections of spheres that face away from one another. The line through the center of the lens perpendicular to the surfaces, the *optical axis,* points toward a star, a body so far away that all its rays are effectively parallel to one another. The ray of starlight directed along the optical axis is not refracted. But a ray hitting the lens away from the center strikes at an angle to the perpendicular to the lens's surface (line *a*) and is bent in the direction of the optical axis. When this ray leaves the glass it is now bent *away* from the perpendicular to the lens's surface (line *b*), again toward the optical axis.

All the rays falling on the lens, otherwise known as the **objective,** are refracted to cross or *focus* at the **focal point** at a distance *F*, the **focal length,** from the lens. If a piece of paper is placed at the focal point, you will see a point of light, the

Figure 9.7   *The great nineteenth-century refractor of the Yerkes Observatory of the University of Chicago, located in Williams Bay, Wisconsin, has a lens 40 inches (about 1 m) in diameter.*

image of the star. Substitute a photographic plate for the paper and you can take a picture of it. You now have a **refracting telescope,** or *refractor,* which acts just like an ordinary camera.

The brightness of the star's image (the amount of energy falling into it in joules/s) depends on the telescope's **light-gathering power,** the amount of light that the lens can capture, which in turn depends on the area of the lens or on the square of its aperture, *D*. If you double the size of the lens, the image becomes four times brighter, and you are able to record stars that are four times fainter than you could before. Telescopes, like the "40-inch" at the Yerkes Observatory in Figure 9.7, are even named by aperture.

Every objective lens has a *field of view* over which it can collect light. In Figure 9.8, the telescope's optical axis is pointed between two stars A and B. The light from each falls onto the objective at a slightly different angle so that the images are also off the optical axis. The images now fall on a **focal plane** that is perpendicular to the optical axis at the focal point. A piece of film placed there will register the images of both stars at the same time and will record a picture of the sky. In Figure 9.8, star A is

actually above star B. But in the image in the focal plane, star B is above star A. The astronomical telescope inverts all images, up for down, left for right. The less the curvature on the lens, the longer the focal length and the greater the separation of the stars (*S*) in the focal plane in millimeters.

Next, assume that the telescope points toward an extended object (one that does not appear as a point) such as a planet with its edges at A and B. Now *S* is the size of the image. *S* will be directly proportional to focal length, and the area of the image will be proportional to S^2 or to F^2. The *surface brightness* of the image (the flux passing through it in joules/mm²/s) then decreases as F^2 increases, but it also increases according to aperture squared, or D^2. The apparent brightness, or visibility, of an extended object therefore depends on $(D/F)^2$, or on $1/f^2$ where $f = F/D$ is called the *focal ratio.* The smaller the focal ratio, the brighter the image, and the shorter the exposure time you need when you take a picture with your telescope or camera.

If you wish to look at an object through the telescope, you can place an **eyepiece** behind the focus (at a distance equal to the eyepiece's focal length) to make the rays parallel again so that they can be focused onto the retina of the eye (see Figure 9.6). The eyepiece acts on the image made by the objective and increases the angular separations of stars and the angular sizes of extended objects, thereby making them look bigger. (The field of view, however, is still inverted.) The degree of magnification, or the **magnifying power,** is the ratio of the focal length of the objective divided by the focal length of the eyepiece, or

$$M = F_{\text{objective}}/F_{\text{eyepiece}}.$$

It is easy to change the power of a telescope just by switching eyepieces.

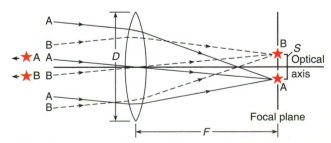

Figure 9.8   *Two stars, A and B, both at great distance, are viewed with the telescope. The stars lie S' mm apart on either side of the optical axis. Their images will fall on the focal plane that passes through the focal point perpendicular to the optical axis. The field of view is reversed in the focal plane, up for down and left for right.*

Figure 9.9 *Refraction in the atmosphere smears starlight into a seeing disk that may be as much as several seconds of arc across.*

Figure 9.10 *The objective mirror of the Palomar 200-inch (5-m) telescope is at the bottom of the open tube and the secondary system is at the top. The secondary mirror can be replaced with a cage in which the astronomer can ride and observe at prime focus.*

More important is the telescope's **resolving power,** which is the ability of a telescope to see fine detail or to reveal the separation between two sources of light that are close together. A single slit or aperture will produce a diffraction pattern just like the double slit of Figures 8.7 and 8.9. As a round aperture, a telescope objective will produce a central circular bright spot or **diffraction disk** of specific angular size surrounded by diffraction rings. If two stars are so close that their diffraction disks overlap, they will be seen as one image. The larger the aperture, the smaller the size of the diffraction disk. The ability of the telescope to resolve detail then improves along with the aperture.

The resolving power of an optical telescope is given in seconds of arc by the ratio $13/D$, where D is the aperture in centimeters. A 10-cm telescope can resolve two stars about 1.3" of arc apart, but no closer; a 20-cm, 0.75" apart. However, the resolving power is limited in practice for most telescopes by the twinkling of stars. Variable refraction in differ-

ent layers of the Earth's atmosphere makes a stellar image appear to jump around, smearing the image of a star into a **seeing disk** (Figure 9.9) that is typically a second of arc or so across and much larger than the diffraction disk.

Refractors suffer from some serious problems. The light refracted by the objective is also dispersed. Red rays have a longer focal length than the blue rays, producing *chromatic aberration* and a smeared colored image. The problem can be partially corrected with a compound *achromatic lens* that consists of simple lenses made with different kinds of glass. The effect of chromatic aberration can be further reduced with large focal length: look at the size of the Yerkes 40-inch in Figure 9.7. In addition, ordinary glass will not pass ultraviolet light, restricting observation at shorter wavelengths unless very expensive optical materials are used. The size of a refracting lens is also limited because the weight of the glass causes it to sag, distorting the lens and reducing the quality of the image.

9.3
The Reflecting Telescope

The solution to the refractor's difficulties was initiated by none other than Isaac Newton, who created the first **reflecting telescope** (Figure 9.10; see also Figure 7.1), which uses a curved mirror rather than a curved lens to focus the light. Since the angle of reflection is independent of wavelength, there is no chromatic aberration; ultraviolet light is not absorbed. Mirrors are also supported from the back, allowing them to be made much larger than lenses.

A spherical mirror surface (Figure 9.11a) will not reflect parallel light rays to a unique focal point. Those that bounce from the outer zones away from the center will focus closer to the mirror than will those that reflect from the inner zones. This *spherical aberration* will cause the image to be blurred. The effect can be corrected by making the mirror surface into a *paraboloid* (Figure 9.11b), the surface generated when a parabola is spun about its axis. Then all the rays that arrive parallel to the optical axis will come to a single point. Other than the way in which the light is collected, the reflector and refractor behave similarly, and all the terms defined above have the same meanings.

The focal point of a mirror, called the *prime focus* (Figure 9.12a), is in front of the light-collecting surface rather than behind it. Therefore, the act of viewing the sky will block incoming light. The prime focus can still be used, but only if the mirror is significantly larger than the astronomer or the device used to detect the light. The effect of the obstruction is only to cut the brightness of the observed image by 10% or so.

Newton solved the problem by setting a small flat *secondary mirror* in the converging beam at a 45° angle to the big, or *primary*, mirror. The *Newtonian focus* (Figure 9.12b) is at the side of the telescope. The *Cassegrain focus* (Figure 9.12c), however, uses a curved secondary mirror to extend the focal length and to send the light directly back through a hole in the primary. A long focal length can then be put in a short tube, allowing for a smaller building to house the instrument and a significant savings in cost. Additional mirrors provide useful variations on the Cassegrain theme. The *Nasmyth focus* (Figure 9.12d) adds a tilted flat mirror at one of the points on which the telescope rotates on its mounting so as to send the light to a convenient fixed position. Additional mirrors can direct the light to a *coudé focus* that is usually buried in the basement of the observatory building, where massive analyzing equipment can be installed. Many reflectors have interchangeable mirrors that allow them to be operated at a variety of foci.

The reflector began to reach maturity in the hands of William Herschel, who constructed metal

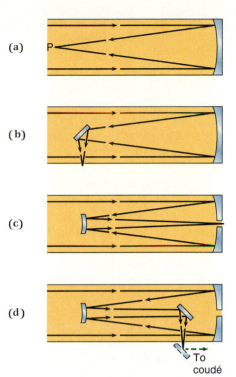

Figure 9.12   *In* **(a),** *the light arrives at the prime focus P without obstruction and in* **(b)** *it is sent by a flat secondary mirror to the Newtonian focus at the side of the instrument. The Cassegrain focus in* **(c)** *uses a curved secondary to extend the focal length and shoot the light through a hole in the large primary mirror. The Nasmyth focus in* **(d)** *uses a third mirror to send the Cassegrain's light again to the side. With additional mirrors light can be sent to a permanent coudé focus.*

mirrors as large as 48 inches across. Telescope mirrors are now made from glass or modern ceramics. The glass is ground into a paraboloid, and then a thin film of highly reflective aluminum is deposited on the curved surface within a vacuum chamber. The first of the great reflectors of this century was the 100-inch (2.3 m) built atop Mount Wilson in southern California in 1917. This was the largest telescope in the world until the Palomar 200-inch or 5-m (see Figure 9.10) went into operation in 1947. With the exception of a 6-m telescope built in the Soviet Union (which has proved less effective), the Palomar telescope reigned as king for 45 years. Its light-gathering power is staggering. A star seen through the eyepiece would be nearly a million times brighter than if viewed with the naked eye.

Over the past few decades, a number of telescopes were built in the 2- to 4-m range, but only recently has significant progress been made in aperture. Several 8-m instruments are now being planned or are under construction in both hemispheres of the Earth. To increase light-gathering power even more, astronomers are using multiple-

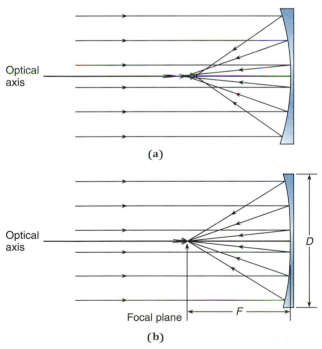

Figure 9.11   **(a)** *The top mirror is a section of a sphere. The light that reflects from the outer parts come to a focus before the rays from the inner parts, and images are blurred.* **(b)** *A mirror shaped like a paraboloid will bring all the rays to a focus at the same point.*

mirror systems. The prime example is the mammoth 10-m Keck telescope (Figure 9.13), now in service on Mauna Kea in Hawaii, 4,200 m above sea level. The Keck, the world's largest optical instrument, features 36 hexagonal 1.8-m mirrors. Each is a piece of a giant computer-controlled paraboloidal surface that has four times the light-gathering power of Palomar. The most ambitious project is the European Southern Observatory's Very Large Telescope, to be located in Chile. It will consist of four 8-m instruments that can be combined to provide a light-gathering power equivalent to a single 16.4-m mirror, 2.6 times that of the Keck. Table 9.1 provides a list of largest telescopes.

An image formed by a paraboloidal mirror will come to a perfect focus only on the optical axis. The result is that the effective field of view of a standard reflector is severely limited: that of the Palomar 5-m is only a quarter of a degree wide. An effective solution to the problem was invented in the 1930s by Bernhard Schmidt, an optician at the Hamburg Observatory in Germany. The **Schmidt telescope** (Figure 9.14) uses a combination of a spherical primary and a thin refracting lens to bring the light to a focus over a very wide field of view. The 48-inch Schmidt on Palomar can take photographs 6° across! Schmidts frequently work as survey instruments, finding objects of interest for the big telescopes to examine. The Palomar Schmidt and its southern hemisphere cousins have surveyed the entire sky to faint limits (see Figure 4.14).

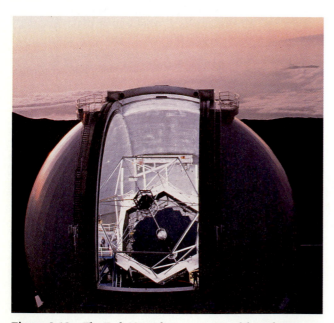

Figure 9.13   *The Keck 10-m telescope consists of three dozen separate mirrors, each 1.8 m across, arranged over a paraboloidal surface.*

TABLE 9.1
The Largest Telescopes

Refractors

40-inch (1-meter), Yerkes Observatory, University of Chicago, Williams Bay, Wisconsin, U.S.A.

36-inch (0.9-meter), Lick Observatory, University of California, Mount Hamilton, California, U.S.A.

Reflectors

16-meter Very Large Telescope optical array (four 8-meter telescopes),[a] European Southern Observatory, Cerro Paranal, Chile

11.8-meter, Large Binocular Telescope, University of Arizona, Arcetri (Italy) Astrophysical Observatory, Mt. Graham, Arizona[a]

11-meter spectrographic telescope, McDonald Observatory, University of Texas and Penn. State, Mt. Locke, Texas[a]

10-meter, Keck Telescope, California Institute of Technology and the University of California, Mauna Kea, Hawaii.

10-meter Keck II, Mauna Kea, Hawaii[a]

8.3-meter Subaru Telescope, National Astronomy Observatory of Japan, Mauna Kea, Hawaii[a]

8-meter Gemini Telescopes, U.S., United Kingdom, Canada, Chile, Brazil, and Argentina, two instruments, Mauna Kea, Hawaii, and Cerro Pachon, Chile[a]

6.5-meter Magellan Project, Las Campanas Observatory, Carnegie Institution (US), Las Campanas, Chile[a]

6-meter, Caucasus Mountains, Russia

5-meter Hale Observatory, Palomar Mountain, California, U.S.A.

4.5-meter equivalent, Multiple Mirror Telescope, University of Arizona and the Smithsonian Center for Astrophysics, Mount Hopkins, Arizona, U.S.A.[b]

4.2-meter Herschel Telescope, Royal Greenwich Observatory, Canary Islands, Spain

4-meter telescopes, National Optical Astronomy Observatories, Kitt Peak, Arizona, U.S.A., and Cerro Tololo Interamerican Observatory, Chile

4-meter Anglo-Australian telescope, Siding Spring, Australia

Radio

8,000-kilometer Very Long Baseline Array (VLBA), National Radio Astronomy Observatory; a VLBI that spans the distance from the Virgin Islands to Hawaii

36-kilometer Very Large Array (VLA) Interferometer, National Radio Astronomy Observatory, Socorro, New Mexico, U.S.A.

20-kilometer Westerbork Radio Synthesis Observatory Interferometer, Westerbork, The Netherlands

300-meter single fixed dish, Arecibo Observatory, Arecibo, Puerto Rico

100-meter single dish, Max Planck Institute for Radio Astronomy, Bonn, Germany

100-meter single dish, National Radio Astronomy Observatory, Green Bank, West Virginia, U.S.A.[a]

[a]Planned or under construction.
[b]To be replaced by a single 6-meter primary.

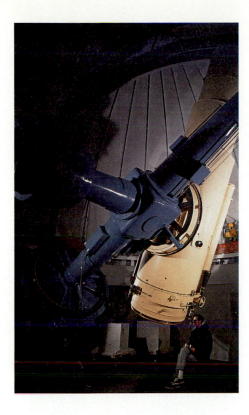

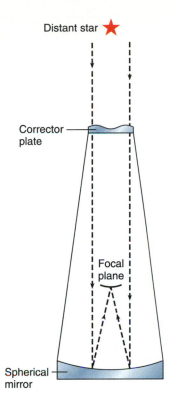

Figure 9.14   *A Schmidt telescope uses a complex correcting plate (exaggerated here) coupled to a spherical mirror to allow wide-angle photography. The primary is larger than the correcting plate so that the focal plane receives uniform illumination.*

9.4
Observatories

The astronomical telescope rotates on two axes mounted perpendicular to each other. The classic design is the *equatorial mount* (Figure 9.15), in which one axis points to the north (or south) celestial pole. When the telescope is rotated about this *polar axis* it will trace out a parallel of declination and follow a star along its daily path. The polar axis is geared to a sidereal clock, allowing the telescope to follow the stars automatically as they move on their daily paths. The *declination axis* is mounted on the polar axis and moves the telescope north and south. Each axis has a graduated circle attached to it so that the instrument can be pointed in hour angle (derived from right ascension, α, and sidereal time) and declination, δ. When the telescope is set, the object to be examined then appears in a low-power finder telescope mounted alongside the main instrument. When the object is centered in the finder, it appears in the bigger telescope. The Yerkes refractor in Figure 9.7 is built this way. Other telescopes use somewhat different designs (see Figure 9.10), but the principle is the same.

Most modern research instruments are controlled by computer. The astronomer enters α, δ, and the epoch of precession on a keyboard. The

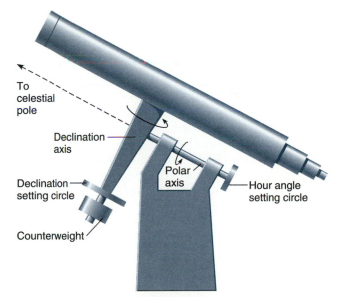

Figure 9.15   *In the classic equatorial mount, the polar axis points to the celestial pole. The telescope can be rotated on it east and west in hour angle. The declination axis is attached to it and swings the telescope north and south. Each axis has a setting circle marked with coordinates. The polar axis is attached to a sidereal clock that can guide the telescope across the sky to follow the daily path of a star. The declination axis also holds a counterweight to balance the heavy instrument across its point of rotation.*

BACKGROUND **9.1** Do It Yourself

Astronomy is a remarkably accessible science: you need only go outside and look. Even simple binoculars will provide marvelous views. Binoculars are described by a pair of numbers, for example, 7 × 35, the first giving the magnifying power, the second the aperture in millimeters. For nighttime use, 7 × 50 is excellent. Small telescopes, which are simple and affordable, are better. The minimum is the 60-mm (2.4-inch) refractor, commonly available in department stores and discount houses. Most amateur telescopes come with a variety of eyepieces, allowing the power to be changed. As a general rule, always use the lowest power needed, which will yield the widest field of view and the brightest images. The maximum that can generally be supported is about 50 power per inch of aperture.

Classic Newtonian reflectors (Figure 9.16) will give the greatest aperture for the money and are readily available from several national mail-order firms. The minimum usable mirror diameter is 3 inches (75 to 80 mm). The bigger the aperture you can afford, the better: a 6-inch will make stars seem four times brighter than a 3-inch. At a more sophisticated (and expensive) level, you might opt for a Cassegrain design, in which a long focal length is folded into a short tube. Cassegrains are therefore relatively easy to transport into a dark site. They start as small as 3 inches, and there is really not much of an upper limit except that imposed by the checkbook.

Astronomical observing is a learned skill that requires practice. Use your new telescope first in the daytime to align the low-power finder. When you do you will become aware of one of the astronomical telescope's most confounding properties: the view is inverted. Do not try to "erect" the image with a special eyepiece—you will only

be introducing another lens that absorbs precious light. At night you will discover another problem: the telescope also magnifies the apparent speed of the Earth's (or the sky's) rotation. Therefore, to follow an object on its daily path, you have to move the telescope constantly, and, because of the field inversion, apparently in the "wrong" direction! Have patience, you get used to it. If you have invested enough, you may have a sidereal clock drive that will follow a star for you, but only if you have remembered to align the telescope's polar axis on the celestial pole. That means making a latitude adjustment and another adjustment at night to point the axis north (or south if you live in the southern hemisphere). A great deal of help—and fun—can be had from your local amateur astronomy club.

Figure 9.16   *The simple, easy-to-use Newtonian reflector is the amateur's workhorse.*

computer then corrects for precession and the aberration of light to find the object's precise coordinates, determines the hour angle from an internal sidereal clock, and commands precision motors to point the telescope. The finder is replaced by a video camera that looks through the main telescope and relays the scene to a television set in a separate control room (Figure 9.17). The speed of the computer allows us to dispense with the equatorial mount. If the telescope is aligned along the direction of gravity—to the zenith and horizon—mechan-

ical stability is increased and costs reduced. Equatorial coordinates entered into the computer are converted into altitude and azimuth and the telescope pointed as before. The Keck telescope in Figure 9.13 has such an *alt-azimuth* mounting.

A telescope is an outdoor instrument that must look directly into the open sky. No warmed air can be allowed to degrade seeing conditions. The most common housing is a dome that keeps out weather during the day and wind and stray lights at night. The telescope points through a shutter (see Figures

9.10 and 9.13) that is opened in the evening and closed in the morning. The dome is on wheels, so that the shutter can be directed toward any part of the sky.

Until this century, telescopes were commonly placed for convenience near the astronomer. Today, they are ideally situated on remote, dark, desert mountaintops that have minimal cloud cover and are as high as possible to reduce atmospheric effects. From the top of a tall peak you see the stars with wonderful clarity, and on an especially good night they seem to stare at you, hardly twinkling at all.

9.5
Analysis and Instrumentation

Until the late nineteenth century, the human eye was the chief astronomical detector. The development of photography, with its ability for creating permanent records, revolutionized the science. An astronomical camera is simply a plate-holder attached to the telescope's focus. During the time the plate is exposed, it will record photons, allowing the astronomer to build up images even of things that cannot be directly seen. Except for the Sun, Moon, and planets, celestial sources of light are quite dim, and astronomical exposure times are measured not in fractions of a second but in minutes and hours.

The photographic plate is a wonderful device for storing data, but it has some severe deficiencies. Its *quantum efficiency*—the ratio of the number of photons recorded to the number striking the detector—is at best only about 1 or 2%, and is much lower in the red and infrared. Worse, photography

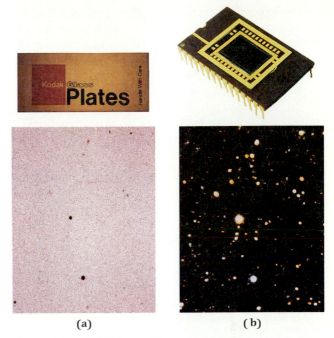

(a)   **(b)**

Figure 9.18   *A CCD image (b) can record far more stars in a specific exposure time than a photograph (a). The chip itself (shown above the CCD image) is only a few centimeters across and is read by computer onto a video screen.*

is not well suited for precise measurements of image brightness. The solution to these problems was initiated in 1913 with the development of the *photoelectric cell,* which uses the photoelectric effect (see Section 8.2.3). Electrons knocked loose by incoming photons can be directed by an external electric field onto a plate, where they set up an electric current that is directly proportional to the image's brightness. The quantum efficiency of the resulting **photoelectric photometer** approaches 40%, and the device also allows precise measurements of stellar magnitudes.

This development pointed the way to electronic imaging, which has almost completely replaced photography. The most effective apparatus, developed since the 1970s, is the **charge-coupled device,** or CCD (Figure 9.18). A CCD chip is a light-sensitive surface divided into rows and columns of individual cells called *pixels.* Each pixel can build up an electric charge that is proportional to the amount of light that falls on it. A computer records the charge of each pixel and stores the data on magnetic tape or disk. The data are in numerical, or digital, form, and they can be reconstructed by computer to produce an image on a video terminal (which can then be photographed for reproduction). The brightness of every position in the picture can also be accurately measured. CCDs have quantum efficiencies that can approach 90%,

Figure 9.17   *The Kitt Peak 4-m telescope is operated from a control room that contains video displays showing telescope position, the field of view, and information about the telescope's condition.*

so the exposure times needed to produce images are vastly shorter than they are with photographic plates. CCDs, however, are not without their problems: they cannot yet be made very large, so they are restricted to small fields of view, and their ultraviolet efficiencies are low.

In a photograph or CCD image, a star will still appear as a fuzzy ball because of our turbulent atmosphere. There are two ways around the prob-

lem. One, extremely expensive, is to take the instrument into space. The other is to use advanced electromechanical systems to counteract the twinkling. It is possible to build a telescope that can actually follow the twinkling movement by feeding the jumping pattern to a set of motors that controls the optics. The resulting image appears as it would if less air were present (Figure 9.19), and we can in principle reach the telescope's theoretical resolving power.

Photography, CCD imaging, and photoelectric photometry allow us to observe the structures of extended objects and to measure the magnitudes and positions of stars. To tell what stars are made of, however, we need to spread the light into its spectrum with the **spectrograph** (Figure 9.20). A narrow slit or aperture is placed at the telescope's focal plane. Light passes through it to a *collimating mirror* that makes the incoming rays parallel and sends them onto a diffraction grating that disperses them. All rays of the same color are now parallel to one another, but each color is going in a slightly different direction. The spectrum is then focused by a *camera mirror* onto the focal plane, where it can be pho-

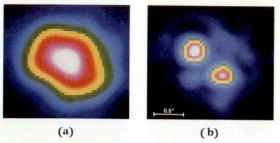

(a)   **(b)**

Figure 9.19   *Extraordinary resolution can be achieved with an adaptive optics system.* **(a)** *An ordinary CCD image records two stars only 0.38 seconds of arc apart made with a seeing disk 0.8 seconds of arc across.* **(b)** *The adaptive optics system is now turned on, improving the resolution to 0.22 seconds.*

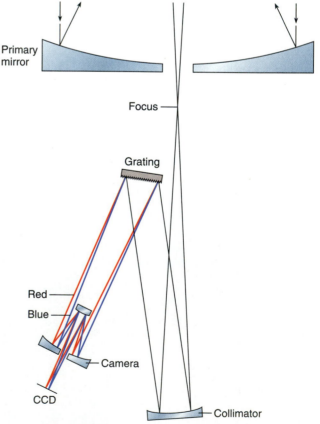

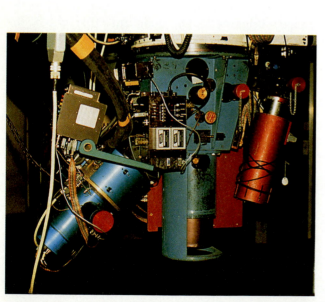

Figure 9.20   *This spectrograph is attached to the Cassegrain focus of the 2.3-m telescope of the University of Arizona; its light path is diagrammed on the right. The light comes out of the telescope's focal plane and is collimated, that is, the rays made parallel. The grating disperses the radiation and the camera mirror focuses it onto a detector, now usually a CCD.*

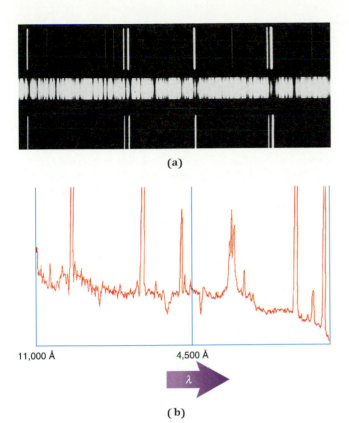

(a)

11,000 Å   4,500 Å

λ

(b)

Figure 9.21   **(a)** *A small section of the absorption spectrum of Betelgeuse, recorded with a photographic plate, is flanked by iron emission lines produced in the spectrograph. Their precisely known wavelengths allow those of the absorption lines to be measured.* **(b)** *A CCD detector can make a graphical spectrogram, here one of a cloud of interstellar gas. Its emission lines appear as spikes on top of a continuum.*

tographed (Figure 9.21a) or recorded by a CCD to create a **spectrogram.** CCD recordings could be reconstructed to look like photographs, but instead they are almost always displayed graphically, with the brightness of the spectrum plotted against wavelength (Figure 9.21b). (The visual version of the spectograph, used to illustrate Kirchhoff's laws in Figure 8.17, is called a *spectroscope.*) Most spectrographs have a variety of interchangeable cameras and gratings that allow the degree of dispersion as well as the observed wavelength region to be changed.

Within the spectrograph is a device for generating an emission-line *comparison spectrum* from electrically vaporized iron or electrically excited helium, neon, or argon that can be directed into the spectrograph. The precise wavelengths of the emission lines are known from laboratory measurements, and they serve to calibrate the resulting spectrum to allow the measurement of accurate stellar wavelengths. In a photographic spectrogram, such as the one in Figure 9.21a, twin comparison spectra are placed alongside the stellar spectrum. If

a CCD is used, a separate image of the emission spectrum generated by the spectrograph's calibration device allows the assignment of specific wavelengths to specific pixels.

9.6
The Invisible Universe

Many critical astronomical processes radiate little or nothing in the optical spectrum. Without the ability to analyze electromagnetic radiation in other wavelength domains, we would be unaware of these processes and the bodies associated with them. Modern astronomy would be impossible.

9.6.1 The Infrared

In the late 1700s, William Herschel (Figure 9.22) opened the first nonoptical domain when he used thermometers to measure the energies of different portions of the solar spectrum. To his surprise, he found that the thermometers would respond beyond the visible red end. He had discovered infrared radiation. However, only recently, in this age of electronics, have sensitive detectors and imaging devices become available. Several large telescopes are now even dedicated to IR observations.

The Earth's atmosphere is not very transparent in the infrared (see Figure 8.10) because of strong absorption lines belonging to atmospheric carbon dioxide and water vapor. As a result, infrared astronomers will go to great lengths to reduce the

Figure 9.22   *Sir William Herschel, in many ways the founder of modern observational astronomy, ponders a rising thermometer that is placed just off the red end of the spectrum. Invisible infrared rays are heating it. Outside the window is Herschel's famed 48-inch reflecting telescope, at the time the largest ever built.*

BACKGROUND 9.2   Astronomical Research

You are an observational astronomer and have a brilliant idea for a research project. How do you pursue it? You cannot just run to a telescope and start working: someone else is already using it. Moreover, research can be expensive. You need funds to travel to the telescope, to run or use the computers needed to examine the data, and possibly to hire an assistant—commonly a graduate student—to help you.

The first step might be to write a research proposal to the National Science Foundation or to NASA that outlines the nature of the work and presents a budget. It will be sent out for competitive peer review to be read and evaluated by other astronomers. If they judge your idea worthwhile and workable, you might then find yourself with a research grant that will last one or more years.

A NASA proposal will incorporate the use of a spacecraft. If you plan to observe from a ground-based observatory, however, you must also apply for telescope time, which is commonly allocated in four- or six-month blocks. So again you write a proposal that outlines what instruments you need to use, what objects you want to observe, and how much observing time you will require. All the proposals are then reviewed by members of a telescope allocation committee, who decide which are the most important. You might then be granted a few nights on specific dates, and you travel to the observatory a day or so ahead to learn (or relearn) the complex electronic instrumentation. If the weather is good and the instruments all work, you fly home with magnetic tapes that contain all your data.

Data acquisition is only the first step. You and your assistant must spend the next few days, weeks, or even months in data reduction, the process whereby the raw numbers from the observations are converted into useful information. You might, for example, have to correct your data for the absorbing effects of the Earth's atmosphere, which depend on wavelength, distortions produced by the telescope itself, or motions of the Earth that affect velocity measurements made via the Doppler effect. Finally, the reduced information must be interpreted in terms of some physical model by you or by a theoretician whom you might invite to participate in your project. Data from only a few nights' observation with a large modern instrument can easily provide enough work for a year. (Of course, you might equally well be the theoretical astronomer. In that case you might apply for massive computer time to implement your new theoretical ideas.)

Finally, when you have learned something about the objects you have observed (or have developed a successful theoretical model), you write a description of your research, including the procedures and the results, and send the paper to the editor of a scientific journal. The editor sends it to an expert referee who judges the quality of the scientific procedures and the merits of the paper's content. The referee's report will be sent to you, giving you an opportunity either to revise or to argue your points. Eventually, perhaps two or more years after you had your idea, you see the results in print. The time lag is so great that usually you will be working on several ideas and projects at one time and may be faced with a number of observing runs (or meetings with colleagues and collaborators) within the next year, all on different projects, always keeping you busy.

---

amount of air over their heads. Two IR observatories are located high on Mauna Kea with the Keck 10-m. To observe in the submillimeter domain where infrared and radio merge, there is even an observatory at the south pole, which has some of the driest air on Earth. The Kuiper Airborne Observatory (Figure 9.23), a telescope mounted in a Lockheed C-141 Starlifter, can climb to 12,000 meters, half again as high as Mount Everest. At that height the amount of water vapor overhead is reduced by some 99%.

9.6.2 Radio Telescopes

Radio waves from space were also discovered by accident. In 1933, Karl Jansky, an engineer for Bell Laboratories, was assigned to find the source of interference that was disturbing transatlantic radio communication. He therefore built a large radio antenna in New Jersey (Figure 9.24). An **antenna** is merely one or a set of metal wires or rods. Because a radio wave is an electromagnetic signal, it will produce an electric current in the antenna that can be amplified, detected, and measured. An

Figure 9.23 *The Kuiper Airborne Observatory (KAO) carries a 0.9-meter telescope to 12,000 meters, above most of the Earth's water vapor, for a clearer view in the infrared.*

Figure 9.24 *The first radio telescope was constructed by Karl Jansky in 1933 to track down radio interference. The source turned out to be the Milky Way. The telescope was no more than a large directional antenna.*

antenna can be designed to have greater sensitivity in one direction than in another; that is, it can be pointed, even if crudely. Such a device is the simplest form of **radio telescope.** Jansky saw that the interference had a period of 23 hours 56 minutes, the length of the sidereal day, which meant that the signals had to be coming from outside the Earth.

Professional astronomers of the time did not know what to do with the discovery. Instead, in 1936 it fell to an avid amateur scientist named Grote Reber to build the first reflecting radio telescope in his Wheaton, Illinois, backyard from spare and homemade parts. To increase the amount of radiation collected and to make the telescope more directional, he used a 28-foot-wide parabolic reflector, a metal radio mirror, to focus the radio waves onto the antenna (Figure 9.25). His telescope was steerable over the sky. With it, Reber made the first crude map of celestial radio sources.

The new science was advanced by the development of radar systems during World War II. Today's astronomers have available dozens of fully steerable

instruments with diameters up to 100 m, the size of a typical sports field. The largest radio telescope is a fixed instrument *300* meters in diameter built into a basin in Arecibo, Puerto Rico (Figure 9.26).

Radio telescopes "see" radio radiation just as optical ones see light. They "listen" to nothing. There is no signal riding the wave that allows you to hear anything with an ordinary radio. The purpose of the radio telescope is to measure the amount of radio radiation coming from celestial objects. The instrument can be made to scan over a source to map it and to produce computer-generated radio images that look very much like photographs made with optical telescopes. Some radio telescopes even have spectrographs that can be tuned to specific frequencies to search for absorption or emission lines.

Artificial radio waves generated by ordinary radio transmitters hinder observation and present a severe problem. As a result, certain important radio bands have been reserved by worldwide agreement exclusively for the use of radio astronomers. Fortunately, much of the radio spectrum is unaffected by the Earth's atmosphere, so that radio observatories can be built in valleys to screen them from stray

Figure 9.25 *Grote Reber built the first paraboloidal reflecting radio telescope. The 28-foot dish focused radio radiation from space onto the antenna at the top.*

Figure 9.26   *The Arecibo Observatory radio telescope is 300 meters in diameter. The antenna that actually detects the radiation is supported at the prime focus by wires slung from three towers. It is a zenith instrument that can be pointed somewhat by moving the antenna. It is also a powerful radar that is used to study the surfaces of the nearby planets.*

artificial emissions. Only instruments designed for microwave observations need to be at high altitudes.

9.6.3 Radio Resolution and Interferometers

Like optical telescopes, radio telescopes produce diffraction patterns. The size of an interference pattern is proportional to wavelength, as seen in Figure 9.4. As the central diffraction disk gets bigger, the resolving power decreases. The resolving power, already known to be proportional to $1/D$, is now seen to be proportional to λ/D, and radio waves are thousands to millions of times longer than optical waves. A 100-m radio telescope operating at a wavelength of 10 cm has a resolution of only four minutes of arc, worse than the unaided human eye!

A solution would be to build bigger telescopes, but the practical engineering limit has already been reached. We must take a cleverer approach. Instead of building a kilometer-wide radio telescope, we build *two* radio telescopes separated by an arbitrary distance, and point them in the same direction (Figure 9.27). The electrical signals from the twin instruments are linked together and we observe their sum. The effect is analogous to that of a double slit that produces a diffraction pattern. We therefore see interference fringes produced by the radio waves from space. With this radio **interferometer** we can reconstruct an image with a resolution equivalent to that of a single dish with a diameter equal to the separation between the two individual telescopes.

However, a two-telescope interferometer still provides limited information on the structure of a source. It is better to use three or more, or even to

fill in the area of a hypothetical single dish by constructing an array of instruments. The largest of these is the Very Large Array (VLA) of the National Radio Astronomy Observatory (Figure 9.28) in New Mexico. Its 27 radio telescopes are spread in a Y, with arms 21 km long. The effect is that of a single-dish instrument 36 km in diameter, roughly the size of the beltway around Washington, D.C.! With it, astronomers can achieve a maximum resolution

Figure 9.27   *The twin telescopes of the Owens Valley Radio Observatory point skyward as they observe a celestial interference pattern from which a radio picture of the sky can be constructed.*

Figure 9.28   *Twenty-seven 25-m radio telescopes of the Very Large Array spread over 26 miles of the New Mexican desert. The telescopes can be moved on railroad tracks to change the interferometer diameter and resolving power.*

TABLE 9.2
Some Astronomical Spacecraft[a]

Name	Launch, Status	Function[b]	Region or Wavelength	Countries
High Energy Astronomical Observatory (HEAO2, or Einstein Observatory)	1971–72	I, S	X ray	U.S.A.
International Ultraviolet Explorer (IUE)	1976, active	S	UV: 1,100–3,000 Å	U.S.A., U.K., ESA[c]
Infrared Astronomical Satellite (IRAS)	1983	I, S	IR: 12, 20, 50, 100 microns[d]	U.S.A.
Cosmic Background Explorer (COBE)	1989–91	I, S	microwave radio	U.S.A.
Röntgen Satellite (ROSAT)	1990, active	I, S	X ray	Germany, U.S.A.
Hubble Space Telescope (HST)	1990, active	I, S	UV through IR	U.S.A.
ASTRO	1990	I, S	UV	U.S.A.
Compton Gamma Ray Observatory	1991	I, S	gamma ray	U.S.A., ESA
Extreme Ultraviolet Explorer (EUVE)	1992	I, S	UV: 70–760Å	U.S.A.
Advanced X-ray Facility (AXAF)	planned	I, S	X ray	U.S.A.
Space Infrared Telescope (SIRTF)	planned	I, S	IR	U.S.A.

[a]For planetary probes, see Part III.
[b]I denotes imaging capability; S, spectral capability.
[c]European Space Agency, a consortium of European countries.
[d]A micron (μ) is 10^{-6} m = 10^{-3} mm = 10^4 Å; 1 μ = 10,000 Å.

of a tenth of a second of arc, better than any ground-based optical telescope.

At greater sizes, direct linkage of the interferometer components becomes impractical. So we separate the instruments entirely, synchronize the observations with precise clocks, and mix the data by computer instead of by direct cabling. The concept is called a **very long baseline interferometer,** or VLBI, and there is no limit to its size. We can make VLBIs as large as the United States itself. Telescopes have even been separated by nearly the diameter of the entire Earth to produce an astonishing resolution of a ten-thousandth of a second of arc. Someday we will have a link between the Earth and the Moon.

9.7
Space Observatories

The best view of the Universe is from space, from where we can see not only the entire infrared, but also all of the ultraviolet, X-ray, and even gamma-ray spectral regions that are blocked completely by our atmosphere. The first attempts to observe from space were made with rockets, but these could climb above the air for only short periods of time. For detailed observing, telescopes are placed in Earth-orbiting satellites.

A remarkable number of astronomical spacecraft have been flown. They come in two broad varieties: those that travel directly to the bodies of the Solar System and those that are placed in orbit around the Earth and are designed more for general observation. The first type will be examined in Part III; several in the latter category are listed in Table 9.2.

In principle, a space telescope is just like one on Earth. The kinds of detectors used depend on wavelength: specialized radiation detectors in the X-ray domain, video or CCD-type systems in the optical. There are some major differences, however, between space- and ground-based instruments. An orbiting space telescope orients itself by locking onto bright stars and is stabilized and pointed by gyroscopes (rapidly rotating wheels that resist turning) and small rocket motors. The craft must also have an on-board power supply, usually large panels with chemical cells that turn sunlight into electricity. All observing is done by remote control and radio command from the ground, and all the data are returned via a radio link. When its fuel supply is exhausted, or too many motors break down, the satellite dies. Space telescopes such as the International Ultraviolet Explorer (IUE) (Figure 9.29a), and the Infrared Astronomical Satellite (IRAS) (Figure 9.29b), and the rest in Table 9.2 have provided us with unparalleled views of the Universe.

The most versatile of astronomical satellites is the Hubble Space Telescope (HST) (Figure 9.30). Conceived decades ago, this 2.4-m (90-inch) tele-

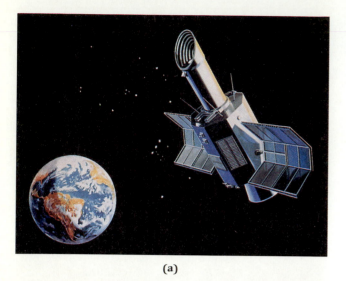

(a)

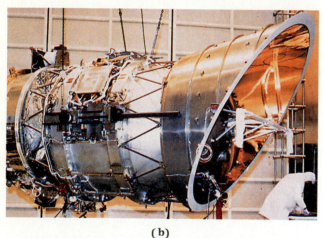

(b)

Figure 9.29   **(a)** *The International Ultraviolet Explorer (IUE) observes from geosynchronous orbit.* **(b)** *The Infrared Astronomical Satellite (IRAS) is shown under construction.*

Figure 9.30   *The Hubble Space Telescope (HST), a complete orbiting observatory, is launched into space from the bay of the space shuttle. It is able to make images as well as to do photometry and spectroscopy from the deep ultraviolet to the infrared.*

scope finally flew in 1990. The surface of the primary mirror is so smooth that if it were expanded to the size of the Atlantic Ocean, the "waves" would only be a few centimeters high. Unfortunately, an error made in testing the mirror caused it to be ground and polished to the wrong shape, degrading the focus. Computer-processing of the images allows most of the error to be corrected but still leaves the telescope operating at low efficiency.

The HST can observe with high resolution from the ultraviolet into the near infrared. It contains two spectrographs, two cameras, and a photoelectric photometer with a rapid response time. A resolving power as small as 0.07 seconds of arc can be achieved with its faint-object camera. One of the major wonders of the telescope is its pointing ability: its guidance system is designed to maintain the instrument in a specific direction to within 0.1 seconds of arc for a protracted period.

The great advantage of the Hubble over other space telescopes is that it was designed to be serviced in orbit by crews of the space shuttle. A correcting lens will be installed to fix the problem with the primary mirror (causing the removal of the high-speed photometer), and the instruments will be replaced by newer models. The corresponding disadvantage is that the HST had to be placed in low orbit around the Earth so that it could be launched and reached by the shuttle, which cannot go more than about 500 km high. Because the telescope is so close to the Earth, the orbital period is only 90 minutes. The motion is so fast that celestial objects rise and set every 45 minutes. Consequently, HST cannot look at anything for very long at any one time, and so its efficiency is considerably reduced. Nevertheless, it is another instrument for the coming century. With it and others we have finally broken our science loose of the constricting Earth, and in a sense have taken our telescopes to the stars themselves.

KEY CONCEPTS

Antenna: That part of a radio excited electrically by radiation.

Aperture (D): The diameter of a telescope objective.

Charge-coupled device (CCD): An electronic imaging device.

Diffraction disk: The central fringe of a circular diffraction pattern.

Dispersion: The spreading of radiation into a spectrum by refraction or diffraction.

Eyepiece: A lens attached to a telescope that makes light rays parallel so they can be viewed by the eye.

Focal length (F): The distance from a focusing lens or mirror to the focal point.

Focal point/plane: The point on the optical axis, and the plane perpendicular to the axis, where images are focused.

Image: The depiction of an object in the focal plane.

Interferometer: Two or more linked radio telescopes that improve resolution by detecting interference of radio waves.

Lens: A curved piece of glass that can focus radiation.

Objective: The main telescope lens or mirror.

Photoelectric photometer: A device that uses the photoelectric effect to measure the brightnesses of celestial objects.

Radio telescope: A telescope that collects and detects radio radiation.

Reflecting telescope: A telescope that focuses radiation with a mirror (usually a paraboloid) and that has a variety of focal positions.

Reflection: The return of radiation from a surface at an angle equal to the angle of incidence.

Refracting telescope: A telescope that uses a lens to refract light to a focus.

Refraction: The bending of the path of radiation as it goes from one substance to another.

Schmidt telescope: A reflecting telescope with a refracting correcting lens that provides a wide field of view.

Seeing disk: The apparent disk of a star, produced by variable refraction in the atmosphere.

Spectrogram: A photographic or graphical recording of a spectrum.

Spectrograph: A device that separates electromagnetic radiation by wavelength.

Telescope: A device for gathering and focusing electromagnetic radiation.

Very long baseline interferometer (VLBI): An interferometer with separate telescopes synchronized by clocks.

KEY RELATIONSHIPS

Light-gathering power:

The degree to which a telescope collects electromagnetic radiation; proportional to the square of the diameter of the objective.

Magnifying power:

The amount by which a telescope eyepiece multiplies the apparent angular diameter of an object;

$$M = F_{objective}/F_{eyepiece}$$

Resolving power:

The ability of a telescope to discern a separation between objects; proportional to λ/D; for optical telescopes: R.P. = 13"/D cm

EXERCISES

Comparisons

1. List the possible foci of a reflector and give an advantage or a disadvantage of each.
2. Rate the advantages of a reflector relative to a refractor.
3. Distinguish between equatorial and alt-azimuth telescope mountings.
4. Compare the advantages and disadvantages of a photographic plate and a CCD.
5. Compare the characteristics and purpose of the paraboloidal reflectors used for optical and for radio telescopes.
6. Why do single-dish radio telescopes have much worse resolving power than do optical telescopes?
7. Distinguish between an ordinary radio interferometer and a very long baseline interferometer.

Numerical Problems

8. How many times fainter can you see with a 20-cm telescope than with a 10-cm telescope?
9. What is the light-gathering power of the largest optical reflector relative to the largest refractor?
10. What is the resolving power of (a) a 20-cm telescope; (b) a 2-m telescope?
11. What are the focal lengths of the eyepieces that you would need to produce 100 and 200 power with a telescope of 240-cm focal length?
12. What is the resolving power of a 10-m radio telescope working at a wavelength of 10 cm?
13. At a wavelength of 10 cm, what is the resolving power of a very long baseline interferometer if one element were on the Earth and the other on the Moon?

Thought and Discussion

14. Show in a drawing the approximate path of a light ray that is both reflected *and* refracted from a glass surface.
15. Why will a penny dropped into a glass of water appear to float if looked at through the water's surface?
16. How does the degree of refraction change with wavelength?
17. Why do telescopes show diffraction patterns, and what do the patterns have to do with resolving power?
18. Why is the theoretical resolution of a 2- or 3-m telescope not achieved in practice?
19. How can chromatic aberration be reduced?
20. What is the shape of most telescope mirrors? Why?
21. How does the Cassegrain telescope get a long focal length in a short tube?
22. Why is the Schmidt telescope system needed?
23. At what latitudes is an equatorial telescope mounting the same as an alt-azimuth mounting?

24. Why are mirrors rather than lenses used in spectrographs?

25. What are the functions of the following spectrograph components: **(a)** collimator; **(b)** grating; **(c)** camera mirror?

26. For what two reasons are space telescopes necessary?

Research Problem

27. Select a particular telescope: optical, radio, or one in space. Research the history of the device from inception through development to implementation. Find out why it was needed and why it has its particular design characteristics.

Activities

28. Measure the aperture of a pair of binoculars in millimeters. Cover one lens completely and look through the other at a group of stars like the Pleiades. Count the number of stars you can see. Then place a cover over the lens with a circular hole cut to half the lens's size. Count the number of stars you see again. Cut the aperture even further and count again to see the effect of lens diameter.

29. Use a prism to cast a spectrum of sunlight on a wall. Do the same thing with a grating; use a compact audio disk. Draw the order of colors and the span over which you see red and blue.

Scientific Writing

30. The U.S. space program is often criticized as too expensive. Write a justification of the program based on its value to astronomy.

31. Your school plans to purchase a small telescope for student use. Decide what kind of instrument you would like and describe it and its capabilities to the school's chief administrator.

10 The Earth
11 The Moon
12 Hot Worlds
13 Intriguing Mars
14 Magnificent Jupiter
15 Beautiful Saturn
16 Outer Worlds
17 Planetary Creation and
    Its Debris

Planetary Astronomy

PART

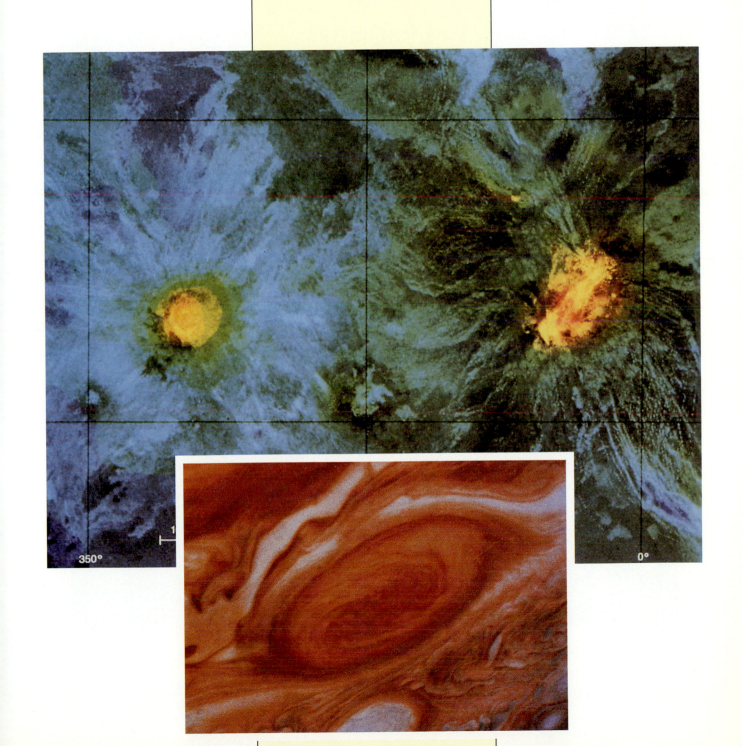

350°

0°

10

The Earth

*The physical
structure
and
atmosphere of
our home*

*Clear air, mountains, liquid water, and life
characterize our Earth, rendering it different
from all the other planets
of the Solar System.*

TABLE 10.1
A Profile of the Earth

Planetary Data			Atmosphere (surface)		
equatorial radius	6,378 km		pressure	$10,330 \text{ kg/m}^2 = 1,033 \text{ g/cm}^2$	
polar radius	6,358 km		mean temperature	10°C	
mass	5.97×10^{24} kg				
mean density	5.5 g/cm^3		**Composition by Number**		
uncompressed density	4.5 g/cm^3		N_2	77%	
escape velocity	11.2 km/s		O_2	21%	
age	4.6×10^9 years		Ar	1%	
core temperature	7,000 K		H_2O	1%[a]	
core pressure	3.6 megabars		CO_2	0.03%	

Layers	% Mass	Radii (km)	Layers	Height (km)	Temperature (°C)
solid inner core	1.7	0–1,220	troposphere	0–12	20 to –60
liquid outer core	30.8	1,220–3,480	stratosphere	12–50	–60 to 0
mantle	67.0	3,480–6,360	mesosphere	50–90	0 to –100
crust	0.7	6,360–6,370	thermosphere[b]	>90	–100 to >1,000

Planetary Features

Cloud cover: Considerable and variable water-vapor clouds.

Craters (meteor)[c]: A few obvious, most eroded and difficult to find.

Basins: Ocean basins produced by plate tectonics.

Mountains: Tectonic uplifts caused by plate motion.

Life on the Surface

Plate motion: 16 crustal plates in constant motion, producing rifts, mountains, and volcanoes; surface age varies from zero to 3.8 billion years.

Volcanoes: Considerable volcanic action caused by subduction of plates in addition to plume-type volcanism.

Water: Considerable liquid water in ocean basins, solid (frozen) water at the poles, vaporized water in the atmosphere.

[a]Variable.
[b]Nearly coincident with the ionosphere.
[c]See Chapter 11.

The Earth is unique within the Solar System. Three-fifths of our world is covered with liquid water, its atmosphere is abundant in oxygen, its surface is mostly young and constantly changing, and it teems with life. All these properties are connected, making the Earth a massive synergistic system. In fact, each planet of the Solar System is unique, each a different experiment of nature. The Earth provides an entry to their study. In turn, they illuminate the nature of our home.

10.1
Age of the Earth

Of all the physical properties of the Earth (summarized in Table 10.1), perhaps the most profound is its age, found from **radioactive dating.** The Earth's surface is covered with hard, solid rock. Several kinds of rock contain radioactive elements that decay into lighter daughter elements (see Section 8.1.3): ^{238}U (uranium), for example, will turn into ^{206}Pb (lead). When a rock is created by solidification from the molten state, an initial ratio of ^{238}U to ^{206}Pb is sealed within (Figure 10.1). The original fraction of ^{206}Pb is known from rocks that do not contain ^{238}U. The ratio decreases steadily at a rate given by the uranium's half-life. The current measured ratio then gives the rock's age. The rocks of the Earth show a huge range of ages, from literally yesterday to an astounding 4.2 billion years. The Earth must be at least as old as its most ancient rocks. The ratio of ^{232}Th (thorium) to its daughter product ^{208}Pb yields the same answer.

The oldest rocks of the Moon go back even further, 4.3 to 4.5 billion years, and the record is held by **meteorites.** Interplanetary space is full of debris, rocks and chunks of metal ranging from mere grains of sand to bodies many kilometers

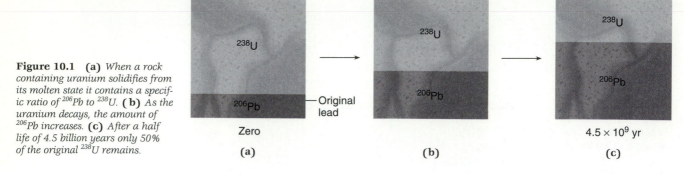

Figure 10.1 **(a)** *When a rock containing uranium solidifies from its molten state it contains a specific ratio of ^{206}Pb to ^{238}U.* **(b)** *As the uranium decays, the amount of ^{206}Pb increases.* **(c)** *After a half life of 4.5 billion years only 50% of the original ^{238}U remains.*

across. We occasionally see them as *meteors* when they penetrate the Earth's atmosphere, their glow giving rise to their popular name of "shooting stars." Called meteorites if they hit the ground, they will be addressed in detail in Chapter 17. Almost all those measured are about 4.5 to 4.6 billion years old. Since there is good evidence that everything in the Solar System was formed at about the same time, that figure is identified as the age of the Earth—the time that has elapsed since it was born out of a swirling cloud of interstellar gases that also gave birth to the Sun.

10.2
The Active Surface

The Earth's surface is dynamic. We feel it shift below us and watch great explosions spew liquid rock into the air. Most of its external layer is young in spite of the ancient origin of the planet as a whole. New rocks are constantly generated and old ones destroyed as the Earth cyclically renews itself.

10.2.1 *Land, Sea, and Mountains*

The rocky surface of the Earth is divided into two broad areas. The higher **continents,** or dry land

Figure 10.2 *Rock layers are buckled by powerful forces from below.*

masses, occupy about 40% of the planet's surface. The remaining 60% is comprised of basins that are filled with water to form **oceans.** The continents contain the oldest rocks, attesting to their relative permanency, whereas the seabed is nowhere more than 200 million years old, revealing that it is constantly renewed and destroyed.

Rocks consist of a variety of *minerals* that can be categorized by their chemistry, their crystal structure, and the way they were formed. Most are **silicates,** a broad class of minerals made of silicon and oxygen often in combination with a variety of metals. The most common near the continental surfaces are *quartz,* made of silicon dioxide (SiO_2), and a variety of *feldspars,* a more complex class of silicate that incorporates light metals like aluminum, potassium, and calcium. Also common are iron- and magnesium-bearing silicates.

Quartz also has a specific crystal structure and is among the loveliest of rocks. The most common continental surface rocks, however, are several varieties of *granite,* characterized by coarse crystals produced by slow cooling from a once-molten state. Granites are made largely of different kinds of feldspars and quartz. By contrast, the **basalts** have a fine-grained structure in which the crystals can scarcely be seen. They are made chiefly of calcium-bearing feldspar and heavy-metal silicates. Although basalts are found on land, they dominate in the ocean beds. The lighter rocks that make the bulk of the continents sit atop this relatively heavy basaltic material rather like puffs of marshmallow in a cup of cocoa.

Rocks are also classified as *igneous, sedimentary,* and *metamorphic.* The three kinds can be transformed into one another by the *rock cycle.* Igneous rocks like granite and basalt solidify directly from liquid **magma** (or **lava**) ejected from the hot terrestrial interior. Sedimentary rocks such as sandstone and limestone (a form of carbon-containing *carbonate* rock) are created when rocks erode into

Figure 10.3   *A map of the globe made by satellite radar shows organized mountain ranges (red) ringing the Pacific Ocean and within southern Asia. More mountain chains lie on the ocean floor, the longest running down the middle of the Atlantic basin. The distribution of earthquakes is similar.*

finely divided material that settles into sea or lake beds and is compressed into stratified layers that overlay the bedrock below. Metamorphic rocks, of which there are many kinds (including marble), are produced by chemical changes in sedimentary and igneous rocks when they are subjected to great weight and heat. Some of the metamorphic rocks are lifted back to the surface by powerful forces where they erode again; others, liquified by great heat far below the terrestrial surface, are reborn as igneous rocks. The result is a process that endlessly repeats itself. Various processes erode and crush rock. When mixed with by-products of plant and animal life, crushed rock is called *soil.*

The most obvious continental features are *mountains* (see the chapter opening photo). They are constantly being worn away in part by rainfall, their eroded rocks and soil carried away to the oceans to become sedimentary rock. Some process, then, must keep building new ones. Cliff faces (Figure 10.2) disclose that many mountains are built of layers of sedimentary and metamorphic rocks that are often buckled, contorted, even stood on end, demonstrating the awesome forces at work.

Mountains are organized into groups and long chains (Figure 10.3), the greatest of which ring the Pacific Ocean basin and include the Rockies of North America and the Andes of South America. Another huge group lies within central Asia, of which Everest, 8.85 km above sea level, is the culmination. Mountains are not confined to the continents, but lie under the ocean as well. Among the most prominent is the Mid-Atlantic Ridge, which runs down the entire spine of the Atlantic Ocean.

10.2.2 Surface Activity

Mountain-building proceeds at an unnoticeable pace over millions of years. Much more rapid and dramatic are **earthquakes,** shocks produced when the Earth's upper layer, or **crust,** slips and suddenly moves, causing the ground to vibrate like a drumhead. Tiny, unfelt shocks occur every day; rare, huge rumbles can demolish cities (Figure 10.4). Although no place is immune, earthquakes are concentrated along mountain chains, both on land and at sea (see Figure 10.3), showing the close relationship between the forces that build mountains and the shaking of the crust.

Figure 10.4   *The vibrations of the 1989 San Francisco earthquake produced enormous damage. The quake occurred when part of the Earth's crust slipped near Santa Cruz, California.*

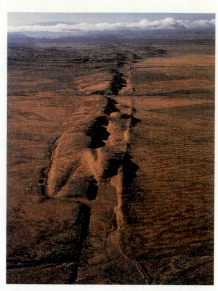

Figure 10.5   *The San Andreas fault of California is clearly visible as a seam in the Earth.*

Figure 10.6   **(a)** *Mount St. Helens erupts in 1982.* **(b)** *Lava pours from the Arenal volcano in Costa Rica.* **(c)** *Crater Lake fills the caldera of an ancient volcano in Oregon.*

(a)

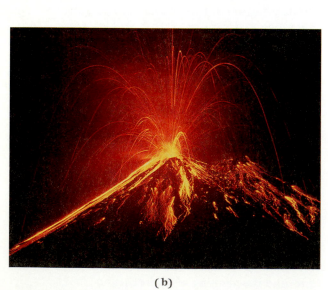

(b)

(c)

Figure 10.7   *The "Big Island" of Hawaii, topped by Mauna Loa, is a shield volcano.*

Running up the coast of California like a tailor's seam is the San Andreas **fault** (Figure 10.5), along which two separate sections of the crust are sliding past each other, the western side moving northward. The joint is fixed so tight, however, that the movement cannot be smooth. When the forces that cause the motion become greater than the bonds that hold the sides together, the land masses suddenly slip by several meters, sending fearsome vibrations out for many kilometers. California and many other places in the world are laced with these faults, any of which can slip at any time. Earthquakes can also occur deep within the crust, many kilometers down. One of these events in 1817 near New Madrid, Missouri, made the Mississippi River run backward and church bells chime in Boston.

Volcanoes (Figure 10.6a) are less common than earthquakes but are no less devastating. They are usually mountains from which erupt hot gases, solid materials called *pyroclastics* that consist of rocks, finely divided ash, and magma (Figure 10.6b). Most volcanoes do not erupt continuously. Some have intervals of many years, even centuries, between active periods. Others are truly dead, remnants of activity long gone. A volcanic peak often

contains a pit or *crater*, formed when mass was removed in an eruption, or a *caldera* (Figure 10.6c) that developed when the mountain peak slumped into an underground chamber emptied of its molten lava.

There are three kinds of volcanoes. *Cinder cone volcanoes* are produced by thick, viscous, lava that erupts explosively, throwing ash and rocks for great distances. They have steep sides caused by falling pyroclastic debris. In contrast, the magma of a *shield volcano* is runny, so it flows easily. Over millennia, the outflow creates a broad mountain with a low slope. Shield volcanoes are not associated with mountain ranges. The "Big Island" of Hawaii (Figure 10.7), which contains inactive Mauna Kea with its observatories and the very active Mauna Loa and Kilauea, is the best example. Measured from base (the seabed) to top, it is 11 km high and the tallest mountain possible on Earth. If you placed more mass on top of it, the weight would make the mountain flow out at the bottom. Intermediate between these two kinds are the more common, classic *composite volcanoes* like Mount St. Helens (Figure 10.6a), Mount Shasta in California, and Fujiyama in Japan. Composite volcanoes can be very dangerous. In 1883, the entire island of Krakatau (in Indonesia) was destroyed when several cubic kilometers of rock were blown into the air. The resulting dust changed world climate and produced spectacular sunsets for years.

Volcanoes are found beneath the sea as well as on land. Cinder-cone and composite volcanoes are distributed around the Earth in a manner similar to mountains and earthquakes, suggesting that their origins are related. Their concentration on the Pacific rim has given it the name "ring of fire."

In spite of their prominence, volcanoes are far from the only source of magma. Huge amounts have periodically flooded from fissures in the crust to produce, not mountains, but vast, thick *lava plateaus.* Much of Washington, Oregon, and Idaho is covered by a lava plateau well over 2 km thick that was formed by multiple ejections over a period of 70 million years. Among the largest is the Deccan plateau that covers most of southern India. The ocean bed beneath the sediments is a vast basaltic lava plain that has issued from undersea crustal fissures commonly associated with ridges like the Mid-Atlantic Ridge.

To understand the origins of these surface structures, we must probe deep within the planet, to its very core.

10.3
The Earth's Structure

What is inside the Earth? Our ability to look directly is limited: the deepest mines go down a mere two kilometers, the deepest wells only a dozen. We must probe instead by indirect but no less reliable means.

The first clue to the Earth's structure is its average density (see Section 8.3.2), found by dividing its mass by its volume, which is $(4/3)\pi$ times the radius cubed. The result is 5.5 g/cm^3. The rocks of the Earth are heavy, and compression raises the density; when this effect is taken into account, the uncompressed density drops to 4.5 g/cm^3. However, the densities of surface rocks and volcanic basalts are typically between 2.6 and 3.0 g/cm^3. Something very heavy, with a density well above 5 g/cm^3, must lie inside to compensate for the light stuff at the surface. Such a high density implies a heavy metal: the most common is iron, followed by nickel.

Since the metal is heavy and rock is light, and since the interior of the Earth is hot (as we know from volcanoes), we might expect the Earth to have to have undergone **differentiation,** a process by which the heavy metal sinks to the center under the force of gravity and the lighter rock floats to the top. The Earth should then have an iron **core** that contains around a third of the total terrestrial mass.

A finer probe of the Earth is provided by the destructive power of earthquakes. A quake acts much like the clapper of a bell, causing the entire Earth to ring with **earthquake** (*seismic*) **waves** that penetrate the globe. There are two different useful kinds. The *primary wave* (P wave) is a compressional wave that vibrates the rock back and forth along the direction of motion (Figure 10.8a). It is essentially a sound wave, but with a frequency too low for the human ear. The *secondary wave* (S wave) is a shear, or transverse, wave that vibrates the rock perpendicular to the wave's motion, much like a wave in a pool of water (Figure 10.8b). Both kinds are detected by *seismographs* (Figure 10.9).

The two kinds of waves move at different speeds through the Earth. The delay in their arrival times tells how far they have traveled from their point of origin to the seismograph. Their speeds

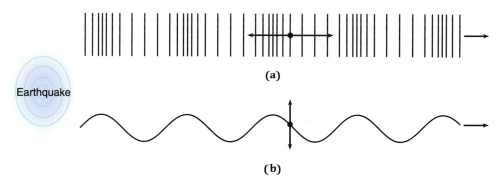

Figure 10.8 **(a)** *The particles through which a primary (P) or compressional earthquake wave moves shake back and forth along the direction of motion to produce waves of high-density compression and low-density rarefaction.* **(b)** *A secondary or shear (S) wave vibrates the particles up and down perpendicular to the wave's motion, much like a water wave.*

(a)

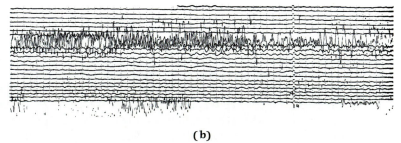

(b)

Figure 10.9 **(a)** *A seismograph records earthquake waves from around the world. The vibrations can be transmitted to a pen that draws a line on a revolving drum.* **(b)** *The resulting chart, or seismogram, here records an earthquake that struck San Fernando, California, on February 9, 1971.*

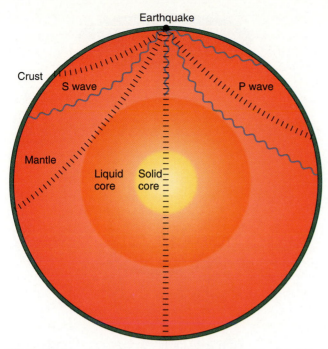

Figure 10.10   *An earthquake sends S and P waves traveling into the rocky mantle toward the iron core. The S waves are stopped by liquid. Opposite the site of the quake is a huge shadow zone in which no S waves appear, allowing us to deduce both the existence and size of a liquid core. The waves are affected by density and temperature variations and are reflected and refracted by other discontinuities.*

also depend on the materials, densities, and temperatures encountered along the paths and, like light waves, they are reflected and refracted at discontinuities. Most important, S waves will not pass through liquid. Analysis of the strengths and arrival times of the waves of large numbers of quakes have allowed geophysicists to build up a picture of the terrestrial interior. We now know where the various layers are and, with the aid of theory and laboratory measurement, even know how temperature and density change all the way to the center.

The S waves are never detected by seismographs on the side of the Earth opposite the quake (Figure 10.10). There is in fact a sizable shadow in which they do not appear. We therefore know that the inside of our planet has a fluid core and can even measure its size. Refraction of the waves shows that the chemical composition also changes at the boundary of the fluid. The core comprises almost exactly one-third of the terrestrial mass and extends 3,500 km from the center, somewhat over half the Earth's radius. It consists of a mixture of iron, followed by nickel, a variety of lesser metals, oxygen, and sulphur.

Conditions within the core (Figure 10.11) can be defined in terms of temperature, density, and **pressure.** Pressure represents the degree of com-

pression, or the outward force that a gas, liquid, or solid would apply per square meter to a wall separating it from a vacuum. At the core's outer edge, the temperature is 5,000 K and the pressure is nearly 1.5 million *bars* (1.5 megabars), where a bar represents standard surface atmospheric pressure. That pressure is high enough to compress iron from its normal density of 7.3 g/cm^3 to over 10 g/cm^3.

The state of a substance depends on the balance between temperature and pressure. Increasing temperature makes atoms move faster and can turn a solid to a liquid and a solid or liquid to a gas. Increasing pressure acts oppositely. In the outer part of the core, the balance is such that the iron is molten and flows like water, a massive metal ocean. Temperature and pressure both rise toward the center under the crushing force of gravity. At a distance of 1,200 km from the center (5,200 km below the surface) the pressure, now more than 3 megabars, wins out over temperature, and the iron actually freezes or turns solid. At the very center, the temperature is nearly 7,000 K, hotter than the surface of the Sun! This solid, frozen nucleus contains 5% of the core's total mass and about 2% of the mass of the whole Earth.

Remarkably, the core makes its presence known at the surface. The liquid iron's temperature decreases outward and is in a constant state of circulation through **convection** (Figure 10.12). If the bottom of a fluid is heated, it expands and becomes less dense. It can then rise, like a helium balloon in the air. But as it moves upward it loses heat, becomes denser, contracts, and falls. Combination with the Coriolis effect produced by the Earth's rotation (see Section 2.7) then establishes a more complex circulation pattern. Iron is an electrical conductor—that is, it allows the free movement of electrons among atoms—so the circulation of the iron generates a magnetic field.

In its simplest form (Figure 10.13), the Earth's field is a **dipole** because it has two *magnetic poles,* one called north, the other south. The *magnetic axis,* the line that connects the poles, is tilted relative to the rotation axis by about 17°. Similar magnetic poles repel, opposite ones attract. A magnetized compass needle points toward the magnetic poles, giving the approximate direction to geographic north.

Surrounding the metallic netherworld of the core is a thick layer of rock that extends nearly to the surface; called the **mantle** (see Figures 10.10 and 10.11), it accounts for about two-thirds of the Earth's mass and is entirely distinct from the core. At the core-mantle boundary, the temperature

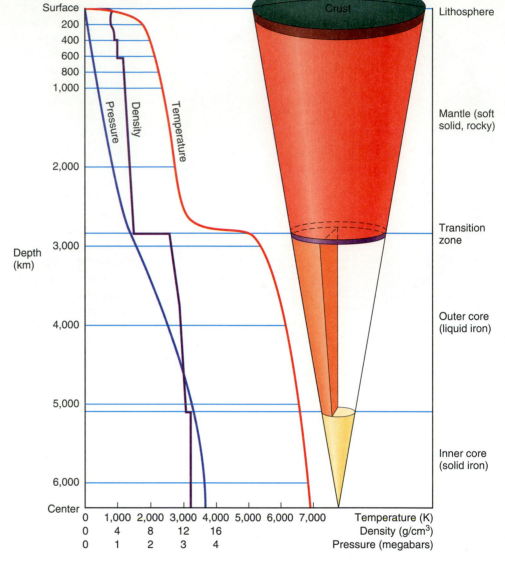

Figure 10.11 *A diagram of the layering of the Earth is accompanied here by graphs that show how pressure, density, and temperature change as depth increases. Between the iron core and the mantle is a particularly sharp discontinuity in density and temperature. At the top of the Earth is the solid crust; it is firmly attached to the top part of the mantle, and with it makes the lithosphere.*

drops by some 2,000 K and the density falls from about 10 g/cm³ to 6 g/cm³. The mantle, made of heavy iron and magnesium-bearing silicates, is the source of the volcanic lavas.

At a depth of about 100 km, the mantle separates into two parts that can slide along each other. The upper one is the **lithosphere.** The topmost part of the lithosphere is the Earth's crust, which consists of light rock that has literally floated upward. On the very top is the lightest rock, the granitic continents. The thickness of the crust is highly variable. It is only 10 km or so thick beneath the seas, whereas it reaches downward 60 or more km beneath the continental land masses.

As we will see, the Earth was (at least partially) molten at the time of its formation. If it is 5 billion years old, why is the interior still hot? It takes

heat to melt ice. Conversely, when water freezes, it must give up the heat. The major source of heat in the Earth is the steady freezing of the liquid core as it slowly solidifies; additional heat is produced by the decay of radioactive elements. The crust is an excellent insulator and dissipates heat only slowly, most of it escaping through volcanoes and other vents. The result is a planet with a very hot interior.

10.4
Continental Drift and Plate Tectonics

A map of the world (Figure 10.14) shows that the eastern coastlines of North and South America fit the curves of western Europe and Africa rather like parts of a jigsaw puzzle, circumstantial evidence

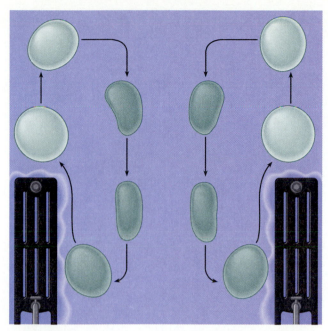

Figure 10.12 *Twin radiators heat a room. The air near them heats, expands, and ascends to the ceiling. There it cools, becomes denser, and falls. The result is a pair of convection currents that make the air circulate.*

that the land masses were once joined. More solid evidence abounds. You can fit the Americas with Europe and Africa in such a way that similar rock structures straddle the boundaries, the Appalachian Mountains of the eastern United States continuing into Norway. When an igneous rock solidifies, the iron oxides within it magnetize in the direction of the local terrestrial magnetic field. Fossil magnetic field directions on the two sets of continents join each other, just like the rocks. At one time, these great continents were part of a single land mass. Now they are separated by 5,000 km!

The Earth's lithosphere is actually divided into 16 separate **plates** (Figure 10.14) that are in continuous motion relative to one another, resulting in global **continental drift.** This geologic restlessness is caused by convection within the upper part of the mantle. The part of the mantle below the lithosphere is a solid, but one so hot that it acts like a soft plastic that can flow. Since the interior of the Earth is hot, the upper mantle will slowly turn over in great, circulating currents. Rising material, forcing its way to the surface, cracks the crust, generally beneath the oceans where the crust is thinnest. The result is a vast network of suboceanic ridges that girdle the globe and form one kind of plate boundary. All along these ridges, we can see evidence of volcanic activity (Figure 10.15), lava pouring out from the mantle making new crust. The entire island country of Iceland, at the northern end of

the Mid-Atlantic Ridge, was made this way. As the lava from the vents in a ridge accumulates as new crust, it forces the opposing plates to move away from the plate boundary, carrying the seabeds and continents with them (Figure 10.16).

The evidence for such motion is overwhelming. The youngest rocks of the Earth are those next to the suboceanic ridges. Toward the east and west they get older, reaching a maximum of about 200 million years near the continents. Magnetic fields frozen into the solid sea floor tell the same story. The seabed is divided into strips that more or less parallel the ridges. Within each of them, the direction of the frozen field is the same, but alternates from strip to strip (Figure 10.17). We can explain this phenomenon only if the Earth's magnetic field periodically reverses, switching its south pole for its north, and if the seabed is at the same time moving outward. From the ages of the rocks, we find that the field switches erratically with an average interval of about a million years. The reversals are somehow caused by (possibly chaotic) oscillations in the liquid core.

We can actually *see* the drift taking place. Observation of celestial sources made with very long baseline interferometers (Section 9.6.3) yields the distances between the individual radio telescopes to within a wavelength of the radiation

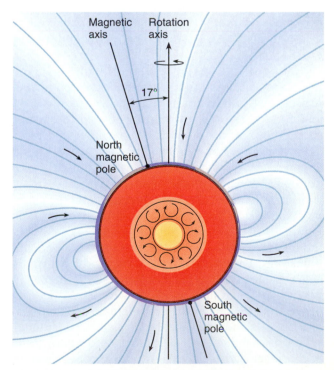

Figure 10.13 *The Earth's dipole magnetic field is derived from terrestrial rotation and core convection, but is tilted by about 17°. As a result, the north magnetic pole is in northern Canada.*

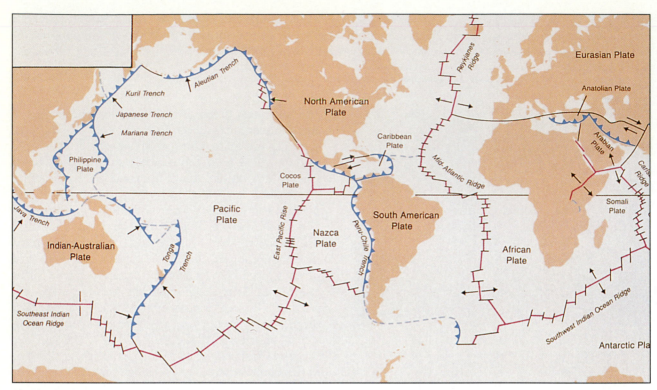

Figure 10.14 *A map of the world shows that the outlines of the Americas fit against those of Europe and Africa, indicating that the land masses were once united. The continents float on 16 crustal plates that are in constant motion (arrows) relative to one another. Compare with Figure 10.3.*

observed. As a result, we can measure the separations between points in America and Europe to about a centimeter. The distance is observed to increase at a rate of about two centimeters per year.

Many of the major geologic features of the Earth are caused by this plate motion or **plate tectonics,** as can be seen by comparing Figures 10.14 and 10.3. There are no spaces between the plates, so the boundaries at which they meet and push against each other can be the scene of awesome violence. If a plate holding a seabed runs into one holding a continent, the seafloor can dive downward in a process called **subduction,** and the continental edge is lifted into a mountain chain (see Figure 10.16). A shallow sea that had collected sediments may now be raised, exposing sandstone and limestone. The Pacific Ocean, attacked from many sides, is shrinking, and mountains ring the basin. Where the ocean floor subducts we find the deepest ocean trenches, like those off the east coast of Asia, and great island arcs like Japan and the Aleutians. When the subducted seabed heats and melts, some of the magma rises. Where it pops through the surface, it creates chains of volcanoes, and we see the rock cycle in action. The plates may also meet at a fault like the great San Andreas Fault in California,

Figure 10.15 *Lava flows and hot springs abound in the neighborhood of suboceanic ridges. This "black smoker" is a stream of hot, sulfurous water blowing out of the Earth 5 km below the ocean surface. (Parts of the device that took the picture are in the foreground.)*

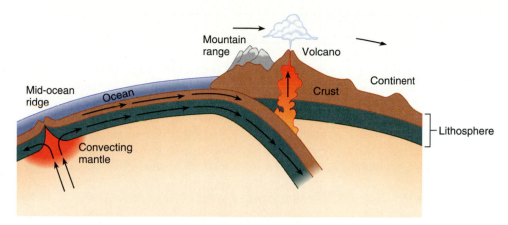

Figure 10.16 *Upwelling mantle material forces its way through the surface, creating a mountain range or ridge and pushing the seabed—with its continents—aside. The oceanic crust and lithosphere are subducted beneath a continental plate, the collision raising mountains on the continent. As the subducted material heats, molten matter drifts toward the surface to produce a volcano.*

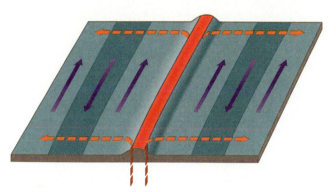

Figure 10.17 *The Earth's magnetic field periodically reverses its direction. As the new rocks push away from the suboceanic ridge (dashed arrows), they carry with them the direction of the field (solid arrows) at the time they solidified. We therefore observe strips in which the field directions alternate.*

where the Pacific and North American Plates slide past one another. When the plates suddenly slip, earthquakes rattle the ground. Mountains are also raised by direct collision between continental land masses. The towering Himalayas are buckled crust created when the subcontinent of India crashed ever so slowly into Asia.

Plate boundaries are not the only sites of rising mantle material. It also comes up in great **plumes** that may go down as far as the core-mantle boundary. As the magma emerges from the crust at the top of the plume, it may produce a lava plateau or create an isolated shield volcano like Hawaii. The islands of the Hawaiian chain get smaller toward the northwest. The plate on which the islands sit is moving, but the plume is fixed. The island on top of the plume grows until plate motion carries it away, after which the island gradually erodes as it is replaced by another. The result is a scar, or plume track, in the crust.

Since we know how the plates are moving, we can line up rock and other structures to tell how various land masses were once joined. About 200

million years ago, the continents had been assembled into one major land mass called *Pangaea,* from the Greek for "one Earth" (Figure 10.18). They separated under the force of the convecting mantle, the Americas breaking away from present Europe and Africa, the rupture creating the Atlantic Ocean. The appearance of the globe will continue to change: North America will one day probably collide with Asia.

10.5
The Atmosphere

Surrounding our planet is a layer of gas, the **atmosphere,** a product of volcanic action and interactions with the Sun, sea, and life. Although the atmosphere seems deep and thick to us, it is actually thin and fragile, the breathable part a mere 10 km high. Chemically unique among planetary atmospheres, it is the only one capable of supporting living things as we know them.

10.5.1 Composition and Structure

The air is a mixture of gases, and consists of 77% nitrogen, N_2 (counted according to the number of molecules per cubic meter), 21% oxygen, O_2, 1%

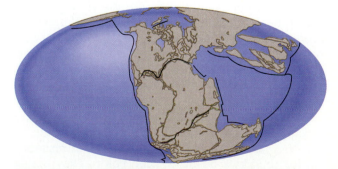

Figure 10.18 *The world of 250 million years ago featured a supercontinent called Pangaea. It cracked open in several places to produce the continents of today.*

BACKGROUND **10.1** Changes in the Earth's Atmosphere

Our planet's thin atmosphere is affected by a variety of natural and artificial processes. Two gases are of great importance to human life, carbon dioxide (CO_2) and ozone (O_3). CO_2 is largely responsible for keeping the Earth warm, but too much of it would lead to thermal disaster. The amount of CO_2 in the atmosphere depends on the balance between the rates of its creation and destruction, which involve absorption by green plants and the oceans, the rock cycle, and emission by animal life. There is great concern that this balance is being upset by the rapid burning of fossil fuels. These consist of plant life that long ago locked up vast amounts of carbon. Today's measurements show an increase in CO_2 from times past. If we do produce too much CO_2, the atmosphere will warm, perhaps to the point of endangering our own existence.

Yet there are many uncertainties. High CO_2 levels have been produced naturally between periods of glaciation and by extreme volcanic activity. We are still uncertain as to how the Earth's various cycles work and to what degree the oceans can clean the atmosphere. But uncertainty should not mean complacency. If the CO_2 content gets too high, it could be irreversible.

Ozone in the high stratosphere prevents ultraviolet rays from reaching the ground. This gas is destroyed by chlorine and fluorine. Many compounds involving these elements are or were used in industry and in the home (refrigerants like freon and propellant chemicals in aerosol cans are examples, though they have now generally been banned). As these molecules are released, they float upward and are broken down by sunlight. The free chlorine and fluorine atoms then attack the ozone.

There have been strong indications that the ozone layer is thinning. In particular, a hole in the layer that develops for natural reasons each summer over Antarctica has been getting larger every year. It is not entirely clear that manufactured chemicals are responsible, but it is certain that they have potential for doing serious damage. We depend on our thin atmosphere for life, and we must take care of it before it is too late to repair the damage.

---

atomic argon, and 1% water vapor. The last is highly variable as a result of rainfall and evaporation from the oceans. Mixed in is 0.033% carbon dioxide (CO_2) and a vast number of trace gases that include helium, neon, krypton, and methane, all of which are measured in only a few parts per million. Nitrogen, which provides most of the atmospheric pressure, is an inert gas (that is, it does not readily react chemically). Oxygen, however, is chemically active, allowing the chemical burning, or oxidation, of food to provide heat and energy for breathing animals. The water and the CO_2 provide an insulating blanket around the Earth through the **greenhouse effect** (Figure 10.19). Sunlight heats the ground with high-energy radiation. The Earth, however, behaves as a blackbody that radiates in the infrared, in which water and CO_2 have myriad opaque spectrum lines. The heat has difficulty escaping, and the terrestrial temperature is some 35 K higher than it would be without these gases.

The molecules of the atmosphere are responsible for the sky's beautiful blue color. They *scatter* sunlight as photons bounce off them in random directions. A short-wave (blue) photon is much more likely to be scattered than a long-wave (red)

photon, so blue photons enter our eyes from all over the sky, causing it to take on a cool blue hue that is independent of the color of the Sun. Correspondingly, the Sun appears a bit redder than it would without an atmosphere, an effect particularly noticeable at sunrise and sunset, when the path that sunlight must traverse is 38 times the thickness of air directly overhead (Figure 10.20a). The

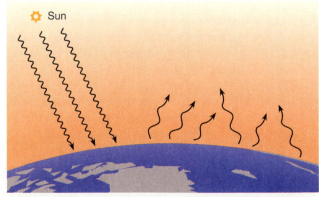

Figure 10.19   *The Earth is warmed in part by the greenhouse effect. Short-wave, high-energy radiation from the Sun heats the ground, but long-wave, infrared radiation is trapped by water and carbon dioxide in the atmosphere.*

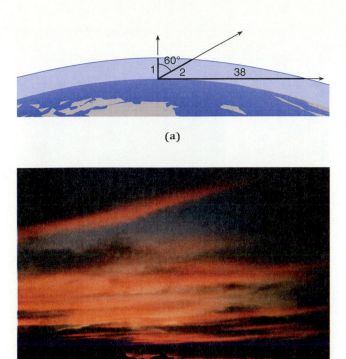

(a)

(b)

Figure 10.20 **(a)** *The thickness of the atmosphere through which one must look to see outside depends on the zenith angle. At a zenith angle of 60°, sunlight must penetrate twice as much air as it would were it at the zenith; at sunset, 38 times as much. (The lengths are distorted in the diagram.)* **(b)** *The reddened setting Sun can produce beautiful sunsets or sunrises.*

Sun can then be quite red, and sunlight reflected from clouds sometimes fills the sky with glorious colors (Figure 10.20b).

Air is surprisingly heavy. Over every square meter of Earth is a mass of 10,330 kg (14.7 pounds per square inch). The densest part lies at the surface, the pressure halving every 5 km in elevation and dropping by a factor of 10 for every 20 km. At the top of Mount Everest, the pressure is about a third that at sea level and there is marginally enough oxygen to sustain a human being (most mountaineers attempting ascent use bottled oxygen). At 100 km the pressure is down to 10^{-5} its value at sea level. There is no place where the atmosphere ends: it just keeps thinning until it blends with the gases of interplanetary space. Even at 200 km there is enough air to produce sufficient friction to degrade the orbit of a spacecraft.

The atmosphere has several distinct layers (Figure 10.21 and Table 10.1). That closest to the surface, the **troposphere,** is only 10 to 15 km thick, but contains the tallest mountains and 90% of the air. The troposphere is characterized by temperatures that drop steadily as elevation increases from an average of around 20°C at the surface to

about –60°C at the top. Most weather phenomena occur in the troposphere. On top of it lies the **stratosphere,** where temperature *rises* with altitude as a result of the formation of ozone (O_3). Ozone absorbs solar ultraviolet light, a process that both heats the layer and keeps these harmful rays from penetrating to the ground. Commercial jet aircraft commonly fly in the transition zone between the two layers, and military craft can range far into the stratosphere.

At an altitude of about 50 km, the temperature has climbed close to 0°C, and then in the **mesosphere,** it begins again to drop, approaching –100°C at an altitude of 85 km. The air is now extremely thin, only a ten-thousandth its pressure at the ground. Here, at the entrance to the **thermosphere,** molecules are broken into atoms, and the atoms become ionized under the action of high-energy solar photons. Since radiation is being absorbed, the temperature again climbs. At 200 km it reaches close to 1,000 K (700°C). From the black-body rules (see Section 8.3.1), you might expect the sky to be bright. But the thermosphere is much too thin to be a blackbody. Its temperature refers only to the speeds of the atoms, not to the amount of radiation the gas gives off, and is therefore called a **kinetic temperature.**

The ionized air within the thermosphere and upper mesosphere, commonly referred to as the *ionosphere,* is used for short-wave radio communication. Radio signals of certain frequencies will be refracted back to the ground around the curvature of the Earth and reflected up again. Successive refractions and surface reflections bounce the signals all the way around the world, allowing you to listen to news and music from Africa, Asia, South America, and Europe.

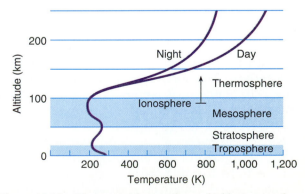

Figure 10.21 *The terrestrial atmosphere is highly structured. Four major zones—the troposphere, stratosphere, mesosphere, and thermosphere—lie atop one another, each defined by whether the temperature is increasing or decreasing with height. The thermosphere and the top of the mesosphere also constitute the ionosphere, where atoms and molecules are ionized by sunlight. Temperatures in the thermosphere change from night to day.*

BACKGROUND 10.2 Optics and the Earth's Atmosphere

The interplay of sunlight and the atmosphere can produce stunning visual effects. Chief among them is the *rainbow* (Figure 10.22). An afternoon thunderstorm flees to the east, and the Sun, low on the horizon, illuminates the still-falling drops. As a ray of sunlight enters each drop it is refracted and dispersed, reflected off the back, and then dispersed again as it exits. The combined spectra of all the drops produce a circular colored arc with red on the outside and violet or blue on the inside. It always has a radius of 42° centered on a point below the horizon opposite the Sun. A double reflection inside the drop will give a fainter secondary bow, with a radius of 51°, in which the colors are reversed. Supernumerary bows seen just below the primary rainbow are caused by interference of light within the raindrops.

Then look for a day with high, wispy cirrus clouds. If the weather is cold, they may be made of ice crystals in the form of hexagonal prisms. When sunlight is refracted through them they can produce wonderful displays of colored rings around the Sun, the principal one having a radius of 22° (Figure 10.23). This solar *halo* will commonly be brightened into *mock suns* or *sundogs* that lie on a line through the Sun parallel to the horizon.

At night, moonlight playing through the same light clouds produces a colored ring around the Moon, with white or blue touching the edge of the lunar disk and red around the outside. This *corona* is caused by diffraction of light waves as they pass by the water drops or ice

Figure 10.22 *A rainbow is produced by refraction and dispersion inside millions of falling raindrops. In the inset, white light falls from the Sun onto one of them and is dispersed, reflected, and dispersed again into the viewer's eye. The combined effect of all the drops makes the dispersed sunlight seem to come from a circle that lies 42° around the point opposite the Sun. The outer bow is produced by double reflection in each drop and is 51° in radius. A bright rainbow will also have several pink supernumerary bows just inside it.*

10.5.2 Weather and Climate

There are few topics of daily interest greater than the **weather,** the short-term changes in the atmosphere. The long-range patterns constitute a location's **climate.** The driving force for both is solar heating. Sunlight is concentrated most strongly at the equator. Convection causes the warmed equatorial air to rise and to expand toward the north and south, where it cools, falls to the ground, and returns. The resulting closed circulation pattern is called a **Hadley cell** (Figure 10.24a).

The atmosphere is also subject to the Coriolis effect. The returning air in the Hadley cell is closer to the Earth's axis than the rising air near the equa-

tor, and is consequently moving more slowly. The returning air then appears to be deflected westward, resulting in the easterly tropical trade winds. In the temperate zones of the Earth (between the tropics and the arctic regions), Hadley circulation is ineffective, and the Coriolis effect causes surface winds to blow generally (though not always) in the other direction, away from the equator and from the west. Toward the poles, the directions again reverse.

Because of the rapid change in temperature toward the poles within the temperate zones, atmospheric circulation breaks down into large-scale, turbulent cells or eddies that produce periodic pressure variations. In the northern hemisphere,

BACKGROUND **10.2** Continued

crystals in the clouds. Sometimes multiple fringes can be seen. A bright Moon can also produce halos and the lunar analogue to sundogs, *moondogs.*

Numerous other atmospheric effects await the aware and patient viewer. When the Sun is setting or rising, it is not round. Remember that, in general, *the Sun should not be viewed without adequate protection, as it is so bright that it can permanently damage your eyes.* However, if it is very hazy and the Sun is right on the horizon and *comfortable* to look at, it is momentarily safe. The atmosphere, though thin, is a refracting medium. As a result, everything in the sky is refracted slightly upward, the degree of refraction increasing as we look closer to the horizon. The effect is generally subtle but, at the horizon, amounts to half a degree. Light from the lower edge of the solar disk strikes the atmosphere at an angle greater than that from the upper edge and therefore is refracted upward more strongly. The result is that the Sun looks squashed. If there are air layers in which the temperature suddenly changes, they will produce anomalous refractive effects and the top of the rising or setting Sun may appear rippled.

If the atmosphere refracts light, it must also disperse it. A view of a bright star near the horizon through a telescope shows its image stretched out in a short line perpendicular to the horizon, with blue or green on top and red at the bottom. This effect produces a major problem for an astronomer who may be trying to observe the spectrum with a spectrograph.

Sunlight is also dispersed. When the Sun is setting, it actually consists of overlapping colored images. The air is so thick near the horizon that the blue and violet light is scattered out; green is the shortest wavelength that gets through. Just as the upper edge of the Sun disappears, the last thing to be seen is an intense green rim that, if conditions are right, results in a sudden *green flash.* The event is commonly seen in Pacific Ocean sunsets.

Figure 10.23   *Ice crystals in high clouds produce a stunning array of colored halos and arcs around the Sun. The bright, common 22° halo carries sundogs at the same altitude as the Sun. On top of the 22° halo is a brightened portion called the upper tangential arc. Many more circles and arcs are potentially observable.*

the Coriolis effect makes the air circulate counterclockwise around low-pressure cells, forming systems called **cyclones** (Figure 10.24b). There may be half a dozen or more of these drifting eastward around the globe at any given time. They, and their associated pressure changes, are responsible for the weather fronts commonly observed passing across the United States. Air circulating around a high-pressure cell creates an **anticyclone,** which rotates clockwise. In the southern hemisphere, the rotation directions are reversed.

Sunlight on the oceans evaporates water. The vapor rises with the convecting air and is blown across the Earth by winds. When it cools, the vapor condenses into clouds and, sometimes, rain. Rain that falls on land runs back into the ocean as streams and rivers, carrying the debris of erosion with it, steadily increasing the salinity of the sea, and completing the *water cycle.* By this process of natural distillation, the water cycle yields the fresh water living things need.

Weather is an example of a chaotic system (Section 7.7). As a result, forecasting for specific locations can be difficult and over long periods may turn out to be impossible. Climate is better understood, since it involves long-range circulation patterns that tend strongly to repeat one another and that respond to the changes in solar heating brought

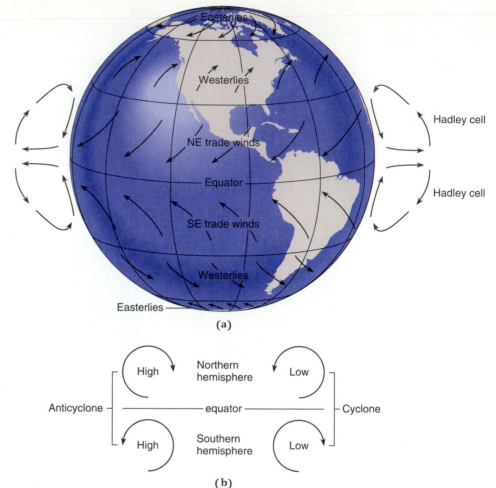

Figure 10.24 **(a)** *Wind patterns in the atmosphere are determined by circulation cells and the Coriolis effect. The equatorial Hadley cell drives air north. On its return south, it is deflected to the west (that is, it blows from the east). In the temperate zones, the wind blows in the opposite direction.* **(b)** *Cyclones are winds that blow around cells of low pressure. In the northern hemisphere, they rotate counterclockwise; in the southern, clockwise. Anticyclones blow around high-pressure cells, and their winds are oppositely directed.*

about by the tilt of the Earth's axis: the seasons. However, even climate has its complicating factors. For example, an 11-year solar activity cycle causes changes in the amounts of ultraviolet and X-ray radiation falling on the Earth. Moreover, over tens of thousands of years, slow perturbations in the Earth's orbit combine to alter the average terrestrial temperature. These may be the cause of the *glaciers,* vast sheets of polar ice that push halfway to the equator to create periodic ice ages. How the long-term variations in the brightness of the Sun add to these slow climatic fluctuations is still a mystery.

10.5.3 *The Origin of the Atmosphere*

The earliest atmosphere of the Earth was very different from what we breathe today. Formed after our planet had solidified from its molten state, it likely consisted of water, carbon dioxide, and nitrogen. After the surface and the air had sufficiently cooled, the water would have precipitated into oceans, leaving CO_2 and N_2 behind. There the story might have ended, except for two extraordinary

features of the Earth: liquid water and the development of life.

Single-celled life forms go back an extraordinarily long time, nearly 3.6 billion years, arising only a billion years after the planet was created. Some, at least, survived by photosynthesis, a process in which carbon dioxide is converted to oxygen under the action of sunlight, allowing one gas to be exchanged for the other. The evolution of life into multicelled organisms hastened the process of oxygenating the atmosphere. The death of living creatures places carbon into soil, then into rocks. When the rocks erode, some of the carbon flows into the sea and is used by small marine animals to make shells. Eventually they die and fall to the seafloor, where under pressure they become carbonate sedimentary rocks such as limestone. The seabeds are subducted under continents, and the carbon disappears into the Earth's mantle. Some of the carbon is liberated directly from continental rocks as CO_2, while some finds its way back to the surface through volcanoes. The result of this cycle, in which much of the Earth's carbon is tied up in rock, is a drastic decrease in atmospheric CO_2. If all this absorbed carbon were released, our planet would have a dominantly CO_2 atmosphere with a pressure of about 80 bars.

Oxygen is highly reactive, and initially much of it was taken out of the atmosphere by the oxidation of metals in rocks. By about 2.5 billion years ago, however, the rocks could hold no more; free oxygen began to be a significant presence, allowing for the development of life forms that could use it for heat and energy.

Our planet is at just the right distance from the Sun to have liquid water. Otherwise, as we will see clearly from examination of other planets, these processes—and our own evolution—would not have occurred.

10.6
The Magnetosphere

The Earth's magnetic field, created in its liquid core, extends well into interplanetary space. The simple tilted dipole of Figure 10.13, however, is severely distorted by the action of the Sun. The Sun loses a tiny part of its mass each year through the **solar wind.** The outflowing matter is in the form of a **plasma,** a gas that consists of ions (in this case mostly protons) and electrons. At the Earth, the wind has a density of about 10 to 100 particles per cubic centimeter and is moving at some 400 km/s. The Sun's magnetic field, also a rough dipole, is carried away from the Sun by the charged particles of the wind. Solar rotation causes the field to wrap into a spiral, and at the distance of the Earth, the magnetic field lines are almost perpendicular to the line connecting the two bodies (Figure 10.25).

The Earth interacts with the solar wind something like a boat plowing upstream. The Earth's field is squashed inward, and the wind flows around the planet like a wave spreading out from the bow. The surface where the solar magnetic field and the wind first meet the Earth's field is actually a curved **shock wave.** The particles cannot get out of the way fast enough so they pile up, producing a sudden increase in pressure. This *bow shock* occurs about 15 Earth radii out. The wind and the magnetic field then spread out to encompass the planet and to push the terrestrial field into a huge *magnetotail* that stretches 1,000 Earth radii downstream, in the direction away from the Sun. The entire structure is called the **magnetosphere.**

A fraction of the particles find their way inside the Earth's magnetic field and become trapped in two doughnut-shaped zones called the *Van Allen radiation belts.* The inner one is about 1.5 Earth radii from the terrestrial center, the outer one about 4. These two belts were among the initial discoveries made by the first artificial satellites sent aloft by the United States in 1957. An astronaut orbiting close to the Earth, within about 1,000 km of the surface, will be inside the inner belt and protected from the energetic particles in the solar wind. The environment outside the belts is much more dangerous.

The magnetosphere is responsible for one of nature's loveliest sights, the **aurora:** *borealis* in the northern hemisphere, *australis* in the southern hemisphere (also called the northern and southern

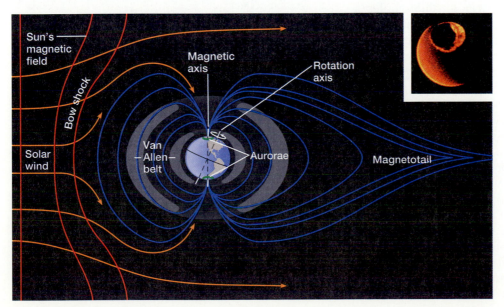

Figure 10.25 *The simple dipole of the terrestrial magnetic field (Figure 10.13) is compressed on the sunward side by the solar wind. The shock wave where the solar wind hits the Earth's field is actually about 15 Earth radii upstream. Opposite the Sun, the magnetic field is stretched into a long tail that extends 1,000 Earth radii away. The solar magnetic field, coupled to the moving wind, is bent around the Earth. Some electrons and protons from the wind become trapped in the two Van Allen belts. The magnetic and electrical interactions form huge current rings around the magnetic poles, causing the upper atmosphere to glow (inset).*

Figure 10.26 *A spectacular display of the aurora borealis, generated by huge electrical currents that ionize the upper atmosphere, hovers over the landscape.*

lights). A display (Figure 10.26) may begin quietly, with no more than a large patch of diffuse red or green light. From there it can grow to develop streamers that look like giant searchlight beams with pulses of radiation moving along them, or may take the form of huge draperies hanging in the sky. An hour or so later it all fades away, leaving only the perpetual stars.

The flow of the solar wind across the Earth's magnetic field injects particles into the Earth's upper atmosphere and sets up gigantic electrical current rings about 20° across. This electrical energy ionizes the upper atmosphere and makes it glow as free electrons are recaptured by ions (inset, Figure 10.25). The rings' energy output of a trillion joules/s is greater than that used by the entire United States.

The aurora is most commonly seen beneath the current rings, which in the northern hemisphere means in Alaska and northern Canada. The solar wind varies considerably in intensity, and the magnetic interactions and the flowing currents are highly unstable. Solar activity, which can produce great magnetic explosions on the Sun, can intensify the solar wind and produce disturbances within it that expand the current rings and extend the aurora toward the equator, allowing much of the population in lower latitudes to see it. This magnetic activity can cause compass needles to swing awry and, at its most intense, can induce currents in power lines that produce major power blackouts. The aurora powerfully illustrates the extended environment of the Earth and its relation with the brilliant Sun, which dominates not only the Earth but also the other planets, to which we now turn.

KEY CONCEPTS

Atmosphere: The gases that surround the Earth; layered into the **troposphere, stratosphere, mesosphere,** and **thermosphere** according to how the temperature changes with height.

Aurora: Lights in the upper atmosphere caused by the interaction between the terrestrial and solar magnetic fields and the solar wind; borealis in the north, australis in the south.

Basalts: Fine-grained metal-bearing silicates (as opposed to coarse-grained granites).

Climate: Long-term conditions and changes in Earth's atmosphere.

Continental drift: The motion of continents across the mantle.

Continents: Raised portions of the Earth made of light rock.

Convection: The up-and-down circulation of a heated fluid.

Core: The metallic (nickel-iron) core of the Earth.

Crust: The thin, light, top layer of the Earth.

Cyclone: Motion of air around a low-pressure zone, counterclockwise in the northern hemisphere, clockwise in the southern; an **anticyclone** rotates oppositely around a high-pressure zone.

Differentiation: The separation of a planet's interior into layers of different composition.

Dipole: A field with two poles.

Earthquake waves: Compressional and transverse waves sent through the Earth by the shocks of an earthquake.

Earthquakes: Vibrations caused by the slippage of the Earth's crust along faults in the Earth's crust.

Fault: A separation in the Earth's (or a planet's) crust.

Greenhouse effect: The process by which carbon dioxide and water vapor in the Earth's atmosphere trap radiated heat.

Hadley cell: An atmospheric circulation cell that brings warm air from the equator to higher latitudes and cool air back.

Kinetic temperature: Temperature defined by atomic velocities.

Lithosphere: A moving rock layer that consists of the crust and the top part of the mantle; 10 km thick under the ocean, 60 km or more under the land.

Magma: Liquid rock, or **lava.**

Magnetosphere: The structure around the Earth filled with its magnetic field and particles trapped from the solar wind.

Mantle: The thick layer of hot rock that surrounds the Earth's core.

Meteorites: Rocky or metallic debris from space that hits the ground.

Oceans: Basins filled with water.

Plasma: A gas that consists of ions and electrons.

Plates/Plate tectonics: The divisions of the Earth's crust and the process by which plates are moved over the mantle.

Plumes: Rising columns of hot mantle material that break through the crust and produce shield volcanoes and volcanic floods.

Pressure: The outward force of compressed matter per unit area.

Radioactive dating: Dating of rocks and other objects by establishing the ratio of the amount of the daughter product of radioactive decay to that of the parent element.

Shock wave: A wave of sudden increase in pressure.

Silicate rocks: Rocks (including granites and basalts) made of compounds that contain silicon, oxygen, and a variety of metals.

Solar wind: A thin, ionized gas blowing from the Sun.

Subduction: One crustal plate diving beneath another.

Weather: Short-term variations in the Earth's atmosphere.

Van Allen belts: Doughnut-shaped zones around the Earth filled with high-energy particles.

Volcanoes: Vents in the Earth's crust through which molten rock (lava), ash, and gas escape.

EXERCISES

Comparisons

1. What are the differences between granite and basalt?
2. Make a table to compare the properties of the Earth's mantle with those of the outer part of the core.
3. What are the principal differences between a cinder-cone volcano and a shield volcano?
4. Compare the origins of igneous, sedimentary, and metamorphic rocks.
5. How do continents differ from ocean basins in density and thickness?
6. Compare the behaviors of S-type and P-type earthquake waves.
7. Compare the worldwide distributions of mountain ranges, earthquakes, and volcanoes.
8. How did the creation of the Himalaya Mountains differ from that of the Andes?
9. Compare the way in which temperature changes with height in the troposphere, stratosphere, mesosphere, and thermosphere.
10. How do cyclones differ from anticyclones?

Numerical Problems

11. What is the age of the rock in Figure 10.1b?
12. If you could shrink the Earth to half its present diameter but maintain the mass, what would be its average density?
13. How high would you have to fly in a balloon to reach a point where the pressure of the atmosphere is one-eighth of its surface value?
14. California and China are separated by about 5,000 km. At the present rate of continental drift, how long will it take before the two collide?

Thought and Discussion

15. How can the "Big Island" of Hawaii be the tallest mountain on Earth when Mount Everest is the highest point above sea level?
16. Why can no mountain be taller than Hawaii from base to top?
17. Name some cities that would *not* receive shear waves if there were an earthquake in Brazil.
18. What layer of the Earth produces the lava that comes from volcanoes?
19. On what basis do we surmise that the core of the Earth is made of iron?
20. Why is the center of the Earth so hot?
21. Why is the Earth differentiated?
22. What cycles are involved in our active Earth? How are they related to one another?
23. Why is the mantle of the Earth in a flowable plastic state?
24. What is the structure of the lithosphere, and why is the layer significant?
25. What evidence suggests that part of the Earth's core is liquid?
26. What would be the consequences if the Earth's liquid core suddenly turned solid?
27. Describe three kinds of plate boundaries, the joints at which crustal plates meet.
28. What is the evidence for continental drift?
29. What process causes continental drift?
30. How does carbon dioxide keep heat trapped near the surface of the Earth?
31. What is ozone, and why is it important?
32. What is the Earth's magnetotail, and how is it formed?
33. How do we know that the magnetic field of the Earth occasionally switches directions?
34. Where are the Van Allen radiation belts? With what are they filled?
35. Why is the aurora borealis less common over New York City than over Fairbanks, Alaska?

Research Problems

36. The theory of plate tectonics met with considerable resistance during the twentieth century. When did this theory originate and when and why was it finally accepted?
37. Investigate and outline the development of the discovery of atmospheric ozone depletion and the progress that has been made in halting it.

Activities

38. Take a compass outdoors at night and find north. How does the direction differ from that found from Polaris? Knowing that the north magnetic pole is on Prince of Wales Island in Canada, estimate from a globe the difference you would expect. Compare the two numbers. Where on Earth would the difference be greatest?

39. Over the course of the school term, keep a log of the atmospheric effects you observe, including weather fronts, atmospheric optical effects, and aurorae. Comment on their origins.

40. Over the course of the school term, keep track of earthquake reports that you see in the newspapers. Plot the earthquake locations on a world map and comment on the relation between these locations and other geographic features. Do the same for volcanoes.

Scientific Writing

41. A brilliant aurora appears tonight. You are on the staff of your school newspaper and must write a 200-word article (one typed page) about its nature and origin in terms your readers can understand. You cannot use technical vocabulary without defining terms.

42. A nonscientist claims that the Earth is unchanging. Refute this claim in an article for the scientifically literate.

11

The Moon

The nature and significance of our nearest neighbor

The crew of Apollo 8 looks back at the Earth over the cratered surface of the Moon.

We return to the Moon, no longer as Earthbound sky watchers (Chapter 5) but as visitors who will examine the physical nature of another world. The Moon is the only body in the Solar System other than the Earth and (properly filtered) Sun on which we can see features with the unaided eye, shadings of light and dark that have long inspired deep curiosity. Our closest celestial companion has a profound effect on the Earth, providing nightly illumination, producing the tides, and gradually lengthening our day. It now offers scientific illumination as well, its surface yielding insight on the distant past. If we know the Moon, we can better know the Earth.

11.1
The Moon and the Earth

Most planetary satellites are small, inconsequential relative to their more massive parents. There are two exceptions: the lone satellites of Pluto and the Earth. The Earth-Moon pair could in fact reasonably be described as a double planet, the Moon so large and close as to have notable gravitational consequences for its bigger companion.

11.1.1 Lunar Properties

The mean distance to the Moon was found long ago by traditional parallax measures to be 384,400 km, with typical apogee and perigee distances of 405,500 and 363,300 km respectively. Its orbit is distorted from an ellipse by the gravity of the Sun, making extreme apogee and perigee respectively somewhat larger and smaller than average. At its average distance, the Moon subtends an angle of 15.56 minutes of arc. Since the angle in seconds of

Figure 11.1   *A laser beam takes off for the Moon from the McDonald Observatory in Texas. It will return in 2.5 seconds after reflection from mirrors placed on the Moon by Apollo astronauts.*

arc is $206,265R/D$ (see MathHelp 3.1), where R and D are respectively the radius and distance of the Moon, we find the physical radius to be 1,738 km (0.272 that of Earth). Astonishing accuracy in distance measurement can now be achieved as a result of the Apollo lunar landing program, in which the astronauts left banks of mirrors behind to reflect laser beams (Figure 11.1). We can time the round trip of a signal and, from the speed of light (accounting for its slower passage through the Earth's atmosphere), determine the lunar distance at any time to within a meter.

The mass of the Moon can be determined from its gravitational effect on the Earth. The motion of the Earth about the pair's center of mass (see Section 7.5) produces a parallax effect that causes the nearby planets to appear to shift back and forth by several seconds of arc. Location of the center of mass allows us to determine the ratio of the lunar and terrestrial masses. As early as 1757 the French mathematician Alexis Clairaut showed that the lunar mass was less than 2% that of the Earth. From the application of Kepler's generalized third law to spacecraft in orbit about the Moon (see Sections 7.4 and 7.5) we find a precise modern value of 7.3×10^{22} kg, or 0.0123 (1/81.3) that of the Earth (M_{Earth}).

The acceleration of gravity at the lunar surface depends on M/R^2, so it is 0.166 that of Earth. As a result, your weight on the Moon would be only about one-sixth that at home. If you did not have to wear a heavy space suit, you could jump six times as high, and for a good astronaut basketball game, the hoop would have to be set 60 feet up! The escape velocity, which is proportional to $\sqrt{(M/R)}$, is 0.213 that of Earth, or 2.4 km/s, making it relatively easy for astronauts to lift off. The mean lunar density is only 3.34 g/cm^3, just over half that of Earth. Lunar data are summarized in Table 11.1.

11.1.2 The Tides

There is a place in Nova Scotia where you could tie your rowboat to a pier and come back six hours later to find it dangling 15 m in the air. Everywhere along the seacoast, water periodically rises and falls with the **tides,** flowing first up and then down the slope of the beach between *high tide* and *low tide* (Figure 11.2). Captains of sailing ships would ordinarily float their craft out to sea on the ebbing tide, storms that occur near high tide can bring disaster to coastal communities, and many plants and animals require the steady wash of water within the tidal zones. The interval between two successive high tides is 12^h25^m, equal to that between succes-

TABLE 11.1
A Profile of the Moon

	Planetary Data	Atmosphere
mean distance	384,400 km[a]	none
mean radius	1,738 km	
mass	7.3×10^{22} kg	
mean density.	3.34 g/cm^3	
uncompressed density	3.3 g/cm^3	
gravity	0.17 g$_{Earth}$	
escape velocity	2.4 km/s	
magnetic field	fossil only	
rotation period[b]	27.3 days	
axial inclination	5.7°	

Layers	% Mass	Radii (km)
core	4[c]	400[c]
mantle	84	400–1,670
crust	12	1,670–1,740

Planetary Features

Basins: Large number of giant impact basins from a few hundred km to over 3,000 km wide; those on farside multiringed, those on nearside filled with maria that obscure multiring effect.

Craters: Range from hundreds of km across to pits in rocks; vast number overlap in highlands; far fewer in maria; there are some young craters with large ejecta blankets and rays.

Highlands: 3.8 or more billion years old; heavily cratered by the late heavy bombardment.

Maria: Volcanic plains between 3 and 3.8 billion years old produced by lava that flooded impact basins after the end of the late heavy bombardment and that continued to about a billion years ago; relatively few craters; most on nearside.

Mountains: Basin walls and ejecta blankets; no tectonic mountains.

Plates and plate motion: Single continuous plate; no motion.

Surface temperature: 383 K maximum daytime, 103 K minimum nighttime.

Volcanoes: Few ancient shield volcanoes and domes, none active.

Water: None.

[a]Varies from 406,000 km at mean apogee to 363,000 km at perigee.

[b]Synchronously locked with revolution period about Earth; rotation period relative to the Sun is 29.5 days.

[c]Upper limit; the core could be much smaller.

sive transits of the Moon across the meridian (first above the horizon, then below), demonstrating the lunar link.

Once Newton formulated the theory of gravity, he immediately saw that the tides are caused by a *differential* gravitational pull. Figure 11.3a shows that the lunar gravity is strongest on the side of the Earth closest to the Moon, less strong at the center, and weakest on the side away from the Moon. Subtract the arrow at the center (the average pull) from the other arrows. The result, shown in Figure 11.3b, is a stretching effect in the lunar direction. The oceans flow toward the line connecting the Earth to the Moon, producing two opposing *tidal bulges,* with shallower water in the perpendicular direction. Any coastal point on the rotating Earth will then pass twice through high and low water between two successive meridian transits of the Moon above the horizon.

The ocean is rotating with the Earth, and the water takes time to adjust to the changing lunar pull. Therefore, the tidal bulges will be shifted for-

(a)

(b)

Figure 11.2   *The high tide at Cutler, Maine* **(a)**, *is several meters above the low tide* **(b)** *that takes place six hours later.*

ward in the direction of rotation (Figure 11.3c). A person on the beach will typically see the Moon pass across the meridian a few hours before high tide. Although the size of the tidal bulge in the open ocean is only about a meter, it can be greatly exaggerated at the shore. The degree of exaggeration and the time of high or low tide depends critically upon the topography of the coastline. The tide is not confined to the oceans. The Earth's solid body is also deformed in a tide that points toward the Moon with a bulge about 30 cm high.

The Sun produces tides with about half the effect of the Moon. Lunar and solar tides are independent. When the Sun and Moon are on a line, at new and full lunar phase, the two tides combine to produce the highest and lowest tides of the month, the *spring tides.* When the Moon is in its quarter phases, the solar tide partially fills in the lunar tide to produce *neap tides,* which have a lower range between high and low water. The strength of a tide depends on the inverse cube of the distance of the attracting body. At perigee, the Moon is 11% closer to the Earth than it is at apogee, and the tide will be $1.11^3 = 1.37$ times higher. The tides will therefore generally be highest (and lowest) when perigee coincides with new or full phase; they will be even more extreme if these events coincide with perihelion near January 2. Extreme low tides often reveal sights like old wrecks that usually lie beneath the sea.

The gravity of the Moon pulls back on the tidal bulge and applies a subtle but steady brake to the rotating Earth, causing our day to lengthen by 0.0015 seconds per century; at one time our day must have been much shorter than it is now. Superimposed on the braking effect are other forces caused by the tides in the Earth's solid body, atmospheric tides, and interactions between the Earth's fluid core and mantle. These can either slow the Earth or temporarily speed it up. As the length of the day changes, we must alter our clocks to keep pace. For many scientific purposes, however, it is necessary to have day and a second (1/86,400 of a day) of constant length. The conflict is resolved by a variation on Universal Time (see Section 3.4.2) called **Coordinated Universal Time** (UTC), in which the length of the second does not vary. As UTC and ordinary UT get out of synchronization, the world's timekeepers add or subtract a *leap second* to UTC once or twice a year to bring them back together. UTC is the world standard, the kind of time broadcast on radio station WWV.

The tidal effect of the more massive Earth on the Moon is much greater than the effect of the Moon on the Earth. Solid tides—the Moon has no oceans—have dissipated enough rotational energy to cause the Moon to keep one side pointing toward us, prohibiting our view of most of the farside. (We can actually see 59% of the lunar surface from Earth, largely because the variable speed of the Moon in its eccentric orbit makes it appear to wobble back and forth.) The Moon is therefore tidally locked onto the Earth, and is in **synchronous rotation,** with equal rotation and revolution periods. Such locking is common in close orbiting bodies. The Moon does rotate on its axis relative to the *Sun,* however, turning once in a synodic month. As a result, someone on the Moon would see the Earth hang steadily in the sky, but the Sun would slowly rise, move along a daily path, and set. To an approximation, the Earth-Moon system is isolated and its total angular momentum is roughly constant. A prime example of the conservation of angular momentum (see Section 7.4), the slowing Earth transfers angular momentum to the Moon (through the gravitational pull of the tidal bulge), causing the Moon to move farther away from us by about three centimeters a year.

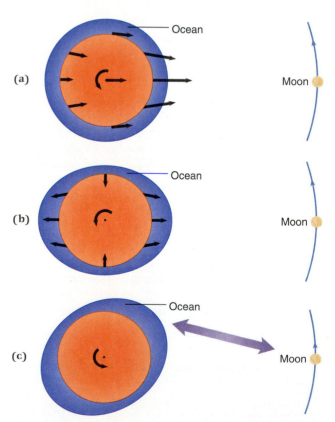

Figure 11.3 *The depth of the ocean is greatly exaggerated.* **(a)** *The arrows show the relative strength of the lunar gravitational pull on Earth's oceans at different distances from the Moon.* **(b)** *The force at the Earth's center is subtracted from all the arrows. What remains is the differential force, which causes water to flow toward the line connecting the Earth and the Moon.* **(c)** *As a result of the Earth's rotation and the time it takes the water to flow, the tidal bulge leads the Moon.*

11.2
Surface Features

We had our first look at the real nature of the Moon in 1609 when Galileo turned his primitive telescope to it and was startled to see craters and mountains. Astronomers have been examining it ever since with progressively more powerful instruments, with flybys and orbiters, and finally—in one of the grandest of human achievements—with direct visits.

11.2.1 *The Global View*

The most obvious property of the Moon may be the clarity of its features (Figure 11.4). The Moon has neither water nor an atmosphere to produce clouds or haze. Its gravity is too low to hold any form of air. The Moon turns only once relative to the Sun in a synodic month of 29.5 days, so the interval between lunar sunrise and sunset is a bit over two weeks. After baking in sunlight for several days the surface rocks can reach a temperature of 110°C (383 K); at night, they cool to –170°C (103 K) in the absence of a heat-retaining atmosphere.

Even with the naked eye you see that the lunar surface is divided into light and dark areas, the dusky features making the face of the "man in the Moon" and other whimsical characters. The bright areas, with albedos (percentage reflectivities: see Section 5.2) of about 15% are called the **lunar highlands.** They are covered with pits, or **craters,** of all sizes that completely cover the ground. The dark markings have long been known as **maria,** Latin for "seas" (the plural of *mare*). They carry old, fanciful names such as Mare Serenitatis ("Sea of Serenity") having nothing to do with their real natures. The maria are made of smooth dark rock with an albedo of about 8%, are set into depressions or basins in the lunar crust, and contain relatively few craters. Several are roughly circular in outline and are ringed by lunar mountain ranges.

11.2.2 *Exploration*

From Earth we can see only a little over half the Moon, and our turbulent atmosphere limits the resolution to about a kilometer—a very poor view indeed. Study of the Moon changed forever in September 1959 when the Soviet Union flew its *Luna 3* spacecraft around the then-mysterious farside. Five years later, the United States succeeded in its Ranger rocket program. *Ranger 7* had a television camera in its nose cone and was designed for a "hard"—that is, a crash—landing. Just before impact, it sent back images of craters and boulders only a few meters in size. By then, the Soviet Union

and the United States were in a race to conquer the Moon, to land on and explore it.

Hard landings provide limited information. Better machines were needed. The enormously successful Lunar Orbiters launched by the United States in 1966 and 1967 photographed nearly the entire lunar surface with a resolution of only a few tens of meters; the spectacular pictures (developed automatically) were televised back to Earth. Meanwhile, the Surveyors made soft intact landings under rocket power to analyze the surface material, returning the results by radio.

These craft paved the way for the carefully planned Apollo missions (Figure 11.5), six of which made successful soft landings, the first on July 20, 1969. While in orbit around the Moon, the visitors also took large numbers of extraordinary high-resolution photographs. The astronauts who set down on the lunar surface walked the landing sites, collected rocks and soil for return to Earth, and left scientific equipment—seismographs and reflecting mirrors for example—behind. In the later missions, they even had electric-powered cars that could roam over several kilometers, considerably increasing the scope of the exploration. In all, the missions returned 380 kg of lunar material to Earth, a treasury of information that will be studied well into the next century. In the meantime, Soviet scientists continued to survey the surface with automated imaging craft and landed three Luna robot craft that together returned another precious 300 grams from other regions of the Moon. These brave explorations, combined with previous and continuing Earth-based studies, have finally revealed the nature of the lunar surface and something of its remarkable history.

11.2.3 *Craters*

A typical crater is a depression in the ground circled by an elevated wall (Figure 11.6). Craters are best seen when they are near the **terminator,** the sunrise or sunset line on a body orbiting the Sun, where elevated areas cast long shadows and stand out in strong relief.

Early astronomers thought the lunar craters were volcanoes like those on Earth. Most volcanoes, however, are mountains with a crater or caldera on top. Instead of being *on* mountains, however, the larger lunar craters commonly have mountain peaks *in* them. The craters are more easily explained as impact phenomena (Figure 11.7). They are holes in the lunar ground dug by collisions with meteorites (see Section 10.1).

Any rock dropped from a great height must land at a speed at least equal to the escape velocity.

(a)

Figure 11.4 **(a)** *Photographs of the first (right) and last (left) quarters of the Moon show the rugged highlands and the smooth maria. Names are given on the accompanying map.* **(b)** *The lunar maps show the maria in blue and the highlands in orange; craters are indicated in violet, major ejecta blankets in green, rays in yellow, and mountain ranges in red. Linear structures are cross-hatched in black. The landing sites of the Apollo craft are shown by red dots and mission numbers. North is at the top. Invert the map when looking through a telescope.*

1. Langrenus
2. Geminus
3. Macrobius
4. Stevinus
5. Rheita
6. Metius
7. Fabricius
8. Franklin
9. Cepheus
10. Endymion
11. Atlas
12. Hercules
13. Fracastorius
14. Piccolomini
15. Posidonius
16. Theophilus
17. Cyrillus
18. Catharina
19. Julius Caesar
20. Eudoxus
21. Aristoteles
22. Vlacq
23. Zagut
24. Pitiscus
25. Nearchus
26. Barocius
27. Maurolycus
28. Gemma Fricius
29. Abulfeda
30. Godin
31. Agrippa
32. Cuvier
33. Heraclitus
34. Licetus
35. Faraday
36. Stöfler
37. Fernelius
38. Aliacensis
39. Werner
40. Apianus
41. Bohenenberger
42. Parrot

43. Albategnius
44. Hipparchus
45. Horrocks
46. Rhaeticus
47. Manilius
48. Autolycus
49. Aristillus
50. Cassini
51. Archytas
52. Bond
53. Walter
54. Regiomantanus
55. Purbach
56. Lexell
57. Arzachel
58. Alpatragius
59. Alphonsus
60. Ptolemaus
61. Herschel
62. Pallas
63. Archimedes
64. Plato
65. Maginus
66. Clavius
67. Blancanus
68. Scheiner
69. Tycho
70. Longomontanus
71. Wilhelm
72. Bulliadus
73. Guericke
74. Parry
75. Eratosthenes
76. Timocharis
77. Copernicus
78. Pytheas
79. Lambert
80. Euler
81. Gassendi
82. Kepler
83. Aristarchus
84. Mairan

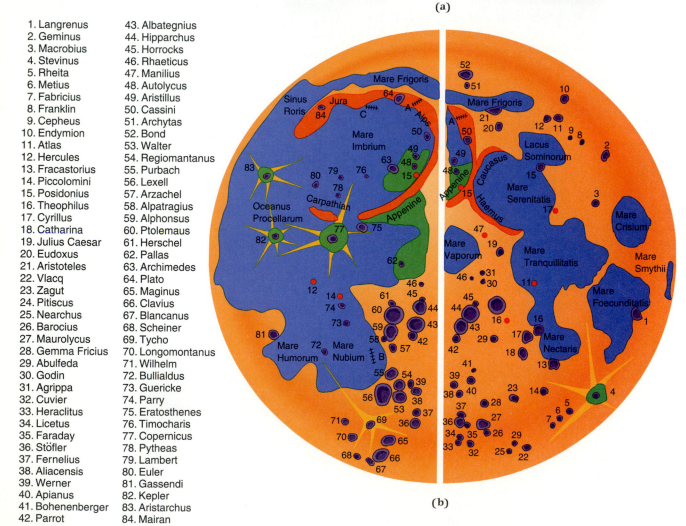

(b)

A. Alpine Valley
B. Straight Wall
C. Straight Range

BACKGROUND **11.1** *Apollo* to the Moon

It is easy these days to be casual about the grand adventure that sent astronauts to the Moon 25 years ago. The Moon ships were amazing vessels that evolved from the earlier Earth-orbiting Mercury and Gemini spacecraft and that in turn developed into our present-day space shuttle. The Saturn-V rockets (Figure 11.5) that launched the *Apollo*s were the most powerful ever built. They stood nearly 100 meters tall, the height of a 30-story building. Almost the entire length and weight consisted of three rocket stages and fuel, the first stage of which was constructed of five enormous engines, each 2 m across. The astronauts rode aboard a small capsule at the top. People hundreds of kilometers away who witnessed the only night launch claimed that it looked like the Sun coming over the horizon.

The first launches were tests in Earth orbit. *Apollo 8* made the initial trip to the Moon in 1968, launched in an elliptical orbit around the Earth with the Moon near the craft's apogee. Then the engineers broke with the Earth, placing *Apollo 10* into orbit around the Moon. Finally, as a billion people listened in, *Apollo 11* eased its way onto Mare Tranquillitatis on July 20, 1969, and Neil Armstrong took the first step onto an alien planet.

Three astronauts participated in each mission. The Moon ship had two parts, command and lunar modules that were linked together shortly after launch. Once in lunar orbit, two astronauts crawled from the command into the lunar module, leaving the third behind in orbit to conduct remote sensing and photography (see Figure 11.5). The pair in the lunar module separated from the mother ship, fired rockets to slow it down, and settled to the surface.

The lunar module was a self-contained vehicle with its own rocket and launch pad; when the exploring astronauts finished, they lifted off the Moon's surface, leaving the launch pad behind. Accelerating to lunar orbit, they linked back to the command module and crawled inside. The lunar module was jettisoned, in some cases steered to a crash landing on the Moon to create an artificial moonquake, allowing the exploration of the interior by means of lunar seismographs left behind. The command vehicle then returned to Earth.

Six Apollo missions landed on the Moon. Although they were primarily oriented toward politics and engineering, the scientific return proved incomparable. But when the government and the public lost interest, the program was terminated. There are no serious plans to go back in the near future.

Figure 11.5   (**a**) *Ever so slowly, the* Saturn V *rocket carrying* Apollo 8 *lifts off for the Moon as it stands on a tail of fire over 200 meters long.* (**b**) *The astronaut aboard the* Apollo 10 *command module takes a picture of the orbiting lunar lander.* (**c**) *The two aboard the lunar module return the favor.*

(a)

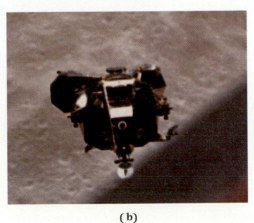

(b)

(c)

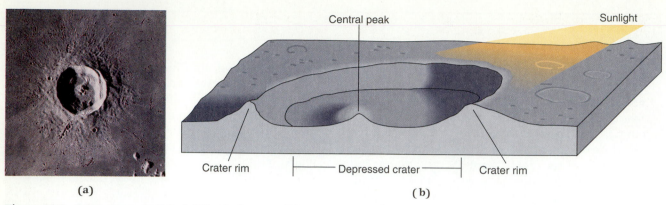

Figure 11.6 (**a**) *Lunar crater Euler is 28 km in diameter.* (**b**) *A cross section shows a crater to be a depression surrounded by a circular elevated wall. Craters this large commonly have central mountain peaks. Shadows cast by slanting sunlight make the crater visible.*

Moreover, interplanetary debris moves on orbits about the Sun with velocities of some 30 km/s, so collision speeds can be much higher. A 10-meter-diameter body has a mass of 10^7 kg, and at 10 km/s carries an astounding energy of 10^{15} joules, equivalent to the explosive power of a nuclear bomb. When it hits the ground, it stops. Some of the energy is transformed into heat that vaporizes the impacting body. Most of the energy, however, goes into violent shock waves that penetrate and compress the lunar rock. Subsequent decompression expels a vast amount of fractured and melted debris, excavating a hole with a diameter some ten times that of the impacting body (depending on its speed) and a depth about one-fifth the crater's width. Rock is fractured to a much greater depth.

Some of the ejected material piles up to produce a circular crater wall with a height roughly equal to the crater's depth. The rest surrounds the new crater in a large **ejecta blanket** (Figures 11.7 and 11.8). Rubble and resolidified melted rock also fall back inside, partially filling the crater. If the force of the blow is sufficiently large to create a crater more than about 20 km in diameter, the force with which the crater floor rebounds from the shock is great enough to produce a central mountain peak, which then drags the crater walls down in a series of terraces (alternatively, slumping walls might push up the central peak). Big pieces blown out of a large crater will dig new pits of their own and produce numerous **secondary craters** that surround the primary. Streamers of the rocky debris can arc across the lunar surface, creating strings of secondary craters and white **rays** (Figures 11.8 and 11.9); Tycho's rays extend for a thousand kilometers and cover nearly an entire lunar hemisphere. Some rocks, blasted off the Moon, have even made their way to Earth (Figure 11.10).

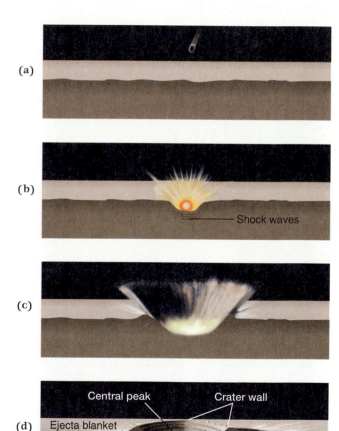

Figure 11.7 (**a**) *A large meteoric body speeds toward the lunar surface.* (**b**) *It strikes and begins to vaporize as it sends powerful shock waves into the rock.* (**c**) *Decompression of the shock ejects vast amounts of broken rock to create the crater walls and the ejecta blanket.* (**d**) *The crater floor is now covered with rubble and solidified melted rock. If the impact is forceful enough, rebound of the crater floor pushes up a central peak and creates terraces in the walls.*

(a)

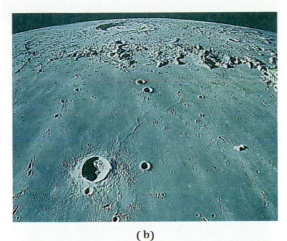

(b)

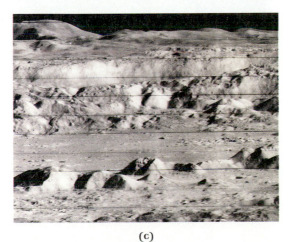

(c)

Figure 11.9   *Ray systems are most visible near full Moon. The brightest of them come from the crater Tycho (85 km in diameter), seen toward the south pole. Copernicus, Aristarchus, and Kepler also have extensive ray systems.*

Figure 11.10   *This 3-cm-wide meteorite, found on Earth, closely matches the samples brought back by Apollo astronauts. It was probably blasted out of the Moon by an immense meteoric impact.*

Figure 11.8   *Three views of Copernicus (a 90-km-wide crater on the southern edge of Mare Imbrium) display all the properties of a massive impact.* **(a)** *An overhead photograph shows the elevated walls and the extensive ejecta blanket. Outside that are numerous secondary craters, and inside is a central peak.* **(b)** *We look over Mare Imbrium and the Carpathian Mountains at Copernicus on the horizon. Elongated secondary craters string away from the impact, as do the rays. The crater Pytheas, in the foreground, also displays a prominent ejecta blanket and some of its own secondary craters.* **(c)** *Looking directly into Copernicus we view the central peak, the interior filled with rubble and impact melts, and the terraced walls.*

11.3
The Lunar Highlands

Tens of thousands of craters can be seen from Earth in the lunar highlands. Figure 11.11 displays an immense range of sizes and qualities. Clavius (near the bottom and identified on the map of Figure 11.4) is 120 km across. The number of craters goes up by a factor of 100 for every decrease of a factor of 10 in diameter, reflecting the distribution in the diameters of the impacting bodies. Hundreds of craters are seen in Figure 11.11 at the limit of resolution. Close-ups by spacecraft and observations by astronauts show them ever smaller (Figure 11.12), ranging down to tiny pits in rocks.

The entire surface of the highlands is cratered; the devastation of the original surface is complete. The Moon can be probed by the detection of moonquakes, using the small network of seismographs the Apollo crews left behind. We find the average thickness of the crust (the lighter rocks) to be about 70 km. Geologists estimate that the force of the immense bombardment that created the jumble of craters has crushed the top half.

Craters pile on craters, showing that they must have been created over a period of time. The relative ages can be determined by **stratigraphy,** a dating technique that uses the way in which various features lie on top of one another. Some craters are

Figure 11.12   *This view of the lunar highlands near Sinus Medii (in the center of the lunar disk) was taken by an orbiting Apollo 10 astronaut. It shows thousands of tiny craters superimposed on a heavily battered terrain. The oldest craters have been nearly destroyed by the younger ones; the smallest ones are only 50 m in diameter.*

Figure 11.13   *A lunar breccia is a mixture of various kinds of rocks that have been fused together under the high pressure and temperature produced by a meteoric impact. The process makes a rock that looks like broken concrete.*

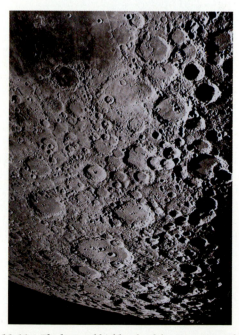

Figure 11.11   *The battered highlands of the Moon toward the south pole show an extraordinarily rugged terrain, with craters ranging to only a kilometer across. By contrast, Mare Nubium, up and to the left, is quite smooth.*

so badly beaten up (see Figures 11.11 and 11.12) that they are only barely recognizable: these must be the oldest. Even earlier craters must have been entirely obliterated. Craters with few superimposed meteoric strikes must be the youngest.

Stratigraphy provides relative ages; absolute ages are determined by radioactive dating of rocks. The majority of the highland rocks are **breccias** (Figure 11.13), which were fused together from smaller smashed particles under the force of meteoric impact. Other *impact melts* are stones that solidified directly from earlier rock rendered molten in the collisions. These are all old, with ages between 3.8 and 4 billion years. The heavy cratering must have taken place over a relatively short interval of less than 700 million years, from the time of the origin of the Solar System to roughly 3.8 billion years ago. This period is therefore known as the time of the *heavy bombardment;* the tail end of it, which produced the craters we see, is therefore called the **late heavy bombardment.**

The highland rocks consist largely of calcium- and aluminum-rich feldspar to the exclusion of other varieties and heavier silicates. Intact crystals of varying compositions can be assigned crude ages of 4.3 to 4.5 billion years, and the large quantities of this feldspar (some 20 km thick) show that the Moon (like the Earth) was at one time covered with a liquid magma ocean at least 400 km deep, one that may even have involved the entire interior. However, the variety of rocks found suggests different sources of magma and a complex history not fully understood.

11.4
The Maria

By comparison with the densely cratered highlands, the maria look peaceful. Within them, the crater density is so low that undamaged surface can be seen (Figures 11.14 and 11.15). The circular outlines of most maria (see also Figures 11.4 and 11.9) indicate that they are contained by enormous impact craters called **impact basins.** The lunar mountain ranges are actually the basin walls, towering as much as 9 km (roughly the height of Everest) above the basin floors. The Imbrium basin's enormous ejecta blanket, the Fra Mauro Formation, surrounds the basin, and is particularly well exposed east of Copernicus (see Figure 11.4).

The color and reflectivity of the dark mare rock suggests it is solidified lava that has at least partially filled the basins. Flow marks are clearly visible in Figure 11.15. Many craters in and around the maria also look as if they have been flooded with the same material (Figure 11.16). Evidence for volcanism is also provided by long, sinuous valleys called **rilles** that look like river beds (Figure 11.17) but are actually channels in which molten lava once flowed. Similar features are seen on Earth. Around the edges of the maria are deep, rectangular, ditchlike cracks called **grabens,** and inside the maria are interior folds or **wrinkle ridges** (Figures 11.15 and 11.16) that look like a tablecloth pushed in from the sides.

Rock samples (Figure 11.18) brought back by the Apollo astronauts show that the mare material consists of fine-grained volcanic basalt similar to,

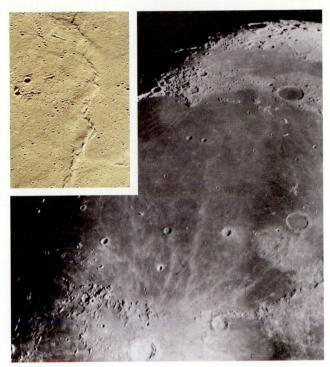

Figure 11.14   *The Imbrium basin presents a fairly smooth surface to the Earthbound camera. Copernicus is at the bottom. A close-up of Sinus Medii (inset) taken by Apollo 10 shows that even under close scrutiny the maria look relatively unbattered (compare with Figure 11.12).*

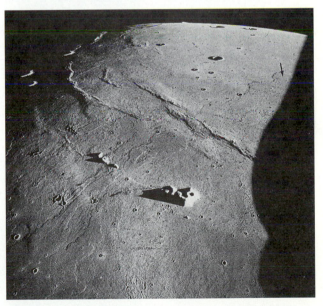

Figure 11.15   *A close-up of the central Imbrium basin shows numerous small craters pocking an enormous lava field, flow marks (arrows), and large wrinkle ridges.*

Figure 11.16 *Mare Humorum is a basin flooded with lava, some of which has flooded nearby craters as well. Graben and wrinkle ridges parallel to the maria walls are the result of slumping after solidification.*

Figure 11.17 **(a)** *Hadley Rille, photographed by the Apollo 15 crew, snakes across the floor of a 240-km-wide section of eastern Mare Imbrium. The arrow shows the landing site.* **(b)** *The character of this channel is more clearly seen in the closeup.* **(c)** *Astronaut James Irwin stands at the rille's edge.*

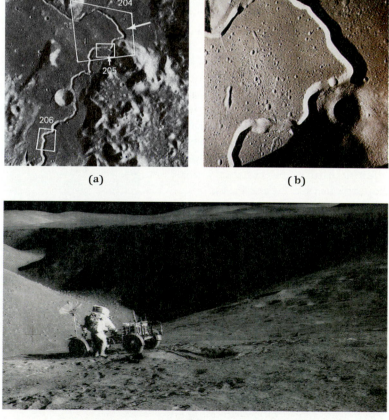

(a)

(b)

(c)

yet tantalizingly different from, that found on the Earth. The frozen lavas contain calcium-rich feldspar, like the rocks of the highlands, but are also abundant in heavier iron-rich silicates. **Volatile** elements—which solidify only at low temperature and melt (or turn to gas) easily—are greatly depleted; there is no water at all. On the other hand, the rocks are relatively rich in **refractory** elements (those that melt at high temperature) like titanium.

There are no Earth-style tectonic processes on the Moon. There are no lunar continents or continental drift, no subduction and remelting, and no plate boundary mountain ranges. Neither are there active terrestrial-style volcanoes. We see only some structures that look like ancient shield volcanoes and a few domes that apparently have been created by upward-pushing lava. The maria were not produced by magma from volcanoes, but are volcanic floods reminiscent of the kinds that made the thick plateaus of the American northwest and southern India (see Section 10.2.2).

The basins that contain the maria are huge—the Imbrium basin is at least 1,100 km in diameter. The impacting bodies must have been tens or even a hundred or more kilometers across. The force was apparently sufficient to fracture the lunar crust in a basin's interior. Hot magma from the lunar mantle subsequently worked its way up through the cracks (or forced its way up by itself) and spread across part or even all of the basin floor in a series of flows. The solidified lava loading the basins slumped toward the center, pushing up the interior wrinkle ridges (see Figure 11.16). It pulled the rock at the edges of the maria apart, creating the grabens and weakening the crust at the basin rims. More magma could then escape through the weakened crust to run into the basin through the sinuous rilles.

Stratigraphy allows the relative dating of the basins and mare flows, and radioactive dating of the mare rocks allows absolute limits on true ages. Among the oldest is the huge Procellarum Basin, which formed before the heavy bombardment ended. Its original circular outline—an astounding 3,400 km in diameter—can be seen as Oceanus Procellarum and Mare Frigoris (see Figure 10.4). The basin actually extends to the east as far as Mare Tranquillitatis and to the south beyond Mare Nubium. There are vague suggestions of even older basins. The other major nearside basins (those whose maria are named in Figure 11.5) were created in a series of impacts that took place between

Figure 11.18   *A basaltic rock brought back to Earth from a lunar mare is made of calcium-rich feldspar and iron-rich silicates.*

about 3.8 and 3.9 billion years ago, beginning with Nectaris and ending with Imbrium and Orientale (just barely seen at the lunar limb). These wiped out the highlands that once existed there and destroyed and hid much of the Procellarum basin.

The craters we see within the maria were created after the basins were formed and filled with lava. Some of them have flooded floors, and a few are but ghostly rings (see Figure 11.11); these had to be formed *after* the creation of the basins but *before* the lava filled them in. Therefore there must have been a significant interval between the time of the great impacts and the eruption of the mare magma from below. The bulk of the mare lavas erupted between 3.2 and 3.8 billion years ago, but the flows continued at an ever-decreasing rate until roughly a billion years ago. Since even the oldest mare surfaces are lightly cratered, the heavy bombardment must have ended fairly quickly. Over the last 3.1 billion years, the cratering rate seems to have proceeded at an almost constant pace; the crater density on the plains of the maria is more or less consistent with the current rate at which meteorites are now known to fall on the Earth. The rays from the great craters Copernicus, Aristarchus, Tycho, and others lie over almost everything else, showing them to be relatively young. Tycho may be only 100 million years old.

The absolutely dated lunar chronology (Figure 11.19) provides a key for studying other planets and their satellites. If the surface of another body is saturated with craters, we know it must have suffered through the late heavy bombardment and must be old. If the surface is not saturated, we can estimate its age from the crater density and the cratering rate calculated from the age and crater density of the

BACKGROUND 11.2   A Day on the Moon

It seems likely that one day we will return to the Moon to set up permanent bases. Lacking air, our satellite is an impressive observatory site, and the farside is shielded from interfering radio radiation from Earth. What would a lunar day be like?

Pick a spot—for example, Mare Vaporium, near the center of the lunar disk as seen from Earth. The Sun would first appear when the Moon is in our first quarter. It would be night even minutes before sunrise, since there is no air to scatter sunlight and create twilight. The landscape would still be relatively bright, however, because of the reflected light from the third-quarter Earth. Your first hint of impending sunrise would be a glow from the solar corona, seen on Earth only during a solar eclipse. Then the tip of a mountain or crater wall to the west would suddenly light up as it caught the first direct solar rays. Since the lunar day is 29.5 Earth days long, the Sun would creep up at 1/29 the pace it does on Earth; at the Earth's equator

it takes the Sun two minutes to vault the horizon, so on the Moon it would take about an hour.

Very slowly the temperature of the surface rocks would climb, an increase that would be of little consequence to any residents, since most would be living underground to shield them from the effects of both temperature and the solar wind. It would actually be rather dangerous to spend too much time on the surface because of this high-speed particle radiation. Only a meter or so down, however, the Moon's temperature remains quite stable and no particle radiation can penetrate.

A week later, it is finally noon and the "new" Earth has disappeared. Another week, and the Sun begins to set. It turns dark as soon as the solar disk disappears, except for the now-brightening Earth and a few peaks that catch the last of the solar rays. Now you settle down for two weeks of long, cold lunar night. This pattern would be repeated day after lunar day, as it has for billions of years.

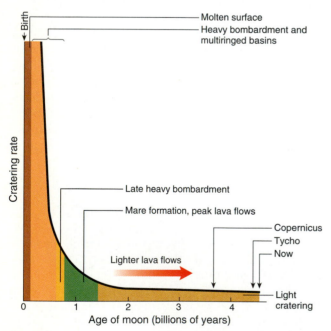

Figure 11.19 *A rough chronology for the Moon shows that heavy cratering and basin formation took place very early in lunar history. Since the Moon was 1.4 billion years old—3.1 billion years ago—the surface has been relatively quiet, with light cratering and diminishing lava flows.*

lunar maria (assuming that the rates apply elsewhere). The Moon long served as a calendar for humanity; now it becomes one for the Solar System.

In spite of the apparent extent of the flows, they cover only 17% of the lunar surface and are typically only a few hundred meters thick. The total amount of magma extruded during the 700 million year peak is really quite small, amounting to no more than that produced by Earth's Vesuvius, although the eruptions that made the rilles could be extremely vigorous for brief periods.

A major surprise of the space program was that almost all the maria are on the nearside, the side of the Moon facing the Earth. The once-mysterious farside (Figure 11.20) is covered with craters and several large **multiringed impact basins** like the one that contains Mare Orientale (Figure 11.21). These rings may have been caused upon impact by traveling shock waves that froze in place, but more likely were produced by the slumping of the lunar crust around a once-smaller basin. All the large basins on the Moon should initially have been multiringed. The difference between the two lunar sides is that dark mare lava filled the nearside basins, covering the inner rings; except for a few like Mare Orientale, the basins on the farside

Figure 11.20   *The Apollo 16 crew took this picture of the lunar farside, which shows a distinct lack of maria. Mare Crisium, prominent in nearside photos, appears at far left. The other two maria at the left are Mare Smythii and Mare Marginis, both barely visible from Earth.*

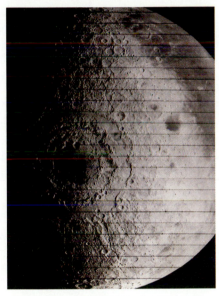

Figure 11.21   *Mare Orientale can just be glimpsed from Earth. From above, this spectacular feature is clearly a giant crater, a multiringed basin partially flooded with mare lava.*

remained empty. The lunar crust is thinner on the nearside of the Moon than on the farside, probably as a result of the huge Procellarum impact. Therefore, it was easier for the lava to reach the surface.

On the maria, smooth, gritty plains stretch into the distance (Figure 11.22). The surface, like that of the highlands, is covered with a pulverized soil, or **regolith,** that contains no organic or biological material. It consists of a fine, glassy grit made over billions of years as micrometeoric impacts shocked and ground up the original surface, the resulting heat turning the glassy crystals dark (Figure 11.23). The regolith is a few meters deep in the younger

maria, but may reach down tens of meters in the older, more beaten highlands.

Debris splashed from a new crater can dig through the regolith and expose lighter-colored material, forming the rays. But as the bright streaks are exposed to meteoric infall, they too turn dark, and the rays disappear; we therefore see rays only around young craters.

11.5
The Lunar Interior

Like the Earth, the Moon has a core, mantle, and crust (Figure 11.24), but the proportions are different. The low average lunar density of 3.3 g/cm^3 suggests that any iron core—if it exists at all—is

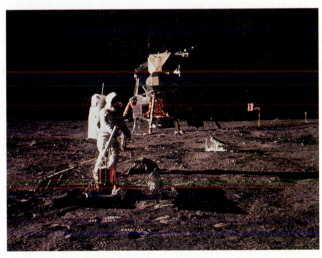

Figure 11.22   *Edwin Aldrin, Jr., of Apollo 11, deploys a scientific experiment package in Mare Tranquillitatis; note the bulky life-support system Aldrin had to carry. The lunar module sits on its launch pad in the background. The sky is black, and the "soil," or regolith, is a fine grit produced by micrometeoric bombardment.*

Figure 11.23   *This bootprint left by an Apollo 11 crew member shows the structure of the fine lunar soil. Unless this area is struck by a meteorite, the print will remain for a billion years or more—a testimony to the Moon's unchanging surface.*

very small. Unlike the Earth, the Moon has no dipole magnetic field, so the core must be solid throughout. However, the lunar rocks *are* magnetized. At the time of their cooling billions of years ago, the Moon may have possessed a magnetic field and a hot, fluid core. The sum of the evidence, however, makes the core smaller than 400 km in radius and less than 2 to 4% of the total lunar mass. Surrounding the core is the rocky mantle, which contains around 84% of the lunar mass. We know little about it. However, since the mantle is the obvious source of the mare lavas, it must have a relatively low volatile content and contain heavier minerals than the highland crust, those rich in iron and magnesium (see Section 11.4). On top is the beaten, lighter crust, some 70 km thick.

The differentiation was produced while the young Moon was covered with the liquid magma ocean (see Section 11.3). The lightest rock floated upward to create the lunar highlands, and the heavier rock sank into the mantle; a portion then rose later as the maria lavas. The preservation of the history of the early days of the Moon now begins to tell us something of how our own Earth developed, the record of which was long ago wiped away.

The crust then solidified and was crushed by the heavy bombardment of the first 700 million years of the lunar lifetime to perhaps half its depth. On top lie the craters of the late heavy bombardment, the lava-filled basins, and a veneer several meters thick of the pulverized regolith that has accumulated over the past 3 or 4 billion years. The lunar mass is so low that not even the decay of radioactive elements could keep the interior warm for long. As a result, the total magma output, while obvious even to the naked eye, is really quite small, and the Moon now presents us with no volcanic and little seismic activity.

11.6
Impacts and the Earth

The Earth ought to have been struck by meteoric bodies at least as frequently as the neighboring Moon: Where are the Earth's craters? Even though the oldest rocks on the Earth are 4.2 billion years old, the majority solidified after the late heavy bombardment and the creation of the great lunar basins. Most of the craters that formed later have been erased by intense tectonic activity and by ice, wind, and water erosion. Our early record is entirely gone.

The Moon's many rayed craters, however, are geologically relatively young, so the Earth's surface

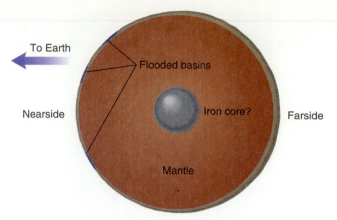

Figure 11.24   *A cross section of the Moon shows a small iron core, a thick, rocky mantle, and a lower-density crust that is pulverized to about half its depth. The nearside crust is thinner than that on the farside. The core may be considerably smaller than represented here.*

should still show some identifiable impact craters. It does. The most famous is Meteor Crater near Winslow Arizona (Figure 11.25a), which was produced by an iron meteorite that struck about 50,000 years ago. Other craters are harder to find as they have been severely eroded, filled in, or covered with vegetation. When viewed from above, however, their circular outlines give them away (Figure 11.25b). Even miners and oil-well drillers occasonally locate craters when they find fractured rock that could have been produced only by violent collisions. Such events are not just prehistoric; as recently as 1908, trees near Tunguska in Siberia were flattened over an area 75 km across by the impact of a small body from space.

These collisions, however, pale beside one that may have taken place 65 million years ago. At that time, the Earth was dominated by the dinosaurs, whose skeletons were preserved as fossils in layers of what is now rock. Their era is dated by stratigraphy combined with radioactivity. Then with geologic suddenness these huge creatures—and some 95% of all other species—disappeared and mammals took over. Arguments have raged among scientists for decades about what created that moment, known as the *KT boundary*, which divides the Cretaceous (the word starts with a *K* in German) geologic time period from the Tertiary.

A clue is provided by meteorites found on Earth, which are relatively rich in the element iridium. All over the planet, layers of clay deposited at the time of the KT boundary are also iridium-rich, suggesting that the Earth was struck by a massive projectile from space perhaps 5 km across. The resulting impact would have raised huge clouds of dust and, if it hit the ocean, steam. The resulting change in climate could have killed large numbers of life forms. The most likely candidate is a huge impact feature at the northern edge of the Yucatán peninsula in Mexico (Figure 11.26) that dates to the KT event. Although other means of extinction, such as massive volcanism, are still viable, such catastrophic impacts have to occur. We see them on the Moon, and they must have had devastating effects on the Earth, even if one was not responsible for the demise of the dinosaurs. Indeed, it is now widely believed that life could not develop on Earth until basin-forming impacts, like those that created the lunar basins, had ceased some 3.8 billion years ago. Our study of the lunar globe, therefore, leads us back to Earth to help us understand our home.

11.7
The Origin of the Moon

The last of the lunar mysteries involves the Moon's origin. Other large satellites of the Solar System revolve in the planes of their planets' equators. Ours, however, orbits close to the ecliptic plane, demonstrating that it is a body created by the Solar System at large. This simple evidence argues against a fission hypothesis, whereby the Earth simply divided in two as it was being formed. (A **hypothesis** is a theoretical idea presented for testing by experiment or observation.) If it had, the Moon would have been in the terrestrial equatorial plane. The same argument might be presented against a co-accretion hypothesis, which holds that

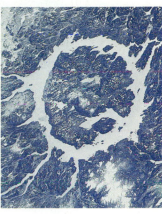

(a)   (b)

Figure 11.25 **(a)** *Meteor Crater in Arizona displays the characteristics associated with lunar craters, including an upraised rim, a deep depression over a kilometer across, and an ejecta blanket of meteoric debris.* **(b)** *Lake Manicouagan in Canada, 75 km across, bears a sharp resemblance to a meteor crater.*

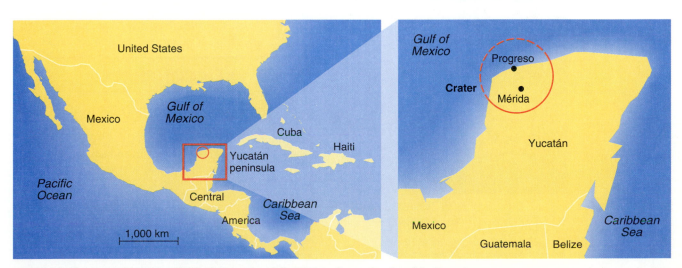

Figure 11.26 *A meteoric impact that may have caused the KT event lies at the edge of the Yucatán peninsula of Mexico.*

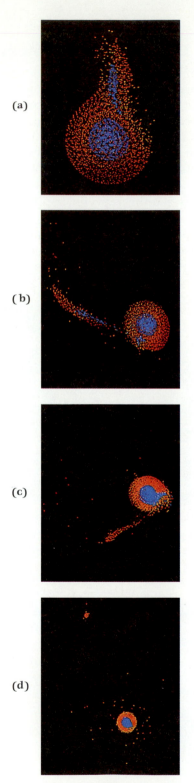

(a)

(b)

(c)

(d)

Figure 11.27 *A computer model shows how the Moon could have been formed by collision with the early Earth:* **(a)** *An independent, differentiated body hits the Earth.* **(b** *and* **c)** *The cores have nearly merged and portions of the mantles fly away into orbit.* **(d)** *This gasified material consolidates into the current Moon, which has little core.*

the Moon and Earth developed simultaneously from the same cloud of matter. In that case, moreover, we would expect the two bodies to have the same iron content. Although the lunar regolith can be rich in iron, the Moon as a whole is highly deficient in the metal because it has little if any iron core.

Other hypotheses accommodate the lunar orbit. One is that our satellite was a separate body orbiting the Sun and was captured from space by the Earth. That idea is not very plausible. A primitive, independent Moon would have approached the Earth on a hyperbolic orbit. An external force, perhaps produced by a third body, would be needed to reduce the relative speeds, allowing capture into an elliptical path. None can be identified.

The hypothesis on which most scientists now tend to agree involves a massive collision (Figure 11.27). At a time near the formation of the Solar System, almost 4.6 billion years ago, an independent body roughly the size of Mars is thought to have crashed into the early Earth. Both were already differentiated and had large iron cores at their centers. The colliding body's core merged with that of the Earth upon impact. Part of the Earth's mantle was ripped away and, with that of the colliding body, was vaporized and sent into terrestrial orbit. The Moon then accreted from this ring of gas and dust. The hypothesis explains the low lunar iron content, the Moon's deficiency of volatile elements (dissipated into space), and the orientation of its orbit. Computer modeling confirms that the event could have taken place, as does evidence for even more violent collisions in the Solar System that will be apparent as we explore the other planets.

KEY CONCEPTS

Breccia: A rock that consists of fused pieces of smaller rocks.

Coordinated Universal Time (UTC): A hybrid time formulated with a constant second and adjusted to keep up with the changing rotation period of the Earth.

Craters: Pits caused by impacts.

Ejecta/Ejecta blanket: The debris thrown out of a crater by the impact that made it/the blanket that this debris forms around the crater.

Grabens: Cracks caused by the pulling of surface rock.

Hypothesis: An idea tested by experiment or observation.

Impact basins: Large impact craters, sometimes filled with dark lava.

Late heavy bombardment: The last period of meteorite fall before 3.8 billion years ago; produced lunar highland cratering.

Lunar highlands: Original lunar crust crushed by heavy cratering during the late heavy bombardment.

Maria: Dark areas made of lava flows that fill many impact basins.

Multiringed impact basins: Impact basins with multiple concentric walls.

Rays: Bright lines that emanate from young craters; caused by secondary impacts that expose light-colored rock.

Refractory (elements or compounds): Those that melt or boil only under high temperature.

Regolith: Lunar (or planetary) soil produced by constant pulverization by meteorites.

Rilles: Channels caused by running lava.

Secondary craters: Craters caused by rocks ejected from impact sites.

Stratigraphy: A means of relative dating by observing how one feature lies on top of another.

Synchronous rotation: Identical rotation and revolution periods.

Terminator: The line on a celestial body that separates night from day.

Tide: A periodic flow of water caused by the differential gravity of the Moon and Sun; generalized to mean any distortion caused by differential gravity.

Volatile (elements or compounds): Those elements or compounds that melt or boil at low temperatures (including water).

Wrinkle ridges: Ridges caused by the compression of surface rocks.

EXERCISES

Comparisons

1. What is the difference between neap and spring tides?
2. How does the rotation of the Moon differ from that of the Earth?
3. Compare the ages and major features of the lunar highlands and the maria.
4. Compare the masses of the cores of the Moon and Earth.
5. How do grabens, wrinkle ridges, and sinuous rilles differ from one another?
6. How does the lunar regolith differ from terrestrial soil?
7. Compare the chemical composition of the lunar surface rocks with those of the Earth, and then compare the average compositions of the whole bodies.

Numerical Problems

8. Show that the surface gravity of the Moon and the escape velocity at the lunar surface are respectively 0.166 that of Earth and 2.4 km/s.
9. Show that the density of the Moon is 3.3 g/cm^3.
10. Show why the interval between high and low tide is 12^{h}25^m.
11. The Moon will someday have a synodic period of 35 days. What will be the interval between successive high tides on Earth?
12. Figures 11.12 and 11.17b respectively represent areas approximately 40 and 20 km across. Count the number of craters per unit area bigger than 0.5 km in each. What can you conclude?
13. How long is the shadow cast by a lunar mountain 5 km high when the Sun has an altitude of 45°? What is the length of the shadow in seconds of arc as viewed from Earth? What is the shadow length of the mountain when the Sun is overhead?
14. Craters A and B are respectively 2 and 0.5 km across. Roughly how deep are they relative to the surrounding plane? About how large were the meteorites that made them?

Thought and Discussion

15. Where are the lunar mountain ranges? How were they formed?
16. What lunar features are associated with ejecta blankets?
17. What are lunar rays made of? Why are most craters without rays?
18. Why do we believe that maria are associated with impact craters?
19. Why do we believe that maria are volcanic?
20. What data support the several hypotheses of lunar formation?
21. How can we determine relative and absolute ages of lunar features?
22. Why is the collision theory of lunar formation consistent with the small lunar core?
23. Why are there so few meteorite craters on Earth?
24. How could you tell whether a depression on Earth is a meteor crater?
25. Why are more multiringed craters visible on the farside of the Moon than on the nearside?

Research Problem

26. Examine a variety of astronomy texts that have appeared over the last century to see how the theory of the formation of lunar craters has evolved.

Activities

27. Draw a naked-eye sketch of the full Moon, showing the location of the maria. Compare your sketch with a lunar map. What was the smallest feature you could see?

28. If you have access to a telescope, select a mare and locate the smallest crater you can find. How big is it?

29. Select a photograph in this chapter and, with the aid of the lunar map in Figure 11.5, determine and list the relative ages of a few craters and other features you can identify.

Scientific Writing

30. You are hired by a telescope manufacturer to write part of an instruction manual. Convey to a beginner who has just purchased a telescope the excitement to be found in viewing the Moon, explaining some of the features to be seen and their significance.

31. NASA has decided to mount a major initiative to study the Moon employing a variety of spacecraft and by making landings. You are enthusiastic about the project, but Congress is not. Write a letter to your representative that explains the importance of a study of the Moon with special emphasis on the Moon's significance for learning about the Earth.

12

Hot Worlds

The contrasting natures of the two terrestrial planets Mercury and Venus

The artist has drawn Pioneer Venus Orbiter *flying above the clouds that cover Venus.*

We might logically assume that Venus and Mercury, both inside the Earth's orbit, would have some physical similarities. Instead, they are as different as the Earth and the Moon. In some ways, Mercury more resembles the Moon, while Venus makes a close couple with Earth. The Moon then teaches us not only about the Earth but about Mercury as well, and the contrast of Venus with Earth lets us know our own planet better. Thus we do not study the planets singly, but jointly, in a broader science of **comparative planetology** that shows all the bodies of the Solar System to be linked together.

12.1
The View from Earth

Although Mercury and Venus are relatively close to the Earth, they were long shrouded in mystery. Venus (Figure 12.1a) is covered with clouds and therefore has a high albedo, 72%. In contrast, Mercury's dark surface (albedo 5.5%) is accessible (Figure 12.1b), since there are no clouds at all. But Mercury is so close to the Sun that at best, in twilight, the Earth's atmosphere permits only a murky view of the tiny disk. Nevertheless, we can still derive significant information about these planets from traditional astronomical techniques.

The relative distance from Earth to any body in the Solar System in AU is easily found from its orbital location. The absolute distance in kilometers to an orbiting body then gives the number of kilometers per AU, that is, the distance of the Sun from the Earth, and the scale of the Solar System. Such absolute distances were first measured in the early 1700s by parallax, here the difference in position as viewed from two locations on Earth (see Figure 5.2). In this century, astronomers used parallaxes

of close-passing asteroids (a class of interplanetary debris concentrated, but not confined, to orbits between Mars and Jupiter). If an asteroid with a distance of 0.01 AU is found to be 1.5 million km away, the length of the AU is 1.5/0.01 or 150 million km. The semimajor axes of the orbits of Mercury and Venus, already known in AU, were then found to be 57.9 and 108 million km. From the angular diameters and distances of Mercury and Venus, we find their respective physical diameters to be about 4,900 and 12,000 km, or 0.38 and 0.95 that of Earth. Mercury is 40% larger than the Moon, and Venus is almost a twin of our planet. (These and other characteristics of Mercury and Venus are summarized in Tables 12.1 and 12.2.)

Neither Mercury nor Venus has any natural satellites from which to find masses by applying Kepler's generalized third law. Instead, respective masses of 0.055 and 0.81 that of Earth were originally found by measuring the gravitational perturbations that the two planets exert on each other and on the Earth. Extremely high precision has been attained from spacecraft orbiting Venus and passing close to Mercury. From the planets' masses and radii we find average densities of 5.4 g/cm^3 for Mercury and 5.3 g/cm^3 for Venus. Analogy with the Earth, which has a similar average density of 5.5 g/cm^3, suggests that the two inferior planets are both differentiated and have iron cores.

Like all planets, Venus and Mercury shine in the optical spectrum by reflected sunlight. In the radio spectrum and in the far infrared, however, they emit energy because they are blackbodies warmed by the Sun. Furthermore, in the case of Venus, this long-wave radiation comes from the actual surface and can penetrate the clouds. In principle, we measure the flux of radiation (F) (see Section 8.2.3) from the planet as it arrives at the Earth (Figure 12.2). A sphere around the planet with a radius equal to its distance from the Earth, D, has a surface area of $4\pi D^2$ square meters. Each square meter must pass flux F, so the total energy passing through the sphere per second is $4\pi D^2 F$. That much energy must leave the planet per second, and is therefore its luminosity, L.

We know that $L = 4\pi r^2 \sigma T^4$ (see Section 8.3.1), where T is the effective blackbody temperature and r is the planetary radius, so $T^4 = L/4\pi r^2\sigma$. We therefore can calculate the planet's temperature. We could also determine the peak of the blackbody curve and use the Wien law, or measure the flux at a single radio wavelength (F_λ), assume a blackbody, and use the more complex Planck equation.

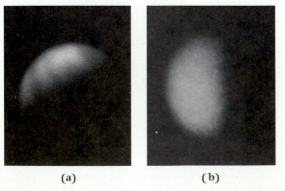

(a)   (b)

Figure 12.1   **(a)** *An ultraviolet photograph of Venus shows vague markings in an obscuring cloud deck.* **(b)** *Because of atmospheric turbulence, Mercury presents only a featureless disk from Earth.*

TABLE 12.1
A Profile of Mercury

Planetary Data

distance from Sun	0.38 AU
radius	2,439 km = 0.38 R_{Earth}
mass	3.3×10^{23} kg = 0.055 M_{Earth}
mean density	5.4 g/cm^3
uncompressed density	5.3 g/cm^3
gravity	0.38 g_{Earth}
escape velocity	4.3 km/s
surface temperature	700 K (days) to 88 K (nights)

Layers	% Mass	Radii (km)
iron core	60	0–1,750
mantle plus crust	40	1,750–2,440

Atmosphere

pressure 10^{-12} Earth

Composition by Number

sodium and potassium	97%
helium	3%
hydrogen and oxygen	(trace)

Other

magnetic field	0.01 Earth
rotation	58.65 Earth days
axial inclination	0°

Planetary Features

Cloud cover: None.

Craters: Numerous and much like the Moon's, except that craters are situated on intercrater plains; lower relief and smaller ejecta blankets caused by higher gravity; most of surface ancient, nearly 4 billion years old.

Maria: Several volcanically flooded multiringed basins.

Origin: Early collision may have blown away much of mantle.

Plates and plate motion: None.

Scarps: Younger than most craters; produced by shrinkage of planet.

Surface temperature: 700 K daytime maximum, 88 K night minimum, 348 K average; cold ice caps at rotation poles; hot poles near equator.

Water: Ice caps may be water, otherwise very little.

Volcanoes: None known.

TABLE 12.2
A Profile of Venus

Planetary Data

distance from Sun	0.72 AU
radius	6,051 km; 0.95 R_{Earth}
mass	4.87×10^{24} kg = 0.81 M_{Earth}
mean density	5.3 g/cm^3
gravity	0.90 g_{Earth}
escape velocity	10.4 km/s
surface temperature	740 K

Layers	% Mass	Radii (km)
iron core	30?[a]	0–3,000?
mantle plus crust	70?	3,000–6,050?

Atmosphere

pressure   98 Earth

Composition by Number

carbon dioxide	96%
nitrogen	4%
water, sulfur dioxide, argon, carbon monoxide	(traces)

Layers	Height (km)
troposphere	0–100 km
thermosphere[b]	>100 km

Other

magnetic field	none
rotation	243 earth days; retrograde
axial inclination	177°

Planetary Features

Craters: Lobate ejecta blankets; multiple craters caused by the breaking up of impacting body in atmosphere; dark areas crushed by atmospheric shock waves.

Clouds: Solid cloud cover about 50 km high; sulfuric acid.

Mountains: All major mountains in Ishtar Terra; probably produced by mantle downwelling.

Tectonics: Small compared to Earth; most features volcanic; some faults; jumbled tesserae; no plates or plate motion.

Volcanism: Most landforms volcanic uplifts; huge number of shield volcanoes; coronae; surface heavily repaved with volcanic flows, most occurring in a global upheaval half a billion years ago.

Water: None on surface; very little in atmosphere.

[a]The interior structure is uncertain; core probably solid.
[b]Disappears at night.

Since Venus is 40% closer to the Sun than is Earth, and by the inverse-square law receives twice as much solar radiation, it should be hotter than Earth. However, the high temperature, first measured in the late 1950s and now known to be 740 K (470°C), exceeded all expectations. A block of lead placed on Venusian ground would melt. Moreover, the daytime and nighttime temperatures were found to be the same, implying insulation by a thick atmosphere.

Mercury's temperature is almost as high. It has no insulating atmosphere, but is even closer to the Sun. The lack of atmosphere makes the daytime and nighttime temperatures quite different, up to 700 K (430°C) on the sunlit side and down to a chilly minimum of 88 K (–185°C) on the dark side. The surface rocks and regolith are excellent insulators, however, and only a meter below the surface the temperature is maintained at a more comfortable average of 348 K or 75°C.

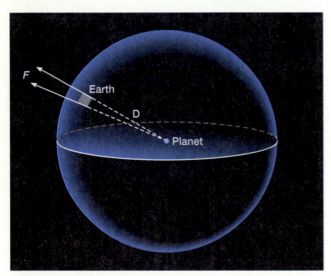

Figure 12.2 *The Earth is on the surface of a sphere of radius D centered on a planet. F is the flux of radiation from the planet, in joules/m²/s, at the Earth. The luminosity of the planet, L, is the flux times the sphere's surface area, or 4pD²F.*

12.2
Modern Observation

We have uncovered some of the secrets held by Mercury and Venus with two techniques, **radar** (*ra*dio *d*irection *a*nd *r*ange) and spaceflight. Radar, developed principally by the British in World War II to detect and warn of incoming enemy bombers, uses artificially produced signals sent to and reflected from the body being examined. An astronomical radar system is a radio telescope that can both receive and transmit. A radio signal at a specific wavelength is sent toward a target planet (Figure 12.3). A small portion of the signal bounces off the body and is returned. The time it takes the radio beam to travel to the target and back at the speed of light gives the distance. If the target is moving toward or away from the antenna, the wavelength of the returned signal will be Doppler-shifted, allowing the radial velocity to be found. (The same technique is used by police radar.) A body moving through space can therefore be tracked with high precision.

Radar provides precise planetary distances and is the best method available for calibrating the AU and for determining the distance of the Sun. It is also used to measure the rotation rates of planets. The planet in Figure 12.3 is rotating, its axis directed perpendicular to the page. The top half in the drawing is approaching the radar relative to the center and the lower half is receding. As a result, different parts of the planet produce slightly different Doppler shifts. The spread in returned wavelengths gives the rotation speed. It is likely that a real planet has a tilted axis. In that case, the observations will give the component of the rotation speed directed along the line of sight. Observations of the planet over its orbit, however, allow the determination of both the direction of the rotation axis and the true rotation speed. The speed and the diameter then give the rotation period.

Figure 12.3 *A radar telescope sends a radio beam (in violet) to bounce off a planet (dashed waves) that is rotating with its axis perpendicular to the page. The distance is found from the round-trip travel time of the signal, and the radial velocity of the planet can be found from the Doppler shift. We can also determine the rotation period from the spread in the wavelengths of the returned signal.*

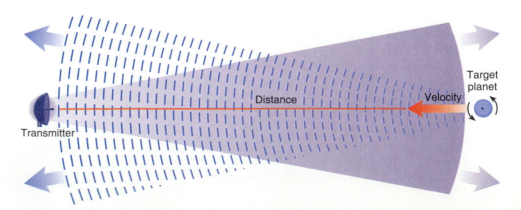

(a)

Figure 12.4 **(a)** *Magellan was launched from the shuttle bay in 1989; the blue Earth is seen below.* **(b)** *It has an altimeter that measures the altitude of the ground directly below it and a side-looking radar that maps the surface by selecting combinations of time delays and Doppler shifts.*

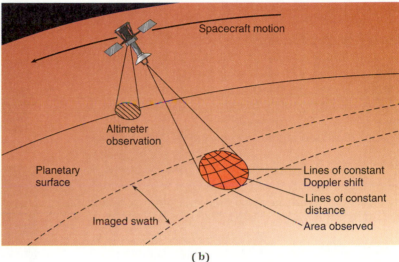

(b)

Different parts of the spherical planet in Figure 12.3 are at different distances from the transmitter, so there will be a spread in time over which a burst of radiation is returned. Each point on a rotating planet's surface will produce a unique combination of Doppler shift in the reflected signal and the time interval between transmission and reception. Planetary astronomers can therefore discriminate different locations on the surface and can make maps of radio reflectivity, allowing us to "see" below the Venusian clouds. The most powerful of all radars is the 300-m Arecibo telescope (see Figure 9.26), which has been used to map part of the Venusian surface with a resolution of about 100 km.

An Earthbound telescope, however, is no match for a direct visit by a spacecraft. The United States' *Mariner 10,* placed in solar orbit, flew past Mercury three times in 1974 and 1975 and imaged 45% of its surface. Venus has been much more popular. *Mariner 2* began exploration in 1962 and *Mariner 10* obtained optical images en route to Mercury. The Soviet Union began sending probes in 1965. Several of its *Venera* craft measured properties of the atmosphere, landed on the surface, and sent back images from the ground. In 1979, the United States' *Pioneer Venus Probe Carrier* also dropped sensors into the atmosphere, and in the 1980s the Soviet Union launched two *Vega* craft that placed floating helium balloons.

Four spacecraft have carried radars for high-resolution mapping. *Pioneer Venus Orbiter* was placed in orbit about the planet in 1979 and yielded maps with a resolution of some 10 km. Six years later, the Soviet Union responded with *Venera 15* and *Venera 16,* which gave even higher resolution. Finally, the United States sent *Magellan* (Figure 12.4) to observe the planet between 1990 and 1993. It resolved features as small as 100 m across and provided a spectacular view of an alien and forbidding surface.

12.3
Mercury

Even with modern techniques, we still do not have a good understanding of the planet closest to the Sun. Mercury has been visited briefly by only one spacecraft, and radar from Earth gives poor resolution. Nevertheless, what we do know provides a fascinating view of a strange body and reveals something about the way in which the planets were formed.

12.3.1 Rotation

With a semimajor axis of only 0.4 AU, the tide raised in Mercury by the Sun is stronger than the combined luni-solar tide at the Earth. Mercury was therefore once believed to keep one face always pointed to the Sun, just as the Moon does to Earth. Radar observations made in 1965 shattered that view when Mercury was observed to have a counterclockwise sidereal rotation period of 59 days, two-thirds of its 88-day sidereal year (Figure 12.5). Its axis is almost perfectly perpendicular to its orbital plane.

If the planet rotates once in two-thirds of its sidereal year, it spins 1.5 times relative to the stars in 1 sidereal year. In Figure 12.5, position *a* initially faces the Sun and position *b* faces away. In one year, *a* and *b* are reversed, and in two years, *a* again faces the Sun. The solar day on Mercury, the interval between successive passages of the Sun across the meridian, is therefore two sidereal years, or $88 \times 2 = 176$ Earth days.

Mercury's odd rotation is still caused by tidal locking. It is the result of the planet's high orbital eccentricity, which produces a large variation in the tidal force. The controlling factor is the high strength of the tide at perihelion, the point at which the planet moves the fastest. The rotation period is close to what the orbital period *would* be were the orbit circular with a radius equal to the perihelion distance.

12.3.2 The Atmosphere

Because of its low gravity and high temperature, Mercury was originally expected to have no atmosphere. However, spectrometers aboard *Mariner 10* found evidence for helium as well as for hydrogen and oxygen. The surface pressure is a mere 10^{-12}

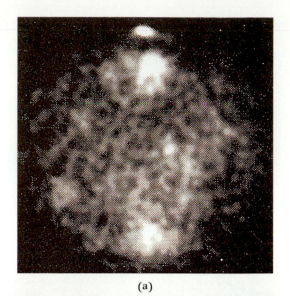

(a)

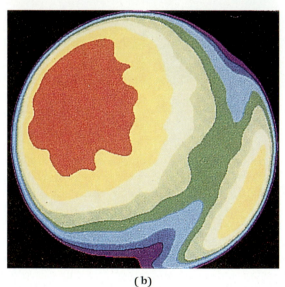

(b)

Figure 12.6   (**a**) *A radar map of Mercury shows a bright patch of ice near the north pole.* (**b**) *A temperature map of the surface of Mercury made with the Very Large Array in New Mexico shows a hot pole (red).*

bar, that is, 10^{-12} that of Earth's atmosphere. Some of this thin gas is likely captured from the solar wind and escapes into space soon after it is caught.

Earth-based spectroscopy also shows absorption lines of sodium and potassium. These elements are the dominant constituents of the atmosphere and are relatively permanent. They do not come from the solar wind but from the planet itself. Mercury has a magnetic field that is about 1% the strength of the Earth's. Ions from the solar wind ride the field lines down toward the surface. Because the atmosphere is so tenuous, the ions can slam into the ground and liberate (or **sputter**) these heavy atoms from surface rocks. (Alternatively, atmospheric

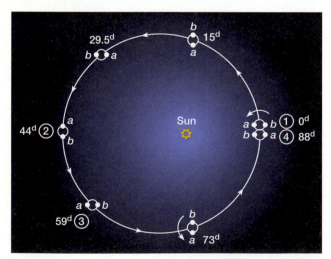

Figure 12.5   *Mercury starts off at perihelion (position 1) with point a at the subsolar point and b opposite. Half a Mercurian year later, at aphelion (position 2), the Sun is just setting at point a and rising at b. At position 3 the planet has finished one sidereal rotation. After return to perihelion, position 4, the planet has spun 1.5 times, and the opposite point b is at the subsolar point. After another year and 1.5 more rotations, point a is back beneath the Sun.*

atoms may have escaped from the planet's interior.) Sodium and potassium are abundant, yet heavy enough to be trapped by the light gravity. The atmosphere of Mercury may not amount to much, but it is there.

12.3.3 Thermal Poles

The cold poles of the Earth support massive ice-caps, but Mercury would seem too close to the Sun to have any such features. However, radar observations show bright reflective spots at the poles that are probably water ice (Figure 12.6a). There, the Sun is at the horizon, and the interiors of craters would be permanently shadowed. Calculations suggest a temperature as low as 125 K (–148°C). Liquid water cannot exist on Mercury: the low atmospheric pressure would allow water molecules to leave the water's surface so quickly that the fluid would evaporate instantly. Instead, some of the ice evaporates directly to a gas, or **sublimes;** under the action of intense ultraviolet sunlight, the gas then breaks down into hydrogen and oxygen, which it contributes to the atmosphere. The ice must be buried under a layer of protective regolith; otherwise it would all sublime away. The ice may be resupplied by water locked up in impacting bodies.

Mercury also has **hot poles** near its equator (Figure 12.6b) that are about 50 K warmer than the average temperature. At perihelion, the planet is 1.5 times closer to the Sun than at aphelion. Because of the inverse-square law, it receives 1.5^2 or 2.3 times as much energy when it is closest than when it is farthest. Consequently, the two opposite spots on the planet that face the Sun at perihelion (*a* and *b* in Figure 12.5) receive more heat than anywhere else. Because of the good surface insulation, these two spots stay warmer than their surroundings during the entire year.

12.3.4 The Surface

Mariner 10 imaged nearly half of Mercury's surface with a resolution of about a kilometer (Figure 12.7), and in local areas achieved resolutions of 100 m. Most of the features identified are named after artists, musicians, and authors. Craters are everywhere. Mercury apparently suffered through the late heavy bombardment, and the surface must be roughly 3.8 to 4 billion years old.

Mercury's surface, like that of the Moon, is differentiated into highlands, lava basins, and plains. The greatest basin is Caloris (Figure 12.7b), a multiringed impact basin 1,300 km across that looks like the Moon's Mare Orientale. The collision that produced it was so violent that shock waves went all the way around the planet and converged at the opposite side, where they jumbled the surface into an extraordinary rugged terrain. Lava subsequently flooded the basin's center. The regolith of the planet is probably similar to that on the Moon, made up of rock that has been battered and ground up by billions of years of bombardment.

For all the similarity with the Moon, however, there are some major differences. The volcanic plains are not dark like the lunar maria but have

(a)

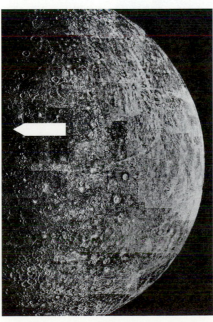

(b)

Figure 12.7 *Two quadrants of Mercury observed by* Mariner 10 *during* **(a)** *approach and* **(b)** *recession show a proliferation of craters that make the planet look much like the surface of the Moon. Caloris, the largest multiringed impact basin known in the Solar System, is indicated by the arrow.*

BACKGROUND **12.1** A Day on Mercury

It is highly unlikely that anyone will set foot on Mercury in the foreseeable future. Living conditions would be formidable, not only because of the searing heat of the Sun, but also because of the solar wind, which is energetic enough to create the bulk of the sparse Mercurian atmosphere.

Yet the idea of living there provides an interesting thought experiment. The Sun would average $1\frac{1}{4}°$ across, 2.5 times greater than its angular diameter seen from Earth. The Sun's angular size would also noticeably change as Mercury passed from perihelion to aphelion, from 1.5° to 1°. The solar day, 176 Earth days long, would proceed with agonizing slowness. At that pace, the Sun would move along its daily path at an average rate of only 2° per Earth day, or a mere five minutes of arc per Earth hour. Watching a sunset or sunrise would be an exercise in extreme boredom, the whole event lasting on the average 18 Earth hours.

Stranger still, the apparent daily solar motion would depend strongly on Mercurian longitude. As Mercury approaches perihelion and begins to move faster, the Sun's east-to-west daily motion would slow, and for a short time, the Sun would actually appear to go backward. Think now of living at a point midway between *a* and *b* in Figure 12.5 where the Sun rises or sets near perihelion. There is a specific longitude where at sunrise, the solar disk would make an appearance, reverse direction and set in the east, and then come back up again!

Other than that bizarre event, the day would proceed much as on the Moon, but with surface temperatures climbing to the lead-melting point. After an equally long sunset, the nighttime temperature would plunge to –185°C. Of course, you could live underground where the temperature could be maintained fairly comfortably with a little air conditioning, but then you would see nothing at all. It would not be a pleasant place to live, or even to visit.

Figure 12.8 *The craters of Mercury are flatter than those on the Moon because of higher gravity. A scarp (arrow) 500 km long and up to 3 km high has broken the prominent crater in the center.*

albedos similar to the highlands'. They are also more beaten up, indicating that the flows are older than those on the Moon. In the highlands, craters do not fill the entire surface, but are set onto extensive **intercrater plains.** These seem to represent volcanic flows that took place earlier than those on the lunar surface. The Mercurian surface is also more extensively covered with ray systems. A close look at the craters (Figure 12.8) shows that they appear less rugged, with lower walls than their lunar counterparts and smaller ejecta blankets. In the higher gravity, the ejecta do not go so far, and the craters slump to larger diameters.

The most unusual surface features are **scarps** (see Figure 12.8), huge cliffs that can stretch for 500 km. They appear to hve been produced when Mercury shrank by a couple of kilometers as it cooled from the high temperatures generated during its formation, forcing the planet's crust to buckle and slip. Because the scarps cross the craters, the contraction had to occur sometime in the last 4 billion years or so, after the heavy bombardment ended.

12.3.5 *Interior and Origin*

Mercury's magnetic field, though weak, implies the existence of at least a partially molten iron core. Because of the planet's small size, the core should

Figure 12.9 *Mercury's hot iron core is relatively the largest in the Solar System.*

have cooled below the melting point of pure iron. There must be some impurities like sulphur present that lower the melting point and keep some of the core in the liquid state. An iron core is also supported by Mercury's average density of 5.4 g/cm^3, since the interior has to be dense to compensate the low density of the rock at and near the surface. Because Mercury is small, its uncompressed density (see Section 10.3) is still a hefty 5.3 g/cm^3, well above Earth's value of 4.4 g/cm^3. The planet actually has the highest uncompressed density of any body in the Solar System. It must consequently have the largest iron core, occupying some 60% of its mass and over 70% of its radius (Figure 12.9). Surrounding the iron core is a rocky mantle only 700 km thick. The depth of the lighter surface crust is unknown.

Mercury's proximity to the Sun likely played a strong role in its strange construction, as intense solar heat would have removed a good portion of the lighter volatile elements during the planet's formation, leaving a higher proportion of metals. Still, theory suggests that there is too much iron. One hypothesis posits a violent collision between a primitive Mercury (which had already differentiated to an iron core and mantle) and another body in the same kind of event that may have formed the Earth's Moon (Figure 12.10). The impact blew away much of Mercury's original mantle, leaving it with a relatively larger iron center. The newly transformed planet then remelted (perhaps as a result of heat generated by the decay of radioactive elements) and differentiated again, sending dirty iron to the center. Mercury was subsequently scarred by

countless blows from the debris of the early Solar System and clobbered with larger bodies to produce multiringed basins. At various times it released molten rock from its interior, contracted slightly, and finally collected a smattering of new craters.

Collision theories may seem like arbitrary speculation, but look again at the mighty Caloris basin and at the lunar Procellarum basin. Collisions not that much greater than the ones that caused those features could have knocked the planet apart. Like the Moon, Mercury provides a window to the early Solar System and to the violent conditions under which the Earth itself formed.

12.4
Venus

Venus seems a near-twin of Earth. However, Venus's smaller distance to the Sun, slightly lower mass, and probably other factors have conspired to produce a planet that in many ways is profoundly different.

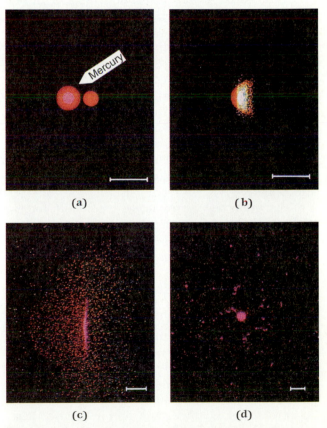

Figure 12.10 *In a computer simulation, a body approaches primitive Mercury* **(a)**, *which at that time had a thicker mantle than it does today. Upon collision* **(b)**, *the two merged, and much of the lighter rock was blown away* **(c)**. *The remaining iron-rich material reassembled in* **(d)**.

12.4.1 *Rotation*

Movement of the dusky cloud patterns that surround Venus suggests a rotation period of four Earth days in the retrograde, or clockwise, direction. Radar results obtained in 1962, however, amazed everyone. Venus's sidereal rotation period is indeed backward, but it is a ponderous 243 days. As a result, the planet's solar day is 117 Earth days long. The axial inclination is 177.3° (a value over 90° indicates reverse rotation), only 2.7° from the perpendicular to the orbital plane. Though the origin of the odd rotation remains a mystery, it seems possible that a collision of the kind hypothesized for the creation of the Moon and for the removal of part of Mercury's mantle may have been responsible.

12.4.2 *The Atmosphere*

The atmosphere of Venus contrasts starkly with the atmospheres of Mercury and the Earth. The high, uniform surface temperature of 740 K indicates that a dense layer of insulating gas surrounds the planet. From Earth, strong absorption lines of carbon dioxide were seen in the reflected spectrum even in the 1930s.

Various probes dropped by spacecraft into the Venusian air, however, have provided a superb measure of atmospheric composition and structure. The pressure at the surface is *98 bars,* and it is almost pure (96%) CO_2. Nearly 4% is nitrogen, and the remainder is a complex mixture of water vapor, sulfur dioxide, argon, carbon monoxide, neon, and other chemicals. The amount of water is extremely small, only a ten-thousandth that in the Earth's atmosphere.

The great CO_2 content is responsible for the high surface temperature through a greenhouse effect (Section 10.5.1) gone mad. The small amount of CO_2 (and the water) in our air raises the Earth's temperature by about 35 K above the blackbody temperature that would be produced by solar heating alone. But on Venus, which has 300,000 times as much atmospheric CO_2 to block outgoing heat, the temperature rockets to 740 K, nearly 500 K hotter than it would be with no atmosphere.

Remarkably, the two planets have similar total amounts of volatile elements like carbon, oxygen, and nitrogen. On Earth, almost all the carbon is tied up in rock, notably limestone, and in decayed vegetation in the form of oil and coal. On Venus, it is all in the air. The reason for the atmospheric difference probably lies in the planet's proximity to the Sun, which is responsible for a **runaway greenhouse effect.** There are neither oceans nor any chemical processes, including life, that can lock

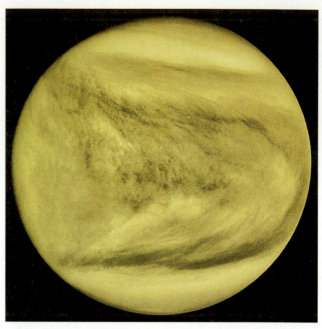

Figure 12.11 *Venus posed for* Mariner 10 *in 1979. A banded structure converges toward the equator. The dark, V-shaped area in the center can be seen from Earth (see Figure 12.1).*

carbon into the surface. When Venus was young, its surface may have been relatively cool (although hotter than Earth's) and may have held a substantial amount of liquid water. Water is an important greenhouse gas. The nearby Sun would have gradually evaporated the water. The resulting greater atmospheric temperature in turn evaporated even more water and allowed less carbon to bind into rocks. With more water and CO_2 in the atmosphere, the planet's temperature kept climbing. The water vapor floated upward, where it was split by solar ultraviolet light into its constituent hydrogen and oxygen. The hydrogen escaped, and the water disappeared forever. As we add CO_2 to our own atmosphere, Venus teaches us a sobering lesson.

The clouds, which surround Venus in a continuous sheet (Figure 12.11), were revealed by ground-based spectroscopy to be sulfuric acid droplets, not water. Spacecraft probes showed that they lie about 50 km above the ground (Figure 12.12), much higher than the water clouds of Earth. Clouds will condense only when air is saturated with vapor. A gas holds less vapor when it is cold than when it is warm. Because the Venusian surface is so much hotter that the Earth's, the low condensation temperature is achieved only at great altitude. The clouds can actually rain sulfuric acid, but the drops evaporate long before they hit the ground.

The planet's cloud cover is a result of the greenhouse effect and the intense surface heat.

Since there is little water, sulfur dioxide in the atmosphere cannot become tied up in surface rocks, so it climbs upward where **photochemical reactions** (chemistry aided by sunlight) turn it into the acid. The same effect happens in large cities on sunny days: in a sense, Venus is covered with an incredibly thick smog. The clouds are highly reflective and let little sunlight penetrate to the surface. Without them, Venus would be even hotter.

Because Venus has little free atmospheric oxygen, there is no significant ozone and therefore no stratosphere in which temperature climbs with height as on Earth. Just over 100 km above the surface lies an odd variable thermosphere whose temperature rises with altitude during the day under the action of sunlight but disappears at night, the temperature stabilizing at about 150 K. Venus has no measurable magnetic field, so there is neither a magnetosphere nor an aurora.

There is little weather on the surface of Venus. The slow rotation produces a rather constant wind pattern that blows in the direction of planetary rotation (east to west) at a speed of only 3 or 4 km/hr. Even so, the winds are strong enough to push surface soils around. The heat and dryness preclude any rain. There are no significant seasons,

as Venus's orbit is almost circular and the obliquity of the ecliptic is a mere 2.7°. Near the cloud layer, however, the wind velocity increases to a fierce 300 km/hr because of solar heating. The winds blow the whole upper atmosphere around the planet in only four Earth days, giving optical observers a false impression of the rotation period. Solar heating also produces convection and a single simple Hadley cell (see Section 10.5.2) that carries heat from the equator toward the poles. The combination of atmospheric rotation and north-south circulation give rise to the V-shaped cloud pattern seen in the Pioneer images (see Figure 12.11).

12.4.3 Surface Topography

Figure 12.13 shows a radar map of the Venusian surface with low areas colored blue and green and high ones yellow and red, in order of elevation. We might at first imagine ocean waters lapping against continental shores. Then we realize there *is* no water. The elevation reference level is not set by oceans but by the average radius (refined by radar) of 6,051 km. From bottom to top, the variation in surface elevation is about 13 km, similar to that on Earth and consistent with the similar gravities of the two worlds. The extremes are much less common on Venus, however. Unlike the Earth, which is divided into distinct high and low areas, Venus's surface is rather strongly concentrated at one level, two-thirds in broad, rolling plains. The highlands cover only about 10% of surface area (as opposed to the 40% occupied by terrestrial continents). The lowlands make up the difference (about 25%), the deepest of them dropping nearly 3 km below the reference level.

There are only two large high regions, one near the north pole called Ishtar Terra ("land of Ishtar"), about the size of the continental United States, and another, comparable to half of Africa, called Aphrodite Terra. (Venusian features are generally named after women.) Ishtar Terra contains the only mountain ranges, which ring a broad plateau known as Lakshmi Planum. To its east rise the towering Maxwell Montes (from the plural of the Latin *mons*, "mountains"), which climb 10 km above the reference level, higher than Everest. Along the broad equatorial belt are several other high areas that include Beta Regio (*regio* means "region"), Alpha Regio, Eistla Regio (between Beta and Aphrodite), and Atla Regio at the eastern end of Aphrodite.

To what degree is the planet like Earth? Are the terrae real continents? Are there tectonic plates? To answer the questions, we need to examine the surface in greater detail.

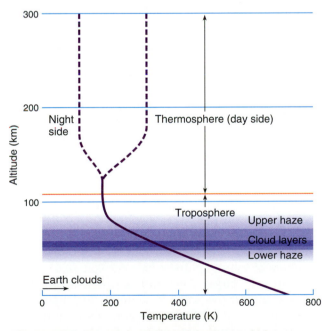

Figure 12.12   *Venus's troposphere, the region in which temperature drops with altitude, is deep, passing through the sulfuric acid clouds at 50 km. There is no stratosphere. On the daytime side, the temperature climbs above an altitude of 100 km, producing a thermosphere. On the nighttime side, however, temperature falls and the thermosphere disappears. Adapted from "The Atmosphere of Venus," by Gerald Schubert and Curt Covey from* Scientific American, *July 1981, page 68. Copyright © 1981 by Scientific American, Inc. All rights reserved.*

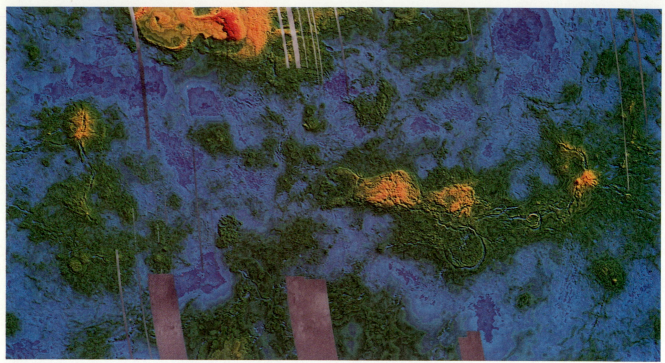

Figure 12.13   *The surface of Venus is between 70° N and 70° S latitude mapped here by Magellan radar. Altitudes are color coded: blue, 2 km below the mean planetary radius of 6,051 km; green, near the mean radius; yellow, 3 to 4 km high; and red, at a maximum of 11 km high. The blank strips are regions with no data. The flat map distorts the sizes of land masses toward the poles: Ishtar Terra is actually smaller than Aphrodite Terra.*

12.4.4 Surface Activity

We can see nothing of Venus through the clouds. With the exception of lander images, everything we know about planetary surface structure comes from radar. The radar images that follow record radio reflectivity; in general, smooth areas are dark and bright areas are rough.

Venus displays a remarkable variety of volcanic landforms. They are broadly distributed across the surface and do not concentrate along anything that might resemble a plate boundary. Most of the equatorial highlands, including much of Aphrodite Terra, appear to be volcanically induced large rises and domes, pushed upward by upwelling mantle rock driven by convection. There may be as many as 100,000 volcanoes, most of them small shield volcanoes typically 10 km in diameter and a few hundred meters high, concentrated into extensive fields in the lowlands or rolling plains. Larger shield volcanoes, a few of which have reached great size, are concentrated more toward the highland rises. Sif Mons (Figure 12.14), in Eistla Regio, is a beautiful shield volcano that rises 2 km above the surrounding plain and is 50 km across. Its neighbor, Gula Mons, is a kilometer higher. Lava flows extend outward from each for 500 km. The dark material near the top of Maat Mons in Atla Regio

(Figure 12.15) suggests relatively recent lava flows (though no absolute dating is available).

Shield volcanoes are formed from relatively runny lava (see Section 10.2.2). Venus also displays numerous steep-sided domes (Figure 12.16) that may be produced by more viscous magma. Unlike Earth, however, Venus has little in the way of pyroclastic deposits caused by explosive lava. Instead, it has several volcanic forms that are not present on Earth at all. The most prominent are the circular **coronae** (Figure 12.17), which range from 200 to over 1,000 km across. They are broad rises with faulted and fractured centers surrounded by concentric fractures all within a peripheral moat, and are commonly associated with volcanic flows. The volcanic rises and shield volcanoes attest to enormous numbers of mantle plumes (see Section 10.4), some of them very large. The coronae are also produced by plumes, but the amount of upwelling matter (or the time over which the plume was active) was apparently not enough to create a true volcanic rise.

The majority of volcanic outpourings, however, come not from volcanoes but from flooding that emanates from fissures (Figure 12.18). Up to 80% of the planet seems to have been paved and repaved by volcanic outflow of one kind or another. We see rilles, or channels, up to 2 km wide and an

Figure 12.14   *Twin volcanos, Sif (left) and Gula Montes (right), imaged by Magellan radar, tower 2 and 3 km above the terrain. They are separated by 700 km. Lava flows extend from one to the other.*

Figure 12.15   *Radar data can be reconstructed by computer to show how volcanos such as Maat Mons might look as seen from near the planet's surface. The image (500 km wide) is exaggerated vertically by a factor of 20, and is colored to match the effect of heavily filtered sunlight. Dark rock near the top has been produced by recent volcanic flows. A 25-km-diameter meteor crater lies in front of the mountain.*

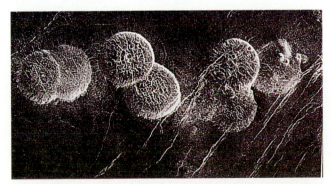

Figure 12.16   *A group of steep-sided domes in Alpha Regio (called the "English Muffins"), each 25 km across, is made of more viscous lava than that associated with shield volcanoes.*

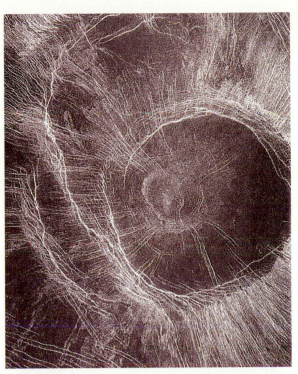

Figure 12.17   *A 200-km-diameter corona in the plains southwest of Beta Regio has been pushed upward by a lava plume. The stretching of the rock faulted and fractured the center, and produced concentric fractures out to a surrounding moat.*

Figure 12.18   *An immense volume of lava has flowed from the left through a breach in a series of mountainous ridges that run from top to bottom.*

astounding 3,400 km long, through which lava once flowed (Figure 12.19). The atmospheric pressure on Venus is so high that gas in volcanic magma cannot escape as easily as it can on Earth. As a result, Venusian lava has a lower density and can rise more easily to the surface. The change in atmospheric pressure with elevation is so extreme that it causes changes in the amount of the trapped gas.

Figure 12.19   *A lava channel 2 km wide runs for 300 km from the top of the picture to the bottom into the volcanic plain Sedna Planitia.*

Volcanic flooding is therefore concentrated toward the lowlands, where atmospheric pressure is highest. The large volcanic constructs tend toward highland areas, where the magma is denser.

The Soviet *Venera* landers (Figure 12.20) gave an eerie picture of the volcanic surface, revealing a dry desolate flat land that stretches toward the horizon. One scene in Figure 12.19 is bouldered, and the other consists of flat rocks and fine, loose regolith. The surface rocks are basaltic, varying in detailed composition from one place to another, much as we find on Earth.

Venus is also pocked with about 900 known meteor craters (Figure 12.21). Like those on Mercury and the Moon, the larger ones have central peaks and the largest are multiringed. In other ways, however, their appearance is quite different from those seen on the other planets as the result of Venus's thick atmosphere. Ejecta blankets are considerably larger than one would expect from the sizes of the craters, and often end in petallike lobes, a phenomenon alien to the Moon or Mercury. The high temperature of the atmosphere heats and softens the surface rock, which allows it to melt more easily in the impacts and to flow for large distances. It is also possible that rubble blasted out of the craters entrains, or encloses, air, which acts as a kind of lubricant and causes some of the ejecta to flow.

The Venusian atmosphere is so thick that meteoric bodies can be crushed as they penetrate, the air acting like a thick, viscous wall. Smaller bodies

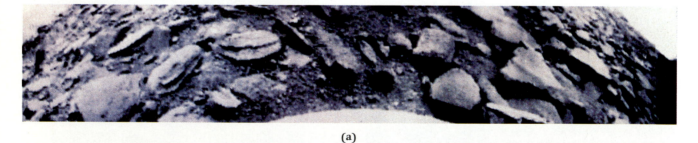

(a)

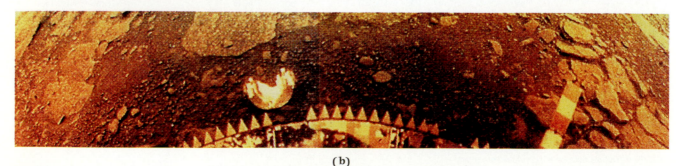

(b)

Figure 12.20   **(a)** *Venera 9 set down to the east of Beta Regio on a slope to a deep plain and imaged a panorama of boulders, some a meter across. In the upper corners we look to a distant horizon.* **(b)** *Venera 13 landed to the south of Beta Regio in a rolling plain and revealed flat rocks and a basaltic regolith. The sawtooth ring is a part of the lander and the odd-looking plate on the ground is an ejected lens cap.*

Figure 12.21   *A view across the plains region Lavinia Planitia in the southern hemisphere shows three meteor craters, each with central peaks. Howe, in the foreground, is 37 km across, and Aglaonice at upper right, 63 km.*

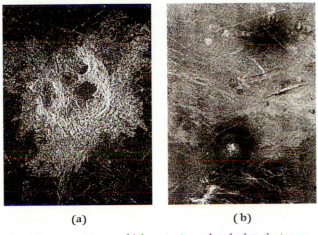

(a)   (b)

Figure 12.22   **(a)** *A multiple crater is produced when the incoming body breaks apart in the thick atmosphere.* **(b)** *The crater at lower left is surrounded by a dark ring, the result of an atmospheric pressure wave. The 50-km-wide dark splotch at upper right is the result of an atmospheric shock wave produced by a body that was completely destroyed.*

are annihilated long before they strike the ground. As a result, there is a significant deficiency of craters smaller than 35 km across and none smaller than about 2 km. Although crushed, the debris from the larger incoming meteoric bodies stay together to make a single hole. Smaller bodies, however, break into several pieces that spread to produce odd-looking multiple impacts (Figure 12.22a). More impressive yet are dark rings around several craters (Figure 12.22b). They are probably caused by atmospheric shock waves that precede the impacting body and smash the surface over a large area. In some cases, no craters are seen at all, only dark splotches on the ground, shadowy testimony to awful violence.

The craters tell us something of Venus's history. There is no record of the late heavy bombardment. These ancient craters were likely destroyed by volcanic resurfacing. From the number of craters and the known cratering rate on the Moon, we find that Venus's craters have been accumulating for only about half a billion years. In that time, they have not been significantly altered by volcanic action. They are distributed across the Venusian surface at random and very few have been flooded with lava. This observation strongly suggests that Venus underwent a catastrophic global volcanic upheaval about half a billion years ago that wiped away all of the older craters, and that there has been very little volcanic activity since. Such an event is consistent with the long lava channels that imply enormous outflows.

Venus has no long ridges that span the globe as they do beneath terrestrial seas, nor are there volcano chains of the Hawaiian type that show plate motion over a plume. No evidence exists for sliding and colliding crustal plates. The low areas are not ocean beds, and the highlands are not really continents. Venus, more like the Moon and Mercury, consists of one crustal plate. Instead, tectonic activity is caused almost entirely by large-scale convection of mantle rock and by smaller plumes. Upwelling currents produce crustal spreading, resulting in cracks or faults of the kind that cut through Aphrodite Terra and Beta Regio (Figure 12.23). These are similar to rift-type faults in the

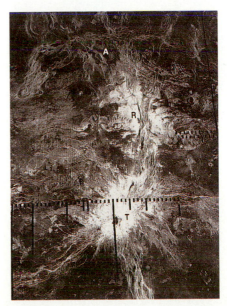

Figure 12.23   *Beta Regio is a large volcanic rise near the equator. It is split by Devana Chasma, a great rift valley. Tessera landforms are indicated by the letter A. R and T are two volcanoes, Rhea and Theia Montes. The black lines that cut across the lower part of the image represent gaps in the radar data*

BACKGROUND 12.2 A Day on Venus

With a pressure suit and an intelligent choice of location—perhaps a pole, where the sunlight slants obliquely—Mercury might be at least livable for an astronaut. There is even water available from the ice caps. One really wonders about Venus. It seems unlikely that human beings will ever walk the planet. The conditions—the heat and atmospheric pressure—are too severe. Nevertheless, radar images and Venera pictures show a fascinating landscape. What would it be like to stand in Lakshmi Planum and gaze at Maxwell Montes off in the distance, or to climb Rhea Mons in Beta Regio to look into the rift valley, or even to try to walk across the tesserae? You would certainly have a long enough period of daylight, 58 Earth days, to make any journey you wished.

Unlike days on the Moon or Mercury, however, your day would start with a long twilight because of slow rotation and sunlight scattering from the high, thick atmosphere. Of course no stars would ever be visible at night, and come daylight all you would see would be a high, probably featureless, cloud deck. To your initial confusion, the sky would begin to brighten in the *west* because of Venus's retrograde spin. It would never get very light outdoors and would proba-

bly appear somewhat as it does at home under a very heavy thunderstorm cloud deck, the kind that brews tornados. The sunlight that got through would be heavily absorbed and would have an orange cast to it (see, for example, Figure 12.15), giving the landscape an eerie burnished glow.

As the long day proceeds, you would have to be careful never to venture out without a protective suit. Not only could you not breathe, but you would literally be cooked in the searing heat, comparable to the temperature of an oven in a self-cleaning cycle. You would not have much weather to worry about, however; the forecast would be the same day after day, hot and dry, with a slight wind from the east. For all the evidence of volcanic and meteoric activity on the planet, you would never likely experience any. The craters have had half a billion years to accumulate, and geologic processes are slower than they are on Earth.

As the Sun sets in the east, it would gradually get darker but it would never cool off. The air is too good an insulator. You would probably want to leave as quickly as you could to go home to the wet, blue, cool Earth.

Figure 12.24 *Bright, complex, ridged terrain, or tesserae, run from the upper left to center, and are surrounded by younger lava flows. At bottom left is a partially flooded impact crater.*

Earth's continental blocks, exemplified by the Great Rift in eastern Africa. Smaller cracks—grabens—can extend for hundreds of kilometers. Downwelling currents produce crustal compression and ridges, and the combination of the two over time can create complicated terrain with ridges crossing over grabens. Extensive systems of ridges a few kilometers wide and a few hundred meters high cross much of the plains and lowlands. Much of the highlands is covered with **tesserae** (Figure 12.24), wildly jumbled and rugged terrain alien to Earth and caused by repeated episodes of tectonic activity. Aphrodite Terra also displays some evidence for subduction of crust.

The small fraction of surface covered by mountains is consistent with the lack of plate motion. What then produced the towering Maxwell Montes in Ishtar Terra (Figure 12.25)? Perhaps the whole region has been elevated by an underthrusting crustal block. However, the mountains are oddly distributed about Lakshmi Planum, a vast volcanic plateau 5 km high. It was flooded by a pair of vents now seen as immense calderas, one 3 km deep.

Figure 12.25 *Ishtar Terra consists of a vast volcanic plateau, Lakshmi Planum, ringed to the west, north, and east by ridged mountains. Blue then violet represent the lowest areas, yellow then red the highest. The plateau contains a pair of giant calderas called Colette and Sacajewea. The Maxwell Montes lie to the right.*

Ishtar Terra was probably produced by a combination of upwelling and downwelling mantle currents, its towering mountains compressed and raised by a downwelling flow. Ishtar is more like a terrestrial continent than any other landform on Venus. The mountains, which contain slopes greater than 30°, must be young. Otherwise, the steep sides, which are softened by the high atmospheric temperature, would have collapsed under their own weight.

12.4.5 The Interior

The interior of the Venus remains more mysterious than its surface. The average density indicates a metallic core similar in size to the Earth's, constituting roughly a third the planet's mass and half the radius. From the mass of the planet, the core ought to be molten, like Mercury's and part of the Earth's, but there is no way of probing it, as we have no seismographs on the surface. The lack of a magnetic field, however, strongly suggests that the interior is solid.

Why should Venus, with a diameter and mass nearly equal to those of the Earth, be so different from our planet, with a solid instead of a partially fluid core, and a surface dominated by volcanism instead of tectonic plates? Remarkably, the answer may lie in the atmosphere. Its extreme temperature heats, softens, and weakens not only the crustal rock but lithospheric rock as well. As a result, the lithosphere is only a few tens of kilometers thick, thinner than its terrestrial counterpart. Because of the change in rock characteristics relative to Earth, the lithosphere has limited horizontal movement, enough to produce some tectonic effects, but not enough to produce plates and plate motion. Instead, the convection currents of the mantle produce volcanic rises and features associated with plumes. The weakness and thinness of the outer layers make the planet far more subject to volcanic flooding than the Earth, apparently resulting in periodic catastrophic upheavals in which Venus is globally repaved. The loss of heat could then have cooled the core to the point of solidification. Between such episodes, Venus's internal heat is lost by simple **conduction** (heat transfer from atom to atom), by plume-type volcanism, and small-scale crustal rifting and spreading.

Thus we see that two planets, neighbors with such seemingly similar outward attributes, can be very different. In a sense, we have begun to experiment on the planets, to see how changes in conditions can lead to other changes in characteristics, and by doing so, we better understand and appreciate our own world.

KEY CONCEPTS

Comparative planetology: The science of comparing planets and their varied conditions, using one body to learn about another.

Conduction: Heat transfer from atom to atom.

Coronae: Large, volcanically uplifted areas with fractured centers, surrounded by concentric fractures.

Hot poles: Hot regions on Mercury that alternately face the Sun at perihelion.

Intercrater plains: Volcanic flows on Mercury that took place during the late heavy bombardment.

Photochemical reactions: Chemical reactions that take place under the action of sunlight.

Radar: An active observational technique in which radio waves are reflected from a body to determine its distance, speed, and surface features.

Runaway greenhouse effect: A process in which high temperature produces more atmospheric carbon dioxide, which produces higher temperature, and so on.

Scarp: A steep cliff or fault.

Sputter: To remove atoms from a material by impact with high-speed atoms or ions.

Sublime: To pass directly from the solid to the gaseous state.

Tesserae: Jumbled, disrupted volcanic terrain on Venus.

EXERCISES

Comparisons

1. Compare the surface ages of Mercury, Venus, and the Earth. Why do they differ?
2. Compare the rotations of the Earth, Venus, and Mercury.
3. Compare the basins of Mercury with those of the Moon.
4. List the differences found among the craters of the Earth, the Moon, Mercury, and Venus.
5. Compare Venus's shield volcanoes with its coronae.
6. What are the differences between the collision theories for the Moon and Mercury?
7. Compare the atmospheres of the Moon and Mercury.
8. What do the atmospheres of the Earth and Venus have in common?
9. Compare the magnetic fields of Mercury, Venus, and the Earth. What do they tell us about the planetary cores?

Numerical Problems

10. How long does it take a radar signal to reach and return from **(a)** an airplane 20 km away; **(b)** the Moon; **(c)** Venus, if the planet is at inferior conjunction?
11. How much stronger are the solar tides at Mercury's perihelion than at aphelion?
12. If the radio (and infrared) luminosity of Venus were only a quarter of what is now observed, what would be the temperature of the surface?
13. What is the mass of Mercury's iron core compared with that expected for Venus?

Thought and Discussion

14. How can we find the masses of Venus and Mercury?
15. At what point in Mercury's orbit would the Sun appear to stand still or even go backward in the sky? Why?

16. How can Mercury have ice at its poles when its surface can get so hot?
17. Why does Mercury have two permanent hot poles?
18. What is the origin of Mercury's scarps?
19. Why do we believe that Mercury has the largest iron core of any planet in the Solar System?
20. How do we know that Venus's atmosphere is made of carbon dioxide?
21. Why are the clouds of Venus made of sulfuric acid and not of water?
22. How have upwelling and downwelling mantle material affected Venus's surface?
23. Why is there a deficiency of small craters on Venus?
24. What is the origin of dark rings around some of Venus's craters?
25. Why do we believe that Venus's surface was globally repaved with volcanic magma about half a billion years ago?
26. What has been the effect of Venus's atmosphere on the nature of the planetary surface and the core?

Research Problem

27. Before radio astronomy and radar revealed the nature of Venus, astronomers speculated about its surface conditions. Using library materials, examine the nature of the some of the speculations made over the last century and comment on why the astronomers making them were misled.

Activities

28. Compile a summary of spacecraft that have visited Venus and Mercury and give the functions of each.
29. Using photocopied library materials, such as *National Geographic* magazine, compile a booklet showing surface features of the Earth (volcanoes and the like) that have counterparts on Mercury, Venus, and the Moon. Under each, comment on how they differ from one planet to the other.

Scientific Writing

30. You are asked to write an article for a technical engineering journal about the influence of modern technology on science. You choose radar as an example. Explain to your readers how radar devices and techniques have helped us learn about Venus.
31. As a feature article in your local newspaper, write about a hypothetical trip to Venus, taking the reader through the clouds, down to the surface, and then home. Convey the excitement of seeing a new world for the first time.

13

Intriguing Mars

The nature of the planet most like the Earth

*Viking 1 lets us watch
a Martian sunset
bring night to a new world.*

The ancient mythologies of the two planets flanking the Earth represent the extremes of human passion, love and war. This pair of worlds indeed provides great contrast, but the reverse of that suggested by their names. Venus is a hellish place where no human being is ever likely to tread, whereas Mars is comparatively benign, even inviting to human exploration, with many features that we associate with home.

13.1
The View from Earth

Unlike that of Venus, the Martian surface is easily seen (Figure 13.1), resulting in a rich but strange observational history. The best time to view the planet is at a favorable opposition (see Section 6.2.1), when the angular diameter climbs to 25 seconds of arc. From angular diameter and distance, we find that Mars has a radius of 3,400 km (refined by spacecraft to 3,394 km), intermediate in size between the Earth and the Moon. Its mass of 6.4×10^{23} kg (10.7% that of Earth and about double Mercury's) is easy to determine from the orbits of its two tiny moons and Kepler's generalized third law. The first surprise is Mars's light average density of 3.9 g/cm³ (the uncompressed density is 3.8 g/cm³), intermediate between those of the Moon and the uncompressed Earth. Analogy with the Earth (see Section 10.3) suggests that Mars's core must be smaller or less dense than those of the

Figure 13.1 *From Earth, we see permanent dark markings on the Martian surface. The one at the center is called Syrtis Major; to the lower right is the large basin Hellas, and at the upper left, haze blankets a polar cap.*

TABLE 13.1
A Profile of Mars

Planetary Data

distance from Sun	1.52 AU
radius	3,394 km = 0.53 R_{Earth}
mass	6.4×10^{23} kg = 0.11 M_{Earth}
mean density	3.9 g/cm³
uncompressed density	3.8 g/cm³
escape velocity	5.0 km/s
gravity	0.38 g_{Earth}

Layers	*% Mass*	*Radii (km)*
iron core	15?	0–1,200
mantle	80?	1,200–3,300
crust	5?	3,300–3,394

Atmosphere

pressure	0.007 bar

Composition by Number

CO_2	95%
N_2	2.8%
Ar	1.6%
O_2	0.13%
CO	0.07%
H_2O	0.03%

Layers	*Height (km)*
troposphere	0–100 km
thermosphere	>100 km

Other

magnetic field	$<10^{-4}$ Earth's
rotation period	24^h36^m
axial inclination	24°

Planetary Features

Basins: Two major multiringed impact basins in southern hemisphere.

Bulges: Two large volcanic bulges in northern hemisphere.

Cloud cover: Wisps of water clouds over high mountains; dry-ice clouds over poles.

Craters: Heavy cratering in southern hemisphere from late heavy bombardment; light cratering in the northern.

Faults: One great fault, Valles Marineris, caused by volcanic uplifting.

Mountains: Volcanic; none tectonic.

Plates and plate motion: One plate; no horizontal crustal movement; vertical movement produces bulges.

Polar caps: Permanent polar caps of dry and water ice in the south, water ice in the north; seasonal polar caps of dry ice.

Surface air temperature: Highly variable and dependent on latitude; 240 K (–33°C) maximum under full sunlight to 170 K (–103°C) at night; ground temperature can exceed 0°C.

Volcanoes: Several large shield volcanoes in northern hemisphere; several ancient patera (collapsed volcanoes).

Water: Vapor in atmosphere, ice in polar caps and below surface; once flowed as liquid.

other terrestrial planets. These and other properties are summarized in Table 13.1.

Aside from its obvious orange-red color, Mars displays a variety of more or less permanent surface features. From the apparent movement of dark markings (Figures 13.2a and b), we find an Earth-like rotation period of 24^h37^m. Moreover, the axial inclination is 25°, only 1.5° greater than Earth's. Mars therefore has seasons similar to ours, but with two significant differences. First, the Martian sidereal orbital period is 1.88 years, so each season is almost twice as long as on Earth. Second, because of Mars's high orbital eccentricity ($e = 0.09$), it is 20% closer to the Sun and receives 45% more sun-

light at perihelion than at aphelion. Aphelion occurs during southern winter and perihelion during southern summer. The result is that the southern hemisphere has significantly hotter, shorter summers than the northern and colder, longer winters, quite unlike Earth.

The planet is enveloped by a thin, somewhat hazy atmosphere. Absorption lines of carbon dioxide were found in the 1940s and those of water vapor in the 1960s. Small, bright clouds are also occasionally seen. During the southern-hemisphere summer, a huge dust storm can obscure the planet's surface.

Surrounding the rotation poles are caps of ice that seem similar to those of Earth (Figures 13.2c and d). They change dramatically with the seasons and, as they grow, are veiled by a thin haze. Because of the hemispheric imbalance of the seasons, the southern polar cap experiences wider variations in size than does the northern. At maximum, it extends almost halfway to the equator (about 55°S latitude), and the northern one shrinks. Half a revolution later the northern cap grows, but since northern winter occurs around perihelion, it does not advance as far. The dark markings also vary with the seasons (see Figures 13.2c and d) and change their shapes after major dust storms. Their average appearance stays about the same, however, allowing us to name them.

These features pale beside another that brought Mars lasting public fame. In 1877, the Italian astronomer Giovanni Schiaparelli observed a network of fine lines that criss-crossed the planet (Figure 13.3a) and named them *canali*, or "channels." Unfortunately, the word was simply translated into English as **canals.** And who builds canals? Why, people. Thus began a prolonged, sometimes frus-

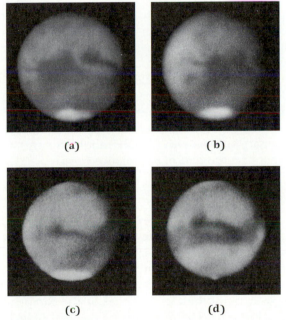

Figure 13.2 **(a and b)** *Mars rotates on its axis, the dark markings moving to the right. The polar cap shrinks and the dark markings expand between late spring* **(c)** *and summer* **(d)**.

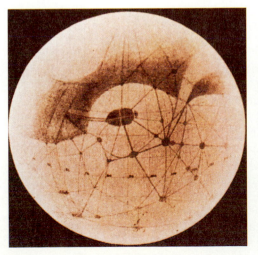
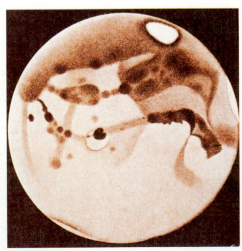

(a) (b)

Figure 13.3 **(a)** *Percival Lowell's 1894 drawing of Mars shows an interconnected set of canals.* **(b)** *The French astronomer E. M. Antoniadi resolved them into spots. The dark, curved line at left center is actually a great chasm.*

BACKGROUND 13.1 Martian Canals and the Scientific Method

The episode of the Martian canals teaches a lesson on how science should, and usually does, operate. The essential assumption of science is that nature is not capricious; experiments or observations must be *repeatable*. Only Lowell and a few others could "see" the canals; those who could not were told that their equipment or their eyes were inferior. The French astronomer E. M. Antoniadi, who considered the canals false apparitions, used some of the best telescopes available in his day and could not see them. Lowell strained credulity when he attempted to explain why larger telescopes were less able to detect delicate features than smaller ones. This kind of explanation is usually a clue that something is amiss.

The history of science is replete with parallel instances, one a topic of recent headlines. The Sun is powered by thermonuclear fusion, which creates heavy elements from light ones with the release of vast amounts of energy. On Earth, the process has been used to produce hydrogen-bomb explosions. Scientists have tried with limit-

ed success to generate nonexplosive fusion for power production, because it requires enormously high temperatures. In 1989, a pair of physicists working at the University of Utah apparently discovered a way of fusing elements at room temperature using a catalyst (a substance that aids a reaction without being changed by it). They immediately announced the results of their "cold fusion" experiments, and scientists everywhere took note. If true, here was a potential source of cheap, reliable power. Unfortunately, few scientists could repeat the fusion reaction. So many failed, in fact, that it became clear that cold fusion does not exist, at least not on the scale suggested by the first experiment.

The moral of these tales is that nature is reliable. The scientific process, which depends on experimental repeatability, operates to check phenomena that are observed only by particular fortunate or "sensitive" individuals. Scientists cannot become part of the science itself.

---

trating, sometimes ridiculous Martian mania. Martians, it was imagined, constructed the canals to transport water from the poles into dry desert regions, allowing the dark "oases" to expand and blossom with thick vegetation during the summer.

The Martians' cause was taken up in the 1890s by the American astronomer Percival Lowell, who established an observatory in Arizona to study Mars and who fervently believed in the canals until his death in 1916. For decades, an argument raged over their existence. Only some astronomers could see them; others resolved the canals into patterns of dots (Figure 13.3b). Under less than excellent conditions, these dots appeared as lines, producing the effect of canals. A far less glamorous explanation for the changes in the dark areas held that they are caused by wind patterns that shift dust around the globe. Belief in the canals and the possibility of life was so strong, however, that some observers claimed the dark areas to be green; one astronomer even announced the discovery of the spectral signature of chlorophyll, a molecule involved in photosynthesis by plant life. Martian madness reached its apex in 1939 with the radio dramatization of H. G. Wells's story *The War of the*

Worlds. Orson Welles convinced millions of listeners they were actually listening to a news broadcast of a Martian invasion.

The best telescopic view of Mars from Earth, however, is no better than that of the Moon with the naked eye. The discovery of the real planet, in truth more interesting than anyone imagined, had to wait for the close-up views made possible by the space age.

13.2
Flights to Mars

The first craft to arrive at Mars were simple flyby missions, *Mariner 4* in 1965 and *Mariners 6* and *7* in 1969. At the same time, the USSR sent *Mars 1* and *2.* These Mariners, however, were severely hindered by their inability to make images of the whole planet. The true nature of Mars was finally revealed by *Mariner 9,* which went into orbit in 1971 (frustratingly, during a major dust storm) and eventually examined the surface in great detail with a resolution of about a kilometer.

The real Martian heroes were the Viking craft (Figure 13.4), both of which entered Martian orbit in 1976 and produced images of nearly the entire globe with resolutions that approached only 20 m. They also carried landers that descended to two different surface locations. The machines arrived with a formidable battery of testing equipment. Each had two televisionlike cameras that could pan around the horizon, devices for measuring the chemistry of the atmosphere, weather stations, and seismometers. They also had arms 3 m long that could reach out, take scoops of dirt, and drop them into automated laboratories designed to test for signs of life. *Viking 2* worked until 1980 and *Viking 1* until 1982, both recording daily and seasonal variations in the Martian air and surface. They were the among the most successful space experiments ever launched.

It is imperative for advances in planetary science that we return to Mars. The *Mars Observer*, launched in 1992, was designed to assess the Martian surface composition, explore the structure and composition of the atmosphere, and image the surface with resolutions as high as 1.5 m. Sadly, communications with the craft were lost in 1993 just as it was to have gone into orbit. We do not know when we will go back.

13.3
The Martian Atmosphere

Observation by spacecraft showed the atmospheric pressure to be a mere 0.007 bar and to consist of 95% CO_2, 3% nitrogen, and 2% argon. Mars's atmosphere is remarkably like that of Venus, but one ten-thousandth as dense. If we include the lithosphere of the Earth and add up the total inventory of carbon and nitrogen, all three planets are similar,

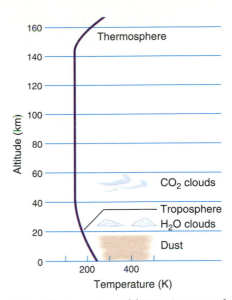

Figure 13.5   *The temperature of the Martian atmosphere declines from an equatorial surface value through the troposphere until it levels off above 30 km. The temperature does not begin to climb again until the thermosphere, well over 100 km above the surface.*

showing both unity in their formation and development and the effects of liquid water and life in shaping the Earth's atmospheric evolution. The Martian air also contains 0.03% water. That, however, is some 30 times less than we find in the Earth's atmosphere. If all the water in the Martian atmosphere could precipitate out, it would make a planetary ocean a mere millimeter deep.

Since Mars is farther from the Sun than Earth and has a thinner atmosphere, it is colder. Though the surface air temperature under a summer noon Sun cannot reach much above 240 K (–33°C), the ground temperature can exceed 0°C, the freezing point of water on Earth. However, the air pressure on Mars is too low to allow liquid water to exist. The CO_2 in the Martian atmosphere produces a greenhouse effect as it does on Earth, but it is much less effective as an insulator, raising the air temperature by only about 5 K. Consequently, temperatures drop sharply to 170 K (–103°C) or below at night; the water vapor in the air can then turn to fog and coat the surface rocks with frost.

The atmospheric structure (Figure 13.5) is more like that of Venus than Earth. The troposphere extends to a great height, its temperature steadily declining from the surface value of 240 K (–33°C) to about 150 K (–123°C) at 30 km altitude. At an altitude of about 150 km we encounter a thermosphere warmed by the absorption of solar ultraviolet radiation. The Martian magnetic field is extremely weak, 10^{-4} that of the Earth or smaller, so there is no significant magnetosphere.

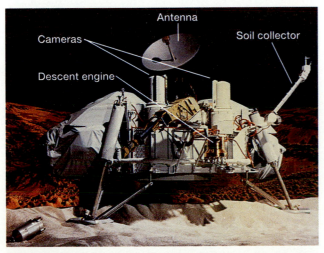

Figure 13.4   *A working model of* Viking 1 *poses on Earth before its August 20, 1975, launch.*

At sufficiently low temperatures, carbon dioxide will freeze. On Earth, CO_2 becomes **dry ice** at $-79°C$. Martian conditions, therefore, create two different kinds of clouds. Icy water clouds form at altitudes of about 20–30 km, over tall volcanic summits. Dry-ice clouds, commonly seen in polar areas, are two and a half times as high. During the winter, carbon dioxide precipitates in the form of dry ice to expand the white polar caps. So much CO_2 is removed that the atmospheric pressure drops by 20% above the cooling pole (pressure variations on Earth rarely exceed about 3%).

The weather on Mars is highly predictable and varies with the seasons. The atmospheric circulation pattern is somewhat similar to that of the Earth, with an equatorial Hadley cell. On Mars, however, it blows *across* the equator from one tropic to another rather than north and south *from* the equator. At higher latitudes, the Martian atmosphere exhibits cyclonic patterns similar to our own. Except during dust storms, winds tend to be benign, with speeds of several kilometers per hour.

13.4
The Beaten Volcanic Surface

Before the days of spaceflight, Mars was commonly thought to be much like Earth. The first close-up views by spacecraft destroyed that image. More comprehensive examination, however, showed that Mars indeed bears some resemblance to our home planet, but in ways no one expected.

13.4.1 Topography

The major Martian features revealed by spacecraft are mapped in Figure 13.6. Surprisingly, few agree with the light and dark markings seen from Earth. Like the Moon, Mars has two distinctly dissimilar hemispheres (Figures 13.6 and 13.7), though for different reasons. The line of division is roughly a great circle inclined about 30° to the equator. The southern hemisphere consists largely of heavily cratered highlands, typically 2 or 3 km above an arbitrary reference level defined by the average atmospheric pressure at the planet's surface. Within the southern highlands are two vast impact basins, Hellas Planitia (see Figure 13.1) and Argyre Planitia.

Lightly cratered volcanic plains near the reference level dominate much of the northern hemisphere. Set within the northern sector are two huge rises or bulges. The one to the east, Elysium Planitia (Figure 13.7), climbs to an elevation of 5 km and is topped by a pair of volcanoes, with another lower down the slopes. The Tharsis bulge, to the west, is twice as high, more massive, and contains a trio of volcanoes along a 1,000-km-long ridge at its top. To the northwest of these is one of the most massive geologic structures in the Solar System, Olympus Mons.

Near the margin of the two regions, to the east of the Tharsis bulge, is a spectacular canyon complex, Valles Marineris, whose name honors the Mariner flights. It stretches east to west for nearly 5,000 km, roughly the length of the continental United States. It is so big it is visible to terrestrial visual astronomers (see Figure 13.3b), as are the volcanoes of the Tharsis bulge.

13.4.2 Structure and Development

The density of craters in the southern highlands testifies that Mars, like Mercury and the Moon, suffered through the late heavy bombardment and that this portion of the surface must be ancient. For so many craters to persist so long, the average rate of erosion must be small, showing that Mars has had neither thick atmosphere nor liquid water over extended periods of time. Moreover, the land has not been destroyed by subduction caused by continental drift, implying that the crust is one solid plate. Yet some activity has occurred. Surface features have been degraded and there is a reduction in the proportion of smaller craters relative to that found on the Moon, so some erosion has taken place. And, as on Mercury, there are intercrater plains, suggesting large volcanic flows that roughly coincided with the end of the late heavy bombardment.

Stratigraphy shows that the two large multi-ringed impact basins of the southern hemisphere, Hellas and Argyre, are among the oldest features on the planet and were formed during the heavy bombardment. They are spectacular features, with rubble strewn for hundreds of kilometers (Figure 13.8). Hellas is 4 km below the reference level and 6 km below the surrounding cratered plain. The basins, like the smaller craters, appear to have been subject to some erosion. They are partially filled with lava and bright Martian dust.

If the craters of the south speak of great age, the volcanoes of the north tell of relative youth, as their lava flows show few impact craters. Their sizes are astonishing. Olympus Mons (Figure 13.9), far down the northwest slope of the Tharsis bulge, towers 27 km above the reference level and 25 km from base to top, three times the equivalent height of the island of Hawaii. The top is so high that bright clouds visible from Earth form around its summit. The volcano, nearly 600 km across,

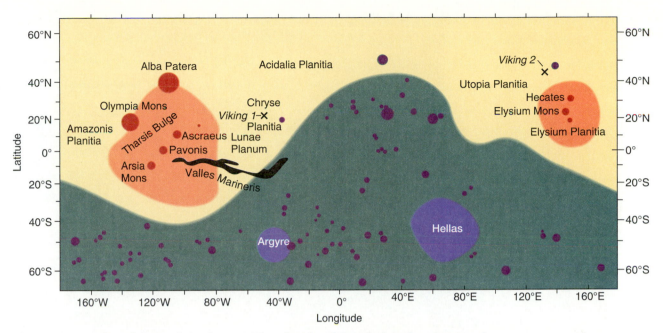

Figure 13.6   *A color-coded map of Mars between 65° north and south latitude shows how the cratered uplands of the southern hemisphere (green) are separated from the volcanic plains of the north (yellow). The two volcanic bulges, Tharsis and Elysium, are orange, and volcanoes are red. A variety of the larger craters are indicated by violet, and the two prominent impact basins by blue. The* Viking 1 *and* Viking 2 *landing sites are also labeled.*

Figure 13.7   *A mosaic created from* Viking 1 *pictures shows the dichotomy between the cratered southern highlands and the northern volcanic plains and bulges. Above the curved dark area at left center is the bulge of Elysium Planitia with its three volcanoes. The white area at upper right consists of water clouds that have formed about Olympus Mons, which is off the picture. At the bottom are the seasonal CO_2 snows of the south polar cap.*

Figure 13.8   *The Argyre basin, a massive multiringed impact crater 800 km across partially filled with dust and lava, occupies the lower left quadrant of the picture. The upper part of the basin rim has been destroyed by another large impact.*

Figure 13.9 *Olympus Mons, the largest volcano in the Solar System, towers 27 km into the thin Martian air and sits on the northwest slope of Tharsis bulge. At its top is a 90-km-wide caldera. The drawing compares the volcano with the much smaller Hawaiian chain.*

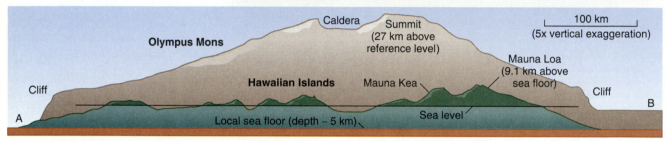

would cover the state of New Mexico or most of New England. Its low slope (only 4°) clearly shows that it is a shield volcano. Lava flows have run down its sides. The flows overlap, younger ones covering older ones, providing evidence for the growth of the structure. At its top is a vast group of calderas 90 km across; at its base is a cliff that stands 6 km tall.

Shield volcanoes on Earth are limited in mass because plate motion eventually removes them from the upwelling mantle plumes that generate them (see Section 10.4). The enormous mass of Olympus Mons and the absence of a plume track reveal that such plate motion does not exist on Mars, a finding consistent with the continued existence of craters from the late heavy bombardment. Olympus Mons has sat in that one spot, getting ever bigger, for much of Mars's history, its great elevation reflecting the planet's lower gravity (see Section 10.2.2). The small number of craters on its slopes demonstrates that lava flowed relatively recently, perhaps only a half a billion years ago.

No less impressive are the three shield volcanoes that top the Tharsis bulge (Figure 13.10). None is as high above its base as Olympus, but the bulge is so big that the volcanic peaks are at the same level as their bigger brother. They lie on a straight line and are obviously connected by some large fault that shows all of Tharsis to be a volcanic structure, a huge mound pushed upward by the pressure of hot magma from below.

Somewhat like Earth, Mars exhibits various types of volcanoes. The pair at the top of the somewhat older Elysium Planitia (see Figure 13.7) peak only about 7 km above the reference level and may have had a more explosive nature than the four in Tharsis, as there is evidence for ash deposits and pyroclastic flows. In addition, there are several huge **pateras** (from the Latin word for "saucer"), structures unknown on Earth, that appear to be ancient collapsed shield volcanoes (Figure 13.11). Alba Patera, on the north slope of Tharsis, is 1,600 km across, wider than Olympus though only one-fourth as high.

The plains of Mars are volcanic and have diverse histories, showing that the planet has endured many episodes of violent volcanism. The simplest plains lie within the confines of the two vast bulges of the northern hemisphere. Lightly cratered, they are therefore of relatively recent origin. Flow patterns are clearly associated with the upward push that made the bulges. The region east of Tharsis, however, is exceedingly complex. It contains two major plains regions, Lunae Planum ("plateau of the Moon") and Chryse Planitia ("plains of gold"). Lunae Planum is a rather heavily cratered raised area. Stratigraphic analysis and crater density shows that it is one of the oldest plains on the planet, forming shortly after the late heavy bombardment but before the great volcanoes. Like most of the others, it is covered with wrinkle ridges, demonstrating compression. Chryse Planitia was formed later, roughly during the raising of Olympus Mons. The far northern plains are also complex, are intermediate in age between Lunae Planum and Chryse Planitia, and seem to have suffered from the effects of underground ice.

Mars may be one plate, but there is still spectacular evidence for ancient tectonic activity. No words can quite describe Valles Marineris (see Figure 13.10), a collective term for a number of huge interrelated canyons that stretch over 5,000 km of Martian terrain. These are fault canyons, caused by an expanded crack in the thick Martian crust. The valley complex developed at about the time of the formation of Lunae Planum as a result of the raising of the Tharsis bulge. Its character changes considerably along its course. At the western end is Noctis Labyrinthus ("labyrinth of the night"), a complex system of grabens (Figure 13.12a) that coalesces

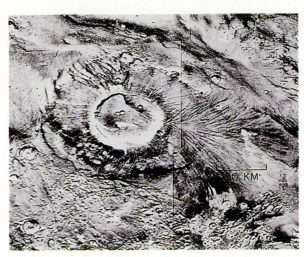

Figure 13.11   *Apollinarsis Patera is a collapsed shield volcano about 150 km across near the Martian equator south of the Elysium bulge.*

(a)

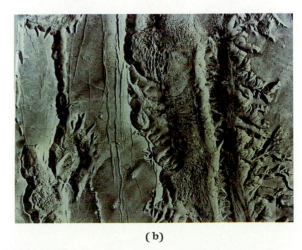

(b)

Figure 13.12   **(a)** *Early morning fog has formed in Noctis Labyrinthus, a grabens network at the far western end of Valles Marineris.* **(b)** *Chasms 40-km-wide to the east of Noctis Labyrinthus display high cliffs and rubble-filled interiors caused by landslides.*

Figure 13.10   *The three volcanoes that top the Tharsis bulge are seen to the left. Valles Marineris, which consists of several individual canyons up to 9 km deep in the middle, stretches eastward from the bulge. Above the eastern end are the complex plains of Lunae Planum and Chryse Planitia.*

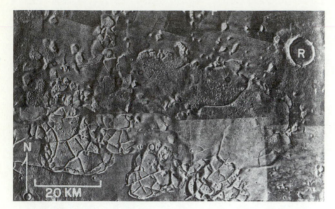

Figure 13.13   *This chaotic terrain lies within the Elysium bulge. Enormously complex, it includes fractured zones and broad slumped areas filled with jumbled blocks of rock.*

Figure 13.14   *Dry runoff channels in the Martian highlands show a branched structure quite similar to terrestrial river systems.*

Figure 13.15   *You look onto the floor of what may once have been a massive lake that filled Candor Chasma. The dark surface consists of sediments.*

into parallel canyons that continually deepen toward the east. They have rubbled floors caused by enormous landslides that have widened the chasms (Figure 13.12b). The central region is the broadest (up to 700 km wide), the deepest, and the most complex, with several canyons branching to the north. At the eastern end, the great valley opens out into a peculiar landform called **chaotic terrain** (Figure 13.13), characterized by a highly irregular assembly of large blocks of rock and fractures.

Relative ages are fairly secure. The features discussed (and there are a great number of others) were formed in the following approximate order: multiringed basins along with the heavy bombardment and highland volcanism; Lunae Planum and Valles Marineris; beginning of Tharsis and Elysium volcanism and the covering of the northern plains; beginning of Olympus Mons; Chryse Planitia; plains west of Olympus; cessation of Olympus volcanism. Absolute ages, however, are highly uncertain. The heavy bombardment ended roughly 3.8 billion years ago and the most recent lava flows may be only half a billion years old.

13.4.3 Water

Water, a hallmark of Earth, sustains life. Mercury and Venus have very little water and the Moon none at all. Though Mars at first seems arid, there is abundant evidence that its surface once held water,

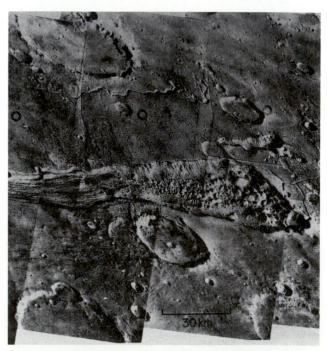

Figure 13.16   *Ravi Vallis seems to erupt from the Martian surface. It once contained water that flowed into Chryse Planitia.*

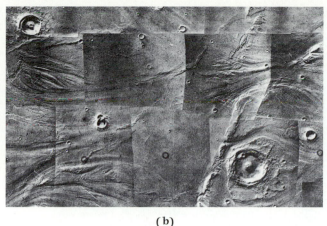

(a)   (b)

Figure 13.17 *Chryse Planitia shows abundant evidence for catastrophic flooding.* **(a)** *Craters stand high above land scoured by a massive outflow. The teardrop shapes show that the water flowed from the area at the bottom of the picture.* **(b)** *Running water has cut ridges into the plain.*

vast amounts of it. The ancient cratered highlands of the south contain dry **runoff channels** that look like common terrestrial riverbeds, with branching tributaries flowing toward a main stream (Figure 13.14). Because they have not produced much erosion, the runoff channels could not have been active for long, and since they are related to the oldest areas of Mars, conditions for running water could have been right for only a short period of time.

More dramatic are younger features associated with the Tharsis bulge and Valles Marineris. The floor of Candor Chasma, the northern canyon in the center of Marineris, is dark with layers of sediment (Figure 13.15). To the east, the canyon walls look scrubbed, and where the canyon opens up at its eastern end, there are clear effects of erosion. Candor Chasma was apparently filled with a huge lake whose containing walls suddenly fell, allowing the water to flow out of the canyon and scour the land as it went.

More dramatic yet are great, dry riverbeds called **outflow channels** that concentrate between Valles Marineris and Chryse Planitia. Unlike terrestrial rivers or the runoff channels, which accumulate from a series of tributaries into a broader stream, they arise seemingly from nowhere, directly from slumped chaotic terrain (Figure 13.16). These were not quiet streams, but massive outpourings of water that had thousands of times the flow rates of the greatest rivers on Earth. The running water massively eroded the land (Figure 13.17), the scars showing that the water ran from Tharsis into the plains.

Direct proof of standing or running water is provided by a few terrestrial meteorites called SNCs (Figure 13.18a), the initials standing for three subcategories. These contain gases similar to those in the Martian atmosphere and were apparently blasted off the planet by meteoric impact. When sliced and analyzed, they reveal clays that appear to have been soaked in water for centuries (Figure 13.18b).

Surface water did not just run, but also froze. The southern highlands appear to have been eroded by extensive glaciers (Figure 13.19). Ice is still

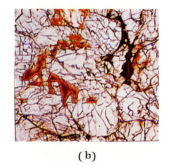

(a)   (b)

Figure 13.18 **(a)** *This SNC meteorite was probably blown off Mars's surface by a meteorite impact.* **(b)** *Another SNC has been sliced to reveal orange-brown, water-soaked clays.*

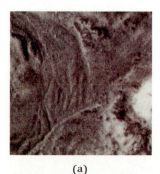
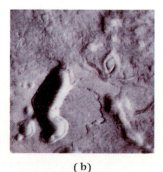

(a)   (b)

Figure 13.19 *Some landforms on Mars are similar to those on Earth produced by glaciers and include* **(a)** *erosion from a melting glacier;* **(b)** *complex terrain produced by massive disintegration of ice. Each scene is about 40 km across.*

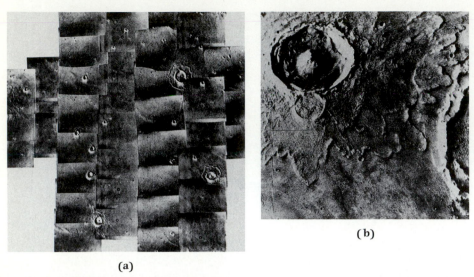

(a)

(b)

Figure 13.20   **(a)** *A mosaic of Viking images of an area 350 km wide in Chryse Planitia east of the Tharsis bulge shows a mass of rampart craters: they look as if a handful of rocks had been thrown into thick mud.* **(b)** *The lobate ejecta blanket surrounding the 18-km-wide crater Yuty is seen in detail.*

present as a constituent of the polar caps, to be discussed below.

Where did the running water come from and where did it go? Meteorite craters provide a critical clue. The ejecta blankets of meteor craters frequently have petallike lobate flow patterns ending in low cliffs that give them the name **rampart craters** (Figure 13.20). They can be explained by ice locked into the ground as permafrost. At impact, the ice melted and the ejecta flowed from the point of explosion in a homely, quick-freezing mudflow. The abundance of polar ice also makes higher-latitude craters take on a softer, almost out-of-focus appearance; the ice-logged soil is weaker, allowing the walls of the craters to slump and round over time.

Planetary scientists have sketched a fascinating picture of the planet's watery history. Like Earth (and maybe Venus), Mars was endowed with an initial supply of water. Since the planet is farther from the Sun than Earth and colder, water saturated the surface rocks as ice. Heat associated with the impact cratering of the heavy bombardment and the ancient volcanism that made the intercrater plains of the southern highlands could have melted the ground ice and caused the water to flow and cut the runoff channels. Water may even have entered an early atmosphere to reappear as precipitation. By the end of the late heavy bombardment, it had sunk back as permafrost, permeating the ground to a depth as great as 3 km.

Large-scale volcanism associated with the raising of the Tharsis bulge and the Tharsis volcanoes

again liquefied the water, except on a much greater scale: this time it gushed from the ground to create the runoff channels and chaotic terrain. Carbon dioxide released from the rocks in the process (as well as water vapor) could then have generated a temporary thick greenhouse atmosphere with a surface pressure of 2 bars or more that warmed the planet and melted the rest of the ice. As the water poured into Chryse Planitia and the northern plains, we believe it to have pooled into a large ocean—*Oceanus Borealis* (Figure 13.21)—some 1,700 m deep. The result was a warm, wet climate. Precipitation could then have created the glaciers of the southern highlands.

The water subsequently cycled back into the rock. When the driving heat from volcanism terminated, the water froze. The atmosphere disappeared into the rock with the water and dissipated into space. There may have been several such episodes triggered by periodic volcanism. Eventually, volcanic activity died away, leaving the planet cold and arid, the water locked underground forever.

13.4.4 *Viking and the Surface*

The pictures taken by the Viking landers are so clear that we feel we are looking through a window, preparing to step into the landscape. *Viking 1* landed in Chryse Planitia (Figure 13.22). The camera shows a nearly flat horizon and a surface spread with fine sand or dust that covers and surrounds thousands of rocks, some quite large. *Viking 2* (Figure 13.23) set down nearly half a planet

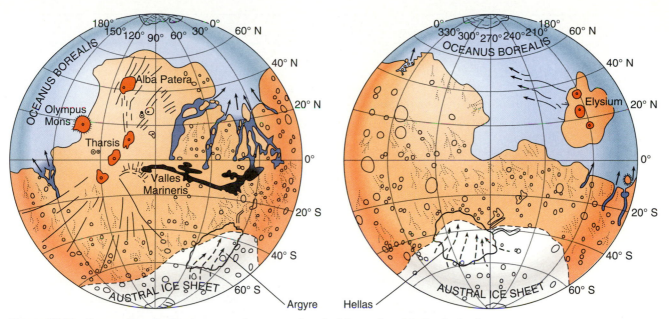

Figure 13.21   *Oceanus Borealis (blue) may once have covered much of the northern Martian lowlands with water supplied through the outflow channels (dark blue), while an ice sheet (white) covered the southern highlands. Major volcanoes are in dark red and the larger impact basins are outlined and labelled.*

Figure 13.22   Viking 1 *took this 100°-wide view of Chryse Planitia, revealing a bleak landscape filled with rocks and drifted sandy soil.*

Figure 13.23   Viking 2 *images show a rockier terrain than did* Viking 1, *with less drifted sand.*

away, on Utopia Planitia at the edge of the bulge of Elysium Planitia. Here we see more rocks and less sand.

The automated laboratories aboard the craft showed that the Martian regolith is rich in iron and is hydrated (infused with water), rather like a ter-restrial clay (consistent with the properties of SNC meteorites). The planet's red color is caused by the reaction of this iron with atmospheric oxygen—the god of war turns out to have rusty armor. The sur-face has, moreover, become crusted and compact-ed into a kind of hardpan. Both Viking craft

Your spacecraft takes you to a location near the equator of the red planet during the southern-hemisphere summer. A day there would certainly be more pleasant than one on Venus. You could not go outside without protective garments, however: the air is too thin to support human life, and there is no oxygen to breathe. Furthermore, the Sun shines almost unimpeded by the atmosphere, so lethal ultraviolet rays reach the ground. Nevertheless, you could get out of your spacecraft and easily walk around. Other than the need for a spacesuit, a Martian day might seem rather like one spent in a high, very dry, cold desert rather like central Antarctica.

At sunrise, preceded by a weak twilight, you see frost or maybe a little fog. After sunrise, these watery effects would rapidly dissipate. You might extract some ice from underground to bring back to the ship for drinking water. It is a cloudless day, though airborne dust tints the air a bit pink. The wind blows gently from the south, and by noon the ground temperature has risen nearly to the freezing point of water. As you walk, the regolith crunches under your feet, and you occasionally kick up little puffs of dust. The wind rises, you see a few dust devils, small whirlwinds, playing off in the distance.

The Sun proceeds on its daily path, as at home, and near sunset you return to the ship to turn up the electric heat and prepare for a temperature drop of 100°C. While you anticipate the next day of exploration, Deimos shines faintly, like a slowly moving star. Perhaps Phobos will rise in the west, making you long for the more familiar Moon of home.

---

apparently landed in the ejecta fields of meteorite impacts that had fractured the rocks out of the solid volcanic surface. The rocks are heavily pitted, either as a result of exploding gas bubbles or wind abrasion.

The sand grains are very fine, only a thousandth of a millimeter or so in diameter. Slow winds cause the sand drifts seen near the large boulder at the left edge of the panorama in Figure 13.22. Wind storms can easily raise the fine material and cover the entire planet in a dust storm. Much of the dust settles in the northern hemi-

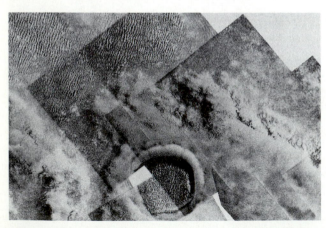

Figure 13.24 *The sand dunes of the northern polar region consist of fine material from the rest of the planet deposited by dust storms.*

sphere near the edge of the polar cap, where it has created vast dune fields (Figure 13.24). This dust—or rather the lack of it—produces the planet's dark markings as atmospheric circulation patterns blow the sandy soil free of the dark volcanic rocks.

13.4.5 The Poles

As winter comes on in each hemisphere, broad **seasonal polar caps** form when carbon dioxide "snow" precipitates out of the clear atmosphere. Beneath these are **permanent polar caps** that remain during the summers; they are very different from each other. The southern one (Figure 13.25a) is about 500 km across, the northern (Figure 13.25b) about twice that size. Even though the southern-hemisphere summer is warmer than the northern, the permanent south polar cap consists of dry ice overlying water ice and the northern one of only water ice, the reverse of what we would expect. The reason seems to be the ubiquitous dust. Because of the orbital eccentricity, the dust blows most fiercely in southern summer and is deposited in the northern ice. Dust is an efficient absorber of sunlight, and it will cause the northern cap to be become so warm in the summer that all its dry ice sublimes away, leaving only frozen water.

Both permanent caps are cut with deep valleys,

the result of warming by sunlight. The northern one shows **layered deposits** of alternating dust and ice, which appear to represent long-term climatic changes (inset, Figure 13.25b). The orbital cycles of Mars are much more extreme than Earth's. Coupled with periodic changes in the obliquity of the Martian ecliptic, they make climatic cycles similar to those that produced the terrestrial ice ages, with periods of 200,000 years or so. The polar caps preserve the record. Someday they will tell us a great deal of the history of the planet and may help explain some of the mystery still surrounding the Earth's glacial cycles.

13.5
The Interior

The interior of Mars remains mysterious. From the planet's average density (and the low density near the surface), it has an iron core that gravity measurements suggest is contaminated with some other material, probably sulfur. The size of the core is relatively similar to that of the Earth. The absence of a magnetic field implies that the core has solidified, the result of relatively rapid cooling in a small planetary mass. The mantle is basaltic like our own, and surface conditions suggest a thick crust perhaps 100 km deep.

Tectonic activity has been limited to the volcanic northern hemisphere. The activity has been mostly vertical, with no horizontal plate motion. Giant mantle plumes produced two massive bulges and a huge crustal crack (Valles Marineris) as well as several smaller ones.

13.6
Life

Only a billion years after our planet's formation, shortly after the end of the late heavy bombardment, life had begun on Earth (see Section 10.5.3). Should Mars not also have been a rich breeding ground? The planet once had liquid water, a fundamental ingredient needed for life as we know it, and a significant atmosphere. Surface conditions may once have been much more Earthlike than they are today. Did life ever emerge there? Is it active today?

The Viking cameras showed no forms of life. There are no obvious plants or animals in Figure 13.22 or 13.23, nor evidence that any ever existed. The landscape looks desolate and forbidding. However, this sterile appearance does not preclude microscopic life, which still dominates on Earth and was the precursor of larger life forms. To evaluate Mars's possible biological history, it is necessary to examine the soil, or regolith. Terrestrial life is based on organic molecules (see Section 8.1.5), those built from carbon atoms. The Viking landers carried a set of sophisticated experiments designed to search for signs of active microscopic life and for organic matter in the soil that might announce that life had once existed.

Shortly after a Viking craft settled to the Martian surface, its long arm (Figures 13.4 and 13.26) was activated to scoop up numerous soil samples. In one experiment, the dusty regolith was heated and the resulting gases examined with a mass spectrometer, a device that can deduce the chemical

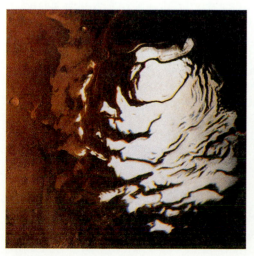
(a)

(b)

Figure 13.25 (a) The southern permanent polar cap is made of frozen carbon dioxide and water ice. (b) The northern cap is composed solely of water ice. Both are cut by deep valleys. The valley wall in the northern cap seen in the inset is layered with sand and ice.

composition of a gas by the weights of the molecules that compose it. The Vikings found no organic molecules of any kind. The regolith seems to be free of them to less than one part in a billion.

The regolith was also tested with various nutrient baths designed to feed any life forms that might be present. A gas detector then searched for simple by-products of biological activity, like oxygen or carbon dioxide. In other experiments, the regolith was exposed to baths and gases containing radioactive ^{14}C. The regolith and the gas above it were then examined by a radioactivity detector to see whether any of the ^{14}C had been incorporated or emitted, which might have indicated its consumption by microbes.

For a brief and exciting moment it looked as if Viking had found signs of biological activity. On closer examination, however, it became clear that the source of the results was not life but a chemically active regolith oddly rich in peroxides (compounds containing oxygen atoms bonded together). The peroxides are created by the action of ultraviolet sunlight, a process not possible on Earth, where atmospheric ozone screens out these rays. The eventual conclusion was that neither Viking craft had found any evidence for life.

Although the regolith was sampled at only two locations, it is fairly certain that the samples are representative, since the fierce dust storms of Mars should homogenize the surface. Nevertheless, researchers still have some reservations. We need to sample deep within the ice-saturated subsoil and within sediments produced by standing water. Ideally, we should bring extensive samples back to Earth for exhaustive testing.

Why did life apparently fail to start on Mars? Because of its low mass and low gravity, Mars could not keep a thick atmosphere for long, and there was insufficient tectonic activity over the planet's

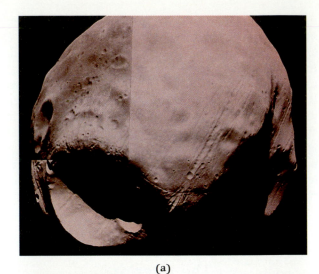

(a)

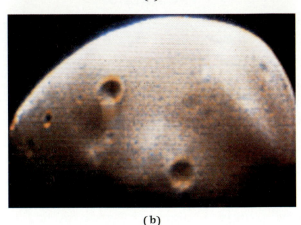

(b)

Figure 13.27   *The Martian moons, Phobos* **(a)** *and Deimos* **(b)**, *are heavily cratered. The large crater on Phobos is 10 km across. The fractures radiating away from it were probably created by the impact.*

history to resupply the gas from the interior. As volcanic activity wound down, the carbon stayed in the rocks, the greenhouse effect diminished, most of the atmosphere escaped for good, and the planet turned into a giant refrigerator. Closer to the Sun, Venus became a runaway oven. And here we are in the middle, with our relatively high mass (which keeps plate motion going) and just the right amount of solar heating. Our neighbors are teaching us something about our own precious world.

13.7
The Martian Moons

Mercury and Venus have no satellites. The Earth's is so big and important we devoted a whole chapter to it. In contrast, Mars has a pair of odd little moons (Figure 13.27). Phobos and Deimos (from the Greek for "fear" and "panic," the war god's atten-

Figure 13.26   *The scoop on* Viking 1 *dug several trenches in the soil before feeding samples into a variety of experiments designed to search for life.*

dants) were discovered at the U.S. Naval Observatory by Asaph Hall in 1877. Difficult to see from Earth, these bodies have a curious history. Their existence was first suggested by Kepler on the basis of the numerical progression of satellites outward from the Sun and subsequently popularized by Jonathan Swift in *Gulliver's Travels* 150 years before their actual discovery.

Both satellites orbit close to the planet, counterclockwise in its equatorial plane, and in synchronous rotation. Deimos, the outer one, is 23,460 km from the planet's center and has a sidereal period of only 30^h18^m. Thus it would appear to move slowly on a daily path to the west, taking nearly 5.3 Martian days between successive moonrises. Phobos is considerably closer, 9,378 km from the Martian center and 6,000 km from the surface. Its sidereal period is 7^h39^m, less than the Martian rotation period, and it therefore rises in the west and sets in the east.

Phobos and Deimos were probably not formed with Mars, but were most likely captured from the **asteroid belt,** a zone of rocky and metallic debris that lies mostly between Mars and Jupiter (to be examined in detail in Chapter 17). Such a capture requires an external force to slow the incoming body (see Section 11.7); friction with a once-thick atmosphere may have done the job. The satellites would have been captured while in the ecliptic plane, and over the years the rotation-induced equatorial bulge of Mars would have dragged them into its equatorial plane.

The satellites are too small for their gravity to shape them into spheres, and they remain irregular blocks. Phobos is roughly 20 × 28 km across, little Deimos 12 × 16 km. They are darker than our Moon, with albedos of only 6%, and both are heavily cratered. Phobos has a remarkable crater 10 km in diameter, about half the size of the satellite itself.

TABLE 13.2
The Moons of Mars

Moon	Radius (km)	Distance from Planetary Center (km)	Orbital Period
Phobos	10 × 14	9,378	7^h39^m

Satellite Properties

Small, rocky, and heavily cratered; probably captured asteroids.

A little larger, and the moon would probably have broken in half, providing further support for the hypotheses of planetary structure and properties that involve catastrophe. We will find more as we enter the realm of the Jovian planets.

KEY CONCEPTS

Asteroid belt: A zone of debris that lies mostly between the orbits of Mars and Jupiter.

Canals: Illusory linear features on Mars caused when the observer's eye links real but actually separate features.

Chaotic terrain: A landform on Mars with an irregular assembly of large blocks of rock and fractures.

Dry ice: Frozen carbon dioxide.

Layered deposits: Layers of ice and dust in the Martian polar caps.

Outflow channels: Large channels created by eruptions or releases of vast amounts of water.

Pateras: Ancient collapsed shield volcanoes.

Permanent polar caps: Polar caps that never dissipate, consisting of dry ice (and water ice) in the south, water ice in the north.

Rampart craters: Craters, unique to Mars, whose ejecta have flow patterns that end in cliffs.

Runoff channels: Ancient dry riverbeds, evidence of low-volume water flows.

Seasonal polar caps: Dry-ice polar caps that extend well toward the equator in winter.

EXERCISES

Comparisons

1. Compare the rotational characteristics of the Earth and Mars.
2. Compare the sizes and compositions of the Martian seasonal and permanent polar caps.
3. How and why do the permanent Martian polar caps of the north and south differ?
4. Compare the atmospheres of Mars, Earth, and Venus.
5. How do the volcanoes of Mars differ from those of the Earth?
6. Distinguish between outflow channels and runoff channels. What do they tell us of early Martian conditions? How do they differ from the canals?

Numerical Problems

7. What is the average flux of sunlight on Mars relative to that on Earth?
8. What is the latitude of the Martian arctic circle?

Thought and Discussion

9. What might Mars be like after half a precessional cycle?

10. Why does Mars lack a magnetosphere?

11. How do we know there is water vapor in Mars's atmosphere?

12. Can it rain on Mars? Explain your answer.

13. Why are the largest Martian volcanoes atop bulges?

14. What is the evidence for tectonic activity on Mars?

15. Water was once abundant on Mars. How do we know?

16. Why do the dark markings on Mars change with the seasons?

17. How do we know that the southern highlands are older than the northern plains?

18. What is meant by "chaotic terrain"? With what is it associated?

19. What is the origin of the layered deposits in Mars's north polar cap?

20. What conditions on Mars are lethal to human beings?

21. What do we know about conditions in the Martian core? On what evidence?

Research Problem

22. Read a science-fiction novel about Mars. List the conditions the author adopts and evaluate them in the light of modern knowledge.

Activities

23. Photocopy Figures 13.1, 13.7, 13.8, and 13.10. Label the prominent features and identify as many as you can, using the map in Figure 13.6.

24. If Mars is near opposition, use a telescope to sketch the surface features. Particularly note the polar caps and their sizes.

Scientific Writing

25. A local astrologer has published a pamphlet that announces disaster because Mars is coming into opposition with the Sun. Counter with another pamphlet that explains why a Martian opposition is actually a positive event that will tell us more about both the red planet and the Earth.

26. A friend who has read Percival Lowell's old ideas about life on Mars has become convinced that Martians exist. Write a letter giving counterevidence, from the past and present, that shows why Lowell was wrong.

14

Magnificent Jupiter

The giant of the planetary system is more closely related to the Sun than to the Earth

Spinning eddies accent Jupiter's turbulent, colorful, cloudy atmosphere.

Huge Jupiter is entirely different from the five little worlds near the Sun. Its characteristics are in some ways more solar than terrestrial, and with its retinue of satellites, it behaves something like a miniature Solar System.

14.1
The View from Earth

Jupiter is one of the finest telescopic sights of the sky, displaying a large disk—47 seconds of arc across at opposition—from which we derive an astonishing equatorial radius of 71,500 km, 11.2 times that of Earth. (Jovian characteristics are summarized in Table 14.1.) The giant planet is noticeably oblate, with a 7% smaller polar radius of 66,900

km. The surface displays a series of stripes parallel to the equator, with lighter **zones** separated by darker **belts** (Figure 14.1). Though variable in brightness and width, the belts and zones are permanent features. A modestly bright *equatorial zone* roughly 15° wide is bordered by dark *equatorial belts,* and beyond them are bright *tropical zones* at about 30° north and south latitude. This alternating pattern becomes increasingly complex toward the poles.

The horizontal stripes are constructed of intricate networks of clouds at different temperatures and, probably, different altitudes. A great deal of subtle color can be seen, with reds, browns, and even blues intermixed with white. Within the belts and zones lie a variety of white and brown ovals.

TABLE 14.1
A Profile of Jupiter

Planetary Data		*Atmosphere*	
distance from the Sun	5.2 AU	***Composition by Number*[a]**	
equatorial radius	71,492 km = 11.2 Earth radii	hydrogen	95.5%
polar radius	66,854 km	helium	4.4%
mass	1.90×10^{27} kg = 318 Earth masses	methane (CH_4)	0.09%
mean density	1.33 g/cm^3	ammonia (NH_3)	0.02%
surface (cloud deck) gravity	2.4 g$_{Earth}$	water	0.01%
escape velocity (cloud deck)	59.6 km/s	ethane (C_2H_6)	trace
rotational period (interior)	$9^h55^m30^s$	acetylene (C_2H_2)	trace
axial inclination	3.1°	phosphine (PH_3)	trace
magnetic field strength	14 times that of Earth	other gases	trace
tilt of magnetic field axis	9.6°		
effective temperature	124 K		

Layers		% Mass	Radii (km)
rock and ice core		4	0–13,000
central temperature	25,000 K		
central density	50 megabars		
liquid metallic hydrogen mantle			13,000–50,000
liquid molecular hydrogen mantle			50,000–68,000[b]
gaseous atmosphere			68,000–69,000[c]

Planetary Features

Cloud cover: Apparent surface consists of deep clouds only; no solid visible surface.

Magnetosphere: 2-million-km radius; highly distorted by solar wind; contains extended magnetodisk.

Surface features: Dark belts and bright zones parallel to equator; many dark ovals; Great Red Spot in southern hemisphere; highly turbulent.

Temperature: Effective temperature, 124 K, is 15 K hotter than expected because of internal heat caused by contraction.

Winds: Direction correlates with belt-zone pattern; 350 km/s prograde wind in equatorial zone; winds reverse direction within equatorial belts, becoming retrograde.

[a]Abundances of atoms and molecules are relative to the number of hydrogen atoms, not hydrogen molecules.
[b]Liquid mantle blends gradually into gaseous atmosphere.
[c]Mean radius.

North tropical zone
North equatorial belt
Equatorial zone
South equatorial belt
South tropical zone

Figure 14.1 *Jupiter displays alternating sets of bright zones and dark belts. Latitudes are shown at left. The Great Red Spot is to the left at about –23°*

By far the most prominent of them is the **Great Red Spot** (GRS), a reddish region twice the size of Earth that lies in the southern hemisphere at a latitude of about 25°S.

Jupiter's remarkably short rotation period, derived from cloud markings, depends on latitude. Within the bright equatorial zone, the planet spins every $9^h50^m30^s$, an interval called **System I.** Well outside this region, the rotation period is $9^h55^m41^s$, known as **System II.** Jupiter is therefore in **differential rotation,** the gases of its atmosphere moving past one another at high speed. The planet's spin axis has a tilt of only 3° relative to the orbital perpendicular; as a result, there are no significant seasonal changes during the 12-year sidereal orbital period.

Even binoculars will show you Jupiter's four bright moons, discovered by Galileo in 1609. Their positions change continuously as they orbit the planet, presenting an ever-entertaining sight. By applying Kepler's generalized third law to these **Galilean satellites,** we find Jupiter's mass to be 1.90×10^{27} kg, 318 times that of the Earth, almost a thousandth that of the Sun, and twice that of all the other planets put together! The resulting density of 1.3 g/cm³, only 30% greater than that of water, is dramatically different from the densities of the terrestrial planets. The immediate implication is that Jupiter is structured much differently from these smaller bodies and is made of much lighter materials.

In keeping with its other superlatives, Jupiter—with the exception of the Sun—is the strongest radio source in the sky. The radio data are divided into three broad domains. At the shortest wavelengths, those of a few centimeters and into the infrared, Jupiter behaves like a blackbody, producing **thermal radiation** because of its warmth. Its effective temperature, first derived from the radio luminosity and the average radius (later, more ac-

curately, by spacecraft) is a chilly 124 K (–146°C). That value may not seem unreasonable, given Jupiter's great distance from the Sun, but sunlight should heat the planet to a blackbody temperature of only 109 K, 15 K lower. Because the luminosity of a blackbody is proportional to T^4, the planet radiates nearly $(124/109)^4$ or 1.7 times as much energy as it gets from the Sun. Consequently, it must have its own powerful internal heat source.

At longer wavelengths, Jupiter produces **nonthermal radiation** generated by processes that have nothing to do with blackbodies or heat. In the decimeter (tens of centimeters) wavelength range, the planet emits **synchrotron radiation** caused by electrons moving near the speed of light and trapped in magnetic fields. Since an electron is a charged particle, a magnetic field will continuously deflect it (Figure 14.2a), accelerating it into a spiral path around a magnetic field line and causing it to radiate in the direction of its motion. Theory shows that the total energy radiated by the electron gas will be proportional to an inverse power of the frequency. The spectral-energy distribution (Figure 14.2b) will then look very different from that of a blackbody, allowing us to distinguish immediately between the synchrotron and thermal mechanisms. The existence of synchrotron radiation clearly means that Jupiter has a powerful magnetic field and a radiation belt at least somewhat analogous to the radiation belts around the Earth.

At wavelengths of tens of meters, we observe **decameter bursts** of nonthermal radiation produced by a process more complex than the synchrotron mechanism. The bursts vary regularly with a **System III** period of $9^h55^m30^s$, close to the System II rotation period. The nonthermal radiation is caused by the Jovian magnetic field, which is produced by and anchored in the planet's deep core. Therefore, System III is the rotation period of

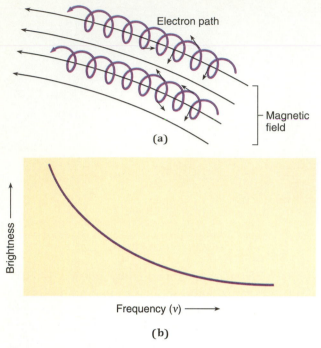

Electron path

Magnetic field

(a)

Brightness

Frequency (ν)

(b)

Figure 14.2 **(a)** *A high-speed electron caught in a magnetic field will be deflected into a continuous spiral path. As it accelerates, it will radiate energy in the direction of its motion.* **(b)** *The sum of the radiation of all electrons, which are moving at different speeds, has a spectrum that depends on a power of the frequency.*

the interior and of the vast mass of the planet. The shorter equatorial period is produced by winds that blow the clouds from west to east at a fierce 350 km/hr. The winds tend to move more slowly at higher latitudes, giving the effect of a longer period.

14.2
Spacecraft

To understand Jupiter properly, we must move in closer. The first spacecraft exploration was made by *Pioneer 10,* launched in 1972, which flew past the planet in 1974. *Pioneer 11* followed a year later, and Jupiter's gravity sent it on to Saturn (see Section 7.7). These craft had relatively crude imaging devices and detectors with which to explore Jupiter's magnetic field.

The real heroes of Jovian exploration, if not of the whole U.S. space program, were the twin Voyagers (Figure 14.3). *Voyager 2* was launched on August 20, 1977, and *Voyager 1* two weeks later. However, *Voyager 1* was the first to arrive at Jupiter, on March 5, 1979; *Voyager 2* passed by on

Figure 14.3 **(a)** *The orbits of the twin Voyagers took them past the Jovian planets into interstellar space.* **(b)** *The craft, dominated by its large communications antenna, also carries guiding rocket motors, a boom with a plutonium dioxide power pack to generate electricity from the heat of radioactive decay, a long rod that carries magnetometers, and a camera boom that also carries particle detectors and spectrometers.*

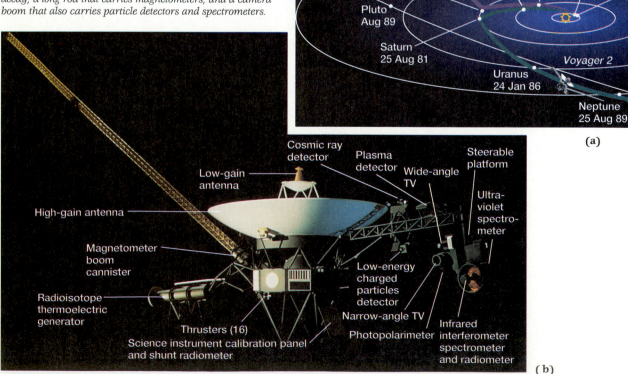

July 9 of the same year. Jovian gravity then propelled them both past Saturn in 1980 and 1981. *Voyager 1* was programmed to pass Saturn's largest satellite, Titan, at which point the craft was ejected by the planet from the plane of the Solar System. *Voyager 2,* however, used Saturn's gravity for a trip to Uranus that took over four years. Uranus then shot it to Neptune, where it arrived on August 25, 1989, almost exactly a dozen years after launch. Along with the Pioneers, they have left the planetary system behind, never to return.

The Voyagers' 11 experiments (Figure 14.3b) compiled an astonishing record of discovery. The public became most familiar with the wide- and narrow-angle television cameras, which sent back extraordinary pictures. Equally important were devices that detected and measured magnetic fields and charged particles, as well as various spectrometers (digital spectrographs). All the information acquired was transmitted to large radio telescopes on Earth. The Voyagers' plutonium power packs are expected to work until about the year 2020 as the hardy spacecraft continue to explore the immediate environment outside the planetary orbits. The spectacular results to come are a tribute to the hundreds of scientists and engineers who worked so long on the project.

14.3
Jupiter's Interior

No one is ever likely to penetrate Jupiter very deeply: even a short way below the clouds conditions are lethal to both human and robotic probes. We must determine what is inside by what we see on the outside. The low density of 1.3 g/cm^3 indicates that the planet is made mostly of hydrogen, the lightest and most common element in the Universe. In support, spectroscopy shows abundant atmospheric molecular hydrogen (H_2).

One of the chief tools in the study of the interior is the concept of **hydrostatic equilibrium.** Within a body, pressure increases toward the center because of the increasing weight of the overlying layers. In the thin layer of matter in Figure 14.4, the pressure at the bottom is greater than at the top, providing an outward shove that exactly compensates the inward pull caused by the layer's weight. Consequently, the layer maintains a stable position. This principle helps us build mathematical **models** of planets, with which we calculate the temperature, pressure, and state of matter at each point necessary for the layers to be self-supporting.

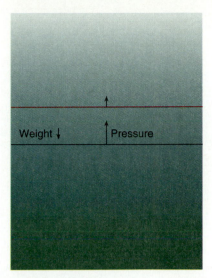

Figure 14.4   *In hydrostatic equilibrium, the outward push caused by the pressure difference across a layer will support that layer's weight, keeping it from rising or falling.*

For Jupiter to have its observed size and mass, it must be mostly hydrogen with about 8% helium (by number of atoms), roughly the same composition as the Sun. The model also shows that temperature and pressure climb rapidly within Jupiter, reaching 25,000 K and 50 megabars at the center, respectively nearly 4 and over 13 times the values found inside the Earth.

Like the Earth, Jupiter is a layered body, although the details of its structure are very different (Figure 14.5). A spinning body's oblateness depends on both its rotational speed and how its matter is concentrated. Detailed analysis of Jupiter's shape shows that its mass is strongly condensed to the center and that it has a core consisting of 12 or so Earth masses of **"rock"** and **"ice"** (about 13,000 km in radius). In this context, these terms do not have their normal meanings. Rock here refers to common heavier elements like silicon, oxygen, and iron; ice is a generic term for lighter compounds like water, methane, and ammonia, and the word is used even if these substances are in the liquid state. If Jupiter were closer to the Sun, solar heat would have driven away the light hydrogen and helium and the remaining core might resemble our own planet, though it would be vastly larger.

Surrounding the core is a mantle made of hydrogen and helium in unusual states. At the low temperatures near the visible surface, hydrogen is in its gaseous molecular form; as we proceed inward, the pressure (calculated from the model) climbs so high that near a depth of about 1,000 km, the molecular hydrogen gradually begins to liquefy

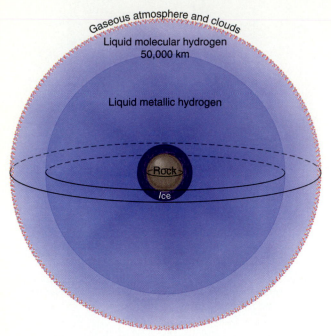

Figure 14.5 *A model of Jupiter shows a hot core of heavy elements called "rock" (silicon, iron, and others) and "ice" (lighter materials solidified or liquefied despite high temperature). Surrounding the core is an inner mantle of liquid metallic hydrogen, encased by an outer mantle of liquid molecular hydrogen. About 8% of the atoms are helium. The liquid H$_2$ gradually merges with the gaseous atmosphere we see*

(there is no real "surface"). At a depth of about 20,000 km (about a quarter of the way to the center), the pressure is so great that we expect the liquid hydrogen to behave like a metal and conduct electricity. Such high pressures have never been achieved in the laboratory, however, so the theory remains uncertain.

Unlike the Earth, which is heated internally by core crystallization and radioactive decay, Jupiter is still hot from its formation 4.5 billion years ago. As it cools, it contracts and slowly releases its internal energy, which explains why it is 15 K hotter than we would expect for a blackbody of the same size.

14.4
The Atmosphere

The Jovian atmosphere is a wonderful natural laboratory in which we can see bizarre analogies of terrestrial weather phenomena—meteorology carried to remarkable extremes, giving us deeper insight into our own world.

14.4.1 Cloud Temperatures and Winds

Since Jupiter has no solid surface, astronomers use as a reference point an atmospheric pressure of 100 millibars, which is the pressure at a level slightly

above the cloud tops and close to the surface as perceived by eye (Figure 14.6). Measurements by the Voyagers show that temperature rises with increasing depth from a minimum of 100 K near the 100-millibar level to over 300 K at a level 125 km farther down.

The optical appearance of the belts and zones depends not on temperature but on reflective properties. In the infrared the belts are bright and the zones are dark, suggesting that the belts are warmer than the zones and may be gaps in the upper cloud deck through which we can see the lower cloud layers. Atmospheric colors have been tentatively associated with different depths and temperatures. Red clouds appear to be the highest, with white, brown, and finally blue clouds each about 20 km progressively farther down.

Voyager observations of cloud movements compared to the System III (interior) rotation period yield wind velocities (Figure 14.7). Winds blow either east or west but display little northerly or southerly flow. The equatorial zone is a broad avenue of Jovian air moving from west to east or *prograde* (in the direction of rotation), with a fairly steady velocity of 350 km/hr relative to that of the underlying planet—explaining why the System I rotation period is five minutes shorter than System III. In the north and south equatorial belts the winds reverse directions, their speed decreasing with increasing latitude until the winds are blowing east-to-west (retrograde). The atmosphere is **shearing,** as winds at different latitudes move past one another at different speeds. The phenomenon may

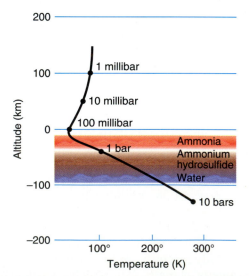

Figure 14.6 *Temperature drops with increasing altitude in the Jovian atmosphere, as it does on Earth. Near the 100-millibar level, it suddenly increases in a manner reminiscent of the Earth's thermosphere. Clouds of different color and composition lie at different depths.*

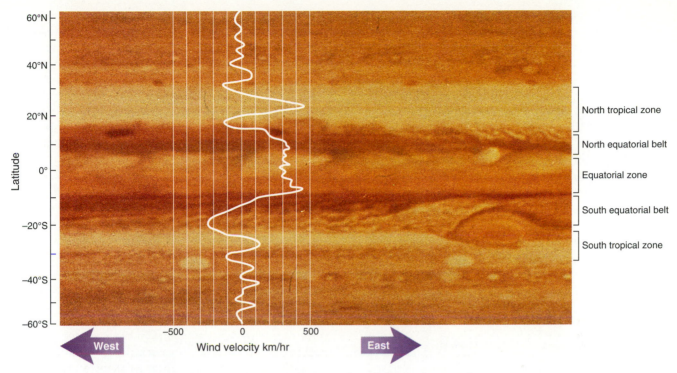

Figure 14.7   *Computer processing allows a Voyager image of Jupiter to be spread into a flat map; the equator runs from left to right. The graph shows the speed of the winds relative to the System III rotation period. Note the complexity of the clouds and the numerous ovals.*

clear clouds from the upper atmosphere, allowing us to look to the lower clouds below. Farther from the equator the winds reverse again, with distinct differences between the two hemispheres. Oddly, the greatest wind speed is in a narrow, dark region in the north tropical zone. Toward the poles the reversals become smaller, correlating with the less obvious distinctions between zones and belts.

Wind shear is associated with beautiful, turbulent patterns in which individual swirling eddies (see the chapter opening image) come and go, the atmosphere always changing its appearance. The tempestuous Jovian air, like that of our own planet, is electrically charged. When the Voyagers flew around the planet and looked back at the night side, it was illuminated by mammoth lightning bolts.

Two competing (but perhaps complementary) theories attempt to explain Jupiter's appearance. One holds that Jupiter's meteorology is like Earth's, only sped up: the cyclonic patterns are stretched into lines parallel to the equator by the planet's high rotation speed (Figure 14.8a). An alternative suggests that the interior of the planet is constructed of concentric cylinders (Figure 14.8b) rotating at different speeds relative to one another. The belts are their outward manifestations, formed where the ends of the cylinders are cut off at the planet's sur-

face. In either case, the lack of atmospheric movement to the north and south is confounding. Jupiter is much more strongly heated by the Sun at the equator than it is at the poles, yet the temperatures of the two regions are nearly the same. Whatever mechanism transports heat to the poles works below the visible limits imposed by the clouds.

Above the top cloud deck, the atmosphere thins and the temperature climbs. Ultimately, a great thermosphere—an ionosphere—heated and ionized by solar ultraviolet radiation extends 3,000 km upward from the visible surface.

14.4.2 *Chemical Composition*

As expected, spectroscopy shows that the atmosphere is dominated by hydrogen and helium. Although helium has no readily available absorption lines, its abundance can be measured by the effect it has on the hydrogen lines as a result of collisions between helium atoms and hydrogen molecules. The ratio of helium to hydrogen (again, by number of atoms, not molecules) is about half that derived from the model of the planet, only 0.044. Much of the helium, which is heavier than hydrogen, has sunk under the influence of Jupiter's strong gravity.

Jupiter's atmosphere has an extraordinarily rich chemistry. Absorption lines of both methane

(CH$_4$) and ammonia (NH$_3$) were found in the 1930s. Modern spectroscopy (Figure 14.9) reveals traces of other **hydrocarbons** (hydrogen-carbon molecules) including ethane, acetylene, and a dollop of propane (C$_3$H$_8$), made by sunlight acting on methane; the process also produces the temperature inversion above the clouds. Water and two unusual chemicals are also observed, phosphine (PH$_3$), and, of all things, germane (GeH$_4$), a compound like methane in which germanium has been substituted for carbon.

These chemicals apparently condense to produce clouds at different temperatures, hence at different altitudes. The highest are almost certainly made of ammonia crystals (see Figure 14.6). Theoretical calculations indicate that the next layer down ought to be ammonium hydrosulfide

crystals (NH$_4$SH), made by the interaction of ammonia and hydrogen sulfide (H$_2$S). Confusingly, however, there is no atmospheric sulfur. At a lower level, we expect familiar frozen water clouds. All these clouds should be white, not at all like Jupiter's rather colorful disposition. There must be many trace materials, based perhaps on sulfur or phosphorus, or even complex organic compounds, that make the reds, browns, and blues that we see.

14.4.3 Ovals and the Great Red Spot

Everywhere within the Jovian atmosphere we find strange ovals, some white, others dark. The dark features (see Figure 14.7) appear to be holes through which we are seeing lower layers. They are observed exclusively at a latitude of about 14° in

Figure 14.8 (a) *A model of Jupiter's atmosphere derived by speeding up a model of the Earth generally recreates Jupiter's appearance.* (b) *In another model, cloud belts and zones are produced by the tops of rotating cylinders that extend all the way through the planet*

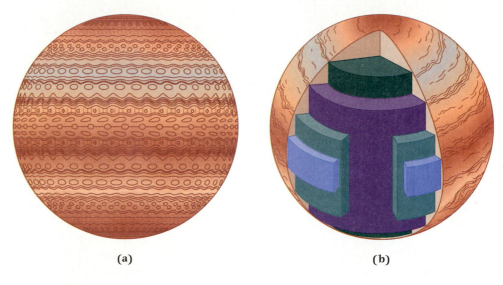

(a)   (b)

Figure 14.9 *The infrared spectrum of Jupiter, taken by Voyager, shows the absorption lines of a variety of chemical compounds, including ammonia, methane, water, a methane molecule with a deuterium atom (C$_2$H$_3$D), and germane (GeH$_4$).*

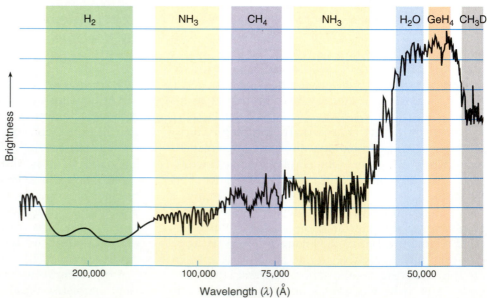

Figure 14.10   *The Great Red Spot of Jupiter, an atmospheric anticyclone, is associated with enormous turbulence as winds blow past it*

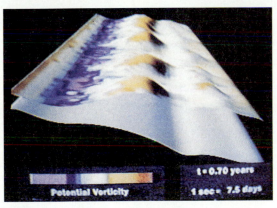

Figure 14.11   *A computer model of the Great Red Spot shows it developing from rising columns of gas that roll in the boundary between a belt and a zone.*

the north equatorial belt. White ovals are seen only in the southern hemisphere. No one knows why there is such clear discrimination.

None of these features, however, can compare with the Great Red Spot (Figure 14.10), discovered by Giovanni Cassini in 1665. Although highly variable in appearance, we know it has been present for over 330 years. It is immense, 26,000 km by 14,000 km, big enough to hold two Earths side by side, and is apparently trapped between the south equatorial belt and the south tropical zone in a giant wind shear. As a result, the GRS is forced to roll like a great ball bearing counterclockwise with a spin period of about six days. We can actually see the winds blow past it, diverging around the large feature.

The spin direction of the GRS is anticyclonic (see Section 10.5.2), showing that the spot is a

region of high pressure. It slowly drifts back and forth both in longitude and latitude, as if floating in the atmosphere below. Its reddish color and dark appearance in the infrared show that it is cold and probably elevated above its surroundings. Computer modeling (Figure 14.11) shows that rising Jovian "air" will in fact form elevated domes that can stabilize between the oppositely blowing belt and zone. The energy that keeps the GRS going must be supplied continuously from below by the planet's internal heat. Why there is only one spot, and none in the northern hemisphere, is a mystery.

14.5
The Magnetosphere

Of all Jupiter's features, perhaps the most intriguing is the immense magnetosphere (Figure 14.12) caused by the powerful magnetic field generated in Jupiter's fluid, electrically conducting interior by convection and rotation. To a loose approximation, the field is a dipole passing not quite through the center of the planet. Like the Earth's, the field is tilted by about 10°, but it is oppositely directed, with magnetic north near the north rotation pole. (The Earth's north magnetic pole is technically a south pole since the north pole of a compass needle points toward it.) Because the Earth's magnetic field switches directions periodically, the difference in magnetic orientation between Jupiter and the Earth is of no significance. Electrons trapped in this field close to Jupiter produce the nonthermal decimeter emission.

The magnetic field is subject to two influences, one from the Sun, the other from the innermost Galilean satellite, Io. The satellite is covered with sulfur-spewing volcanoes. Some of the volcanic material escapes into space and is ionized. These ions are trapped in the magnetic field along with those that enter from the solar wind. Since the field rotates with the planet in less than 10 hours, the ions move much faster than the satellite, which orbits in 1.8 days. As the magnetic field whips past Io and slams ions into it, other atoms (including those of sodium) are sputtered from the surface. The result is a doughnut-shaped ring or **torus** filled with an energetic plasma of ions and electrons coincident with the satellite's orbit, the **Io plasma torus** (Figures 14.12 and 14.13).

In part as a result of Jupiter's fast rotation, the plasma spins itself out into a huge **magnetodisk** that lies roughly parallel to the Jovian magnetic equator. Because of the tilt of the field axis relative to the rotation axis, the magnetodisk takes on a

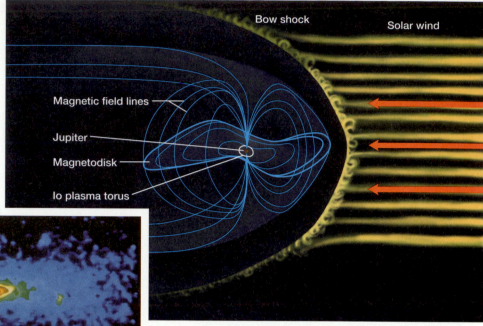

Figure 14.12 *The solar wind hits and distorts Jupiter's magnetic field, producing a bow shock. Sputtering from Io creates a plasma torus in Io's orbit. Rotation produces the magnetodisk in the center. From Earth we see a variety of phenomena (inset), including a neutral sodium cloud spreading throughout a disk 6° wide.*

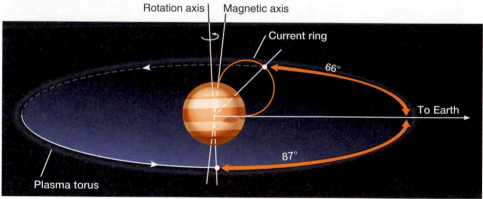

Figure 14.13 *Io is less than six Jovian radii from the planet, deep inside the powerful magnetosphere. Ions released from its surface spread into a plasma torus. A current ring associated with the decameter bursts connects Io to Jupiter. The bursts are strongest when Io is 87° ahead of the position in which it crosses (transits) Jupiter as viewed from Earth, and again 66° after transit.*

wavy form as the planet spins. Like the Earth's magnetosphere, Jupiter's magnetic environment is highly distorted by the impact of the solar wind, which produces a bow shock (see Section 10.6) and blows the field into a long magnetotail that stretches beyond Saturn's orbit five AU away!

The Jovian magnetosphere is enormous, extending well past the outermost Galilean satellite, Callisto, more than 2 million km away. From Earth, the magnetosphere with its embedded magnetodisk is one of the largest structures in the sky, several times the angular width of the full Moon (see inset, Figure 14.12). A vast amount of energy is involved: at its most intense, the density of the particle radiation is thousands of times the fatal human dose.

Figure 14.14 *Voyager imaged Europa and orange Io as they transitted the face of the planet. Comparable to the Earth's Moon in diameter, they put the size of the giant planet and even the Red Spot in vivid perspective.*

Io is responsible for other dramatic effects. The swiftly rotating magnetic field electrifies the satellite as it whips past, causing a ring of electrical current to flow between Io and Jupiter (see Figure 14.13). This ring appears to be responsible for the decameter bursts, which take place close to the planet's atmosphere and are strongest when the satellite is 87° east (as viewed from Earth) and 66° west of the line that connects the Earth with Jupiter. However, the mechanism that generates the bursts is far from understood.

14.6
Jupiter's Satellites

In the outer Solar System, satellites take on much greater significance than they do among the terrestrial planets. Jupiter commands four large satellites. Mixed with them are a dozen small bodies, increasing Jupiter's total known retinue to 16.

14.6.1 *The Galilean Satellites*

Io, Europa (both seen in Figure 14.14), Ganymede, and Callisto were all discovered by Galileo. They make up a system (Table 14.2 and Figure 14.15) that stretches from Io, 400,000 km from Jupiter (5.9 Jupiter radii, R_J) to Callisto at 1.9 million km (26.3 R_J). Io orbits in a mere 1.8 days, whereas more distant Callisto takes 16.9 days to make a circuit.

The Galilean satellites have a number of common characteristics. They all move counterclockwise along nearly circular orbits in the plane of Jupiter's equator. Tides induced by Jupiter have made all four synchronous rotators, so each keeps the same face pointed toward the planet at all times. Radius measures, difficult from the Earth, are best made by passing spacecraft. They range from 1,570 km (just smaller than the Moon) for Europa to 2,630 km (8% larger than Mercury) for Ganymede. The determination of masses is harder, requiring the observation of mutual gravitational

TABLE 14.2
The Jovian Satellites

Name	Distance (10^6 km)	(R_J)[a]	Period (days)	$i°$	e	Radius (km)	Mass (10^{21}kg)	Density (g/cm^3)
Metis	0.1280	1.79	0.295	0	0.00	20		
Adrastea	0.1290	1.80	0.298	0	0	12 × 8		
Amalthea	0.1813	2.54	0.498	0.4	0.00	135 × 75		
Thebe	0.2219	3.11	0.675	0.8	0.01	50		
Io	0.4216	5.89	1.769	0.04	0.00	1,815	89.4	3.57
Europa	0.6709	9.38	3.551	0.47	0.01	1,569	48	2.97
Ganymede	1.070	15.0	7.155	0.19	0.00	2,631	148	1.94
Callisto	1.883	26.3	16.689	0.28	0.01	2,400	108	1.86
Leda	11.09	155	239	27	0.15	8		
Himalia	11.48	161	251	28	0.16	90		
Lysithea	11.72	164	259	29	0.11	20		
Elara	11.74	164	259	28	0.21	40		
Ananke	21.2	297	631 r[b]	147	0.17	15		
Carme	22.6	316	692 r	163	0.21	22		
Pasiphae	23.5	329	735 r	147	0.38	35		
Sinope	23.7	332	758 r	153	0.28	20		

Satellite Features

Galileans: Outer two have silicate cores, deep icy mantles, surfaces of dirty ice; inner two higher density, fewer volatiles, and smaller icy mantles because of proximity to Jupiter; Io highly volcanic (sulfur) as a result of tidal heating; surface ages of satellites increase with distance as tidal heating declines.

Outer satellites: All small and rocky; two groups of four; outer group retrograde; probably shattered captured bodies.

Inner satellites: Three small, Amalthea intermediate size; all rocky.

[a]Jupiter radii.
[b]Retrograde orbit.

(a)

Figure 14.15   (a) *The Galilean satellites (exaggerated in size by a factor of 6) are shown at their maximum distances from Jupiter. The four small inner satellites and the main ring are also indicated. On this scale, the two sets of outer satellites respectively lie about 0.6 and 1.2 meters away from Jupiter.* **(b)** *The Galilean satellites are drawn to scale relative to the limb of the planet*

(b)

BACKGROUND 14.1 Jupiter and Science

This extraordinary planet has had an equally extraordinary scientific history. With the invention of the telescope, Jupiter and its family immediately showed Galileo that Copernicus was probably right. Here were tiny satellites, swinging around night after night; if Jupiter could influence other bodies and keep them in orbit, so could the Sun.

Lesser known is Jupiter's contribution to basic physics, which shows the powerful interdependence of the sciences. As Jupiter's Galilean satellites orbit, they are *occulted* by the planet— that is, they are occasionally hidden as they swing behind it. They are also invisible from Earth when they *transit,* or cross in front of the planetary disk. If Jupiter is not at opposition, its shadow will project to the side (Figure 14.16), and occasionally one will see a satellite go into eclipse.

In 1675, Olaus Römer, an astronomer from Denmark working at the Paris Observatory, made a detailed study of these eclipses. He found that when the Earth was approaching Jupiter (at A in Figure 14.16), the interval between eclipses was shorter than when the Earth was receding (B). The difference is caused by the finite velocity of light.

Observe an eclipse at a_1. At a_2, you are closer to Jupiter. The light does not have to travel as far, and the eclipse occurs earlier than would be expected. Then observe again at b_1. Now you are

moving away and the light must travel farther, so the eclipse arrives at b_2 late. From the difference in timings between A and B, Römer found that it must take light 11 minutes to cross the gap between the Sun and the Earth (by modern measurement the figure is 8 minutes). At about the same time, the absolute scale of the Solar System was first established by parallax, so the distance to the Sun and Jupiter could be found. The Sun-Earth distance divided by the light-travel time between the two bodies yields the speed of light, which Römer had found to within about 30% of its true value. With this observational experiment, Römer also provided the first concrete proof that the Earth is moving and in solar orbit.

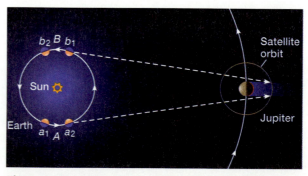

Figure 14.16 *From points A and B in the Earth's orbit, we can see Jupiter's shadow and watch satellites being eclipsed. Because the Earth is moving and because of the finite speed of light, the eclipse intervals between a_1 and a_2 and between b_1 and b_2 will not be the same.*

perturbations and perturbations experienced by spacecraft. Masses range from 0.65 that of the Moon for Europa, the smallest, to double lunar for Ganymede.

The satellites' densities offer the first surprise. The inner two, Io and Europa, have densities of 3.57 and 2.97 g/cm³ respectively, comparable to that of the Moon. They are probably made of similar materials, silicate rock perhaps. However, the densities of the outer two, Ganymede and Callisto, are significantly lower, only 1.94 and 1.86 g/cm³. They are too light to be made just of rock. About half their mass is probably frozen water—*real* ice.

Images taken by the Voyagers show that the four satellites are in fact wildly divergent, the differences and surface ages correlating with distance from Jupiter. Callisto (Figure 14.17), the most distant, has the most ancient surface and is heavily covered with craters from the late heavy bombardment. Its largest feature, Valhalla, is a multiringed basin. Infrared spectra show absorption bands produced by ice. The craters are flattened, consistent with an icy exterior; at the low temperature of the surface, 150 K, ice has great strength but can still flow slowly, leading to crater deformation. The ice is far from pure. The satellite's surface is dark—common among bodies of the outer Solar System—with an albedo of only 0.2. It seems to be coated or mixed with nonreflective carbon compounds. Callisto is probably differentiated, its ice surrounding a core of rock (inset, Figure 14.17).

Ganymede's surface (Figure 14.18) is marginally younger. Dark areas on the satellite are similar to Callisto's, heavily cratered and probably just as old. Ganymede has lighter regions that are less heavily cratered and younger, however, perhaps about the age of the lunar maria. The surfaces of the light areas are heavily covered with grooves that look like grabens (inset, Figure 14.18) and suggest that the body has been slightly pulled apart. Infrared spectra again indicate ice, consistent with the low density. Ganymede's internal structure is probably similar to Callisto's.

The two inner satellites are under the powerful influence of Jupiter. Europa (Figure 14.19) is bright (albedo 0.6) and glazed with relatively clean ice. The smoothest body in the Solar System, with features no more than 50 m high, it looks like a giant skating rink. Europa's high density, however, implies that the ice is only about 25 km deep and surrounds a large rocky core (inset, Figure 14.19). There are no large craters and the surface may be as young as 100 million years. The activity that has smoothed the surface must be associated with heat,

Figure 14.17   *Callisto, an ancient body covered with impact craters, displays a huge multiringed basin called Valhalla. The satellite is differentiated (inset) into a rocky core and icy mantle.*

Dirty ice crust
Water or ice mantle
Rocky core
Large basins
Fresh craters expose ice

Figure 14.18   *Ganymede, the largest satellite in the Solar System, has old, heavily cratered dark regions and (inset) younger, lighter regions with grooved terrain.*

which may have liquefied the water beneath the icy skin. Europa may have oceans; its surface may *be* an ocean frozen at the top.

By all odds, however, Io reigns supreme in the annals of weirdness, perhaps even in the entire Solar System. It is covered with more than 200 volcanoes over 50 km across, a few of which are active and spew not ordinary lava but *sulfur,* which gives the satellite its characteristic orange color (Figure 14.20). Eight eruptions were seen during the *Voyager 1* flyby; six were still going four months later (and two more had erupted) when *Voyager 2* made its pass. Activity is still observed from Earth. There are no impact craters, as the surface is continuously being repaved with sulfurous ejecta. The sulfur is expelled in great plumes (Figures 14.21a) that spray sulfur dioxide in fountains hundreds of kilometers high; in addition, molten sulfur seems to gush from the vents, running across the surface in rivers that are hundreds of kilometers long (Figure 14.21b). There may even be molten sulfur lakes. Io is the most volcanic body in the Solar System and like nothing else we know.

Io's volcanoes were actually predicted and explained before the Voyager flybys. The period of Europa is almost exactly double that of Io, and Ganymede's is double Europa's. Every time Ganymede makes a circuit, Europa and Io are in the same place. Io is therefore subject to a constant gravitational effect called a **resonance** that acts to pull the satellite about, and hence its orbital eccentricity and distance from Jupiter are continuously changing by small amounts.

Io is so close to Jupiter that it is subject to powerful tides. The strength of a tide depends on the inverse of the cube of the distance between the attracting bodies (see Section 11.1.2). As Io changes its semimajor axis, it is flexed and squeezed; the interior heats, melts, and pours through the surface. The interior probably consists of silicates, which are surrounded by a sulfur-rich crust (inset, Figure 14.20). There is even some evidence for silicate volcanoes. All the water boiled away aeons ago. Even though Europa is farther from Jupiter, a resonance seems to have something of the same effect on that satellite and may supply the heat that smooths the surface. One weak plume has been discovered.

The progression of satellite characteristics with distance from Jupiter is reminiscent of the variation of planetary properties with distance from the Sun. The terrestrial planets, made of rock and iron, have relatively little in the way of volatiles (substances that evaporate at low temperatures). On the other hand, Jupiter is made *mostly* of volatile hydrogen. The Sun long ago evaporated all the light stuff from the terrestrials, whittling them down to their heaviest substances. Jupiter probably did the same to the Galilean satellites. When the planet was young its great heat may have driven much of the ice away from Io and Europa, but vast amounts of it

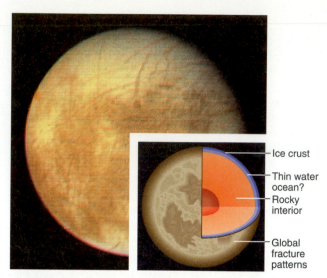

Figure 14.19 *Europa has a young, lightly cratered, smooth icy surface. It seems (inset) to have a rocky interior topped by a thin icy crust above a liquid ocean.*

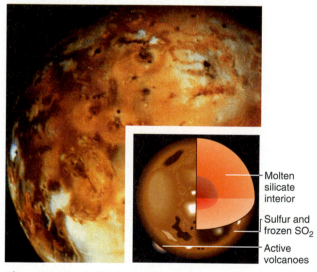

Figure 14.20 *A global view of one hemisphere of Io shows many volcanic calderas. A molten silicate core seems to be topped with a sulfurous crust (inset).*

remained on more distant Ganymede and Callisto. Tidal heating subsequently caused the progression of surface ages with distance from the planet.

14.6.2 *The Inner and Outer Satellites*

The dozen other satellites that belong to Jupiter are very different from the Galileans. Eight lie at great distances outside the orbit of Callisto. They divide into two groups of four, one averaging 11.5 million km from Jupiter, the other twice as far at about 22.5 million km (2° as viewed from Earth). All are in eccentric orbits, with *e* averaging 0.2. The set

BACKGROUND **14.2** A Visit to Jupiter

A trip to Jupiter sounds interesting, but would you really want to go? The conditions make the worst of the terrestrial planets seem benign. The journey alone is arduous. Like the Voyager craft, you would cruise across interplanetary space for almost two years. That is survivable; after all, the crews of tiny sailing vessels did much the same regularly in the seventeenth and eighteenth centuries.

Arriving at Jupiter, where would you land? There is no accessible solid surface. Progressing inward, you would encounter a slushy beginning of a molecular hydrogen ocean, but no person or machine could actually survive even part of the passage. There is a spot among the clouds where you could drift happily in a balloon at a comfortable temperature and pressure, although attempting to breathe the hydrogen/hydrocarbon atmosphere would quickly prove fatal. Instead, you go on to the satellites.

Io would be a fascinating stop. You land on the side facing the giant planet, which appears 20° across in the sky, larger than the Big Dipper seen from Earth. You can rather easily watch the huge planet rotate. You orbit it in less than two days, the stars rising and setting much as they do on Earth, while immense Jupiter hovers stationary in your sky because of tidal locking. In the distance you may see a geyser of sulfur dioxide spewing straight up, like the stream from a giant firehose, so high it disappears. You may even overlook the banks of a molten sulfur river or lake. To appreciate the eerie beauty of the place, however, you would have to be swathed in heavy shielding to protect you from energetic particles. It is doubtful, in fact, that you could take shielding heavy enough.

Proceed to the other satellites and Jupiter retreats into the distance. From bright, shiny Europa it has shrunk to 12° across, and from grooved Ganymede to 8°. Even there, the particle radiation field would probably be deadly. On ancient Callisto, the magnetosphere weakens to a point where you could survive, but for long-term safety you would need to withdraw to one of the tiny outer satellites, where the planet you want to study is about the size of the full Moon in our own heavens. Better to stay home and send a robot.

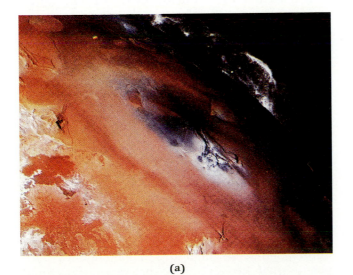
(a)

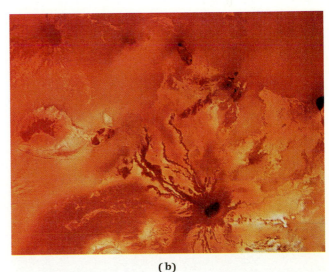
(b)

Figure 14.21 (a) *Pele, a sulfur volcano, seen near the center of the global picture in Figure 14.20, sends a giant plume far above the horizon of Io.* (b) *Rivers of once-molten sulfur seem to have run from the collapsed volcano Ra Patera.*

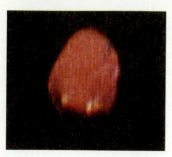

Figure 14.22   *Little Amalthea is probably colored by sulfur spewn from Io.*

TABLE 14.3
Jupiter's Rings

Name	Distance (10³ km)	(R_J)	Width (km)	Thickness (km)	Mass 10⁻⁹M_J[a]
Halo	100–123	1.40–1.72	22,800	20,000	—
Main	123–129	1.72–1.81	6,400	<30	10⁻¹⁴
Gossamer	129–214	1.81–3	850,000	—	—

Ring Features

Thin ring enmeshed in two extended structures, halo toward the planet and the gossamer ring outward.

[a]Jupiter masses.

closer to Jupiter revolves counterclockwise in about 250 days. The outermost satellites, however, orbit backward (retrograde), taking nearly two years to complete a circuit. These eight moons have the appearance of captured asteroids, like the satellites of Mars (see Section 13.7), and are approximately the same size, typically 20 km across. We have no close-up images of any of them. The similarity of the moons within the two groups suggests that they are fragments of a pair of larger bodies.

The four small objects inside the orbit of Io (see Figure 14.15) are quite different. Two are highly inclined to Jupiter's equator but all are in circular orbits, a result of Jupiter's strong gravity. Except for Amalthea (Figure 14.22), found by E. E. Barnard in 1892, they are about the same size as those in the outermost groups. Irregular Amalthea, however, is larger, with radii of 135 and 75 km. Its reddish surface is believed to come from sulfurous material

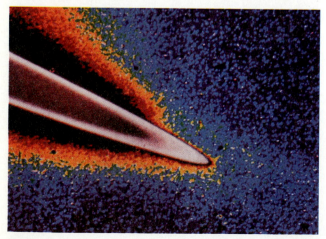

Figure 14.23   *In this false-colored image, Jupiter's main ring, only 30 km thick, is white (it is actually almost black). A fainter halo (here, red) begins at the inner edge of the main ring and spreads toward the planet to the north and south. The gossamer ring (here, blue) stretches outward.*

blown off Io. These inner satellites are probably not captured asteroids, but are still likely to be fragments from collisions among larger bodies.

14.6.3 The Rings

Everyone knows that Saturn is surrounded by **rings** of matter. No one suspected any around Jupiter. They are invisible from Earth, so it was a great surprise when *Voyager* found them in Jupiter's equatorial plane (Figure 14.23 and Table 14.3). The main ring (see Figure 14.15) is narrow and dark. At a distance of 123,000 to 129,000 km from the planet's center, it encompasses the two innermost small satellites, Metis and Adrastea, which inhabit a pair of lanes within it. Although over 250,000 km across, it is less than 30 km thick, and is made of countless dust particles a few thousandths of a millimeter across that orbit the planet. The individual motes are so small that electromagnetic and particle radiation either destroys them or kicks them out of orbit. The result is two thicker components to the ring system, a *halo* that extends inward toward Jupiter, and the broad *gossamer ring* that stretches outward to a great distance. The dust must be replaced continuously, from the innermost satellites or possibly from as-yet-undiscovered bodies.

The Voyagers not only taught us a great deal about Jupiter, but also forced us to ask new questions. Flying toward the planet now is *Galileo*, a sophisticated spacecraft that will not just zip past but will go into orbit about the planet. Most important, in 1995 it will drop a probe that will radio back information about conditions in and chemical compositions of Jupiter's atmosphere and its clouds (Figure 14.24).

Jupiter only opens the door to the wonders to come. The differences among the Jovian planets are as great as those among the terrestrial planets, as we will see upon departing for Saturn.

Figure 14.24 Galileo's *probe is scheduled to parachute into the atmosphere of Jupiter in 1995.*

Io plasma torus: A ring of ions and electrons in Io's orbit.

Magnetodisk: An extended disk of plasma in Jupiter's magnetosphere.

Models: Mathematical descriptions of physical systems (how temperature, density and pressure change with depth, for example) that allow predictions of observations.

Nonthermal radiation: Radiation produced by processes that are not the result of heat and cannot be related to temperature.

Resonance: A gravitational phenomenon in which one orbiting body produces a large gravitational perturbation in another because the two have periods that are simple multiples of each other.

Ring: In the context of the Jovian planets, an encompassing belt of dust and debris.

Rock and **ice:** In the context of the Jovian planets, respectively a mixture of heavier materials like silicates and iron and a mixture of lighter volatiles like water, methane, and ammonia.

Shearing: The act of one part of a substance moving past another at a different speed.

Synchrotron radiation: Radiation produced by fast electrons spiraling in a magnetic field.

Systems I, II, and **III:** The three major rotation periods of Jupiter, respectively equatorial, higher latitude, and internal.

Thermal radiation: Radiation produced as a result of heat, for example, blackbody radiation.

Torus: A doughnut-shaped ring.

KEY CONCEPTS

Belts and **zones:** Respectively, strips of darker, lower altitude, higher temperature Jovian clouds, and strips of lighter, higher altitude, lower temperature clouds.

Decameter bursts: Bursts of radio radiation from Jupiter apparently caused by the action of a current ring that connects the planet to Io.

Differential rotation: Rotation in which different parts of a body rotate with different speeds, resulting in different periods.

Galilean satellites: Jupiter's four largest satellites, discovered by Galileo.

Great Red Spot (GRS): A huge, reddish zone below Jupiter's south equatorial belt; physically, a high-pressure anticyclone.

Hydrocarbons: Chemical compounds based on hydrogen and carbon.

Hydrostatic equilibrium: A condition in which the upward push of pressure is balanced by the downward pull of gravity.

EXERCISES

Comparisons

1. Compare the origins of centimeter, decimeter, and decameter radio radiation from Jupiter.
2. What are the periods and origins of Jupiter's rotation systems?
3. How does synchrotron radiation differ from thermal radiation?
4. Compare the heights and temperatures of zones and belts.
5. Compare the appearance of Jupiter's northern and southern hemispheres.
6. How do the Jovian clouds differ at different altitudes?
7. How do Io and Europa contrast with Ganymede and Callisto?
8. How does Amalthea stand out from the other three satellites inside Io's orbit? How do the four outermost Jovian satellites differ from the next four in?
9. Compare the crater forms and densities on the surfaces of the Galilean satellites.

Numerical Problems

10. Calculate Jupiter's mass from the orbital parameters of two of the Galilean satellites; show all your work. Are the masses identical? Account for any difference.
11. How many Jupiter days constitute Jupiter's year for System I and System III?
12. The Great Red Spot rotates around its center relative to the surrounding clouds. What is the speed of rotation at its edge? (Assume it is circular.)
13. Give the period of a hypothetical satellite that is in orbital resonance with Callisto.

Thought and Discussion

14. How does hydrostatic equilibrium keep Jupiter from expanding or contracting?
15. Where does Jupiter's internal energy come from? Does your answer violate the concept of hydrostatic equilibrium?
16. What phenomenon produces synchrotron radiation?
17. Draw a line representing the radio spectrum and indicate wavelengths. Where does Jupiter emit thermal radiation? Synchrotron radiation? Bursts?
18. Why is the common term "gas giant" a misnomer for Jupiter?
19. Where do you expect Jupiter's magnetic field to be generated?
20. What phenomena are associated with Jupiter's magnetosphere?
21. How is matter from Io incorporated into Jupiter's magnetodisk?
22. How do Jupiter's wind directions and velocities correlate with the planet's zones and belts?
23. How and why is Jupiter so different from terrestrial planets?

24. If you could strip all the hydrogen and helium from Jupiter, what would the remaining body be like? In what ways would it differ from Earth?
25. What is meant by an orbital resonance?
26. What features make Io unique?
27. Why do the internal and external characteristics of the Galilean satellites change with distance from Jupiter?

Research Problem

28. Galileo discovered the four big satellites of Jupiter. Who discovered the others? When? With what equipment?

Activities

29. Use a telescope to make observations of Jupiter. Make a sketch of the cloud-belt pattern, including the location of the Great Red Spot. Chart the motions of the satellites until you can identify them. Estimate their periods and their distances from Jupiter.
30. Draw to scale the orbits of Jupiter's outer satellites (relative to those of the Galileans) and show their orbital inclinations.

Scientific Writing

31. For a magazine devoted to preserving environmental conditions on Earth, write an article that explains how observing and understanding the nature of Jupiter may aid us in gaining knowledge of our own atmosphere.
32. You are an interviewer on a nationally syndicated television program. A famous planetary astronomer is to be your guest. Write a script containing 20 questions that will illuminate the subject of Jupiter for your audience.

15

Beautiful Saturn

The second largest world and its spectacular company of rings and satellites

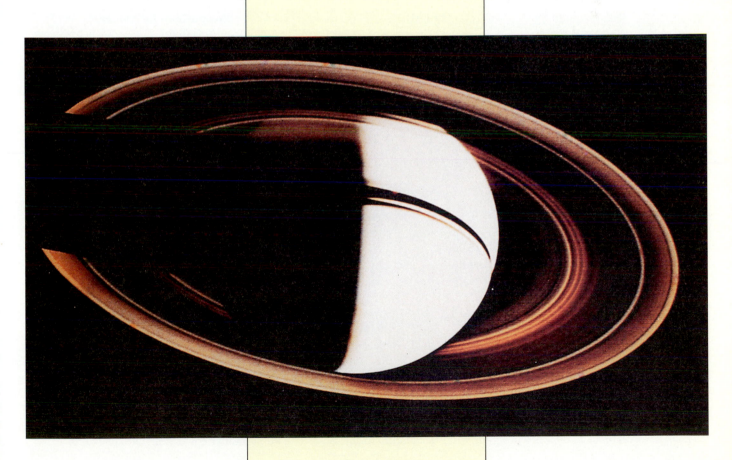

Voyager 2 left Saturn
with this spectacular side view
that shows the huge planet
throwing its broad shadow
onto its rings.

North tropical zone
North equatorial belt
Equatorial zone

Cassini division
C ring
B ring
A ring

Figure 15.1 *This excellent Earth-based photograph of Saturn provides a map of visible features that includes three separate rings surrounding a dusky disk marked with parallel belts and zones.*

TABLE 15.1
A Profile of Saturn

Planetary Data		*Atmosphere*	
distance from Sun	9.54 AU	***Composition by Number*** [a]	
equatorial radius	60,268 km = 9.45 Earth radii	hydrogen	97%
polar radius	54,362 km	helium	1.5%
mass	5.69×10^{26} kg = 95.2 Earth masses	methane (CH_4)	0.1%
density	0.69 g/cm^3	ammonia (NH_3)	0.015%
surface (cloud deck) gravity	0.93 g$_{Earth}$	ethane (C_2H_6)	trace
escape velocity (cloud deck)	35.5 km/s	acetylene (C_2H_2)	trace
rotation period (interior)	$10^h40^m30^s$	phosphine (PH_3)	trace
axial inclination	26.7°	carbon monoxide (CO)	trace
magnetic field strength	0.71 times that of Earth	hydrogen cyanide (HCN)	trace
tilt of magnetic field axis	0°	other gases	trace
effective temperature	95 K		

Layers		*% Mass*	*Radii (km)*
rock and ice core		13	0–16,000
central temperature	15,000 K		
liquid metallic hydrogen mantle			16,000–31,000
liquid molecular hydrogen mantle			31,000–59,000 [b]
gaseous atmosphere			59,000–60,000 [b]

Planetary Features

Cloud cover: Deep clouds only; no solid visible surface.

Magnetosphere: Much smaller and weaker than Jupiter's.

Surface features: Dark belts and bright zones parallel to equator (less prominent than Jupiter's); Great White Spot in northern hemisphere is seasonal storm.

Temperature: Effective temperature, 95 K, is 13 K hotter than expected as a result of contraction and liquid helium separation.

Winds: 1,700 km/s prograde wind in equatorial zone; otherwise no particular correlation with belt-zone pattern.

[a]Abundances of atoms and molecules are relative to the number of hydrogen atoms, not hydrogen molecules.
[b]Based on the equatorial radius.

Nothing represents astronomy quite so vividly as an image of Saturn. Embraced by graceful rings, it is one of the loveliest sights to see in a telescope. The planet bears a strong resemblance to its mighty relative Jupiter, yet there are profound distinctions caused by differences in mass, composition, and distance from the Sun.

15.1
The View from Earth

Saturn is twice as far from the Sun as Jupiter, 10 AU, and its apparent disk is only 20 seconds of arc across at opposition. As a result, Earth-based studies are somewhat limited. Not until *Pioneer 11* and the Voyagers paid their visits did we begin to get a true—and astounding—look at this remarkable planet. Nevertheless, the view from here is stunning and provides a foundation for modern investigations.

Saturn's banded cloud pattern (Figure 15.1) is like Jupiter's, but much fainter, and the equatorial zone is about twice as wide. The angular diameter and distance give an equatorial radius of 60,268 km, making it the second largest of the planets, 84% the size of Jupiter and 9.45 times that of Earth. More oblate than any other planet in the Solar System (Figure 15.2b), its polar radius is 10% smaller (54,362 km). Saturn's properties are summarized in Table 15.1.

The chief cause of the polar flattening is rapid spin. The rotation period of the equatorial clouds is $10^h13^m59^s$, an interval slightly greater than Jupiter's and analogous to its System I. Saturn, however, is a relatively faint radio source because of its weak magnetic field. Only when Voyager examined the planet from close up did mission scientists detect a variable radio source that was rotating with the magnetic field and the planetary interior that produces it. The true period is notably longer, $10^h40^m30^s$, and therefore equatorial winds must be fierce, blowing much more strongly than they do on Jupiter.

Saturn's glory, however, is its rings. Stretching 274,000 km across, a bit more than twice the diameter of the planetary disk and 44 seconds of arc at opposition, they can easily be seen in the smallest of astronomical telescopes. They were first viewed in 1610 by Galileo, but he thought they were knoblike appendages. It was not until 1659 that the Dutch astronomer Christiaan Huygens realized their true nature. They are so big that if they sur-

rounded Mars instead of Saturn, at favorable opposition they would be half the angular size of the full Moon.

There are three principal rings, A, B, and C, lettered inward (see Figure 15.1). The outer A ring is separated from the bright and most obvious B ring by a dark band called the **Cassini division** (discovered by Giovanni Cassini in 1675), which is conspicuous in a small telescope. The inner C ring, known also as the *crepe ring* for its light and delicate appearance, was not found until 1850. The rings sit exactly above Saturn's rotational equator and cast a prominent shadow on the clouds below. The whole system is tilted through an angle of 26.7° (Figure 15.2a). When the axis of rotation is tilted toward the Earth (as it was in 1991), we have a fine view of the north side of the ring surface, and when it is tilted away, we see the south side. Twice during the 29.5-year sidereal orbital period, the axis is seen tilted to the side. The rings are then presented to us edge-on (Figure 15.2b), whereupon they nearly disappear from view, demonstrating their extreme thinness.

Even a casual telescopic observer will notice an eighth-magnitude satellite, Titan, orbiting the planet every two weeks. If you have access to a larger instrument, another 6 of Saturn's 18 known satellites will pop into view, looking like tiny moths

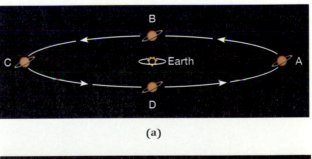

(a)

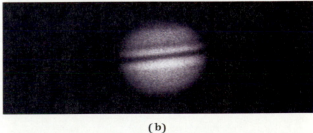

(b)

Figure 15.2 (a) *The observer on Earth sees the northern face of the rings at position A. Seven and a half years later, at B, the rings appear edge-on; after another quarter revolution, at C, the southern face fully presents itself. At D, the Earth is again in the plane of the rings.* (b) *Saturn's rings appear edge-on, detected only by the shadow they throw across the equatorial zone. Note Saturn's oblateness.*

around a candle flame. The orbits of the satellites and Kepler's generalized third law yield a planetary mass of 5.69×10^{26} kg, 95.2 times that of Earth but only 30% that of Jupiter. The resulting density is a mere 0.69 g/cm^3, below that of water, half that of Jupiter, and the lowest in the Solar System.

15.2
Planetary Structure

Comparison between Saturn and Jupiter raises two related puzzles. First, if Saturn is so large, how can its mass be so small; that is, why is its density so low? The answer is that gravitational compression causes the radius to change only slowly with mass. If you could add a large amount of mass to Saturn, you would also increase its gravity. Instead of a significant expansion in radius, there would be an increase in density. Triple the mass and the radius goes up by only about 20%, whereupon you have another Jupiter, which is nearly the largest size physically allowed. After a point, increasing the mass actually would cause the planet to *shrink*. A

mathematical model of Saturn reveals that its mass and radius are consistent with a gross composition of 92% hydrogen and 8% helium (by number of atoms), close to the Jovian and solar values.

Second, if Saturn turns slower on its axis than Jupiter, why is it more flattened at the poles? The principal reason is its lower density, which allows the planet to bulge more at the equator. For Saturn to be so oblate, however, it must also be less centrally condensed than Jupiter, with a larger "rock" and "ice" core that contains the inner quarter of its radius (about 16,000 km) and 13% of its mass (Figure 15.3). Because of lower internal pressure, the inner mantle of liquid metallic molecular hydrogen is notably thinner than Jupiter's, stretching only about halfway (31,000 km) out from the center. The upper part of the mantle, nonmetallic liquid H$_2$, extends most of the rest of the way to the surface. At the top is a gaseous molecular hydrogen atmosphere thicker than Jupiter's, the result of lower gravitational compression.

From infrared measurements we find an effective temperature of 95 K and a radiant flux 1.8 times greater than expected from solar heating

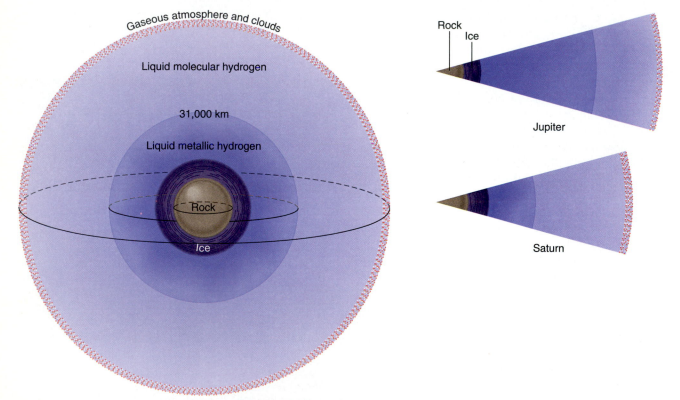

Figure 15.3   *Saturn's rock and ice core is larger than Jupiter's, occupying roughly the inner 25% of the radius. The metallic hydrogen layer, on the other hand, is smaller, extending only about 50% of the way out. The internal structures of the two planets are compared in the wedges.*

alone, compared to 1.7 for Jupiter (see Section 14.1.3). Saturn may have another source of heat in addition to gravitational compression. Given the temperature and pressure of Saturn's deep interior, helium should condense into droplets. As the heavier helium falls inward, it releases additional gravitational energy: Saturn may glow in the infrared in part because of a deep helium rain. This process is apparently just beginning within Jupiter.

Saturn's magnetic field has a strength only 70% that of the Earth and a mere 5% of Jupiter's, consistent with the smaller inner mantle of circulating metallic hydrogen that generates it. The north magnetic pole is up, but Saturn's axis oddly aligns almost exactly with its rotational axis. Since the field is generated by internal convective motion in combination with planetary rotation, that may seem logical, but it is not. Tilted fields like those found in the Earth and Jupiter are in fact expected theoretically; why Saturn's is exactly aligned is something of a mystery. As a result of the weaker magnetic field, Saturn's magnetosphere is one-third the size of Jupiter's and more compressed by the solar wind. Otherwise, it behaves similarly, sputtering atoms from the satellites to fill the radiation belts. The magnetosphere generates radio radiation but no synchrotron emission, only bursts at long wavelengths.

15.3
The Atmosphere

The atmosphere of Saturn is similar to that of Jupiter, but Saturn's atmospheric helium-to-hydrogen ratio (by number of atoms) is only 0.015, far below the Jovian value of 0.044 (see Section 14.4.2). Saturn's helium seems to have sunk below the atmosphere under the influence of gravity much more quickly than Jupiter's. Although the two planets have similar abundances of methane and ammonia, the amount of *gaseous* ammonia in Saturn's atmosphere is minimal because the low temperature freezes almost all of it into clouds. The same seems to have happened for water, which is not seen at all. Abundances of the minor hydrocarbons are lower than in Jupiter's atmosphere, as they are produced by the photochemical action of sunlight, which is much weaker at Saturn's distance from the Sun.

The most immediately obvious difference between Saturn and Jupiter is Saturn's smoother

Figure 15.4 *A Voyager view of Saturn shows a weakly banded structure of zones and belts. The Cassini division is dark and prominent between rings A and B, as is the narrow Encke division in the A ring.*

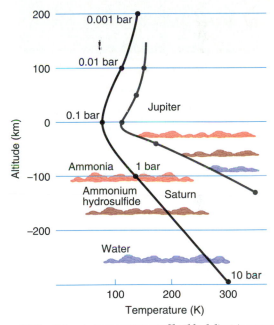

Figure 15.5 *Saturn's temperature profile (black line) is compared with Jupiter's (gray line). The visibly accessible atmosphere of Saturn is much deeper than that of Jupiter, the average temperature much colder, and the clouds at lower levels.*

atmospheric texture (Figure 15.4). Saturn's lower gravity causes its atmosphere to be less compressed than Jupiter's, and the rate at which temperature climbs with increasing depth is considerably lower (Figure 15.5). As a result, the clouds (again, made from layers of ammonia, ammonium hydrosulfide, and water) are formed farther down and the contrasts are less visible. The belts are again warmer and deeper than the zones.

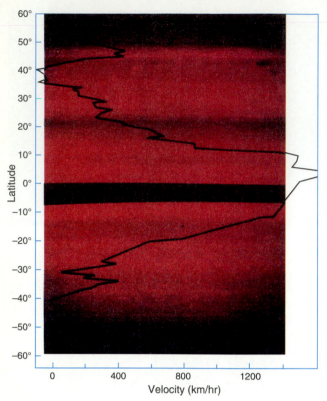

Figure 15.6 *Saturn's broad equatorial wind blows from west to east at 1,700 km/hr. Another narrow, rapid jet blows to the east at about 45°N; otherwise there is little correlation with belts and zones.*

Perhaps the biggest surprise is Saturn's violent winds (Figure 15.6). Almost the entire equatorial zone blows from west to east (prograde) at astonishing speeds relative to the internal rotation of over 1,700 km/hr. There are few east-to-west (retrograde) winds. Little correlation exists between wind speeds and the belt-zone structure except for a peculiar high-speed wind in a narrow zone at a high northern latitude. This departure from Jupiter's pattern is a mystery.

There are a number of small, turbulent ovals set within the clouds, but no feature analogous to Jupiter's Great Red Spot. Saturn's claim to stormy fame is its **Great White Spot** (GWS) (Figure 15.7). Unlike the GRS, the GWS is not permanent, appearing about every 30 years and lasting only a few months. It is produced by seasonal changes that take place within about one Earth year of the onset of northern-hemisphere summer (when the northern side of the ring plane is tilted toward us). The GWS, which first appears as a bright oval in the equatorial zone near 4°N, seems to be caused by a huge bubble of ammonia gas that rises from beneath the clouds under the warmth of the summer Sun. As it reaches an altitude of 250 km above

the cloud tops, the gas crystallizes into bright, reflective clouds that move northward where wind shears stretch them around the planet. The GWS is a brilliant feature, visible in small astronomical telescopes. The last event was witnessed in 1990. The next chance you will have to see one will be around 2020.

15.4
The Rings

In spite of the complexity of Saturn's disk, attention always returns to the rings. A total of seven are now known (Figure 15.8 and Table 15.2). They exhibit a great range of properties, from the enormously broad outer E ring to the remarkably narrow F ring. In addition, the A ring is split by a very narrow dark band found in the nineteenth century called the **Encke division** (after the German astronomer J. F. Encke, though most likely discovered by James Keeler at Lick Observatory in 1888). It can also be seen in Figure 15.4.

Before the Voyager flights, astronomers had an accurate, though quite incomplete, knowledge of the rings' natures. The rings' periodic disappearances from view showed them to be very thin. In 1857 the Scottish physicist James Clerk Maxwell demonstrated on mechanical grounds that if the rings were solid, their revolution about the planet

Figure 15.7 *The Hubble Space Telescope captured the detail of Saturn's Great White Spot in 1990 after it had already been blown a good way around the planet by wind shear. The upper edge is sculpted where the spot runs into a region of lower wind velocity.*

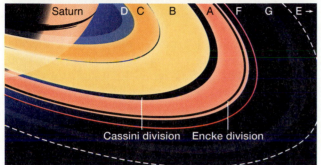

Figure 15.8 *Saturn has seven major named rings. A, B (separated by the Cassini division), C, and narrow F are seen in this Voyager image and colored in the inset map. D, G, and E are indicated on the map but are not evident in the image. The Encke division is dark and clear in the A ring. Each ring divides into hundreds of ringlets. Note that Saturn can easily be seen through the rings.*

would tear them apart; they must, therefore, consist of finely divided material. Forty years later, Keeler provided confirmation when he discovered from their Doppler shifts that the outer ring revolves more slowly than the inner one, in accord with Kepler's laws. Indeed, stars can be seen through the rings, establishing that they are made of billions of orbiting moonlets.

No one, however, was prepared for the spectacular views from the Voyagers. Each main ring is divided into hundreds of small **ringlets** (see Figure 15.8), each only a few hundred kilometers across, and these are subdivided into still smaller ringlets (Figure 15.9) a mere 20 or so kilometers wide. Ringlets even fill the Cassini division, which is not empty at all, just a region with substantially fewer particles.

More astonishing, the Voyagers showed the three classical rings to be no more than a few tens of meters thick, the size of a ten-story building. The C ring and the Cassini division may be thinner yet. At 270,000 km across, the ring system is the thinnest astronomical structure known! Only in the outer E and G rings does the thickness increase to the order of a thousand kilometers.

As a result of constant collisions, which disorganize their motions, the particles making up the

TABLE 15.2
Saturn's Rings

Name	Distance 10^3 km	Distance R_S[a]	Width (km)	Thickness (km)	Mass M_S[b]
D	67–75	1.11–1.24	7,500		
C	74–92	1.24–1.52	17,500		2×10^{-9}
B	92–118	1.52–1.95	25,500	0.1	50×10^{-9}
Cassini division	118–122	1.95–2.02	4,700		1×10^{-9c}
A	122–137	2.02–2.27	14,600	0.1	10×10^{-9}
Encke division	137	2.26	35		
F	140	2.34	30–500		
G	165–174	2.75–2.88	8,000	100–1,000	10^{-20}
E	180–480	3–8	300,000	1,000	

Ring Features

Three main broad rings only a few tens of m thick; several fainter ones both broad and narrow; narrow F ring shepherded by satellites; Cassini division between A and B rings caused by satellite resonance; Encke division in A ring swept clean by satellite; thousands of ringlets compose main rings, which also have spokes and spiral waves; rings composed of dust and ice-coated rocks a few cm to a few m across.

[a]Saturn radii.
[b]Saturn masses.
[c]Even though a gap, the Cassini division contains ring particles.

Figure 15.9 *A high-resolution image of the B ring shows the ringlets are subdivided into yet more ringlets.*

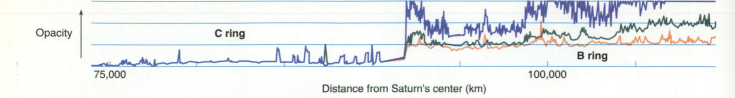

Figure 15.11   *A computer simulation of a volume only 3 m square from within a ring shows hundreds of dirty iceballs, most only a few centimeters across.*

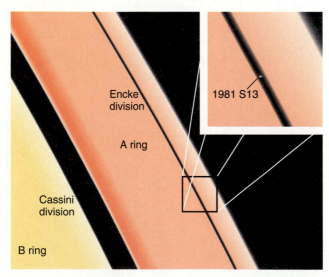

Figure 15.12   *Satellite 1981 S13 glides along the Encke division, clearing it of ring particles.*

rings must be several times smaller than the ring thicknesses. A good determination of the dimensions of these fragments was made when the Voyager team watched first a star and then the spacecraft radio signals dim when they were viewed through the rings (Figure 15.10), thereby obtaining a measure of the rings' **opacities**—the degrees to which they block light—in the optical and at two radio frequencies. The C ring is very transparent, the A ring intermediate, and the B ring quite opaque. The ringlets inside the broad Cassini division are prominent, while the Encke division is almost entirely free of matter.

Ring opacities depend on particles whose minimum sizes are roughly similar to the wavelengths of observation. The optical curve is therefore sensitive to particles of all sizes, from dust on up, whereas the radio opacities respond only to particles a few centimeters across. The measurements reveal a distribution of particle diameters from approximately 1 cm to perhaps 10 m, into which fine dust must be mixed (Figure 15.11). (The same conclusion is drawn from radar and radio observations made from Earth.) The particles are only a few centimeters apart. No spacecraft could survive a journey through the rings.

Particle sizes change across the system. In the C ring and the Cassini division, the curves of Figure 15.10 merge, meaning that these zones contain considerably less dust than either the A or B rings. The much thicker E and G rings, however, must be made of fine dust to have spread out so far from the main ring plane. Infrared observations show that the ring particles are made partly of water ice or, at least, are coated with it, making the rings bright and reflective.

Why are the main rings so thin? The particles are so close to one another that they collide constantly. A particle may therefore attain sufficient energy to be launched out of the plane into an inclined orbit. When it recrosses the plane, it will invariably collide with another iceball, and the subsequent collision lowers its inclination and returns it once again to the plane. Individual moonlets cannot escape, and the result is a ring of extreme thinness.

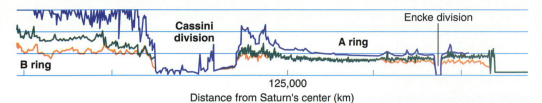

B ring

Cassini division

A ring

Encke division

125,000
Distance from Saturn's center (km)

Figure 15.10 *The structure of the ring system is shown at three wavelengths, the blue line representing the Voyager measurements of ring opacity made in the optical and the other two determinations made at radio frequencies.*

The ring structures present enormous challenges to the study of orbital mechanics. The broad Cassini division has long been known to be caused by a perturbing resonance (see Section 14.6.1) with the satellite Mimas. The period of each particle at the inner edge of the division is exactly half that of the satellite, so ring particles that wander into the Cassini division tend to be kicked back out of it. By contrast, the Encke division is directly swept free of particles by a tiny moon (Figure 15.12). However, the origins of the myriad ringlets are not known. The breaks do not fit any resonances, and there are too many of them. Small, unseen embedded satellites a kilometer or so across may be involved.

Many other phenomena invite explanation. The strikingly narrow F ring is stranded or braided (Figure 15.13). Such a skinny feature should diffuse outward with time, but two **shepherd satellites** (inset, Figure 15.13) flank it and keep the particles in line. More mysterious are the **spokes,** dark structures that fall across the surfaces of the rings (Figure 15.14). They are probably caused by fine dust above the ring plane that is electrically charged and consequently affected by Saturn's rapidly rotating magnetic field. Finally, some orbital resonances with satellites apparently produce bunching of particles. The gravity of the bunches causes further bunching, and the shearing of the differential orbital revolution (the periods are shorter closer to the planet) spreads the bunches into spiraling **density waves.**

Two broad hypotheses—recent and ancient—have been proposed for the origin of the rings. The rings are inside a radius surrounding Saturn called the **Roche limit.** Within it, the tides (see Section 11.1.2) are strong enough to tear a fluid body apart. The effect on a *solid* body depends on its distance, size, and strength. If it is large, weak, and sufficiently close to its parent planet, the stretching force induced by the tide can be greater than the mechanical and gravitational forces holding it together, and it will be broken into pieces. The rings may therefore be relatively recent relics of the breakup of one or more large bodies, perhaps a moon or two that strayed too close, or a large comet that passed too near the planet and was frag-

Figure 15.13   *The narrow, braided F ring is kept together by the shepherd satellites shown in the inset.*

Figure 15.14   *Broad, spokelike features are seen on the ring surfaces.*

mented. (Comets are icy bodies that enter the Solar System from far outside the orbit of Pluto; they will be examined in detail in Chapter 17.) Continued impacts among the fragments would then have reduced them to small particles and dust. A few remaining pieces—tiny satellites—could help produce the rings' structures. Alternatively, they may be the remnants of an ancient dusty disk that developed around Saturn during the planet's formation. Satellites coalesced in the outer part of the disk; within the Roche limit, however, tides would have been too strong to allow a satellite to accumulate from the smaller particles.

The current view leans toward recent origin. If the rings are indeed old, their surfaces should have become coated with dust, rendering them dark instead of bright. Furthermore, old rings ought to have diffused into one another. However, the extremely sharp divisions between the rings suggest forces that maintain their integrity over long intervals. No one yet knows where the rings came from, nor how long they will last.

15.5
Saturn's Satellites

The only planet with bright rings, Saturn also (perhaps not by coincidence) has the largest number of known satellites, a total of 18; they include a diverse gaggle of bodies that share certain similarities, yet present some striking differences (Table 15.3). There is one Galilean-type satellite, the huge moon Titan. With a radius of 2,575 km, it is just smaller than Ganymede, the largest satellite in the Solar System. The other satellites range in radius

TABLE 15.3
The Saturnian Satellites

Name	Distance (10^6 km)	(R_s)	Period (days)	$i°$	e	Radius (km)	Mass (10^{21} kg)	Density (g/cm^{-3})
1981 S13[a,b]	0.137	2.26	0.60			20		
Atlas[b]	0.1376	2.28	0.60	0	0	20 × 15		
Prometheus[b]	0.1380	2.31	0.611	0	0.00	70 × 40		
Pandora[b]	0.1417	2.35	0.63	0	0.00	55 × 35		
Epimetheus	0.1514	2.51	0.69	0.34	0.01	10 × 50		
Janus	0.1515	2.51	0.70	0.14	0.01	110 × 80		
Mimas	0.1855	3.08	0.94	1.53	0.02	195	0.038	1.17
Enceladus	0.2380	3.95	1.37	0.02	0.00	250	0.084	1.24
Tethys	0.2947	4.89	1.89	1.09	0.00	525	0.755	1.26
Telesto[b]	0.2947	4.89	1.89	0	0	12		
Calypso[b]	0.2947	4.89	1.89	0	0	15 × 10		
Dione	0.3774	6.26	2.74	0.02	0.00	560	1.05	1.44
Helene	0.3774	6.26	2.74	0.2	0.01	18 × 15		
Rhea	0.5270	8.75	4.52	0.35	0.00	765	2.49	1.33
Titan	1.222	20.3	15.95	0.33	0.03	2,575	135	1.88
Hyperion	1.481	24.6	21.28	0.43	0.10	175 × 100		
Iapetus	3.561	59.1	79.33	14.7	0.03	720	1.88	1.21
Phoebe	12.952	215	550.5	175[c]	0.16	110		

Satellite Features

Titan: The one large, Galilean-type moon; thick, opaque nitrogen atmosphere denser than Earth's; hydrocarbons.

Smaller satellites: Varying properties; several satellites inside Titan's orbit resurfaced, though Mimas is not; icy Enceladus brightest body in solar system; Dione cratered on one side, wispy terrain on the other; Rhea smooth on one side, wispy on the other; Hyperion irregular; Iapetus bright on one side, dark on the other.

[a]Temporary designation.
[b]Discovered by Voyager team members.
[c]Retrograde and captured.

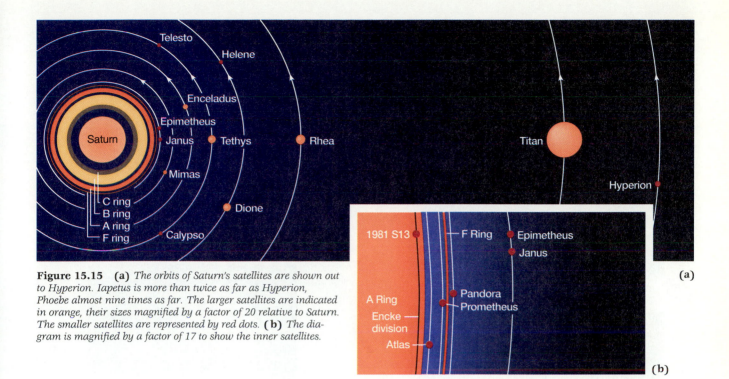

Figure 15.15 **(a)** *The orbits of Saturn's satellites are shown out to Hyperion. Iapetus is more than twice as far as Hyperion, Phoebe almost nine times as far. The larger satellites are indicated in orange, their sizes magnified by a factor of 20 relative to Saturn. The smaller satellites are represented by red dots.* **(b)** *The diagram is magnified by a factor of 17 to show the inner satellites.*

from Rhea (765 km) down to blocks of ice and rock only a few tens of kilometers across.

Orbital radii (Figure 15.15) range from 2.26 Saturn radii (R_S) for the satellite that sweeps the Encke division (1981 S13) through Titan at 20 R_S to Phoebe at 215 R_S. Except for the distant satellites Phoebe and Hyperion, the orbits are all nearly circular; except for Iapetus and Phoebe, they all lie close to the ring plane. Phoebe is also the only Saturnian moon with retrograde revolution. On average, satellite radii increase outward from Saturn to Titan, then decrease.

There are some odd orbital relationships. Janus and Epimetheus nearly share an orbit and exchange places, switching orbits each time they approach one another. Telesto and Calypso *do* share an orbit with Tethys, one 60° ahead, the other 60° behind. These 60° positions, called **Lagrangian points** after the eighteenth-century mathematician Joseph Lagrange, are gravitationally stable. Once a smaller body establishes itself at one of a larger body's Lagrangian points, gravitational perturbations act to keep it there. Helene also orbits at Dione's forward Lagrangian point. All the satellites but Hyperion and Phoebe rotate in synchrony with their orbital periods.

Masses and densities have been determined for seven of the satellites. Since Titan's density is similar to those of the outer two Galilean satellites of Jupiter, it is probably about half water ice. The densities of the others are significantly lower. They may be as much as 65% water ice, the remainder likely being some kind of silicate rock. Unlike Jupiter's brood, Saturn's satellites demonstrate no relationship between density and distance from the planet. Apparently the much-smaller Saturn never got so hot during its formation that the heat affected the amount of water either acquired or retained by its inner moons.

A tour outward from Saturn shows that each of the larger satellites has unique properties. Mimas (Figure 15.16a) is a battered body with an ancient surface much like that of Callisto. Cold, strong ice has preserved the craters. Mimas has an enormous crater 130 km in diameter—two-thirds of the satellite's radius. A larger impact might have broken the moon apart. Next door is Enceladus (Figure 15.16b). One would expect it to mimic Mimas, but nature is not that simple. It is the brightest body in the Solar System, with an albedo near 100%, and parts of it seem to have been paved in smooth, fresh ice by eruptions from watery volcanoes. Enceladus is in the middle of the diffuse E ring and may be its source.

The next three moons display evidence of ancient geologic activity and of resurfacing, although nowhere near the level of that seen on Enceladus. Tethys (Figure 15.17a) is heavily cratered and has a broad canyon that may be the result of expansion. The leading side of Dione (Fig-

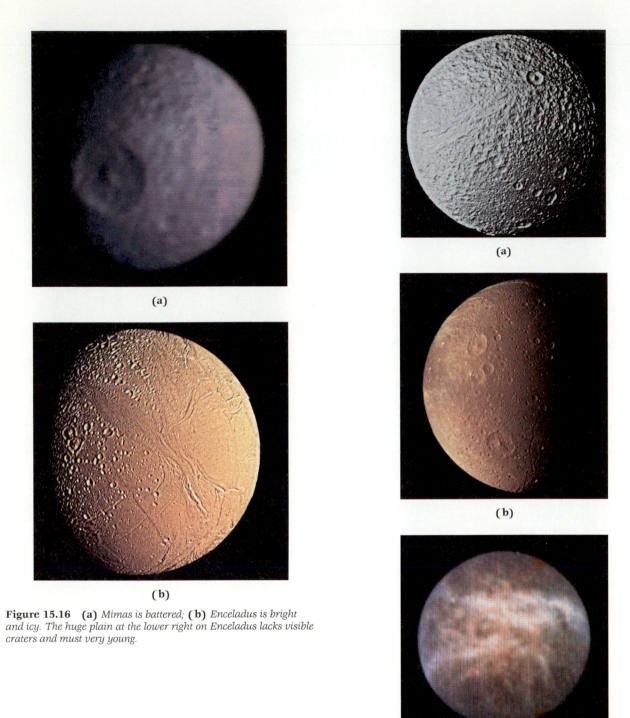

Figure 15.16 **(a)** *Mimas is battered;* **(b)** *Enceladus is bright and icy. The huge plain at the lower right on Enceladus lacks visible craters and must very young.*

Figure 15.17 *Tethys* **(a)** *and Dione* **(b)** *are both cratered. Tethys has a long, wide valley extending over much of its surface. Rhea* **(c)** *displays a wispy terrain much like that seen on the other side of Dione.*

BACKGROUND 15.1 Saturn and the Joy of Science

Like little else, examination of Saturn by the Voyager mission experts demonstrates that science is not a dry, dispassionate study but an exciting exploration into the unknown. Try to imagine what it must have been like for the people at the video consoles as *Voyager 1* approached its destination. The words of Bradford Smith, the Voyager imaging team leader, express it best: "As the final days of the approach phase were upon us, anticipation grew. The satellites were finally being seen as more than just individual disks and there was something very peculiar about the bright rings—they were daily showing more and more structure, far more than could be accounted for by simple satellite resonance theory. We were preparing for an exciting encounter, but even bigger surprises were to come."

Later he notes, "Mimas seemed to be more normal—except for a giant impact crater that is more than a third the diameter of Mimas itself. As the image of Mimas and this absurd crater appeared on our television monitors, there was a sense of *déjà vu*. Of course! It was George Lucas's 'Death Star!'" (Look at Figure 15.16a.)

And still later: ". . . our attention was fixed on the F ring. Resolution was improving rapidly, and the apparent decreasing width of the ring kept pace. . . . those of us watching the monitors were stunned. If . . . we had become somewhat jaded to the unpredictability of the outer solar system, our sense of astonishment was brought back in an instant. . . . Staring back at us from the television monitors were three individual strands, each approximately 20 km wide and separated by a few tens of km; they appeared to be knotted, kinked, and braided. To me, it was the most improbable picture yet sent back by either Voyager spacecraft."

---

ure 15.17b), that which faces in the direction of its orbit, is heavily cratered, but the trailing side displays a so-called wispy terrain that suggests resurfacing. Rhea (Figure 15.17c) shows wisps on its trailing side, but the leading side is much smoother than Dione's, perhaps the result of resurfacing. The variations in the characteristics of these five satellites are probably caused by ices with low melting points and by different levels of tidal heating, each related to differences in size and distance from Saturn.

For now, skip over Titan. Hyperion (Figure 15.18a), just outside Titan's orbit, is the largest irregular body known in the Solar System. It is distinguished by chaotic rotation caused by gravitational perturbations and may actually have no rotation period. Not to be outdone, distant Iapetus (Figure 15.18b) has two different faces, a fact noticed by its discoverer, Cassini, in the 1600s. The leading hemisphere is dark, the trailing one 10 times brighter. Spectroscopy from Earth suggests that the dark material is similar to a tar sand (sand laced with thick hydrocarbon tar). No one understands how Iapetus developed these conditions: perhaps Phoebe is responsible. The backward orbit of that moon plainly reveals it to be a captured asteroid or comet. As we will see in Chapter 17, small bodies in the outer solar system tend to be very dark. Dusty matter loosed from Phoebe in collisions with meteoric bodies will spiral inward, and Iapetus may sweep it up. More likely, the tarry stuff has come from inside Iapetus itself.

Now return to Titan, the most intriguing of the satellites. Methane bands were seen in its spectrum even in the 1940s, suggesting that it might have a fairly thick atmosphere. *Voyager 1* revealed a featureless yellowish haze so dense that nothing could be seen of the surface (Figure 15.19a). Models constructed from Voyager data reveal that Titan's air is mostly—perhaps 80% or more—molecular nitrogen. The remainder is a mixture in unknown proportions of methane (CH_4) and maybe argon. The surface pressure is 1.5 bar, greater than that on Earth, but not too great for a human being to survive. Comfort is not assured, however, as the temperature is a bone-chilling 95 K and all the oxygen is trapped in ice. Temperature drops with altitude, and the haze is formed by condensation of methane near minimum temperature, about 35 km up (Figure 15.19b). Chemical and photochemical reactions produce a remarkable array of hydrocarbons that

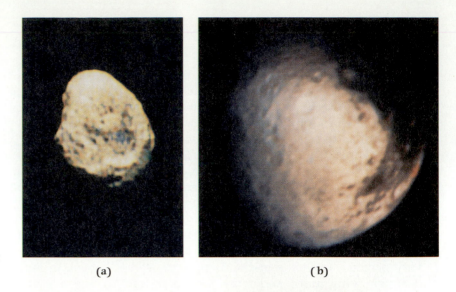

(a)

(b)

Figure 15.18 **(a)** *Hyperion is highly irregular.* **(b)** *Iapetus is dark on the leading side and bright on the trailing.*

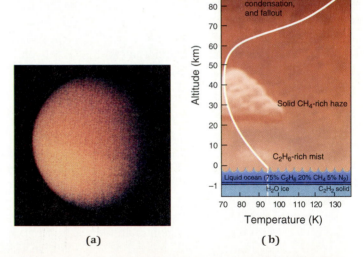

(a)

(b)

Figure 15.19 **(a)** *Titan displays only a thick atmosphere;* **(b)** *the graph shows its structure. Clouds form where the temperature reaches a minimum.* **(c)** *The artist imagines a scene of a cold ethane ocean*

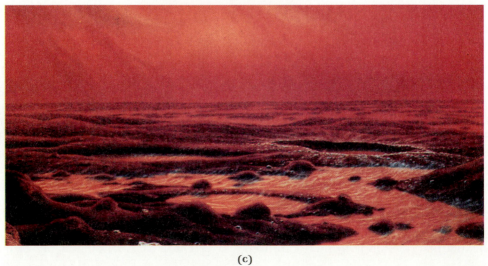

(c)

include ethane, propane, ethylene, and methylacetylene (C_3H_4). So much ethane has been generated that Titan may have ethane seas (Figure 15.19c). The composition of Titan's atmosphere is largely a product of its distance from the Sun and the resulting low temperature. If we could warm the satellite and sublime the ice, oxygen freed by sunlight from the water vapor would convert the methane to carbon dioxide, making the satellite's atmosphere more like that of Mars.

Why should Titan have such a thick atmosphere when Ganymede, of comparable size, has none? The most likely hypothesis rests on the fact that cold ice can hold a great deal of gas. Because Titan is farther from the Sun than Ganymede, and because it is the satellite of a smaller, cooler planet, it was the colder of the two while the satellites were being formed. Titan's ice therefore retained its gas, whereas Ganymede's did not. Subsequent heating then released the gas to make Titan's atmosphere. Moreover, bodies falling toward Jupiter attain higher velocities than those falling toward Saturn. One striking Ganymede could do much more damage than one striking Titan—perhaps enough to have stripped Ganymede of any atmosphere it once possessed.

The Voyagers are long gone, leaving a superb legacy of surprises and puzzles, mysteries that will be difficult to solve from Earth. We need to go back. Currently under construction by NASA and the European Space Agency (ESA), a new explorer named *Cassini* should arrive to orbit the ringed, planet early in the next century, carrying a probe to penetrate the atmosphere of Titan, allowing us to add yet another pearl to the string of worlds that we have directly explored.

KEY CONCEPTS

Cassini division: A resonance gap that divides Saturn's A and B rings.

Density waves: Gravitational accumulations of particles that spread into a spiral pattern as a result of orbital revolution.

Encke division: A narrow gap in the outer part of the A ring produced by a satellite that sweeps up ring material.

Great white spot (GWS): A seasonal northern-hemisphere storm on Saturn that extends around the entire planet.

Lagrangian points: Points of orbital stability 60° ahead of and behind a planet or satellite.

Opacity: The degree to which a substance blocks electromagnetic radiation; the opposite of *transparency.*

Ringlets: Small rings a few hundred km wide that make up Saturn's big rings.

Roche limit: A limit surrounding a planet within which a fluid body would be torn apart by tides.

Shepherd satellites: Small satellites that organize and preserve narrow rings.

Spokes: Dark structures in Saturn's rings, possibly caused by electrical effects.

EXERCISES

Comparisons

1. Compare Saturn's winds with those of Jupiter.
2. Compare the amount of helium in Saturn's atmosphere with that in Jupiter's. Why is there a difference?
3. How is the origin of Saturn's internal heat different from that of Jupiter's?
4. Compare the different natures and origins of the Cassini division and the Encke division.
5. Compare theories of the origins of Saturn's rings.

Numerical Problems

6. If you were to lay Saturn's main rings on a field 100 m in diameter, what would be their thickness?
7. In approximately what years of the next half-century will we see Saturn's rings on edge?

Thought and Discussion

8. Why is Saturn more flattened at the poles than Jupiter?
9. Why are Saturn's cloud belts less visible than Jupiter's?
10. Why is Saturn's density so low?
11. When and where does Saturn's Great White Spot occur?
12. How can you tell that Saturn's rings are not a solid sheet?
13. What evidence is there that Saturn's rings consist of centimeter- or meter-sized particles?
14. What keeps Saturn's F ring in place?
15. What is the likely origin of the spokes seen in Saturn's rings?
16. What is the significance of the Roche limit?
17. How can Telesto and Calypso be in the same orbit as Tethys?
18. Why cannot the surface of Titan be seen?
19. Why does Titan have such a thick atmosphere?
20. Why do we think Phoebe is a captured body?

21. What is the relation between the density of a Saturnian moon and its distance from the planet?

Research Problem

22. Discovery of Saturn's D ring was claimed from Earth, though Voyager images show it could not have been seen. The E ring is commonly credited to Voyager, but it was actually discovered from Earth. Who claimed discovery of the D ring? Who discovered the E ring? In each case, what were the date, site, and circumstances of discovery?

Activity

23. Examine Saturn through a telescope and make a drawing. From the known angular and physical diameters of Saturn's disk, estimate the angular and physical diameters of the rings and of the Cassini division, and the distance of Titan from the planet.

Scientific Writing

24. For your astronomy club's bulletin, write a timetable for Saturn over the next orbital revolution. Describe how the aspects of the planet will change, what phenomena can be viewed over the course of the orbit, and their scientific significance.

25. A high school science teacher writes you a letter asking how to make a scale model of the planet Saturn. Write a return letter in which you describe in some detail how to do it. Pick a reasonable scale for Saturn (say, a baseball), and tell where all the parts should be placed. The object is to provide a vivid demonstration for the students.

16

Outer Worlds

Uranus,
Neptune, and Pluto
float in the
cold depths at the
end of the
planetary system

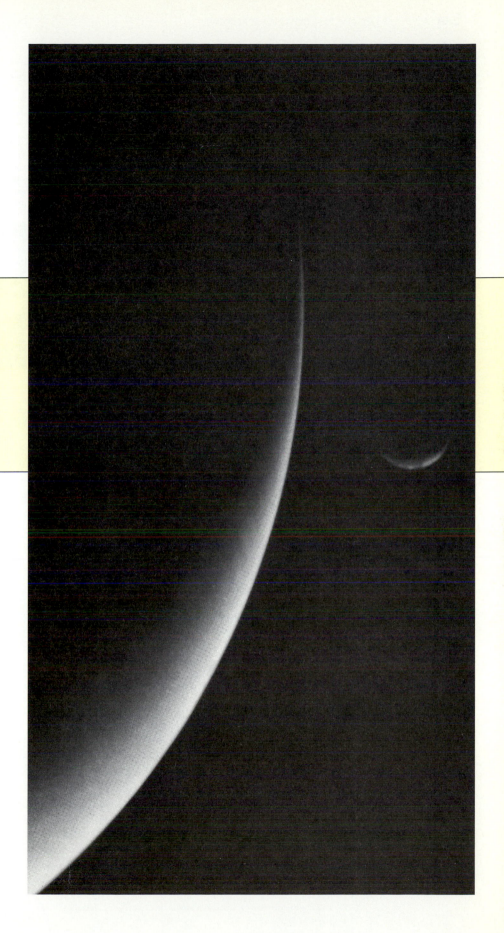

Voyager says goodbye
to the planetary system
as it looks back on Neptune
and its moon Triton.

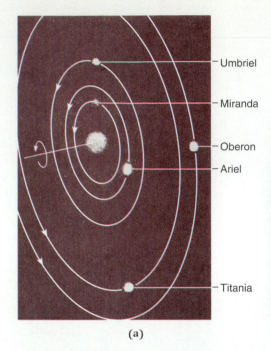

- Umbriel
- Miranda
- Oberon
- Ariel
- Titania

(a)

Figure 16.1   *The outer worlds present a poor sight from Earth.* **(a)** *Uranus's small disk is surrounded by five satellites. The poles of their orbits currently point nearly at Earth, a result of Uranus's 98° axial tilt.* **(b)** *Neptune is seen with its satellite Triton.* **(c)** *Pluto is indistinguishable from a star*

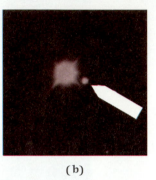

(b)

(c)

TABLE 16.1
A Profile of Uranus

Planetary Data		Gaseous Atmosphere	
mean distance from the Sun	19.19 AU	***Composition by Number*** [a]	
equatorial radius	25,559 km = 4.01 Earth radii	hydrogen	90%
polar radius	24,974 km	helium	9%
mass	8.68×10^{25} kg = 14.5 Earth masses	methane (CH_4)	1%
density	1.29 g/cm^3	other hydrocarbons	trace
surface (cloud deck) gravity	0.79 g$_{Earth}$		
escape velocity (cloud deck)	21.3 km/s		
rotation period (interior)	17^h14^m		
axial inclination	97.9°[b]		
magnetic field strength	0.74 times that of Earth		
tilt of magnetic field axis	58.6°[c]		
effective temperature	59 K		

Layers	***Radii (km)***
rocky core	0–7,500
possible liquid water mantle	7,500–25,000

Planetary Features

Atmosphere: High helium relative to value expected on the basis of settling; planet has lost hydrogen.
Cloud cover: No clouds; apparent surface consists of hydrocarbon haze.
Magnetosphere: Large but weak.
Surface features: Almost featureless; near-invisible bands, no belts or zones; a few barely visible clouds near the equator.
Temperature: Effective temperature 59 K, as expected from solar heating; little or no internal heat source.
Winds: Pattern the reverse of Jupiter's and Saturn's; retrograde at least 350 km/hr near equator; prograde at 600 km/hr at higher latitudes.

[a]Relative to the number of hydrogen atoms, not hydrogen molecules.
[b]Retrograde rotation.
[c]Offset by 30% from center.

We now reach out to three cold worlds that, with their icy refrigerated companions, further mystify and delight as well as illuminate the constitution and origin of the planetary system. Uranus and Neptune are vaguely intermediate in construction between Jupiter and the terrestrial planets, and Pluto may be the prototype of a new class of object altogether.

16.1
The View from Earth

Our Earthly view of these far worlds is poor indeed. Uranus (Figure 16.1a) presents a tiny disk just four seconds of arc across, and no surface detail has ever been seen from Earth. Neptune (Figure 16.1b), with a disk half as large, is even harder to examine, although visual observers have reported some structure, and photographs taken at the wavelength of a methane absorption band show a dark

belt near the equator. The two planets are very nearly twins, with equatorial radii of 25,600 km and 24,800 km respectively (as refined by *Voyager 2* measurements), four times as big as Earth (see Tables 16.1 and 16.2). Pluto (Figure 16.1c) is little more than a dot against the starry sky, its angular diameter a mere 0.08 seconds of arc.

Uranus and Neptune fit nicely into the broad class of Jovian planets. Their dense atmospheres are loaded with methane. Pre-Voyager measurements and estimates gave rotation periods less than that of Earth. Uranus has five medium-sized moons and Neptune has a big one that fits the class of Galileans (see Figure 16.1). The satellites' orbits (and the perturbations they induced in the Voyager flight paths) yield masses of 14.5 Earth masses for Uranus and 17.1 for Neptune—large, but between those of the Earth and Jupiter. The densities are low, 1.29 and 1.64 g/cm^3 respectively, indicating that they are made of light materials.

TABLE 16.2
A Profile of Neptune

Planetary Data		Gaseous Atmosphere	
		Composition by Number[a]	
mean distance from the Sun	30.06 AU	hydrogen	87%
equatorial radius	24,764 km = 3.88 Earth radii	helium[b]	12%
polar radius	24,343 km	methane (CH_4)	0.5%
mass	1.02×10^{26} kg = 17.1 Earth masses		
density	1.64 g/cm^3		
surface (cloud deck) gravity	1.12 g_{Earth}		
escape velocity (cloud deck)	23.3 km/s		
rotation period (interior)	16^h07^m		
axial inclination	29.6°		
magnetic field strength	0.43 times that of Earth		
tilt of magnetic field axis	46.8°[c]		
effective temperature	59 K		
Layers	**Radii (km)**		
rocky core	0–7,000		
possible liquid water mantle	7,000–24,000		

Planetary Features

Atmosphere: Greater helium and more hydrogen loss than Uranus.
Cloud cover: Several layers amidst a deep blue hydrocarbon haze.
Magnetosphere: Large but empty because of distance from Sun.
Surface features: Great Dark Spot and D2 storm systems; two areas of high, bright cirrus.
Temperature: Effective temperature, 59 K, is 7 K above solar heating, implying internal heat source.
Winds: Like Uranus, pattern the reverse of Jupiter's and Saturn's; retrograde at 1,700 km/hr in equatorial region; prograde at higher latitudes.

[a]Relative to the number of hydrogen atoms, not hydrogen molecules.
[b]Uncertain, but a high percentage, like that found on Uranus.
[c]Offset 55% from center.

Now, however, comes the first of several surprises. Uranus is tipped on its side, its rotation axis almost in the planet's orbital plane. The pole around which the planet rotates counterclockwise is actually tilted by 98° relative to the northern orbital perpendicular (see Figure 16.1a) and points slightly south rather than north. Like Venus, Uranus therefore rotates backward. This southern pole now points nearly toward the Earth. There is no known cause for this anomaly, but we suspect that Uranus was knocked over in a giant collision with another body sometime in the early days of the Solar System. Neptune spins more normally, with a tilt of 30° to the orbital perpendicular. However, its big satellite, Triton, while moving along a nicely circular orbit, *revolves* backward, the only large moon in the Solar System to do so.

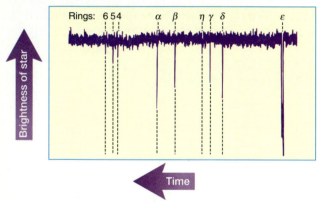

Figure 16.2 *The graph shows brightness variations undergone by a star just before Uranus covered it as seen from Earth. The sudden drops were caused by narrow planetary rings that occulted the star. The major rings were named with Greek letters α through ε; the weaker rings were named later.*

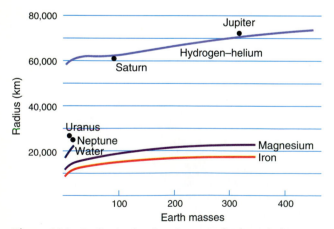

Figure 16.3 *Radius is plotted against mass for theoretical nonrotating planets of different compositions. The top curve shows what is expected for a liquid mixture of 90% hydrogen and 10% helium. It fits well with Saturn and Jupiter, but not with Uranus and Neptune, which are much closer to the curves for liquid water and all-metal planets (the lower curves).*

A closer look extends the list of oddities. In March 1977, several groups of astronomers set out to watch Uranus pass in front of a star. Since the planet's angular speed in seconds of arc per second of time was known, the duration of the **occultation** (the concealment, or covering, of one body by another) would give Uranus's precise angular diameter, and the way in which the light dimmed would provide data for studying the Uranian atmosphere. What was found illustrates the way science often works—you look for one thing and find another. The star began to wink in and out before the planet reached it (Figure 16.2), indicating that something was getting in the way. After Uranus moved off the star, the events were repeated in reverse. The only reasonable conclusion is that Uranus has *rings*. But unlike Saturn's, they are very narrow and optically invisible. Nine were eventually found before the Voyager flyby.

Then in 1979, Voyager identified rings around Jupiter. If three of the Jovian planets have rings, why not a fourth? The same occultation technique was tried with Neptune. Dips were indeed seen in the stellar light as the planet approached but not as it receded. Later work showed that sometimes the rings were there, sometimes not, indicating the existence of incomplete **ring arcs,** matter concentrated into particular segments of a circular orbiting path. Rings around Jovian planets are not unusual, but rather the norm.

Better understanding of Uranus and Neptune required the scrutiny beautifully provided by *Voyager 2*. Results from this extraordinary spacecraft and from sophisticated Earth-based observation, combined with theoretical analysis, now give us an idea of the natures of these distant planets. Neptune and Triton, in turn, provide insight into lonely Pluto, the only planet in the Solar System that has not yet been visited by a product of human creativity and engineering.

16.2
Interiors of Uranus and Neptune

These planets at first look like Jupiter and Saturn, but looks are deceiving. Because of gravitational compression, the radius of a hydrogen-helium Jovian planet changes only slowly with mass (see Section 15.2). Although Saturn has a mass less than a third that of Jupiter, its radius is still 84% as large. If Uranus and Neptune were made of hydrogen and helium, we would expect them to be only slightly smaller than Saturn and nearly 60,000 km in radius

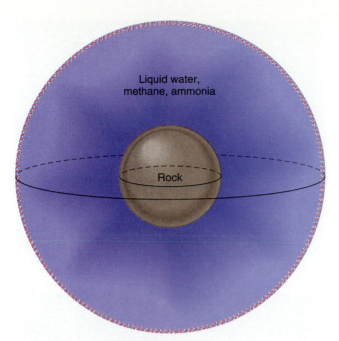

Figure 16.4   *Modeling of Uranus and Neptune shows them to have similar internal structures: rocky cores possibly surrounded by deep mantles of liquid water (mixed with methane and ammonia), topped off by cloudy hydrogen atmospheres.*

(Figure 16.3). Yet each is only about 45% Saturn's size. They must therefore be made largely of denser material that causes them to contract gravitationally, most likely water and "rock" (see Section 14.3). Their average densities, which are higher than those of Jupiter or Saturn, support this conclusion. The equatorial bulges also indicate a lower central concentration of mass.

One detailed model (Figure 16.4) suggests a rocky core surrounded by a deep mantle of liquid water (mixed with much smaller amounts of methane and ammonia), topped by a cloudy atmosphere of light hydrogen and helium. The core might also be surrounded by a dense, gaseous mixture of water and hydrogen that is again overlaid with clouds. If you could strip most of the hydrogen and helium from Jupiter, the remainder might look at least something like the twin outer planets.

The mystery surrounding Uranus and Neptune deepens when we look at two additional characteristics. Like Jupiter and Saturn, Neptune (with an effective temperature of 59.3 K) radiates 2.6 times the amount of energy expected from the distant Sun. Uranus's effective temperature, however, is nearly identical to Neptune's, 59.1 K, even though it gets more solar heat. In spite of the close similarity between the two planets, Uranus has no measurable flow of internal heat at all. Perhaps the event that knocked the planet askew served to chill its interior.

Voyager 2 found that the magnetic-field strengths of Uranus and Neptune are not unusual, respectively about 74 and 43% that of the Earth, compared to 71% for Saturn. The peculiarity is in the placement of their magnetic dipoles (Figure 16.5). Jupiter's magnetic axis, like the Earth's, is tilted by 10° against the rotation axis, and Saturn's is aligned; however, Uranus's is tilted by an astonishing 59°. Moreover, the dipole is not centered in the planet, but is 30% of the way to the planetary surface. Neptune's magnetic dipole is tilted by 47°, not quite so much; but it is even more offset, centering itself 55% of the way out toward the surface. The strange fields enhance the similarity of Uranus and Neptune and are probably related to the planets' internal constructions, but no one knows how or why.

As odd as they are, the fields finally allowed the accurate measurement of the internal, or System III, rotation period, which is 17^h14^m for Uranus and 16^h7^m for Neptune, notably longer than the Jovian and Saturnian periods.

16.3
Atmospheres

Although these outer two planets have internal structures quite different from those of the two big Jovians, their atmospheres are grossly similar. Yet even these hold surprises and are not at all like we expected before Voyager.

16.3.1 *Uranus*

Voyager scientists were crestfallen as they watched the image of the planet loom larger and clearer prior to encounter. There was at first nothing to see, just an attractive blue-green disk (Figure 16.6a). Only when the color is enormously enriched by computer (Figure 16.6b) can we discern a pattern of stripes running parallel to the equator. No real belts or zones can be identified. Immense contrast enhancement eventually also showed some whitish clouds near the equator that moved along with rotation (Figure 16.6c).

Yet the atmosphere is still of considerable interest. The helium abundance (the ratio of helium to hydrogen) is about 0.09 by number of atoms, similar to that found in the Sun but about double that in Jupiter's atmosphere and six times that in Saturn's. Jupiter's and Saturn's low helium abundances are caused by gravitational settling, a process that should also be active in Uranus's atmosphere. Uranus's original presettling helium abundance

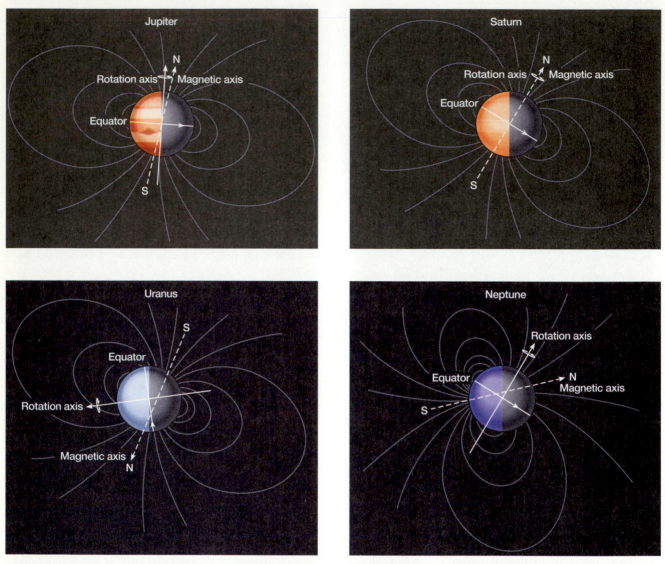

Figure 16.5   *The magnetic fields of Uranus and Neptune are oriented quite differently from those of Jupiter and Saturn.*

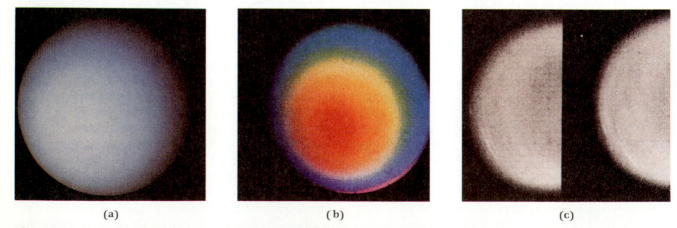

(a)   (b)   (c)

Figure 16.6   **(a)** *A hypothetical astronaut aboard* Voyager 2 *would have seen Uranus as a beautifully colored blue-green sphere, with no features to mar its surface.* **(b)** *Only with great color enhancement can bands of clouds be seen encircling the south pole and parallel to the equator, which lies near the edge of the disk.* **(c)** *Contrast enhancement shows white clouds near the equator that rotate with the planet.*

Figure 16.7 *Unlike Uranus, Neptune shows a wealth of detail, including familiar cloud belts and oval storms.*

ought therefore to have been *greater* than solar. The planet seems to have lost some of its hydrogen as a result of its low gravity, consistent with the interior construction outlined above.

Also consistent with this helium enrichment is a high carbon content, demonstrated by a methane abundance of 1% (relative to the number of hydrogen atoms), more than ten times Jovian. This compound gives the planet its color, since methane fiercely absorbs red sunlight, leaving only blue to be reflected. Like the previous two Jovian planets we have examined, the temperature of the Uranian atmosphere drops to a minimum near 0.1 bar. The temperature is so low that methane can freeze out and produce haze and clouds, which together with the inevitable hydrocarbon smog (ethane, acetylene, and the like) hide the belt-and-zone pattern and the cloud decks of ammonia and water that presumably exist below. The result is a bland, almost featureless appearance.

High-contrast images showed enough clouds to establish that wind speeds are intermediate between those of Jupiter and Saturn, reaching a maximum of 600 km/hr (relative to the interior rotation speed) at a latitude of 50° to 60°S in the prograde direction. Most surprising is a retrograde wind blowing at 350 km/hr near the equator, just the reverse of what we see on Jupiter and Saturn. No oval storms have been located. Uranus should experience extreme seasonal changes as the Sun alternately illuminates one pole, then the other. We would therefore expect a strong latitudinal temper-

ature difference and at least something of a north-south wind flow. The equatorial orientation of Uranus's cloud belts and the winds are consequently surprising. The weather patterns of the Jovian planets are apparently controlled by rotation, and axial tilt is of little consequence.

Because of the relatively low gravity, Uranus's thermosphere extends far from the planet. At an altitude of 6,000 km above the cloud tops, the temperature reaches 750–800 K, too high to attribute to the weak sunlight. The reason is unknown. Farther out, we encounter the magnetosphere, which has the familiar Sun-facing bow shock and rearward magnetotail. The magnetosphere contains radiation belts like Earth's, but because of the distance from the Sun, they are relatively weak. The high angle between the field and the rotation axis causes the magnetosphere to gyrate in space like a giant corkscrew.

16.3.2 *Neptune*

Given Uranus's appearance, expectations were low as Voyager approached Neptune in 1989. The planets, however, invariably surprise. Neptune's atmosphere is a rich blue covered with belts, immense storms, and bright clouds (Figure 16.7). Nevertheless, the atmospheres of the two planets are similar. Neptune's helium abundance of 12% (by number of atoms), even higher than Uranus's, again implies significant loss of hydrogen, and the high methane content of 0.5% (about half that found in Uranus's atmosphere) is again responsible for the lovely color. Neptune's temperature profile drops to a minimum (a chilly 50 K) at the 0.1 bar reference level (Figure 16.8), then shoots up toward a high value—750–800 K—in the extensive outer thermosphere. The terrible cold causes the upper cloud deck to be made of methane ice, and the action of sunlight on methane gas produces high hazes of ethane, acetylene, and other hydrocarbons.

Neptune's winds are also a variation on the Uranian theme. They are retrograde (east-to-west) near the equator and prograde (west-to-east) at high southerly latitudes. The reversal from the pattern exhibited by Jupiter and Saturn is a total mystery and probably relates to quite different internal structures. Neptune's equatorial wind speed is huge, 1,700 km/hr, comparable to that of Saturn.

The glory of Neptune lies in its storms and clouds. The planet is dominated by a huge, dark oval at a latitude of 20°S, the **Great Dark Spot** (GDS) (Figure 16.9). The GDS is about the size of the Earth, or about half the physical size of Jupiter's Great Red Spot; however, relative to the size of each planet, the GDS is half again as large. Time-lapse imaging shows that it spins counter-clockwise with a 16-day period. Since it is in the southern hemisphere, it is an anticyclone and consequently (like the GRS) a high-pressure storm system. We have no idea how old it is, but reasoning from the GRS we might think of it as permanent. In spite of the apparent similarity of the two storms, their placements are different. The GRS is trapped and rolled between winds that blow in opposite directions. The GDS, however, is in a band of steady retrograde winds that flow to the west. It is thought to be the top of a great plume of upwardly convecting Neptunian air that has a very deep but unknown origin.

The southern edge of the GDS is bordered by a line of bright clouds called the Bright Companion. These high clouds form roughly 100 km above the surface apparent when winds are sent aloft after hitting the high-pressure zone, rather like the clouds that form on the windward sides of high terrestrial mountains. Below the GDS, near 40°S, is another area of bright clouds not associated with any storm ovals: called the Scooter, it appears similar to the bright clouds seen in Uranus's atmosphere (see Figure 16.6c) and may be the tops of convective plumes or the result of waves in the atmosphere. Farther down, at 55°S, is Dark Spot 2 (D2), another storm oval. This one is more like Jupiter's GRS in that it does seem to roll between easterly and westerly winds. Wisps of high clouds

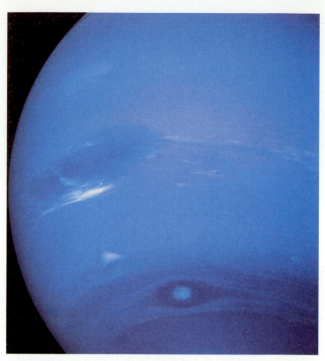

Figure 16.9   *A close-up of Neptune shows the major cloud and storm systems. At left center is the Great Dark Spot. Surrounding its lower edge is the Bright Companion, a zone of high, wispy clouds. Just below that, at 40°S, is the Scooter, another set of clouds; toward the bottom, at 55°S, is D2.*

Figure 16.10   *High cloud streaks some 200 km long throw shadows on the thick cloud deck 50 km below. This image was taken at closest approach, the resolution a mere 11 km*

(Figure 16.10), seen in both the northern and southern hemispheres, throw shadows on the thick cloud deck 50 km below, giving a unique aspect to the Neptunian atmosphere. It is hard to keep up with all the features, which wander around the planet with different periods because of the variation of wind speed with latitude.

Neptune shows the same kind of magnetosphere as Uranus, but weaker, smaller, and emptier of particles, since it is farther from the Sun.

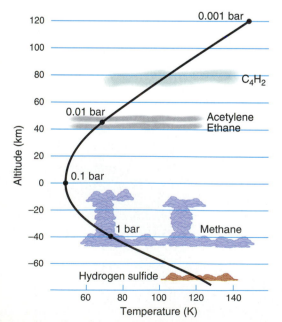

Figure 16.8   *Neptune's atmospheric temperature profile looks much like those of the other Jovian planets. The dominant clouds are made of methane. Rising columns may produce clouds and storm systems. Below them may be clouds of hydrogen sulfide, and above are high, smoggy hazes of various hydrocarbons.*

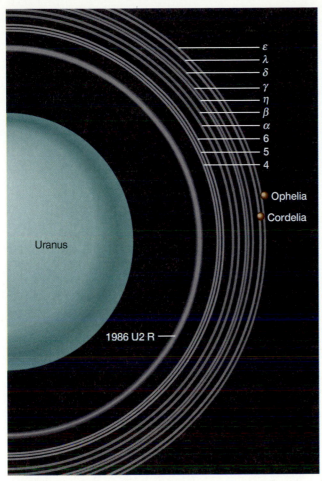

Figure 16.11 *Except for 1986 U 2R, the widths of Uranus's rings and the diameters of the two inner satellites (which shepherd the ε ring) are drawn 50 times larger than the orbital and planetary scale.*

16.4
Rings

The positions of the nine rings revealed by occultation measurements in Uranus's equatorial plane and those of two more discovered by Voyager are mapped in Figure 16.11 (see Table 16.3). All are within two Uranian radii (R_U) of the planet's center and within the Roche limit. Except for the ε ring, which is up to 100 km wide, and a broad sheet close to Uranus, they are only a few km wide (Figure 16.12), reminiscent of Saturn's narrow F ring.

Uranus's rings are very dark; with albedos of only about 5%, they would be invisible to an astronaut. The particles that comprise them are almost black and are believed to be covered with carbon compounds, either original or created by the action of sunlight on methane ice. As Voyager passed behind the rings, the dimming and flickering of its radio signal allowed an assessment of ring constitu-

tion and structure. The ring particles range in size from a few tens of centimeters to a few meters (see Section 15.4), and the big ε ring exhibits the same kind of ringlet structure possessed by the Saturnian system (inset, Figure 16.12). Like Saturn's F ring (see Section 15.4 and Figure 15.13), the ε ring is held in place—shepherded—by a pair of tiny satellites, Cordelia and Ophelia, each only 15 km in radius. No shepherding satellites were found for the other rings, but we suspect they must be there to keep the rings from diffusing outward.

Coarse rocks scatter light in all directions, but very small particles have a strong tendency to scatter light forward in the direction in which the light was initially traveling. When Voyager looked back at the rings with the Sun behind them, it detected a

	Distance		Width	Thickness
TABLE 16.3				
The Rings of Uranus				
Name	(10^3 km)	(R_U)	(km)	(km)
1986 U2 R	38	1.49	2,500	
6	41.8	1.63	1.3	0.1
5	42.2	1.65	2.3	0.1
4	42.6	1.67	2.3	0.1
α	44.7	1.75	7.1	0.1
β	45.7	1.79	7.1	0.1
η	47.2	1.85	0.2	0.1
γ	47.6	1.86	1.4	0.1
δ	48.3	1.89	3.9	0.1
λ	50.0	1.96	1.2	0.1
ε	51.1	2.00	20–100	< 0.15

Ring Features

Most are narrow and very dark; dust pervades ring plane but is relatively absent from rings.

Figure 16.12 *Voyager captured the narrow rings that sweep around Uranus. The detail in the ε ring (inset) was reconstructed by computer from observation of the flickerings of occulted stars.*

Figure 16.13 *Looking at the rings from behind, toward the Sun, Voyager showed the ring plane to be filled with a highly structured sheet of fine dust. The short streaks are trailed images of stars.*

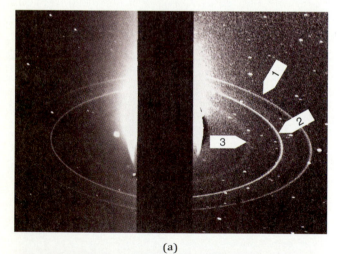

(a)

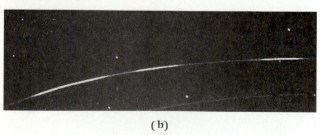

(b)

Figure 16.14 **(a)** *A composite of two images shows three of Neptune's rings encircling the planet (which is behind the vertical bar). The rings are numbered inward 1 through 3; ring 4, which is between rings 1 and 2, is not seen here.* **(b)** *Ring 1 shows arcs in which particles are consolidated. The two halves of the image in (a) were made about an hour and a half apart. Because of orbital movement, the arcs are not seen.*

TABLE 16.4
The Rings of Neptune

Name	Distance (10^3 km)	(R_N)	Width (km)
1989 N3 R	42–45	1.7–1.8	2,500
1989 N2 R	53.2	2.15	15
1989 N4 R	53.2–59.1	2.15–2.4	5,800
1989 N1 R	62.9	2.54	<50

Ring Features

Dusty rings in a dust sheet; particles accumulate into arcs in outer ring.

hundred bands of fine dust (Figure 16.13) that pervade the ring plane. The main rings, however, were no longer as visible, showing that the dust is *less* confined within them.

Because of Uranus's low gravity, its hydrogen atmosphere is distended and extends past the rings. The particles must therefore be subject to atmospheric drag, which will make them spiral into the planet. The major ring particles should last no more than 10 or 100 million years, and the fine dust should disappear in far less time. Therefore, the rings need to be resupplied. We think they are a way station for the smashed-up debris of collisions, particularly between comets and Uranus's satellites. The rocks become trapped by shepherding satellites, collide, and produce dust; small grains quickly swept from the rings by the atmospheric gas produce the dust sheet. All the particles inevitably wind up colliding with Uranus.

Voyager confirmed the Neptune occultation experiments and found four separate rings enmeshed in a dust sheet (Figure 16.14a and Table 16.4) within 2.5 Neptune radii (R_N) of the planet's center. The brightest two (1989 N1 R and 1989 N2 R, rings 1 and 2, numbered inward) are only a few tens of kilometers wide. The tiny satellites Despina and Galatea shepherd the inner one. A ring interior to these (1989 N3 R, ring 3) is bright but broader, about 2,500 km wide. Bordering the outside of ring 2 is a 6,000-km-wide band (1989 N4 R) that extends halfway to ring 1. Neptune's rings are brighter in forward-scattered light than those of Uranus, showing that they are filled with dust. In addition, the material in the outer ring is not distributed evenly, but is clumped into short arcs (Figure 16.14b) by a gravitational resonance with the satellite Galatea, which is just inside it.

CHAPTER 16   OUTER WORLDS   **277**

16.5
Satellites

Jupiter and Saturn are rich in accompanying satellites, and Uranus and Neptune follow the pattern. But each set has unique qualities. Voyager again taught us not to make generalizations.

16.5.1 The Moons of Uranus

The five largest satellites of Uranus (see Figures 16.1 and 16.15 and Table 16.5)—Oberon, Titania, Umbriel, Ariel, and Miranda—range between 760 km and 235 in radius. Ten other small bodies, the largest (Puck) only 75 km in radius, were discovered by *Voyager 2*. The noticeable correlation between the sizes of the satellites and their distances from the planet remind us of Saturn's system. All 15 rotate synchronously with their orbital periods, and all are within Uranus's magnetosphere.

The ten little moons are irregular and very dark, with albedos of only about 5%. They are probably covered with organic carbon compounds and are likely to be a source of the ring particles. The bigger five have densities averaging 20% greater than Saturn's intermediate satellites, indicating a constitution of more rock and less ice, about 50% of each. They are brighter than the inner satellites, with albedos averaging about 30%. These larger bodies are surely differentiated, with rock on the inside and bright ice on the outside.

Figure 16.15   *The 15 moons of Uranus are magnified in size by a factor of 40 relative to their orbital sizes and to Uranus.*

TABLE 16.5
The Satellites of Uranus

Name	Distance $(10^6$ km$)$	(R_U)	Period (days)	$i°$	e	Radius (km)	Mass $(10^{21}$ kg$)$	Density (g/cm^3)
Cordelia	49.75	1.95	0.35	0.1	0	15		
Ophelia	53.76	2.10	0.38	0.2	0.0	15		
Bianca	59.16	2.31	0.44	0.2	0	20		
Cressida	61.77	2.42	0.46	0.0	0	35		
Desdemona	62.66	2.45	0.47	0.2	0	30		
Juliet	64.36	2.52	0.49	0.1	0	40		
Portia	66.1	2.59	0.51	0.1	0	55		
Rosalind	69.93	2.74	0.56	0.3	0	30		
Belinda	75.26	2.94	0.62	0.0	0	35		
Puck	86.01	3.37	0.76	0.3	0	75		
Miranda	129.8	5.08	1.41	3.4	0.0	235	0.069	1.35
Ariel	191.2	7.48	2.52	0.0	0.0	580	1.26	1.66
Umbriel	266.0	10.4	4.14	0.0	0.0	585	1.33	1.51
Titania	435.8	17.1	8.71	0.0	0.0	790	3.48	1.68
Oberon	582.6	22.8	13.46	0.0	0.0	760	3.03	1.58

Satellite Features

Sizes generally increase outward from Uranus; inner two shepherd the ε ring; small satellites, very dark, may provide material for dark rings; larger five are brighter; among outer three, Oberon and Umbriel have old, cratered surfaces, but Titania is younger and icy; inner two have been reworked and may have been broken apart, Miranda as often as five times; rockier than Saturn's intermediate satellites.

 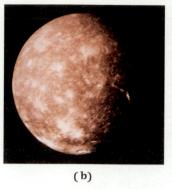 

(a)   **(b)**   **(c)**

Figure 16.16 **(a)** *Oberon is heavily cratered.* **(b)** *Titania has a younger surface and grabens.* **(c)** *Umbriel, again, is old with rayless craters. (The three images are not on the same scale.)*

(a)

(b)

Figure 16.17 **(a)** *Ariel, the brightest of Uranus's moons, is partially covered with bright ice and displays great, deep valleys.* **(b)** *Miranda has a trio of heavily grooved ovals associated with grabens close to 10 km deep. (The two images are not on the same scale.)*

The larger satellites exhibit a variety of surfaces. Oberon (Figure 16.16a), the most distant, is saturated with craters, its ancient surface a survivor of the great bombardment period of the early Solar System. Though nearby and of nearly the same size, Titania (Figure 16.16b) is different. Its surface displays grabens and evidence that it has been extensively resurfaced by the extrusion of water ice. Umbriel (Figure 16.16c), the darkest, also has an old surface.

The inner two satellites have been subject to considerable activity. Bright Ariel (Figure 16.17a), with an albedo of 40%, has far fewer craters than the others and gigantic rift valleys tens of kilometers deep. Its surface has been repaved by icy volcanism. It may have been heated by tidal resonances (see Section 14.6.1) no longer in effect because of orbital changes. Miranda's cratered and faulted surface (Figure 16.17b) features three roughly oval structures some 200 km across, outlined by a series of deep, concentric grooves and associated with grabens up to 10 km deep (Figure 16.18). This moon appears to have been shattered into pieces—perhaps as many as five times—that subsequently reassembled (the same might have happened once to Ariel). The odd ovals could be lingering frozen evidence of heavier rock that sank when the satellite differentiated, or perhaps they are bubbles of ice that rose from below, possibly the tops of giant, icy convection cells. Miranda has also been partially repaved with ice, and may, along with Ariel, have been tidally heated long ago.

16.5.2 The Moons of Neptune

Now for something different. Neptune, too, has a family of moons (Table 16.6). Five small ones under 100 km in radius are bunched near the planet, within or near the rings. Farther out, about double the ring distance, is larger Proteus, some 200 km in radius. These six satellites are dark, like the

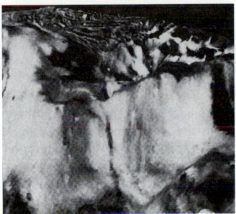

Figure 16.18 *Voyager's data allowed computer reconstruction of views across Miranda's grooved terrain. The cliffs (exaggerated here) are the greatest straight drop in the Solar System*

small Uranian moons, and are probably surfaced with some kind of carbon-based material. Proteus has a counterpart discovered from Earth, Nereid, that is 5.5 million km from the planet and has an orbital period of nearly an Earth year. Its high orbital eccentricity of 0.75, the record for satellites, suggests that it is a captured body.

Finally, Triton (Figure 16.19), 1,350 km in radius, is in a league with the Jovian Galileans and Titan. Unlike the other large satellites, however, its orbit is not in the plane of the parent planet's equator but is tilted by 157°, meaning that it revolves backward. These odd properties suggest that Triton was *also* once independent, captured from interplanetary space, perhaps by collision with another body or by encounter with a once-thicker Neptunian atmosphere. This conclusion is consistent with the lack of large Galilean-type satellites orbiting Uranus: neither planet formed any on its own, as did Jupiter and Saturn.

Triton is truly distinctive. Its density is high for an outer satellite, 2.07 g/cm^3 (exceeded only by Io and Europa), indicating that a large fraction—some 75%—is rock. Thus we see a continuance of the reversal of the rock-to-ice ratio with distance from the Sun first indicated by the Uranian satellites. The maximum water content seems to occur at Saturn. Triton also has an atmosphere. Although its surface pressure is only 2 × 10^{-5} bar, the atmosphere can produce high hazes. It is composed

	Distance		Period				Radius	Mass	Density
Name	*(10^6 km)*	*(R_N)*	*(days)*	*i°*	*e*		*(km)*	*(10^{22} kg)*	*(g/cm^3)*
Naiad[a]	48.0	1.9	0.29	0	0		30		
Thalassa[a]	50.0	2.0	0.31	4.5	0		40		
Despina[a]	52.5	2.1	0.33	0	0		90		
Galatea[a]	62.0	2.5	0.43	0	0		75		
1989 N2[b]	73.6	2.9	0.55	0	0		95		
Proteus[a]	117.6	4.7	1.21	0	0		200		
Triton	354.8	14.3	5.88	157[c]	0.0		1,350	2.14	2.07
Nereid	5,513	223	360.2	29	0.75		170		

TABLE 16.6
The Satellites of Neptune

Satellite Features

Five small, two intermediate, and one big satellite; two of inner six shepherd a ring; outer two, including Triton, probably captured; Triton has nitrogen-methane atmosphere, effective temperature 38 K, large frozen nitrogen-methane polar caps, and terrain that features grabens, basins, and nitrogen geysers; Nereid in highly elliptical orbit.

[a]Discovered by Voyager.
[b]Temporary name.
[c]Retrograde.

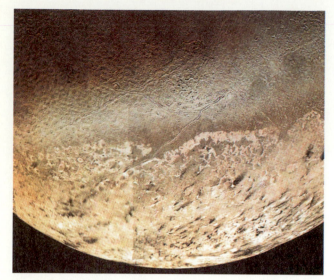

Figure 16.19 *Triton displays a remarkable variety of landforms reminiscent of the terrestrial planets. At the bottom is the huge south polar cap. To the north are vast areas of cantaloupe terrain covered with filled grabens.*

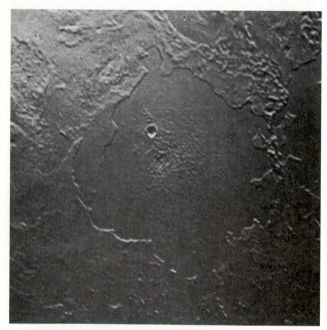

Figure 16.20 *The cantaloupe terrain of Triton contains basins that have been flooded with water-methane-ammonia "magma" from below.*

almost entirely of nitrogen with about a hundredth of a percent of methane. The body is so cold, only 38 K, that the nitrogen and methane freeze and fall to the surface as frost and snow, producing a vast polar cap that extends to a latitude of 25°. As the cap sublimes to gas under the action of sunlight, it augments the atmospheric pressure, just as the sublimation of the CO_2 cap does on Mars. Within the polar cap are geysers spewing nitrogen into the thin air and forming elongated clouds when they are blown downwind.

The equatorial region displays cratered areas and a strange mottled surface known as **cantaloupe terrain.** The long streaks seen in Figure 16.19 are grabens that reveal geologic activity. Close-ups show basins (Figure 16.20) that, like the grabens, have been flooded by volcanism, the "lava" an extruded mush of water, methane, and ammonia. Triton is probably highly structured, with a water-ice crust atop a liquid mixture of water, methane, and ammonia, which in turn rides on a rocky core. Triton's unexpected activity indicates internal heating, possibly caused by tides that changed and flexed the satellite during capture.

This body is like nothing we have seen before. Remarkably, as we leave Neptune behind, we encounter another.

16.6
Pluto

Since its discovery by Clyde Tombaugh in 1930, Pluto has had the distinction of being the last planet. Although it appears as little more than a point from the Earth, we have long known that it is not another Jovian. It is too small, and a periodic varia-

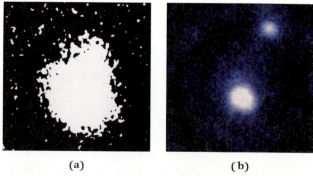

(a)   (b)

Figure 16.21 **(a)** *Pluto's satellite Charon was discovered as a blip on the planet's upper right edge. Turbulence in the Earth's atmosphere makes the images fuzzy and does not allow their easy resolution.* **(b)** *Pluto and Charon are nicely resolved by the Hubble Space Telescope.*

TABLE 16.7
A Profile of Pluto

Planetary Data		Gaseous Atmosphere	
mean distance from the Sun	39.53 AU	pressure	10^{-6} bar
equatorial radius = polar radius	1,190 km = 0.19 Earth radii	**Satellite Data**	
mass	1.3×10^{22} kg = 0.0022 Earth masses		
density	1.8 g/cm^3	**Charon**	
surface gravity	0.04 g_{Earth}	distance from Pluto:	19,640 km
escape velocity	1.1 km/s	radius:	590 km
rotation period (interior)	6.39 days	mass:	1.1×10^{21} kg
axial inclination	122.5°^{a}	density:	1.2 g/cm^3
magnetic field strength	not known		
tilt of magnetic field axis	not known		
temperature	40 K		

Layers	% Mass	Radii (km)
rocky core	60–75	0–900?
icy mantle	40–25	900–1,190

Planetary Features

Not a Jovian, but similar to Triton; nitrogen–carbon monoxide–methane atmosphere, large frozen N–CO–CH$_4$ polar caps; wide temperature range as a result of high eccentricity; atmosphere probably all freezes out at aphelion; with large moon, Charon, a double planet; Charon has water ice.

^{a}Retrograde rotation.

tion in its brightness shows that it is spinning slowly, with a period of 6.4 days. What, then, is it?

Some of the curtain of mystery was lifted in 1978 with the discovery that Pluto has a satellite (Figure 16.21a). Charon (in Roman mythology, the ferryman who transported the dead into Pluto's domain) was immediately seen to be large relative to its planet (see Table 16.7). Its orbit was first determined from fuzzy direct images and later by a specialized optical technique called **speckle interferometry,** in which the astronomer takes a large number of very short exposure images that freeze the turbulence induced by the Earth's atmosphere (see Section 9.2). These images are then combined by computer, permitting the reconstruction of a refined image similar to what would be seen were we taking the picture from space, which the Hubble Space Telescope can now do (Figure 16.21b).

Charon's orbital period is 6.4 days, the same as Pluto's rotation period. Each is tidally locked onto, and continually points the same face toward, the other. The orbit has a radius of 19,640 km and is highly tilted, Charon moving retrograde with a high orbital inclination of 122°. Because of the tidal locking, Pluto must be turning retrograde as well.

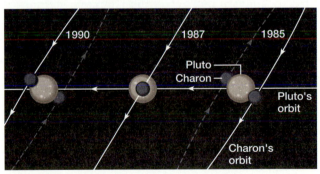

Figure 16.22 *As Pluto creeps slowly around the Sun, the angle at which Charon's orbit is presented to the Earth changes. In 1985, Charon and Pluto partially eclipsed and occulted each other. In 1987, the events were total and central, and by 1990 were partial again.*

Kepler's generalized third law applied to the mutual orbits of Pluto and Charon as determined with the Hubble Space Telescope shows that Pluto has a mass only 0.0022 times that of the Earth, a mere 18% that of the Earth's Moon. Charon's mass is 8% that of Pluto. The system has greater claim to be a double planet than even our own.

The timing of Charon's discovery could not have been better. Between 1985 and 1991, Pluto's

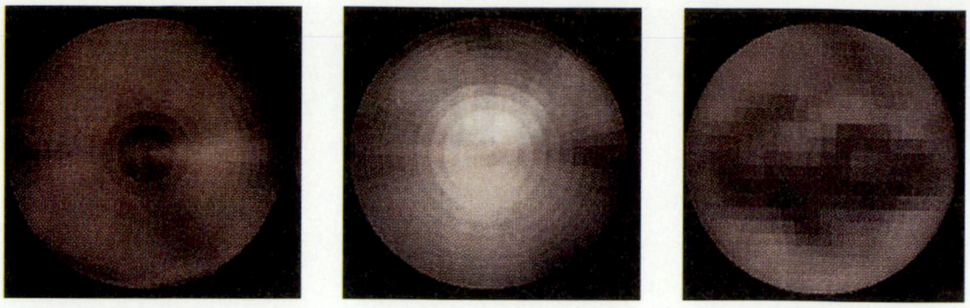

Figure 16.23 *Three images of Pluto constructed by computer from occultation data show* **(a)** *the north polar region,* **(b)** *the south polar region with its ice cap, and* **(c)** *an equatorial view.*

(a)   (b)   (c)

equatorial plane and the plane of Charon's orbit were directed at the Earth, which made the bodies cross in front of one another in a long series of mutual transits and occultations (Figure 16.22). These events provided a wealth of information. If the orbit is known, so are the orbital velocities. The durations of the transits and occultations allow the computation of the radii of the two bodies, even though neither is seen as a disk. Such measurements coupled with occultations of stars give a radius of 1,190 km for Pluto, less than half that of the previous minimum record holder, Mercury, and only two-thirds the size of our Moon. Charon has a radius of just 590 km. With a density of about 1.8 g/cm^3 (and possibly higher), Pluto is at least 60% rock, again more than the Uranian satellites or even Triton. Charon's density of 1.2 g/cm^3, however, indicates a much higher fraction of water ice, similar to that found in the Saturnian satellites.

The spectrum of Pluto shows absorptions of gaseous methane and of nitrogen and carbon monoxide ices. The little planet has an atmosphere with a surface pressure near 10^{-6} bar in which N$_2$ must dominate. Pluto's average albedo is very high, about 50%. At a temperature of around 40 K, some of the atmosphere freezes to the surface as a nitrogen–carbon monoxide–methane snow.

Pluto's orbit is quite eccentric (see Section 6.1). The planet is now near perihelion, where it is closer to the Sun than Neptune and where it would be expected to be brightest. Yet as Pluto approached perihelion it actually dimmed, which indicates an extensive polar cap. Fifty years ago, the planet directed its bright, icy cap toward us. Since then, it has turned its pole away so that we look increasingly at the less-reflective equatorial region. Moreover, as sunlight warmed the ground, some of the ice evaporated, leaving a smaller amount of bright snow; Pluto therefore became fainter even though more sunlight was falling on it. At the same time,

the atmospheric density has gone up. Thin as it is, Pluto's atmosphere seems to suffer the ultimate in variation: near planetary perihelion it is thickest, and at frigid aphelion the atmosphere all falls to the ground and disappears.

As Charon cuts across Pluto, the dimming is not constant, indicating variations in Pluto's surface brightness. The transit data, combined with brightness variations caused by rotation, allow the construction of a map of reflectivity that actually shows the bright south polar cap and a darker, mottled equator (Figure 16.23).

Charon is somewhat different. The surface appears to be uniform in brightness (with an albedo of about 38%) and its spectrum shows no methane absorptions. Instead, there are spectral features from water ice, consistent with its low density and high internal water content. Theory suggests that Charon lost its methane as a result of its lower gravity, leaving the heavier water behind.

We wish we could actually see Pluto's surface. The trajectories of the Voyagers never took them near the planet, and there are only indefinite plans for a mission to it. But maybe, in a way, we do know what Pluto looks like. It bears an uncanny resemblance to Triton. The two are nearly the same size, mass, and density, with similar atmospheres and polar caps. Furthermore, the bodies have comparable orbits around the Sun. If Triton were free of Neptune, which it probably once was, it might look very much like Pluto. So if you want to see Pluto, at least to an approximation, look back at Figures 16.19 and 16.20.

An old theory asserted that Pluto is an escaped satellite of Neptune, and that the separation event in some way was responsible for Triton's retrograde orbit. However, orbital analysis shows that Pluto has probably never been close to Neptune. Planetary scientists now think that Pluto and Triton represent a kind of body different from the other

BACKGROUND **16.1** Life at the End of the Solar System

Earlier, we spent a day on blistering Mercury, which is tucked so close to the Sun that lead would melt on its surface. What is the other extreme like?

Travel now to an outpost on Pluto—the end of our planetary system, where we stare out into the vast depths of interstellar space. The trip alone is daunting. It took Voyager, with accelerating gravitational assists by Jupiter, Saturn, and Uranus, a dozen years to make the journey to Neptune.

Finally, you arrive on this dim, cold world at a time when it is at its average distance from the Sun, 40 AU. Put on your pressure suit and climb out of your craft. A strange Sun shines overhead in a black sky. There is no apparent disk! To your eye, it appears only as a blinding star. Look away, and a huge, dim, quarter moon more than 3 degrees across hangs over the horizon. Crunching across a methane snow frozen to only a few tens of degrees above absolute zero, you arrive at your cabin, where you can finally breathe oxygen without your space suit.

Your day is a long one. Very slowly the Sun moves across the sky. It was noon when you arrived; 1.5 Earth days later, the distant star has finally set. Although the looming moon steadily goes through its phases, it stays motionless in the sky, the captive of mutual synchronous rotation. Night sets in and the temperature drops even further. At least your view of the stars is good.

Time to call Earth. Pick up the phone and start talking. No one answers. You talk for about an hour and a half, telling the mission commanders or perhaps your family what it is like 6 billion kilometers from home. You hang up. Your "Hello, I'm fine" has *just reached the orbit of Neptune*. It will arrive at Earth five hours after you began speaking. Six or seven hours after that, your phone rings.

Gaze outward into space, away from the Sun. There are no planets going through their familiar oppositions. Look toward the Sun. All planets are inferior. With luck you might glimpse Jupiter at greatest elongation a mere 7° from the daylight star. Earth is almost lost. In your loneliness, you pull out your telescope for a glimpse of home. There it is (Figure 16.24), a fragile blue dot against the frigid sky. It's going to be a long year.

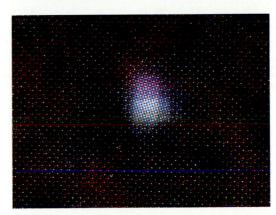

Figure 16.24 *Earth is seen from Neptune in one of* Voyager *2's last images as it left the planetary system for deep space.*

planets and their satellites. Perhaps there are many more out there, well beyond the orbit of the last Jovian, Neptune. Maybe we are staring out at a pair of the few remaining large **planetesimals** (the word inaccurately meaning "infinitesimal planets"), primitive, solid bodies that gave birth to the planets themselves. In the next chapter, we examine more evidence for the existence of planetesimals, and finally put the whole Solar System together.

KEY CONCEPTS

Cantaloupe terrain: A landform toward Triton's equator filled with grabens and basins.

Great Dark Spot (GDS): A dark, oval, high-pressure storm in Neptune's southern hemisphere, probably caused by a convective plume.

Occultation: An event in which one body hides or crosses in front of another.

Planetesimals: Primitive bodies of the Solar System that preceded the formation of the planets.

Ring arcs: Segments of a planetary ring enhanced by concentration of matter.

Speckle interferometry: An observing technique designed to produce high resolution; it uses computer-combined images taken at very short exposures to overcome the effects of turbulence in the Earth's atmosphere.

EXERCISES

Comparisons

1. Graphically compare the relative angular diameters of Jupiter, Saturn, Uranus, Neptune, and Pluto at opposition to show the relative ease of visibility of surface features.
2. Contrast the Great Dark Spot, the Great Red Spot, and the Great White Spot. How are they similar? How do they differ?
3. Compare the winds and magnetic fields of Uranus and Neptune with those of Jupiter and Saturn.
4. Why do the colors of Uranus and Neptune differ from those of Jupiter and Saturn?
5. Contrast the rings of Uranus and Neptune, and compare these with the rings of Saturn.
6. List similarities and differences between Triton and Pluto.
7. Why might the atmospheric helium abundances of Jupiter and Saturn differ from those of Uranus and Neptune?
8. How does the internal heat of Uranus differ from Neptune's?

Numerical Problems

9. What are the latitudes of Uranus's tropics and arctic circle?
10. About what year will Pluto and Charon again go into transit and occultation?
11. About when will Uranus's north pole be viewed face-on from the Earth?

Thought and Discussion

12. How are Uranus and Neptune similar to the cores of Jupiter and Saturn?
13. Why should the Jovian planets spin faster than the terrestrial planets, and why does Pluto spin so slowly?
14. Why do we know little about Uranus's northern hemisphere?
15. Why do the particles of Uranus's rings have short lifetimes?
16. What makes us suspect that Miranda was broken apart?
17. Why do we think Triton was once free, and why is Pluto no longer thought to be a former satellite of Neptune?
18. Why are Uranus's smaller inner satellites dark and its outer ones brighter?
19. How can the rings of Neptune and Uranus be so narrow?
20. What is cantaloupe terrain? On what body do you find it?
21. Name the most geologically active bodies of the outer Solar System? What could be responsible for their activity?

Research Problem

22. Explore: pre-Voyager observations of surface features on Uranus and Neptune; early, false preoccultation reports on rings of Neptune; early observations of the angular diameter of Pluto that gave a false idea of its size and mass.

Activities

23. Compare the Jovian planets. Number them 1 through 4 and plot their planetary radii, masses, densities, effective temperatures, ratios of internal to external heating, rotation periods, number of satellites over 1,000-km radius, number of satellites between 100- and 1,000-km radius, and number of small satellites.
24. At opposition, Uranus is visible to the naked eye, and is an easy sight in binoculars. *Sky and Telescope* and *Astronomy* magazines annually publish maps showing its path. Use them to find the planet. Sketch its position among the stars.

Scientific Writing

25. Write an essay for a textbook (like Background Essay 16.1) on what life would be like on Triton.
26. You are a public relations officer with NASA, whose leadership plans to send a spacecraft to explore Pluto. Write a press release to convince the public that such exploration is a good idea.

17

Planetary Creation and Its Debris

Asteroids, comets, meteors, and the creation of the Solar System

Comet West spreads its great tail above the morning horizon in 1976.

How did the 9 planets and their 61 known satellites arise? What gave us the gift of Earth? Ancient heavy cratering shows that there were once an astonishing number of small bodies—planetesimals—in orbit about the Sun, so many that we believe the planets were assembled from them. If so, the scraps of planetary creation ought to be all around us, providing the clues we need to look back through time to the beginning of the Solar System, to glimpse, however dimly, the construction of our home.

17.1
Meteorites

From over the nighttime horizon comes a brilliant ball of light, leaving a long trail behind it (Figure 17.1), like a torch thrown across the sky. Sparks shoot from its leading edge, and in a second or two it explodes in a colorful blast. You have just seen a **meteor** (from the Greek *meteoron,* "a thing in the air") of a special kind called a **fireball.** It was caused by a **meteoroid** (a piece of interplanetary debris) perhaps about the size of a softball, speeding along on a solar orbit at 20 to 40 km/s and colliding with the Earth. At an altitude of some 75 km, interaction with the atmosphere made the meteoroid heat up and ionize the surrounding air. Recombination of ions and electrons then produced the visible event. What appeared as sparks were generated as the meteoroid's outer surface melted and broke away. At its great speed, the meteoroid

Figure 17.2   *A woman in Peekskill, New York, examines the trunk of her car, which has just been bashed in by a 12-kg stony meteorite.*

could fly from Atlanta to Chicago in under a minute and might be seen over a several-state area.

The great majority of meteoroids are small, their number increasing rapidly with decreasing size. These vaporize in the air or disintegrate in a small explosion. If sufficiently large, however, perhaps the size of a marble, one may survive its passage through the atmosphere and strike the ground to become a **meteorite.** A rare large meteoroid a few meters across might even shatter the ground on impact to produce a crater (see Sections 11.2.3 and 11.6).

Meteorites are surprisingly common. Some 800 with masses of at least 10 g fall on the Earth every day, most of them landing in the oceans or on uninhabited land. Nevertheless, that rate produces an average of one strike/km^2 every 1,800 years. In your lifetime, one should drop within roughly 5 km of you (Figure 17.2). Occasionally a large meteoroid will break up upon entering the atmosphere and produce a rain of hundreds of fragments over a few hundred square kilometers, as occurred in Murchison, Australia, and Allende, Mexico, both in 1969. Given the infall rate, hundreds could accumulate within each square kilometer before the weather erodes them away. Unfortunately, most are usually hard to recognize and are buried rather quickly. Nevertheless, many museums have large collections.

The simplest classification scheme for meteorites includes **stones, irons,** and the intermediate **stony irons** (Table 17.1). The irons (Figure 17.3a) are the most easily identified and the most common to find. No other natural process will place a chunk of raw, unmanufactured iron in a farmer's field. They range in size from tiny scraps to bodies a few meters across with masses of several metric tons. They are always *alloys* (combinations of metals) with a nickel content of at least 5 to 10%, and possibly much higher. If sliced, polished, and

Figure 17.1   *A brilliant fireball meteor flashes past the distant stars.*

TABLE 17.1
Types of Meteorites

Type	Subtype	Frequency	Composition	Formation
Stones	carbonaceous chondrites	5%	water, carbon, silicates, metals	primitive
	chondrites	81%	silicates	metamorphic
	achondrites	8%	silicates	igneous
Stony irons		1%	50% silicates, 50% free metal	differentiated
Irons		5%	90% iron, 10% nickel	differentiated

Figure 17.3 **(a)** *An iron meteorite from Canyon Diablo in Arizona is about 40 cm across and weighs a quarter-ton.* **(b)** *Large crystals are seen when an iron meteorite is cut, polished, and etched with acid.*

Figure 17.4 **(a)** *This carbonaceous chondrite taken from the 1969 Allende fall, has been broken to show the interior.* **(b)** *Millimeter-sized chondrules are set in a dark, grainy silicate matrix.*

etched with acid, they show beautiful crystal structures (Figure 17.3b), indicating that they have undergone a long, slow period of cooling. The irons are similar in composition to what we might find at the Earth's core and thus seem to have come from inside large differentiated bodies. The meteorite that created Meteor Crater in Arizona (see Section 11.7) was an iron.

Stones are far more common than irons but seem rare because they are so hard to distinguish from ordinary rocks. (This bias against discovering stones is an example of **observational selection,** a major problem in astronomy.) In the Antarctic, where the stones are preserved and recognizable in glaciers, they constitute about 94% of the total meteorites found. The stones, which are primarily silicates, include about 10% metal combined into chemical compounds, and are similar in composition to the rock of the terrestrial mantle.

Eighty-six percent of the stones are **chondrites** that consist of a dark, fine-grained silicate rock matrix with small spherical inclusions called **chondrules.** They look as if they have been melted and resolidified, and many are brecciated (that is, they appear to have been smashed apart and welded back together). However, an important 5%, the **carbonaceous chondrites** (Figure 17.4), show no such effects and are abundant in carbon and water. They have not been altered and appear to be the true primitives of the Solar System. The **achondrites** constitute the remainder of the stones, contain no chondrules (neither do some carbonaceous chondrites, somewhat confusing the issue), and have been altered by heat. Over 40% of them, the **eucrites,** resemble the mantle of the Earth in that they are made of basalt and look as if they have undergone considerable volcanic processing.

The third general kind of meteorite, the stony irons, constitute only about 1% of all meteorites.

They are made of silicates intermixed with roughly 50% free iron.

The ages, orbits, and compositions of the meteorites provide clues to their origins. With the exception of those that have arrived from the Moon or Mars (see Figures 11.10 and 13.18), they are all about 4.5 billion years old. The Allende carbonaceous chondrite, dated at 4.57 billion years, is the oldest object we know. Since it is primitive and unaltered, we identify its age with that of the Solar System itself. The meteorites seem to be remnants of the system's formation. In a few instances, we have been able to reconstruct the orbits of recovered meteorites from the paths of their meteors (Figure 17.5). All were highly elliptical, with perihelia near the Earth's orbit (hence the collision) and aphelia between the orbits of Mars and Jupiter. They appear to originate from the zone between these two planets.

Perhaps the most extraordinary feature of meteorites is their trace constituents. Certain isotope ratios within them are similar to those found in exploding stars and in the atmospheres of stars that are losing matter back into space. Some meteorites contain ^{26}Mg, the daughter product of highly radioactive ^{26}Al, which has a half-life of only 720,000 years; other meteorites have been found to contain tiny bits of carbon crystallized by high tem-

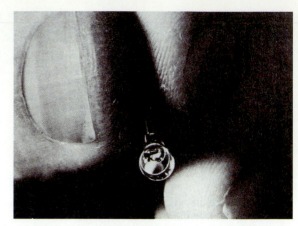

Figure 17.6   *Diamond dust, extracted from a carbonaceous chondrite, testifies to shock waves in the matter that made the Earth.*

perature and pressure (Figure 17.6)—diamonds—a trillion per gram. Their presence can be explained by shock waves from exploding stars acting on interstellar carbon. These discoveries allow us to go back in time *before* the origin of the Solar System and explore where its raw material came from. Most remarkably, the chondrites contain complex organic compounds, including amino acids, the building blocks of proteins and of life. The acids are not terrestrial contaminants, but were formed—in ways that we do not yet understand—within the meteoroids or their parent bodies. We do not yet know where the first molecules that could have turned into life came from, but these ancient stones have some of the complex chemicals already in place. The meteorites provide a link between the stars and the Earth and may illuminate the origin of life itself.

17.2
Asteroids

In 1772, an obscure mathematics professor named Johann Titius discovered a curious sequence in the distances of the planets from the Sun. Popularized by Johann Bode, the pattern led to the discovery of a new class of objects.

17.2.1 Discovery

A modern version of **Bode's law** begins with a geometric progression of numbers that starts at 0, goes to 3, then doubles repeatedly. Below each number write a 4:

0	3	6	12	24	48	96	192	384	768
4	4	4	4	4	4	4	4	4	4

Now add the columns, divide by 10, and compare the results to the actual distances of the planets from the Sun in AU:

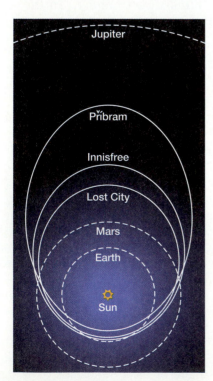

Figure 17.5   *The elliptical orbits (solid curves) of the meteoroids responsible for three named meteorite falls carried the original rocks out well past the orbit of Mars. (The orbits are given the same orientations for purposes of comparison.)*

	Mercury	Venus	Earth
prediction	0.4	0.7	1.0
distance	0.4	0.7	1.0

	Mars		Jupiter	Saturn
prediction	1.6	2.8	5.2	10.0
distance	1.5	—	5.2	9.5

	Uranus	Neptune	Pluto
prediction	19.6	38.8	77.2
distance	19.2	30.1	39.5

Figure 17.7   *Eros moves while its picture is being taken, instantly identifying itself as an asteroid.*

Given Pluto's odd characteristics, we can probably ignore it, so Neptune is the only outcast.

Bode's law is not a physical law, just a numerical relation, and it may be merely coincidental. Nevertheless, it predicted a planet at 2.8 AU, so astronomers of the late eighteenth century set out to look for it. On January 1, 1801, a new faint body was sighted from a Sicilian observatory by the Italian astronomer Giuseppe Piazzi, who named it Ceres (after the Roman goddess of grain, and the ancient deity of Sicily). The orbit, calculated by the great German mathematician Karl Friedrich Gauss, had a semimajor axis of 2.8 AU.

At seventh magnitude, vastly fainter than Mars or Jupiter, Ceres had to be very small, hardly qualifying as a planet; we now know it has a radius of only 457 km. More disconcerting, Ceres was not alone. A year later, Heinrich Olbers found Pallas also at 2.8 AU, which was followed by Juno (2.7

AU) and Vesta (2.4 AU, and visible to the naked eye). It was beginning to look as if these minor planets or **asteroids** (meaning "starlike") might be pieces of what had once *been* a planet.

By 1872 over 100 were known. They are easy to find. At 2.8 AU, Ceres has a period of 4.7 years, so it moves 0.2 degrees per day or 30 seconds of arc per hour. On a long photographic exposure or CCD image, an asteroid will leave a short streak (Figure 17.7). Over 20,000 have been found and more than 5,000 have known orbits. Several hundred new ones are discovered every year. An asteroid's name is preceded by its discovery number, for example, 1 Ceres, 2 Pallas, or 334 Chicago. The properties of the 15 largest are given in Table 17.2.

We now know that the asteroids never constituted a true planet. What they are is much more interesting.

TABLE 17.2
The Fifteen Largest Asteroids

Name	Radius (km)	a[a] (AU)	P[a] (years)	e[a]	$i°$[a]	Albedo (percent)	Rotation (days)	Class
1 Ceres	457	2.77	4.61	0.10	11	0.10	9.1	C
2 Pallas	261	2.77	4.61	0.18	35	0.14	7.8	C-like
4 Vesta	250	2.36	3.63	0.10	7	0.38	5.3	S-like[b]
10 Hygiea	215	3.14	5.59	0.14	4	0.08	18	C
511 Davida	168	3.18	5.67	0.17	16	0.05	5.1	C
704 Interamnia	167	3.06	5.36	0.08	17	0.06	8.7	neutral[c]
52 Europa	156	3.10	5.46	0.12	7	0.06	5.6	C
15 Eunomia	136	2.66	4.30	0.14	12	0.19	6.1	S
87 Sylvia	136	3.49	6.52	0.05	11	0.04	5.2	C[d]
16 Psyche	132	2.92	5.00	0.10	3	0.10	4.2	M
31 Euphrosyne	124	3.16	5.58	0.10	26	0.07	5.5	C
65 Cybele	123	3.43	6.37	0.13	4	0.06	6.1	C[d]
3 Juno	122	2.67	4.36	0.21	13	0.22	7.2	S
324 Bamberga	121	2.68	4.41	0.29	11	0.06	29.4	C?
107 Camilla	118	3.49	6.50	0.08	10	0.06	4.8	C

[a] a, semimajor axis; P, orbital period; e, orbital eccentricity; i, orbital inclination.
[b] Basaltic achondrite (eucrite).
[c] Flat spectrum, no classification.
[d] Possibly like M.

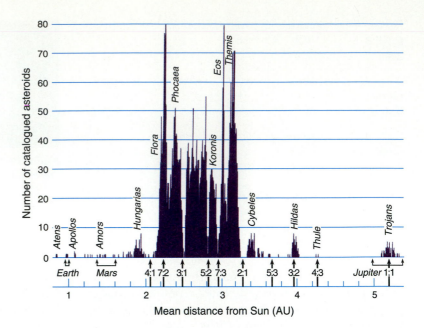

Figure 17.8   *The distribution of asteroid semimajor axes is concentrated strongly within the main belt between 2.1 and 3.2 AU. Resonances with Jupiter that produce the Kirkwood gaps are indicated by period ratios (Jupiter:asteroid) above the bottom scale. The names of several asteroid families are indicated. Aphelion and perihelion distances are shown for the planets.*

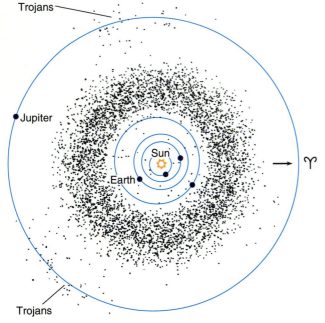

Figure 17.9   *The positions of 5,011 numbered asteroids, plus those of the inner planets and Jupiter, are plotted for May 1, 1992. The Kirkwood gaps are not visible because of the smearing effects of orbital eccentricity. A few Amor-Apollo-Aten asteroids are near or have crossed the orbit of the Earth. The Trojans occupy Jupiter's Lagrangian points. ♈ indicates the direction of the vernal equinox.*

17.2.2 Orbits

The asteroids are distributed between the orbits of Mars and Jupiter, are especially concentrated into a **main belt** between 2.1 and 3.2 AU (Figure 17.8). Most of their orbits have inclinations under 15° (a few tilt as much as 30°) and eccentricities typically about 0.15. A few asteroids, however, have highly eccentric orbits that take them within the orbit of the Earth and even inside that of Mercury. The three meteorites whose orbits are shown in Figure 17.5 clearly came from the asteroid belt. Meteorites are asteroids that have crashed into the Earth.

The distribution of asteroid semimajor axes is characterized by gaps (the **Kirkwood gaps,** discovered by Daniel Kirkwood in 1866) and **families** (see Figure 17.8). There are no bodies with semimajor axes of 3.3 and 2.5 AU. In a snapshot of the momentary positions of asteroids, however (Figure 17.9), the gaps are hidden, since the asteroids constantly pass through these distances on their elliptical orbits. The Kirkwood gaps are the analogues of the Cassini division in Saturn's ring system; they are caused by resonances with Jupiter, whose orbital period is exactly twice that of an asteroid at 3.3 AU and three times that of one at 2.5 AU. There are several other resonances (see Figure 17.8) with a variety of period ratios. The families consist of asteroids with closely similar orbits. Some, like the Floras, consist of broken parts of a once-larger body. Others (the Hildas) are associated with resonances that act to assemble asteroids rather than reject them. The most notable resonance family,

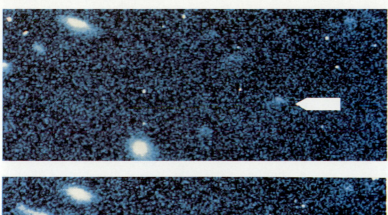

Figure 17.10 *Two CCD images show the movement of distant 1992 QB1 over a few days. The camera tracked the moving body, so stars appear as streaks.*

the Trojans (see Figures 17.8 and 17.9), have the same period as Jupiter and lie in a broad distribution around its Lagrangian points, 60° ahead of and behind the planet.

Three other families are of highly practical interest. The Amors approach the orbit of the Earth, the Apollos cross it, and the Atens have semimajor axes *less* than that of the Earth. Collectively they are known as AAA Objects, or **AAAOs.** There are about 1,000 known Amor-Apollos, but only 80 Atens. All are on elliptical orbits that take them out to the main belt. None has an orbit that intersects that of the Earth, but constant perturbations produced by the planets give all AAAOs the capability of colliding with us. Any one could produce an impact crater and perhaps disaster. It is sobering to think that there are a small number of Atens because the Earth has destroyed most of them in collisions! Meteorites like those in Figure 17.5 are just small AAAOs that are unrecognized until they hit.

Gravitational perturbations should prevent an extensive asteroid belt between Jupiter and Saturn. Only a few asteroidal bodies are known beyond Jupiter: dark 944 Hidalgo, which ranges from the main belt nearly to Saturn; 2060 Chiron between Saturn and Uranus (both have radii of about 90 km); 5154 Pholus near Uranus; 1992 QB1 (Figure 17.10), a 100-km-radius body with a semimajor axis of about 41 AU, just beyond the average distance of Pluto; 1993 FW, a similar body that may be even farther away; and a couple of others.

17.2.3 Physical Properties

Of the physical characteristics of asteroids, rotation is the easiest to assess. Most are irregular, and as they tumble in space, present different cross-sections to our view. Rotation periods can be determined from the resulting variations in brightness.

Close-up imaging by spacecraft provide the best radii, but few objects have been so viewed. *Viking* and other craft imaged the asteroidal moons of Mars (see Figure 13.27), and while on its way to Jupiter *Galileo* took a picture of 951 Gaspra (Figure 17.11a). The best Earth-based view is a radar image of 4179 Toutatis made at Arecibo (Figure 17.11b) when the (apparently double) asteroid passed 3.6 million km from Earth. Angular radii (and even shapes) can also be assessed by speckle interferometry (see Section 16.6) and occultations of stars. Combination of angular size with distance then yields physical dimensions. Once the radii are known, optical albedos can be derived by comparing the observed brightnesses to those calculated for perfectly reflecting surfaces. The most general technique for determining radius involves the combination of optical and infrared photometry. The amount of emitted infrared radiation depends upon surface area and temperature (see Section 8.3.2). The temperature depends on how much solar energy the asteroid absorbs, and consequently on its distance from the Sun and on its albedo. The optical brightness depends on size and albedo but not temperature, so its comparison with the infrared

(a)

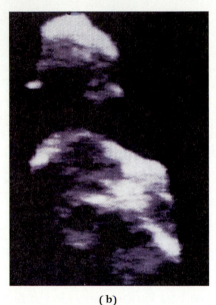

(b)

Figure 17.11 (a) *The* Galileo *spacecraft reveals the beaten surface of 951 Gaspra, which measures about 20 × 12 × 11 km. The different colors indicate a variety of materials. (b) Radar imaging shows 4179 Toutatis to be a double asteroid, two rocks each 3 or 4 km across in (or near) contact with one another.*

data yields radius and albedo simultaneously. The sizes, shapes, and albedos of a selection of asteroids are shown relative to their distances from the Sun in Figure 17.12.

Figure 17.13 shows the **size distribution** of asteroids, defined as the way their number varies with their radii. Large ones are rare, and the number increases rapidly with decreasing size. Although the data are not complete below about 20 km, a reasonable extrapolation of the curve shows that there must be vast numbers of asteroids in the neighborhood of a kilometer in diameter and even more at smaller radii. The number keeps increasing as we descend through boulders, rocks, and pebbles, consistent with the size distribution of

meteorites and just what we would expect if asteroids are products of collisions. They have been grinding one another down from larger bodies for billions of years. The collisional activity is vividly illustrated by the irregular shapes and battered surfaces of 951 Gaspra (see Figure 17.11a) and the Martian moons.

Masses of asteroids are hard to find but can be evaluated for a few by the perturbations they exert on Mars (as assessed by radar observations) and on one another. Ceres, the largest, has a mass of only 1.2×10^{21} kg, or 2×10^{-4} the mass of the Earth. Pallas and Vesta have masses about 20% of this value. The resulting densities of Ceres and Pallas are about 2.7 g/cm^3, typical of rock. That of 10 Hygeia is lower, 2.1 g/cm^3. Vesta is unusual, about 3.6 g/cm^3, showing that more than rock is involved in its structure.

From the average density and the distribution in radii in Figure 17.13, we can make an estimate of the total combined mass of all asteroids. From the way that the curve bends over at upper left, we know that most of the mass in the asteroid belt is concentrated in the larger objects. As numerous as they are, little asteroids do not count for much. Ceres is estimated to possess about a quarter of the total mass. All the asteroids combined would make a mass of about 5×10^{21} kg, 8×10^{-4} the mass of the Earth: they would not constitute anything close to a respectable planet.

Remarkably, we can relate the different meteorite classes to different kinds of asteroids. Asteroids are classified on the basis of their albedos, colors, and spectra, which contain absorption features that give clues to the nature of the surface minerals. Class C asteroids are dark, their albedos in the neighborhood of only 5%. Because of the similarity of their spectra, the C asteroids are almost certainly the progenitors of the carbonaceous chondrites. S asteroids are over twice as bright. Their spectra fit those of the ordinary chondrites and, unfortunately, those of the stony irons as well, making discrimination difficult. The M asteroids have characteristics quite similar to those of the iron meteorites. The P and D asteroids have very low albedos and are quite reddish. Finally, bright Vesta defines its own class and has a spectrum almost identical to the stony basaltic eucrites, which are almost certainly pieces broken from it. You can look into the sky at Vesta while holding some of it in your hand!

17.2.4 Origins of Asteroids and Meteorites

The physical conditions of asteroids and their relations to the meteorites allow us to reconstruct something of the asteroids' origins. Like the car-

(a)

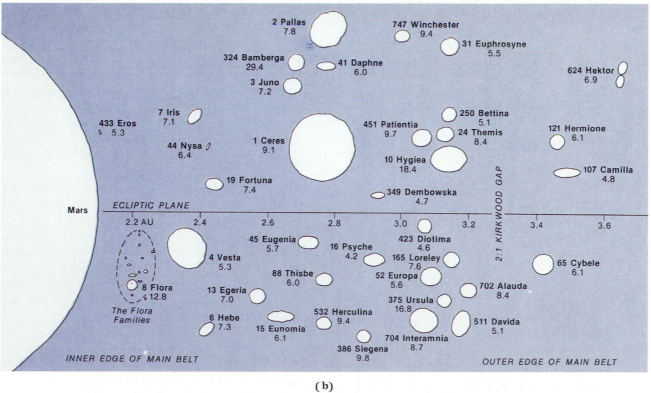

(b)

Figure 17.12 *The sizes of the larger asteroids are compared to Mars.* **(a)** *The painting shows colors and albedos.* **(b)** *The drawing gives names, distances from the Sun, and albedos (by percentage). The distance scale is enormously compressed relative to the scale of sizes; inclinations are indicated by the positions relative to the ecliptic plane.*

bonaceous chondrites, C type asteroids must be primitive and essentially unchanged over the aeons, whereas the S and M asteroids must be the remnants of differentiated bodies that have been smashed to pieces. The iron meteorites are probably the cores of differentiated asteroids and the achondrites are the mantles; the stony irons may come from the boundary layers. The relative velocities among asteroids are not all that high. Following a collision, some of the debris will be scattered to form new, smaller asteroids, but much will fall back to the parent body, leaving its surface a pile of fusing rubble. When another collision knocks this outer crust away, the result is a brecciated asteroid, which if it hits the Earth will be a brecciated meteorite.

We do not know why some asteroids differentiated while others did not. Moreover, the parent body or bodies of the present asteroids had so little mass that they should have cooled very quickly after formation, leaving them little time to differentiate. Where did the heat come from? One suggestion is that it was produced by the rapid decay of ^{26}Al, which was once abundant in meteorites.

A clue to these mysteries may be found in the spatial distribution of the asteroids (Figures 17.12 and 17.14): the farther from the Sun, the more primitive the body. The S asteroids are closer than the C. The dark red P and D, which are presumed to be covered with an organic carbon residue and to be primitive, are more distant yet. Dark 5145 Pho-

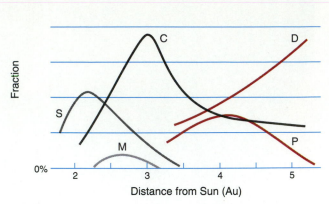

Figure 17.14 *The rough distributions of different classes of asteroids show that the differentiated ones are closer to the Sun than the dark, primitive ones.*

lus, one of the most distant asteroids known, is even redder. The Sun may have played a role in heating the closer asteroids, possibly through intense magnetic fields generated when it was very young and just forming. This mechanism and the decay of ^{26}Al may have shared in the heating process.

We now believe the asteroids to be remnants of planetesimals, the original small bodies out of which the planets formed. Several small planetesimals probably began to develop in the space between Mars and Jupiter. Those near Jupiter may have been kicked by the huge planet into eccentric orbits toward the Sun and in turn gravitationally stirred other planetesimals, prompting them to collide. Each collision produced more objects, causing further collisions, and the asteroid belt ground itself down. The few remaining large bodies like Ceres were lucky to avoid great hits.

Jupiter continues the process: resonances, particularly that caused by a 3:1 ratio of orbital periods, force asteroids into elliptical orbits. Some of these asteroids come close to the Sun and become AAAOs. Planetary perturbations continue to modify their orbits until they collide with one of the terrestrial planets. Those that hit the Earth scream through the sky as meteors before coming to rest as meteorites.

This process has apparently given us a very distorted view of the relative numbers of different kinds of asteroids. Jupiter is selectively throwing ordinary chondrites at us, whereas the majority of the asteroids are carbonaceous chondrites, primitive stones that only rarely reach the Earth.

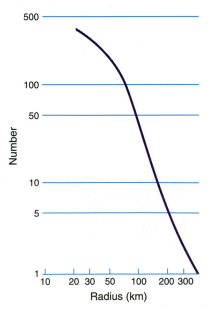

Figure 17.13 *The curve shows the number of asteroids larger than a given radius.*

Figure 17.15   *The gaseous coma of Halley's Comet surrounds a tiny nucleus, indicated by the black dot (on this scale, the nucleus would be only a thousandth the size of the dot). The plasma tail streams out from the coma away from the Sun in nearly a straight line. The dust tail fans away to the left.*

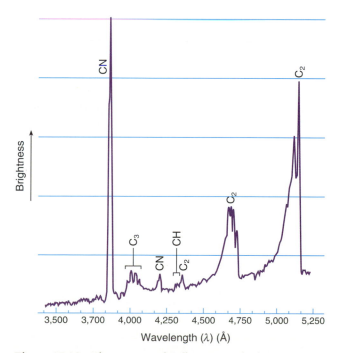

Figure 17.16   *The spectrum of Halley's Comet displays emission bands of numerous molecules, including C_2, C_3, CH, and CN (cyanogen).*

17.3
Comets

If Saturn is the favorite icon of astronomy, a **comet** is a close second. The sight of a great comet (Figure 17.15), its long gossamer tail streaming a quarter of the way across the sky, will bring multitudes outdoors to look. Over the centuries, comets have been viewed as messengers of doom and destruction. In fact, life on Earth may have been impossible without them.

17.3.1 *Appearance and Orbits*

Comets are commonly confused with meteors. Meteors are atmospheric phenomena, while comets are typically as distant as planets and move at comparable speeds. They therefore shift only a few minutes or degrees of arc from night to night as they slowly work their way through the background of the stars.

A typical comet has three basic parts. At the core is a dark **nucleus,** the solid interplanetary body itself. Cometary nuclei are so small that none has ever been resolved from Earth. Surrounding the nucleus is a large, resolved halo of light called the *head* or **coma** that can be a degree or more across and over 100,000 km in diameter. The spectrum of the coma consists of molecular emission lines (Figure 17.16), showing it to be a gas.

Streaming out of the coma is a **tail** that may be several degrees and millions of kilometers—even up to an AU—long. Comets have two kinds of tails. One is narrow and points nearly straight back from the Sun. Spectroscopy reveals emission lines of ionized gas (Figure 17.16), so it is called the *gas, ion,* or **plasma tail.** It tends to be blue because of emissions of CO^+ in the blue part of the spectrum. The other kind is generally curved and fan-shaped. Spectroscopy shows the spectrum of sunlight scattered from small grains; this **dust tail** is yellowish, the color of sunlight. Perspective effects can make it seem to point in any direction. Which of the two kinds of tail is the brighter depends on the particular comet; sometimes one or the other even appears to be missing. The two together can combine to make a stunning sight.

Comet orbits are highly elliptical, and many have perihelion points inside the orbit of the Earth, some greatly so. Comets are easily seen only when they are within the confines of the terrestrial planets. There are two broadly different types of comet. The **short-period comets** (Figure 17.17) have

Figure 17.17 *The short-period comet Giacobini-Zinner has a period of 6.6 years and sports a short but bright plasma tail.*

Figure 17.18 *Comet Ikeya-Seki, called a sungrazer because it came within a few hundred thousand km of the solar surface, had a brilliant tail dozens of degrees long.*

orbital periods less than about 200 years (Table 17.3a). The shortest known is Encke's Comet, which goes about the Sun in only 3.3 years. These move generally prograde, counterclockwise about the Sun, and inclinations tend to be low, under 35°. Aphelion points range across the planetary system, but are typically near the orbit of Jupiter. Short-period comets are by definition seen frequently, often at every passage near perihelion, but most are dim and have modest tails. Their orbits are well known, and their returns into the inner Solar System can be predicted, as first shown by Edmund Halley (see Section 7.6).

The **long-period comets** (Table 17.3b and Figure 17.18) have periods over 200 years, most far longer. The elliptical orbits of some are so eccentric that in the vicinity of the Earth they cannot be discriminated from parabolas, so their periods cannot be measured and are estimated to be in excess of a million years. Long-period comets have no respect for the ecliptic. As many proceed retrograde as prograde, and their orbital inclinations are random. You are just as likely to find one in the Big Dipper as in Leo. The majority of the known comets are of the long-period variety.

Unlike Halley's Comet (named for the astronomer who predicted its orbit) most comets are named after their discoverers. Long-period comets are random events, and the majority are found by dedicated amateur observers. Some two dozen or more comets are seen each year (including a few known periodic ones), most of them faint and

TABLE 17.3
Selected Comets

Name	Year of discovery	q(AU)[a]	e	i°	P(years)	Characteristic
Short-period						
Biela	1772	0.86	0.76	13	6.6	vanished after 1851
Encke	1786	0.34	0.85	12	3.3	shortest period
Giacobini-Zinner	1900	1.03	0.71	32	6.6	visited by spacecraft
Halley's	—	0.59	0.97	162	76	brightest
Long-period						
De Chéseaux's	1744				—[b]	six dust tails
Donati's	1858	0.58	0.996	117	—	
Morehouse	1908	0.95	1.0	140	—	very dusty
Great January	1910				—	rivaled Halley's
Arend-Roland	1956	0.32	1.0	120	—	sunward spike
Ikeya-Seki	1965	0.008	1.0	142	800	sungrazer
Bennett	1970	0.54	0.996	90	—	
West	1976	0.20	1.0	43	—	nucleus broke into four pieces

[a]Perihelion distance.
[b]No listing means the period is very long.

BACKGROUND 17.1 Great Comets and Halley Madness

Perhaps once a generation we will witness a great comet, one that dwarfs the dozens of others that pass by the Earth every year. Great comets can provide spectacular interludes in our perpetual searches of the heavens. Several have already been depicted here. Comet West of 1976 (the chapter-opening figure) and 1965's Ikeya-Seki (Figure 17.18) are among the brighter ever seen. However, even these beauties are surpassed by some that appeared in the nineteenth century (certainly enhanced by the darker skies of the time). The great comet of 1843 had a tail 300 million km long, and the central coma of the great comet of 1881 (Figure 17.19a) was visible in broad daylight. Certainly one of the more spectacular was De Chéseaux's, which had six

separate dust tails (Figure 17.19b). We never know when another great comet will appear, as the events are truly random, but almost certainly one or more will arrive near the Earth in your lifetime. It is well worth making a trip to the dark countryside to see and admire it.

Halley's Comet is in a class by itself. Although it has a short period, it is so bright that it clearly deserves to be called a great comet. Its period is roughly the length of a human lifetime, which gives most of us an opportunity to see it once and a favored few the chance to view it twice. It has been recorded throughout much of written history and appears in the Bayeux Tapestry (Figure 17.20a), which chronicles the Nor-

(continued on page 298)

(a)

(b)

Figure 17.19
(a) *The great comet of 1881, Tebbutt's Comet, was so bright that its central coma was seen in daylight.* **(b)** *The six tails of De Chéseaux's Comet rose above the morning horizon in 1744.*

(a)

(b)

Figure 17.20
(a) *Halley's Comet is seen embroidered as a heavenly sign into the Bayeux Tapestry, a pictorial history of the Norman conquest.* **(b)** *In a religious painting by the Florentine master Giotto it probably represents the star of Bethlehem.*

man invasion of Britain in 1066, and in Giotto's *Adoration of the Magi,* painted in 1303 (Figure 17.20b).

In our own time, Halley's has produced a brand of popular mania. Its 1910 return was one of the showiest ever, as the comet came into perihelion and was brightest when it was close to the Earth. The comet was used widely in advertising, and apparently inspired a thwarted blood sacrifice of a woman in Oklahoma by a band of fanatics. More intriguing, astronomers predicted that on the night of May 18–19, 1910, the Earth would actually pass through the tail. Shortly before, the deadly gas cyanogen had been discovered in the comet's spectrum, so there was widespread fear that life on Earth might be destroyed.

On that night some people packed their windows and doorways with wet rags to keep out the fumes, while sophisticates in New York held rooftop parties to celebrate the end of the world. Entrepreneurs got rich selling comet hats and comet pills to ward off the evil effects. Nothing happened. The tail of a comet is so vacuous that no effects were seen at all.

The 1985 encounter with Halley's Comet, one of the most disappointing in its two-millennium recorded history, was seen well only from the southern hemisphere. Yet the hype was enormous and sold a great many telescopes to people who had little idea of how to use them. No doubt we can expect more of the same when this comet returns in 2061.

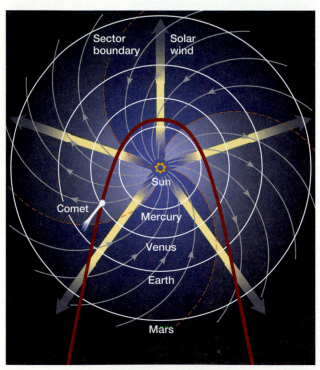

Figure 17.21   *As the solar magnetic field is dragged into the planetary system by the solar wind, the Sun's rotation winds it into a spiral. The polarity of the field reverses at the four sector boundaries. The comet's gas tail is formed by the interaction of the wind and field with the comet's ionized coma.*

below naked-eye visibility. They are formally designated by letters in the order in which they were discovered within a given year and then by Roman numerals in order of perihelion passage. Comet Arend-Roland (named after its co-discoverers) is also 1956h and 1957 III. Short-period comets are designated by P/, so Halley's becomes P/Halley.

17.3.2 *Physical Nature*

In 1950 Fred Whipple of Harvard proposed the basic model of a comet still in use, the *icy conglomerate* or *dirty snowball.* Comet nuclei, it states, are made of a variety of ices, mostly water, into which are mixed large quantities of small solids ranging in size from a few millimeters down through grains of dust less than a micron (10^{-3} mm) across. As a comet approaches the Sun, it absorbs progressively more solar energy, begins to heat, and its ice starts to sublime to a gas. The comet's gravity is low, and the gas expands into space under the force of its own pressure. Dust particles are carried away with the gas; together they form the coma. The force of the ejection is strong enough to produce a rocket effect that slightly changes the comet's orbit, sometimes making it difficult to chart its future course.

Ultraviolet sunlight breaks down the water, causes chemical reactions that make new compounds, partly ionizes the gas, and excites atomic and molecular electrons into outer orbits. When the electrons cascade back downward, they produce the emission spectrum of Figure 17.16. Simple diatomic (two-atom) molecules like CH, CN, C_2, and CO dominate the mixture, but there are also molecules like CH_3CN and bound strings of formaldehyde (H_2CO) molecules. Mixed in are also a number of free elements. Hydrogen freed in the reactions spreads out around the coma in a cloud that can be a million or more kilometers across.

A comet's plasma tail is a creature of the solar wind (see Section 10.6 and Chapter 18) and the Sun's dipole magnetic field. Since the solar wind is ionized, it interacts with the field, dragging it outward from the Sun. The Sun is rotating, however, with a period of about a month. The result is that the outbound magnetic field lines wrap into a spiral, much like sprays of water from a rotating water sprinkler (Figure 17.21). Some of the field lines come out of the Sun and others return, the two directions separated by *sector boundaries*.

As the comet plows into the magnetic field lines, they are captured by the electrically charged gases. The field wraps around the comet like a wind sock and stretches the tail away from the Sun (Figure 17.22a). The solar wind flows around the ionized and magnetized comet, creating a bow shock of the sort associated with the planets. When the comet crosses a sector boundary (where the magnetic field suddenly reverses direction), the tail is disconnected from the head (Figures 17.22b and 17.23). The old tail flies away, and the comet grows a new one. Comets therefore function as probes by

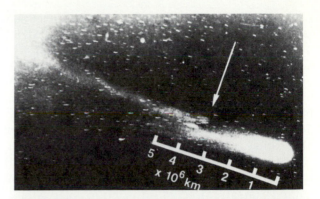
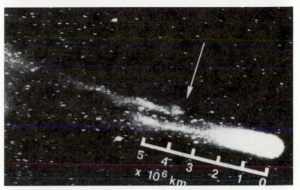
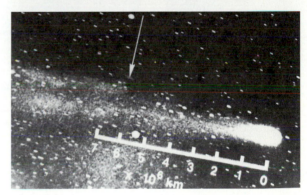

Figure 17.23 *In its 1910 return, Comet Halley displayed a classic disconnection event over a period of hours.*

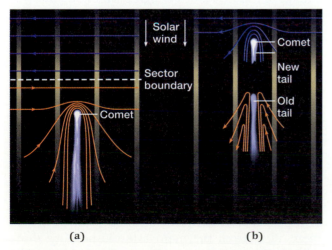

(a)   (b)

Figure 17.22 *In* **(a),** *the comet captures the solar magnetic field to create the plasma tail, bending the field (orange lines) into a bow shock. In* **(b),** *the comet has crossed a sector boundary, where the field lines (blue) run in the other direction. The tail then disconnects and a new one grows in its place.*

which we can examine the Sun's wind and magnetic field.

As the comet's ices sublime, dust is released and pushed away from the Sun by the pressure of the intense electromagnetic radiation. Since the particles have considerable mass (at least as compared to the molecules of the gas), they lag behind the comet in its orbit, and the dust tail thereby spreads outward into a great fan.

A comet's tiny nucleus is so small it can be studied only by spacecraft. In 1984 and 1985, an armada of spaceships was launched to rendezvous with the returning Halley's Comet. Two from the Soviet Union, two from Japan, and one sent by the

European Space Agency (ESA) met the comet as it crossed the ecliptic plane during the week of March 8 to March 14, 1985. The twin Japanese craft (*Suisei* and *Sakigake*) were designed to measure properties of the magnetic field and solar wind 10^5 to 10^7 km upstream from the comet and in its magnetosphere. The Soviet *Vega* ships, carrying a variety of detectors including spectrometers, dust detectors, and cameras, passed within 10^4 km. ESA's *Giotto* craft (named for the artist of the painting in Figure 17.20b) carried an imaging camera. In addition, the United States redirected an orbiting solar satellite equipped with particle and magnetic field detectors, renamed the *International Cometary Explorer* (*ICE*), to pass 0.2 AU to the sunward side of Halley's on March 25. In September 1985 *ICE* had passed a mere 8,000 km from the nucleus of Comet Giacobini-Zinner (see Figure 17.17), its instruments confirming the theory of the tail's formation.

 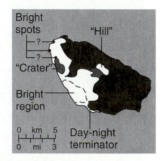

Figure 17.24   *The nucleus of Halley's Comet is revealed in this composite of pictures taken by ESA's Giotto spacecraft. Features are identified at right. The bright sunward side is to the left. The terms "hill" and "crater" are identifiers only and their natures are uncertain. Gas is spewing from jets or geysers under the action of sunlight.*

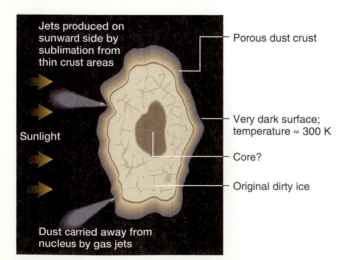

Figure 17.25   *A model of Halley's Comet suggests a rocky core surrounded by dirty ice topped by a dust crust. Gas and dust stream out of the crust in jets.*

Passing only 600 km from Halley's nucleus, *Giotto* obtained a detailed view (Figure 17.24) even as its camera was being destroyed by dust particles hitting it at 70 km/s. The nucleus has the shape of a giant peanut, 8×16 km across. It is almost black, one of the darkest bodies known in the Solar System, with an albedo of only 3%. On the daylight side is a dark area thought to be a crater of some sort, and a small "hill" on the nighttime side seems to catch sunlight. The most prominent features are geysers of gas with entrained dust that spray from bright spots on the sunlit half. Halley's nucleus rotates with a period of about six days. As it is warmed by the Sun, the subsurface ice sublimes and shoots off into space, forming the streamers of dust seen to come from cometary nuclei.

From the orbital changes resulting from the rocket effect and the rate at which mass was observed to be lost, the comet's mass could be estimated at about 10^{14} kg, indicating that its density may be as low as 0.25 g/cm^3. The surface seems to be extremely porous and fragile. There may be a rocky core at the center (Figure 17.25) surrounded by undisturbed dirty ice. Some of the dust from the jets piles up on the surface, resulting in crustal thickening.

Comet nuclei are ephemeral objects, typically losing 1% of their mass at each solar passage; the exact amount depends on the comet's size, the fraction of volatile compounds, and degree of crustal insulation. The long-period comets are the brighter because their nuclei have lost the least matter. With the exception of Halley's, the short-period comets are faint. After enough perihelion passages, a comet's nucleus will lose all its volatiles and either disintegrate or reveal itself. Comet West of 1976 broke into four pieces as it went around the Sun. Very gradually they will move apart and, in the distant future, return like a line of ducklings, though separated by many years. Then the Sun will go to work on these. Short-period comets go dark much faster. Biela's Comet, discovered in 1772, broke in two at its 1846 perihelion passage, was seen at its next return in 1852, and then vanished. Even Jupiter can break up comets (Figure 17.26).

If a comet's nucleus does not completely disintegrate, it may look like an Apollo asteroid. There are too many AAAOs for Jupiter to have kicked them all in from the main asteroid belt; we think that roughly half the AAAOs are dead comets. The issue is even more confused by the asteroid 2060 Chiron. As it approached perihelion, beyond Saturn, it generated a gaseous cloud reminiscent of cometary activity (Figure 17.27). This so-called asteroid may be a comet nucleus or something akin to it, one with a diameter of 180 km. The neat distinction between comets and asteroids is now blurred. Both are planetesimals, and there may be a continuum of behavior between the two.

17.3.3 Meteors, Meteor Showers, and the Zodiacal Light

Standing outdoors at three in the morning, facing in the direction of the Earth's orbital motion, you will typically see at least three meteors an hour. Unlike the rare fireballs, most will be small and faint, quickly extinguishing themselves in the upper atmosphere. Several times a year you can see a **meteor shower** (Figure 17.28 and Table 17.4) that might provide more than a meteor a minute. If you trace the paths of these shower meteors backward, you will see that they all appear to emanate from a particular point in the sky called the **radiant.** Meteor showers are named after the constellation that contains their radiant; for example, the Perseids of August 12 come from Perseus.

The meteoroids that produce a particular shower all orbit together in a stream around the Sun (Figure 17.29a). The railroad tracks in Figure 17.29b are actually parallel to one another, but look as if they diverge from a point on the horizon. To a good approximation, the little meteoroids are moving along parallel tracks (Figure 17.29c). When they hit the Earth's atmosphere and become visible, they seem to emanate from the shower's radiant. The radiant for a particular shower depends on the relative directions of the motion of the Earth and the meteoroid stream.

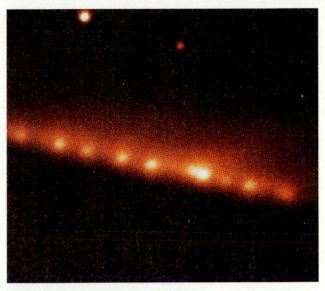

Figure 17.26   *Comet Shoemaker-Levy apparently passed too close to Jupiter in 1992 and broke into at least 17 pieces (of which a dozen are visible here).*

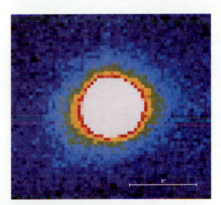

Figure 17.27   *Is 2060 Chiron a comet or an asteroid? Here it has developed a small coma as it is warmed by the Sun.*

Figure 17.28   *Several meteors radiating from the constellation Perseus flash across the sky.*

TABLE 17.4
Prominent Meteor Showers[a]

Name	Dates[b]	Radiant α	Radiant δ	Hourly rate[c]	Comet
Quadrantids[d]	January 4	15^h16^m	$+49°$	100	?
Lyrids	April 20	18^h08^m	$+33°$	10	1861 I
Eta Aquarids	May 5	22^h24^m	$-01°$	30	Halley's
N. Delta Aquarids	July 29	22^h36^m	$-17°$	15	?
Alpha Capricornids	August 1	20^h32^m	$-08°$	10	?
Perseids	August 12	03^h12^m	$+57°$	80	Swift-Tuttle[e]
S. Delta Aquarids	August 13	22^h56^m	$+02°$	10	?
Orionids	October 21	06^h20^m	$+16°$	20	Halley's
Leonids[f]	November 17	10^h12^m	$+22°$	10	1866 I
Geminids	December 13	07^h32^m	$+33°$	80	3200 Phaethon[g]

[a]Selected from about 100 known annual showers.

[b]Peak; meteors can usually be seen on many days to either side.

[c]For a dark sky; number can vary considerably.

[d]Named after the defunct constellation Quadrans.

[e]1862 III.

[f]Source of spectacular meteor storms every 33 years.

[g]Classified originally as an asteroid; probably a defunct comet.

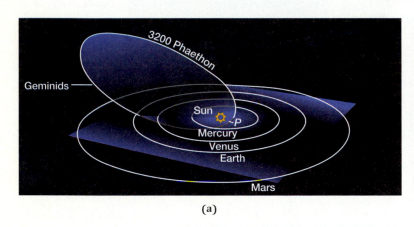

(a)

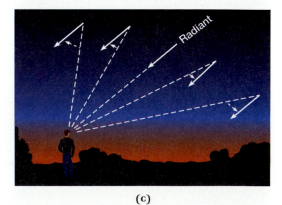

(b)

Figure 17.29   **(a)** *A stream of meteoroids follows the path of the defunct comet 3200 Phaethon (once thought to be an asteroid). When the Earth crosses the orbit, we see the Geminid meteor shower (P represents perihelion).* **(b)** *Parallel railroad tracks appear to come from a point in the distance.* **(c)** *Meteoroids on parallel tracks (solid arrows) hit the atmosphere and as we look at them (dashed arrows), appear to us to be coming from a point or radiant in the sky.*

(c)

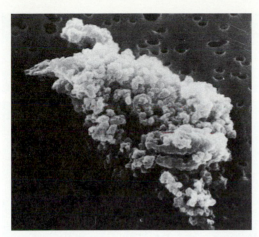

Figure 17.30 *Small particles only a few microns across, picked up by high-flying airplanes, are suspected of being bits of comet fluff of the kind that produce meteors.*

Biela's Comet, which disappeared after 1852, should have returned in 1859, 1865, and 1872. Instead, on the night of November 27, 1872, it was replaced by an intense meteor shower, the now-defunct Andromedids, demonstrating an intimate link between meteors and comets. The connection was confirmed by Giovanni Schiaparelli in 1866 when he found that the orbit of the Perseid meteor stream was the same as that of Comet Swift-Tuttle, 1862 III. As comets' nuclei disintegrate under the power of solar heating, they leave behind a trail of debris, probably surface dust compacted into fragile, rocklike structures (Figure 17.30). Gradually, the orbits of the particles diverge into a band. When the Earth crosses the orbital path of the stream, we see a shower.

The particles may be unevenly distributed along the comet's orbit, and there may be a large assembly of them moving around the Sun. The annual shower may generate only a few meteors per hour, but if the Earth collides with the dense cloud, a **meteor storm** (Figure 17.31) will ensue. The most famous example, the Leonids of November 17, ordinarily do not produce many meteors. But every 33 years—the orbital period of the parent comet—the Earth and the dense collection of debris collide and a lucky viewer may see one of the most awe-inspiring sights that nature can provide, thousands of meteors per hour, hundreds per second, falling from the sky. Great displays of Leonids occurred in 1799, 1833, and 1866. Then Jupiter perturbed them away, but they returned in 1966 with another spectacular show; 1999 is not too far away. The Andromedids were such a storm. The Draconids, a product of P/Giacobini-Zinner, did the same in 1933 and 1946 and were never seen again.

However distributed, the tiny particles gradually spread out from the original orbital path and lose sight of their siblings. These are the sources of the sporadic meteors, the ones that do not belong to showers. On any clear, moonless night you can see the result of a comet's demise.

The fine debris of a comet that is ejected to form the dust tail also spreads out into the Solar System, where it combines with dust produced by asteroid collisions. The asteroids and the short-period comets, those ripe for destruction, concentrate toward the ecliptic. The dust scatters sunlight and produces a faint conical glow, the **zodiacal light,** which rises from the horizon (Figure 17.32) along the ecliptic after dark or before dawn. At the antisolar point—180° from the Sun—it brightens as efficient backscattering (light bounced off in the reverse direction) produces the **gegenschein** (German for "counterglow"). The gegenschein is a dim presence, hard to find, and a real challenge to the naked-eye observer.

All this dust is temporary. The scattering of radiation produces a drag effect on the particles and makes them spiral into the Sun. They must then constantly be replaced by comet disintegration and asteroid collisions.

Figure 17.31 *An old woodcut gives a spectacular impression of the Leonid storm of November 13, 1833.*

17.3.4 The Sources of Comets

If comets are constantly disintegrating and have presumably been doing so for 4.5 billion years, where do the new ones come from? In 1950, the Dutch astronomer Jan Oort suggested there must be a huge pool of comets, now called the **Oort comet cloud** (Figure 17.33), that extends far beyond the limits of the planetary system. Several trillion comets must be contained within a thick disk 75,000 AU in radius, and another trillion or more within a vast sphere that extends 150,000 AU outward, halfway to the nearest star. The vast majority of these comets are on orbits that never take them close to the planetary system, and they are therefore without comas and tails. Over the ages, however, their orbits are disturbed by passing stars and massive clouds of interstellar gas, which will send a few plunging inward on a near-collision course with the Sun (a few have actually been seen to hit). Since they come from so far away, such comets have orbital periods so long that they will likely never be seen again, and since the Oort cloud is so spread out, the orbital inclinations will be randomly distributed.

Some long-period comets may become the short-period variety by being perturbed into smaller orbits by Jupiter; however, there are simply too many for this to be the only source. Most of the short-period comets must be supplied by a second reservoir called the **Kuiper belt** (after Gerard Kuiper, who suggested it in 1950) that is aligned with the ecliptic plane and may be only about 200

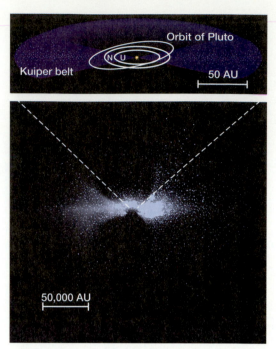

Figure 17.33 *Comets are distributed in the great Oort cloud, which may extend halfway to α Centauri. The inset above shows the Kuiper belt outside the orbit of Neptune.*

AU in radius (inset, Figure 17.33). As a result, the orbits of the short-period comets have low inclinations. We cannot see members of the Oort cloud: they are too far away and too faint. However, the Kuiper belt is potentially accessible, and may be represented by 1992 QB1 and 1993 FW (see Section 17.2.2), the most distant known members of the Solar System. Perturbations by the giant planets eject some of these comets from the Solar System and move others inward toward the Sun, to even smaller orbits, where with periods of only a few years, they are quickly destroyed or turned into bodies that look like asteroids.

The number of comets is huge, but their combined mass is not. From estimated masses, all the comets rolled together would perhaps make a hundred Earths.

17.4
The Creation of the Solar System

Strong evidence suggests that asteroids, comets, and perhaps bodies like Pluto and Triton are the remains of the planetary building blocks, the planetesimals. The asteroids lie in the ecliptic plane and go around the Sun counterclockwise, as does (more or less) one family of comets. The positions

Figure 17.32 *The zodiacal light, sunlight scattered from interplanetary dust, is seen along the ecliptic after evening twilight or before dawn.*

of the asteroids fit into the zone between Mars and Jupiter where Bode's progression of distances suggests a planet should lie. If not for Jupiter's gravitational influence, these bodies might have assembled into one. The colors and compositions of comet nuclei and asteroids correlate with distance from the Sun, as they do for the planets and their satellites. The inner asteroids are bright and rocky, the outer ones dark and icy. The distant comets seem to be the darkest of all, made mostly out of ice mixed into and coated with carbon and silicates.

It is no simple matter to reconstruct events that took place 4.5 billion years ago, but there is enough evidence to give us a reasonable idea of how the Solar System was assembled. The Sun appears to have been made from a contracting cloud of interstellar gas that was rich in molecules and infused with fine dust. The details of this process and the formation of stars in general will be discussed in Chapter 23. We consider here only that the cloud, which was rotating and producing the Sun at its center, was subject to the law of conservation of angular momentum. As it squeezed down, it spun faster, flattening into a disk called the **solar nebula** (Figure 17.34a). This remnant of solar formation was the birthplace of the planets. The disk became the ecliptic plane. All the planetary motions, the

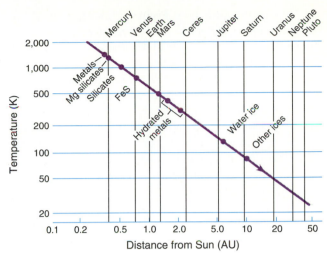

Figure 17.35   *Temperature in the early solar nebula is plotted against distance from the Sun. Different substances condense at different distances (indicated by dots) as the temperature drops.*

counterclockwise orbits and spins, represent the rotation direction of the pre-planetary disk.

Within this disk, the original interstellar dust was subject to a variety of forces and influences. Among the most important controlling factors was temperature. It was very low, in the tens of degrees K, at the outermost fringes. Toward the inner Solar System it climbed sharply, near 2,000 K close to the forming Sun where the dust would have been vaporized. The dust was in physical and chemical equilibrium with the surrounding gas, which means that as the solar nebula cooled, the gas could condense onto the original dust grains to build larger particles of matter. Some molecules—refractory ones (see Section 11.4) like metallic oxides and silicates—enter and leave gas phase at high temperature (Figure 17.35) and could condense into and onto solid grains everywhere in the nebula, even near the Sun. Others—volatiles like water, methane, and ammonia—will stay in the gas phase at higher temperatures; these could condense only at great distances from the Sun, beyond several AU. The composition of the raw material out of which the planets were made therefore depended on how far from the Sun they were formed. The realm of what would become the terrestrial planets was filled with growing grains made of metal and rock; the domain of the developing Jovians contained metal, rock, and vast amounts of water and other ices. The differences in ice content found among the outer satellites and Pluto may reflect changes in the solar nebula's chemical composition with distance from the Sun.

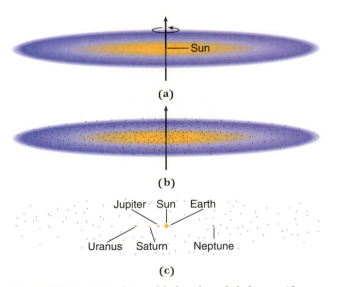

Figure 17.34   **(a)** *In this model, the solar nebula began with a disk around the newly forming Sun.* **(b)** *Planetesimals formed in the disk as a result of dust accretion and gas condensation.* **(c)** *The planetesimals eventually accreted into planets and planetary cores as the solar nebula, dissipating under the action of the new Sun, was accreted into the outer layers of the Jovian planets. Some planetesimals remained in the disk outside the planetary system; those inside that were not swept up were ejected or thrown into the Oort cloud.*

Within the disk (see Figure 17.34a), everything was moving in the same direction at velocities governed by Kepler's laws of motion. At a specific radius from the Sun, all the grains had nearly the same speed and moved slowly relative to one another. As a result, when two bumped together they did not necessarily break apart, but could fuse. As the grains were growing fatter by gas condensation, they also enlarged by accretion of other particles. The effect snowballed as their increasing sizes allowed them to capture still more particles. When they became big enough, gravity began to play a role, and the largest ones grew at the expense of the smaller. Within perhaps 100,000 years, the dust had congealed into trillions of small bodies, the planetesimals (Figure 17.34b). The warmer ones near the center of the budding Solar System were rocky, the colder ones farther away were icy. During all this activity, the new, hot Sun was driving the gas out of the inner Solar System, ending gas condensation altogether.

The planetesimals continued to collide and grow. The larger ones dominated because of their gravitational pulls and began to consume the others. In the inner Solar System, four bodies finally won out, slowly becoming rocky Mercury, Venus, Earth, and Mars. In the outer Solar System, the accretion of planetesimals made the cores of the great planets Jupiter, Saturn, Uranus, and Neptune (Figure 17.34c). The gravitational pulls of the growing planets were so high that the impacting planetesimals imparted enormous energy upon collision and the new planets melted, allowing them to differentiate. In the outer Solar System, where the solar nebula was still more or less intact, the great gravity of the cores began to sweep the hydrogen and helium into vast envelopes, completing the formation of the Jovian planets. These behaved much like the Sun, in that the remaining gas circulated about them in a disk. Out of the disks came the Jovian satellite families, miniature analogues of the Solar System.

The sizes of the planets depended on the distribution of mass within the solar nebula while they were being created. The new Sun probably swept much of the gas from its immediate neighborhood, and as a result, the planets get bigger as distance from it increases. Beyond Jupiter, the density of gas in the original solar nebula probably dropped with distance, so the outer planets could not grow as large. Mars turned out small because Jupiter's gravity ejected planetesimals from the region in which the planet was being formed.

After some 10^8 years, there was a Sun and a retinue of eight cooling planets. The terrestrials began to develop solid crusts, then mantles. Solidification and radioactive decay have helped keep their interiors hot throughout the aeons. Trillions of planetesimals were left to a variety of fates. A large fraction was swept up by the planets and their satellites. By that time the planetary surfaces had turned solid and the bombarding meteoroids produced permanent craters; we still see the record of this extraordinary bombardment laid down on the surface of the Moon and other bodies. The giant planets agitated the icy planetesimals, and huge numbers of them made it into the inner Solar System. They crashed into the Earth and other terrestrial planets, bringing supplies of volatiles that had previously been driven away, particularly water and perhaps even carbon. It is possible that they supplied enough water even to fill the oceans, providing the necessary conditions for life.

The planetesimals between Mars and Jupiter were stirred up by Jupiter's gravity, and collisions among them smashed many to produce the current asteroids. Jupiter hurled some, including vast numbers of distant, icy planetesimals—the comets—out of the Solar System altogether. Uranus and Neptune, with smaller gravities, threw cometary planetesimals into the Oort cloud. Planetesimals beyond the planetary system stayed more or less in place as comets in the Kuiper belt. A few really large planetesimals, such as Pluto and Triton, lurk at the outer fringes of the Solar System and may be representative of many more, similar bodies.

Though the pace of planetary and Solar System evolution slowed dramatically, it never ceased. The planets continued to cool; the terrestrial planets developed atmospheres from their interiors and—along with the satellites of the Jovian planets—established unique volcanic and tectonic characteristics that on Earth and Venus are still highly active. And when we stand outdoors on a dark night, we can still watch a few remaining planetesimals orbit the Sun, or even slice through the atmosphere.

We seem to have explained everything neatly. Do not be deceived. A great many mysteries and uncertainties remain, and much of this satisfying picture may be wrong. For example, the Jovian planets could have formed from discrete rings of gas surrounding the Sun, rather than by conglomeration of planetesimals. The qualitative explanation provided here is a mere outline. Any genuine proof must be mathematical; that is, we must be able to use the equations of physics to construct models that establish whether this chain of events could actually take place. Such models are actually semi-successful, indicating that we are on the right track. Some models have even predicted the formation of the terrestrial planets in about the right positions. But an enormous number of details have not been explained: no wonder, since we have not yet explored this amazing system of ours in complete detail and do not even have a full census of its contents. Whatever the final picture assembled, one result was a great blue ball, the third planet from the Sun, the birthplace of life. From it we look out on the rest of our home, the Solar System.

KEY CONCEPTS

AAAOs: Asteroids in the Amor, Apollo, or Aten families, all of which come close to the Earth's orbit.

Asteroids: Small bodies that orbit mostly between Mars and Jupiter; concentrated in a **main belt** between 2.1 and 3.2 AU; C, S, and M types relate to meteorite classes.

Bode's law: A numerical progression (not a true law) that matches the distances of most planets from the Sun.

Comet: A fragile interplanetary body with a **nucleus** of dirty ice that is heated by the Sun to produce a surrounding **coma** and **tails** of gas and dust; **short-period comets** have periods less than 200 years and come from the Kuiper belt; **long-period comets** have random orbits longer than 200 years and come from the Oort cloud.

Dust tail: The diffuse, fan-shaped tail of a comet caused by sunlight scattered from dust released by the nucleus.

Families: Groups of asteroids that result from the breakup of a parent body or from assembly by resonances.

Fireball: A brilliant meteor that can produce a meteorite.

Gegenschein: Brightening of the zodiacal light in the direction opposite the Sun.

Kirkwood gaps: Gaps in the distribution of the semimajor axes of asteroid orbits caused by resonances with Jupiter.

Kuiper belt: A disk-shaped reservoir outside the orbit of Neptune that produces the short-period comets.

Meteor: A bright tube of ionized atmosphere caused by the passage of a meteoroid.

Meteor shower: A shower of meteors that emanates from a **radiant;** caused when the Earth passes near a comet's orbit.

Meteor storm: An intense meteor shower caused by a concentration of cometary debris.

Meteorites: Meteoroids that land on the Earth; classified as **stones, irons,** and **stony irons; stones** are subclassified as **chondrites** (those with **chondrules**), **carbonaceous chondrites** (primitive chondrites with a high carbon content), and **achondrites** (those with no chondrules). Many achondrites are basaltic **eucrites** from 4 Vesta.

Meteoroid: A piece of interplanetary debris that produces a meteor when it enters the Earth's atmosphere.

Observational selection: An effect in which data preferentially selected by some often-unidentified means distort statistical findings.

Oort comet cloud: A reservoir over 100,000 AU in radius that contains trillions of comet nuclei; the origin of the long-period comets.

Plasma tail: A comet's ionized gas tail; points away from the Sun and is structured by the Sun's wind and magnetic field.

Size distribution: The way in which the number of a set of objects varies with their sizes.

Solar nebula: The disk of gas and dust around the forming Sun, out of which grew the planetesimals and the planets.

Zodiacal light: A band of light in the zodiac caused by sunlight scattered from cometary and asteroidal dust.

EXERCISES

Comparisons

1. Distinguish meteors, meteoroids, and meteorites.
2. Distinguish between ordinary and carbonaceous chondrites.
3. What is the difference between the Apollo and the Aten asteroids?
4. Distinguish S, C, and M asteroids.
5. What is the difference in formation between the iron and the stony-iron meteorites?
6. What chiefly distinguishes P and D asteroids from the others?
7. What are the differences between short- and long-period comets?
8. What are the differences in the structures of the Kuiper belt and the Oort comet cloud?
9. How do a meteor shower and a meteor storm differ?
10. How does the zodiacal light relate to the gegenschein?

Numerical Problems

11. What would be the period and distance from the Sun of an asteroid at the 7:3 Kirkwood gap?
12. Assume that Jupiter is at the vernal equinox. What are the approximate right ascensions and declinations of the centers of distributions of the two sets of Trojan asteroids? What is the angular extent of one set of the Trojan asteroids as viewed from Earth?

13. Estimate the total mass of the comets, making all your assumptions clear.
14. Presuming that Neptune is an aberration, where would you expect to find the next planet beyond Neptune according to Bode's law? What would be its orbital period? How bright would you expect it to be, relative to Neptune? Guess its size, and incorporate your estimate into the brightness calculation.

Thought and Discussion

15. What are chondrules?
16. Why do we think that iron meteorites were once part of the interiors of larger bodies?
17. What evidence is there for interstellar material in asteroids?
18. What significance can be attached to asteroid families?
19. What heating mechanisms could have differentiated some of the asteroids? Which of the mechanisms might explain why the more distant asteroids are the more primitive?
20. What controversies involve Chiron and Phaethon? How do these two bodies confuse the distinction between asteroids and comets?
21. How do we know that asteroids rotate?
22. How was the Oort cloud filled with comets?
23. Why do comets lack tails when they are at Saturn's distance from the Sun?
24. What causes comets' tails to disconnect?
25. Why do the meteors in a shower appear to come from a radiant?
26. Why are the distant planetary satellites icy, whereas the terrestrial planets are rocky?
27. How were dust particles able to accumulate into planets in the early days of the Solar System?

Research Problems

28. Assume that the meteoroids in Figure 17.5 came from their aphelia. What kind of asteroids and meteorites would they be? How certain are your results?
29. If the *Circulars* of the International Astronomical Union are available in your library, make a list of the comets that have passed perihelion in the last three years. What is the ratio of the numbers of short- to long-period comets? How does the number of comets vary with brightness? What does that tell you about the size distribution of comets?

Activities

30. Plot the albedos of the 15 brightest asteroids against their distances from the Sun. What is the significance of your results?
31. Plot the densities or the ratios of rock to ice within the Jovian satellites and Pluto against distance from the Sun. What are the possible origins of the differences?

Scientific Writing

32. Someone sends you samples of rock and iron, claiming that they are meteorites. Write a letter to your correspondent. Take either side of the issue, and explain the reasons for your opinion.
33. You have been invited to give a lunch talk to your local Rotary Club, and you choose as your topic the origin of the Earth. Write the script for a 20-minute talk explaining how the Earth formed. Your audience has no background in astronomy and will be unfamiliar with technical terms.

18 The Sun
19 The Stars
20 Stellar Groupings: Doubles,
     Multiples, and Clusters
21 Unstable Stars

Stellar Astronomy

PART IV

18

The Sun

*The central engine
of the
Solar System
and an example
of a star*

A total eclipse draws attention to the Sun.

Our study of the planets is still missing the link needed to make them fully comprehensible. They are the by-products of the formation of the central body of the Solar System, the Sun. Viewed from another star, the Sun would totally dominate the system, and (with our current technology) the planets would be invisible.

The Sun is one of some 200 billion stars in our Galaxy. It is near the middle of the ranges of the various stellar properties and is the only star on which we can see detailed activity. The Sun thus provides a basic model for understanding general stellar behavior and it is the passageway that takes us from our neighborhood to the vast city of stars and the world of galaxies beyond.

18.1
Basic Properties

The first step in examining the Sun is to establish and summarize its remarkable characteristics (Table 18.1). From the solar angular radius of 16 minutes of arc and the distance, we find the physical radius to be 6.96×10^5 km, 109 times that of the Earth (Figure 18.1) and about 0.005 AU. From the radius and a mass of 1.99×10^{30} kg (see Section

7.5), we find the average density to be only 1.4 g/cm^3, similar to that of Jupiter. However, if you could add considerable mass to Jupiter, its fluid interior would compress (see Section 15.2), and the density would greatly increase. The low average

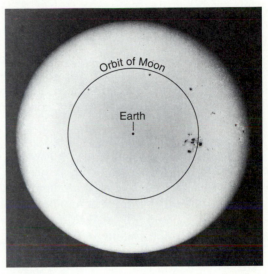

Figure 18.1 *The Sun's diameter is 109 times that of Earth (represented by the dot in the center) and more than 3.5 times the distance between the Earth and the Moon.*

TABLE 18.1
A Profile of the Sun

Data

radius	6.96×10^5 km
mass	1.99×10^{30} kg
average density	1.4 g/cm^3
gravity	27.9 g$_{Earth}$
escape velocity	618 km/s
general magnetic field	about five times that of Earth
rotation	
equatorial period	25.4 days
inclination to ecliptic	7°15'
effective temperature	5,780 K
solar constant (at Earth)	1,368 joules/m^2/s
luminosity	3.83×10^{26} joules/s

Layer[a]	% Mass	Radii (km)	Temperature (K)	Pressure (bar)	Density (g/cm^3)
core	40	0–2×10^5	1.5×10^7	1.5×10^{11}	160
radiative envelope	59	2×10^5 to 5×10^5	4×10^6	6×10^8	2
convective envelope	1	5×10^5 to 6.96×10^5	1×10^6	1×10^6	0.01
photosphere	—	6.96×10^5	5,780	0.1	10^{-9}
chromosphere	—	—	10,000	low	low

[a]Temperature, pressure, and density refer to the center of the core and to a point midway through the other layers. The photosphere is only a few hundred km thick and the chromosphere a few thousand km thick. The masses of the outer three layers are insignificant. The radius of the corona is ill defined and extends many solar radii from the photosphere, eventually merging with the solar wind.

density of the Sun therefore demonstrates that it is gaseous throughout.

Unlike the planets, the Sun does not shine by reflected light but is entirely self-luminous. The luminosity is expressed as the flux of solar radiation at the Earth, a quantity called the **solar constant.** Sophisticated devices aboard spacecraft, which can detect radiation that does not penetrate the Earth's atmosphere, measure a flux of 1,368 joules/m²/s. From the distance of the Sun (see Section 12.1), we find an astounding solar luminosity of 3.85×10^{26} joules/s. It would cost $\$7 \times 10^{18}$—the gross national product of the United States for 7 million years—for your local power company to run the Sun for a mere second.

Combination of radius, luminosity, and the Stefan-Boltzmann law yields an effective surface temperature of 5,780 K. From the Wien law, the peak of the blackbody curve falls at about 5,020 Å, in the green part of the spectrum. The combined radiation at all wavelengths makes a shaft of sunlight appear a soft yellow-white to the eye.

From the rates of passage of dark spots on the solar surface, we find that the Sun is a differential rotator, taking 25 days to turn near the equator and closer to 30 days near the poles. The rotation is in part responsible for a general, global, dipole magnetic field that aligns with the rotation axis and has a strength about five times the Earth's.

Astronomers have long divided solar phenomena into two parts: the **quiet Sun,** a collection of features that are always present, and the **active Sun,** which includes features that come and go and are associated with intense magnetic activity.

18.2
Special Tools

The Sun is so bright that telescopes and observational devices used to study it have specialized designs. The McMath-Pierce solar telescope at Kitt Peak (Figure 18.2a), for example, bears little resemblance to the telescopes studied in Chapter 9. It is capable of creating a high-resolution solar image 76 cm across (Figure 18.2b). Sunlight can also be sent to a spectrograph that can produce a solar spectrum with a resolution of better than a hundredth of an Ångstrom. The vacuum telescopes at Kitt Peak and at Sacramento Peak in New Mexico are actually evacuated to remove the degrading optical effects of turbulent air.

The Sun has been a major target for spacecraft designed to observe in the ultraviolet and X-ray parts of the spectrum, where we can study high-energy processes. Early rocket flights were followed in 1962 with a series of Orbiting Solar Observatories, the program terminating in 1971 with *OSO 7*. These craft could return images of the Sun in the light of a variety of ultraviolet emission lines. One of the more ambitious programs involved *Skylab,* a 36-m-long manned laboratory that orbited Earth from 1973 to 1979, when its orbit decayed and the ship broke up in the atmosphere. *Skylab* was equipped for UV and X-ray imaging and spectroscopy as well as for high-resolution optical imaging. It was followed in 1980 by *Solar Maximum Mission,* which carried spectrographs and imaging devices that could observe from the optical to the gamma-ray region of the spectrum.

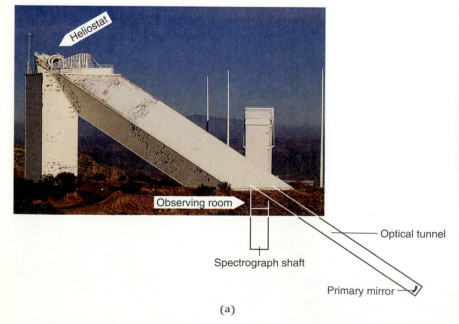

(a)

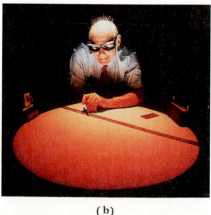

(b)

Figure 18.2 **(a)** *The McMath-Pierce solar telescope has rotating twin mirrors at the top that project sunlight down the inclined tube to a primary mirror. The light is then returned by a fourth mirror to the observing room* **(b),** *where the projected image is so bright the astronomer needs sunglasses.*

The newest entry in the Sun's space fleet, *Ulysses,* was sent in 1990 to Jupiter, whose gravity threw it out of the ecliptic plane in 1992. In 1994 it will pass under the south solar pole and in 1995 back over the north pole to provide different and critical views of the workings of solar rotation, the magnetic field, and the production of the solar wind.

18.3
The Quiet Sun

The quiet Sun consists of three basic components and exhibits a variety of phenomena. As we will see, it is not really so quiet.

18.3.1 *The Thin Grainy Photosphere*

In spite of its gaseous nature, the brilliant apparent solar surface, called the **photosphere** (the "sphere of light"), has a razor-sharp edge, or **limb.** As we will see, 92% of the atoms in the outer part of the Sun are hydrogen. The hydrogen is neutral, but many metal atoms are ionized and contribute electrons to the gas. A hydrogen atom can pick up one of these electrons and become a negative hydrogen ion (H⁻). This second electron is easily ripped away by absorbing any photon with a wavelength less than 16,400 Å. There are so many H⁻ ions that radiation is strongly impeded, making the solar gases highly opaque. As a result, the photosphere, the layer of the Sun from which light escapes directly into space, is only a few hundred km thick. From Earth, this distance subtends less than a second of arc, so the photospheric limb looks sharp and well defined.

The small but real depth to the photosphere is still sufficient, however, to produce **limb darkening.** The Sun is not uniformly bright across its surface (Figure 18.3a), but becomes steadily darker away from the center. The effect can be explained only if temperature increases with depth. If we look (through instruments that offer protection to the eye) at the center of the apparent solar disk (Figure 18.3b), our vision penetrates to a certain depth within the gaseous photosphere. If we look at the limb, our sightline probes through about the same thickness of gas. But since we are now no longer looking along the perpendicular to the surface, we do not see as deep. If the Sun's temperature increases inward, the radiation from the limb must come on the average from cooler layers than it does when the apparent solar disk is viewed at the center. Lower temperature means a lower flux of radiation, so the limb appears relatively darker than the center and, in response to the Wien law,

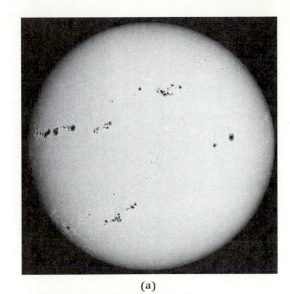

(a)

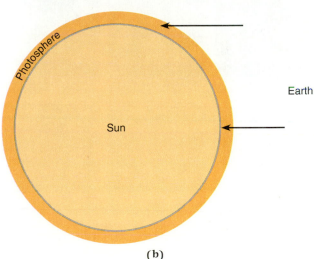

(b)

Figure 18.3   (a) *The solar photosphere, covered with spots, is darkened toward the limb.* **(b)** *When we look at the center of the photosphere, our vision penetrates to deeper, hotter, and brighter gases than when we look at the limb. The thickness of the photosphere is greatly exaggerated for clarity; on this scale, it would be only 0.01 mm thick.*

redder. Measurement of limb darkening is in fact used to find the photospheric **temperature gradient,** the rate at which temperature changes with distance (here, depth into the Sun).

A close look at the photosphere shows that its texture is not smooth but grainy. The solar surface is covered by millions of relatively tiny **granules** (Figure 18.4a), bright cells about a second or two of arc across—about 700 km, the size of Texas—separated by about their own widths and set into a darker background. As we watch, they continuously change their appearance (Figure 18.4b). Each typically lasts for only a few minutes before it disappears and is replaced by another.

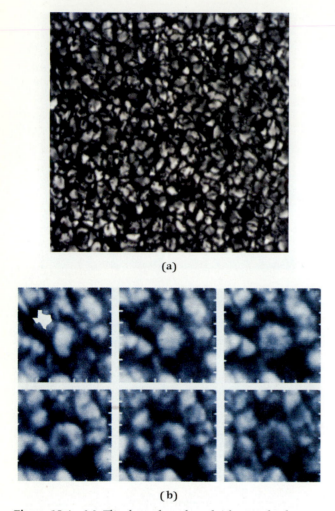

(a)

(b)

Figure 18.4 **(a)** *The photosphere shows bright granules that are the tops of rising convection cells.* **(b)** *A sequence of panels made at four-minute intervals during the* Spacelab 2 *shuttle mission shows the granules changing with time. An outline of Texas provides a scale.*

The granules' natures are revealed by their spectra. In Figure 18.5a, each horizontal strip corresponds to a particular point on the Sun, bright strips to granules, dark ones to the granules' background. The spectra of the granules are Doppler-shifted to shorter wavelengths and the background spectra to longer, showing that the granules are rising and that the darker matter is falling (Figure 18.5b). The granules are thus seen to be the tops of giant convection cells that bring heat to the surface from below. They then radiate their energy, cool, darken, and fall. Enclosing the granules is an even larger pattern of supergranules that produce horizontal motions.

Also superimposed on the granulation is a complex up-and-down pattern of **solar oscillations** that has a period of about five minutes. Large areas of granulation first rise at a speed of about 200 m/s while other areas descend at the same speed (Fig-

ure 18.6). The range of motion is small, at most only a few tens of kilometers. The entire Sun is ringing in space like a huge bell, its vibrations driven by the mass motions of convection. No bell, however, rings with just one tone, but has many vibrations of different frequencies called overtones. Careful observation reveals that superimposed on the already complicated pattern of Figure 18.6 are hundreds of overtone oscillations with periods ranging from 4.5 to 6.7 minutes.

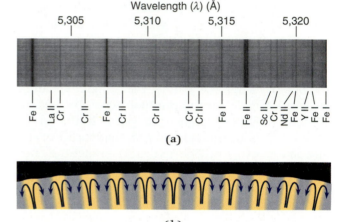

(a)

(b)

Figure 18.5 **(a)** *The wiggly appearance of solar absorption lines is caused by Doppler shifts produced by rising and falling granules. Roman numeral I denotes a line of a neutral element, II that of a singly ionized element, and so on.* **(b)** *The granules are the tops of rising convection cells.*

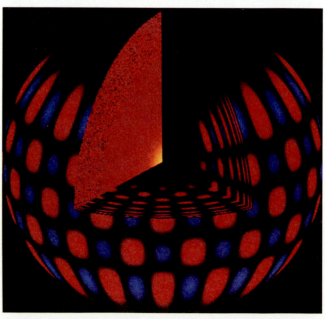

Figure 18.6 *The Sun pulsates with a principal period of about five minutes, with some parts of the surface moving inward (red) and others moving outward (blue) at the same time. After 2.5 minutes the directions will reverse.*

18.3.2 The Solar Spectrum and Chemical Composition

The solar spectrum (Figure 18.7a) displays a great number of absorption lines in a classic example of Kirchhoff's third law (see Section 8.3.2). In the simplest sense, the lower, hotter layers of the photosphere generate a continuous spectrum. The upper layers constitute a gas of lower temperature and density and superimpose the absorptions. (Reality is more complex. The continuum and lines are actually formed in much the same regions. The photosphere is more opaque at the wavelengths of the absorption lines, where photons have difficulty escaping. Therefore when we look at an absorption line, our vision does not penetrate as deep, we see only to cooler layers, and the line is dark.) The

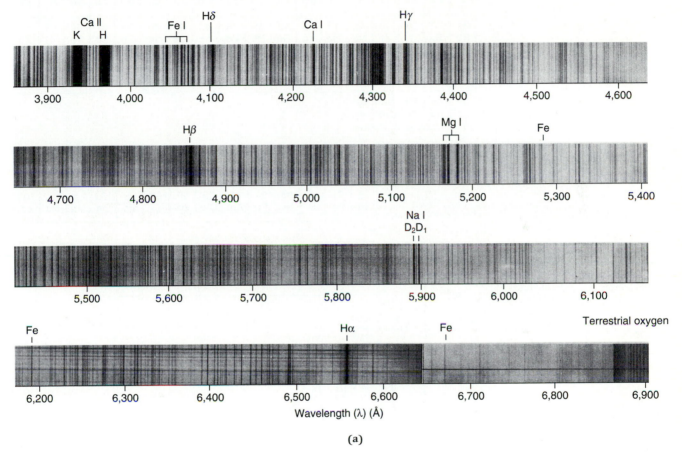

(a)

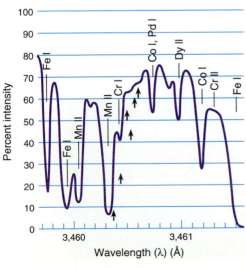

(b)

Figure 18.7 **(a)** *The optical solar spectrum is dominated by hundreds of absorption lines (single letters are original Fraunhofer designations).* **(b)** *An extremely spread out spectrum that shows nine lines in just one Ångstrom of the ultraviolet. Arrows show the expected positions of lines of the rare element rhenium (Re), which are too weak to be detected.*

Figure 18.8 *The periodic table arranges the chemical elements by increasing atomic number according to their chemical properties. The two rows at the bottom are inserted as indicated. Elements identified in the Sun are shown in color according to the solar layer or region in which they are found.*

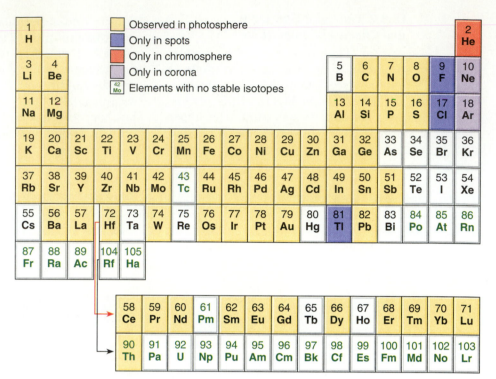

Observed in photosphere
Only in spots
Only in chromosphere
Only in corona
Elements with no stable isotopes

Figure 18.9 *The abundances of the elements in the Sun are plotted according to powers of ten against atomic number. The gaps indicate elements that have yet to be observed. Lithium and beryllium (atomic numbers 3 and 4) are rare; iron, nickel, and cobalt are relatively abundant.*

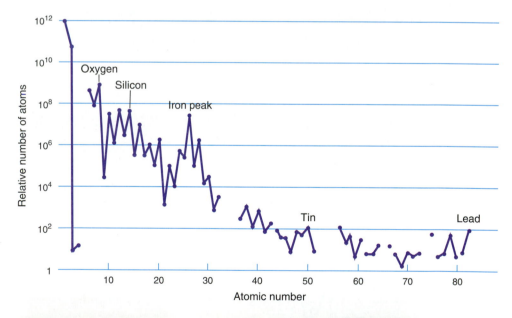

Figure 18.10 **(a)** *The red rim of the chromosphere surrounds the eclipsed Sun.* **(b)** *The narrow chromosphere spreads out into an emission spectrum.*

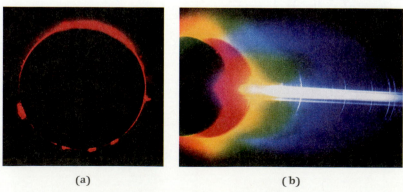

(a)   (b)

lines of hydrogen, neutral sodium (Na I), and ionized calcium (Ca II) are the most prominent (I represents the spectrum of a neutral atom, II that of a singly ionized atom, and so on). The majority of the other lines are produced by common metals such as iron, chromium, titanium, manganese, and vanadium (Figures 18.5a and 18.7b). Some 70% of the natural elements (Figure 18.8) are observed in the solar gases, and we are confident they are all in fact present. Those not seen simply have such low abundances that their weak absorptions are swamped by other lines (see Figure 18.7b).

The strongest lines in the solar spectrum are those of Ca II at 3,933 and 3,968 Å, called H and K by the pioneer of spectroscopy, Joseph von Fraunhofer (see Section 8.3.2). If we naïvely relate elemental abundance directly to line strength, there would seem to be more calcium than hydrogen. However, for hydrogen to produce Balmer absorption lines, the atoms must have electrons in the second energy level (see Section 8.3.3 and Figure 8.18). Within the gas, electrons are shuffled among the levels by atomic collisions. The higher the temperature, the more energetic the collisions, and the greater the percentage of hydrogen atoms with elevated or excited electrons. At the temperature of the solar photosphere, only one atom out of every 10^{10} will have an excited electron, and as a result, very little solar hydrogen is capable of absorbing the Balmer lines. The Fraunhofer lines of ionized calcium, however, arise from the ground state, so nearly every ion is capable of producing the lines. When we take the different absorption efficiencies into account, we see that calcium is actually about 10^8 times *less* abundant than hydrogen. Similar arguments made for other atoms show that the Sun is made overwhelmingly of hydrogen and that metals, like calcium, have low abundances.

The composition of the Sun is shown graphically in Figure 18.9 (results for helium and for some other elements are derived from solar layers that are discussed below). We find that the solar photosphere consists of 92% hydrogen (by number of atoms) and about 8% helium. *All the other atoms constitute no more than about 0.1%.* Generally, the greater the atomic number of the element, the less there is of it. This rule has some interesting exceptions, however. First, lithium, beryllium, and boron (between helium and carbon) are extremely scarce. Then, there is a rise around iron called the *iron peak* that also includes cobalt and nickel. Finally, elements with even atomic numbers are somewhat more abundant than their neighbors with odd num-

bers. These patterns provide critical clues to the ways in which the elements were created in the birth of the Universe and in stars, subjects to be addressed in later chapters.

Above atomic number 2, the relative abundances of the atoms are roughly similar to what we find in the Earth and in stony meteorites. The Earth, the terrestrial planets, and the planetesimals are a distillate of the material out of which the Sun was made.

18.3.3 The Solar Chromosphere

At the instant the Sun becomes fully eclipsed, the edge of the darkened Moon is surrounded by a thin, red glow, the **chromosphere,** the "sphere of color" (Figure 18.10a). Lying above the photosphere, this layer is a few thousand km thick and produces so little radiation that it cannot be seen outside of an eclipse or without special instrumentation. Its spectrum (Figure 18.10b) consists of emission lines, showing it to be a gas under low pressure. The element helium was discovered by its chromospheric emission lines, hence its name, after Helios, Greek god of the Sun.

Temperature steadily declines upward in the photosphere to a minimum of about 4,500 K at the top (Figure 18.11). Then the spectrum shows us that as we enter the chromosphere, temperature begins to *rise,* increasing first to about 8,000 K and then to over 10,000 K. From the Stefan-Boltzmann law, we might expect the hotter chromosphere to be brighter than the photosphere. Because of the chromosphere's low density, however, it is not a blackbody, and relative to the photosphere cannot emit much radiation. The temperature, like that of the terrestrial thermosphere, is a kinetic temperature that describes the speeds of the atoms and ions in the gas, not the gas's luminosity.

Like the photosphere, the chromosphere quite literally bubbles with energy. Photograph the Sun with a narrow filter, one confined to the wavelength of the ionized calcium line at 3,933 Å or the hydrogen Hα line at 6,563 Å. Since the chromosphere produces emission lines at these wavelengths, we can take its picture in the form of a **spectroheliogram** (Figure 18.12) with little trace of the underlying photosphere. Projecting from the chromosphere's base are millions of tiny needles of rising gas, or **spicules** (see Figures 18.11 and 18.12). Like the granules, they last no more than a few minutes before they vanish and are replaced by others.

The chromosphere is thought to be heated by photospheric convection. As the gases churn over,

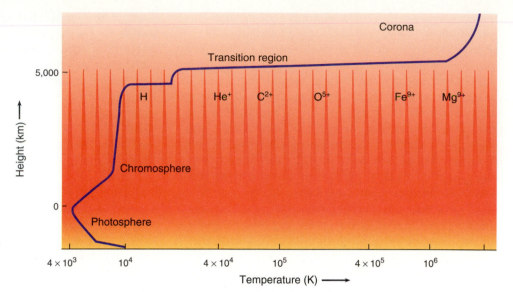

Figure 18.11 *Temperature drops within the photosphere to about 4,500 K at the top of the layer, then rises in the chromosphere. At the top of the chromosphere, a few thousand km above the photosphere, temperature rapidly rises in a transition region to over 2 million K in the corona. Observed ions are shown along the curve. Needlelike spicules project upward from the bottom of the chromosphere into the corona.*

they create sound waves that propagate outward into the lower-density gas; there the speed of sound is lower, and the waves become shock waves (see Section 10.6). The shocks dissipate their energy into the surrounding gas, heating it and producing the spicules. Convection also generates waves and turbulence in the solar magnetic fields, which may deposit their electromagnetic energy into the chromosphere. A complete theory, however, still eludes us.

Figure 18.12 *A spectroheliogram made in the light of Hα shows the chromosphere and the upward-projecting spicules.*

18.3.4 The Corona

The chromosphere is a transition to the outer halo of the Sun, the **corona** (Latin for "crown"), seen in Figure 5.16 and in the chapter-opening photograph. Superimposed upon a continuous spectrum (produced by photospheric light scattered from electrons) are emission lines of metals like iron ionized up to 16 times. For atomic collisions to strip so many electrons from an atom requires temperatures over 10^6 K, a conclusion supported by radio observations. Radio waves interact strongly with an ionized gas, and at any density there is a critical frequency below which they are blocked. High-frequency radio emission, above 50,000 MHz, emanates from the photosphere and penetrates the corona. However, low-frequency radio waves emitted by the photosphere are turned back by the corona; they can escape only if emitted by the corona itself. The intensity of low-frequency radio emission yields a temperature above 1 million K. The lower the frequency, the higher we look, allowing us to examine the corona's structure.

The division between the chromosphere and corona is not sharp. At the top of the chromosphere is a transition region where the temperature rapidly ascends from around 20,000 K to over 1 million K as it climbs to a coronal average of 2 million K, hot enough to produce copious X rays. Yet, again, the density is so low that the corona is far from being a blackbody. It is therefore optically dim, producing only about as much light as the full Moon.

The corona's feeble glow is largely hidden by sunlight scattered by the Earth's atmosphere. To see the corona properly, we must either wait for an eclipse or go into space (where the X rays may also be observed). Nevertheless, its inner portions can be viewed with a *coronagraph* (invented by Bernard Lyot of the Paris Observatory in 1930), a telescope that contains a disk that fits over (and blocks) the image of the photosphere and is placed on a high mountain peak to minimize the effects of atmospheric scattering.

The corona is too energetic to be heated by shock waves. It is instead a creature of solar magnetism, which is not a feature of the quiet Sun so much as the active Sun, to which we now turn for a better understanding of solar workings.

18.4
The Active Sun

As the Sun plies its daily path, we pay it little heed except to note its warmth. We are largely unaware of the seething turmoil of its surface or its great magnetic storms and wracking explosions. Only when we see an aurora, or our radio or television broadcasting goes awry, or our electrical power shuts down, might we have direct contact with the active Sun.

18.4.1 Sunspots and the Magnetic Activity Cycle

Spots on the Sun were recorded by the Chinese some 2,000 years ago. Then, in 1610, Galileo found them again. **Sunspots** (Figures 18.3 and 18.13a) are darkened areas of the Sun. They tend strongly to appear in pairs, and the pairs in groups or **centers of activity.** Their sizes range from expanded spaces between the granules to huge systems 20,000 km or more in diameter, much larger than the Earth (see Figure 18.1).

Sunspots have two distinct parts: dark, central **umbras** (which here have nothing to do with shadows except darkness) and surrounding lighter **penumbras.** With central temperatures of about 4,500 K, the umbras radiate 35% as much light as the photosphere and appear dark only by contrast. They have no granulation and are actually flat depressions a few hundred km deep in the visible photospheric surface. The penumbras display radial lines and slope upward, connecting the umbras with the unspotted surface. The spots are not permanent features. They are born as small dark spaces between the granules and then grow. They can live from a few hours to weeks, depending mostly on the sizes they attain.

The most significant feature of a spot is an intense magnetic field, typically 5,000 or so times the strength of Earth's. The field is in a loop that

leaves the Sun at one spot of a pair and reenters at the other (Figure 18.13b). The radial lines in the penumbras trace out the direction of the field lines. The spots are creations of the magnetic field, which somehow (as yet there is no comprehensive theory) acts as a giant refrigerator, inhibiting convection and the flow of energy. Immediately surrounding the photospheric spots are bright areas or *faculae* (Latin for "little torches"), easily seen near the solar limb in Figure 18.3, that represent energy blocked by the spots. Spectroheliograms (Figure 18.14) show the chromosphere to be bright above the spots rather than dark, a manifestation of the magnetic energy. The magnetic loops have finite lifetimes. As they grow and then disintegrate, so do the spots.

Solar magnetic fields are detected and measured by means of the **Zeeman effect.** An orbiting electron is a tiny electric current that generates a magnetic field. An external magnetic field will interact with the electron's field and will make the electron's orbit precess about its axis. The energy

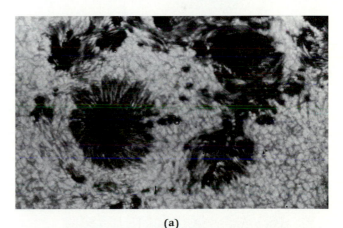

(a)

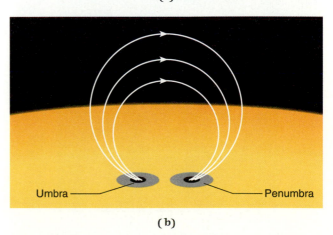

(b)

Figure 18.13   **(a)** *A huge sunspot group shows the individual spots to have dark umbras surrounded by lighter penumbras.* **(b)** *A typical sunspot pair is created by a magnetic field that loops out of the Sun and cools the photospheric gases.*

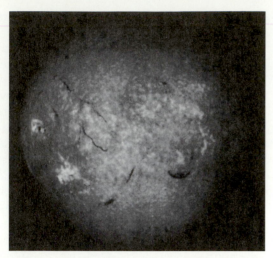

Figure 18.14 *A spectroheliogram shows active regions to be bright in the chromosphere. Sunspots are not visible. The long dark streaks are called filaments.*

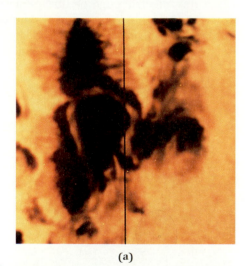

(a)

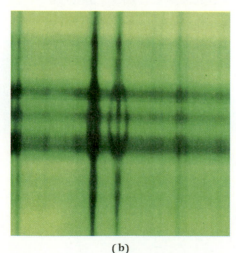

(b)

Figure 18.15 **(a)** *The slit of a spectrograph is placed across a sunspot.* **(b)** *The dark horizontal strips are the spectra of the spots. The absorption lines from the unspotted photosphere are single but are split within the spot by the Zeeman effect.*

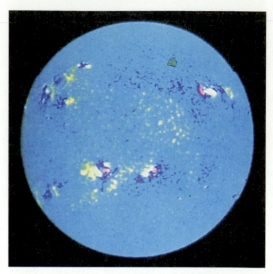

Figure 18.16 *The strengths and directions of solar magnetic fields are determined by the Zeeman effect, from which we can produce a magnetogram, a magnetic map. Dark blue areas show a north magnetic direction, yellow south. Magnetic directions of spot pairs are the same in one hemisphere, the opposite in the other.*

associated with the precession is quantized and can take on only specific values. As a result, the electron's energy levels can be split into two or more parts. Spectrum lines created by the jumps between energy levels will also be split (Figure 18.15). The stronger the magnetic field strength, the greater the splitting.

When we use the Zeeman effect to map the global solar magnetic fields (Figure 18.16), we see that the leading spots—those ahead in the direction of rotation—of one hemisphere have the same magnetic direction (north or south), and the following spots the opposite magnetic direction. In the other hemisphere, the magnetic directions are reversed. The leader spots are generally slightly closer to the equator than the followers, showing that the underlying magnetic field lines are all tilted relative to parallels of solar latitude.

In 1851 the German astronomer Heinrich Schwabe discovered that the number of sunspots on the solar surface varies with an approximately 11-year period originally called the **sunspot cycle** (Figure 18.17a). As a new cycle begins, a few spots will appear in both hemispheres at solar latitudes near 45°. The number of spots slowly grows with time, and their average northerly and southerly latitudes creep toward the solar equator (Figure 18.17b). The number reaches a maximum when the average latitude is about 10° to 15° north and south; thereafter, the count diminishes to near zero. As one cycle begins to run out, a new one starts at high latitudes, and for a time the two overlap.

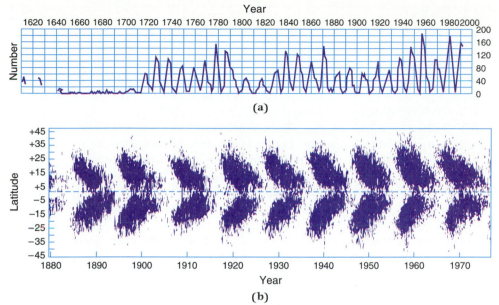

(a)

(b)

Figure 18.17 **(a)** *The sunspot cycle is shown for 1615–1990; the average monthly number of spots is given on the vertical axis. The variation from one maximum to the next is quite large, and the cycle seems to have been nearly absent between 1645 and 1710.* **(b)** *A more detailed* butterfly diagram *shows how the positions and numbers of spots change with solar latitude. At any time, the spots occupy a large latitudinal range.*

Perhaps more remarkable is the behavior of the magnetic fields. During one cycle the magnetic directions stay the same, the leading spots having the same magnetic direction as the pole of their resident hemisphere (Figure 18.18a). In the next cycle, however, they are reversed (Figure 18.18b), as are those of the poles (the whole solar dipole flipping over). Eleven or so years later, all the magnetic directions reverse again. The 11-year sunspot cycle is really a 22-year **magnetic activity cycle.**

Associated with the cycle is a set of subtle currents of photospheric gas that flow easterly (prograde) and westerly (retrograde) at a mere 10 km/hr relative to the steady rotation (Figure 18.19). The flows have slight tilts relative to the solar equator; they begin near the rotation poles

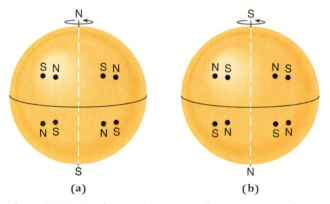

(a)   **(b)**

Figure 18.18 *Leader spots always have the same magnetic direction as the pole of the hemisphere that contains them. During one 11-year cycle* **(a)**, *the magnetic directions stay the same. During the next* **(b)**, *all the magnetic directions are reversed.*

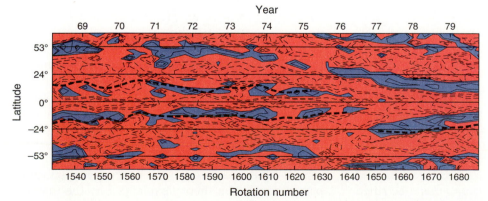

Figure 18.19 *The blue areas show regions of 10-km/hr prograde gas flows. The flows take 22 years to drift to the equator and they pick up the spots along the way.*

At the maximum of the solar activity cycle, the Sun produces copious quantities of high-energy radiation, and it is no surprise that there are long-term climatic effects. We see the familiar 11-year pattern in tree rings, and the cycle has been linked to periodic droughts in the central United States. Perhaps more important, the exact length of the cycle (which varies between about 9.5 and 12 years) has been closely tied to the average terrestrial temperature. As the period goes below 11 years (as it has been doing lately), the Earth's climate warms, confusing measures of global warming by the greenhouse effect. The most dramatic consequence of the solar cycle was the demise of *Skylab* in 1979. The 1979–80 maximum was one of the biggest ever recorded. The increased solar flux expanded the outer atmosphere and increased the drag on the spacecraft, which came crashing to Earth before a shuttle launch could be prepared to boost it into a higher orbit.

There may be more profound effects. There was little discussion of sunspots following their discovery by Galileo. In 1893 the English astronomer E. W. Maunder (who created the concept of the butterfly diagram of Figure 18.17b) concluded that between about 1645 and 1715 sunspots and the cycle effectively disappeared (see Figure 18.17a). During this period,

Europe was plunged into what is called the little ice age. Settlements were frozen out of Greenland, and rivers in what is now the southern United States froze in winter.

There is no proof of a cause-and-effect relation between the two events. However, we can recreate something of a historical record by looking at how radioactive ^{14}C varies with time in plant life. This form of carbon is created when the atmosphere is struck by *cosmic rays,* high-energy particles from beyond the Solar System. Cosmic rays are kept from the inner Solar System by the solar wind, so there are fewer of them near maximum and consequently less ^{14}C is absorbed in living things. This isotope of carbon is useful for radioactive dating for a period of about 50,000 years. As a result, we can determine the strength of solar activity over time. We can also look at old climate patterns by examining the growth rates of trees through tree rings. Cool weather indeed correlates with long-term low activity.

We do not know how the correlation operates and why the solar cycle disappears. If we are to maintain ourselves here on Earth and protect ourselves against the next time the cycle vanishes—which it almost certainly will—we had better find the reasons.

---

and contain magnetized zones but no spots. The currents slowly drift to the equator, taking the full 22-year period. As they pass 45° latitude, the spots begin to develop and are carried with the flows in the equatorial direction. At the same time, new flows develop near the poles. Consequently, there are *two* 11-year cycles going on *at the same time,* one at high latitudes (with no spots), one at low (with spots).

18.4.2 The Origin of the Solar Cycle

In spite of decades of observation and theoretical analysis, there is still no adequate theory for the source of the solar magnetic cycle. At its heart must lie the Sun's general magnetic field. The gases in the interior of the Sun are ionized. Convection and motion stirred up by rotation cause this electrified

matter to circulate, creating a **solar dynamo** that generates the dipole magnetic field in much the same way that planetary fields are produced.

There are presently two popular hypotheses that incorporate this field. The older (Figure 18.20a) begins with an ordered dipole whose field lines are locked into the ionized gases of the surface. However, since the Sun rotates faster at low latitudes than at high, the field lines become stretched out along the equator and eventually spiral around the Sun, which, with convection, twists and concentrates them into dense ropes. The intense magnetism produces a pressure that makes the magnetic ropes more buoyant, and they float upward with the convection cells. When a magnetic rope is squeezed through the surface, it produces a pair of spots. The theory explains why the magnetic directions of spot pairs must be the same in one hemisphere and opposite those in the other. As the magnetic field becomes more wound up, activity increases. Finally, after about 11 years, the tightly wrapped field breaks apart and the number of spots decreases. The field then reorders itself, but with reversed directions.

The other hypothesis involves the drifts of gas of Figure 18.19. They may result from huge tubes of gas that begin rolling at high latitudes and creep to low (Figure 18.20b). As they roll against one another, they concentrate the magnetic field. After 11 years, when they pass about 45° latitude, the rollers break up and produce the spots. After 11 more years, they reach the equator and dissipate. A complete theory may have to take both pictures—and other hypotheses as yet undefined—into account.

18.4.3 Other Active Phenomena and the Solar Wind

The sunspots are the most obvious manifestation of the magnetic cycle. There are many others. The solar limb in Figure 18.10a displays several sheets of chromospheric gas, or **prominences,** that can extend tens to hundreds of thousands of km into the realm of the corona. When seen against the chromosphere (see Figure 18.14), they absorb light headed toward the observer, appear dark, and are known as **filaments.**

The prominence in Figure 18.21a floats in a serene arch 100,000 km high, clearly outlining a magnetic loop of the kind shown in Figure 18.13b. There are most likely sunspots below. Somehow the magnetic field and strong electrical currents that flow within the prominence insulate it from the surrounding hot coronal gas. Other prominences condense directly from the corona and rain matter down onto the photosphere, and still others (Figures 18.21b and 18.21c) will suddenly erupt into space. Since prominences are magnetic phenomena, their number varies with the solar cycle.

Spectroheliograms occasionally reveal a much more violent phenomenon, a sudden explosive brightening or **flare** in the space between a spot pair (Figure 18.22a). Flares range from minute flickerings to events that can encompass nearly a whole activity center; some are so brilliant they can be seen in ordinary white light. With temperatures up to 20 million K and perhaps much greater, most

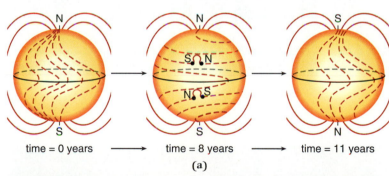

time = 0 years ⟶ time = 8 years ⟶ time = 11 years

(a)

Figure 18.20 **(a)** *The diagrams illustrate two hypotheses for the origin of the solar cycle. At the start of an 11-year cycle the Sun's magnetic field is a dipole. The field lines are frozen into the solar gases and differential rotation causes them to be wrapped around the Sun. They become concentrated and pop through the surface to produce centers of activity and magnetically opposite spot pairs. When the field lines become thoroughly tangled, they break down and reorder with reversed direction.* **(b)** *Within the Sun lie several parallel rolling tubes that creep toward the equator from the poles over a 22-year period and break up into separate cells. Spots appear where the rolling tubes have sufficiently concentrated the magnetic field.*

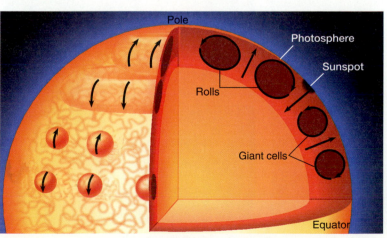

(b)

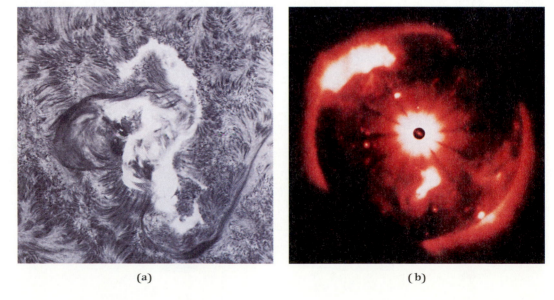

Figure 18.21 **(a)** *A quiescent, cool prominence can hang in the corona for hours or even days as it traces out a magnetic field loop.* **(b)** *An eruptive prominence took off into space in March 1969.* **(c)** *Radio images show the gas departing at nearly 300 km/s.*

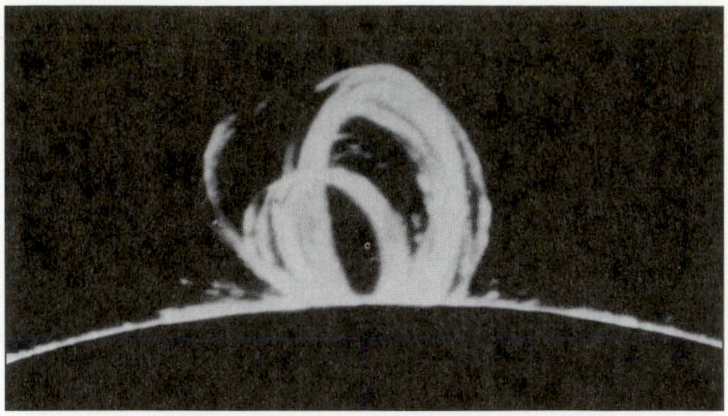

(a)

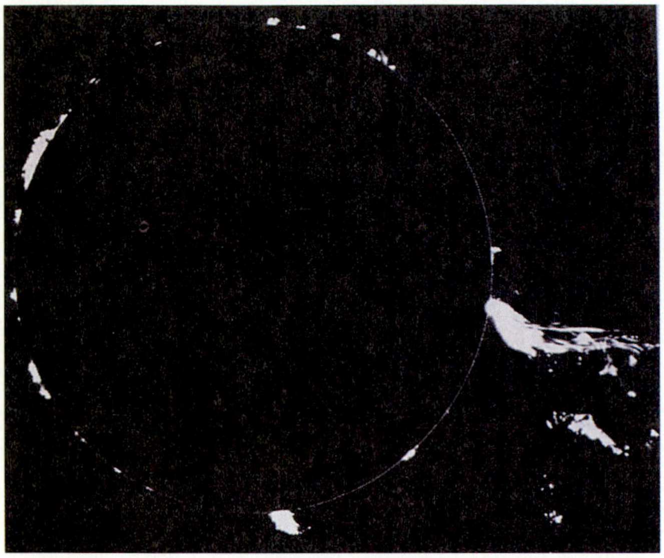

(b)

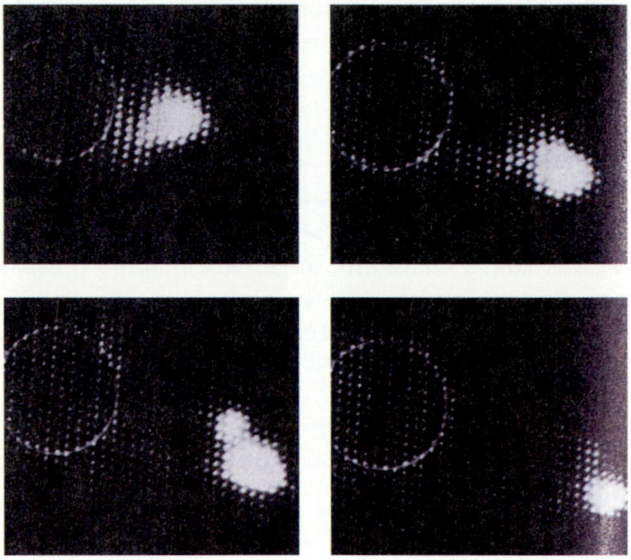

(c)

Figure 18.22 **(a)** *A huge solar flare illuminates the chromosphere above a center of activity.* **(b)** *A long X-ray exposure taken from* Skylab *shows the bright corona. At the center, a short X-ray exposure reveals a flare.*

(a)

(b)

of their energy is emitted in the X-ray spectrum (Figure 18.22b). A large flare can liberate 10^{24} joules/s, 1% of the solar luminosity. Flares actually occur in the corona and are caused by instabilities in the loops that arch from the photosphere between spot pairs (Figure 18.23). If magnetic lines become too twisted and tangled, energy is liberated in a giant spark that causes electrons and protons to accelerate upward and downward along field lines. When the downward ones hit the chromosphere, they heat it, producing X rays and the optically visible flare and evaporating gas from the chromosphere, driving it into the corona. The outbound electrons, accelerated to nearly one-third the velocity of light, produce bursts of radio radiation as they pass through the coronal gases and then speed outward through the Solar System.

The corona exhibits great magnetic structure (Figure 18.24a). **Coronal loops** and long **coronal streamers** lie along the magnetic fields outlined in Figure 18.23. Confinement of the coronal gas to regions above centers of activity is best seen in X-ray images (Figure 18.24b). Between these concentrations lie distinct **coronal holes** that contain little of the hot X-ray-emitting gas. When the X-ray-bright regions are at the solar limb, they create the optically visible loops and streamers. The X-ray-bright areas are further connected by fainter large-scale loops to create a "hairy ball" (Figure 18.25) that has arches of various sizes connecting different solar active areas.

The corona's structure continuously changes in response to the creation and destruction of the magnetic loops that form the centers of activity. On occasion, parts of the corona will simply take off

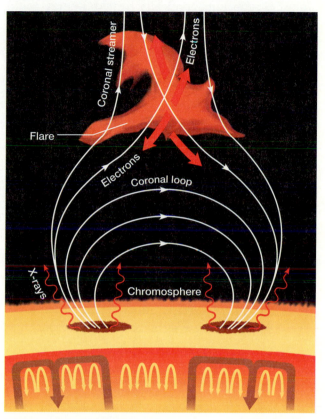

Figure 18.23   *If the field lines from a magnetic loop that create a spot pair become too twisted and tangled, they short-circuit and release their energy in a solar flare. Liberated high-speed electrons generate X rays when they are stopped. The structure of the field defines coronal loops and streamers seen in Figure 18.24.*

Figure 18.24   **(a)** *The optical corona is broken into loops and streamers extending far into space.* **(b)** *X-ray images show coronal holes and bright spots above active regions on the Sun.*

(a)

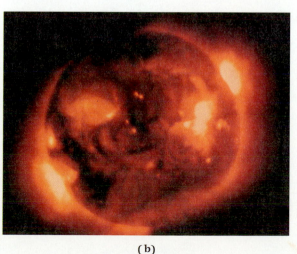

(b)

into space in **coronal transients** (Figure 18.26) that seem to be produced when the magnetic field is released at the foot of a loop. As a result, the corona is constantly dissipating and has to be replaced from the photosphere. The corona is almost certainly magnetically heated. Energy could be riding upward from below on waves that propagate along field lines. Heat may also be generated by myriad small microflares.

Observations of the density and speed of the solar wind at the Earth (about 10 protons and electrons per cubic centimeter rushing at some 500–700 km/s) show the Sun is losing matter at a rate of about 2×10^{17} kg (about 10^{-13} of itself) per year. The wind is controlled by the corona. However, it does not stream from the bright coronal patches

and loops: they are confined by magnetic fields. Instead, most of it pours outward from the coronal holes, where it meets little resistance. The wind is somehow accelerated from the surface of the Sun by these same magnetic fields in a process whose origin is still highly uncertain.

As the number of confining magnetic loops increases during the solar cycle, the wind diminishes and is concentrated more toward higher latitudes. At the same time, the wind is sporadically modulated and enhanced by particles and shock waves generated by solar flares. The particles take about three days to arrive at the Earth, and when they slam into the terrestrial magnetosphere, they produce magnetic storms and aurorae that are also intimately related to the solar cycle.

18.5
The Solar Interior

Whatever the final explanation of the solar cycle, it is clear that it begins deep within the solar interior. Nothing can be understood about the Sun without knowledge of solar structure and of the processes that generate its enormous luminosity over such a long lifetime.

18.5.1 *The Sun's Source of Energy*

No kind of ordinary combustion could keep the Sun luminous for long. Even if it were made of gasoline, it would burn only a few thousand years. In the 1800s, the British physicist William Thomson (Lord

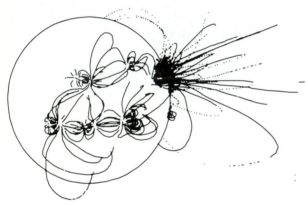

Figure 18.25 *The "hairy ball" model of solar magnetism shows loops of various sizes, some connecting spots within a group, others connecting spot groups, still others extending far into space.*

Figure 18.26 *A spectacular composite image shows both an eruptive prominence and a coronal transient. The latter appears as an arch of coronal gas surrounding the prominence*

Kelvin) and the German Hermann von Helmholtz thought they had the answer. Since the Sun is a gas, it must contract under the force of its own gravity, the squeeze raising the interior temperature. The hot gas then radiates. The Sun would need to contract about 20 m per year to produce the observed power of 4×10^{26} joules/s; at that rate it could live for 10^8 years. However, the science of geology showed that the Earth's rocks are far older than that. Although gravity is the force that raises the interior temperature, it cannot have kept the Sun illuminated for the 4.6-billion-year age of the Solar System.

In 1920, the British astrophysicist Sir Arthur Eddington realized that solar hydrogen might provide the key to the puzzle. From Einstein's relation $E = Mc^2$, the Sun could be run by the conversion into energy every second of 2×10^9 kg of mass. The helium atom was known to be 0.7% lighter than four times the weight of hydrogen. Eddington surmised that if four hydrogen atoms or protons are converted, or fused, into a helium atom, the missing mass might appear as energy. In the late 1920s Cecilia Payne-Gaposchkin of Harvard led the way in showing that stars are made mostly of hydrogen. Therefore, nearly 0.7% of the solar mass might become available for the **fusion** of hydrogen into helium, enough to enable the Sun to live for 10^{11} years.

But how do the protons merge? The solution was provided by the development of quantum mechanics and the discovery of the neutron (Section 8.1.1), the **positron,** and the **neutrino.** The positron is a positively charged electron, an example of **antimatter,** which is normal matter with charges (among other properties) reversed. The neutrino is a massless (or nearly so) neutral particle produced in atomic reactions that carries energy at or near the speed of light. Matter is almost transparent to neutrinos, and typically it would take a block of lead a quarter of the way from here to the nearest star to stop one. By 1938, Cornell's Hans Bethe and Charles Critchfield had consolidated the discoveries into the **proton-proton (p-p) chain** (Figure 18.27).

The first step in the chain is to bring two protons so close together that the strong (nuclear) force can overcome the electrostatic repulsion created by like positive charges (see Section 8.1.2). One way to accomplish this act is to throw the particles together at great speed by heating the gas to extremely high temperature, as gravity does in the solar center. The process is therefore known as **thermonuclear fusion** or **nuclear burning.**

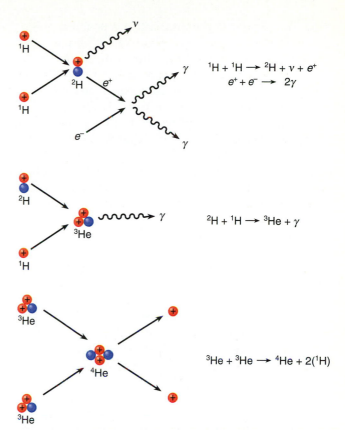

$$^1H + {}^1H \longrightarrow {}^2H + \nu + e^+$$
$$e^+ + e^- \longrightarrow 2\gamma$$

$$^2H + {}^1H \longrightarrow {}^3He + \gamma$$

$$^3He + {}^3He \longrightarrow {}^4He + 2({}^1H)$$

Figure 18.27   *The proton-proton chain begins with the merger of two protons (red), one converting to a neutron (blue), to form a deuterium atom, a positron (e^+), and a neutrino (ν). The ejected positron hits an electron to create a pair of gamma rays (γ). The deuterium immediately encounters another proton, which makes light helium (^3He) and another gamma ray. Two ^3He atoms then collide to produce ^4He with two protons left over.*

However, the electromagnetic repulsion of two protons is so great that even at the solar center, speeds are not high enough. They must be aided by another process. A subatomic particle also behaves like a wave (see Section 8.3.3) that drops off quickly to either side of a maximum. In a loose sense, the proton (that is, the particle-like aspect of the proton's behavior) can be found anywhere along its wave with a likelihood that depends on the wave's height, and it can jump from one place in the wave to another. At temperatures above about 7 million K, the fastest protons can approach closely enough so that if one jumps along its wave, it can appear near enough to the other to be grabbed by the short-range strong force and can tunnel though the electric barrier.

However, even the strong force cannot keep two protons together. For two protons to fuse, one must lose its positive charge and decay into a neutron with the ejection of a positron and a neutrino, turning the atom into deuterium, ^2H. The neutrino flies directly out of the Sun, but the positron quickly encounters a negative electron. When normal matter and antimatter come into contact they annihilate each another and produce a pair of gamma rays. The deuterium nucleus then immediately absorbs another proton (Figure 18.27b), which makes light helium (^{3}He) and another gamma ray. Two ^{3}He nuclei finally collide and make normal ^{4}He (Figure 18.27c). The two protons left over are ejected back into the surrounding gas. The result is that four protons are turned into one helium atom and energy is released in the form of gamma rays and neutrinos.

We have produced rapid fusion reactions on Earth to make hydrogen bombs, but the bombs use heavier hydrogen isotopes, for which reactions are quick. The Sun's fusion process is highly controlled by the first step of the p-p chain, which is very slow. Physicists and engineers are trying to control hydrogen fusion in the laboratory as a means of producing clean and abundant power, but the technology still seems years away.

18.5.2 *The Solar Model*

Once we know how the Sun's energy is generated, we can find how long the Sun will live. First, however, we must construct a **solar model** that will give temperatures, densities, pressures at all points within the interior. Its calculation requires the simultaneous solution of several equations, a task well suited to high-speed computers. First, we adopt the principle of hydrostatic equilibrium (see Section 14.3). Second, we apply the **perfect-gas law,** which relates the pressure of an ordinary gas, P, to the density of atoms (N) per unit volume and the temperature:

$$P = NkT,$$

where k is a physical constant. For example, if you double the temperature without changing the density, the pressure also doubles. Third is the equation that governs dependency of the rate of nuclear fusion on temperature and density (or pressure). Fourth, we have to consider the rate at which energy flows out of the solar interior. The solar gases are highly opaque, and a newly created gamma ray flies only a fraction of a centimeter before it is absorbed and re-emitted by an atom. The energy works its way outward randomly, moving first forward then backward, but on the average each re-emission takes place farther from the center, where the temperature is lower. From the blackbody laws, the wavelengths of the emitted radiation must

lengthen, but since the total energy stays the same, more photons must be emitted to compensate. What starts off as a single deadly gamma ray in the center is then emitted from the photosphere a million years later as hundreds or thousands of optical photons.

From the solar model (Figure 18.28), we find that the central density is 160 g/cm^3, ten times denser than lead, but the temperature is so high— 15 million K—that matter is still in the gaseous state. The temperature is high enough (above roughly 7 million K) to sustain thermonuclear fusion within the inner 30% of the solar radius (2.5% of the volume), a region known as the solar **core** (Figure 18.29). The compression is so great that the core includes *40%* of the Sun's mass. At the calculated fusion rate (allowing for changes in structure as the fuel supply diminishes), there was originally enough hydrogen fuel in the core to last 10 billion years. Since the Solar System is about 5 billion years old, the Sun is roughly halfway through its allotted lifetime.

Surrounding the solar core is the **envelope,** where the temperature is too cool to allow nuclear

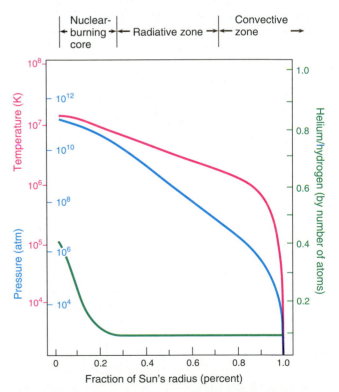

Figure 18.28   *The standard solar model shows how temperature, pressure relative to the Earth's atmosphere at sea level (bars), and the helium-to-hydrogen ratio change with distance outward from the center. The composition changes only in the nuclear-burning core.*

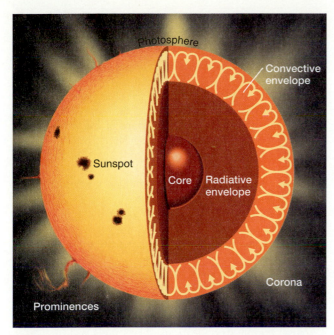

Figure 18.29 *A cutaway model of the Sun shows the core, the radiative and convective envelopes, and the surrounding photosphere, chromosphere, and corona.*

burning. The envelope provides the insulating blanket needed to keep the core hot and transmits energy from the interior to the outside. From the model, we find that it is divided into two parts. In the inner portion, the gas is quiet and energy moves by radiation (as it does in the core). However, as the surface is approached, the temperature gradient becomes progressively steeper, forcing convection currents to arise. The whole outer envelope moves energy by convection, the result of which can be seen in the various aspects of granulation.

The convection layer can be probed accurately with solar seismology, the analysis of the solar oscillations (see Section 18.3.1). A bell's natural vibration frequencies depend on its size, composition, and shape. The same is true of the Sun. From the observed solar vibrations we find that convection extends about 30% of the way into the Sun (see Figure 18.29), that the helium content agrees with that derived from other methods, and that the convective envelope maintains the same pattern of differential rotation found at the surface.

18.5.3 Neutrino Trouble

The only proper way of checking the theory of the solar interior is to look directly into the solar center. Electromagnetic radiation, however, escapes to us from the solar surface. Our only direct access to

the core is through neutrinos, which fly out unimpeded. An amazing 66 billion should pass through every square centimeter of the Earth (and of you) every second. A neutrino detector could allow us to assess the neutrino flux at the Earth, from which we could find the neutrino generation rate in the Sun and the rates of thermonuclear reactions.

It is difficult to capture neutrinos—even the Earth is like the purest glass to them. Fortunately, they can initiate a variety of nuclear reactions. There is, for example, a small but real chance that ^{37}Cl can absorb a neutrino to become radioactive ^{37}Ar. The first solar neutrino detector, which is still working after 25 years, is a tank of 100,000 gallons of the cleaning fluid perchloroethylene, C_2Cl_4 (Figure 18.30). To shield it from high-energy particles from space, it is buried 1.5 km deep in the Homestake gold mine in Lead, South Dakota. About a quarter of the chlorine atoms are ^{37}Cl. Every two weeks the argon is swept out with helium gas. As the argon decays it produces radiation, allowing the number of atoms to be counted. Laboratory measurements of the probability of neutrino capture then give the neutrino flux at the Earth. The experiment has consistently measured only a quarter of the expected neutrinos.

A second detector, built in the Kamioka zinc mine in Japan, detects flashes of light when elec-

Figure 18.30 *This neutrino detector, built by Ray Davis and his group in the Homestake gold mine in South Dakota, is a tank filled with perchloroethylene cleaning fluid*

trons in water molecules are struck by high-energy neutrinos. It counts only half the expected number, but unlike the chlorine detector, is directional, clearly showing the neutrinos to be coming from the Sun.

Unfortunately, these experiments do not measure the neutrinos produced in the main p-p chain, which have low energies, but higher-energy neutrinos made in several secondary reactions, some of which involve the temporary manufacture of beryllium and boron. Two recent detectors, a joint U.S.-Russian venture in the Caucasus Mountains (SAGE) and a joint European experiment in Italy (Gallex), use the element gallium, counting neutrinos when they convert ^{71}Ga into radioactive germanium, ^{71}Ge. These instruments can detect all the neutrinos, including those of the main p-p reaction. Together they find about half the expected number, but with a large experimental uncertainty.

The deficiency is confounding and confusing. Numerous explanations have been put forward, none of them fully satisfactory. For example, we can explain the lack of neutrinos in the chlorine experiment by lowering the temperature at the solar center, but then the discrepancy with the Kamioka detector is worsened. Another explanation involves not the Sun, but the neutrinos themselves. Matter is complex. The proton and electron actually represent only one of three known, successively heavier, *families of matter.* The heavier version of the electron is called the muon and the heaviest the tau particle. Each is associated with its own kind of neutrino, and each neutrino is paired with an antineutrino (which spins in the opposite direction from the "normal" neutrino). Of the six kinds, neutrino telescopes can detect only normal electron neutrinos. Theory tells us that if neutrinos have a small amount of mass, they can switch back and forth from one kind to another through interactions with electrons in the solar interior and cannot change back after they leave the Sun. Perhaps a combination of various explanations will work. We will probably not know the answer until the observations have been refined and new detectors built.

Whatever the final outcome, the Sun may be telling us something valuable about the nature of matter, as well as helping us to understand our terrestrial environment and the natures of the billions of other stars in the sky, to which we now lift our sight.

KEY CONCEPTS

Active Sun: The collective phenomena of the solar magnetic cycle.

Antimatter: Particles with reversed charges; a **positron** is a positive electron.

Centers of activity: Magnetic regions of the solar surface that contains sunspots and associated phenomena.

Chromosphere: The hot, transparent layer just above the photosphere.

Core: The inner 30% of the solar radius in which nuclear reactions are produced.

Corona: The hot (2 million K), low-density layer above the chromosphere.

Coronal holes: Gaps in the corona where there is little hot gas.

Coronal loops and **streamers:** Gas in the inner corona confined by magnetic loops above centers of activity and streams of coronal gas above the loops that extend far into space.

Coronal transients: Ejections of coronal gas.

Envelope: The thick layer of the Sun that blankets the core and transmits solar energy.

Flares: Sudden releases of magnetic energy in the corona that brighten the chromosphere.

Granules: Bright cells in the photosphere, about a second of arc across, caused by convection.

Limb darkening: The darkening of the surface of the photosphere toward its edge, or **limb;** caused by the transparency of the photosphere and the inwardly increasing temperature.

Magnetic activity cycle: The 22-year cycle in the amount of solar magnetic activity, which produces the 11-year **sunspot cycle.**

Neutrino: A massless (or nearly massless) particle that carries energy at or near the speed of light.

Photosphere: The bright apparent surface of the Sun.

Prominences: Arches of cool chromospheric gas confined by magnetic fields; seen as **filaments** against the chromosphere.

Proton-proton (p-p) **chain:** A fusion reaction that turns four atoms of hydrogen into one atom of helium.

Quiet Sun: The constant phenomena of the Sun, which do not take large part in the solar magnetic cycle.

Solar constant: The flux of solar radiation at the Earth.

Solar dynamo: The combination of rotation and convection that produces the solar magnetic field.

Solar model: A mathematical reconstruction of the Sun that gives the temperature, density, and composition at all points.

Solar oscillations: Multiple movements or vibrations of the solar surface.

Spectroheliogram: A picture of the chromosphere made by taking a photograph in the light of a chromospheric emission line.

Spicules: Needlelike projections of gas at the top of the chromosphere.

Sunspots: Cool, dark solar regions, consisting of central **umbras** surrounded by lighter **penumbras,** and caused by intense magnetic fields that inhibit the flow of energy.

Temperature gradient: The rate at which temperature varies with distance.

Thermonuclear fusion (nuclear burning): The process of energy generation by the combination of lighter atoms into heavier atoms.

Zeeman effect: The magnetic splitting of spectrum lines.

KEY RELATIONSHIP

Perfect gas law:

$$P = NkT \; (N = \text{number of particles/cm}^3, \\ k \text{ is a constant}).$$

EXERCISES

Comparisons

1. What is the distinction between the quiet Sun and the active Sun?
2. Compare the temperatures of the photosphere, chromosphere, and corona.
3. How do sunspot umbras differ from penumbras? How do prominences differ from filaments?
4. Compare the heating mechanisms of the corona, chromosphere, and photosphere.
5. What are the similarities and differences between electrons and positrons, and between neutrinos and photons?
6. What is the difference between the solar core and the solar envelope?

Numerical Problems

7. What is the solar rotational velocity at the equator? What is the maximum Doppler shift for the sodium lines at 5,893 Å?
8. When do you expect the next solar maximum?
9. The element mercury is not found in solar spectra. On the basis of the abundances of other elements, estimate the ratio of the number of mercury atoms to the number of hydrogen atoms.
10. Estimate the length of one of the filaments shown in Figure 18.14 and the maximum altitude of the prominence in Figure 18.21. Why might your estimates be too low?
11. What mass of antimatter per day would you need to run a power plant that produces energy at the rate of 100 million joules per second?
12. What mass of hydrogen would the power plant in Question 11 use if it ran on hydrogen fusion?

Thought and Discussion

13. Why is the Sun redder at the limb than at the center?
14. Temperature rises with depth in the photosphere. If instead it fell, what would be the effect on limb darkening?
15. Why does the limb of the Sun appear so sharp?
16. If energy passed through the outer solar envelope by radiation instead of by convection, how might the photosphere look?
17. What are solar granules and what causes them?
18. How can we use solar oscillations to learn about the Sun?
19. Why are the ionized calcium lines are so strong in the Sun relative to those produced by more abundant hydrogen?
20. How do we know the temperature of the corona?
21. Why does the chromosphere have emission lines and the photosphere absorption lines?
22. What is a spectroheliogram and what part of the Sun do we observe by making one?
23. How do the magnetic fields of sunspots alternate from one cycle to the next?
24. How are magnetic fields detected and measured on the Sun?
25. How do coronal flares produce the observed chromospheric flares?
26. What are the phenomena associated with solar flares, including those that affect the Earth?
27. Why are prominences bright at the solar limb, but appear dark when seen against the chromosphere?
28. How do the corona and solar activity relate to the solar wind?
29. Describe the loop connections in the "hairy ball" model of the solar magnetic field.
30. What are the steps of the proton-proton chain?
31. What is antimatter?
32. Why are we not killed by the gamma rays emitted in the solar core?
33. If the convective envelope occupies 30% of the solar radius, why does it constitute only about 1% of the solar mass?
34. You reside on a world orbiting another star. How might you sense the existence of an activity cycle on our Sun even though that body appears to you only as a star and you cannot see its surface?

Research Problem

35. Using data found in popular astronomy magazines, extend the sunspot cycle given in Figure 18.17a to the present.

Activity

36. Use a telescope to observe the Sun (*by projection of the image only*) over a period of two weeks. Sketch the positions and structures of sunspots and how their appearance changes from limb to center. Why are the faculae more visible when the spots are near the limb? Why are spots elongated near the limb?

Scientific Writing

37. For a nature magazine read by people who are not scientifically trained, describe what we do *not* understand about the Sun. Examine particularly the neutrino problem and give a sense of the depth of our ignorance.

38. For a children's magazine, take a hypothetical ride to the Sun straight into its core. Describe the sights along the way from the corona into the center and convey the awesome nature of the journey.

19

The Stars

*The fundamental
observed properties
of the stars*

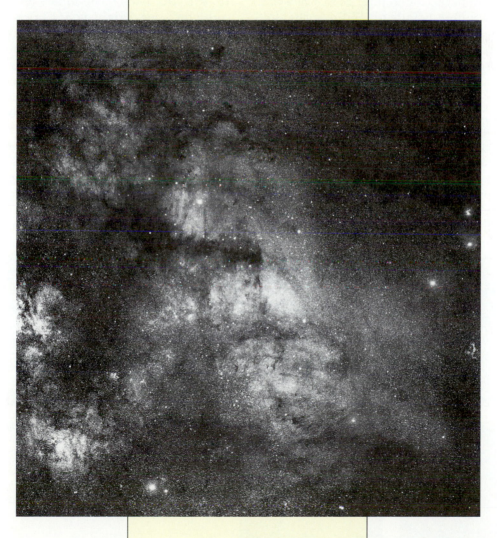

*A million stars
shine within the Milky Way in
Scorpius and Ophiuchus.*

When the Sun sets, we encounter the stars and venture into deep space. Reasoning from the Sun, we define them as bodies that derive, have derived, or will derive their energy from thermonuclear fusion. In turn—as studies of planets aid us in knowing the Earth—investigations of stars help us learn more about our own star.

19.1

Distance

Few stellar properties are as significant as distance. Without it, we can learn little of the physical nature of a star. The fundamental method of determining distance involves measurement of a star's parallax, a concept introduced in Section 3.6 to demonstrate that the Earth revolves. The stars are so far away that parallaxes were not found until 1837, when the German astronomer Friedrich Wilhelm Bessel turned his telescope and his practiced eye to 61 Cygni. This star is actually a double, the two components gravitationally bound and moving through space together. Over the course of a year, he saw the pair swing back and forth through an angle of two-thirds of a second of arc as the Earth went around the Sun. This tiny angle is that subtended by a nickel seen face-on at a distance of 6.5 km.

Parallax translates easily into distance. Nearby stars A and B in Figure 19.1a are in the ecliptic plane. Both appear to move back and forth over a six-month period against the background of the distant stars (Figure 19.1b). The **parallax** (p), is defined as half the star's total angular shift. Parallax is inversely proportional to distance. Star B is double the distance of star A, and its parallax, p_B, is half that of A, p_A. The distance of a star, d, is measured in **parsecs** (pc), where

$$d = 1/p.$$

The parallax of 61 Cyg is 1/3"; therefore the distance is $1/(1/3)$, or 3, pc.

Place star A at a distance of one parsec. Its parallax and the angular size of the Earth's semimajor axis (a_E) as seen from the star are both 1". The parallax angle in radians (see MathHelp 3.2) is a_E/d where both are measured in AU. Since one radian contains 206,265 seconds of arc, $a_E/d = 1/206,265$ radian. The distance d of one parsec must then be 206,265 astronomical units. Since the AU is 1.50×10^8 km, the parsec contains 3.09×10^{13} km. The commonly used **light-year** (ly) is the distance a beam of light (or a photon) will travel in a year at 299,793 km/s. Since there are 3.16×10^7 seconds per year, the length of a light-year is 9.47×10^{12} km, and the parsec is also equal to 3.26 ly. The closest star is Proxima Centauri, a telescopic companion to α Centauri, which has a parallax of 0.772" and a distance of 1.30 pc (4.22 ly). Data on 61 Cyg, α Cen, and all the stars within 4 pc are given in Table 19.1.

Good parallaxes require several years of observation and careful statistical analysis of the data. No measurement can be made with absolute precision; all measurements have inherent errors. If 100 people measure the length of a rope, each will get a

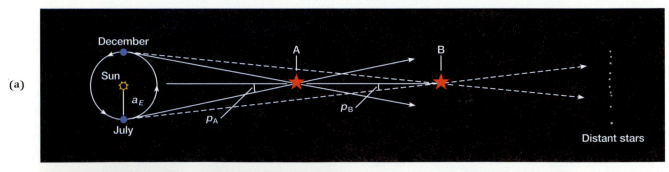

(a)

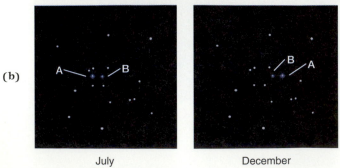

(b)

July   December

Figure 19.1   **(a)** *Two stars, one twice the distance of the other, display parallax shifts. The astronomical parallax (p) is half the total angular shift. It is also the angle subtended by the semimajor axis of the Earth's orbit. Doubling the distance from star A to star B halves the parallax.* **(b)** *Simulated photographs of stars A and B made six months apart show how they change their positions against the background stars, which are assumed to be infinitely far away.*

TABLE 19.1   The Closest Stars[a]

Name	α (2000) h	m	δ °	'	Distance (pc)	μ (''/yr)	v_t (km/s)	v_r (km/s)	v_s (km/s)	Magnitude V	M_V	Spectral Class
Proxima Cen	14	30	-62	41	1.30	3.68	23	-23	33	11.0	15.4	M5 Ve[b]
α Cen A	14	33	-60	50	1.33	3.68	23	-25	34	-0.01	4.87	G2 V
α Cen B								-21	31	1.33	6.21	K1 V
Barnard's Star	17	57	+04	33	1.83	10.31	88	-108	139	9.54	13.25	M5 V
Wolf 359	10	56	+07	03	2.39	4.71	52	13	54	13.53	16.68	M8 V
Lalande 21185	11	04	+36	02	2.53	4.78	56	-84	101	7.50	10.49	M2 V
Sirius A	06	45	-16	43	2.63	1.33	17	-8	19	-1.46	1.42	A1 V
Sirius B	06	45	-16	43	2.63	1.33	17	-8	19	8.68	11.56	DA
Luyten 726-8 A	01	38	-17	58	2.68	3.36	43	29	52	12.45	15.27	M5 V
Luyten 726-8 B[c]								32	54	12.95	15.77	M6 V
Ross 154	18	50	-23	49	2.92	0.72	10	-4	11	10.6	13.3	M4 V
Ross 248	23	42	+44	12	3.16	1.59	24	-81	84	12.29	14.80	M6 V
ε Eri	03	33	-09	27	3.27	0.97	15	16	22	3.73	6.14	K2 V
Ross 128	11	48	+00	49	3.34	1.37	22	-13	26	11.10	13.50	M5 V
Luyten 789-6	22	39	-15	20	3.41	3.26	51	-60	79	12.18	14.60	M7 V
ε Ind	22	03	-56	47	3.46	4.70	77	-40	87	4.69	7.00	K4 V
Cin 2456 A	18	43	+59	37	3.47	2.30	38	0	38	8.90	11.15	M4 V
Cin 2456 B								10	39	9.69	11.94	M5 V
61 Cyg A	21	07	+38	45	3.48	5.25	85	-64	106	5.21	7.55	K5 V
61 Cyg B						5.17	83	-64	105	6.03	8.37	K7 V
Procyon A	07	39	+05	13	3.50	1.25	20	-3	21	0.38	2.71	F5 IV
Procyon B										11.7	14.0	DA
Lacaille 9352	23	06	-35	52	3.52	6.90	117	10	117	7.36	9.59	M2 V
τ Cet	01	44	-15	56	3.54	1.92	32	-16	36	3.50	5.79	G8 V
Gr 34 A	00	18	+44	61	3.56	2.90	49	13	51	8.07	10.32	M1 V
Gr 34 B								20	53	11.04	13.29	M6 V
G 051-015[d]	08	29	+26	47	3.63	1.53	24	—	—	14.81	17.03	M6.5e
G 268-135[e]	01	12	-18	04	3.74	1.34	24	—	—	12.05	14.19	M5e
Luyten's Star	07	28	+05	17	3.78	3.74	66	26	71	9.82	11.98	M5 V
Lacaille 8760	21	17	-38	52	3.86	3.46	63	21	66	6.67	8.75	M0 V
Kapteyn's Star	05	11	-44	56	3.87	8.81	162	245	294	8.81	10.85	M0 V
Krüger 60 A	22	28	+57	43	3.97	0.86	16	-26	31	9.85	11.87	M3 V
Krüger 60 B[f]										11.3	13.3	M4 V

[a]Several stars are double; the components (indicated by A and B) are listed separately; the A component is the brighter, B the fainter.
[b]The appended "e" means that emission lines are seen.
[c]UV Ceti.
[d]DX Cnc.
[e]YZ Cet.
[f]DO Cep.

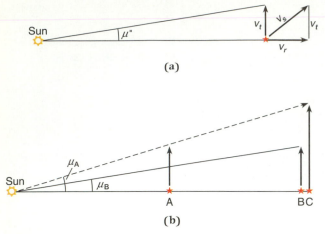

(a)

(b)

Figure 19.2 **(a)** *A star moves through space relative to the Sun with space velocity v_s. This velocity is resolved into the tangential velocity v_t and the radial velocity v_r. The tangential velocity causes the star to appear to move across the celestial sphere at the angular rate of μ seconds of arc per year.* **(b)** *Stars A and B are moving at the same speed. Since B is twice the distance from the Sun as A, the proper motion μ_B is half μ_A. Star C, at the same distance as B, moves twice as fast and consequently has twice the proper motion of B.*

(a)

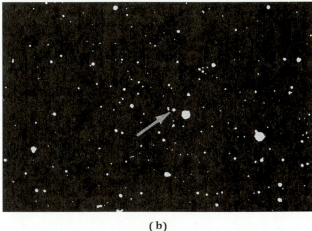

(b)

Figure 19.3 *Barnard's Star (visible only telescopically), caught in 1892 **(a)** and in 1915 **(b)**, sails through Ophiuchus at 10 seconds of arc per year.*

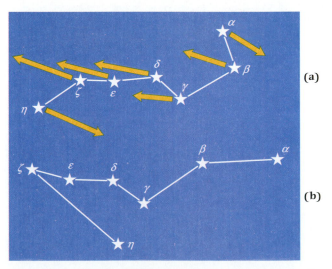

(a)

(b)

Figure 19.4 *The Big Dipper as it appears now **(a)** and as it will appear a quarter million years into the future **(b)**, when the proper motions of the stars (represented by the yellow arrows) will have changed it into a new figure.*

slightly different value. The length of the rope is then defined as the average of the hundred values, and from the spread in individual answers, we derive a formal error. Errors are expressed by a ± attached to the measurement, which gives the range within which the actual value probably falls. For a measurement to be valid or useful, it must be larger than its error: 1,000 ± 2 is quite good; a value of 2 ± 2 is meaningless except that it establishes an upper limit of 4. Errors are often called *noise,* and the ratio of the measurement to its error is called the **signal-to-noise (S/N) ratio.** For a value to be minimally believable, S/N should be at least 2; obviously, the larger the better.

The technology of parallax measurement is in the midst of a revolution. The parallaxes of thousands of stars have been determined by traditional photography, which has a typical error of about ±0.01". A meaningful parallax ought then to be at least 0.02", limiting the resulting distances to a mere 50 pc. Photoelectric devices can now reduce the error to close to 0.001", allowing accurate distance determinations to 250 pc or so. Such measurements are slow and laborious. The process is being vastly speeded up by the European Space Agency's *Hipparcos* satellite, which is automatically measuring the positions and parallaxes of 120,000 stars with an error of 0.002". This vast work is expected to be completed around 1997.

Even with these advances, however, our ability to measure parallaxes is limited to stars within a

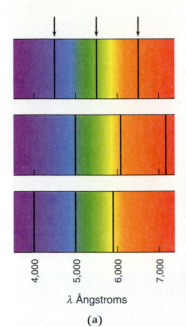

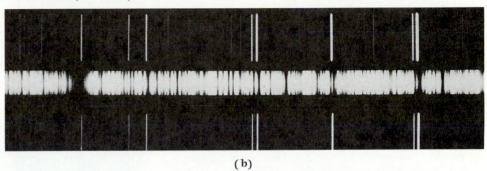

Figure 19.5   (a) *The arrows indicate the laboratory wavelengths of three absorption lines produced by an imaginary astronomical body. On top, the radial velocity is zero and there is no Doppler shift. In the middle, the body is receding at a velocity of a tenth that of light and the lines are shifted to the red by a few hundred Ångstroms; on the bottom, a shift to the blue shows that the body is approaching.* **(b)** *Several iron absorption lines in the spectrum of Arcturus can be identified from their counterparts in the emission comparison spectrum. They exhibit small but measurable Doppler shifts toward the red caused by a recession velocity of a few km/s. (Most of the shift in this case is produced by the orbital motion of the Earth.)*

λ Ångstroms

(a)

(b)

few hundred parsecs. As we will see, they are only the first step in a chain of reasoning that allows us to determine distances as far as we can see.

19.2
Stellar Motions

All stars are in motion. Figure 19.2a shows a star moving relative to the Sun. Its velocity, which must include both speed and direction, is called the **space velocity** (v_s). This motion can be resolved into two mutually perpendicular components: the tangential velocity (v_t) across the line of sight, and the radial velocity (v_r) along the line of sight (see Section 8.4). The three velocities make a right triangle to which we can apply Pythagoras' theorem (the sum of the squares of the lengths of the two perpendicular sides equals the square of the length of the hypotenuse) to create the **velocity equation,**

$$v_t^2 + v_r^2 = v_s^2.$$

A star's motion across the line of sight is seen as an angular change across the celestial sphere (Figure 19.3) called the **proper motion** (μ), measured in seconds of arc per year. Figure 19.2b shows that proper motion depends directly on tangential velocity and inversely on the distance:

$$\mu = v_t/4.74d,$$

where d is in pc, and the constant (4.74) is derived from the mixture of units ("/yr, km/s, and pc).

Nearby stars naturally tend to have big proper motions. The largest known is 10"/yr for dim Barnard's Star (seen in Figure 19.3), only 1.8 pc away. The majority of the naked-eye stars are much farther, many tens and hundreds of parsecs, and the proper motions are correspondingly small. Beyond a thousand or so parsecs (a *kiloparsec*), the motions become difficult, if not impossible, to see. As small as they are, however, over the course of millennia these motions doom the familiar constellations to extinction (Figure 19.4). Knowledge of distance allows us to calculate tangential velocity from proper motion ($v_t = 4.74d\mu$). Typical speeds are 20 to 50 km/s relative to the Sun, although a few (see Table 19.2) top 100 km/s and one (Kapteyn's Star) exceeds 200 km/s.

Radial velocities are found from the Doppler effect (see Section 8.4). We simply measure the wavelengths of a star's absorption lines with the spectrograph (Figure 19.5). The Doppler formula solved for velocity,

$$v_r = \frac{(\lambda_{obs} - \lambda_{rest})}{\lambda_{rest}}c,$$

provides the radial velocity. If the star is receding, the shift is toward longer wavelengths or to the red, λ_{obs} greater than λ_{rest}, and the radial velocity is positive. If the star is approaching, the shift is to the blue, and the radial velocity is negative. Like tangential velocities, radial velocities are typically a few tens of km/s (see Table 19.1), though we find a small number with values greater than 100 km/s.

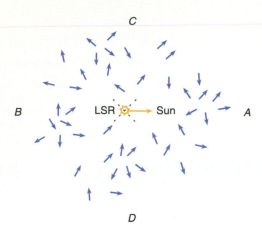

Figure 19.6 *The stars in the Sun's neighborhood have random velocities. From the point of view of their average motion, the local standard of rest (LSR, the dashed cross), there are just as many going in one direction as in any other. The Sun is moving toward A relative to the LSR. The average apparent motions of the stars will reflect the solar motion. The maximum average radial velocities of approach and recession will be in the A and B directions respectively, the maximum proper motions toward C and D.*

We can now combine these with the tangential velocities to derive the space velocities, v_s.

Measurements of motions are made relative to the Sun—but the Sun must be moving, too. In Figure 19.6, local stars within a few hundred pc are shown milling about, moving in random directions with random velocities. The Sun moves through the swarm relative to the stars' average collective motion, which can be identified with a concept called the **local standard of rest** (LSR). From the point of view of the LSR, there are just as many stars approaching as receding, just as many moving to the left as to the right. Now think of standing on the Sun and looking in the direction of solar motion (toward *A* in Figure 19.6). On the average, the stars will appear to be coming toward you at a speed that reflects your velocity. In the opposite direction, toward *B*, the stars will on the average appear to be going away. The average proper motions in the *A* and *B* directions, however, will be zero. To the sides (*C* and *D*), perpendicular to the direction of motion, you would see the reverse: the average proper motions will be at a maximum and to the left, and the average radial velocities will be zero.

From the average motions and velocities of stars all over the sky, we find that the Sun is moving relative to the LSR at a speed of 20 km/s toward right ascension 18^h08^m and declination $+30°$, a point midway between the classical outlines of Hercules and Lyra. The solar motion can then be subtracted from the velocities of the other stars to see how they are moving relative to the LSR.

Now expand your view from the local stars to encompass our **Galaxy** (Figure 19.7a), which contains the Sun, all the stars you see at night, 200 billion others, and vast amounts of interstellar gas. The great majority of its stars are found within a flat disk about 25 kpc across that makes the Milky Way. The disk is surrounded by a sparsely populated spherical halo. The disk stars are called **Population I,** those in the halo **Population II.** Population II extends into a denser **bulge** that surrounds the galactic center. The Galaxy holds itself together by the force of the combined gravity of its members, and all stars have orbits about the center. For any given star, the Galaxy effectively behaves as if all

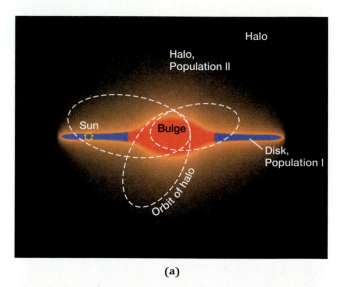

(a)

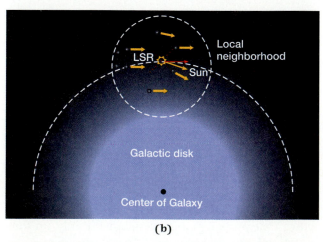

(b)

Figure 19.7 **(a)** *An edge-on view of the Galaxy reveals a disk of Population I stars encompassed by a thinly populated Population II halo that condenses into a thick central bulge. The orbits of halo stars are inclined to the disk and are elliptical.* **(b)** *In the disk the local neighborhood of stars (actually, the LSR) has a circular orbit about the galactic center. The stellar orbits are not identical, however, so the stars slowly drift past one another, resulting in the observed proper motions and radial velocities.*

the mass interior to the stellar orbit were condensed to that center. Therefore, the star and the Galaxy behave as a two-body system and Kepler's first and second laws apply: a stellar orbit is an ellipse with the galactic center at one focus, and the orbital speed depends on distance from the center.

The local standard of rest orbits the galactic center on a circular orbit at a speed of about 220 km/s (Figure 19.7b); the source of this number will be addressed in Chapter 26. However, the individual stellar orbits—the semimajor axes, eccentricities, inclinations to the plane—are all slightly different, and as a result, so are the stellar velocities relative to the LSR and to the Sun. The stars and the Sun therefore slowly drift past one another, which gives them their observed radial velocities and proper motions.

The stars of the Galaxy's halo, those of Population II, are different from the Population I stars of the disk. To be in the halo at all, their orbits must be highly inclined to the galactic plane. Further, the orbital paths are quite elliptical. During the completion of their galactic circuits, they must pass through the galactic plane twice (see Figure 19.7a). Since their paths are so different from ours, we see them moving by us quickly, explaining the existence of the high-speed stars in Table 19.1. (These and others like them are excluded from the determination of solar motion.) The stars of Population II are different in other ways. Watch as we build the case for the distinction between the populations, the accumulated data leading to a theory of galactic origin and evolution.

19.3
Magnitudes

The range of stellar distances is great, and so is that of stellar brightnesses. In 130 B.C., Hipparchus divided the stars into six brightness categories or **magnitudes** (m) that extended from first magnitude for the brightest stars to sixth for the faintest he could see (see Section 4.5). These are now called **apparent magnitudes** because they represent the brightnesses of stars as they appear in the sky. In the nineteenth century, it was found that the first-magnitude stars were roughly 100 times brighter than the sixth. Hipparchus' classification was then placed on an exact scale, in which five magnitude divisions correspond precisely to a factor of 100 in brightness. One magnitude thus corresponds to a brightness ratio of the *fifth root* of a hundred, or 2.512. . . . Two magnitudes correspond to a ratio of 2.512^2, three to 2.512^3, and so on (Table 19.2). To

TABLE 19.2
Magnitudes and Brightness Ratios

Magnitude Divisions	Brightness Ratio
1	2.51
2	6.31
3	15.85
4	39.81
5	100.00

go beyond five divisions, you just combine the ratios: six magnitude divisions are a factor of 100 × 2.512 in brightness, ten magnitudes 100 × 100 = 10,000, and so on.

The scale was set by establishing the average apparent magnitude of a group of faint stars around the north celestial pole as 6.0. The very brightest stars then fall into apparent magnitude 0 and even –1. Venus at its brightest is –4, and the full Moon and Sun are –12 and –27. Stars observable only through the telescope go above 6th: a small instrument allows you to see easily to 10th and a big one able to detect nearly to 30th. With modern photometry (see Section 9.5), it is possible to measure the apparent magnitudes of stars to better than a hundredth of a division. A broad category like second magnitude extends from 1.50 to 2.49. The 40 brightest stars in the sky and their properties are listed in Table 19.3.

Apparent magnitudes are complicated by the range of stellar colors. Color is not just aesthetically appealing, but is a measurable stellar property related to temperature. Most stars behave something like blackbodies. Figure 19.8 shows three blackbody curves with temperatures of 3,400 K, 5,800 K, and 23,000 K, their maxima scaled to 100%. The cool blackbody radiates mostly in the red and will appear reddish to the eye. The intermediate one will seem yellow and the hot one somewhat bluish.

Apparent magnitudes determined by the human eye are called **apparent visual magnitudes,** m_{vis}. The eye is most sensitive to yellow-green light near a wavelength of 5,500 Å. Apparent visual magnitudes are now measured with photoelectric photometers (Section 9.6.1) that isolate the part of the spectrum shown as the yellow area in Figure 19.8 and are called *V*. However, early photographs of the sky (Figure 19.9) effectively recorded blue light in the neighborhood of 4,500 Å (the blue area in Figure 19.8). Blue stars like Rigel will therefore appear brighter photographically than

TABLE 19.3   The 40 Brightest Stars[a,b]

Proper Name	Greek-letter Name	α (2000) h	m	δ °	'	Magnitude[a] (V)	Distance[c] (pc)	M_V	μ ("/yr)	v_t (km/s)	v_r (km/s)	v_s (km/s)	Spectral Class
Sirius	α CMa	06	45	−16	42	−1.46	2.65	1.42	1.32	13	−8	15	A1 V
Canopus	α Car	06	24	−52	41	−0.72	70	−5	0.034	11	21	24	F0 II
Rigel Kentaurus	α Cen A	14	40	−60	50	−0.01	1.33	4.37	3.68	23	−25	34	G2 V
	α Cen B					1.33	1.33	5.71	—		−23	31	K1 V
Arcturus	α Boo	14	16	+19	11	−0.04	10.3	−0.10	2.28	49	−5	49	K1 III
Vega	α Lyr	18	37	+38	47	0.03	7.5	0.65	0.35	12	−14	19	A0 V
Capella	α Aur	05	17	+46	00	0.08	12.5	−0.40	0.43	25	30	39	G5 III + G0 III
Rigel	β Ori	05	15	−08	12	0.12	265	−7	0.004	5	21	21	B8 Ia
Procyon	β CMi	07	39	+05	13	0.38	3.5	2.71	1.25	20	−3	21	F5 IV
Achernar	α Eri	01	38	−57	14	0.46	27	−1.7	0.14	18	16	24	B3 V
Betelgeuse	α Ori	05	55	07	24	0.50	320	−7	0.027	41	21	46	M2 Ia
	β Cen	14	04	−60	22	0.61	95	−3	0.031	14	6	15	B1 III
Altair	α Aql	19	51	+08	53	0.77	5.0	2.30	0.66	16	−26	30	A7 V
Aldebaran	α Tau	04	36	+16	31	0.85	19	−0.49	0.20	18	54	57	K5 III
Antares	α Sco	16	29	−26	26	0.96	190	−5.4	0.024	22	−3	22	M1.5 Ib
Spica	α Vir	13	25	−11	10	0.98	67	−3.2	0.054	17	1	17	B1 V
—	α Cru	12	27	−63	06	1.58	120	−3.8	0.030	17	−11	20	B0.5 IV
						2.09	120	−3.3	0.034	18	−1	18	B1 V
Pollux	β Gem	07	45	+28	02	1.14	10	1.00	0.63	32	3	32	K0 III
Fomalhaut	α PsA	22	58	−29	37	1.16	6	2.02	0.37	12	7	14	A3 V
Deneb	α Cyg	20	41	+45	17	1.25	450	−7.2	0.005	12	−5	13	A2 Ia
—	β Cru	12	48	−59	41	1.25	150	−4.6	0.042	30	16	34	B0.5 III
Regulus	α Leo	10	08	+11	58	1.35	22	−0.38	0.25	26	6	27	B7 V
Adhara	ε CMa	06	59	−28	58	1.50	190	−4.9	0.002	2	27	27	B2 II
Castor	α Gem	07	36	+31	53	1.58	15	0.72	0.20	14	6	15	A1 V
—	γ Cru	12	31	−57	07	1.63	14	−0.5	0.27	18	21	28	M3 III
—	λ Sco	17	34	−37	06	1.63	67	−2.5	0.029	9	−3	9	B2 IV
Bellatrix	γ Ori	05	25	+06	21	1.64	35	−1.08	0.018	3	18	18	B2 III
Elnath	β Tau	05	26	+28	36	1.65	36	−1.13	0.18	30	9	31	B7 IV
Miaplacidus	β Car	09	13	−69	43	1.68	48	−1.73	0.18	41	−5	42	A2 IV
Alnilam	ε Ori	05	36	−01	12	1.70	460	−6.6	0.004	8	26	27	B0 Ia
Al Nair	α Gru	22	08	−46	58	1.74	37	−1.10	0.20	35	12	37	B7 IV
Alioth	ε UMa	12	54	+55	57	1.77	20	0.26	0.11	10	−9	13	A0 V
—	λ Vel	08	10	−47	20	1.78	59	−2.1	0.007	2	35	35	WC8 + O7
Dubhe	α UMa	11	04	+61	45	1.79	26	−0.28	0.14	17	−9	19	K0 III
Mirfak	α Per	03	24	+49	52	1.79	63	−2.2	0.033	10	−2	10	F5 Ib
Wezen	δ CMa	07	08	−26	24	1.84	740	−7.5	0.009	30	34	45	F8 Ia
Kaus Australis	ε Sgr	18	24	−34	23	1.85	43	−1.3	0.13	26	−15	30	B9 III
Alkaid	η UMa	13	48	+49	19	1.86	29	−0.45	0.13	17	−11	21	B3 V
—	ε Car	08	23	−59	31	1.86	60	−2.0	0.029	8	2	8	K3 III + B2 V
Girtab	θ Sco	17	37	−43	00	1.87	37	−1.0	0.016	3	1	3	F1 II
Menkalinen	β Aur	06	00	+44	57	1.90	25	−0.09	0.055	7	−18	19	A2 IV

[a]Compiled principally from the *Bright Star Catalogue*, D. Hoffleit and C. Jaschek, Yale University Observatory, New Haven, 1982.

[b]Several stars in the list are double or multiple. If two components of a double star can be seen through the telescope (α Cen, α Cru), they are listed separately; if they are too close together (Capella, ζ Ori), their individual spectral classes are shown in the last column; if the stars are similar (Spica) or multiple (Castor), the components' characteristics are combined.

[c]Distances above about 50 pc are derived from spectral classes.

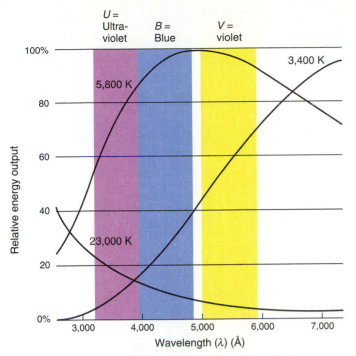

Figure 19.8   *Three blackbody curves are scaled with their maxima to the top of the graph. The three vertical bands correspond to the wavelengths of the three major magnitude systems, visual (V), blue (B), and ultraviolet (U). The coolest blackbody is brighter at V than B while the hottest blackbody is brighter at B.*

visually, and reddish stars like Betelgeuse will appear brighter visually than photographically. Photography led to a new system of **apparent photographic magnitudes** (m_{ptg}), replaced in turn by photoelectrically determined apparent blue magnitudes called *B*.

B and *V* magnitudes are scaled to one another such that they are the same for white stars with temperatures around 9,200 K. We can therefore quantify color by the difference between *B* and *V*, *B* – *V*, called the **color index**. *B* – *V* is slightly negative for hot blue stars (since they have lower *B* magnitudes), and becomes progressively more positive for cooler yellow, orange, and red stars; the full range runs from –0.4 to about +2 (Figure 19.10). The measurement of color index is a quick way of establishing temperature. Magnitude measurements are also commonly made in the ultraviolet (*U*), the red (*R*), the infrared (*I*), and at several longer wavelengths.

However, since these magnitude systems are confined to particular wavelength bands, none of them can express *total* apparent brightness. Cool stars radiate most of their light in the infrared, hot stars emit in the ultraviolet; in both cases, *V* magnitudes will be relatively faint. The **apparent bolometric magnitude,** m_{bol}, is responsive to *all* the

energy from the star, and is set equal to apparent visual magnitude for yellow-white stars near 7,000 K. The difference m_{bol} – *V*, the bolometric correction, also depends strongly on temperature (see Figure 19.10).

Our goal is the measurement of true, or intrinsic, stellar luminosity, the energy radiated by a star in joules/s. The apparent magnitude of a star, however, depends both on its luminosity and on the inverse square of its distance (see Section 8.2.3 and Figure 8.11). To compare the luminosities of stars independently of distance, astronomers use **absolute magnitude** (*M*), defined as the apparent magnitude a star would have if it were placed at a standard distance of 10 pc. If a star is distance *d* away and you move it to 10 pc, it will be $(d/10)^2$ times brighter (or as bright). If *m* is an apparent

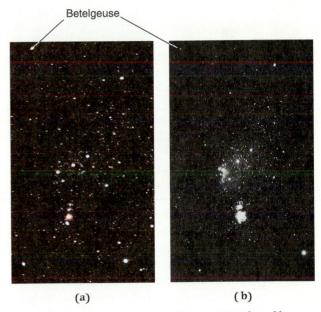

(a)   **(b)**

Figure 19.9   *Orion as it appears to the eye* **(a)** *and to a blue-sensitive photographic plate* **(b)***, on which red Betelgeuse shrinks to third magnitude.*

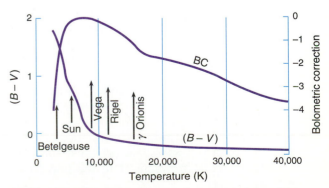

Figure 19.10   *Color index, B – V, is scaled on the left and the bolometric correction (BC) on the right.*

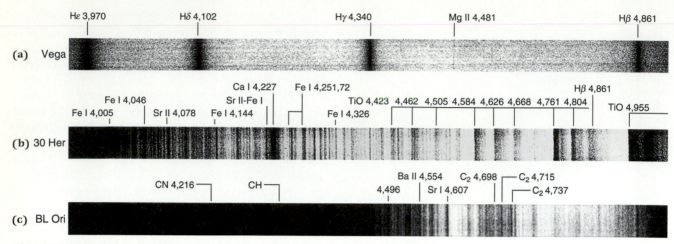

Figure 19.11 *No two stars could have spectra more different than Vega* **(a)** *and 30 Herculis* **(b)***.
Vega has overwhelmingly strong hydrogen lines and a weak Ca II K line, whereas 30 Her has no significant hydrogen lines, a strong line of neutral calcium, and absorption bands of titanium oxide. BL Orionis*
(c) *has strong carbon lines.*

visual magnitude, or *V*, *M* is an absolute visual
magnitude called M_V. Similarly, blue apparent magnitudes (*B*) relate to M_B, m_{bol} to M_{bol}, and so on.

As an example, Deneb (α Cygni) has an apparent visual magnitude of +1.3 and is 450 pc away. If
moved to 10 pc, it becomes $(450/10)^2 = 2,025$
times, or 8.3 magnitudes, brighter. Its absolute visual magnitude is therefore 1.3 − 8.3 = −7, 3 magnitudes brighter than Venus at maximum. Procyon
has *V* = 0.4 and a distance of 3.5 pc. If moved to 10
pc it becomes $(3.5/10)^2 = 0.12$ times as bright, or
2.3 magnitudes fainter, so M_V is 2.7. The Sun, with
an apparent visual magnitude of −27, has a modest
absolute visual magnitude of +4.83, about as bright
as the faintest star in the Little Dipper (see Figure
4.4). Apparent and absolute magnitudes and distance are conveniently related through the **magnitude equation,**

$$M = m + 5 - 5 \log d,$$

where *d* is distance in parsecs (see MathHelp 19.1
for a discussion of logarithms). Careful comparison
of a star's apparent magnitude with a laboratory
blackbody at a known temperature finally allows us
to relate absolute magnitudes to true luminosities
in joules/s.

The range of absolute magnitudes is staggering.
There are a few rare stars with M_V = −10 and M_{bol}
= −12. Others are as faint as M_V = +20. We find
stars 15 magnitudes—a million times—both fainter
and brighter than the Sun. The brightest stars are
30 magnitudes more luminous than the faintest—a
factor of a *trillion*.

19.4
Stellar Spectra
and the Spectral Sequence

Magnitudes give only partial information about a
star. To learn the natures of the stars, we need to
look at their spectra. The first such examinations
puzzled the spectroscopic pioneers of the nineteenth century: only a few stars had spectra matching that of the Sun. Some had powerful hydrogen
absorption lines, others none at all; it looked as if
stars might be chemically different from one another. Several classification schemes were invented to
organize the observations, culminating in one created by Edward C. Pickering at Harvard in 1891. Pickering ordered stellar spectra by letters A through O
according to the strengths of their hydrogen lines,
with A representing the strongest. Sirius and Vega
(Figure 19.11a) went to the top of the list, while
Betelgeuse and 30 Her (Figure 19.11b), with molecular absorption bands of titanium oxide (TiO) and
insignificant hydrogen lines, fell into class M. Class
N (Figure 19.11c) also displays the signatures of
molecules, but those of carbon (principally C_2)
instead of TiO.

Pickering and his assistants, who included
Williamina P. Fleming, Antonia Maury, and Annie
Jump Cannon, classified the stars with *slitless spectrograms* (Figure 19.12). A standard spectrograph
(see Section 9.5) isolates one star at a time with a
narrow slit or aperture. A *slitless spectrograph* is a
large prism placed over a telescope objective. It

MathHelp 19.1 Logarithms

Logarithms (logs) are in constant use in any branch of physics or mathematics. The concept is simple. The logarithm of a number is the power to which 10 must be raised to yield that number. For example, $10^2 = 100$, so the log of 100 is 2; $10^6 = 1,000,000$, so the log of 10^6 is 6. There is no reason why exponents have to be integers: log 10 is 1, and log 20 is 1.3. Tables of logarithms are available in libraries, and many hand calculators are designed to supply them readily.

Logarithms provided a means for simple calculation in the days before computers. If you multiply 10 by 100 you get 1,000, or $10^1 \times 10^2 = 10^3$. The numbers are multiplied, but the exponents, or logarithms, are *added*. To multiply two numbers, all you need do is to look up the logs, add them, and look up the number in the table that corresponds to the sum. Even though they are no longer used much for calculation, logarithms are important parts of many equations, among them the magnitude equation.

allows the spectrum of each star in the field of view to be recorded simultaneously. Because the resulting photograph can be scanned quickly, stars can be classified with speed and accuracy.

As classification proceeded, several of the original categories (for example, E and H) were dropped. Then Fleming and Cannon realized that the B stars, which have He I lines, do not actually fit between A and F but, for continuity in the sequence, are better placed if positioned before A. The O stars, which had then-mysterious lines of ionized helium, also had He I lines, so they had to precede class B. By 1901 the system was in place, the vast majority of stars falling within the seven major classes of the **spectral sequence,** OBAFGKM (Figure 19.13 and Table 19.4). The Sun,

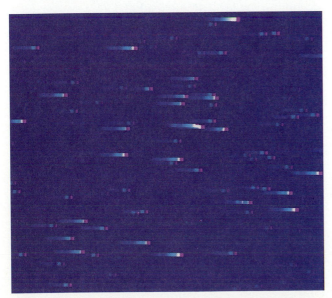

Figure 19.12   *A slitless spectrogram of the Hyades star cluster in Taurus is produced by a prism placed over the telescope's objective lens.*

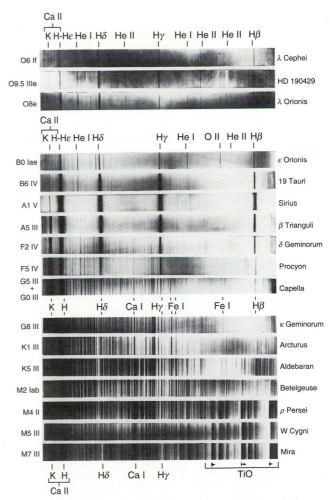

Figure 19.13   *The spectral sequence (OBAFGKM) descends from class O at the top to M at the bottom. The Balmer lines of hydrogen are strongest in the A stars and then decline both upward and downward. Helium lines become prominent in class B and ionized helium in O. Metal lines, particularly Ca II, increase in strength downward from class A, and in class G the neutral calcium line starts to become prominent. The M stars are dominated by bands of titanium oxide and have no hydrogen absorptions. The Roman numerals represent luminosity classes.*

TABLE 19.4
Properties of the Spectral Classes

Class	Characteristic	Color	B – V	Effective Temperature (K)	Examples
O	He II, He I	Bluish	–0.3	28,000–50,000	χ Per, ε Ori
B	He I, H	Blue-white	–0.2	9,900–28,000	Rigel, Spica
A	H	White	0.0	7,400–9,900	Vega, Sirius
F	Metals; H	Yellow-white	0.3	6,000–7,400	Procyon
G	Ca II; metals	Yellow-white	0.7	4,900–6,000	Sun, α Cen A
K	Ca II; Ca I; molecules	Orange	1.2	3,500–4,900	Arcturus
M	TiO; other molecules; Ca I	Reddish	1.4	2,000–3,500	Betelgeuse
R[a]	CN; C	Orange	1.7	3,500–5,400	
S[b]	ZrO; other molecules	Reddish	1.7	2,000–3,500	R Cyg
N[a]	C_2	Red	>2	1,900–3,500	R Lep

[a]Carbon stars.
[b]Mild carbon stars.

Figure 19.14
(a) *The M 6 star V774 Centauri, laden with TiO, is contrasted with the S6 star T Camelopardalis* **(b)**, *which displays ZrO bands.*

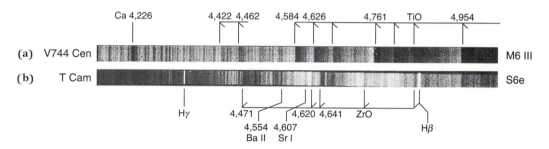

with strong Ca II H and K lines and relatively weak hydrogen, is class G.

Cannon also saw that simple lettering was too crude to express subtle differences, so she decimalized the system. Each class can be broken into ten parts: for example, A0 through A9 is followed by F0 through F9, and so on. In this expanded scheme, the Sun is class G2. By 1924, Cannon's classification of an astonishing 225,300 stars was published in her and Pickering's immense *Henry Draper Catalogue,* named after the man who provided the funds. Stars are commonly called by their HD numbers. Annie Cannon died in 1941; an extension of her work appeared in 1948 with a total of 350,082 stellar classifications. Her name is one of the most revered in all astronomy.

As work proceeded, it was necessary to establish two more categories. Like class N, class R also shows carbon molecules, but not so intense. Together, R and N are now called **carbon stars.** Finally, class S (Figure 19.14), with weaker carbon lines, has properties midway between those of

classes M and N and includes absorption lines of zirconium oxide (ZrO) instead of TiO.

Early in the development of the spectral classification scheme, astronomers realized that the spectral sequence is actually a *temperature* sequence. The classes correlate nicely with blackbody colors and color indices. Class O stars are bluish with temperatures of about 40,000 K, A stars are white and about 9,000 K, the G2 Sun is yellow-white at 5,800 K, and class M stars are reddish and about 3,000 K. The basic properties of the classes are summarized in Table 19.4 and Figure 19.15.

Analysis of stellar spectra in the 1920s, particularly by Harvard's Cecilia Payne-Gaposchkin, demonstrated that in spite of their spectral differences, the photospheres of all the stars of the standard sequence OBAFGKM have similar solar-type chemical compositions (see Figure 18.9). The origin of the sequence has to do not with composition but with temperature-dependent ionization and excitation (Figure 19.16). In class M, the temperature is sufficiently low to allow molecules to form. Neutral

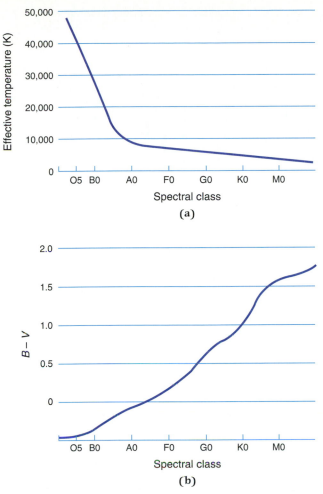

Figure 19.15   *In the two graphs, temperature* **(a)** *and color index* **(b)** *are plotted against spectral class.*

metals like calcium and sodium have powerful lines as well, since they arise from the ground state and no collisional preexcitation is needed (see Section 18.3.2). For warmer stars, through class K, energetic collisions between atoms and molecules cause the molecules to break up, and calcium starts to ionize: TiO disappears, and Ca I and Na I weaken and are replaced by Ca II. The calcium begins to be doubly ionized in class G and Ca II weakens (Ca III is not visible in optical spectra). During the progression, more electrons are collisionally excited into the second energy level of hydrogen. Hydrogen lines become visible in class K and continue to increase in strength all the way to A. The metals become more highly ionized and their lines disappear from optical view (they can be seen in the ultraviolet), making the A spectra appear supremely simple.

Above class A, the hydrogen lines weaken. At 9,000 K the atomic collisions become vigorous enough to begin to produce significant ionization. The number of neutral hydrogen atoms then declines, and hydrogen begins to lose its ability to produce absorption lines. The temperature now becomes so high that high-speed collisions can elevate electrons into the second level of neutral helium, which has double the energy of the second level of hydrogen. He I then becomes strong near 20,000 in the B stars. At higher temperatures, helium ionizes and the He II lines become strong within class O.

Once we understand the origin of the sequence, we can use the theory of atomic excitation and ion-

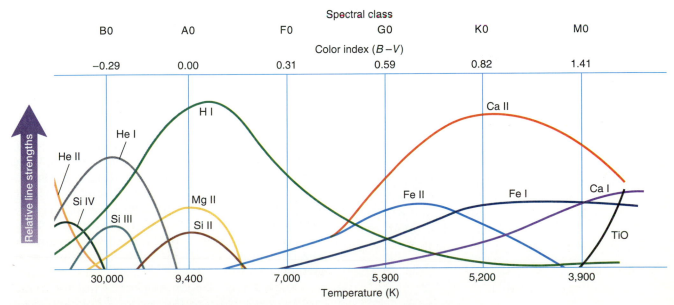

Figure 19.16   *Different atoms, ions, and molecules produce their spectra at different temperatures. Molecules are populous only in class M. Neutral metals are present at low temperatures. Toward higher temperatures, more highly ionized species are created.*

ization to derive stellar temperatures. We need look only at the ratios of the various ions and apply curves similar to those of Figure 19.16.

19.5
The Hertzsprung-Russell Diagram

From the Stefan-Boltzmann law, the energy radiated by a star per unit area per second (the flux) depends on the fourth power of its temperature. It is thus natural to try to correlate spectral class and total stellar luminosity. Such correlations were made independently by the Danish astronomer Ejnar Hertzsprung in 1908 and by the American astronomer Henry Norris Russell of Princeton in 1913. The final result, the **Hertzsprung-Russell** (HR) **diagram,** is the most important single tool of stellar astronomy.

19.5.1 *Giants and Dwarfs*

Russell's original plot of absolute magnitude against spectral class is shown in Figure 19.17. As expected, stars lie nicely along a main band from lower right—low temperature and luminosity—toward upper left. The great surprise was the discovery of some cool (and therefore orange or red) stars that, instead of being dim, are very *bright*. These can be seen in a secondary band that extends up and to the right; they are more apparent in Figure 19.18, a famous later version of the HR diagram that includes many more stars. The luminosity of a spherical blackbody, L, equals $4\pi R^2\sigma T^4$ (see Section 8.3.1), where R is radius. The only way a star can be bright and cool is if it has a large radius, which produces a large surface area. Hertzsprung discriminated the two broad classes of cooler, redder, stars by calling the dimmer ones (those on the main band) **dwarfs** and the brighter ones **giants.** Though the main band is now called the **main sequence,** all its stars are still also called dwarfs. The Sun is therefore a G2 dwarf or G2 main sequence star. The giant stars occupy the *giant branch*.

It is easy to compare stellar diameters with that of the Sun by means of the equation $L = 4\pi R^2\sigma T^4$. Write it twice, once for the star ($\star$) and once for the Sun ($\odot$). Then divide one by the other. The constants cancel out and we are left with

$$\frac{L_\star}{L_\odot} = \frac{R_\star^2 T_\star^4}{R_\odot^2 T_\odot^4},$$

or

$$L_\star/L_\odot = \left(R_\star/R_\odot\right)^2\left(T_\star/T_\odot\right)^4.$$

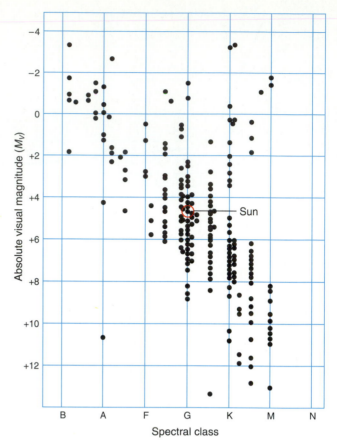

Figure 19.17 *In Henry Norris Russell's original diagram, he plotted stars according to absolute visual magnitude and spectral class. The dwarf sequence (now called the main sequence), which includes the Sun, goes from lower right to upper left. The giant branch scatters out toward the upper right. One white dwarf sits at bottom left.*

Here, $L_\star$, $R_\star$, and $T_\star$, the stellar luminosity, radius, and temperature, are expressed in terms of the solar values. The above equation can be solved for $(R_\star/R_\star)$, whence

$$R_\star/R_\odot = \sqrt{L_\star/L_\odot}\Big/\left(T_\star/T_\odot\right)^2.$$

A red M giant has a temperature of 3,500 K, 0.61 that of the Sun. The giant's absolute visual magnitude is around 0 and the absolute bolometric magnitude is roughly –2, so the star is 7 magnitudes brighter than the Sun, corresponding to a luminosity ratio of about 600. The radius of the star is then $\sqrt{600/0.6^2} = 68$ times larger than the Sun. The solar radius is 0.005 AU. If placed at the Sun, the star would extend to 0.34 AU, almost as large as the orbit of the planet Mercury. The term *giant* is apt indeed.

Remarkably, the immense sizes of the giants are not even close to the record. Sprinkled across the top of the HR diagram in Figure 19.18 are stars even more luminous and larger than the giants: the

supergiants. The brightest shown here have absolute visual magnitudes near –4 (bolometrically –6), 4 magnitudes (40 times) brighter than the red-giant example. A red (class M) supergiant of the same temperature as the above red giant is then $\sqrt{40}$ times, or over 6 times larger. Such a star has a radius of 2 AU, 1.4 times the size of the orbit of Mars. A few rare red supergiants have absolute bolometric magnitudes that approach –10 and sizes comparable to the orbit of Saturn!

Stellar radii are indicated in Figure 19.18 by dashed lines. As we climb the main sequence from the Sun, the stars become larger. An O star at the top of the main sequence has a radius 20 times that of the Sun. (This O star, however, is still considered a dwarf. These colorful terms—dwarf, giant, or supergiant—depend strictly on the specific branch or zone of the HR diagram in which the star is found.) At the bottom of the main sequence, the stars are small: the radius of Proxima Centauri, an M 5 dwarf, is only about a third solar.

Finally, in the lower middle of Figure 19.18 are three dim stars, companions to the stars 40 Eridani, Sirius, and Procyon. The three appear as white stars

Figure 19.19 *Sirius B, smaller than Earth, shines dimly next to its brilliant companion.*

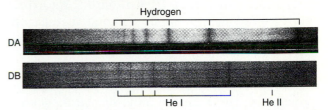

Figure 19.20 *The spectra of a pair of white dwarfs reveal composition differences.* **(a)** *A DA star has absorption lines of hydrogen, but nothing else.* **(b)** *A DB star has helium absorption lines only.*

roughly of class A with temperatures in the neighborhood of 10,000 to 20,000 K. To be both hot and faint (the exact opposite of the giants), they must be very small. Sirius's companion, Sirius B (Figure 19.19), the hottest of the three (whose temperature is now known to be 23,000 K), is only three-quarters the size of Earth. Because of their color, these tiny stars, and others like them, became known as **white dwarfs.** They are commonly placed into two main divisions that reflect composition, not temperature (Figure 19.20). The DA (the "D" distinguishes them from ordinary dwarfs on the main sequence) stars have almost pure hydrogen photospheres, whereas those of the DB variety are almost pure helium. The temperatures of each kind range across most of the HR diagram.

The inferred radii of the larger stars on the HR diagram are confirmed by observation. Although stars are too far away to be seen directly as disks (partly because of blurring by the Earth's atmosphere), there are still ways of measuring their angular radii—which with distances give physical radii. The most important method uses optical interferometry. In 1920 A. A. Michelson (who won a Nobel prize for his experiments with light) attached a long steel beam to the top of the 100-inch telescope on Mount Wilson (Figure 19.21). At each end he placed a moveable mirror, the two acting as a double slit. Light from the pair was reflected down the telescope tube by a pair of fixed mirrors, resulting in an interference pattern that could be seen through the eyepiece. Since a star is

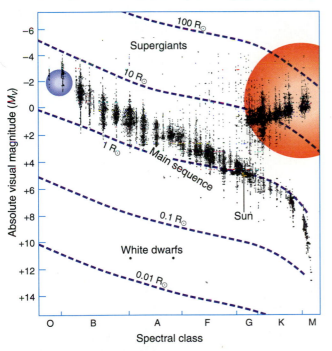

Figure 19.18 *A famous version of the HR diagram, compiled over 60 years ago by W. Gyllenberg of the Lund Observatory in Sweden, shows the giant branch to be very distinct from the main (dwarf) sequence. A few supergiants sprinkle across the top and there are three white dwarfs. The relative sizes of stars in solar units are indicated by the colored circles and by dashed lines. The cool, dim red dwarfs (which, compared to other stellar sizes shown, are about as big as the dots used to represent them) seem to plunge downward because they radiate much of their energy in the invisible infrared and appear too faint to the eye.*

not a point, the pattern was smeared. The outer mirrors were moved to find the separation at which the pattern disappeared, which depends on the star's angular diameter. Michelson and his coworkers found that Betelgeuse has an angular size of 0.047 seconds of arc, consistent with its diameter as estimated from blackbody laws and its distance. More sophisticated interferometry, as well as speckle interferometry (see Section 16.6), allows such measurement for hundreds of stars. Once we know stellar radii, we can determine precise effective temperatures from $T^4 = L/4\pi\sigma R^2$. We expect interferometers to become so good that astronomers should someday be able to *image* stellar surfaces with reasonably good resolution.

The different realms on the HR diagram reflect different stages of **stellar evolution**—changes wrought by age. Like the Sun, all main-sequence stars produce their energy by the fusion of hydrogen into helium. When the hydrogen fuel runs out, main-sequence stars turn into giants and supergiants. The giants ultimately die as white dwarfs. The supergiants die as bodies so strange they cannot even be placed on the HR diagram. The full explanations will wait until Chapters 24 and 25.

19.5.2 Luminosity Classes

One of Pickering's assistants, Antonia Maury, separated hot stars (spectral classes A and B) into categories based on the widths of the spectrum lines. This division was one of the keys that opened the door for Hertzsprung. He found that the average proper motion of narrow-line stars (Figure 19.22a) is much less than that of the broader-line stars (Figure 19.22b), and consequently the narrow-line variety must be much farther away (see Section 19.2). Yet the two kinds were seen to have comparable apparent magnitudes. The narrow-line stars must

be very luminous and very large. Hertzsprung had discovered the supergiants.

Collisions among atoms in a gas minutely jiggle and smear atomic energy levels. As a result, the absorption lines are also smeared and broadened. The denser the gas, the closer the atoms are to one another, the more frequent the collisions, and the broader the lines. The masses of giants and supergiants are spread within an immense volume, and their photospheric densities are vastly lower than those of the dwarfs. As a consequence, the collision rate is lower and the absorption lines are narrow. The broadening effect reaches a maximum among the white dwarfs (Figure 19.22c). These tiny stars have amazingly high densities and the lines are correspondingly broad.

All spectral classes have density-dependent features. He II is seen in absorption in O dwarfs and in

Figure 19.21 *The 20-foot Michelson interferometer had twin mirrors at the ends of a steel beam (M_1 and M_4) placed across the 100-inch telescope. Two more mirrors (M_2 and M_3) sent the light to the objective. Stellar diameters were inferred from the resulting interference patterns.*

Figure 19.22 *Three stars, all about the same temperature, have very different absorption line widths.* **(a)** *Those of the supergiant HR 1040 are very narrow.* **(b)** *The lines of θ Vir, an ordinary main-sequence dwarf, are broader.* **(c)** *The spectrum of the white dwarf 40 Eri B shows immensely broad lines.*

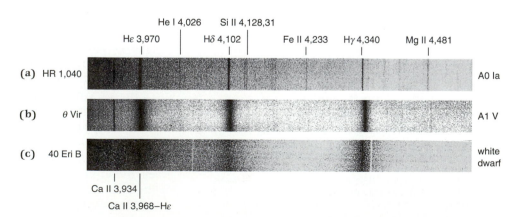

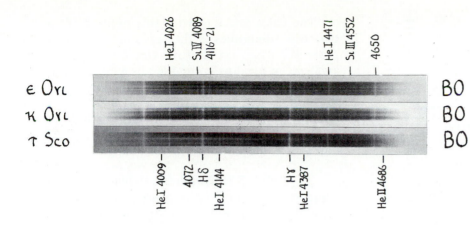

Figure 19.23 *A portion of the MKK atlas shows luminosity criteria for the B stars. Most are subtle. The 4,089 Å line of Si IV is considerably stronger in the supergiant than it is in the dwarf (lower density reduces the interaction between ions and electrons and selectively increases the level of silicon ionization).*

TABLE 19.5
The Luminosity Classes

Class	Type of Stars	Examples
0	Extreme, luminous supergiants; hypergiants	ρ Cas[a]; S Dor
Ia	Luminous supergiants	Betelgeuse, Deneb
Ib	Less luminous supergiants	Antares, Canopus
II	Bright giants	Polaris, θ Lyrae
III	Giants	Aldebaran, Arcturus, Capella
IV	Subgiants	Procyon
V	Main sequence (dwarfs)	Sun, α Cen, Sirius, Vega, 61 Cyg
sd	Subdwarfs	—[b]
D	White dwarfs	Sirius B, Procyon B, 40 Eri B

[a]0-Ia.
[b]All faint and obscure.

emission in O supergiants; in class K, CN molecular absorption is weak in the dwarfs and strong in the giants and supergiants (lower density reduces collisions and enhances molecule formation). We can therefore determine the part of the HR diagram occupied by a star—dwarf, giant, supergiant, or white dwarf—from its spectrum alone.

These criteria were codified in 1943 by William W. Morgan of the University of Chicago, Philip C. Keenan of Ohio State, and Edith Kellman in the famous **MKK** atlas of spectra, in which they also presented spectra of stars to serve as standards for the Pickering/Cannon classification (Figure 19.23). They divided the HR diagram into six **luminosity classes** (Table 19.5 and Figure 19.24). The dwarfs of the main sequence are luminosity class V, the giants are III, and the supergiants I. Because the luminosity range of supergiants is so great, they are broken into subclasses Ia and Ib. Between I and III are bright giants, class II, and between III and V are **subgiants,** class IV. Keenan later added class 0

(zero) that contains the brightest known stars in the Universe; the intermediate case of Ia–0 is seen in Figure 19.24. The white dwarfs are not included and are referred to only as D.

The MKK luminosity classes are appended to the Harvard spectral types. All the stars in Tables 19.1 and 19.3 and in the various figures (like Figure 19.13) are listed with these MKK classes (or as D). Betelgeuse is an M2 Ia star, Arcturus is K1 III, and the Sun is G2 V.

The MKK system allows us to estimate the distance to a star. We take a spectrogram, compare the spectrum to standards in the MKK atlas, find the class, and place the star on the HR diagram to determine the approximate absolute visual magnitude. Since $M = m + 5 - 5 \log d$, and m can be measured,

$$\log d = \frac{(m - M + 5)}{5},$$

giving us the star's distance.

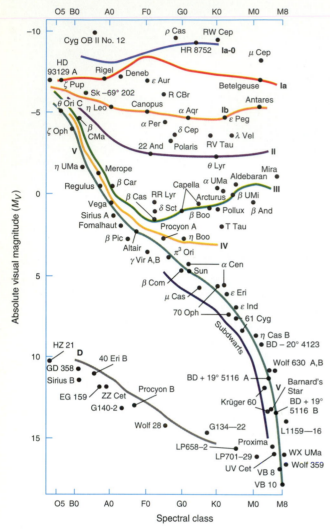

Figure 19.24 *This modern HR diagram shows the loci of the average positions of the luminosity classes (including the subdwarfs) and the locations of about 100 representative stars. Those that fall right on the luminosity loci were placed there deliberately according to their classes. Because of their odd spectra, the white dwarfs are placed by temperature.*

TABLE 19.6
The Fractions of Stars in Different Spectral Classes per Unit Volume of Space[a]

Class	Dwarfs	Giants and Supergiants	
O	4×10^{-7}	F	7×10^{-4}
B	1.4×10^{-3}	G	2.3×10^{-3}
A	7×10^{-3}	K	6×10^{-3}
F	0.036	M	5×10^{-4}
G	0.090		
K	0.11		
M	0.72		

[a]From *Astrophysical Quantities*, C. W. Allen, Athlone Press, London, 1973; white dwarfs are excluded.

The chief problem with this method of **spectroscopic distances** is the calibration of the complete HR diagram: before the method can be applied, we need to know the absolute magnitudes of all the MKK classes. Unfortunately, most of the stars within the current limit of trigonometric parallax are on the main sequence; there are few giants and no supergiants. The problem of calibration will be addressed in the next chapter.

19.5.3 Numerical Distribution

Figure 19.18 shows the densest concentrations of stars among giants and class F dwarfs; M dwarfs appear relatively uncommon. The truth is just the opposite. Our view of the nighttime sky is severely distorted by observational selection (see Section 17.1). The bright stars of the upper main sequence and the giants are easy to see and therefore we count large numbers of them. Over half the 40 brightest stars in Table 19.3 are of class A or B; there is only one K dwarf and no M dwarfs. The M dwarfs are so intrinsically faint that none are even visible to the naked eye.

It is vital for theories of star formation that we know how the numbers of stars change with spectral class. To find the correct distribution, we do not count the observable stars, but those within a restricted volume of space. We must use one large enough to encompass samples of even the rarest kinds, for example a sphere a kiloparsec in radius. Such a complete survey would be extremely difficult, however, since the sphere would encompass roughly a billion stars and because we cannot see

Figure 19.25 *The great nearby Andromeda Galaxy, M 31, 690 kpc away, shows off its blue Population I disk and the red Population II bulge.*

BACKGROUND **19.1** The Discovery of the Stellar Populations*

Scientific discoveries often come about from odd circumstances. Population types were discovered as the result of a war. In 1944 Walter Baade, a staff astronomer at Mount Wilson Observatory, was engaged in studying the magnificent Andromeda Galaxy, M 31 (Figure 19.25), with the 100-inch reflector. Most of the observatory's astronomers were engaged in war-related activities, but he was a German scientist displaced by World War II and had the place much to himself.

Even 50 years ago the lights of Los Angeles had largely destroyed the once-superb sky. But on the nights that Baade observed, the city was blacked out for defense, rendering the sky beautiful once again. He photographed the graceful spiral system through red and blue filters and found that stars of different colors were not distributed in the same way. Bright blue ones appeared sprinkled throughout the galaxy's disk and spiral arms, while red ones sparkled in the thick central region. This central portion is seen from photographs of other galaxies to be a great bulge in the disk, looking like a huge, vaguely spherical blister. Different kinds of stars occupy different positions. To discriminate between them, Baade called the blue disk stars Population I and the red ones of the bulge Population II. Investigations of our own Milky Way Galaxy have revealed the same phenomenon: a red bulge that extends into a red halo and a blue disk. Galaxies with structures like our own all show this phenomenon, providing a key to how they were born, how they will live, and how they will evolve into the future.

*Adapted from *Stars*, J. B. Kaler, New York, *Scientific American Library*, 1991.

the faintest stars a kiloparsec away. In practice, we count the faint stars within smaller volumes and extrapolate to the larger.

What we observe is astonishing. The number of stars climbs steeply as we proceed down the main sequence (Table 19.6). The vast majority, some 70%, are dim M dwarfs. The seemingly common B stars constitute only 0.1% of the total, and rare class O dwarfs a mere 0.00004%! Less than one star in a hundred is a giant, and the supergiants are about as common as O stars.

19.5.4 Galactic Distribution

One of the great revelations of twentieth-century astronomy is the discovery that the different spectral classes are deeply related to galactic location, that is, to population type. The hot stars of the main sequence tend to lie along the plane of the Milky Way. A few, like Regulus in Leo, *appear* off the plane of the Galaxy, but that is an illusion; they are actually nearby and, like the Sun, are still within the galactic disk. Proper placement by distance shows the O and B stars to occupy a plane no more than 200 pc thick over a disk 25 kpc in diameter. The effect is quite noticeable in the low velocities of the bright stars of Table 19.3. They are all going around the galactic center on orbits that are much like the Sun's, mostly confined to the disk. The blue stars of the upper main sequence belong to Population I. Even though Population I contains stars all the way down the main sequence (including the Sun), the disk of our Galaxy, as well as those of others, takes on a bluish color (Figure 19.25).

As seen from Table 19.1, the high-velocity stars—those passing through the disk from the halo—are principally the red dwarfs of the lower main sequence. Direct examination of the halo also shows numerous red giants, but no supergiants. Population II consequently consists at least in part of these high-velocity red stars. The bright central bulge of a galaxy, which is more a part of the halo than the disk, then appears reddish (see Figure 19.25).

The difference between the populations is further illuminated by a group of stars called **subdwarfs** that lies on the HR diagram just to the left of the main sequence, from roughly class F on down (see Figure 19.24), and deviates from the solar norm of chemical composition. These stars are deficient in metals, some having less than a hundredth the solar metallic composition (the record is an astonishing 10^{-4}). The spectral sequence is designed for stars with solar compositions. As a result of their weakened metal lines, the subdwarfs falsely fall into spectral classes that are too warm for their actual temperatures. Metal lines concentrate toward shorter wavelengths, so if metal content and line absorption are reduced, the stars

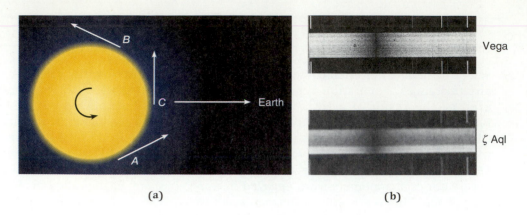

Figure 19.26 (a) *Light from different parts of a spinning star is Doppler-shifted by different amounts (from A to shorter wavelengths, from B to longer, and from C not at all), resulting in smeared absorption lines.* (b) *Slowly rotating Vega has a much narrower Hγ line than does rapidly spinning ζ Aquilae.*

will also appear too blue for their temperatures. Moreover, the lower metal content makes a subdwarf more transparent to outgoing radiation, causing it to be smaller and actually hotter than a star with higher metal content and the same luminosity. The subdwarfs reside in the halo and correlate with high velocity. They are all Population II. The Population I disk stars are more enriched in heavy elements. The result of these observations is a clear relation among main-sequence luminosity, galactic location, and metal content, which provides clues to the evolution of both the stars and the Galaxy as a whole.

19.6
Rotation and Stellar Activity

Rotation is a basic property of all astronomical bodies, and has a strong relation to the HR diagram. Figure 19.26a shows a star spinning in space. Assume it is stationary relative to the Earth. The center (C) of the disk is moving across the line of sight and absorption lines created there will not be Doppler-shifted. However, side A is approaching the observer and its absorptions are Doppler-shifted to shorter wavelengths; side B is receding and *its* lines are shifted to longer wavelengths. The total effect is to smear or broaden the star's spectrum lines.

By comparing the width and shape of an absorption line with that expected for a nonrotating star, the rotation velocity can be found; and if the radius of the star can be estimated, the rotation period can be determined. The effect is easily seen by comparing the spectrum of Vega, which rotates at only 15 km/s, with ζ Aquilae, spinning at 345 km/s (Figure 19.26b).

Unfortunately, the measured rotation speed depends on the tilt of the star's axis relative to the line of sight. If the pole of a rapid rotator is pointed

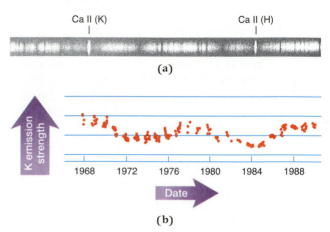

Figure 19.27 (a) *The Ca II H and K lines of 61 Cygni B contain emissions that indicate a chromosphere.* (b) *Long-term changes in the Ca II K emission line demonstrate the existence of a magnetic or star-spot cycle.*

at the observer, there will be no rotational Doppler shifts. However, by looking at large numbers of stars and assuming that the axes are randomly oriented, we can measure the average spin rates for different spectral classes. Stars on the lower main sequence, classes G, K, and M, are generally slow rotators, spinning at one or a few km/s at their equators. At class F5, however, main-sequence rotation speeds jump suddenly to 30 km/s, climb to 100 km/s at F0, and to over 200 km/s among the B stars.

The cause of the relation between main sequence spectral class and rotation is magnetism. As in the Sun (see Section 18.4.3), a combination of rotation and convection can produce stellar magnetic fields. Stellar winds similar to the solar wind drag the magnetic fields outward. The field lines, still tied to the stars like ropes, act as **magnetic brakes** that over billions of years slow the rotations. Dwarfs hotter than about F5 apparently lack the necessary convection zones and have never slowed down from their initially rapid spins.

The solar magnetic field produces chromospheric effects, the corona, and solar activity. We would expect to find the same phenomena associated with other stars. Stellar coronae, indicated by X-ray emissions, are found among solar-type stars. Within the broad Ca II H and K absorptions of the cooler, more slowly rotating dwarfs we also see narrow emission lines that are generated by stellar chromospheres (Figure 19.27a). The H and K emissions of some stars vary periodically (Figure 19.27b), indicating the existence of stellar activity cycles. The Sun, our original example of a star, is anything but unique.

KEY CONCEPTS

Bulge: The thicker part of the halo toward the center of the Galaxy.

Carbon stars: Stars that have at least as much carbon as oxygen.

Dwarfs: Stars of the **main sequence,** the main band of stars that falls from lower right to upper left (luminosity increasing with temperature) on the HR diagram; MKK class V.

Error: The range within which a measured value probably falls.

Galaxy (the): The collection of 200 billion stars and interstellar matter in which we live; shaped like a disk surrounded by a sparsely populated spherical halo.

Giants: Large stars that are brighter than the main sequence for a given temperature; MKK classes II and III.

Hertzsprung-Russell (HR) **diagram:** A plot of stellar magnitudes against spectral types.

Light-year: The distance light travels in a year.

Local Standard of Rest (LSR): The average motion of the stars near the Sun.

Luminosity (MKK) **classes:** The categorization of stars by luminosity and size (dwarf, giant, subgiant, supergiant).

Magnetic brake: The process by which a magnetic field tied to a stellar wind slows a star.

Magnitude: A system for gauging stellar brightness, in which 5 magnitudes correspond to a brightness factor of 100. **Apparent magnitudes** (m) are measures of the brightnesses of stars as they appear in the sky; **visual magnitudes** (V) are measured by a yellow-green detector; **photographic** or blue **magnitudes** (B) are measured by a blue detector; **bolometric magnitudes** represent the total stellar energy; **absolute magnitudes** (M) are the apparent magnitudes that would be seen at a distance of 10 pc.

Parallax: In stellar astronomy, half the total angular shift of a star as the Earth goes around the Sun.

Parsec (pc): The distance at which the Earth's semimajor axis subtends one second of arc; 3.26 light-years.

Population I: The stars of the galactic disk; includes stars of high metal content and luminous blue stars.

Population II: The stars of the galactic halo; includes red stars and stars of lower metal content, but no luminous blue stars.

Proper motion (μ): The angular motion of a star across the line of sight.

Signal-to-noise (S/N) **ratio:** The ratio of the values of observational data to random background "noise" not associated with the source.

Space velocity: The velocity of a star relative to the Sun.

Spectral sequence: The categories of stars organized by temperature (OBAFGKM) and by carbon composition (RNS).

Spectroscopic distance: Distance estimated from absolute magnitude as determined by spectral class.

Stellar evolution: Changes in stars brought about by aging.

Subdwarfs: Low-metal Population II stars to the left of the lower main sequence.

Subgiants: Stars classified between the giants and the main-sequence dwarfs; MKK class IV.

Supergiants: Huge, luminous stars plotted across the top of the HR diagram; MKK class I.

White dwarfs: Small dim stars about the size of Earth plotted across the bottom of the HR diagram; class D; hydrogen rich DA and helium rich DB.

KEY RELATIONSHIPS

Color index:

Photographic minus visual magnitude, or $B - V$.

Distance:

$$D \, (\text{pc}) = 1/p''.$$

Magnitude equation:

$$M = m + 5 - 5 \log d \; (d \text{ is distance in pc}).$$

Proper motion:

$$\mu \, ('' /\text{yr}) = v_t \, (\text{km/s})/4.74d \, (\text{pc}).$$

Stellar luminosity:

$$L_\star / L_\odot = (R_\star / R_\odot)^2 (T_\star / T_\odot)^4.$$

Stellar radius:

$$R_\star / R_\odot = \sqrt{L_\star / L_\odot} \big/ \left(T_\star / T_\odot \right)^2 .$$

Velocity equation:

$$v_t^{\,2} + v_r^{\,2} = v_s^{\,2}.$$

EXERCISES

Comparisons

1. Distinguish between tangential velocity, radial velocity, and proper motion.
2. What are the differences among apparent, absolute, and bolometric magnitudes, and between B and V magnitudes?
3. What are the differences between Populations I and II?
4. Compare the dimensions of dwarfs, giants, supergiants, and white dwarfs.
5. What is the difference between: an M6 V and an M6 III star; an F5 Ia and an F5 Ib star?
6. How does a main-sequence star differ from a subdwarf?
7. What are the differences between DA and DB white dwarfs?

Numerical Problems

8. A star appears to shift back and forth over the year through an angle of 0.05 seconds of arc. How far away is it in parsecs and light-years?
9. What is the parallax of a star 500 pc distant? Can this parallax be observed?
10. If you lived on Jupiter, what would be the parallax of Proxima Centauri? How distant would Proxima be in light-years?
11. The space velocity of a star is 30 km/s and the radial velocity is 12 km/s. What is the tangential velocity?
12. The Hγ line of a star is shifted by 0.05 Å to the red. What is the star's radial velocity?
13. A star has a proper motion of 0.05"/yr, a parallax of 0.03", and a radial velocity of 10 km/s. What is its space velocity?
14. A second magnitude star is how many times brighter than an 11th magnitude star?
15. A star has an apparent visual magnitude of 11.15 and is a kiloparsec away. What is its absolute visual magnitude?
16. What are the apparent color and color index of a star with a temperature of 3,000 K?
17. What is the absolute bolometric magnitude of a star with $M_V = 1.23$ and $T = 3,000$ K? What is the star's spectral class?
18. Compare the strengths of the He I and He II lines and the Si II and Si III lines for an O9 star.
19. An M 2 giant and dwarf have $M_V = 0$ and +10 respectively. How many times bigger than the dwarf is the giant?
20. What is the diameter of a K5 II star relative to the Sun?

21. The spectrum of a star shows it to be an M giant. Its apparent visual magnitude is 10.0. How far away is it?

Thought and Discussion

22. The star 61 Cygni was chosen by Bessel for parallax study because of its proper motion. What was his reasoning?
23. To measure a parallax, the background stars are used as a reference. They must have parallaxes, too. What is the effect of their parallax on the distance measurement of a nearby star?
24. How is the direction of solar motion among the nearby stars determined?
25. What is the spectral class of a star whose spectrum contains: TiO bands; ZrO bands; strong hydrogen lines; He II lines?
26. How do we know that the spectral sequence is a temperature sequence?
27. How can you tell spectroscopically an A dwarf from an A supergiant?
28. How is it possible for a dwarf to be bigger than a giant?
29. What are MKK classes?
30. Where on the HR diagram do we find the most stars and where do we find the fewest?
31. What effect does rotation have on a stellar spectrum?
32. Why do some of the nearby stars have larger velocities relative to the Sun than the more distant bright stars?
33. Why are there no supergiants in the solar neighborhood?

Research Problem

34. Cecilia Payne-Gaposchkin was instrumental in the discoveries of how stars shine. Using library materials, examine her career, list some of her contributions to astronomy, and comment on how they were made.

Activities

35. Construct a map of the sky with right ascension on the horizontal axis and declination on the vertical axis. Use declinations up to 60°N and S. Plot the stars in Tables 19.1 and 19.3. Next to each star place the radial velocity (with the correct sign) in red and the proper motion in blue. How do the radial velocities and proper motions tend to change with right ascension and declination? From your data make a crude estimate of the motion of the Sun through the surrounding stars.
36. Make your own HR diagram from the stars of Tables 19.1 and 19.3. Color each group differently. What profound differences are there between the two sets of stars and why do these differences exist?

37. From Tables 19.1 and 19.3 make a graph of space velocity against spectral class. Use colored symbols to discriminate among dwarfs, giants, and supergiants. Explain the meanings of the correlations.

Scientific Writing

38. A planetarium has a newsletter for its season passholders. Write an article for the interested public on the HR diagram that introduces the concepts of spectral classes, dwarfs, giants, and supergiants (without going into detail about the mechanism of the formation of the spectrum itself).

39. You have been asked to write a chapter in a fourth grade science book on astronomy. For this exercise, write a section on the distances of stars. Introduce the idea of parallax and give a sense of the immense distances involved.

20

Stellar Groupings: Doubles, Multiples, and Clusters

How stars are grouped, and the determination of stellar masses and distances

The open cluster NGC 3293 in Carina sparkles with a variety of stars.

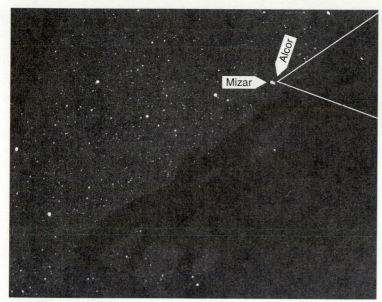

Figure 20.1   *The Big Dipper glides across the slit of the 2.3-meter telescope of the University of Arizona's Steward Observatory. The second star from the end of the handle, Mizar, is gravitationally paired with its companion Alcor. Mizar is itself a double, the white components separated by 14 seconds of arc.*

Stars have a powerful tendency to group, a consequence of the way they are formed. Of the nearest stars in Table 19.2, nearly 60% are members of double or triple systems; as yet undetected companions will push the fraction even higher. Stars also form quadruples, quintuples, and even more complex systems. Collections of larger numbers—clusters—range from a few dozen stars to grand congregations of tens or even hundreds of thousands. These groupings give astronomers the means to derive stellar masses, to probe stellar ages, to test theories of stellar evolution, and to establish the distance scale of the Universe.

20.1
Binary Stars

Binary stars are pairs bound by gravitational attraction, and therefore each partner must orbit the other. There are three different kinds that we will take up one at a time.

20.1.1 *Visual Binaries*

The components of **visual binaries** are farther apart than the seeing limit (about 0.5") imposed by the Earth's turbulent atmosphere, and can therefore be seen individually through the telescope. The first to be discovered was Mizar (Figure 20.1), resolved as a stellar pair about 1650 by Jean Baptiste Riccioli. Tens of thousands of visual binaries are now known; a sampling is presented in Table 20.1 on the next page.

To be certain that neighboring stars are not just in chance alignment (such pairs are called *optical doubles*), we should ideally see orbital motion (Figure 20.2). If the stars are far apart, however, orbital periods can be thousands or even millions of years, rendering orbital movement undetectable. In that case, to be considered binary, the stars must have the same distances and motions. Among such **common proper motion** (CPM) pairs are Mizar and Alcor of Figure 20.1 and Proxima and α Centauri (itself an orbiting binary).

Binary and multiple stars present themselves in a wonderful array of separations, brightness ratios, color differences, and arrangements (Figure 20.3). Some, like γ Leonis, are simple doubles. In other systems, for example ζ Cancri, a third star orbits an

1908

1915

1920

Figure 20.2   *Krüger 60, one of the stars nearest to Earth, is a double whose components show obvious orbital motion over only a dozen years.*

TABLE 20.1
A Sampling of Visual Binaries

Name	α (2000) h	m	δ °	'	Magnitudes (V)	Spectral Classes	Separation "	Position Angle °	Remarks
γ And	02	04	+42	20	2.26–4.84	K3 II–B8 V + A0 V	10	63	Gold, blue
γ Ari	01	54	+19	18	4.75–4.83	A1p–B9 V	7.8	360	Both white
ε Boo	14	45	+27	05	2.70–5.12	K0 II–A2 V	2.9	338	Orange, white
ζ Cnc	08	12	+17	39	5.44–6.20	F9 V–G9 V	0.9	337	1968, $P = 60$ yr
ζ Cnc A–C[a]					6.01	G5 V	5.8	83	Triple star
α Cen	14	40	−60	50	−0.01–1.33	G2 V–K1 V	8.7	8	1946, $P = 81$ yr
α Cru	12	27	−63	06	1.58–2.09	B0 IV–B1 V	4.4	115	A is spectroscopic binary
β Cyg	19	31	+27	58	3.08–5.61	K3 II + B0 V–B8 V	34	54	Gold, blue
ν Dra	17	32	+55	11	4.87–4.88	A4 V–A6 V	62	312	Both white
α Gem	07	35	+31	54	1.58–1.59	A2 V–A1 V	1.8	140	Both spectroscopic binaries
α Gem A–C					9.5		73	164	Spectroscopic binary
γ Leo	10	20	+19	50	2.61–3.80	K1 III–G7 III	4.4	123	
α Lib	14	50	−16	01	2.75–5.80	A3 IV–F4 V	231	314	Naked-eye
ε Lyr	18	44	+39	39			208	173	Naked-eye
ε¹ Lyr	18	44	+39	40	5.06–6.02	A4 V–F1 V	2.8	359	
ε² Lyr	18	44	+39	37	5.37–5.71	A8 V–F0 V	2.2	98	
σ Ori	05	39	−02	36	4.1–5.1	O9.5 V–B0.5 V	0.2	199	Multiple
σ Ori A–C					8.79	A2 V	11	236	
σ Ori A–D					6.62	B2 V	13	84	
σ Ori A–E					6.65	0.95 V	42	61	
θ¹ Ori	05	35	−05	23	6.73–7.96	O7–B0 V	8		The Trapezium; many more components make a small cluster
θ¹ Ori A–C					5.13	O6	12		
θ¹ Ori A–D					6.70	B0.5 V	21		
ζ UMa[b]	13	24	+54	55	2.27–3.95	A1 V–A1 V	14	151	Both spectroscopic binaries
γ Vir	12	42	−01	27	3.65–3.68	F0 V–F0 V	4.7	306	

[a]Multiple star. This row gives information on the C component and its relation to A, as it does for C, D, or E components below.
[b]CPM with Alcor; quintuple star.

inner pair. Epsilon Lyrae is a double-double, two close double stars slowly orbiting one another. Sometimes a fifth star farther out will orbit a central double-double and it may be double too, leading to a sextuple star like Castor.

The measured properties of a binary are *separation* (*s*, in seconds of arc) of the fainter secondary B component relative to the brighter primary A component and *position angle* (*PA*), the angle made by the line between the two stars and the local hour circle (Figure 20.4). These observations can be made visually, with a measuring device attached to the eyepiece, or by photographic or CCD imaging.

A variety of techniques extends the observation of binaries below the conventional visual limit. If a binary with close components is occulted by the Moon, it will wink out in steps, as first one, then the other, member disappears behind the moving lunar limb. Two occultations observed with the Moon approaching at different angles allow the simultaneous determination of both *s* and *PA*. Different types of interferometry (see Sections 16.6 and 19.5.1) and adaptive optics (see Section 9.5 and Figure 9.19) are also beginning to let us use the telescope's full resolving power.

Slowly, the closer pairs will move (see Figure 20.2), the separation and position angle changing with time. After years, or even many decades, enough data may be accumulated to allow the construction of the orbit of star B about star A (Figure 20.5). If the period is centuries long, all we will have is a partial orbit that must be extrapolated to completion.

These observations allow us to find the most important of all stellar parameters, mass (M), through Kepler's laws. The period (P) is simple to determine, but the semimajor axis (a) is not. The observed orbit is the projection of the true orbit onto the plane of the sky, which will be another ellipse. If we draw the major axis of the projected ellipse we see that the primary (A) component is not at its focus, as would be expected from Kepler's first law. From the position of the star relative to the observed focus, the astronomer can find the true orbit, the true semimajor axis in seconds of arc (a''), and the true eccentricity. We now can construct a long, thin triangle (Figure 20.6), in which $a'' = 206{,}265\,a(\mathrm{pc})/d(\mathrm{pc})$ (see MathHelp 3.1). Since $a(\mathrm{pc}) = a(\mathrm{AU})/206{,}265$ (see Section 19.1), $a'' = a(\mathrm{AU})/d(\mathrm{pc})$, or $a(\mathrm{AU}) = a''\,d(\mathrm{pc})$. We now have the semimajor axis of the orbit in astronomical units. Next, divide Kepler's generalized third law (see Section 7.4) written for the binary by that written for the Earth and Sun. The constants ($4\pi^2/G$) divide out, and we find that $P^2 = a^3/(M_A + M_B)$, or

$$(M_A + M_B) = a^3/P^2,$$

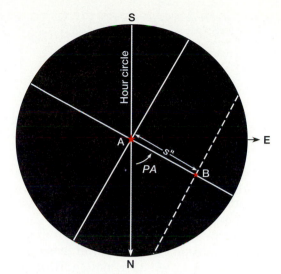

Figure 20.4 *A view through the telescope shows a binary with the brighter A component in the center. The field of view is inverted, north down, east to the right. The crosshairs are set to find the position angle (PA) and the separation (s) of the components.*

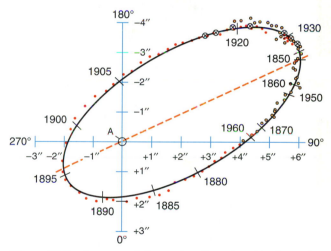

Figure 20.5 *A century of observation allows the orbit of 70 Oph B to be constructed relative to 70 Oph A. The drawn ellipse is the best fit, giving the least difference between the curve and the actual measurements (which contain the inevitable observational errors). Star A, the component of reference for the orbit, is not at the focus of the ellipse because the true orbit is tilted to the plane of the sky.*

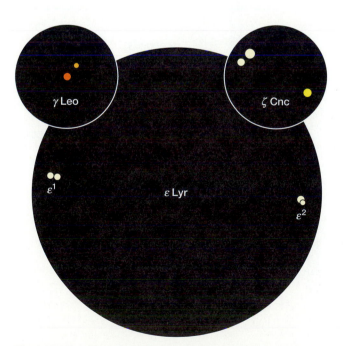

Figure 20.3 *Three visual binaries seen through the eyepiece display different colors, separations, and arrangements. Gamma Leonis is a simple double, ζ Cancri a triple (the outer star going about the pair), and ε Lyrae is a double-double. The last is a naked-eye double that makes a good test of eyesight.*

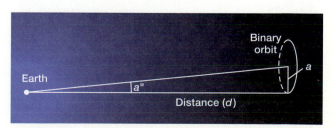

Figure 20.6 *The semimajor axis of a binary star subtends a tiny angle of a'', which equals 206,265 a/d, where a and d are in the same physical units.*

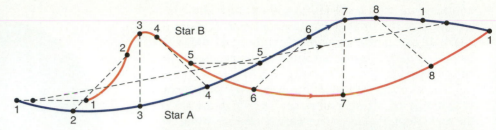

where $(M_A + M_B)$ is the sum of the masses of the two components in solar masses ($M_\odot$), P is the binary's orbital period in years, and a is the orbital semimajor axis in AU.

The determination of individual masses requires the location of the center of mass (see Sections 7.5 and 11.1.1), which is more difficult. In Figure 20.7, the proper motion of the center of mass is a straight line. As the binary moves through space, each star orbits the center of mass and proceeds along a wavy path. The ratio of semimajor axes, a_A/a_B, which is equal to the mass ratio M_B/M_A, is just the ratio of maximum separations from the straight line. Once we have both the sum and the ratio of the masses, we can determine the individual masses. From observations such as these, we find, for example, that α Cen A, a G2 star much like the Sun, has a mass of 1.1 solar masses ($M_\odot$) and that the K1 V B-component measures 0.89 $M_\odot$.

20.1.2 *Spectroscopic Binaries*

If a double star's components cannot be separated by the above techniques, we can sometimes see the system as a **spectroscopic binary.** The observed spectrum will be a combination of the spectra of the two individuals, and if one star does not overwhelm the other, we can see absorption lines from both. If the components are too close to be resolved, their orbital velocities will be high. Assume that identical stars orbit a common center of mass, the center of mass has zero radial velocity, and the orbital plane lies in the line of sight (Figure 20.8a). The stars have equal semimajor axes, a_A and a_B. At time T_1, the stars are moving across the line of sight, there are no Doppler shifts, and the spectrum displays one set of combined absorption lines. After a quarter of an orbit (time T_2), one star is approaching us and the other is receding. The spectra of the two stars are Doppler-shifted in opposite directions, producing doubled absorption lines in the spectrogram (Figure 20.8b).

We measure the radial velocity of each star all around the orbit to determine **velocity curves** (Figure 20.8c), from which we find the period. The

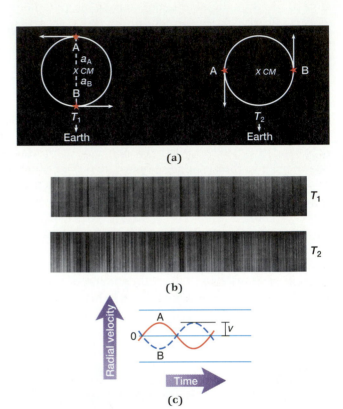

Figure 20.8   **(a)** *A pair of identical stars (A and B) orbit a common center of mass with identical semimajor axes, a_A and a_B. At time T_1 they are moving across the line of sight and there are no Doppler shifts in the spectrum (**b, upper spectrum**). At time T_2 the stars have orbited 90° and now A is moving toward us and B moving away. A's lines are Doppler-shifted to the blue, B's to the red (**b, lower spectrum**), allowing (**c**) the orbital velocities (V) of the stars to be determined.*

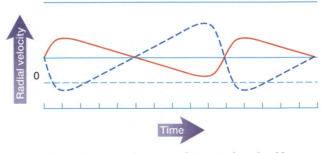

Figure 20.9   *The mass of star A (red) is twice that of B (blue). Star B must then move twice as fast around its orbit as star A and its velocity curve is twice as big. The orbits are eccentric, so the curves are distorted. The center of mass also has a radial velocity that shifts the average of the curves from zero.*

maximum velocity of each star is the orbital velocity (V), which is equal to the circumference of the orbit divided by the period, or $2\pi a/P$. The identical semimajor axes a_A and a_B, are each $PV/2\pi$. Kepler's third law takes a, the semimajor axis of one star relative to the other, or $a = a_A + a_B$. From a and P, the sum of the masses ($M_A + M_B$) is found and each mass is one-half the sum.

Few binaries have such ideal orbits. The center of mass most likely has its own radial velocity relative to the Sun, which shifts the mean velocity away from zero. Generally, the masses will not be the same. The more massive star will be closer to the center of mass and will have a smaller velocity curve than the less massive star. If the orbits are elliptical, the orbital velocities will vary with time, and the velocity curves will be distorted. The final velocity curves then may look something like those in Figure 20.9. Nevertheless, the orbits can be analyzed. The mass ratio is readily found from the ratio of the peaks of the curves relative to the average radial velocity, the degree of distortion yields the orbital eccentricities and orientations, and a is found from a more complex formula involving the ellipse.

The limiting factor in the usefulness of spectroscopic binaries involves the inclination (i) of the orbit relative to the plane of the sky (Figure 20.10). If i is less than 90°, the observed maximum radial velocity will be less than the orbital velocity, our measurement of a will be too low, and we will derive only lower limits to the masses (though the mass ratio can always be found).

20.1.3 *Eclipsing Binaries*

If the binary's orbit is close to the observer's line of sight, we may see it as an **eclipsing binary.** The outstanding example is Algol, β Persei (seen in Figure 4.6). Every 2.8 days, the apparent visual magnitude V drops over a 5-hour interval from 2.1 to 3.8.

As in the case of the solar eclipses, stellar eclipses can be partial, total, or annular (Figure 20.11). The chance of our seeing an eclipse depends on the sizes and separations of the stars. If the members are distant from each other, the orbital inclination i must be almost exactly 90°. The conditions are so stringent that no classical visual binary (one seen without interferometry) has yet been observed to eclipse. Spectroscopic binaries, however, are commonly so close to each other that eclipses can occur with relatively low inclinations.

The primary data for an eclipsing binary are magnitudes plotted against time to make a **light curve** (Figure 20.12). There are many different

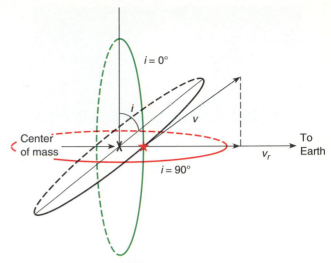

Figure 20.10 *All binary orbits have an inclination (i) relative to the plane of the sky. If i = 90° (red), the orbit lies in the line of sight and we measure the orbital velocity. If i = 0° (green), no orbital velocity can be detected. In the general case (black), the observed maximum radial velocity will be less than the true velocity, and all that can be obtained are lower limits to the individual masses.*

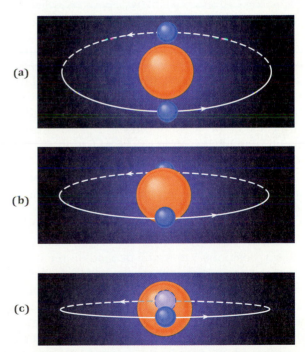

Figure 20.11 *You look at an orbiting binary from Earth. For simplicity, the smaller blue star is shown orbiting the larger orange star. (a) The inclination is low, the stars miss each other, and there is no eclipse. (b) The inclination is higher and once per orbit each star cuts off part of the light of the other, making a partial eclipse. (c) The smaller star can be completely hidden behind the larger, and its eclipse is total. Half a period later, the small star cuts off part of the light of the big one, and the eclipse is annular.*

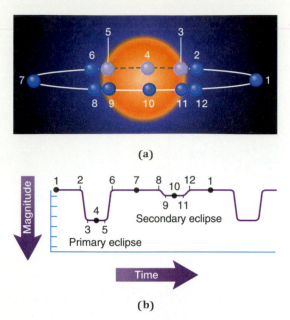

(a)

(b)

Figure 20.12   **(a)** *You look at a small bright star orbiting a large faint one (the center of mass is ignored for simplicity).* **(b)** *The light curve of the system is marked with numbers that correspond to the orbital positions indicated in (a). The primary eclipse lasts from points 2 to 6 and the secondary from 8 to 12.*

kinds of pairings and light curves. A good example involves a hot, bright main-sequence star and a giant. When the dwarf passes behind the larger component (positions 2 to 6) it completely disappears in a deep *primary eclipse.* In between primary eclipses will be a secondary eclipse, when the smaller star blocks some of the radiation from the larger star (8 to 12).

If the system is also a spectroscopic binary we can learn a great deal. Because of the curved limb of the secondary star, the detailed shape of the light curve during the dimming phase (called *ingress*), positions 2 to 3, or during the brightening phase (egress), positions 5 to 6, will depend on the inclination of the orbit. By matching the light curve's shape against theoretical predictions for different inclinations, we can determine *i*. The spectroscopic data then allow us to find the semimajor axes for each orbit and the individual stellar masses. Since we know the orbital speeds, the interval between positions 2 and 5 or 3 and 6 will give (with a correction for the inclination) the diameter of the larger star in km. The ingress or egress duration of the secondary eclipse (positions 8 to 9 or 11 to 12) gives the diameter of the smaller star, also in km. If we know the luminosities, we can derive effective temperatures.

Again, few systems match the idealization. Rotation and tides can distort the stellar shapes from spheres, some of the light from the hotter star may heat the facing hemisphere of the cooler star,

and tides may cause gas streams to flow within the system. As a result, the light curve can be continuously variable outside of eclipse. Moreover, the eclipses may just be partial. Figure 20.13 shows a pair of real light curves. Untangling all the effects can be a Herculean labor. Yet the results—masses, radii, temperatures, shapes, even the degrees of limb darkening and photospheric temperature gradients—are worth all the effort.

If one of the stars is large, the eclipses can be very long. Every 972 days, the K4 II component of Zeta Aurigae (one of Capella's "Kids") hides a B8 V dwarf for 38 days, attesting to a great stellar dimension. VV Cephei, easily seen with the naked eye at magnitude 4.9, consists of an M2 Ia supergiant and an O8 dwarf. As the brilliant, smaller star passes behind the bigger, cool one every two decades, it hides for 1.2 years. From the duration of the eclipse, we find that the supergiant has a radius of 9.4 AU, nearly the size of Saturn's orbit. Epsilon Aurigae, another of the Kids, consists of an F super-

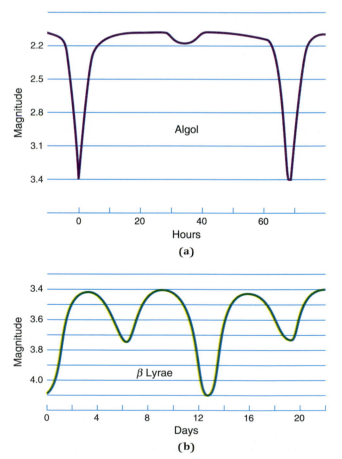

Figure 20.13   *The light curves of two famous eclipsing binaries display complications.* **(a)** *Algol's primary eclipse comes to a sharp point, showing that it is partial.* **(b)** *The light curve of β Lyrae is wildly variable outside eclipse and reveals the effects of distorted stars and flowing streams of gas.*

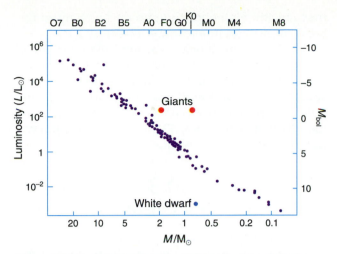

Figure 20.14 *Stellar luminosities (left-hand axis) and absolute bolometric magnitudes (right-hand axis) are plotted against main-sequence masses derived from binary star orbits to show the mass-luminosity relation. Main sequence spectral classes are across the top. Giants and white dwarfs do not fit the relation.*

giant eclipsed every 27 years for an astounding 714 days by an unseen body, probably a huge cloud of dust in which is buried a pair of B stars that make a much closer binary. If you have the patience, the next eclipses of VV Cep and ε Aur will take place in 1997 and 2009 respectively.

20.1.4 *Unseen Companions*

Binary-star components commonly have greatly different magnitudes. When that is the case, the light of one star can overwhelm that of the other, making it difficult, if not impossible, to see. We might then observe what appears to be a single star wobbling drunkenly through space about a constant proper motion path (like one of the curves in Figure 20.7). The classic cases of such **astrometric binaries** (from *astrometry*, the study of star positions) involve the discoveries of the white dwarfs Sirius B and Procyon B. In 1844 F. W. Bessel discovered that neither Sirius nor Procyon was moving in a straight line. He hypothesized that both these great stars, respectively an A1 dwarf and an F5 subgiant, were being shifted by then-invisible "dark stars." The first, the companion to Sirius, was located just 18 years later by Alvan Clarke while testing the new 18-inch lens of Northwestern University's Dearborn Observatory. Just 10" away from its brilliant primary, and faint, only 8th magnitude, it is still extremely difficult to see (see Figure 19.19). Much fainter (11th-magnitude) Procyon B was finally located from Lick Observatory in 1895. Numerous stars display these shallow wobbles, often near the limit of detectability, and it is not always clear whether they are real or not.

The spectroscopic version of the astrometric binary is the **single-lined spectroscopic binary,** which displays only one set of lines that Doppler-shift back and forth as a result of movement around an invisible companion. These are quite common, since if the magnitude difference is at all substantial, the light from the brighter primary swamps the absorption lines produced by the dimmer secondary. Single-lined spectroscopic binaries yield only a complex mathematical function of mass and inclination. Nevertheless, with judicious assumptions made about the primary, we can use the information to assess the properties of the secondary.

20.2
Masses and the
Mass-Luminosity Relation

The most important result of the study of binary stars was the discovery that the absolute magnitude of a main-sequence star depends strictly on its mass. The **mass-luminosity relation** (Figure 20.14) is tightly defined and demonstrates that the main sequence is actually a *mass* sequence that has the stars of greatest mass lying at the top, those with the least at the bottom.

The absolute luminosities of main-sequence stars are remarkably sensitive to mass. On the average, the luminosity is roughly proportional to the 3.5 power of the mass, or

$$L \propto M^{3.5}.$$

One star with twice the mass of another will be *10 times* as bright (between $2^3 = 8$ and $2^4 = 16$). If the mass is multiplied by a factor of 10, luminosity climbs by a factor of 3,000. The exponent, however, itself changes with mass: in the middle of the mass range, it is about 4, and at the low and high ends it drops to 2.

If we reverse the relation, we can estimate the mass of any main-sequence star from its luminosity. Vega's absolute visual magnitude is 0.65 and the bolometric correction is –0.1, so the bolometric magnitude (M_{bol}) = 0.55. From Figure 20.14, we then find a mass close to 2.5 $M_{\odot}$. Proxima Centauri, on the other hand, with M_{bol} = 8.4, has a mass only 45% that of the Sun.

The lowest mass measured for a star is about 0.08 $M_{\odot}$. This value represents the end of the main sequence, below which bodies are not hot enough inside to fuse their hydrogen into helium. The mass of the upper end of the main sequence is harder to ascertain. Figure 20.14 stops at about 24 solar masses and an absolute bolometric magnitude of about −8. Stars at the upper end of the main sequence are exceedingly rare, and all of them happen to be very far away, so far that it is difficult to study binary orbits. However, we can still estimate the masses of the brightest stars by extrapolating the mass-luminosity relation up and to the left (with the aid of theories to be described in Chapter 24) until we reach the magnitudes of the brightest stars. The highest mass appears to be in the neighborhood of 100 to 120 $M_\odot$.

Stars off the main sequence deviate considerably from the mass-luminosity relation and do not define any particular relation of their own because their interior constructions are different from those of main-sequence stars. However, the supergiants clearly have higher masses than do the less luminous giants. They range well into the tens of solar masses, whereas the ordinary giants have masses typically in the neighborhood of 1 to 5 $M_\odot$.

The class of white dwarfs (see Section 19.6.1) serves up the greatest surprise. Orbital analysis of little Sirius B, a star somewhat smaller than Earth, shows it to have a mass of 1.05 $M_\odot$. The Earth is about 100 times smaller than the Sun and contains only a millionth the solar volume. Compression of a solar mass into a volume the size of our planet would produce an average density of *1 million g/cm³*. A sample the size of a golf ball taken from the interior would have a mass of about 35 metric tons (the equivalent of 15 full-sized Cadillacs). Although white dwarf photospheres (where the absorption lines are formed) have lower densities,

they are still high enough to produce the amazingly broad absorption lines seen in Figure 19.22c.

How is such density possible? No experiment in a terrestrial laboratory has come even remotely close. Remember that the atom is mostly empty space, that the proton of a hydrogen atom occupies only one part in 10^{15} of the atoms's volume, so there is plenty of room. The cause of the compression is, again, stellar evolution.

20.3
Open Clusters

Some multiple stars are not hierarchies of doubles like those in Figure 20.3, but small groups of a few stars: σ Orionis and θ^1 Orionis (at the center of Orion's sword) are good examples. When there are more than about a dozen stars, the group is termed a **cluster.** Within a cluster the stars (which themselves can be double or multiple) follow more complex orbits about a common center of mass, each affecting the others. There are two distinct kinds to examine: open and globular.

20.3.1 Structure and Organization

Open clusters are loose, somewhat ragged-looking, systems of a few dozen to a few thousand stars contained in volumes ranging from under 1 pc to roughly 10 pc in radius. Their most defining characteristic is that they are confined to the galactic plane, which contains thousands of them. Consequently, they are strictly Population I.

TABLE 20.2
A Sampling of Bright Open Clusters

Name	Catalogue	Constellation	α (2000) h	m	δ °	′	Diameter °	Distance (pc)	Magnitude[a] (V)
h Per[b]	NGC 869	Perseus	02	19	+57	07	0.5	2,200	7
χ Per[b]	NGC 884	Perseus	02	22	+57	05	0.5	2,200	7
Pleiades	M 45	Taurus	03	48	+24	06	2	138	3
Hyades	—	Taurus	04	26	+15	51	6	45	4
—	M 35	Gemini	06	09	+24	21	0.5	850	8
—	M 41	Canis Major	06	47	−20	44	0.6	640	8
Praesepe	M 44	Cancer	08	40	+20	00	1.6	160	6
Coma Berenices		Coma Berenices	12	25	+26	10	2	90	5
Jewel Box	κ Crucis	Crux	12	53	−60	18	0.15	1,500	7
—	M 6	Scorpius	17	40	−32	12	0.25	490	7
—	M 7	Scorpius	17	54	−34	49	1.3	240	7
—	M 11	Scutum	18	51	−06	18	0.2	1,700	11

[a]Of brightest star.
[b]Together, the Double Cluster in Perseus.

Figure 20.15   *The Pleiades, M 45, the archetype of open clusters, is obvious to the naked eye in Taurus. The wisps are clouds of interstellar dust that scatter starlight.*

Figure 20.16   *The Double Cluster in Perseus (h to the right, χ to the left) can just be seen with the naked eye as a small fuzzy enhancement of the Milky Way. The twin clusters are gravitationally bound and move through space together.*

Open clusters are popular and attractive sights through the telescope (Table 20.2). Several are visible even to the naked eye and make up significant structural and mythological parts of their constellations. The most noticeable is the Pleiades, or Seven Sisters, in Taurus (Figure 20.15), in which nearly a dozen B dwarfs are packed into an area less than a degree across. The telescope shows hundreds of fainter stars. The Hyades (seen in Figure 6.1) makes the head of Taurus, and Coma Berenices forms a whole constellation. In the heart of Cancer, you can find the Praesepe or Beehive cluster, and just below Cassiopeia's chair is a fuzzy spot that turns out to be a pair of open clusters, h and χ Persei, the Double Cluster in Perseus (Figure 20.16). Lucky residents of the southern hemisphere can spot κ Crucis, the Jewel Box (see Figure 1.14), in the Milky Way next to the Southern Cross.

Clusters carry a variety of names. In 1781 a French comet-hunter named Charles Messier compiled a catalogue of 103 "fuzzy objects," since extended to 109. The collection, which contains such nonstellar objects as clusters, glowing interstellar gas clouds, and galaxies, is one of the foundation stones of amateur observing; it is listed in Appendix 2. The Pleiades is M (for Messier) 45 and the Beehive M 44. The most extensive compilation of nonstellar objects was begun by William Herschel and his son, John (who surveyed the southern hemisphere). Their *General Catalogue* was added to and recompiled by J. E. L. Dreyer in 1888 as the *New General Catalogue* (*NGC*), which orders more than 7,000 objects by right ascension: the Jewel Box is also NGC 4755. The *NGC* was later supplemented by the *Index Catalogue* (*IC*), which also goes by right ascension and adds over 5,000 objects.

The looseness of open clusters makes them prone to destruction. The less-massive stars can gravitationally encounter more-massive ones and gain enough energy to be hurled out of the system. The effect is enhanced by tidal forces induced by the Galaxy as a whole, which can stretch a cluster and cause some of the less-massive stars to flee. A cluster then slowly evaporates. Many, if not most, stars were born in clusters and over time escape to wander freely in galactic orbit.

20.3.2 Distances

Clusters provide a means to measure distances across the Galaxy. The **moving-cluster method** provides an important link in the chain. As the Hyades—the best and most studied example— moves through space (Figure 20.17a), the proper motions of its stars closely parallel one another. From the perspective of the observer, however, the

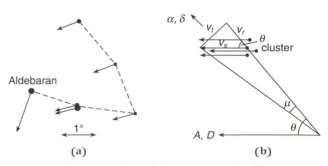

Figure 20.17   **(a)** *If extended, the proper-motion paths of the Hyades stars will converge to a point in space, the position at which the cluster will be found after a sufficiently long period of time (Aldebaran is not part of the cluster).* **(b)** *The angle θ between where the cluster is now (α, δ) and where it will converge in the future (A, D), is also the angle between the radial and space velocities, v_r and v_s, from which the tangential velocity, v_t, and then the distance can be found.*

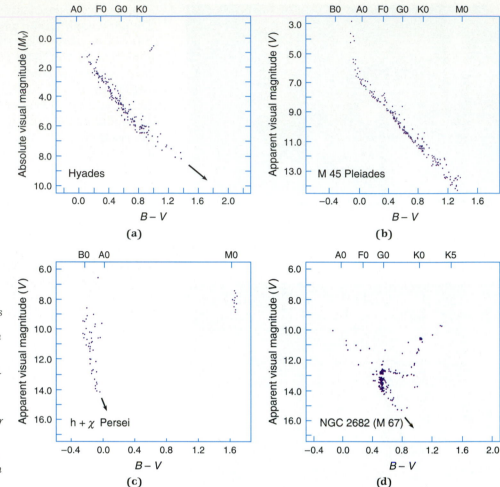

Figure 20.18   *The HR diagrams for* (**a**) *the Hyades,* (**b**) *the Pleiades,* (**c**) *the Double Cluster in Perseus, and* (**d**) *M 67 are all different. That of the Hyades is plotted with absolute visual magnitude, the others with* apparent *visual magnitude. Only h and χ Per have high-mass stars and supergiants. All diagrams use color index, B − V, on the lower axis. Spectral classes are placed across the top. The main sequences of all the clusters continue down through the M stars.*

tracks will seem to converge toward a point. By plotting the converging proper-motion arrows forward, we can locate this convergent point in the sky and determine its right ascension and declination, A and D. The angle θ (Figure 20.17b) between where the cluster is *now* (α and δ) and where it *will be* after an arbitrarily distant time (A and D) is then known.

The angle θ is also the angle between the arrows that represent the radial and space velocities, v_r and v_s. Simple trigonometry allows us to find the tangential velocity, v_t, from the easily observable radial velocity. From Section 19.2, $\mu = v_t/4.74d$, so $d = v_t/4.74\mu$. We have found the distance without parallax. When coupled with the parallaxes of its stars, we find the Hyades distance to be 45 ± 4 pc.

Field stars, those not allied with clusters, are found all over the HR diagram (see Figure 19.18). The HR diagrams of open clusters, however, are different, though varied (Figure 20.18). Since the cluster diagrams commonly use color instead of

spectral class, they are called **color-magnitude diagrams** (spectral classes are indicated here across the tops). The Hyades (Figure 20.18a) has a main sequence that proceeds upward from faint stars, then stops abruptly in class A. There are four K giants, but no upper main sequence and no supergiants. However, the main sequence of the Pleiades (Figure 20.18b) goes into the hot end of class B, and the combined HR diagram of the Double Cluster in Perseus (Figure 20.18c) extends into the O stars. The Double Cluster has no giants, but several *supergiants* instead. M 67 in Cancer (Figure 20.18d) is the opposite of the Double Cluster with a main sequence that stops in class F and has a well-developed giant branch.

The reason for the differences among clusters is age. As we will see, the stars of a cluster are all born at nearly the same time, and as the cluster ages, the high-mass stars die first. A young cluster will have an intact main sequence, but an older cluster will be missing its more-massive stars. The giants in the Hyades developed from the dwarfs of class A, and the M supergiants in h and χ Per are the progenies of O stars.

We now build on the Hyades. Its distance is known with good precision, so its HR diagram can be constructed according to absolute visual magnitude. To this diagram, we add all the nearby parallax stars. As a result, we know the absolute magnitudes for the lower main sequence up through about class A and those of several giants. The other clusters' HR diagrams, however, are constructed with *apparent* visual magnitude, which is possible because all the stars within a cluster are at the same distance from us. Now in your mind lay the HR diagram of the Pleiades (Figure 20.18b) over that of the Hyades (Figure 20.18a), and line up the lower axes. Magnitudes are logarithmic: if you add them together you multiply brightness. As a result, you can slide the diagram of the Pleiades up and down over that of the Hyades until the main sequences fit together (Figure 20.19). On the left-hand axis, you note the apparent magnitude of the Pleiades that corresponds with the absolute magnitude of the Hyades and find the difference, $m - M$ = 5.5. The magnitude equation (see Section 19.3) then yields a distance of 125 pc, almost three times that of the Hyades. You can consequently replot the HR diagram of the Pleiades with absolute mag-

Figure 20.20 *The bright stars in this 5°-wide photograph are members of the loose Monoceros OB 2 association. Near the center is an illuminated gaseous cloud that contains an open cluster.*

nitude and know the luminosities of its B stars. Parallax measurements give 150 pc, so we adopt the average of 138 pc.

This procedure of **main-sequence fitting** can now be applied to the Double Cluster using the extended main sequence of Figure 20.19 to find not only the Double Cluster's distance but also the absolute magnitudes of even brighter main-sequence stars and supergiants. The establishment of the HR diagram is finally completed with a full sampling of open clusters, from M 67 to the youngest known, those with hot O stars. We can therefore calibrate spectral characteristics that depend on luminosity (see Section 19.5.2) and determine the distance of any star whose spectrum we can obtain and classify through the method of spectroscopic distances. We can now wander our Galaxy and know where we are, and can even begin to look into others.

20.4
Associations

A detailed examination of the rare O stars shows that they not only concentrate to the plane of the Milky Way, but also that they (as well as the B stars) congregate into loose groupings called **OB associations** (Figure 20.20). Dozens are known. They are catalogued according to their principal constellation of residence; for example Cygnus OB 2 is the sec-

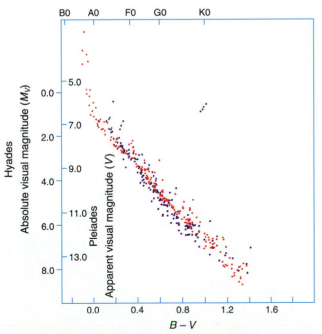

Figure 20.19 *The color-magnitude diagram of the Pleiades (red dots) from Figure 20.18 is superimposed over that of the Hyades (blue dots). The lower axes are aligned and the diagram for the Pleiades has been placed so that its main sequence fits on that of the Hyades. The magnitude difference, $V - M_V = 5.5$, can then be read from the vertical axis and converted to distance.*

ond OB association in Cygnus. Some of the closer ones are quite familiar. The constellation Orion is filled with bright blue O and B stars and is actually one large OB association known as Orion OB 1. Perseus OB 2 contains ξ, ζ, and o Per, and most of the hot stars in Scorpius and Centaurus are part of a massive association.

An association commonly has an open cluster near its center, but unlike a cluster, an association's members are not gravitationally bound to one another. The stars' velocities are too high, and they are all seen to be moving away from a common center, presumably the place of their birth. All associations are therefore rapidly disintegrating. Yet almost all O stars are members of associations, so we know that the stars do not get very far from their birthplaces before they expire. Here is solid proof that massive stars do not live very long, a finding consistent with their lack in most open clusters.

The OB associations also contain red supergiants that have evolved from the O dwarfs. Antares is part of Scorpius OB 2 (along with the B stars δ, β, and τ Sco), and gigantic μ and VV Cephei belong to Cepheus OB 2, which also includes the O stars λ, 14, and 19 Cephei.

20.5
Globular Clusters

Though the open clusters make fine sights through the telescope, they pale beside the great **globular clusters** (Figure 20.21). The largest of these systems are some of the most beautiful objects to be seen in the sky.

20.5.1 *Properties*

Globular clusters are entirely distinct from open clusters. A typical globular contains perhaps 50,000 stars. Their spherical shapes are the natural result of the stars' great combined gravity. The stars are usually heavily concentrated toward the center, the brightness of the image dropping by half only 2 or so parsecs out. The stars then thin outward into space; commonly, some 90% will be found within about 25 pc of the center.

The globulars display considerable variety. Omega Centauri—easily visible to the naked eye—possesses a half million stars within a radius of 30 pc. Several, like the northern hemisphere's M 3, M 5 (Figure 20.22a), and M 13 are not too far behind. Others (like NGC 3201 in Figure 20.22b) are loose organizations with far fewer stars and without much central condensation. The smallest

have radii of only 10 pc or so and only a few thousand stellar members.

The view from inside a globular cluster must be spectacular. The heart of the dense cluster M 15 in Pegasus (Figure 20.23) has 7,000 stars in a spherical volume a mere 0.2 pc across, a seventh the distance between the Sun and α Centauri. The density of stars is a million times greater than we find locally. From within a globular your night sky would be filled with over a thousand first-magnitude stars!

Only 146 globular clusters are known in the Galaxy. Several more are certainly hidden by the dark dust clouds in the Milky Way, but the total number probably does not exceed 200. As viewed from Earth, the clusters concentrate heavily toward the galactic center, with some 60% found within Sagittarius, Scorpius, and the neighboring constellations. Except for this region, however, the globulars do not particularly concentrate toward the Milky Way but are spread out all over the sky. Moreover, the radial velocities tend to be high, a few topping 200 km/s. Both observations indicate that we are looking at members not of Population I but of Population II.

The difference between open and globular clusters is hammered home when we compare their HR (or color-magnitude) diagrams (Figure 20.24). The main sequence of a globular stops at $B - V$ equal to about 0.4 (which corresponds roughly to spectral class F), a pattern similar to that of the open cluster M 67 seen in Figure 20.18d. Both also have distinctive giant branches that curve up and to the right from the main sequence. There are two important differences, however. First, the color-

Figure 20.21 *The greatest of all globular clusters, Omega Centauri, contains half a million stars.*

(a)

(b)

Figure 20.22   (a) *A large globular cluster, M 5 in Serpens, is compared to* (**b**) *a much sparser version, NGC 3201 in Vela.*

magnitude diagrams of nearly all globulars look this way. None has a main sequence that extends below a color index of about 0.4. Second, the globulars display a characteristic never seen in open clusters, a **horizontal branch** that extends to the left from about the midpoint of the giant branch.

Analysis of the spectra of individual stars shows clearly that the Galaxy's globulars are deficient in heavy elements. The range in metal content is quite large, from less than 0.001 solar to about 0.5 solar, but none comes up to that of the Sun. We see then that the main-sequence stars of the globular clusters are subdwarfs like the halo stars discussed in Section 19.5.4, their main sequences truncated at about the same place. The low metal content shifts the globular-cluster main sequences somewhat to the left compared with the Population I main sequence. When corrected for this effect, the sequence stops in class G, at a color index higher (that is, at a cooler spectral class) even than for M 67. The low metal content is also in part responsible for the horizontal branch, making its stars appear bluer than they otherwise would.

20.5.2 Distances and Distribution

With the metal content known, we can correct the position of a globular's main sequence, fit it to the standard main sequence of the combined open clusters, determine the absolute magnitudes of its stars, and find the precious distances. The globulars are very far away, with distances measured in kiloparsecs, consistent with the apparent faintness of objects that contain tens, even hundreds of thousands of stars. The closest is NGC 6397, 2.9 kpc away, and two in the short list of Table 20.3 are 10 kpc or farther. Many are so distant that we cannot even see their main sequences. However, from the nearer and brighter globulars we can establish that the horizontal branches are all close

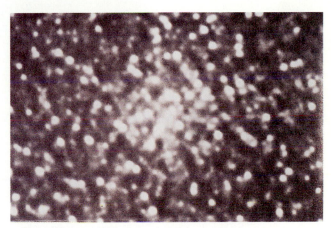

Figure 20.23   *The Hubble Space Telescope resolves the deep core of M 15, revealing stars packed with a density a million times greater than in the solar neighborhood.*

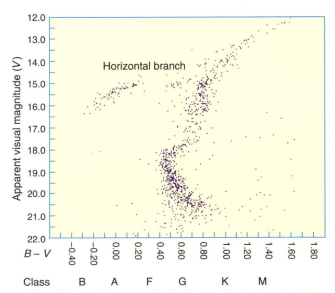

Figure 20.24   *The color-magnitude diagram of the globular cluster M 5 displays a severely truncated main sequence and a distinctive horizontal branch. The spectral classes are given for reference. Since the stars are deficient in metals, the classes are only approximate.*

TABLE 20.3
A Sampling of Bright Globular Clusters

Name	NGC	Constellation	α (2000) h	m	δ °	'	Magnitude (V)[a]	Distance (pc)	Remarks
47 Tuc	104	Tucana	00	24	−72	08	4.0	4,300	Naked-eye
ω Cen	5139	Centaurus	13	26	−47	36	3.7	6,080	Naked-eye; magnificent
M 3	5272	Canes Venatici	13	42	+28	25	6.4	10,000	
M 13	6205	Hercules	16	42	+36	30	5.9	7,400	Barely naked-eye
M 92	6341	Hercules	17	17	+43	08	6.5	7,900	
M 22	6656	Sagittarius	18	36	−23	55	5.1	5,200	Contains planetary nebula
M 15	7078	Pegasus	21	30	+12	09	6.4	11,300	Contains planetary nebula

[a]Total magnitude, considering all stars.

to absolute visual magnitude 0.6, which allows them to be used as distance indicators. The most distant globular known, Pal (Palomar) 2, a 13th-magnitude system in Auriga, is an astounding 150 kpc off, far beyond the outlines of the Galaxy (see Figure 19.7) defined by the distribution of the large majority of its stars.

With the distances known, it is possible to establish the three-dimensional construction of the system of globular clusters. The majority are found to reside in a great Population II halo that surrounds the disk (Figure 20.25). A subset—about 20%—form a thick disk. Its members have higher metal contents, making this group something of an intermediate case between Populations I and II. Like the open clusters, globulars are subject to disruption, though, because of their much greater masses, to a much lesser degree. They are so densely packed with stars that few of them have probably ever broken up completely. Nevertheless, we see that the clusters farther from the galactic center are the larger, as they are less subject to disturbing tides raised by the mass of the galactic disk.

Some 70 years ago, Harlow Shapley, who became a famous Harvard astronomer, made the logical assumption that the center of the system of globulars is coincident with the center of the Galaxy itself. He was then able to use the globular cluster distances and a plot like that in Figure 20.25 to make the first assessment of the galactic center's distance from the Sun, which leads to the scale of the whole Galaxy. The modern value is 8.5 kpc.

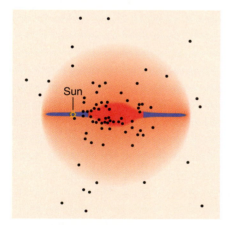

Figure 20.25   *A sample of 80 globular clusters shows their distribution in the Galaxy's halo about the galactic center. Adapted from "Globular Clusters," by Ivan R. King from Scientific American, June 1985, page 80. Copyright © 1985 by Scientific American, Inc. All rights reserved.*

KEY CONCEPTS

Binary stars: Double stars with gravitationally bound components: those of **visual binaries** are detected directly through the telescope; **spectroscopic binaries** are detected by the Doppler shifts of their components; the stars of **eclipsing binaries** eclipse each other; components of **common proper motion** pairs move together; **single-lined spectroscopic binaries** display Doppler-shifted lines of only one component; single observed components of **astrometric binaries** wobble about a common center of mass.

Clusters: Gravitationally bound groups of stars; **open clusters** are loose Population I clusters in the Milky Way; **globular clusters** the rich Population II clusters of the halo.

Color-magnitude diagrams: Plots of color index against magnitude.

Field stars: Stars not in clusters.

Horizontal branch: A branch of roughly constant magnitude extending left from the giant branch in globular cluster HR diagrams.

Light curve: A plot of magnitude against time.

Main-sequence fitting: The means of deriving cluster distances by comparing their main sequences.

Moving-cluster method: A method that allows the measurement of the distance to a cluster from the motions of its stars.

OB associations: Groups of O and B stars in the plane of the Galaxy.

Velocity curve: A plot of radial velocities against time for the components of a spectroscopic binary.

KEY RELATIONSHIP

Mass-luminosity relation:

$$L \propto M^{3.5} \text{ (average) for main-sequence stars.}$$

EXERCISES

Comparisons

1. How do eclipsing binaries and spectroscopic binaries differ?
2. How do stellar associations, globular clusters, and open clusters differ from one another?

Numerical Problems

3. The stars of ξ Boötis orbit one another with a period of 150 years. The semimajor axis of star B about star A is 33 AU. What is the mass of the system?
4. The center of mass of the system in Exercise 3 is found to be 45% of the way from the brighter component to the fainter one. What are the individual masses?
5. A spectroscopic binary has components with circular orbits, inclination of 90° to the plane of the sky, and a period of 20 days. The B component has a maximum velocity of 160 km/s, the A component 80 km/s. What are the semimajor axes of the orbits, and what are the masses of the stars?

Thought and Discussion

6. What are the primary observational data for: **(a)** visual binaries; **(b)** spectroscopic binaries; **(c)** eclipsing binaries?

7. Why is the primary star of a visual binary rarely at the focus of the secondary's *observed* (apparent) orbit?
8. Use Kepler's generalized third law in Section 7.4 to demonstrate that $(M_A + M_B)$ in solar masses = $a^3(\text{AU})/P^2(\text{years})$.
9. The primary component of ξ Boötis is spectral class G8 V, the secondary K4 V. Use the mass-luminosity relation to estimate their masses. How do they compare with those derived in Exercise 4? How might you account for any differences you may find?
10. What is a typical arrangement of components within a sextuple star?
11. Why are the components of spectroscopic binaries closer together than those of visual binaries?
12. Why in general can you find only lower limits to the masses of spectroscopic binary stars?
13. What data are used to derive the radii of the components of an eclipsing binary star?
14. The components of an eclipsing binary have the same temperatures, but A has twice the radius of B. The inclination is 90° and the eccentricity is zero. Sketch the light curve where you plot the brightness of the system (not the magnitude) against time.
15. How can the light of an eclipsing binary vary outside of eclipse?
16. Why are visual binaries not seen to eclipse?
17. What is the basis of the moving-cluster method of distance?
18. Why do some open clusters (like the Praesepe in Cancer) not appear within the Milky Way even though they are Population I?
19. Where do you find the horizontal branch?

Research Problem

20. What were the results of the studies that examined the last eclipses of VV Cephei and ε Aurigae?

Activities

21. Demonstrate why visual binaries are not seen to eclipse. Use two golf balls. Place them six inches apart; you can see that one can cut off the light from the other over a wide range of inclinations. Then place them 50 feet apart, as they might be if they represented a visual double. You can then see that the balls must be exactly aligned for an eclipse to take place.
22. Using binoculars, make drawings of the Pleiades, the Hyades, and Coma Berenices. Estimate the angular diameters of the clusters. From the distances in Table 20.2, what are their physical diameters?
23. Over an academic term, estimate the brightness of Algol or Beta Lyrae and construct a light curve for either of the stars. For references, use α Per ($V = 1.79$) and ρ Per ($V = 3.39$), γ Lyr ($V = 3.24$) and ζ^1 Lyr ($V = 4.36$).

24. Photocopy the HR diagrams of the clusters in Figure 20.18 and derive the distances of M 67 and h and χ Persei. Why might your answers differ from those in Table 20.2?

Scientific Writing

25. You are planning to write a science fiction novel that describes life in a Solar System in which the Sun is binary and Jupiter replaced by a KV dwarf. Write a proposal to a prospective publisher explaining what the system would be like, how it would appear from Earth, and how it would appear from α Centauri.

26. You are a professor of astronomy at your college and are trying to persuade students to take your course. Write a description for an extended course catalogue, concentrating on double stars and clusters. Include their aesthetic qualities as well as their scientific appeal.

21

Unstable Stars

*How stars vary
and return
their mass
to space*

*Cetus, the sea monster,
harbors a monster star in his heart,
the red giant
long-period variable Mira.*

The stars seem immutable, never changing except for their daily and seasonal passages across the sky. Careful observation shows otherwise, revealing **variable stars,** whose magnitudes change with time. In 1572 a new star blazed forth in Cassiopeia. Studied by Tycho, and still called Tycho's Star, it was visible in full daylight. It then faded, and after 16 months disappeared forever. Twenty-four years later, Tycho's friend David Fabricius saw a new third-magnitude star in Cetus, which also vanished. However, Mira, "the amazing one" (Omicron Ceti)—came *back*. To the naked eye, it regularly disappears and reappears with a period of nearly a year. Then in 1784 a more subtle variable star, δ Cephei, which changes by only a magnitude and never disappears, was found. We now know of thousands of these **intrinsic variables,** stars that vary in brightness not because of eclipses but because of processes within the stars themselves.

21.1
Cepheid Variables

Delta Cephei varies between *V* = 3.9 and 5.1 over a 5.37-day period (Figure 21.1) and gave its name to a whole class of **Cepheid variables.** Their light curves are asymmetric, displaying a rapid rise followed by a slow fall; the pattern repeats itself over and over without significant deviation. There are well over 600 of these intriguing stars known in the Galaxy. In addition to δ Cephei, ten others visible to the naked eye are listed in Table 21.1, including one of the most famous stars of the sky, Polaris.

Cepheid periods range from only about a day to as long as 50 days, and there is a broad tendency for the longer-period variables to have larger magnitude ranges, or **amplitudes.** Typically, those with periods of a few days have amplitudes of 0.6 magnitude or so, and those whose periods exceed a month vary by about twice that amount. Cepheids are all F and G supergiants and bright giants. Such luminous stars are rare, and none are close enough for parallax measurement. From their spectral classes, they typically have absolute average magnitudes around –5, ranging from about –2 to –7. When plotted on the HR diagram, they occupy a broad band called the **Cepheid instability strip** (Figure 21.2). As the magnitudes of individual stars vary, so do their spectral classes, typically from F5 at maximum light to about G5 at minimum (see Table 21.1). Spectral variation is caused by a tem-

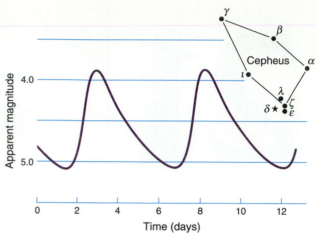

Figure 21.1 *The light curve of δ Cephei repeats its pattern week after week, year after year. The star is located in the inset.*

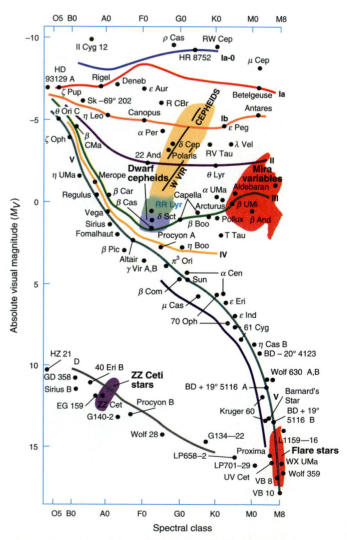

Figure 21.2 *Unstable regions of the HR diagram, outlined in color, range from brilliant Cepheids to huge long-period (Mira) variables to tiny and dim ZZ Ceti white dwarfs.*

TABLE 21.1
Prominent Cepheids

Star	α (2000) h	m	δ °	'	Period (days)	Magnitude (V)	Spectral Class
Polaris	02	24	+89	14	3.97	1.94–2.05	F7 Ib–II
T Vulpeculae	20	51	+28	13	4.44	5.44–6.0	F5 Ib–G0 Ib
FF Aquilae	18	58	+17	21	4.47	5.18–5.6	F5 Ia–F8 Ia
δ Cephei	22	29	+58	23	5.37	3.90–5.09	F5 Ib–G2 Ib
Y Sagittarii	18	21	–18	52	5.40	5.40–6.1	F6 I–G5 I
X Sagittarii	17	47	–29	50	7.01	4.24–4.8	F5 II–G9 II
η Aquilae	19	52	+00	59	7.18	4.08–5.36	F6 Ib–G2 Ib
W Sagittarii	18	04	–29	35	7.59	4.30–5.0	F2 II–G6 II
S Sagittae	19	55	+16	37	8.38	5.28–6.0	F6 Ib–G5 Ib
β Doradus	05	34	–62	30	9.84	4.03–5.07	F6 Ia–G2 Iab
ζ Geminorum	07	04	+20	35	10.15	3.68–4.16	F7 Ib–G3 Ib

perature change (see Figure 19.15) from about 6,300 K to 5,000 K; high temperature occurs at maximum brightness.

The spectra reveal the origin of the variability. During the light cycle, the absorption lines move back and forth because of Doppler shifts (Figure 21.3). The velocities are compatible with **pulsation** (changes in radius), as the visible hemisphere of the star alternately approaches the observer (relative to the star's radial velocity) and then recedes. The velocity curve is nearly a mirror image of the light curve. The star is smallest and begins its outward motion as the radial velocity becomes negative (point A in Figure 21.3), which occurs while the star is in the midst of brightening. Maximum light takes place at maximum expansion speed (point B). As it continues to expand, the star begins to fade and is largest when in the middle of its brightness decline (point C). Minimum light takes place during maximum contraction velocity (point D), and then the star brightens as it approaches its smallest size (again, point A). The change in radius is only about 10% and the star has roughly the same radius at its hottest and coolest. We thus see that the luminosity variation is produced by a combination of changes in both radius and temperature.

Cepheids are fundamental indicators of distance. When the tattered remnants of Ferdinand Magellan's crew returned from their voyage around the world in 1522, they told of two bright patches of light deep in the southern hemisphere that appear to the naked eye like displaced portions of the

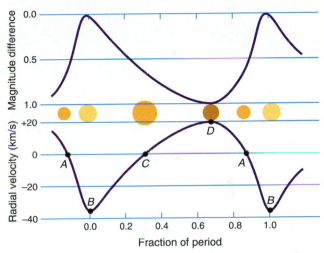

Figure 21.3   *The light curve (top) and radial velocity curve (bottom) of a typical Cepheid are mirror images. Radius changes are shown schematically between the curves. The actual radius variation is only about 10%.*

Milky Way. These **Magellanic Clouds** (Figure 21.4) are small gravitational companions of our own Galaxy. The Large Magellanic Cloud (LMC), some 8° across, is mostly within Dorado (see the star maps in Appendix 1). The Small Magellanic Cloud (SMC), in Tucana, is about half that size.

The Magellanic Clouds are sufficiently close (52 kpc away for the LMC and 58 kpc for the SMC) for us to see that they contain all the different kinds of stars that inhabit our own Galaxy. Because all the stars within each cloud have about the same distance, the LMC and SMC make superb natural labo-

(a)

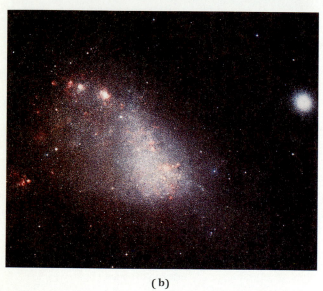

(b)

Figure 21.4   (a) *The Large Magellanic Cloud and* **(b)** *the Small Magellanic Cloud are a pair of irregular galaxies that are close companions of our own Galaxy. The great globular cluster 47 Tucanae appears just to the right of the SMC.*

ratories that allow astronomers to compare relative stellar properties. In 1912 Harvard's Henrietta Leavitt noticed that the longer-period Cepheids in the SMC are also the brighter. When she graphed average apparent magnitude against the logarithm of the period (Figure 21.5), she found the resulting **period-luminosity relation** to be a straight line. The Cepheids in the LMC showed the same phenomenon.

There are a few Cepheids in open clusters in our own Galaxy. From main-sequence fitting (Section 20.3.2), we can find the clusters' distances and the Cepheids' absolute magnitudes. Since we know

the periods of the galactic Cepheids, we can place them on Figure 21.5 and find the absolute magnitudes that correspond to the observed apparent magnitudes; we therefore establish the diagram in terms of absolute magnitude. We can then find the distance of any Cepheid by measuring its period and average apparent magnitude, reading the absolute magnitude from the period-luminosity relation, and applying the magnitude equation. The supergiant Cepheids are so brilliant they can be seen immensely far away, not only in our Galaxy and in the Magellanic Clouds, but in other galaxies as well (Figure 21.6).

Unfortunately, the application of the period-luminosity relation has not been quite so straightforward. There are actually *two* kinds of Cepheids, one for each population type. Cepheids in the Population I galactic disk, now called **classical Cepheids,** are (for a given period) 1.5 magnitudes brighter than the Population II variety, which are known as **W Virginis stars,** after the name of the first one found. Thus there are two parallel period-luminosity relations (Figure 21.7). Before the method can be applied, we must know what kind of star we are observing. Since W Virginis stars are in the Galaxy's halo, they are low in metals, so the varieties can be discriminated by their spectra and compositions.

Cepheids vibrate at natural frequencies or periods that are proportional to the inverse square root

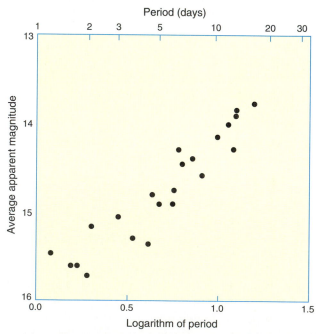

Figure 21.5 *Henrietta Leavitt's original Cepheid period-luminosity diagram shows that the average apparent magnitude changes linearly with the logarithm of the period. The actual periods are shown across the top.*

Figure 21.6   *Although the galaxy M 33 in Triangulum is 720 kpc away, it is still close enough that its Cepheids are easily visible, allowing its distance to be found. Individual Cepheids are indicated by numbers. The stars with lowercase letters are used for magnitude calibration. HS "B" is a luminous blue variable.*

of their average densities ($P \propto 1/\sqrt{\rho}$, where r is density). The larger the star, the lower the density and the longer the period. Larger stars are also more luminous, thereby explaining the period-luminosity relation. Nevertheless, Cepheid pulsation involves only the outer part of the star and a small fraction of the mass. High pressure in deeper layers pushes the outer part of the star outward. As pressure is reduced and the outward push is diminished, gravity drags the outbound layer back, increasing the internal pressure. The cycle thus continuously repeats. If left alone, however, the cycle would die out like an untended child's swing. To keep it going, the cycle must be driven, just as the swing must be pumped to keep it in motion. The principal driver in the star is a layer where helium is becoming ionized some 100,000 km below the stellar surface. When the star compresses, the ionization zone becomes more opaque and absorbs more radiation, causing it to expand and to push the outer layers outward. If the zone is not at the right depth, the star will not pulsate.

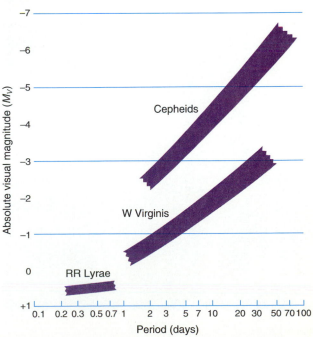

Figure 21.7   *The calibrated period-luminosity relations for classical Cepheids and fainter W Virginis stars are parallel. The curve for Population II RR Lyrae stars extends below that of the W Vir stars.*

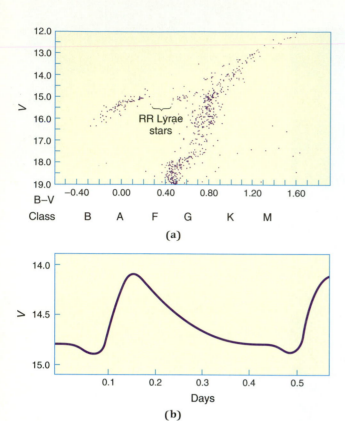

Figure 21.8   (a) *The gap in the horizontal branch of the globular cluster M 5 contains about 100 unplotted RR Lyrae stars.* (b) *The light curve of a typical example shows a period of less than a day and a variation of about half a magnitude. The spectral classes are given for reference. Since the stars are deficient in metals, they are only approximate.*

21.2
Other Regular Pulsators

There are other instability zones on the HR diagram in which stars vary with some precision. Any star found between about spectral classes F5 and A5 within the horizontal branches of globular clusters (Figure 21.8) will vary like a Cepheid. The first such star actually discovered was a field star of the galactic halo, whose HR diagram has its own horizontal branch, called RR Lyrae (located in Figure 21.2), which has given its name to the general class of **RR Lyrae stars** that belong to Population II. (Variable stars are commonly named with single and double Roman letters. The first discovered in a constellation is called R, the second S, and so on down to Z. The series continues with RR . . . RZ, SS . . . SZ, and on to ZZ, then to AA . . . AZ, BA . . . BZ, to QZ, skipping J. The 334 letter combinations are followed by V335 and so forth. RR Lyrae is then the tenth variable to be found in Lyra.) The number of known RR Lyrae stars changes dramatically from one cluster to the next, from zero to over 200; even if seen, they ordinarily are not plotted on the HR diagram because of their variability. The region on the diagram where they would appear is therefore commonly known as the *RR Lyrae gap*.

The RR Lyrae stars are smaller than Cepheids, so their periods are shorter, a day or less, and their amplitudes lower, roughly half a magnitude (Figure 21.8b). They fit nicely onto the period-luminosity

Figure 21.9   (a) *At minimum, Mira is at tenth magnitude and can be seen only by telescope.*
(b) *At maximum, it can nearly dominate its constellation.*

BACKGROUND 21.1 Progress in Science: The Battle for Distance

News media and popular entertainment often make scientists appear to be confused bumblers. Answers to a question might vary from one day to the next, or respected scientists may defend opposite sides of an issue.

The problem is that science is not the neat subject it is supposed to be; it is, in fact, a messy business. Information on a new subject is revealed in fragments that often have little to do with one another, upon which we must, for any progress, immediately build a theory that changes continuously under the barrage of new data. Imagine trying to construct an elephant from a tusk and a tail. The beast you envision will be very different from the one made by someone working from an ear and a foot. Only when all, or nearly all, of the parts are at hand can you create the correct theory, the elephant itself. That takes time—maybe months, maybe years, even centuries.

The process is beautifully illustrated by the Cepheids and our realization of the distance scale of the Universe. In the 1930s and 1940s, astronomers used the period-luminosity relation to find the distance to the Andromeda Galaxy (see Figure 19.27), as well as to others. However, the relation was found long before Baade revealed the existence of two different stellar populations (see Background Essay 19.2). Until then we had no way of knowing that there were two different kinds of Cepheids, and that the two sets of stars were mixed together. As a result, we thought the Cepheids in the other galaxies were fainter and the galaxies closer than they actually are.

The truth was presaged in part by the RR Lyrae stars. When the 200-inch Palomar telescope went into service in 1947, astronomers thought they would be able to see these variables in M 31. However, they were not there. Why should the RR Lyraes in this nearby galaxy be fainter than those in ours? From this and other inconsistencies, around 1950 astronomers discovered the existence of the two kinds of Cepheids. Because the Population I Cepheids—those used to determine the distance to M 31—are intrinsically brighter than those of Population II, the Population I period-luminosity relation had to be adjusted. The measured distances to M 31 and M 33—and of the whole Universe—doubled almost overnight. We now have all the information we need, and the distances of the nearby galaxies relatively are certain, but the battle took almost 40 years.

---

diagram in Figure 21.7 at the end of the curve for the Population II W Virginis stars. Since they are on the horizontal branch, their absolute magnitudes are all about 0.5, making them excellent distance indicators. If we spot an RR Lyrae star by its light curve, we know M_V and can apply the magnitude equation to derive distance.

Below the Cepheid instability strip (see Figure 21.2), we encounter subgiant and bright main-sequence variables of shorter periods and lesser amplitudes called **dwarf Cepheids.** These vary by only a few tenths of a magnitude over an interval of a few hours. Then, between about 10,500 K and 13,000 K among the white dwarfs, we find the **ZZ Ceti stars.** All are hydrogen-rich DA stars (see Section 19.5.1) driven by hydrogen ionization. They do not pulsate simply, like their larger cousins, but are *nonradial oscillators:* some parts of the stars expand outward at the same time that other parts contract inward. Consistent with their small sizes, they vary with periods of only minutes.

21.3
Long-Period (Mira) Variables

Some effort is required to see the variation in a Cepheid. The changes in **long-period** or **Mira variables,** however, are blatant (Figure 21.9). Their periods range between about 100 and 700 days and many are seven magnitudes or more fainter at minimum than at maximum. Miras are not as regular as the Cepheids. The periods are somewhat variable, and so are the amplitudes, the light curves never quite repeating themselves (Figure 21.10). Mira, the prototype of the class, usually brightens to third magnitude, occasionally to second, but sometimes only to fourth. Several Miras that are visible to the naked eye at maximum are listed in Table 21.2.

All Miras are red giants, though they can be of any spectral class, M, R, N, or S (see Section 19.5.1). They can also be of either stellar population, although those of longer period are tied to Popula-

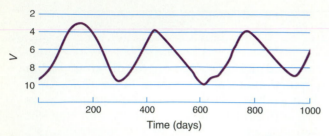

Figure 21.10 *Though periodic, the light curve of Mira is not as regular as the light curves of Cepheids.*

tion I. Mira is the archetype of the M stars. Among the class N carbon stars, R Leporis is a fine example, and χ Cygni well represents the S giants. Their realm on the HR diagram is identified in Figure 21.2. Placement within the zone is a necessary condition for variability, but it is not sufficient. Only a fraction of the stars found there are actually Miras because they must first be in a particular state of evolution (to be defined in Chapter 24). However, the incidence of Miras increases among the cooler giants. Temperatures may be as low as 1,900 K and spectra as cool as M9. As a result, the spectra are extremely complex and have large numbers of molecular bands (Figure 21.11).

Like Cepheids, Miras change their tempera-

tures and spectral classes as they vary. Mira itself, typical of many, goes from M5 III and around 2,800 K to M9 III near 2,000 K. The more luminous stars also take longer to vary. However, the relation is not sufficiently tight to make Miras reliable distance indicators. These cool stars radiate most of their energy in the infrared, where the amplitude is only one or two magnitudes (as it is bolometrically). In the visual, we see only the tail of a very crude blackbody curve. Small changes in temperature shift the curve and produce the large visual amplitudes. The effect is enhanced by large changes in TiO abundance and the strength of its absorptions. From the variations in temperature and bolometric luminosity, we find that the Miras are pulsators, changing their radii by as much as 50% over their long periods.

The pulsation is again driven by deep ionization zones and is powerful enough to produce shock waves. As the waves propagate outward, they cause the hydrogen of the atmosphere to radiate emission lines that develop strongly as the star expands and brightens. As a result, almost all Miras have "e" appended to their spectral classes.

The larger Miras have very low surface gravities, so the shock waves can drive matter from the

TABLE 21.2
Bright Mira Variables

Name	α h	m	δ (2000) °	'	Period (days)	Magnitude (V)	Spectral Class
Mira = *o* Cet	02	19	−03	01	332	2.0–10.0	M5 IIIe–M9 IIIe
R Horologii	02	54	−49	55	404	4.7–14.3	M7 IIIe
R Leporis	04	59	−14	49	432	5.5–11.7	C6 IIe (class N)
U Orionis	05	55	+20	11	372	4.8–12.6	M6.5 IIIe
R Carinae	09	32	−62	45	309	3.9–10.5	M4 IIIe–M8 IIIe
R Leonis	09	47	+11	28	312	4.4–11.3	M8 IIIe
S Carinae	10	09	−61	31	149	4.5–9.9	K5 IIIe–M6 IIIe
R Hydrae	13	29	−23	15	390	4.5–9.5	M7 IIIe
χ Cygni	19	50	+32	54	407	3.3–14.2	S6 IIIe–S10 IIIe
R Aquarii	23	43	−15	20	389	5.8–12.4	M5 IIIe–M8.5 IIIe

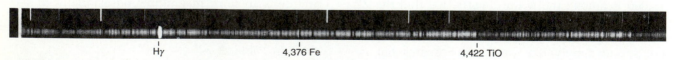

Hγ   4,376 Fe   4,422 TiO

Figure 21.11 *Mira's spectrum displays a huge number of absorption lines, most of which belong to molecular bands. Hydrogen appears not in absorption but in emission.*

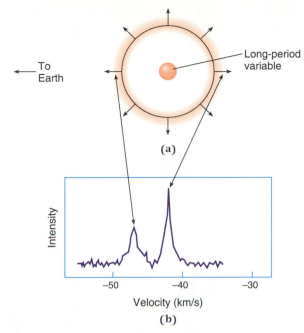

Figure 21.12 (a) *An OH/IR star contains a Mira variable surrounded by an expanding shell that produces* (b) *an OH maser emission line split in two by the Doppler effect.*

star. In the deep cold of space above the stellar surface, more gas condenses into molecules and further into solid particles—tiny grains of **dust**—no more than a fraction of a micron (10^{-3} mm) across. The dust is pushed outward by radiation and drags the gas along with it. The result is a dirty **stellar wind** that can make the stars lose mass at a rate of up to 10^{-5} solar masses per year, 10^8 times the mass-loss rate of the Sun. Some Miras are so encased in dusty clouds that radiation cannot escape directly, and the stars disappear. The starlight heats the dust to a few hundred K, and all we see is the blackbody radiation of the cloud glowing in the infrared.

There are two classes of these encased stars. The envelopes of the oxygen-rich M stars radiate strong emission lines of OH (hydroxyl) at 1,612 mHz in the radio spectrum: these gaseous assemblies are therefore known as **OH/IR stars.** The OH lines from an OH/IR star are actually produced by a natural *maser,* the microwave version of the familiar laser, in which the upper energy levels are excited by infrared emission from the embedded star. These emissions come from an expanding shell of gas (Figure 21.12a) typically 1,000 AU across. The emission from the receding back side is Doppler-shifted to longer wavelengths, the emission from the approaching front side to shorter, and

the lines are split in two (Figure 21.12b). The separation yields typical expansion speeds of about 20 km/s. The OH/IR stars also produce a strong, broad, infrared line that comes from solid silicates, allowing at least an idea of the composition of the dust. The silicates are probably tied up with various metals, as they are in the rocks of the Earth (see Section 10.2.1).

The other class of Miras involves carbon stars. Their dusty expelled shells (Figure 21.13) are rich in carbon-based molecules, some very complex. These shells emit an infrared line associated with silicon carbide, SiC_2. Other observations indicate amorphous carbon in a form similar to graphite.

21.4
Erratic Variables

Many variables are not as regular as Cepheids, RR Lyrae stars, and Miras. Most large giants and supergiants vary to some degree. Huge Betelgeuse, for example, varies over intervals of days, weeks, and months on top of a six-year variation during which it can change its brightness by over half a magnitude. These more-erratic stars are classified into two broad groups, the **semiregulars** and the **irregulars.** Not surprisingly, the semiregulars are somewhat predictable and the irregulars are entirely unpredictable. Other examples of the genres are the supergiant α Herculis (semiregular), Antares (irregular), and the large component of the eclipsing binary VV Cepehi (irregular).

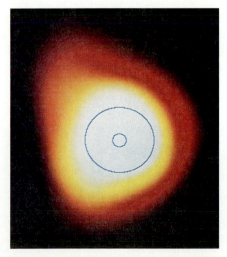

Figure 21.13 *The shell of the carbon-rich Mira IRC +10 216 is imaged in radio radiation produced by carbon monoxide. The circles show the size of the invisible star and the inner edge of the dusty shell.*

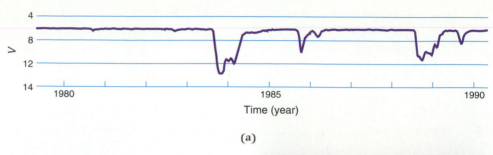

(a)

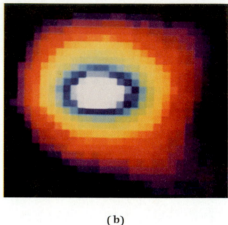

Figure 21.14 *The long-term light curve* (**a**) *of R Coronae Borealis shows that it erratically and suddenly dims. The* Infrared Astronomical Satellite, IRAS, *observed emission from a huge, dusty shell around the star* (**b**) *nearly a third of a degree and 10 pc across, the result of millennia of sporadic ejections of matter.*

(b)

Among the stranger variables are about 20 known **R Coronae Borealis stars,** named for the F giant prototype inside the curve of the Northern Crown. Normally shining at a steady fifth magnitude (Figure 21.14a), it will suddenly disappear from naked-eye view, dropping to perhaps twelfth or even fourteenth. Over an interval of months, the star will slowly and irregularly get brighter as it works its way back to 5th. R CrB is rich in carbon and helium and has little or no hydrogen, just the opposite of what we normally see. The dramatic fluctuations are caused by the condensation of tiny carbon dust grains—effectively, *soot*—that suddenly accumulate and hide the star. Slowly the dust dissi-

pates, and the star becomes visible again. Over time it becomes surrounded by a huge shell of dirty gas (see Figure 21.14b) in another example of extreme mass loss.

The **luminous blue variables** (LBVs) are just as erratic. They are exceedingly rare but so luminous (class 0 or 0–Ia supergiants) that they are easily visible in other galaxies (see Figure 21.6). They exhibit great fluctuations over long periods. One in the galaxy M 33 brightens every 40 years or so; another recently faded by 3.5 magnitudes as its spectrum went from F all the way to M.

A few LBVs are visible in our own Galaxy; they are some of the oddest characters of the sky. In 1945, ρ Cassiopeiae dropped from $V = 4.5$ to sixth magnitude and changed its spectral class from F8 to M3; over the next year it returned to normal. In 1848 Eta Carinae (Figure 21.15a) was the second brightest star in the sky, rivaling Sirius; by 1880 it had faded to eighth magnitude, and it resides at sixth today. Once class F, its optical spectrum now presents only emission lines. The object known as Eta Carinae actually contains brilliant blue supergiants, and the other LBVs probably contain one or more as well. The great changes they exhibit are linked to a thick stellar wind that mimics a cooler absorption spectrum. The expanding gas also con-

(a)   (b)

Figure 21.15 *The luminous blue variable Eta Carinae* (**a**) *is seen in a close-up* (**b**) *to be enmeshed in a cloud of outflowing gas and dust.*

(a)

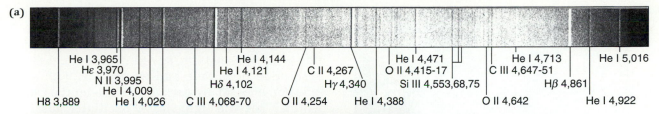

He I 3,965
Hε 3,970
N II 3,995
He I 4,009
H8 3,889   He I 4,026

He I 4,144
He I 4,121
Hδ 4,102
C III 4,068-70

C II 4,267
Hγ 4,340
O II 4,254

He I 4,471
O II 4,415-17
Si III 4,553,68,75
He I 4,388

He I 4,713
C III 4,647-51
Hβ 4,861
O II 4,642

He I 5,016
He I 4,922

(b)

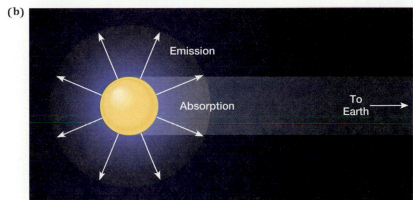

Emission

Absorption

To Earth

Figure 21.16   (a) *The spectrum of P Cygni displays emission lines from mass loss that are flanked to the blue side by absorptions produced by matter flowing toward the observer (**b**).*

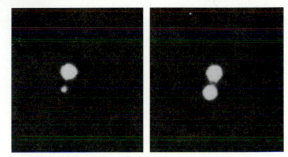

Figure 21.17 *The fainter M4 component of Krüger 60 (DO Cephei) suddenly flares to rival its M3 companion.*

denses into dust that dims and buries the star or stars within (Figure 21.15b).

Mass loss from one of the LBVs, fifth magnitude P Cygni (which reached third magnitude in 1600), is characterized by peculiar spectrum lines. The star's prominent emission lines are flanked to the blue-ward side by strong absorptions (Figure 21.16a). The emissions are produced by the cloud of gas pouring out of the star (Figure 21.16b). The gas coming directly at the observer, however, is seen against the star and produces absorptions that are Doppler-shifted according to the velocity of the wind. These **P Cygni lines** are observed in a wide variety of stars and are extremely useful. The highest observed Doppler shift allows us to measure the wind speed, and from the wind speed and the strengths of the emissions we can find the mass-loss rate, which for P Cygni approaches 10^{-4} M$_\odot$/yr.

From the heights of the HR diagram, plunge to the depths for a look at the lowly but no less fascinating red dwarfs. The fainter M4 component of the binary Krüger 60 (Figure 21.17) will suddenly brighten, or flare, by a magnitude or so, then fade in only a few minutes. The explosive events on these **flare stars** are unpredictable. The flares are visible from the radio spectrum into the X-ray, where they are extremely powerful. They may be similar to the flares seen in the chromosphere and corona of the Sun. The major difference is that the stellar flares involve the release of magnetic energy not just over an active region but over the whole star.

21.5
Eruptive Variables and Interacting Binaries

The flare stars introduce a world of stellar violence. In 1975, a "new star" or **nova** ("new" in Latin) blossomed in northern Cygnus (Figure 21.18). For a few days it rivaled Deneb, then, after a few weeks, slipped from sight. Several novae are discovered each year, most below naked-eye vision, though every decade or two one will reach first or second magnitude. Table 21.3 lists the most prominent ones seen in this century. Nova names use the genetive of the nova's resident constellation followed by the year of the event. If the progenitor star can be identified after the event, it is assigned a variable star designation. Nova Herculis 1934, for example, is related to DQ Herculis.

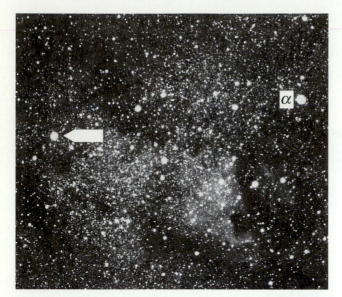

Figure 21.18 *Nova Cygni 1975 reached nearly first magnitude and for a brief time rivalled Deneb (α Cygni).*

TABLE 21.3
Famous Twentieth Century Novae

Nova	Variable Star Name	V_{max}	$M_{V(max)}$[a]
Persei 1901	GK Per	+0.2	−8.5
Aquilae 1918	V 603 Aql	−1.1	−9.2
Pictoris 1925	RR Pic	+1.2	−7.4
Herculis 1934	DQ Her	+1.4	−5.6
Puppis 1942	CP Pup	+0.4	−9.1
Cygni 1975	V 1500 Cyg	+1.8	−9.5

[a]Corrected for dimming by interstellar dust (see Chapter 22).

Novae are divided into two basic classes, fast and slow, that actually represent two ends of a continuum of behavior. A fast nova will jump by a dozen magnitudes or more over a period of only one or two days and will reach an absolute visual magnitude of −8 to −10. The light curve (Figure 21.19) then begins to decay. Fast novae drop through three magnitudes in about a month. Slow novae do not become as bright, reaching absolute visual magnitude −6 or so, but they decay much more slowly, taking maybe six months to decline by three magnitudes. In either case, return to the normal pre-outburst state may take years or even decades.

Nova spectra (Figure 21.20) reveal an expanding cloud of debris around the fading star, the result of an explosive blast. At maximum the stuff is so

thick that all we see are absorption lines. A few days later, the gas thins and starts to produce P Cygni lines, from which we infer expansion velocities of up to 3,000 km/s. After a few weeks, the absorption lines fade and we are left with only emission lines. Many days or weeks after the event, we may see the light curve oscillate, possibly reflecting fluctuations in the rate at which matter had been expelled. Or, the star may fade and then recover as dust grains develop from the gas in the expanding cloud in a manner reminiscent of the R Coronae Borealis stars.

After years or even decades, we see a gaseous cloud expanding around the star (Figure 21.21), a **nova remnant.** We can measure the rate of expansion in seconds of arc per year—that is, we can find the proper motion of the expanding debris relative to the star. The low-density gas of the nova remnant produces emission lines. From their Doppler shifts we learn the rate of expansion in km/s, which we assume to equal the tangential velocity of the expanding cloud (again relative to that of the star). With tangential velocity and proper motion known, the distance is $d = v/4.74\mu$ (see Section 19.2). Nonuniform expansion, however, will produce significant error.

Many years ago, DQ Herculis was discovered to be an eclipsing binary. We have since found that *all* novae are binary; it is what makes them explode. The prenova pair is a combination of a cool main-sequence star and a white dwarf. Surrounding the components of any binary system is a figure-eight-

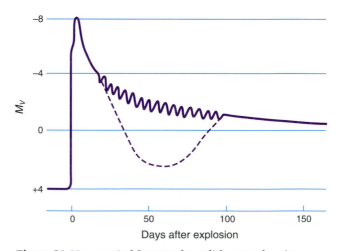

Figure 21.19 *A typical fast nova has a light curve that rises quickly and then decays by some three magnitudes in under a month, taking years to return to normal. As it dims, it may go through a series of oscillations, or the magnitude may drop considerably for a time because of the formation of a dust shell.*

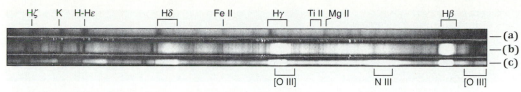

Hζ K H-Hε   Hδ   Fe II   Hγ   Ti II Mg II   Hβ

—(a)
—(b)
—(c)

[O III]   N III   [O III]

Figure 21.20   **(a)** *Spectra of Nova Aquilae 1918 (V603 Aql) at first showed strong absorption lines,* **(b)** *three days later a mixture of violet-shifted absorption and emission lines,* **(c)** *followed after three weeks by emission and fading absorption features; all were the result of the expanding debris of the explosion.*

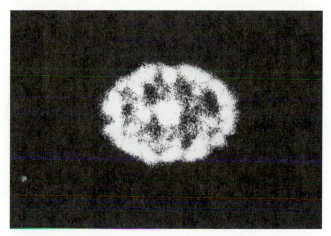

Figure 21.21   *A cloud of debris is seen expanding around DQ Herculis (Nova Her 1934).*

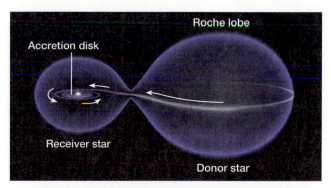

Roche lobe

Accretion disk

Receiver star

Donor star

Figure 21.22   *An equilibrium surface consisting of two Roche lobes surrounds a binary star. If the stars are very close, one (or even both) can fill its lobe and overflow toward the other, frequently into an accretion disk, at the point where the two lobes connect.*

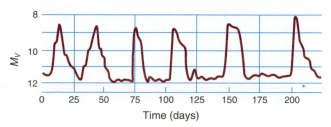

Figure 21.23   *The light curve of SS Cygni shows erratic dwarf nova-eruptions every few weeks.*

shaped surface (Figure 21.22) upon which the inward acceleration of gravity plus the outward acceleration caused by the revolving stars is zero. The part of the surface surrounding each star is called its **Roche lobe.** In a binary destined to become a nova, the components are so close that the main-sequence star becomes tidally distorted and fills its Roche lobe. Matter can then flow from the larger star through the point where the two lobes touch toward the smaller star. Instead of landing directly on the white dwarf, however, the matter first flows into a circulating **accretion disk** from which it is transferred onto the star below. Variations of mass inflow and outflow can cause the light of the system to flicker.

In a white dwarf, nuclear reactions are effectively shut down. Ever so slowly, however, the binary interaction causes fresh hydrogen to build up on the white dwarf's surface. The hydrogen becomes enormously compressed under the huge gravity of the tiny star and is heated to the point of detonation. What we see as the nova is a runaway thermonuclear blast that blows away the top layer of the star. The expanding debris will later be seen as the nova remnant. The interior of the white dwarf is so dense that it is isolated from the surface action and remains unaltered. Once the surface has blown away, the star begins to accumulate new hydrogen all over again, and perhaps a hundred thousand years later the event is repeated.

A nova is an example of a general class of interacting binaries called **cataclysmic variables. Dwarf novae,** like SS Cygni (Figure 21.23), display abrupt and unpredictable brightenings of a few magnitudes every few weeks or months. They seem to be powered not by thermonuclear explosions but by the accretion disks, which suddenly release mass and energy. They may be the first erratic stages of the buildup of a real nova. **Recurrent novae** fall between dwarf and ordinary novae. T Coronae Borealis dramatically brightened in 1866 and 1946 and reached second apparent magnitude. RS Ophiuchi pops off every few decades to reach

Figure 21.24 *The spectrum of the symbiotic star R Aquarii (sandwiched between spectrograph comparison lines) displays emission lines from ionized atoms set against the background of a cool class M continuum.*

fourth. In these objects the donor star is a giant and the receiver an ordinary dwarf.

Symbiotic stars are generally quieter. They are defined by their strange spectra, which combine bright emission lines of highly ionized elements and a cool (usually class M) stellar spectrum (Figure 21.24). A symbiotic is a binary in which a white dwarf is coupled with a giant that fills, or nearly fills, its Roche lobe. As matter flows into the accretion disk, it can compress and heat. Temperatures may become so high that ultraviolet radiation ionizes the gas flowing out of the giant, producing the emission lines. The star constantly flickers as a result of irregularities and instabilities in the gas flows. We will also occasionally see a novalike outburst when fresh hydrogen that has fallen on the white dwarf erupts. CH Cygni, the only symbiotic visible to the naked eye, has been in a state of outburst for years.

21.6
Supernovae

Though Tycho's Star of 1572 was long called a nova, we now know that it is not. Shining within the Milky Way in Cassiopeia, it reached a spectacular apparent magnitude of –4 and rivaled Venus. Even a hundred days after the event, it was as bright as Vega, and it did not fade from sight until 1574. Only 30 years later, in 1604, another spectacular nova graced Ophiuchus. This one, studied by Kepler, outshone Jupiter and reached apparent magnitude –2.5, finally disappearing from view after 14 months.

Studies of galaxies eventually provided the key to understanding these brilliant events. In 1885, a seventh-magnitude nova, S Andromedae, was found in the inner region of the Andromeda Galaxy, M 31. At the time, no one knew that these fuzzy balls and disks of light were actually external stellar systems at great distances. Only when Edwin Hubble found Cepheids in M 31 in 1924 did we realize how far away it really is (the modern value is 690 kpc). The discovery demonstrated that S Andromedae, and similar eruptions that had been seen in other sys-

tems (Figure 21.25), were anything but ordinary. S Andromedae had shone with amazing brilliance, at an absolute magnitude of –17, over 500 times brighter than an ordinary nova and only 10 or so times fainter than the whole galaxy itself! These **supernovae** were clearly among the most violent events in the Universe. From their brilliance and duration there is no doubt that Tycho's and Kepler's stars were supernovae.

Supernovae are rare. In 1054 Chinese astronomers recorded a "guest star" that appeared in Taurus and attained the brightness of Tycho's Star. The Chinese also handed down records of similar explosions in 1006 and 1181. Over the course of this millennium, only five have been seen, for an average of about one every two centuries. None has been seen in our Galaxy since the invention of the telescope, and as a result, we have been forced to study them in other, usually distant, galaxies, where the view is inadequate. With great fascination, then, we watched one erupt in the Large Magellanic Cloud (see Figure 21.4) in 1987. Though 52,000 pc away, it allowed us to examine a supernova in detail, which we will do in Chapter 25.

The modern study of supernovae began in the 1930s when Fritz Zwicky of the Mount Wilson

Figure 21.25 *A supernova gleamed within the galaxy M 63 in Canes Venatici in June 1971. From Earth it appeared at 11th magnitude, but from the distance of the galaxy and the magnitude equation we find that it reached an absolute magnitude of –19.*

Observatory systematically collected data on these events in external galaxies. In the 1940s, Rudolph Minkowski (also of Mount Wilson) classified them into different groups. Type I supernovae have no hydrogen lines in their spectra and were later divided into Types Ia and Ib. Type Ia events uniformly reach an absolute visual magnitude of about –19. At 10 pc, such a supernova would appear in the sky brighter than 500 full moons! After outburst, Type Ia supernovae decay rather quickly (Figure 21.26). Doppler shifts in their spectra show gas exploding outward at a velocity of 10,000 km/s. Type Ib events display strong lines of helium (absent in Type Ia) and are about 1.5 magnitudes fainter at their peaks. Type II supernovae do have hydrogen lines. Their peak luminosities average about a magnitude fainter than those of Type Ia but display a wide range from one to another. Type IIs decay at a slower rate, frequently have plateaus on their light curves in which the magnitude is almost constant for a time, and have expansion velocities about half those of their Type Ia counterparts. Type Ia supernovae are associated with both Populations of a Galaxy (and with galaxies that are exclusively Population II). Types Ib and II are seen exclusively in galactic disks and are allied with Population I.

Supernovae are too bright to be just grand versions of ordinary novae. They must arise from the destruction, or near-destruction, of entire stars. More than one kind of phenomenon has to be involved to produce the different types of supernovae seen among the different population groups. The explosions of Types Ia and II are believed to be driven by the internal collapse of supergiants, and they consequently brighten by eight magnitudes or so. We strongly suspect white dwarfs to be the culprits in the Type Ia events. If so, they brighten by an astonishing 30 magnitudes.

Such explosions ought eventually to be marked by clouds of expanding debris. Just off the eastern horn of Taurus, near ζ Tauri, we find a fuzzy filamentary gas cloud called the Crab Nebula, or M1 (Figure 21.27). It lies at the location of the Chinese "Guest Star" of 1054, and as early as 1921 was suspected of being produced by the event. Over the years, we can watch this **supernova remnant** expand. From their proper motions, the filaments should all have been together about the year 1140, close enough to 1054, given possible accelerations and the inevitable observational errors. From its distance of about 2,000 pc (derived from the radial velocities and proper motions of the filaments), the peak absolute magnitude of the Crab progenitor

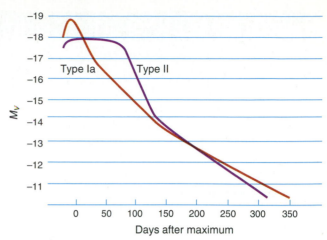

Figure 21.26   *Type II supernovae are fainter than Type I and decay slower.*

Figure 21.27   *The Crab Nebula, M 1, is the remains of the Chinese "Guest Star" of 1054.*

must have been around –16. The hydrogen in the Crab Nebula suggests that the explosion was a Type II (Population I) event. Tycho's and Kepler's stars were probably Type Ia.

Supernovae are destructive blasts that can be seen across much of the Universe and that signal the deaths of stars. What sets of circumstances can lead to these catastrophes? Why do some stars die violently while others quietly produce the modest white dwarfs? In Part V we look at the whole picture of stellar life and evolution, into which the unstable stars (summarized in Table 21.4) and the rest of the stellar citizenry can be placed.

TABLE 21.4
The Characteristics of Variable Stars

Type	Alternate Name	Population	Spectral Class	Amplitude (magnitudes)	Period
Instability Strip and Related Pulsators					
Classical Cepheids	—	I	F, G; I, II	0.5–1.5	1–50 days
W Virginis stars	Population II Cepheids	II	F, G; I, II	0.5–1.5	1–50 days
Dwarf Cepheids	—	I	A; V, IV	Few tenths	Few hours
RR Lyrae stars	Cluster variables	II	F, A; III	Few tenths	<1 day
ZZ Ceti stars	—	I, II	hot white dwarf	Few tenths	Few minutes[a]
Cool Giant and Supergiant Variables					
Long Period Variables	Miras	I, II	M, R, N, S; III	5–10	100–600 days
Semiregular		I, II	K, M; I, II, III	1	10–100 days
Irregular		I, II	K, M; I, II, III	1	—
Other					
R CrB stars[b]	—	I	F; III	5–10	—
Luminous Blue Variables	HS variables	I	O, B; 0, Ia	3–4	Years, decades
Flare stars	—	I, II	M V	1	Erratic
Eruptive					
Novae		I, II	K, M; V + white dwarf	12	10^5 years?
Dwarf novae	SS Cyg, U Gem	I, II	dwarf + white dwarf	Few	Weeks, months
Recurrent novae		I, II	dwarf + giant	Few	Decades
Type Ia Supernovae		I, II	white dwarf + ?[c]	30	—
Types Ib and II Supernovae		I	supergiant	8	—

[a]Nonradial oscillators.
[b]Magnitude suddenly drops.
[c]See Chapter 25.

KEY CONCEPTS

Accretion disk: A disk of gas revolving around a star into which matter flows from a companion.

Amplitude: The range of magnitudes exhibited by a variable star.

Cataclysmic variables: Binary systems that vary as a result of mass exchange.

Cepheid variables: F or G giant or supergiant variables with regular periods; **classical Cepheids** are Population I, **W Virginis stars** are the fainter Population II variety; both fall within the narrow **Cepheid instability strip** in the HR diagram.

Dust: Small, solid grains with a variety of chemical compositions.

Dwarf Cepheids: Short-period Cepheids just above the main sequence.

Dwarf novae: Binaries with irregular, low amplitude outbursts caused by instabilities in the accretion disks.

Flare stars: Main-sequence M stars that display sudden eruptions.

Intrinsic variables: Stars whose brightness varies because of internal processes (as opposed to those that vary by eclipses).

Long-period (Mira) variables: Giant pulsators with periods over about 100 days and large visual-magnitude ranges.

Luminous blue variables (LBVs): Extremely luminous supergiants that sporadically bury themselves in dusty ejecta.

Magellanic Clouds: Two small companion galaxies to our Galaxy.

Nova: A stellar brightening caused by the explosion on the surface of a white dwarf of matter flowing from a main-sequence star.

Nova remnant: The cloud of debris expanding around a nova.

OH/IR stars: Class M Miras buried in dusty shells created by strong stellar winds.

P Cygni lines: Stellar emission lines flanked by blue-shifted absorption lines; caused by mass flowing from a star.

Period-luminosity relation: The correlation between the periods of Cepheids and their absolute magnitudes, which allows derivation of distances.

Pulsation: Change in stellar radius.

R Coronae Borealis stars: Stars that suddenly dim as a result of dust formation.

Recurrent novae: Binaries with repeating novalike outbursts.

Roche lobe: An equilibrium surface around orbiting stars upon which the acceleration is zero.

RR Lyrae stars: Regular short-period pulsators of the horizontal branch.

Semiregular and **irregular variables:** Giant or super-giant pulsators with erratic periods.

Stellar wind: A flow of mass from the surface of a star.

Supernova remnant: The expanding cloud of debris caused by the explosion of a supernova.

Supernovae: Outbursts that involve the destruction or near-destruction of a star; produced by supergiants of Population I or possibly by white dwarfs of either population.

Symbiotic stars: Interacting binaries with both hot and cool spectral characteristics.

Variable stars: Those whose magnitudes change with time.

ZZ Ceti stars: DA white dwarf pulsators, which are non-radial oscillators.

EXERCISES

Comparisons

1. Make a table comparing the characteristics of classical Cepheids and W Virginis stars.
2. How do dwarf and recurrent novae differ?
3. How do RR Lyrae stars differ from classical Cepheids?
4. Contrast the two kinds of Miras that are surrounded by dusty shells.
5. What kinds of variables would you expect to be contained by open clusters and by globular clusters? How do the variable-star contents of open clusters differ from one another? How do the variable-star contents of the globular clusters differ from one another?
6. Compare the binary components of novae, recurrent novae, and symbiotic stars.

Numerical Problems

7. From the period-luminosity relation for classical Cepheids, calculate the distances of the stars in Table 21.1. What are their individual absolute magnitudes?
8. What would you expect the apparent magnitudes of the brightest Cepheids of the Small Magellanic Cloud and M 31 to be?
9. The nova remnant of Nova Persei 1901 is growing at a rate of 0.5" per year. From Doppler shifts of its emission lines it is expanding at 1,100 km/s. How far away is it?
10. How many times farther away can you see a typical supernova than an ordinary nova?

11. The Crab Nebula now has an angular diameter of about 4 minutes of arc and is now expanding at about 1,500 km/s. Estimate its difference. Why might your distance be different from that in the text?

Thought and Discussion

12. What kinds of variables are found in each spectral class?
13. At what points on a Cepheid's velocity curve is the star the brightest and the faintest?
14. What are the Large and Small Magellanic Clouds? How far away are they? How do we know?
15. What is the cause of Cepheid pulsation?
16. If there were no Cepheids in open clusters, how could you still find the absolute magnitudes of the period-luminosity relation?
17. What do the emission lines in the spectra of Mira variables indicate?
18. Why do the optical and infrared light curves of Miras not have the same amplitudes?
19. List the kinds of variable stars you would expect to be able to *observe* in M 31.
20. Of the variable stars observed in our Galaxy, what kinds would you expect to have the smallest average proper motions?
21. What are the chief characteristics of luminous blue variables?
22. What is the probable nature of a flare star's flare?

Research Problem

23. There are many kinds of variables other than those described in this chapter. What are their natures? Where do they fall on the HR diagram?

Activities

24. Construct your own light curve of δ Cephei by watching it with the naked eye or binoculars over several light cycles. Use ζ ($V = 3.35$), ε ($V = 4.19$), and λ ($V = 5.04$) Cephei as comparisons (see Figure 21.1).
25. Look through a set of charts from the American Association of Variable Star Observers (available from the AAVSO, 23 Birch Street, Cambridge, Massachusetts 02138). Pick a star, and, using a telescope, follow it through the school term to construct a partial light curve.

Scientific Writing

26. Add a summary section to this chapter on the distance scale of the Universe. Begin with the nearby stars and tell how astronomers calibrate one distance method with another until they are able to calculate the distances to the nearby galaxies.
27. You have just heard that the National Science Foundation, chief provider of astronomical funding in the United States, is about to cut drastically its funding for variable-star research. Write a letter to the director explaining the value of such research to astronomy.

22 *The Interstellar Medium*
23 *Star Formation*
24 *The Life and Death of Stars*
25 *Catastrophic Evolution*
26 *The Galaxy*

Birth and Death in the Galaxy

PART

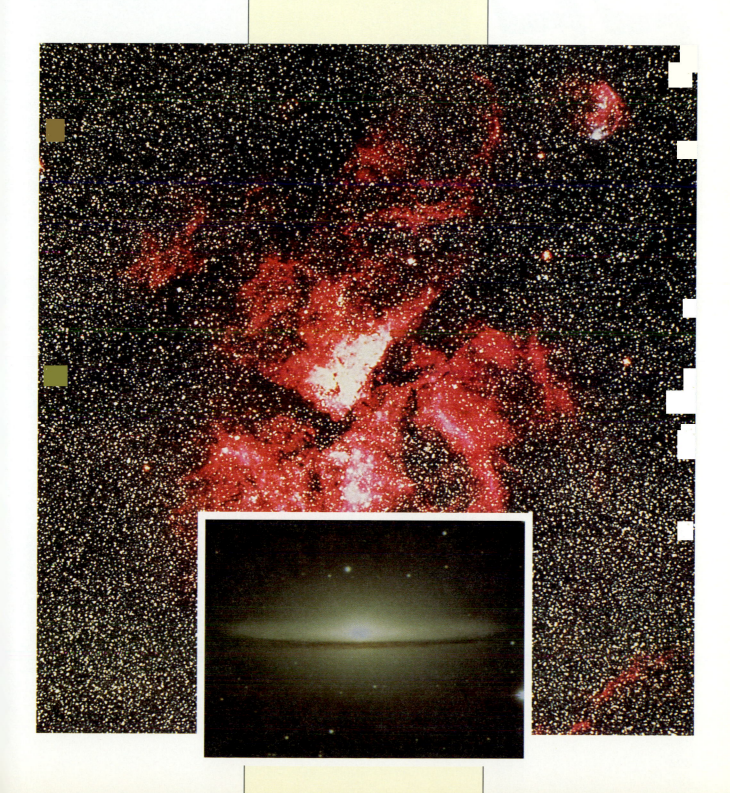

22

The Interstellar Medium

Gas and dust in the immense spaces between the stars

Interstellar gas and dust create the dark Horsehead Nebula near Zeta Orionis.

TABLE 22.1
Prominent Diffuse Nebulae

Popular Name	Constellation	Messier Number	NGC	α (2000) h	m	δ °	'	Distance (pc)	Associated With
—	Triangulum	—	604	01	35	+30	46	720 kpc	M 33, giant H II region
California	Perseus	—	IC499	04	00	+36	00	400	ξ Per
Orion	Orion	42	1976	05	35	−05	25	420	θ¹ Ori
Tarantula[a]	Dorado	—	2070	05	39	−69	12	52 kpc	LMC, giant H II region
Rosette	Monoceros	—	2224	06	37	+05	00	1,500?	open cluster, NGC 2237
Carinae	Carina	—	3324 +3372	10	43	−59	00	2,500?	η Carinae, HD 93129A
Trifid	Sagittarius	20	6514	18	03	−23	02	1,000	—
Lagoon	Sagittarius	8	6523	18	04	−24	23	1,100	open cluster, NGC 6530
Eagle	Serpens	16	6611	18	19	−13	46	2,900	open cluster of same name
Omega	Sagittarius	17	6618	18	21	−16	12	900	open cluster, IC 4707
North America	Cygnus	—	7000	20	45	+30	42	—	—

[a]30 Doradus.

A tour of the Milky Way quickly reveals billows of glowing gas and great clouds of dust that appear as blackened patches against the stellar background. We now take the first step on the path of stellar evolution and look at matter between the stars, the **interstellar medium,** the stuff of which the stars, the planets, and the Sun were created.

22.1
Diffuse Nebulae

The most obvious features of the interstellar medium are bright clouds of illuminated gas called **diffuse nebulae** (*nebula* is Latin for "cloud"). They consist principally of ionized hydrogen gas and are therefore also known as **H II regions.** Table 22.1 provides a list of prominent examples, among them the magnificent Orion Nebula (Figure 22.1) in Orion's sword (easily seen with binoculars) and the spectacular Carina Nebula of the southern hemisphere (Figure 22.2) that contains the luminous blue variable Eta Carinae and the Galaxy's brightest star.

The diffuse nebulae are confined to the plane of the Galaxy, the Milky Way, are always associated with O or hot B stars, and therefore belong to Population I. Because O and B stars have short lives and never get far from their birthplaces (see Section

Figure 22.1   *The Orion Nebula glows because of the ultraviolet radiation of the four stars of the Trapezium at its center.*

20.4), we conclude that the diffuse nebulae are the remnants of the stars' births. These lovely objects exhibit a great range of characteristics, from telescopic showpieces to barely detectable, from nearly circular to ragged and highly irregular in shape. The sizes are astonishing. The stars in the middle of the Orion Nebula yield a spectroscopic distance of 420 pc. From the nebula's angular diameter of

Figure 22.2   *The Carina Nebula, easily seen with the naked eye, contains Eta Carinae (lower arrow) and the ultraluminous star HD 93129A (upper arrow).*

about a degree, the physical diameter is about 8 pc. Vastly larger is the Tarantula Nebula (30 Doradus) in the Large Magellanic Cloud, seen at the upper left in Figure 21.4a. This *giant H II region* is 300 pc across and encompasses whole associations of O and B stars; if placed at the distance of the Orion Nebula, it would fill the entire constellation! At the other extreme, compact nebulae only a fraction of a parsec across stud almost any photograph of the Milky Way.

The diffuse nebulae produce emission-line spectra (Figure 22.3) and are prime examples of Kirchhoff's third law (see Section 8.3.2). The Balmer lines of hydrogen are prominent features, as are lines of neutral helium and many other light elements. Among the strongest lines are those at

5,007 Å and 4,959 Å. Discovered in 1864, they were originally thought to be emitted by an undiscovered element tentatively named "nebulium" (their origins to be discussed below). The emission lines extend from the ultraviolet to the radio and are superimposed on a weak background emission continuum. There are no absorption lines.

These natural artworks fluoresce under the action of the ultraviolet light produced by the hot stars within them. In 1928 the Dutch astronomer Hermann Zanstra described the basic illumination process of **photoionization** (ionization by photons) of hydrogen atoms. Almost all the atoms have their electrons in the lowest energy level, the ground state, $n = 1$ (Figure 22.4a). To become ionized, the atom must absorb a photon with an energy greater than that required to rip an electron from the ground state, which corresponds to the Lyman limit at a wavelength of 912 Å (see Section 8.3.2). The electron then flies away with a kinetic energy and velocity that depend on the energy of the absorbed photon. For a star to produce a significant number of photons with energies greater than the Lyman limit, it must have a temperature greater than 25,000 K, which corresponds to spectral class B1.

The free electron eventually is captured by another proton in a recombination (Figure 22.4b). It can land on any energy level, and as it jumps downward from one energy level to another toward the ground state, it creates **recombination lines** (in the example, Hβ and Lyman α). Recombination lines are also seen from atoms and ions of such relatively abundant elements as helium, oxygen, carbon, and nitrogen. The free electron can also pass close to a proton and be slowed down rather than

Figure 22.3   *The Orion Nebula exhibits emission lines produced by hydrogen, helium, oxygen, and other common elements. Brackets denote forbidden lines; the others are all recombination lines. There is also a faint background continuum produced by free-free radiation and other processes.*

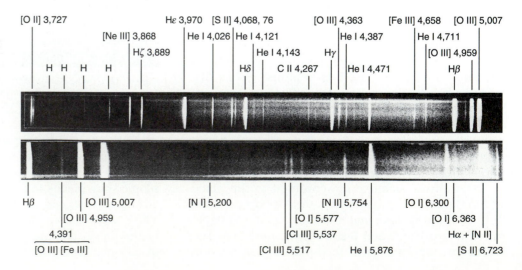

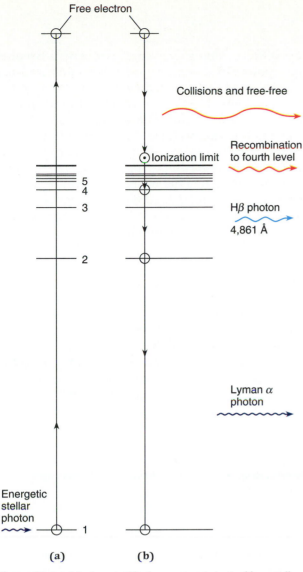

Free electron

Collisions and free-free

Ionization limit

Recombination
to fourth level

5
4

3

Hβ photon

4,861 Å

2

Lyman α
photon

Energetic
stellar
photon

1

(a)   **(b)**

Figure 22.4 **(a)** *A neutral hydrogen atom is ionized by a stellar photon with an energy greater than the Lyman limit. The electron flies away at high velocity.* **(b)** *After losing energy by collisions and free-free radiation, it recombines with a different proton and lands on an energy level (here, the fourth). It can then jump to any lower level with the production of emission lines (here, Hβ and Lyman α).*

captured. The decrease in kinetic energy is accompanied by the release of a photon. The collection of such interactions in the gas produce continuous **free-free radiation** that is especially strong in the radio spectrum.

Ideally, each stellar ultraviolet photon will produce a photoionization. Almost as soon as an electron and proton recombine to form an atom, the atom is re-ionized. As a result, almost all the hydrogen atoms are ionized and the gas is in the form of a plasma (defined in Section 10.6). If the cloud is large enough, a high level of ionization will be maintained outward from the star until all the ultraviolet photons are used up, whereupon the gas almost immediately becomes neutral (Figure 22.5a). The result is a spherical plasma bubble—an H II region—within a neutral cloud called a **Strömgren sphere** after the Swedish-American astronomer Bengt Strömgren, who first developed the idea. The Rosette Nebula (Figure 22.5b) provides a fine natural example. The size of a Strömgren sphere depends on gas density, the number of ionizing stars, and their temperatures. Most diffuse nebulae, however, are more irregular because of nonuniform mass distribution. The Orion Nebula (Figure 22.1) is not a bubble but a blister on the edge of a dark neutral cloud.

The "nebulium" mystery was solved about 1928 by the American astronomer Ira S. Bowen, who mapped the energy level diagram of doubly ionized oxygen (O^{+2}) and discovered level pairs that produce photons at 4,959 Å and 5,007 Å (Figure 22.6). By simplified atomic rules, transitions between these pairs are not allowed, and the lines are therefore called **forbidden lines.** However, according to the complete theory they are indeed allowed, just improbable. Because of the huge number of atoms

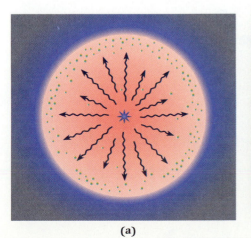

(a)

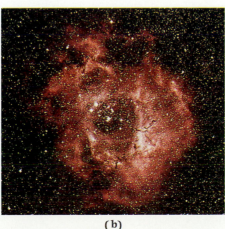

(b)

Figure 22.5 **(a)** *The ideal diffuse nebula is a Strömgren sphere, an H II region set within a neutral cloud. The star ionizes the gas to a radius at which all the ionizing ultraviolet photons are used.* **(b)** *The Rosette Nebula in Monoceros, NGC 2224, fits the picture well.*

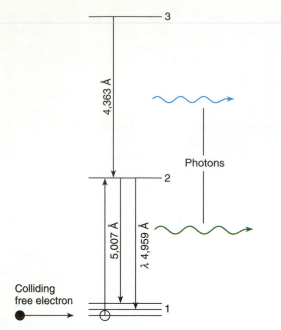

Figure 22.6 *A free electron collides with an electron in the ground level of O^{+2} (which is split into three sublevels) and sends it to higher levels 2 or 3. When the electron jumps back down it can produce forbidden emission lines at 4,363 Å, 4,959 Å, or 5,007 Å, depending on the combination of level pairs.*

in a diffuse nebula, the lines can in fact become quite powerful. Electrons are not excited into the upper levels (from which they descend to make the emission lines) by recombination but by collisions between the ions and energetic free electrons. These and numerous other forbidden lines are indicated in Figure 22.3 by square brackets ([O III] for the O^{+2} lines, [N II] for N$^+$, and so on).

From the strengths of the spectrum lines, we can calculate nebular conditions and compositions. Densities are commonly 100 to 1,000 electrons (and ions) per cubic centimeter, though they may rise to 10^5/cm^3 in compact objects. These values are comparable to the best vacuums that can be produced in laboratories. Yet the nebulae are still bright, testimony to their great masses. That of the Orion Nebula is several hundred solar masses, and of the Tarantula several tens of thousands. Because of its low density, a diffuse nebula, like the solar corona, is not a blackbody. Kinetic temperatures are typically 10,000 K. The compositions are like those of the Sun and the stars: about 92% hydrogen (by number of atoms), 8% helium, and sunlike proportions of other elements.

Figure 22.7 **(a)** *Small sections of two very high dispersion (greatly spread out) spectrograms of ε Orionis show multiple interstellar Ca II K (top) and neutral sodium (Na I) D lines (bottom). We can see evidence for* **(b)** *five separate clouds of interstellar matter along the line of sight, with radial velocities between +3 and +27 km/s.* **(c)** *A graphical spectrogram made by the* International Ultraviolet Explorer *that shows interstellar lines of ionized silicon and aluminum.*

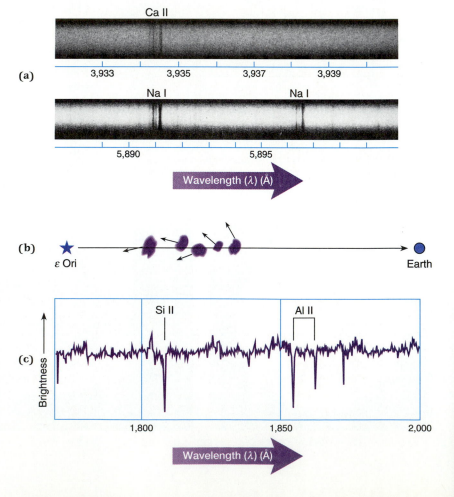

22.2
The General Interstellar Medium

Delta Orionis is a spectroscopic binary whose absorption lines Doppler-shift back and forth over a 5.7-day period. In 1904, however, Johannes Hartmann found that the K line of Ca II did not move. It therefore could not be produced by the stars and had to be created by gas in the space between them and the Sun.

We have since discovered hundreds of **interstellar absorption lines** (Figure 22.7a) that arise from atoms like calcium and sodium in the neutral or ionized states and from simple molecules (CH, CH$^+$, and CN). They are present in the spectra of almost all stars in the galactic disk beyond 100 pc or so, but are absent or weak in the stars of the halo, again showing the interstellar medium to be a Population I phenomenon. The ultraviolet (Figure 22.7c) is especially rich in these interstellar absorptions, displaying lines from sulfur, carbon, and metals such as chromium, manganese, iron, copper, and zinc. The absorption lines shown in Figure 22.7a are split by the Doppler effect, showing that the interstellar medium consists of discrete lumps of gas moving with velocities that differ by a few km/s (Figure 22.7b).

The real spectral realm of interstellar studies, however, is the radio. In 1951 the Americans Harold Ewan and Edward Purcell, acting on a prediction made six years earlier by the Dutch astronomer Hendrik Van de Hulst, discovered a powerful interstellar emission line of neutral hydrogen at a wavelength of 21 cm. Protons and electrons have a property called *spin,* an atomic version of rotation. The magnetic fields associated with the spins cause the particles to interact. The spin axes of the electrons and protons must be aligned, and they are allowed to spin either in the same direction (parallel), or in opposite directions (antiparallel). The result is that the ground state of hydrogen is split in two (Figure 22.8), the parallel state on top.

The electron can be shuffled from one state to the other by collisions with neighboring atoms. Once in the upper state, it can also reverse direction by itself, performing a *spin-flip* wherein it drops to the lower level with the radiation of a 21-cm photon. On the average, a particular electron will wait in the upper state for around 11 million years before it will jump downward. There is *so much* neutral hydrogen in interstellar space, however, that the line builds to great strength and can

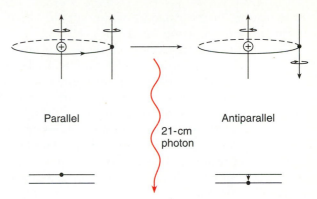

Figure 22.8   *At left, the electron and proton of neutral hydrogen spin parallel and the electron is in the higher of two sublevels of the ground state. The electron then changes its spin direction and jumps to the lower sublevel (right) by emitting a 21-cm photon. On this scale, the second orbital energy level is nearly 4 km away.*

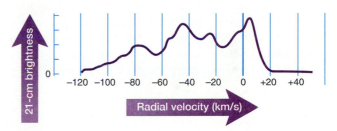

Figure 22.9   *A radio spectrogram in the direction of Cygnus shows several blended components to the 21-cm line of neutral hydrogen that reveal clouds and a general dispersed intercloud medium. The wavelength scale is expressed in terms of velocity through the Doppler shift.*

be seen with modest equipment. A complex **21-cm line** (Figure 22.9) is found along any direction we look within the Galaxy's disk, showing the all-pervading nature of the interstellar gas.

The combination of optical and radio observations reveals cool, discrete, **neutral hydrogen clouds** set within a warmer, lumpy, partially ionized **intercloud medium,** the ionization produced by the light of distant stars. These and other components of the interstellar medium are shown in Figure 22.10. Along any line of sight, there are about 10 clouds per kiloparsec. Typically a few pc across, they have temperatures around 100 K and densities a few tens to a few hundreds of atoms/cm^3. The denser ones are the colder. Masses are several times that of the Sun, and taken together they constitute roughly a quarter of the total mass of the interstellar medium. The warm intercloud gas has a temperature around 8,000 K, but has a much lower density of only a few tenths of an atom/cm^3. It constitutes about half the mass of the interstellar medium. The clouds lie within about

100 to 150 pc of the galactic plane, but the warm gas spreads a few kpc into the halo.

The chemical compositions of these portions of the interstellar medium are oddly different from that of the Sun. The ratios of lighter elements such as oxygen and nitrogen relative to hydrogen are normal. However, others, like silicon and iron, are low by a factor of 10. The depletion increases as the density rises and can reach a factor of a hundred. Something is removing these atoms from the gas and tying them into a less-directly observable form. We will find them later.

Finally, rocket flights in 1968 showed low-energy X rays coming from deep space. Six years later, satellite observations revealed interstellar absorption lines from five-times ionized oxygen (O^{5+}) in the far ultraviolet. Such ionization is incompatible with the other interstellar lines and requires a temperature of at least 300,000 K. The X rays indicate a value closer to a million K. The source, called the **coronal gas** because of its high temperature, has an extremely low density, near 10^{-3} atoms/cm^3.

Such a high temperature cannot be produced by radiation from ordinary stars but almost certainly represents the effects of supernova blast waves. Over the aeons, stellar explosions have interconnected, weaving a tapestry of tunnels within the gaseous medium. Although the coronal gas contains less than 5% of the interstellar medium's mass, it occupies perhaps half the interstellar volume.

22.3
Dust

In any view of the Milky Way you will see what look like holes in the blanket of stars. These gaps are in fact unilluminated gaseous clouds—*dark nebulae*—filled with **interstellar dust,** tiny grains that block the light of the background stars. The smallest and densest of the dark nebulae are called **Bok globules** after Bart Bok, the Dutch-American astronomer who helped reveal their natures. They are typically a few parsecs across and have masses of 10 to 100 times that of the Sun. The most famous

Figure 22.10 *A schematic view of the interstellar medium shows neutral hydrogen clouds (gray) enmeshed in a warm intercloud medium (red), a hot low-density coronal gas (yellow), small, dense, dusty globules (black), and giant molecular clouds (blue) filled with stars and OH masers. Hot stars inside or at the cloud edges create diffuse nebulae (green).*

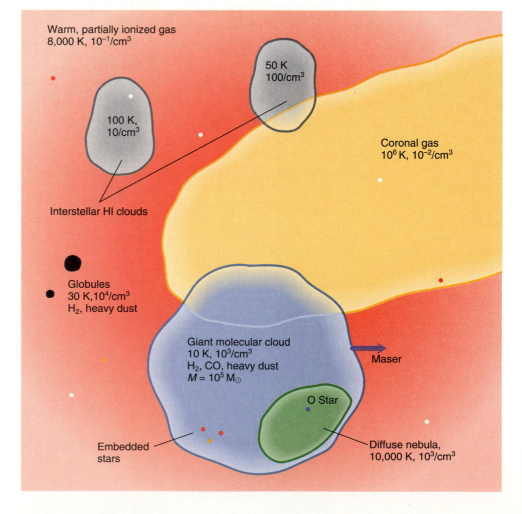

Warm, partially ionized gas
8,000 K, 10^{-1}/cm^3

50 K
100/cm^3

100 K,
10/cm^3

Coronal gas
10^6 K, 10^{-2}/cm^3

Interstellar HI clouds

Globules
30 K, 10^4/cm^3
H$_2$, heavy dust

Giant molecular cloud
10 K, 10^3/cm^3
H$_2$, CO, heavy dust
$M = 10^5\,M_\odot$

Maser

O Star

Embedded
stars

Diffuse nebula,
10,000 K, 10^3/cm^3

TABLE 22.2
Some Famous Dark Nebulae

Popular Name	Constellation	Barnard Number	α (2000) h	m	δ °	'	Angular Size	Distance (pc)	Associated With
Horsehead	Orion	42	05	41	–02	15	4'	100	IC434, ζ Ori
S Nebula	Ophiuchus	72	17	24	–23	34	20'		
Coalsack	Crux	—	12	50	–63	00	5°		
—	Sagittarius	86	18	02	–27	55	3'		
Great Rift[a]	Cygnus, Vulpecula, Aquila, Serpens	—	18 to 20		+ 45 to –10		55°		

[a]Not a single dark nebula but a vast collection of them that runs down the middle of the northern Milky Way.

globules are the Horsehead Nebula (see the chapter-opening photograph) and the 5°-wide Coalsack, seen in Figure 4.7 just southeast of the Southern Cross. These and other dark nebulae are listed in Table 22.2.

The globules are remarkably opaque in the optical spectrum. None of the stars on the other side of Barnard 86 (Figure 22.11a) is visible; their light is totally scattered or absorbed by the dust (Figure 22.11b). Much of the apparent structure of the diffuse nebulae seen in Figures 22.1 and 22.2 is formed by intervening dust. Note also the numerous small, dark features seen against the Rosette in Figure 22.5b.

However, if a cloud is transparent and has one or more stars embedded within it, the dust can scatter the stellar radiation toward the observer to produce a **reflection nebula,** whose spectrum is a faithful reproduction of the embedded stars, absorption lines and all. Such an object can also be produced by a star in front of a dense cloud. The most famous reflection nebula enmeshes the Pleiades (see Figure 20.15), whose hottest star, class B3 Merope, cannot ionize the surrounding gas. Reflection nebulae can often be distinguished simply by color. The main part of the Trifid Nebula in Sagittarius (Figure 22.12) is a Strömgren sphere, and appears red because of Hα radiation. However, the southern extension of the nebula is outside the ionized zone and consists of neutral gas. Its dust scatters light from a nearby B star and is colored blue.

In the 1930s Robert Trumpler of Lick Observatory discovered that the distances of open clusters as determined by main-sequence fitting (see Section 20.3.2) were systematically larger than those estimated from the clusters' angular sizes. He concluded that something was dimming the light of the cluster stars and making them seem too far away.

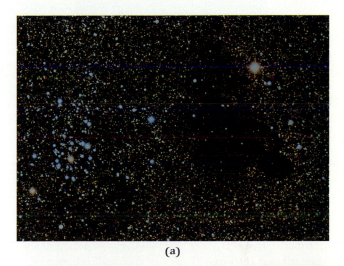

(a)

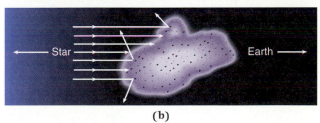

(b)

Figure 22.11   **(a)** *Barnard 86, a Bok globule, is an optically opaque cloud of gas and dust that blocks the light of stars on the other side of it* **(b)***. Note the open cluster next to it.*

The obscuring dust lies not just in the obvious globules, but in the cool clouds and the intercloud medium as well, and is mixed everywhere with the gas. The dust's pervasive nature is revealed by a strip about 20° wide centered on the Milky Way called the **zone of avoidance:** in this region the dust prevents us from optically seeing any external galaxies (Figure 22.13). Other spiral galaxies, those like our own, show the dust as a pronounced dark band lying along their spines (Figure 22.14), vividly demonstrating the Population I nature of the

Figure 22.12 *The Trifid Nebula, M 20, is a diffuse nebula red with the light of Hα radiation from atomic hydrogen. Below it is a blue reflection nebula that scatters starlight.*

interstellar medium. This layer can easily be seen in our own system with the naked eye, manifesting itself as a dark rift that runs through much of the Milky Way (see Figure 1.12).

Since the dust absorbs starlight, it is heated to temperatures of the order of 100 K and radiates at far infrared wavelengths. Gloriously recorded by the *Infrared Astronomical Satellite* (*IRAS*), the dust fills the plane of the Milky Way (Figure 22.15), where it is concentrated into the same clouds that produce the interstellar absorption lines and spreads out into vast sheets of *interstellar cirrus*.

The dust produces one of the most frustrating problems of optical astronomy. Because it absorbs and scatters light, all stars beyond the local neigh-

borhood appear to be fainter and consequently more distant than they actually are. To apply to the methods of main sequence fitting or spectroscopic distance, we must account for this **interstellar absorption.** If starlight is dimmed by A magnitudes, then A must first be subtracted from all observed apparent magnitudes.

On the average, light is dimmed by about one visual magnitude per kiloparsec, but the distribution of the dust is so irregular that an average means little. Fortunately, because scattering and absorption are more efficient at shorter wavelengths than at longer ones, the dust also reddens starlight much as the Earth's atmosphere reddens the Sun at sunset. All stars beyond about 100 pc have a color index, $B - V$, that is greater than it should be for their spectral classes. By comparing the spectral continuum fluxes of similar stars, one reddened by dust, the other not (one far away, the other nearby), we find the **interstellar reddening** to depend roughly on the inverse of the wavelength (Figure 22.16). If the wavelength is halved, about twice as much radiation will be lost. The degree of absorption can be very high at short wavelengths (making ultraviolet observations quite difficult) and goes to zero at long wavelengths, meaning that our view of the Galaxy is clear if we observe in the far infrared or radio spectral domains (Figure 22.17). From the amount of reddening, we can in most cases estimate the amount of absorption, or A, and can therefore determine what the visual magnitude or magnitudes should be without the effect of the dust and thus the true distance of a star or cluster. The relation between reddening and absorption is complex, however,

Figure 22.13 *In this whole-sky map (in which the celestial equator runs horizontally), the blue diamonds represent the diffuse nebulae that outline the Milky Way. The red symbols represent galaxies; they avoid the plane of the Milky Way because of the dust in the galactic disk.*

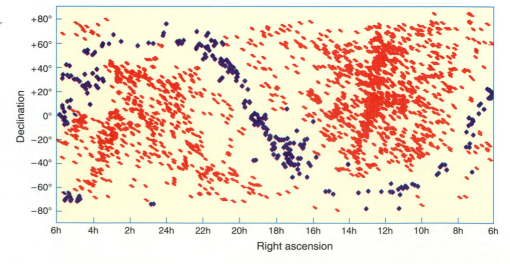

Figure 22.14 *The dust band that runs down the spine of NGC 4565 is similar to that in our own system.*

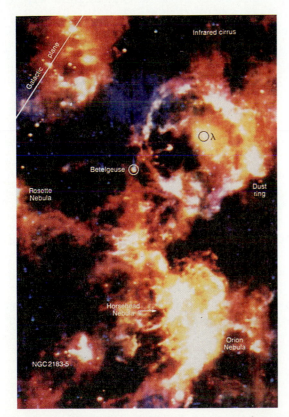

Figure 22.15 *Orion takes on a different appearance when viewed by the IRAS satellite at far infrared wavelengths. All the stars but cool Betelgeuse have disappeared, replaced by masses of heated dust that fill the picture. The dust is concentrated toward the lower half of the constellation and is centered on the Orion Nebula. Toward the top is a huge bubble blown in interstellar space by the luminous O star λ Orionis. Across the top of the picture are thin filaments of dusty interstellar cirrus.*

and the dust introduces an additional degree of uncertainty into the determination of distances, one frequently quite difficult to assess.

From the way in which starlight is reddened and a theoretical knowledge of how light interacts with small particles, astronomers find that the dust grains run from an upper limit of around a quarter of a micron (10^{-3} mm) down to minute particles no more than a few Ångstroms across, not much larger than complex molecules. For every reduction in size of a factor of 2 there are roughly 10 times as many.

The compositions of the grains are similar to those of the particles that surround the dusty Mira variables. A broad absorption feature in the infrared spectra of highly reddened stars reveals the presence of silicates. Laboratory studies also show that the bump at 2,200 Å in the reddening graph shown in Figure 22.16 is caused by solid carbon grains in the form of graphite. Additional carbon is tied into simple unstructured soot. These grains almost certainly originate in Mira winds. Another absorption feature suggests that some 20% of interstellar carbon is in the form of tiny diamonds, created from carbon grains by supernova shock waves. They seem to be similar to those found in primitive meteorites (see Section 17.1), directly linking the stuff of interstellar space with the raw material of the Solar System.

In addition to dimming and reddening, the grains are also responsible for the **polarization** of

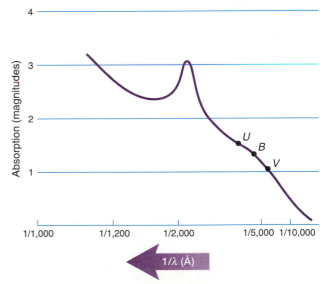

Figure 22.16 *Relative interstellar absorption is plotted against the inverse of wavelength (1/λ). Except for the bump near 2,200 Å caused by interstellar graphite, absorption depends roughly on 1/λ. Adapted from "Interstellar Dust and Extinction" from* Annual Review of Astronomy and Astrophysics, *Volume 28, 1990 by Geoffrey Burbidge, David Lazyer and Allan Sandage, page 37. Copyright © 1990 by Annual Reviews Inc. Reproduced, with permission, from Annual Reviews Inc. and the author.*

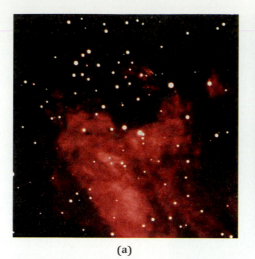

Figure 22.17 **(a)** *An opaque dark cloud within M 17, the Eagle Nebula, appears blank against the optical sky. However, infrared radiation* **(b)** *penetrates the dust and reveals stars buried within.*

(a) **(b)**

starlight. Light (or any kind of radiation) is polarized when a significant fraction of the waves oscillate in a particular direction (Figure 22.18). Light can be polarized by passing it through a molecular filter that allows only a particular direction of oscillation to pass. It is also naturally polarized when it is reflected or scattered. Polarization and its direction can be detected by rotating a polarizing filter in the beam of light to find the direction in which the light is brightest.

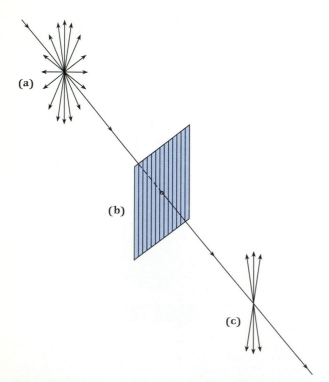

Figure 22.18 *A beam of light proceeds from upper left to lower right.* **(a)** *The light is randomly polarized (or unpolarized) and the waves oscillate in all directions.* **(b)** *It strikes a polarizing filter that allows only a limited range of oscillation planes* **(c)** *to go through.*

We find that distant starlight is usually polarized by a few percent, that is, the grains selectively scatter a few percent more waves that oscillate in one direction than in the other. For the light to be polarized, the grains must be elongated, and more important, must be somewhat aligned with one another to produce the natural filter. The only explanation is that metal atoms within the grains must be affected by a weak magnetic field that runs through the Galaxy, theory showing the long axes of the grains to be partially aligned perpendicular to the field direction.

The silicate and carbon grains from the Mira stars become highly modified in the deep cold of interstellar space (Figure 22.19). They become coated with ices of various kinds and can slowly capture heavier atoms, including iron, removing them from the gas, which explains the observed elemental depletions seen in interstellar clouds.

The degree of interstellar absorption indicates that about 1% of the mass of the interstellar matter is dust. Only about one of the larger grains per cubic meter is needed to cause the observed interstellar absorption and reddening. The dust has an effect only because of the vast distances involved: the path to a star 1 kpc away is 3×10^{18} meters long.

22.4
Interstellar Molecules

Of all the discoveries made about the interstellar medium, the most dramatic may be that of extensive molecular gas. In the optical spectrum, molecular interstellar absorption lines are meager: the only ones known before the 1960s were CH, CN, and CH⁺. Consequently, no one took interstellar chemistry very seriously. However, with the advent of sensitive radio receivers, our concept of interstellar space began to change.

Radio astronomers made their big breakthrough in 1963 with the discovery of the hydroxyl (OH) *radical,* which can generate a powerful maser emission line as it does in OH/IR stars. A radical is an incomplete molecule, one with an unshared electron that needs an atomic partner. OH cannot exist in the pure state on Earth and immediately combines with another atom or molecule to produce a different compound. In interstellar space, however, the density is so low and the atoms so widely separated that OH encounters no likely partner and can survive almost as a permanent species.

Five years after OH was found, radio astronomers discovered emission lines from interstellar water vapor and ammonia (NH_3), and during the next 20 years, dozens of ever more complex molecules (Figure 22.20). As of 1992, 91 different species, including ions, were known (some of them are listed in Table 22.3). Many are organic, based on carbon. Some, like formaldehyde, acetylene,

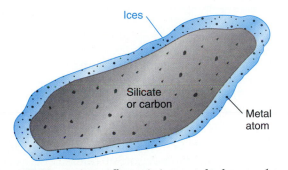

Figure 22.19   *An interstellar grain is expected to have an elongated silicate or carbon core. It is coated with a mantle of water, methane, and ammonia ices in which are embedded atoms of metals and other elements as well as simple and complex molecules.*

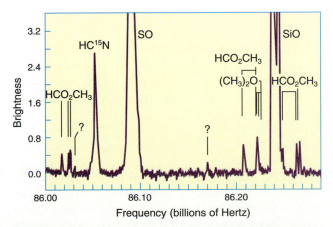

Figure 22.20   *A short segment of the radio spectrum from the Orion Nebula shows a remarkable number of interstellar molecules.*

TABLE 22.3
A Selection of Interstellar Molecules

Two-Atom
CH (methylidine)
C_2 (diatomic carbon)
CN (cyanogen)
CO (carbon monoxide)
CS (carbon monosulfide)
H_2 (molecular hydrogen)
NO (nitric oxide)
NaCl (table salt)
OH (hydroxyl)
SO (sulfur monoxide)

Three-Atom
HCN (hydrogen cyanide)
HCO (formyl radical)
H_2O (water)
H_2S (hydrogen sulfide)
SO_2 (sulfur dioxide)

Four-Atom
HNCO (hydrocyanic acid)
H_2CO (formaldehyde)
HC_2H (acetylene)
NH_3 (ammonia)

Five-Atom
HCOOH (formic acid)
CH_4 (methane)
HC_3N (cyanoacetylene)

Six-Atom
$HCONH_2$ (formamide)
CH_3OH (methyl alcohol)
CH_3CN (methyl cyanide)
C_5H (pentynylidyne)

Seven-Atom
NH_2CH_3 (methylamine)
$HCOCH_3$ (acetaldehyde)
CH_2CHCN (vinyl cyanide)
CH_3C_2H (methylacetylene)

Eight-Atom
$HCOOCH_3$ (methyl formate)

Nine-Atom
CH_3CH_2OH (ethyl alcohol)
$(CH_3)_2O$ (dimethyl ether)
CH_3CH_2CN (ethyl cyanide)
HC_7N (cyanotriacetylene)

Eleven-Atom
HC_9N (cyanotetraacetylene)

Thirteen-Atom
$HC_{11}N$ (cyanopentaacetylene)

and both methyl (wood) and ethyl alcohol (Figure 22.21a), are familiar in everyday life. Others, like CH_9N and $HC_{11}N$ (Figure 22.21b), do not exist on Earth. Many of these were later found in the envelopes that surround carbon-type Mira variables. IRC+10 216 (see Figure 21.13) has nearly 20 different species, including HC_7N and HC_3CN.

Molecules are easily broken apart. They require low temperatures and must be hidden from energetic starlight to survive, so they concentrate in dark, dusty clouds where temperatures may be only a few K. Collision energies are so feeble that for the most part only energy levels just above the ground state (like the one that produces the 21-cm line of hydrogen) can be populated with electrons. As a result, when the electrons jump back to the ground state, the emitted photons have low energies, long wavelengths, and fall in the radio spectrum.

The hydrogen molecule, H_2, is a significant exception. Because of its structure, it has no energy levels that can produce strong radio emission lines and must be observed by its weak infrared emissions and ultraviolet absorption spectrum. Nevertheless, we find that within a dark cloud almost all the hydrogen is tied up in hydrogen molecules—there are few free hydrogen atoms. In Section 22.2, we accounted for three-quarters of the interstellar mass: one-quarter is in neutral hydrogen clouds and half in the intercloud medium. The final quar-

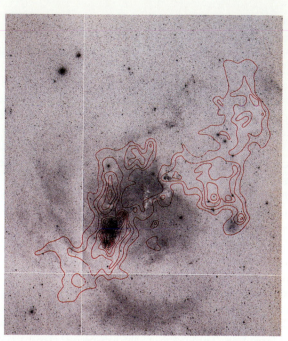

Figure 22.22 *The contours show concentrations of carbon monoxide that outline a giant molecular cloud in Monoceros. They are superimposed on a photograph (presented as a negative) of the Milky Way. Diffuse nebulae are commonly seen at the edges of the clouds, demonstrating recent star birth from the cloud material. This nebula is probably on the forward edge of the cloud.*

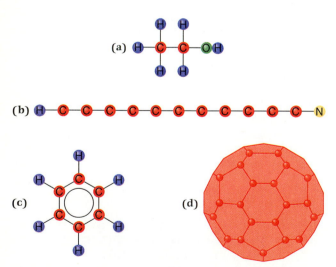

Figure 22.21 **(a)** *Common ethanol, ethyl alcohol (CH_3CH_2OH), is abundant both on Earth and in interstellar space. A pair of carbon atoms form a bond. One has three H atoms connected to it, the other two H atoms and an OH radical.* **(b)** *A weird chain of carbon atoms makes the $HC_{11}N$ molecule.* **(c)** *A benzene ring can build with others to form complex polycyclic aromatic hydrocarbons that may be responsible for certain broad emission lines.* **(d)** *Another possible source of the emission lines is C_{60}, buckminsterfullerene, a soccer-ball-shaped molecule of 60 carbon atoms.*

ter (perhaps even a higher fraction) is H_2, a form that we could not even *detect* until 1970.

The abundances of other molecules are much lower and in line with the compositions of stars. Of them all, carbon monoxide, CO, with powerful lines at millimeter wavelengths, is probably the most important. From comparison of line strengths and excitation mechanisms, we find a fairly constant ratio of about 1 CO molecule for every 10,000 molecules of H_2. CO is therefore a good tracer for the difficult to observe molecular hydrogen. More complex molecules are considerably rarer: there is roughly one acetylene (HC_2H) molecule for every 10^{10} of H_2.

Extensive surveys of CO emission show that it—and the H_2—are concentrated not just in the obvious Bok globules, but also in much more massive systems known as **giant molecular clouds** (GMCs) (Figure 22.22). These clouds are concentrated along the plane of the Milky Way within the optical rifts caused by the dark dust. Some 6,000 GMCs, with a huge range of radii and masses, are known. A typical GMC might be 100 pc across and contain over 200,000 solar masses. GMCs are the most massive discrete structures in the Galaxy, and (since they contain nearly all the H_2) they constitute a quarter or more of the interstellar medium's mass. The GMCs that pass close to the Earth are believed to disturb the Oort comet cloud (see Section 17.3.4) and send new comets into the inner Solar System.

The formation of interstellar molecules involves a variety of reactions and sites. Some are created in the shells of Mira variables by ordinary chemical reactions among atoms in the gas. They stick to the surfaces of dust grains and are quietly wafted into the interstellar medium. Additional reactions take place on the grain surfaces when the molecules are struck by ultraviolet photons from distant stars, are broken apart, and then reform. Within opaque interstellar clouds, hydrogen atoms stick to grains and react to form H_2 molecules, which are then evaporated into space. The H_2 molecules can be ionized by high-speed atomic nuclei called *cosmic rays* and react with themselves to produce ionized triatomic hydrogen, H_3^+. A number of reaction chains involving both molecules and ions can subsequently produce heavier molecules. The process is aided by shock waves driven by winds from hot stars and supernovae. A reasonably complete theory of molecule formation is enormously complex, with a huge number of equations. With the aid of high-speed computers, however, astronomers have been able to make predictions that match the observed molecular abundances reasonably well.

Interstellar chemistry may be far more complicated. A pair of infrared emissions have been tentatively identified as *polycyclic aromatic hydrocarbons* (PAHs). These organic structures, built of *benzene rings* (Figure 22.21c), have strong odors, hence their name. They may be so large and complex that they take on the form of tiny grains, blurring the distinction between the dust and the molecular gas. It has been suggested that 10% of the molecular gas may be tied up in such structures. Another interpretation of the emissions is that they are produced by complex carbon molecules, notably by C_{60} (Figure 22.21d). This molecule is structured like a geodesic dome, and is called buckminsterfullerene (or a bucky ball) in honor of the dome's inventor. Chemistry must also continue on the grain surface, producing other substances that may or may not get kicked off into space. The carbonaceous chondrites (see Section 17.1) are effectively made of interstellar dust and contain amino acids. We have clearly taken but a small step toward understanding the processes that take place deep within the cold, dark globules and molecular clouds.

Our picture of interstellar space (return to Figure 22.10) is now fairly complete: we see cold, giant molecular clouds, dense cold globules, and thinner cool neutral hydrogen clouds embedded in a warm gas tunneled through by hot coronal gas powered by supernova blasts, all of it totaling around 10% of the stellar mass of the Galaxy. The interstellar medium, and the giant molecular clouds in particular, are the sites of star formation, where the cold raw material condenses to form the billions of stars that surround us.

KEY CONCEPTS

Bok globules: Small, opaque, relatively dense, dusty interstellar clouds, typically a few pc across containing 10 to 100 solar masses.

Coronal gas: A very hot gas in the interstellar medium created and heated by supernova blasts.

Diffuse nebulae: Bright clouds of interstellar gas ionized by stars at least as hot as type B1; also called **H II regions.**

Forbidden lines: Emission lines common in diffuse nebulae that build up great strength because of high cloud mass, even though the electron jumps that create them are improbable; they are produced by collisional excitation.

Free-free radiation: Continuous radiation caused by the close passage of electrons and protons.

Giant molecular clouds (GMCs): Massive, dusty clouds (typically 200,000 $M_\odot$ and 100 pc across) made mostly of hydrogen molecules.

Intercloud medium: A thin, warm, partially ionized gas that lies in the space between interstellar clouds.

Interstellar absorption: The dimming of starlight by interstellar dust.

Interstellar absorption lines: Narrow absorption lines superimposed on the spectra of stars, produced by atoms and ions in the interstellar medium.

Interstellar dust: Small, solid grains of silicates or carbon coated with ices and embedded with heavier atoms.

Interstellar medium: The lumpy mixture of gas and dust that pervades the plane of the Galaxy.

Interstellar reddening: The reddening of starlight by interstellar dust; correlated with interstellar absorption.

Neutral hydrogen clouds: Interstellar clouds a few pc across containing a few solar masses of neutral hydrogen gas.

Photoionization: The ionization of gas by energetic radiation.

Polarization: An optical process that makes light waves oscillate in a preferential direction.

Recombination: The capture of an electron by an ion; the process produces **recombination lines** as the recaptured electron cascades to the ground state.

Reflection nebula: A bright cloud created by light scattered by dust from a star too cool to cause ionization (cooler than type B1).

Strömgren sphere: A spherical bubble of ionized gas—a diffuse nebula—(plasma) set within a neutral medium.

21-cm line: A powerful radio line of neutral atomic hydrogen produced in the interstellar medium.

Zone of avoidance: The region around the plane of the Milky Way where galaxies are not seen because of dust absorption.

EXERCISES

Comparisons

1. List the differences between diffuse and reflection nebulae.
2. What is the difference between photoionization and recombination, and between recombination lines and forbidden lines?
3. Compare Bok globules and giant molecular clouds.
4. How does the intercloud medium differ from the coronal gas?
5. Compare the galactic distributions of interstellar dust and interstellar carbon monoxide.

Numerical Problems

6. If you tripled the ionizing ultraviolet luminosity of the illuminating star of a Strömgren sphere, what would you expect to happen to the sphere's radius? What assumption must you make for the estimate to be valid?
7. What is the frequency of the 21-cm line?
8. What is the minimum energy a photon can have and still ionize hydrogen from the ground state?

Thought and Discussion

9. Why is a diffuse nebula also called an H II region?
10. What produces the emission continua of diffuse nebulae in the radio spectrum?
11. Why do spectral classes B1 and B2 divide diffuse nebulae from reflection nebulae?
12. What is described by the temperature of a diffuse nebula?
13. What observations provide evidence for the existence of a general gas in the interstellar medium?

14. What observations provide evidence for the existence of dust in the interstellar medium?
15. Why are certain elements depleted in the interstellar gas?
16. How is the 21-cm line created?
17. What is polarized light?
18. How do we know that dark regions in the Milky Way are really clumps of dust and not gaps in the stellar distribution?
19. Why are we confident that we can use carbon monoxide as a tracer of H_2 in the interstellar medium? Why would ethyl alcohol not make a very good tracer?
20. What is the effect of interstellar dust on the calculation of distances by main-sequence fitting? How does the dust affect the distance scale of the Galaxy?
21. What observations lead us to believe there are two different kinds of interstellar grains? What are they?
22. What evidence leads us to believe that there are magnetic fields in the Galaxy?
23. Where do the different kinds of interstellar grains originate?
24. Why must interstellar molecules be made in dark clouds?
25. What are cosmic rays? What is their role in the formation of interstellar molecules?
26. What are the various components of the interstellar medium? What fraction of the mass does each represent?

Research Problem

27. Find photographs of the diffuse nebulae in Table 22.1. How might the different forms have been created?

Activities

28. Examine the Orion Nebula or the Lagoon Nebula with any telescope you can find. Draw the nebula and compare your drawing with a photograph.
29. On a dark moonless night in the summer, draw your own version of the Milky Way and outline the dark clouds.

Scientific Writing

30. An art magazine wants to do an article on aesthetics in science. Write about the diffuse nebulae, pointing out their beautiful forms and colors and also commenting on the origins of these features.
31. Amateur astronomers are most interested in what they can see with their telescopes. Write a background column for an amateur magazine that describes hidden wonders: interstellar gas, dust, and molecules that cannot be seen without observations made in the infrared and radio.

23

Star Formation

*How stars are born
out of the gas
and dust of
the Milky Way*

*NGC 2264 in Monoceros
is rich in young stars.*

Our Solar System was created some 4.6 billion years ago from a spinning disk of gas and dust. We now know that youthful stars are associated with interstellar matter. As we look into the clouds we can watch stars being made and can learn how the Sun was born.

23.1
Infant Stars

To build a theory of star formation, we must first assemble the data and search for stars that are being created—**protostars** (from the Greek *proto*, "first"). Associated with the dark, dusty clouds of the Milky Way are numerous F, G, K, and M variables called **T Tauri stars** (Figure 23.1a). On the HR diagram they lie up and to the right of the main sequence, among the subgiants. Typical of its class, a T Tauri normally shines at apparent visual magnitude 11, but can unpredictably brighten to 10th or fade to 14th.

T Tauri stars gang together into **T associations,** which are not bound gravitationally and from which the stars are escaping, demonstrating that, like the O and B stars, they have a common and recent origin. In another indication of youth, T Tauri's spectrum (Figure 23.1b) contains a strong absorption line of neutral lithium. This element is rare in the Sun and in most stars because it is easily destroyed by nuclear processes in hot, convecting, stellar envelopes. To have a high lithium abundance, a star must be young.

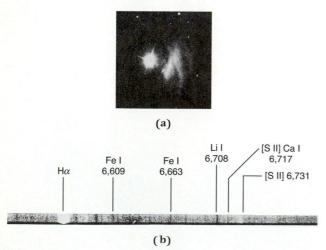

(a)

(b)

Figure 23.1   **(a)** *T Tauri, the prototype of the T Tauri stars, is surrounded by a small nebula associated with its birth.* **(b)** *Its spectrum has emission lines of hydrogen and forbidden emissions of sulfur that reveal a circumstellar cloud and a strong absorption line of lithium that indicates extreme youth.*

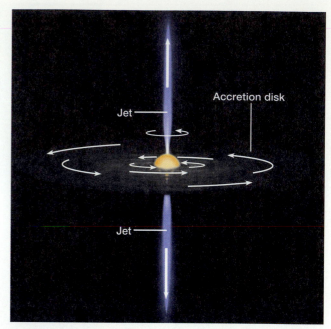

Figure 23.2   *Mass falls from an accretion disk onto a T Tauri star. A wind generated by the star escapes through the poles of the disk as a bipolar flow.*

These variables are exceptionally bright in the infrared, and their spectra display hydrogen and forbidden emission lines, which together indicate clouds of circumstellar gas and heated dust. Their spectra also exhibit P Cygni lines (see Section 21.4) that signify powerful mass outflows. Moreover, we also see *inverse* P Cygni lines (the absorption on the *red* side of the emission) that reveal mass flowing *onto* the stars from the surrounding dust clouds in the aftermath of formation. The accretion rate is erratic and tied to the variability; a star brightens as more matter falls onto its photosphere. X-ray flares and excessive ultraviolet luminosity indicate chromospheric activity that is probably caused by rapid spin and powerful magnetic fields. These stars are so young they have not yet been slowed by magnetic braking (see Section 19.6). We have apparently found real stellar infants, examples of **pre-main-sequence stars,** that will slowly develop into Sun-like dwarfs.

The features of a T Tauri star can be explained by a dusty circumstellar accretion disk that develops from the birth cloud that created the star. Mass rains down from the accretion disk onto the stellar photosphere and helps brighten and activate the chromosphere. For reasons still not understood, the star then develops a powerful wind, perhaps making use of the falling matter. But the wind cannot penetrate the thick accretion disk: its only avenue of escape is through the disk's poles, where the gas is thin, and it therefore blows in a **bipolar flow** of twin opposing jets (Figure 23.2). The presence of such a disk is consistent with the shape of the reflection nebula associated with the T Tauri star R Monocerotis (Figure 23.3). This star is surrounded by a tiny dust halo about a second of arc and 1,300 AU across that is thought to be the actual accretion disk.

The disks and jets are more directly revealed by strange, fuzzy, variable nebulae found in the 1950s by George Herbig of Lick Observatory and Guillermo Haro of the University of Mexico. These **Herbig-Harlo (HH) objects** are always associated with dark clouds and have no obvious sources of illumination. They (or small groups of them) tend to appear in pairs that flank a star (Figure 23.4). This star, often a T Tauri, might be obscured by a thick clot of dust, but can sometimes be revealed by a deep infrared image. Even when we cannot find the stars we are certain they are there. The HH objects are produced when bipolar flows from the star—speeding in excess of 100 km/s—ram into the surrounding interstellar medium and dissipate their energy in hot, bright shock waves.

Figure 23.5 shows a collection of HH objects and their associated bipolar flows. The flow axes

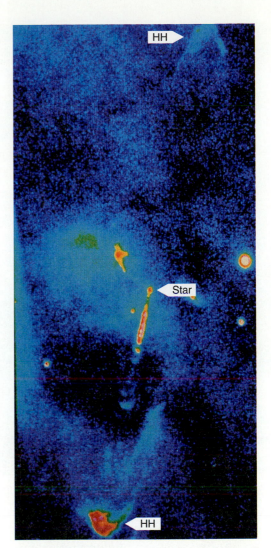

Figure 23.4 *HH 34, in the Orion Molecular cloud about 1° south of the Orion Nebula, consists of two sets of Herbig-Haro (HH) objects. They are shock waves created by jets emanating from the star at center ramming into the surrounding interstellar medium.*

Figure 23.3 *R Monocerotis is at the tip of the fan-shaped reflection nebula NGC 2261. An opaque disk around the star allows illuminating radiation to escape only through its poles. Because the disk is tipped, the "counter-fan" below is dimmed by obscuring dust.*

are almost all aligned with one another as if something in the cloud has choreographed them. Polarization of background starlight reveals an interstellar magnetic field lined up in the same direction. Apparently magnetic fields play a powerful role in stellar development.

To examine star formation further, we must probe deeper into the clouds with infrared and radio observations. Figure 23.6 shows a scene in Taurus with several HH objects. There are no stars between them, and at first they seem unconnected. But with a radio telescope, we see two jets of carbon monoxide issuing from a point in the cloud northeast of them; the western jet causes the HH objects. Molecular hydrogen must accompany the

Figure 23.5   *Six more HH object pairs lie near HH 34, the flow axes of most of them aligned by an interstellar magnetic field running in the same direction. Wisps of gas perpendicular to the flows appear to be aligned by the same field.*

CO, revealing a powerful bipolar flow shooting from an infrared source: a hidden protostar.

Radio observations reveal developing disks as well as jets. The interstellar ammonia molecule radiates its emission lines from particularly dense clouds of gas. Figure 23.7 shows the ammonia (yellow contours) outlining a thick disk of matter. Perpendicular to the dense disk is a bipolar flow of carbon monoxide (blue contours), one jet tilted toward us, the other away. In both cases, the molecules trace the abundant molecular hydrogen.

Molecular clouds are highly fragmented. Figure 23.8 shows the locations of compact knots or **dense cores** of matter typically a tenth of a parsec (20,000 AU) in radius revealed by ammonia radiation. They have densities of a few times 10^4 H_2 molecules/cm^3, masses about that of the Sun, and are very cold, with temperatures of only about 10 K. Dense cores associate with T Tauri stars, and may be the original blobs of matter out of which new stars will be born.

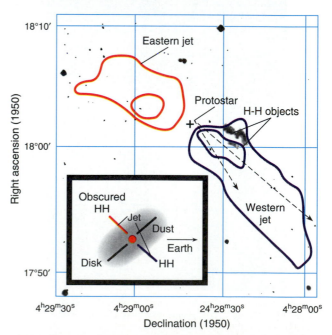

Figure 23.6   *Long jets of molecular gas outlined by CO emission stream from an infrared source (cross) in a dark Taurus cloud. Matter moving away is outlined in red and that flying forward in blue, showing that the flow is tipped to the line of sight (inset). The streamers align with the HH objects to the lower right. Any HH objects to the upper left would be hidden by the intervening dark cloud.*

Figure 23.7 *Yellow curves outlining ammonia radiation show that a thick disk surrounds an infrared star in the dark cloud Lynds 43. A bipolar carbon monoxide flow (blue) runs perpendicular to the disk. The dashed and solid blue lines show receding and approaching matter respectively.*

23.2
The Formation of Solar-Type Stars

We now begin to follow the flow of stellar evolution. Our observations allow us to construct a theoretical pathway that takes lower-mass stars from conception onto the main sequence. Start with a rotating dense core of intensely cold molecular gas and dust. It may be one of many within a giant molecular cloud or by itself within a small Bok globule. The core may have been created when the interstellar medium was compressed by a shock wave from a supernova, by a bubble produced by the wind of a brilliant O star (see Figure 22.15), or by other means to be discussed later.

To form a star, the core must contract. The behavior of a cloud or core depends on the balance of forces within it. Gravity attempts to contract it but gas pressure pushes outward, as does the turbulence stirred up by nearby stars that are forming or already have formed. The dominating factor, however, appears to be the magnetic field that threads its way through the dusty domain (Figure 23.9a). The field strength is locally weak: radio-line Zeeman splitting shows it to be only a few ten-thou-

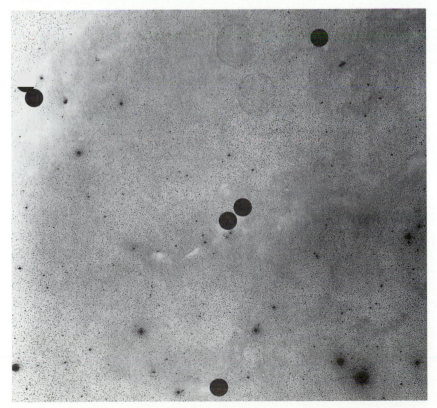

Figure 23.8 *The Taurus-Auriga molecular cloud complex contains dense cores (circles, greatly exaggerated in size) that will ultimately condense to form stars.*

sandths that of Earth. Yet the core is so large that the total amount of magnetic energy involved is huge. A few ions produced by cosmic rays penetrating the core tie themselves to the magnetic field. As the core tries to contract, the ions drag the magnetic field inward, increasing the field strength and providing an outward pressure on the gas. Since the dominant neutral atoms and molecules are always colliding with the ions, they feel the outward pressure too. The result is a stabilizing force that hinders collapse.

Contraction to a star cannot truly begin until the core gets rid of its magnetic field. The neutral atoms and molecules will slowly slide past the ions and fall to the center. As the inside of the core becomes denser it also becomes more neutral, releasing the magnetic field's hold on the bulk of the gas and allowing ever faster contraction. A star is now on its way to forming at the core's center.

The outer part of the core feeds matter into the denser center (Figure 23.9b), the speed of infall

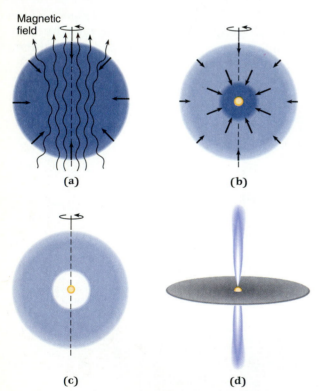

Figure 23.9 (a) *A star first develops as a dense core supported by magnetic pressure, its rotation axis aligned along the field direction.* (b) *As neutral molecules and atoms fall inward, the magnetic field loosens its grip and the core contracts.* (c) *Infalling gas generates a shock wave that heats a protostar a few solar diameters across at the center, creating a clear zone within a surrounding dusty cocoon 10,000 times the size of the Sun.* (d) *Shrinkage makes the surrounding cloud spin faster and develop into a disk, the wind from the star creating a pair of opposing jets.* (*The figures are not to scale.*)

decreasing with increasing radius. The inside of the core gets thicker, the outer more tenuous. The rate of infall exceeds the speed of sound in the cloud, and the incoming gas produces a shock wave that creates heat and a genuine protostar with a radius a few times that of the Sun and a luminosity many times solar. The heat destroys the molecules and the interior dust, clearing a zone out to over 100 solar radii. But the protostar is still surrounded by the outer, cold portion of the dusty cloud, which traps the radiation and converts it to infrared. From Earth we see an infrared source within the molecular cloud, the protostar hidden by a dust shell 10,000 times the size of the Sun, comparable to the size of the planetary system (Figure 23.9c). It may now look something like IRAS 16293-2422 in Figure 23.10.

As the cloud collapses, the conservation of angular momentum causes it to spin faster and faster. Unless angular momentum is removed, the core will tear itself apart before it can contract to stellar dimensions. If the rotation is sufficiently fast, it may indeed do exactly that, and separate into a *pair* of cores, allowing the formation of a binary. A large portion of the angular momentum then goes into orbital motion, permitting the two new cores to continue contraction. These cores may even split again, producing a double-double or even a sextuple star (see Figure 20.3).

A contracting protostar continues to accrete matter, and the temperature rises. After 10^5 to 10^6 years, the interior reaches a temperature of around a million K, and its small amount of deuterium (about one atom out of every 10^5 of ^{1}H) begins to fuse to helium, making the protostar larger. Convection continually sweeps fresh deuterium into the hot center, keeping the nuclear fire burning. As a result of the conservation of angular momentum, the whole assembly of accreting gas flattens into a disk (Figure 23.9d) like the ones around the youthful objects shown in Figures 23.3 through 23.7. A fast wind then blows through the poles of the disk, creating bipolar flows and Herbig-Haro objects. As the star continues to contract, it develops T Tauri characteristics and a strong magnetic field as a result of its rapid spin. The outflowing wind hauls the magnetic field lines outward, producing a drag on the star, removing even more angular momentum and slowing it down. The wind blows the surrounding gas away, dissipating much of the disk. Without the continuous accretion, no more deuterium can fall into the star to be swept into the core. The nuclear fire shuts down and the star dims.

Figure 23.10 *IRAS 16293-2422 (inset) is a cold (20-K) infrared source 20 times more luminous than the Sun, a protostar buried in a dusty cocoon set within the Ophiuchus dark cloud (large photograph). Radio observations of carbon monosulfide show that it is accreting matter. Note the newly formed cluster.*

Theoretical calculations show how the luminosities of protostars change with effective (photospheric) temperature. We can plot graphs of these changes, called **evolutionary tracks,** on the HR diagram (Figure 23.11) and compare them to the locations of observed stars. As the debris of formation clears, our dimming T Tauri star and others with a variety of masses, slowly emerge from their cocoons and become optically visible as they cross the *birthline.* Eventually the star's luminosity stabilizes, and contraction causes its photosphere to heat, and on the evolutionary tracks of Figure 23.11, to move to the left. The remaining accretion disks disperse and the T Tauri activity slowly quiets down.

The steady contraction of a pre–main-sequence star produces heat and considerable luminosity. As the interior temperature passes about 7 million K, protons begin to fuse, initiating the proton-proton chain. This immense new energy source stabilizes the star, stopping the contraction. The star then settles down onto a line on the HR diagram called the **zero-age main sequence** (ZAMS) with a fresh hydrogen supply and the prospect of a long, quiet life.

T Tauri stars become main-sequence stars of classes roughly K though A with masses that range

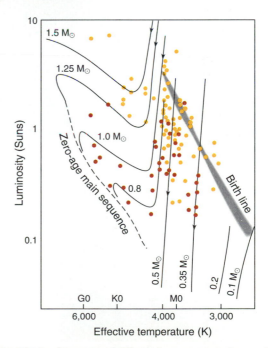

Figure 23.11 *Protostars of different mass contract along evolutionary tracks (black lines) on the HR diagram. They become visible as T Tauri stars (yellow dots) when they cross the birthline. As they shed their envelopes they turn into quieter T Tauri stars (red dots). When they are hot enough to fuse their internal hydrogen, they settle onto the zero-age main sequence (dashed line), their positions depending on mass.*

between about 0.75 and 3 M$_\odot$. The creation of higher-mass B stars is similar, involving higher-mass versions of the T Tauri variables. Stars of lower mass probably form in much the same way, but their pre–main-sequence versions are faint and hard to detect within the dust clouds.

For a star of a solar mass, the process takes about 30 million years. The star's motion gradually takes it away from its birth cloud, and eventually we see our lonely Sun, or perhaps a double like α Centauri, making its way around the Galaxy, its original home long lost.

23.3
Higher-Mass Stars

Curiously, local sites of star formation manufacture no massive O stars. To see how these are made we must go far afield, to distant star-forming clouds that are inherently difficult to study (Figure 23.12). Why different clouds produce different kinds of stars and

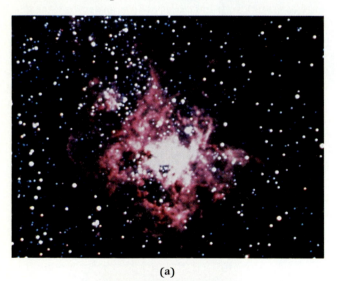

(a)

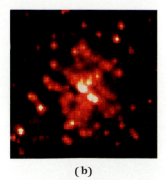

(b)

Figure 23.12   (**a**) *The Tarantula Nebula is a great producer of O stars.* (**b**) *Within the nebula is the object R136a, resolved by the Hubble Space telescope into a cluster of high-mass stars.*

why some produce open clusters and others only loose T and OB associations, are still mysteries. Perhaps only the greatest of the molecular clouds is capable of creating the most massive of stars. Unfortunately, breeders of O stars are so far away that we know little about their production of low-mass stars and how O associations relate to T associations. It is thus difficult to relate the observed numbers of stars within the various spectral classes to the formation rates for stars of different masses.

The formation of O stars is different from that of lower-mass stars. High-mass stars are much hotter than low-mass stars and consume their fuel much more rapidly. All stages, including formation, proceed at a greatly increased rate. An O star is expected to begin hydrogen fusion even before it has stopped accreting matter. Its enormous luminosity hollows out a shell within the birth cloud, which the ultraviolet flux lights up to produce a compact H II region. As the O star develops it sweeps away its surrounding cloud, and its wind compresses the encompassing interstellar medium into a bubble (Figure 23.13). O stars may then beget themselves, the compressing bubbles producing more O stars as the entire surroundings condense in sequential star formation.

The upper limit to stellar mass is unknown. Massive stars are apparently so difficult for nature to make that the chance of any with masses greater than about 120 M$_\odot$ developing within any galaxy is effectively zero.

23.4
Brown Dwarfs

The *lower* limit to the main sequence is imposed by interior temperature. Below 0.08 M$_\odot$, gravitational compression does not generate enough heat to raise the interior above the critical level of about 7 million K required for full hydrogen burning. Yet there seems to be no reason why nature cannot form substellar bodies below 0.08 M$_\odot$. For several years, astronomers have been searching for these **brown dwarfs.** For a time, they should appear bright enough to see, especially in the infrared. Gravitational contraction heats them, producing radiation, and the internal temperatures can exceed the million-degree limit required for the burning of their natural deuterium. There have been a large number of candidates (Figure 23.14), but in spite of great efforts, none has ever been confirmed as a true brown dwarf.

Figure 23.13   *An H II region in the Large Magellanic Cloud called N 70 is actually a giant bubble that the wind of an O star has blown in the interstellar medium.*

Why would nature stop making bodies at the limit at which the internal temperature becomes so low that the proton-proton chain can no longer operate? The answer is important to our understanding of how interstellar clouds produce stars and, as we will see in Chapter 26, to the mass of the Galaxy.

23.5
Planets

We do know, however, that nature makes bodies *much* less massive than stars. In our own Solar System, it made Jupiter and the other planets. Jupiter is not massive enough to initiate thermonuclear fusion, even deuterium fusion, and it is not a brown dwarf. A brown dwarf is expected to be cre-

ated whole as a result of contraction and accretion in a dense cloud. Jupiter and the other planets have been accumulated (at least in part) from a vast number of primitive planetesimals.

We can now finish the story begun in Chapter 17, where we saw that the planetesimals and planets are distributed *in a disk about the Sun.* The disk appears to be the remnant of the original accretion disk that must have formed around the protosun over 4.6 billion years ago. The dust in this disk accumulated to form larger pellets, and finally planetesimals.

We now also see how the inner asteroids may have been heated, the process explaining the observed variation in their properties with distance from the Sun. The strong magnetic field that the Sun had as a T Tauri star likely did at least part of the job. More profoundly, the dust that created the planets came from interstellar space, and initially from the ejected shells of Mira variables. *We are the products of other stars, our Earth a distillate not just of the solar nebula but of the dusty, long-since dissipated interstellar cloud that gave us birth.*

If the Sun created planets from its spinning prestellar disk, other stars might have developed planetary systems too. Even Jupiter, Saturn, and Uranus have their own little sets of "planets," their satellites, further suggesting how common these systems might be. In fact, the process may be another important way in which dense circumstellar disks dissipate. Can we find evidence for extrasolar planets? In 1984 *IRAS* detected the infrared signatures of dusty disks surrounding Vega and Fomalhaut. Others have also been seen. Our own planetary disk is filled with dust from cometary debris and from small particles caused by planetesimal—asteroid—collisions. The Sun has a dusty disk with planets embedded within it. Perhaps Vega and Fomalhaut do, too.

(a)

(b)

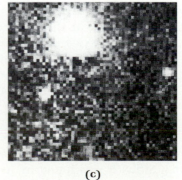

(c)

Figure 23.14   *A possible brown dwarf in the Pleiades is brighter in the infrared* **(a)** *than in the red* (**b**)*, showing that it is cool.* **(c)** *Another brown dwarf candidate hovers near an M dwarf. Both bodies appear to be only faint M stars at the end of the main sequence.*

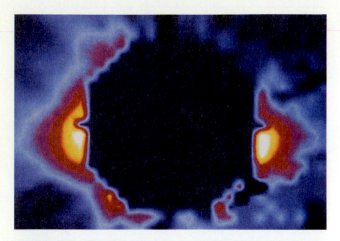

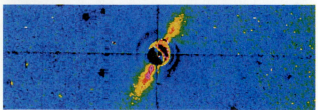

Figure 23.15   *A disk of heated gas surrounds β Pictoris (the star itself is eliminated from the picture) with silicates shown in yellow, ices in red. The extent of the disk (400 AU in radius) is shown below.*

More telling was the discovery in 1985 that β Pictoris is surrounded by an edge-on disk 400 AU in radius, ten times larger than our own planetary disk (Figure 23.15). Our Solar System, however, extends far beyond Pluto and into the Kuiper belt of comets that apparently circulates in a thick disk about the Sun. We have little idea of how far this belt may actually extend. Infrared observations show that the β Pictoris disk contains silicates and ices mixed with carbon, recalling the compositions of comets and asteroids in our own system. Moreover, spectroscopy with the Hubble Space Telescope indicates lumps of matter apparently spiraling in toward the central star. It seems possible that the β Pic disk contains planetesimals and maybe, in its inner portion, real planets.

But can we *see* the planets? Turn the problem around and ask whether someone out there with our instruments could see *our* solar system. From α Centauri, Jupiter would shine at 22d magnitude, an easy sight with modern detectors. However, it would appear at most only 4 seconds of arc from a bright star whose glare would render it invisible. Direct looks at planets orbiting other stars will be exceedingly difficult.

The α Centaurians might, however, use gravitational effects. Recall that Sirius B was found by its

deflection of Sirius A from a straight-line path. Jupiter's mass is only a thousandth the Sun's, effectively placing the center of mass of the Solar System only 0.005 AU—the solar radius—from the solar center. From α Centauri the deflection would be a mere 0.004 seconds of arc, beyond our current detection capability.

A better ploy would be to use radial velocity variations. Since the Sun orbits the center of mass of the Solar System, it is technically a single-lined spectroscopic binary. The circumference of the solar orbit is 0.03 AU ($2\pi \times 0.005$), about 4 million km, which the Sun traverses in Jupiter's sidereal period of 12 years (about 4×10^8 seconds). The solar orbital speed is then 10 m/s, close to the current error for this type of observation. From α Centauri, we would be within a factor of 2 or 3 for detection with a minimum signal-to-noise ratio.

Return now to Earth. It is likely that some stars have more-massive planets that would make them move faster than the Sun. We can also improve the odds of detection by observing more easily deflected lower-mass stars. Several studies have shown such radial velocity wobbles. One candidate is ε Eridani (Figure 23.16), a K2 dwarf 3.3 pc away. The mass of the star can be estimated from its luminosity. With assumptions about the distance of the hypothetical planet from its parent and the inclination of the system, astronomers have suggested a planetary mass between one and five times that of Jupiter.

Yet we must remain uncomfortable with such data. Other phenomena associated with stellar photospheres may well cause slow radial velocity variations. We need to increase the signal-to-noise ratios and to observe for much longer periods of time to see if the variations are smoothly repeated, as would be expected if they were caused by orbiting planets. The totality of the evidence, however, leads us to believe that stellar planets are indeed there, and are a natural part of star formation.

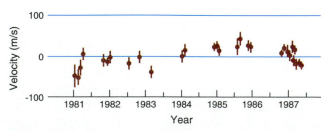

Figure 23.16   *The radial velocity of Epsilon Eridani slowly changes over a period of several years. Does the star have a planet?*

BACKGROUND **23.1** Sunlike Stars

It is only natural to assume that the best chance for finding other civilizations would be to look at stars like the Sun. It is also intriguing to look at such stars simply to see how our Sun might look at a great distance. Table 23.1 gives a short list of the stars that carry Greek letter names or Flamsteed numbers that are most similar to the Sun. All six are classified at or very near G2 V; 16 Cyg A and B are G1.5 and G2.5 respectively. The closest match is 9 Ceti. Note the large spread in absolute magnitude, in part a result of the stellar aging process that will be a focus of the next chapter.

TABLE 23.1 Other Suns			
Name	Apparent Visual Magnitude V	Distance (pc)	Absolute Visual Magnitude M_V
53 Aquarii A	6.57	18	5.35
α Centauri A	−0.01	1.34	4.37
9 Ceti	6.39	20	4.84
ρ Coronae Borealis	5.41	25	3.42
16 Cygni A	5.96	26	3.92
16 Cygni B	6.20	26	4.16

23.6
Life

The ultimate question is not really about other planets but about what may be on them. Before we speculate, however, we must look at our own world.

23.6.1 *Life on Earth*

We have only a few conjectures of how life formed on Earth. But form it did, and quickly, too. In Australia there are primitive fossils of organisms that were alive only a billion years after the planet was created. In a famous 1953 experiment, Stanley Miller and Harold Urey of the University of Chicago passed electric currents simulating terrestrial lightning through a mixture of methane, ammonia, hydrogen, and water that was meant to recreate Earth's early atmosphere. After several days, they recovered a goo of complex molecules that included amino acids, the basis for the formation of life. Such an experiment hardly proves that the seeds of life were initiated in this way, but it certainly provides a basis for speculation. However, it is now believed that the early atmosphere differed from what the experimenters assumed, rendering their test less meaningful.

Another speculation involves meteorites and comets, which contain complex molecules. Since comets were made from the dust that came from interstellar space, it is not unreasonable to think that they also contain the molecules found in molecular clouds and maybe others far more complicated. Perhaps more telling are the amino acids found in meteorites. The comets that crashed into Earth during the heavy bombardment apparently brought our vast stores of water, and perhaps also the original building blocks out of which life eventually developed. It is fascinating to think that not only did the Earth come from interstellar space, but that maybe *we* did, too.

If that was the case, late-arriving planetesimals must also have seeded the Solar System's other planets. However, only the Earth has a near-universal solvent, liquid water, that might have allowed the seeds to grow into more complex molecules, the RNA and DNA that support all of life. Only the Earth was in the right place relative to the Sun—not so far that the water froze, as it did on the outer planets, and not so close that it boiled away, as it did on Venus and Mercury.

If this picture is correct, what happened on Mars? Our neighboring planet once had both water and a thicker atmosphere. If life developed quickly here, why did it not develop there as well? What does that observation say about speculation on the formation of life in general? Or did Martian life actually begin to develop only to die away as the planet turned into a runaway refrigerator? Perhaps we have not looked in the right place to uncover the primitive fossils that may be left.

The origin of life on our planet—and of ourselves—is shrouded in deep mystery. We have only one sample planet on which the formation took place, and it happened in the dim past; most of the

BACKGROUND   **23.2**   **Unidentified Flying Objects**

The Search for Extraterrestrial Intelligence has the great misfortune to be confused in the public mind with unidentified flying objects (UFOs). The two have utterly nothing to do with each other. For centuries, people have been sighting things in the sky—lights and other objects—that they have not understood. Some then make a great leap of logic and propose that, since the objects are unexplained, they must be of extraterrestrial origin and must contain visitors from another planet.

UFOs fall into three categories: natural phenomena that the viewer cannot explain, hoaxes, and delusions. The last of these cannot ever really be "explained," as the evidence is not real. Hoaxes are remarkably easy to fabricate, and many have been perpetrated on an unsuspecting public. Photographs are simply not good evidence. All you need do is take an out-of-focus picture of a flying garbage can lid and you have a picture of a UFO you could probably get published in a supermarket tabloid. Figure 23.17a shows another example, an upside-down banana-split dish hung on a wire.

The natural phenomena taken for UFOs are far more interesting. Venus tops the list. As you drive along a highway it seems to follow you. A jet pilot once tried to shoot it down, and a sheriff and his deputy in Ohio chased it into Pennsylvania, where it disappeared (most likely, the Sun came up). Stars have been mistaken for UFOs, as have been mirages (which are far more common than most people think), oddly shaped clouds, and lights seen through clouds (Figure 23.17b). Fireball meteors, with their long persistent trains, have been mistaken for tubular spaceships. There is absolutely no evidence that any UFO ever seen has anything whatever to do with extraterrestrial visitors.

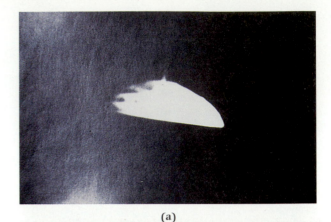

(a)

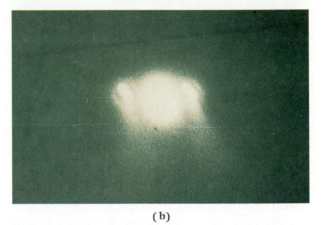

(b)

Figure 23.17   *Photographic evidence of UFOs is not reliable.*
(a) *An upside-down banana-split dish is hung on a wire.*
(b) *The landing lights of an airplane are seen through fog.*

Real star travel is a massively difficult problem. There is not enough energy on Earth to send even a robot craft to the nearest star in a reasonable time, within a human lifetime. The propulsion problem may be solved sometime in the distant future. But if it is, it will have nothing to do with unidentified flying objects, which are creations of the human mind.

---

evidence has been destroyed by erosion and plate tectonics. If we cannot find further evidence in the Solar System, we must look elsewhere, to the stars.

23.6.2 *Life in Deep Space*

Are we alone, or does life exist on some other star's planet? As planets may be a by-product of star formation, is life a natural by-product of planet formation? If it exists, can we detect it? We have not yet

even been able to find extrasolar planets, so we are certainly not going to make observations of plants or insects. The search involves not just life, but *intelligent* life, something similar, however vaguely, to ourselves, something with which we might communicate even though the harboring planets are still invisible.

Before searching, however, we need to investigate the possibilities of life elsewhere. Any speculations involve some variation on the **Drake equation,** initially formulated by astronomer Frank Drake. The probability (P) that a star harbors intelligent life is

$$P = f_p n_h f_l f_i f_t.$$

The first factor, f_p, is the fraction of stars that have planets, and is certainly the best known of the five. Given the number of close binaries, which may preclude planet formation, the value may be 0.1 or even higher. The second, n_h, is the number of planets per star expected to lie in a habitable zone. Around each star there should be a shell in which the planetary temperature would support liquid water. From our own Solar System, the number is 1; if Mars is included, it might be 2. Computer modeling of planet formation indicates that Earthlike planets generally form at distances of about 1 AU from their star.

The other factors are subject only to conjecture: f_l is the fraction of planets in a habitable zone that actually evolves life. In our Solar System, its value is 1 (or, considering Mars, maybe 0.5). But we do not know how life formed: if it was simply by an incredible stroke of luck, the number may approach zero. Then, presuming that life *does* form, what fraction (f_i) develops into *intelligent* life? Again, on Earth it is 1—but over the large scale of the Galaxy or the Universe it may be close to zero. We do not know. Finally, for there to be intelligent life elsewhere for us to contact, we need some idea of how long intelligent life-forms last once they develop. That is expressed by the fraction of planets on which life manages to endure, or the ratio of the lifetime of a civilization to the age of the Galaxy, f_t. Assuming the Galaxy to be 10^{10} years old, and reasoning from the Earth, f_t (so far) is only 10^{-4}. Other intelligent societies may have been around for a billion years, a significant part of the Galaxy's age, and thus f_t could be as high as 0.1.

Multiplying the components gives the fraction of stars in the Galaxy that shelter intelligent life. Taking the highest possible values yields perhaps 10^{-2}, and since there are some 10^{11} stars in the Galaxy, there could be a billion civilizations! The closest one could be on a planet orbiting a neighboring star. But perhaps the chance of evolving intelligent life from single-celled organisms is only one in a billion. Then the final fraction is 10^{-11}, which yields only a single star: our own.

The argument among scientists rages on. There are two opposed camps, one convinced that the upper number is close to correct, the other that there is more truth in the lower. The fact, of course, is that we just do not know. Given the absence of data or even accurate speculation, all we can do—like scientists everywhere and in every discipline—is to go and look.

23.6.3 *The Search for Life*

In spite of the uncertainty, there is clearly a possibility that, because of the enormous number of stars in the Galaxy, the final fraction of the Drake equation is relatively high and somewhere out there other intelligent beings exist. If so, an unknown (and probably very small) fraction of them may have developed a technology like ours that would allow communication. Perhaps they share our wonder and are looking for us. If so, maybe we can make contact.

This subject, the **Search for Extraterrestrial Intelligence** (SETI) has long been taken seriously, and has been endorsed by the International Astronomical Union (the world organization of professional research astronomers), the U.S. National Academy of Sciences, and the National Commission on Space. Because of the immense distances between stars, SETI cannot involve any direct personal contact but communication by radio.

For most of this century, the Earth has been bright in the radio spectrum. We have beamed countless radio and television broadcasts into the cosmos and have announced ourselves with a growing bubble of radiation now approaching 100 light-years in radius, one that encompasses thousands of stars. Maybe some other society is beaming similar radio broadcasts. Or perhaps some civilization is deliberately looking for company by sending specific signals or messages. The task is to look into space to try to find some artificial signal, something periodic that is clearly not part of the cosmic radio-noise background.

The search is daunting, the reward measureless for both science and philosophy. Where do we start in the vastnesses of the sky and the radio spectrum? What stars do we look at, where do we tune our receivers? The first step was taken in 1960 by Drake, whose Project Ozma examined ε Eridani and τ Ceti. He chose radio frequencies near 21 cm on the assumption that if some extraterrestrial engineers were broadcasting an intentional signal, they would place it near an obvious and recognizable spectral reference. He found nothing.

There have since been many, more sophisticated, projects. One at Harvard called META (Megachannel Extraterrestrial Assay) uses a 26-meter radio telescope and a receiver that can scan 8.4 million radio channels at the same time. A second META observatory with two 30-m dishes operates in Argentina. The biggest effort is NASA's 10-year High Resolution Microwave Survey, which employs the 34-m dishes of the Deep Space Network used for tracking space probes, the giant Arecibo radio telescope, and other instruments. It will eventually simultaneously scan 30 million radio channels. Part of this program will involve a survey of the entire visible sky, and another search will concentrate on some 800 nearby solar-type stars. So far, however, with all the effort, there is still nothing.

The lack of data should not yet be discouraging. The Galaxy is enormous, and the signals, especially if they are passive (that is, not produced by a transmitter actively trying to contact someone), may be weak. Is anyone else out there? We will never know if we do not look. Perhaps a century into the future we will have a better answer. So far, however, for all the grand speculation, the search for intelligence in the Universe is a science with but one datum: ourselves.

KEY CONCEPTS

Bipolar flow: Gaseous jets emitted from an object's poles.

Brown dwarfs: Bodies that contract directly from the interstellar medium but with masses below the main sequence cutoff of 0.08 $M_\odot$, too low to allow operation of the proton-proton chain.

Dense cores: Knots of matter in the interstellar medium, observable by ammonia radiation; the first observable step in star formation.

Evolutionary tracks: Graphs of luminosity against effective temperature (or M_V against spectral class) on the HR diagram.

Herbig-Haro (HH) **objects:** Knots of gas in the interstellar medium lit by shock waves from bipolar flows.

Pre–main-sequence stars: Stars that have developed from the interstellar medium but have not yet arrived on the main sequence.

Protostars: Stars in the process of formation.

Search for Extraterrestrial Intelligence (SETI): The search for extraterrestrial civilizations by radio.

T Tauri stars: Active, young, variable pre–main-sequence stars that group into **T associations.**

Zero-age main sequence (ZAMS): The line on the HR diagram outlined by new stars.

KEY RELATIONSHIPS

Drake equation:

$P = f_p n_h f_l f_i f_t$, where P = the probability of a star harboring intelligent life, f_p = the fraction of stars with planets, n_h the number of such planets in a star's habitable zone, f_l the fraction on which life will form, f_i the fraction on which intelligent life develops, and f_t the fraction on which it lasts.

EXERCISES

Comparisons

1. How are T and OB associations **(a)** different; **(b)** similar?
2. How do formations of low- and high-mass stars differ?
3. How do brown dwarfs differ from real stars and from planets?

Numerical Problems

4. A protostar is ten times more luminous than the Sun and has an effective temperature of 100 K. How big is its dust photosphere in AU?
5. What is the maximum deflection in proper motion and the maximum Doppler shift at 5,500 Å of a 0.5 $M_\odot$ star 10 pc away caused by a planet with 1 Jupiter-mass and an orbit 1 AU in radius?
6. What is the approximate apparent magnitude and maximum separation of Earth from the Sun as seen from α Centauri?

Thought and Discussion

7. Why do we think that T Tauri stars are young?
8. What is the evidence for mass-loss in T Tauri stars? For disks and bipolar flows in these and other protostars?
9. How are Herbig-Haro objects produced?
10. How can we tell the direction of a flow observed by radio?
11. What is the observational evidence for dense cores?
12. Why does a dense core not collapse quickly?
13. How can a developing star lose its angular momentum?
14. How does deuterium fusion affect a protostar? What stops the fusion?
15. On the HR diagram, what is the "birthline"?
16. What happens inside a star that settles it onto the main sequence?
17. Why is it impossible to conclude that there is no life elsewhere in the Universe?

Research Problem

18. Brown dwarfs and extrasolar planets have proved extremely elusive. Find instances in which discoveries have been publicized and later withdrawn. What led to the false discovery?

Activities

19. Make a list in order of development of catalogued protostellar objects or pre–main-sequence stars and their properties. Diagram the sequential development of the Sun up to the time it reaches the main sequence and develops planets.

20. Use a telescope to find one or more of the sunlike stars listed in Table 23.1.

Scientific Writing

21. Write an essay in which you solidly link the Sun's planetary system with the disks that are believed to exist about T Tauri stars. Summarize all the evidence from the Solar System that shows why we believe the planets formed in a disk, and summarize the evidence that T Tauri stars have disks.

22. Your local educational television station plans to produce a program that will address the existence of intelligent life in the Universe in the form of a debate between two people. You are hired to write a script that must allow each participant to answer and rebut the arguments of the other.

24

The Life and Death of Stars

*The flow of
stellar lives from the
main sequence
and the quiet deaths
of stars
like the Sun*

*The Dumbbell Nebula
in Vulpecula
wreathes a dying star.*

Stars can live for extraordinarily long times. With a few spectacular exceptions, we can watch the sky our whole lifetimes and will see little change. We can still study stellar evolution, however, by stringing the different kinds of stars together with theory. We now know that the Sun and stars will someday expire, their deaths providing sustenance for new stars, the cycle repeating into the unimaginably distant future.

24.1
The Main Sequence

The evolution of stars is ultimately the result of gravitational contraction. Physical systems try to seek their lowest energy states: a dropped body falls to Earth, an atomic electron descends to the ground level, and stars make themselves as small as possible. The phases of a star's life involve pauses in the shrinkage or the act of shrinkage itself. In the first stage, a dense core contracts to a protostar, then to a T Tauri star, then finally to a main-sequence star. The main sequence is the first pause.

24.1.1 Conditions Along the Main Sequence

The Sun is very stable because the inward pull of gravity is balanced by the outward push of gas pressure. The solar luminosity depends on the internal temperature generated by gravitational compression, and consequently depends on the solar mass. Stars of higher mass have higher internal temperatures and greater luminosities. Simple theory shows that they should brighten in accord with the general form of the mass-luminosity relation ($L \propto M^{3.5}$) as found from binary stars.

Thermonuclear fusion of hydrogen (commonly called nuclear burning) is not immediately responsible for the Sun's luminosity but for the duration over which that luminosity can be sustained. The energy provided by fusion allows the Sun a 10-billion-year pause in its perpetual attempt to contract. Since main-sequence stars are made from the same material, they apparently must also be driven by hydrogen fusion. The main sequence is thus a zone of long-term hydrogen-burning stability.

Solar energy is produced by the proton-proton chain (see Figure 18.27). About the same time this set of reactions was discovered, Hans Bethe realized that hydrogen could also be fused into helium by using carbon as a nuclear catalyst (Figure 24.1). If the temperature is high enough, a proton can penetrate a ^{12}C nucleus to create ^{13}N and a gamma ray. ^{13}N is unstable, decaying into stable ^{13}C when a

proton ejects a positron and becomes a neutron. The positron quickly strikes an electron and creates two more gamma rays. The ^{13}C then absorbs another proton, making stable ^{14}N, which then absorbs yet another proton to create unstable ^{15}O, producing a gamma ray at each step. The ^{15}O decays by positron ejection, becoming stable ^{15}N. When a fourth proton bangs into ^{15}N, the atom ejects a whole helium nucleus and drops back to ^{12}C. By this **carbon cycle,** four atoms of hydrogen are turned into one of helium, energy is created, and the carbon is unchanged.

The carbon cycle is more temperature dependent than the p-p chain. At lower masses, among the K and M dwarfs, it is insignificant, and even in the Sun accounts for only 7% of the nuclear energy. Above 1.8 $M_\odot$, however (near spectral class F0), the carbon cycle dominates, and for the massive O and B stars, the p-p chain can be largely ignored. Once the carbon cycle becomes important, it changes the structure of the core, making energy move by convection rather than by radiation as it does in the solar core (see Section 18.5.2).

At the same time the carbon cycle is becoming important, convection disappears in the envelope, causing energy to move by radiation. The magnetic dynamo diminishes, magnetic braking is no longer

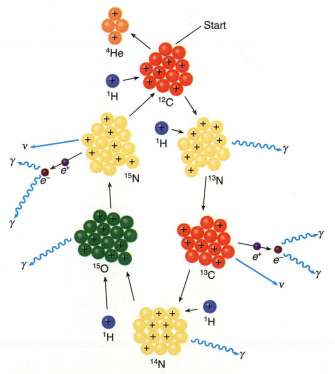

Figure 24.1   *The carbon cycle uses carbon to fuse four protons into a helium nucleus through the creation of isotopes of carbon, nitrogen, and oxygen. Each reaction except the last produces one or two gamma rays.*

effective, and hotter than class F5 or so, main-sequence stars spin faster (see Section 19.6).

At the very top of the main sequence, we find another change in conditions. The hottest O stars—where masses reach into the tens of $M_\odot$—are so extraordinarily luminous that we must even take into account the outward pressure of the stars' own radiation in the construction of interior models.

24.1.2 Stellar Lifetimes

From the rate at which the Sun's hydrogen fuel is consumed, we find the solar main-sequence lifetime to be about 10 billion years. We might expect that more-massive stars, those with larger fuel supplies, should live longer. The truth is the opposite. The lifetime of a fire, whether nuclear or wood, depends on the amount of fuel available divided by the rate at which it is burned. In the simplest picture, the amount of fuel is proportional to the stellar mass, M, and the rate of burning is proportional to the luminosity, L. The lifetime, τ, must then be proportional to M/L. But $L \propto M^{3.5}$, so

$$\tau \propto M/L = M/M^{3.5} = 1/M^{2.5}.$$

Higher-mass stars therefore have dramatically *lower* lifetimes. Vega, 2.5 times more massive than the Sun, will live only one-tenth as long, only a billion years, insufficient time for the development of life as we know it on Earth. On the other hand, the M5 dwarf Proxima Centauri, with half a solar mass, will live ten times as long as the Sun, an amazing 100 billion years.

For accurate assessment of lifetimes over the entire range of the main sequence (Figure 24.2), we need to use the detailed form of the mass-luminosity

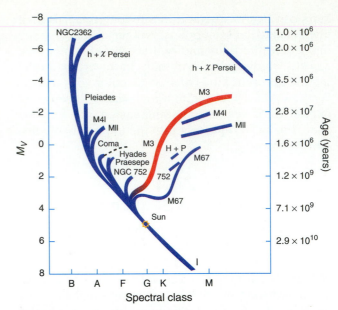

Figure 24.3   *As open clusters age, they lose their main sequences from the top down. Ages that correspond to the top of the main sequence are given at the right. The globular cluster M 3 looks younger than M 67 because the positions of its stars are shifted to the left because of low metal abundance; M 3 is actually considerably older than any of the open clusters.*

relation (see Figure 20.14), not just the average relation. Moreover, only the deep core of a star is hot enough to allow nuclear burning. We therefore also need to factor in the ratio of the *burnable* mass to *total* mass, a ratio that *depends* on total mass. We also have to consider internal structural changes that alter the size and mass of the nuclear-burning core with age.

When these factors are taken into account, we see that a 120-$M_\odot$ O3 star will stay on the main sequence for a mere three million years. At the lower main-sequence mass limit of 0.08 $M_\odot$, stars will live for over a *trillion* years. Below this mass limit, however, a brown dwarf (see Section 23.4) lives only on gravitational contraction and some deuterium-burning. Once formed, it fades relatively quickly.

The differences among the HR diagrams of open and globular clusters (see Figures 20.18 and 20.24) are now understandable. A cluster is born from the interstellar medium with an entire main sequence intact. As time passes, higher-mass stars evolve faster and disappear from the main sequence to become giants and supergiants. The older a cluster, the lower the main sequence's stopping point (Figure 24.3). From theory and the observed point where the main sequence ends, we can derive the cluster's age.

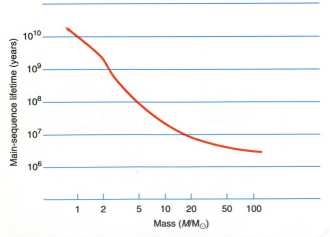

Figure 24.2   *As the masses of stars increase, their main-sequence lifetimes rapidly decrease.*

24.1.3 Evolution on the Main Sequence

A stellar interior is a self-adjusting device. The pressure of a gas depends on temperature and the number of atoms per unit volume, not on their kinds (see Section 18.5.2). As fusion proceeds, four atoms of H are turned into one of He, and the number of atoms drops. To maintain pressure, the gas compresses. Temperature and density (in g/cm^3) climb, and the rate at which the particles fuse offsets the diminishing fuel supply. The main sequence is therefore a zone of stability in the HR diagram.

However, because of the massive internal changes taking place—the abundance of hydrogen decreasing from 90% to zero—the compensation cannot be perfect. As a result, stars undergo small evolutionary changes even as main-sequence dwarfs. Study of any kind of evolution requires detailed calculations of successive models of the star that incorporate changing conditions. The result is an evolutionary track on the HR diagram.

The main-sequence evolution of the Sun provides a fine example (Figure 24.4). Calculations show that at birth our star was about 200 K cooler at the surface, 30% fainter, and nearly 8% smaller than it is now. Over the last 4.6 billion years it has slowly become brighter, larger, and hotter at the surface. These changes will slowly accelerate over the next 4 billion years. By the time the fuel is gone, 5.4 billion years from now, the Sun will be twice as bright and 50% larger.

All main-sequence stars behave in somewhat the same fashion, undergoing slow changes in luminosity and effective temperature with time. As soon as they are born, they begin to "move" (that is, they change graphical position) off the zero-age main sequence either upward or to the right in Figure 24.5. The more massive the star, the greater the changes in external characteristics. As a result, the observed main sequence is not a *line* but a *band* that is narrow at the bottom and flared out toward the top (see Figure 19.18), with the ZAMS along its left-hand edge (see Figures 24.4 and 24.5).

What do these changes mean for life on Earth? In 5 billion years we will receive as much energy as Venus does now, and it is not unreasonable to think that the Earth will be a greenhouse oven long before solar main-sequence life is finished. However, remember that billions of years are involved, compared to which humanity's entire history on the planet is insignificant.

More intriguing is the problem of the past. When life began on Earth 3.5 or so billion years ago,

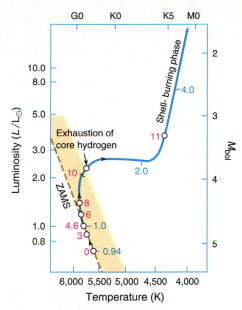

Figure 24.4 *As the Sun ages from the zero-age main sequence, it moves up and a bit to the left on this version of the HR diagram. Ages in billions of years are indicated by red numbers, radii (in current solar units) by blue. The band of the observed main sequence is shaded yellow. When in 5.4 billion years the core hydrogen is gone, nuclear burning will continue in a shell; the Sun will first cool at the surface, then brighten to become a subgiant (blue curve).*

the Sun should have appeared more like it does today from Mars. Did a greater dynamo caused by once-faster solar rotation produce enough chromospheric and flaring activity to warm the Earth? Did a different terrestrial atmosphere provide more insulation? Clearly, the past history—as well as the future—of the Earth are not simple, and for now both are beyond our ability to evaluate.

24.2
Giants and Supergiants

A star waits quietly on the main sequence for perhaps billions of years. Then a critical moment arrives: the nuclear fuel in the core exhausts itself and fusion ceases. What happens next depends almost entirely on the star's mass.

24.2.1 Realms of the Main Sequence

The main sequence can be divided into three parts, the first imposed by the age of the Galaxy. When compensation is made for metal abundance, the globular clusters have the dimmest and coolest main-sequence upper cutoff and must therefore be the Galaxy's oldest systems. Application of theory shows the oldest of them to have ages approaching 17 billion years. No star with a main-sequence lifetime greater than 17 billion years (and above the main-sequence cutoff of 0.08 M$_\odot$) has ever had time to evolve away from the main sequence. This lifetime corresponds to a star of 0.8 solar masses and, in Population I, to spectral class K0. Stars below this limit are said to be on the **lower main sequence** (see Figure 24.5). Since none of them has ever evolved, discussion of the evolution of the lower main sequence is moot.

As we will see, main-sequence stars of up to about 8 solar masses produce white dwarfs. Above 8 M$_\odot$, white dwarfs are not possible and the stars may collapse and explode. This mass corresponds to class B2, near the point that divides reflection and diffuse nebulae. The evolution of the **upper**

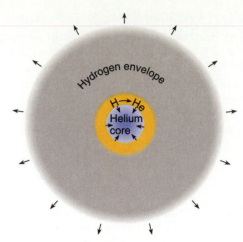

Figure 24.6   *As a star begins its evolution to gianthood, it has a quiet, shrinking helium core surrounded by a shell of burning hydrogen encased in an expanding hydrogen envelope.*

main sequence (above 8 M$_\odot$) is therefore fundamentally different from that of the **intermediate main sequence,** which lies between 0.8 and 8 solar masses. For now, we will consider the evolutionary patterns of both the upper and intermediate main sequences. When the two begin to diverge significantly, we will abandon the upper main sequence and pick up the thread of its tale in the next chapter.

24.2.2 Giants

The cessation of fusion in the solar core 5.4 billion years from now will remove a source of internal support. The nearly pure helium core will finally be able to resume its contraction. An ordinary fire would dim, but because of the huge gravitational energy released, an evolving stellar core becomes *hotter*. The increased heat energy allows hydrogen burning to expand into a shell around the now-quiet core (Figure 24.6), for a time keeping the Sun very much alive.

For about a billion years, the Sun's luminosity will stay roughly constant. However, the new energy of the shrinking core and resulting structural changes will nearly double the size of the outer solar envelope. As a consequence, its surface will cool to class K, near 4,500 K. The nuclear-reaction rates in the shell will then become so intense that the Sun will begin to brighten. As the core—the real heart of our star—heats and condenses, the outer parts continue to balloon outward. The Sun will then take another billion-plus years to climb the **red giant branch** of the HR diagram (see Figures 24.4 and 24.5), first becoming a subgiant and then a true red giant of class M with an effective temperature of only about 3,500 K.

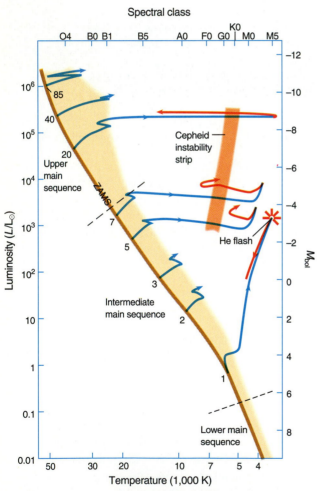

Figure 24.5   *The main sequence is indicated in yellow and the evolutionary tracks for stars becoming giants with dead helium cores in blue. No lower-main-sequence star (0.08 to 0.8 M$_\odot$) has ever had time to evolve off the main sequence. From the intermediate main sequence (0.8 to 8 M$_\odot$), solar-type stars become cooler and brighter as they turn into giants. From the upper main sequence, above 8 M$_\odot$, stars cool at roughly constant luminosity as they become supergiants. The ascents of intermediate-mass stars are terminated by helium burning (shown in red). Solar-mass stars descend the HR diagram, but those of higher mass loop back to the left. As stars cross the instability strip they become Cepheids.*

During this core-contraction phase, the Sun will become 1,000 times more luminous than it is today, climbing to roughly absolute bolometric magnitude –3. The solar radius will increase to about 100 times its current value, and the Sun will extend past the orbit of Mercury, vaporizing the little planet. Imagine such a body in Earth's sky, a glowering orange-red sphere 50° across, stretching halfway from the horizon to the zenith (Figure 24.7).

The present Sun generates a minor wind as it loses mass at a rate of a mere 10^{-13} solar masses per year. As the Sun develops into a giant, greater luminosity and lower surface gravity will produce a stronger wind that will eventually blow millions of times more vigorously. Mass loss now becomes an important factor in evolution. By the time the Sun reaches the top of its giant track in Figure 24.5, it may have lost as much as 20% of its initial matter and will have begun recycling part of itself back into the interstellar medium from which it came.

At the same time, the solar core, with half a solar mass, will shrink to only a few times the size of Earth (Figure 24.8). Its temperature will then be an astonishing 100 million K and its density nearly 1 million g/cm^3. Under these conditions, the atoms are totally ionized, and the free electrons form their own electron gas. The electrons are packed so tightly that the **Pauli exclusion principle,** defined in 1925 by the German physicist Wolfgang Pauli, becomes important. Atomic particles have a variety of quantum mechanical properties, for example, their spins (see Section 23.2). The exclusion principle states that no two identical electrons, protons, or neutrons moving at a specific velocity can occupy a specified minimum volume of space. When the density of a gas becomes so high that these minimum

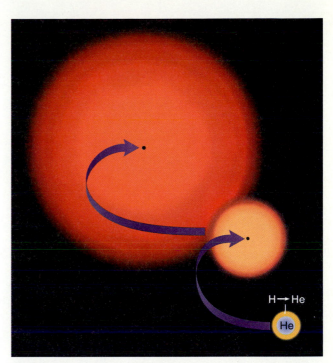

Figure 24.8 *The core of a fully developed giant is tiny compared to the size of the envelope. The dot in the center of the giant on the left is expanded twice at lower right to show the hot, dense helium core (blue) and the surrounding H-burning shell (yellow).*

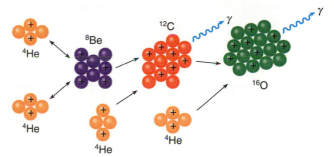

Figure 24.9 *In the triple-alpha process, two ^{4}He nuclei—α particles—collide to produce ^{8}Be, which is struck by another α particle to produce ^{12}C before it falls apart. The carbon nucleus can subsequently capture another α particle to make ^{16}O.*

volumes become filled, the particles of a given velocity can be packed no closer and the gas enters a state of **degeneracy.** We can always add particles to a degenerate gas, but only at ever-higher velocities. *Electron degeneracy* provides powerful outward pressure and support. As the Sun climbs its evolutionary track to become a giant, the density of its now-tiny core becomes high enough to induce partial degeneracy, slowing the contraction.

When the temperature reaches about 100 million K, the inert helium becomes a fuel for a new reaction and begins to fuse into carbon (Figure 24.9). Helium fusion is difficult. Two helium nuclei first collide and meld briefly as ^{8}Be, a highly unstable beryllium isotope that falls apart in only 10^{-16}

Figure 24.7 *The red giant Sun rises over the Earth's parched and blistered landscape.*

seconds. But if the temperature and density are high enough, a third helium nucleus can bang into the ^{8}Be nucleus just before it disintegrates, producing carbon and a gamma ray. In effect, three helium nuclei (or α particles) must come together simultaneously; the process is known as the **triple-**

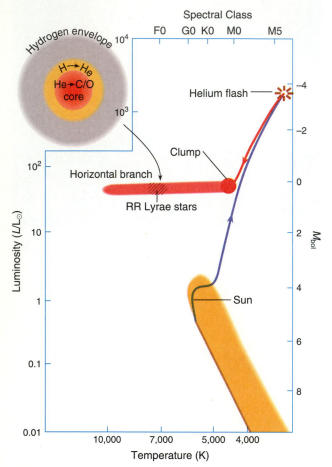

Figure 24.10 *Solar-type stars terminate their ascents of the giant branch at the helium flash and then descend (red curve) to the clump, where the helium-burning core is surrounded by a shell of fusing hydrogen (inset). In globular clusters, stars of about 0.8 M$_\odot$ spread to the left of the clump to form the horizontal branch.*

alpha (3-α) process. An additional α-particle collision can then change some of the carbon to oxygen, with the creation of more energy.

A degenerate gas does not behave according to the perfect gas law, $P = NkT$ (see Section 18.5.2). Instead, pressure is dependent only on density, not temperature. The energy produced by helium ignition does not therefore immediately affect the pressure of the partially degenerate electrons. The core cannot quickly expand, and the temperature, to which the reaction is extremely sensitive, suddenly rises. As a result, the 3-α process ignites violently in a **helium flash.** The huge new energy source causes the core to expand, lifting the state of degeneracy and terminating the ascent of the red giant branch (see Figures 24.5 and 24.10). The future Sun will then reverse direction on its evolutionary track, and will go about halfway back down to reside in an area called the **clump** (because stars tend to accumulate there). Our star will now have reached another pause in its relentless attempt to contract, where it resides for about 10% of its hydrogen-burning lifetime, a K giant with a helium-burning core surrounded by a hydrogen-burning shell (upper inset, Figure 24.10).

More-massive stars behave similarly, though their evolutionary tracks are different in detail (see Figure 24.5). Above about 2 solar masses, the core never becomes degenerate, and helium ignition is quiet instead of explosive.

Stars that exemplify the future of the Sun and other intermediate main-sequence dwarfs are found everywhere, identified by their orange color. Aldebaran, Arcturus, the brightest bowl stars in the Big and Little Dippers (α Ursae Majoris and β Ursae Minoris), the four K giants of the Hyades, and many others are helium-burning giants. As the higher-mass stars, those above about 5 M$_\odot$, move across the HR diagram, they encounter the Cepheid instability strip and start to pulsate. Cepheid variability is a temporary phenomenon. When a Cepheid leaves the strip, its pulsations will cease.

Globular clusters provide more examples of evolutionary action. The giants of a globular are now evolving from dwarfs of about 0.8 solar masses. As the stars fire their helium and come back down from the red giant tip to the clump, they spread out to the left into the horizontal branch because of their low metal content (which makes their photospheres hotter than their Population I counterparts) and differences in mass caused by variations in mass-loss rates. When the evolving stars cross the RR Lyrae gap, they pulsate and vary in brightness (see Section 21.2).

24.2.3 Supergiants

In spite of their great masses and luminosities, the stars of the upper main sequence at first behave similarly to those of the intermediate zone. The 20-$M_\odot$ star in Figure 24.5 begins as a brilliant O7 dwarf, one that may light a diffuse nebula. When its main-sequence lifetime ends it has cooled to class B2, and any remaining birth cloud becomes a reflection nebula. This star tracks to the right at nearly constant luminosity to class M, expanding to a radius of some 5 AU, about 1,000 times that of the Sun, comparable with the orbit of Jupiter (Figure 24.11). Examples of such red supergiant stars are Betelgeuse in Orion and Antares in Scorpius. Given their sizes, mass loss becomes a serious affair; rates approach 10^{-5} solar masses per year, shrinking the star from 25 $M_\odot$ to perhaps 22 $M_\odot$.

The cooling rightward progress will be stopped by quiet helium ignition, which turns the star around and sends it back to the left. It is now down to 15 solar masses and may appear more like Orion's Rigel.

Supergiants, the progeny of stars of the upper main sequence, are fundamentally different from giants, as they will not make white dwarfs. The

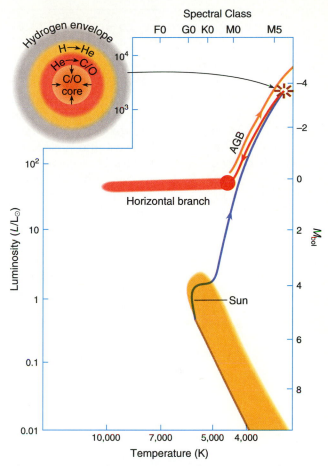

Figure 24.12   *A solar-mass star ascends (orange curve) the asymptotic giant branch (AGB). It now has a dead carbon-oxygen core surrounded by a shell of helium that is fusing into carbon and oxygen, and another of hydrogen burning to helium (inset). Surrounding these is an immense hydrogen envelope that on this scale would fill roughly the average house. At the top of the AGB the star has lost almost all its envelope and has a mass of only about 0.6 $M_\odot$.*

story of supergiants and what they *will* make is reserved for the next chapter. The remainder of this chapter will be devoted to the fate of the intermediate main sequence.

24.3
The Asymptotic Giant Branch and the Degenerate Core

When a giant's helium fuel is consumed and the nuclear fire once again goes out, the core again loses its support and resumes its contraction. Internal compression generates more heat, making the star brighter and larger. It now climbs the giant branch for the *second* time (Figure 24.12). The second ascent is much faster than, and somewhat asymptotic to, the first ascent (asymptotic curves approach but do not meet), so this part of the evo-

Figure 24.11   *The Sun is placed next to the supergiant Mu Cephei, one of the largest stars in the sky. On this scale μ Cep is 5 meters across, the Earth a mere 0.02 mm.*

Throughout the history of science we encounter landmark publications describing work and research that forever change our view of nature. Many are famous, and are included not only in textbooks, but are familiar to the educated public as well. Sterling examples are Copernicus's *De Revolutionibus* of 1543, Galileo's *Dialogue on Two Chief Systems of the World,* Newton's *Principia,* and the papers in which Einstein announced his theories of relativity to the world in 1905 and 1914.

Other benchmark papers are generally known only to the working scientist. One such changed our view of the internal processes of stars. In 1957 four astrophysicists—E. Margaret Burbidge, Geoffrey R. Burbidge, William A. Fowler, and Fred Hoyle, all then of the California Institute of Technology—combined their efforts to write "Synthesis of Elements in Stars," published in *Reviews of Modern Physics.* This immense paper, which has become affectionately known as B²FH, is more than 100 pages long. Here for the first time are set down the basic rules for the creation of the elements past helium. The paper describes not only hydrogen and helium burning, but also α-particle capture that can build even-numbered elements, the s-process, which builds elements all along the periodic table, and the r-process, a mechanism involving rapid neutron capture that can build the heaviest atoms.

Stellar astrophysics was never the same. Theoreticians expanded on the ideas to produce detailed calculations of the expected compositions of stars, which have been largely confirmed by observers. The agreement now found between theory and observation clearly demonstrates that, except for hydrogen and helium, everything we rely on for our lives came from the stars.

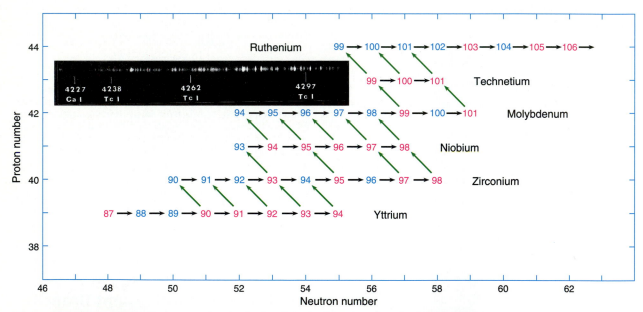

Figure 24.13 *An example of an s-process network of nuclear reactions shows heavier elements being built from lighter. The numbers show various isotopes, the stable ones blue, the unstable red. ^{87}Y captures free neutrons (black arrows), successively becoming ^{88}Y, ^{89}Y, and then unstable ^{90}Y. A ^{90}Y nucleus can capture still another neutron to become ^{91}Y. However, before that happens, one of its neutrons can eject an electron and become a proton, creating ^{90}Zr. Yttrium can keep capturing neutrons (and decaying to zirconium) until ^{94}Y is created, which decays so fast that it all becomes ^{94}Zr. The network moves upward past technetium, a radioactive element seen in the spectra of giant stars (inset).*

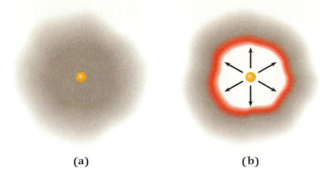

(a)   (b)

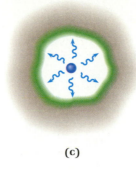

(c)

Figure 24.14 (**a**) *What is left of an AGB star sits inside its expanding cloud of debris.* (**b**) *It develops a fast, hot wind that compresses the inner edge of the cloud and creates a central hole.* (**c**) *When the star becomes hot enough, it illuminates the compressed cloud to form a bright planetary nebula.*

lutionary track is called the **asymptotic giant branch** (AGB). At the center of an AGB star is a dead carbon and oxygen core, the remains of helium burning. It is now surrounded by two shells (inset, Figure 24.12), the inner one fusing helium to carbon and oxygen, the outer fusing hydrogen to helium, the two alternately switching on and off in an increasingly violent fashion. As the star climbs the AGB its changing interior structure causes it to pulsate, and it becomes a long-period or Mira variable (see Section 21.3).

During the AGB ascent, the intense heat causes a variety of nuclear reactions that turn the star into a chemical factory. The most important is the **s-process** (Figure 24.13), in which heavy nuclei up through bismuth are built by the slow capture of neutrons. The AGB star's envelope is in a state of deep convection, as are regions in and near the nuclear-burning shells. As a result, the star's cool atmosphere can become enriched in carbon made by the 3-α process and in elements made by the s-process, including zirconium. At the point where the abundances of carbon and oxygen are about equal, carbon atoms make carbon compounds and the oxygen ties up with the enhanced zirconium to make an S star (see Section 19.4). If the carbon content increases further, the S star becomes a deep-red carbon star of class N. Nitrogen and helium contents also increase.

Direct evidence for these processes is provided by observations of technetium, element number 43, in Mira variables (inset, Figure 24.13). Even the longest-lived isotope, ^{98}Tc, has a half-life of only about 4 million years. There is none in the Earth, yet there it is in stellar atmospheres, a demonstration of nuclear alchemy at work.

The developing AGB stars become larger, brighter, and redder than they ever were on their first ascent of the HR diagram as red giants. The Sun will pass absolute bolometric magnitude –4, and more-massive stars will become even more luminous. (Because much of a cool AGB star's radi-

ation falls in the infrared, however, it will not appear that bright visually.) Mira is now the size of the orbit of Mars (see Section 21.2.1) and the Sun will reach the orbit of the Earth. Spectral classes can extend to M9 and temperatures can drop below 2,000 K. The evolving star has been losing mass since it began its giant phase, and now the rate can reach 10^{-5} M$_\odot$ (more than 3 Earth masses) per year, producing a shroud of gas and dust. The more-massive stars will hide themselves, some becoming bright in the infrared as OH/IR stars.

The core of the star has now shrunk to about the size of the Earth and the temperature has climbed to several hundred million K. We might now expect that the core's carbon/oxygen mixture would ignite and begin to fuse to heavier elements, stopping the contraction and once more stabilizing the star. However, the star has lost so much mass that the interior cannot reach the point of carbon ignition. The Sun will have lost almost half its mass back to space, and a 7-M$_\odot$ star will lose an astounding *80%*. Observational proof is readily found among binary stars. The white dwarf companion to Sirius must have evolved first and therefore must have had the greater mass of the pair, but it is now has only half the mass of Sirius A.

The star's core finally becomes stabilized, not by nuclear burning, but by electron degeneracy, outward pressure that forever prevents any further contraction. The core is on its way to becoming a white dwarf.

24.4
Planetary Nebulae

The star now consists of a degenerate carbon-oxygen core of half a solar mass or more surrounded by hydrogen- or helium-burning shells, which are in turn topped by a light but extensive hydrogen envelope about the size of the present Sun. It is wrapped in an expanding cloud produced by millions of years of mass loss (Figure 24.14a). A fast

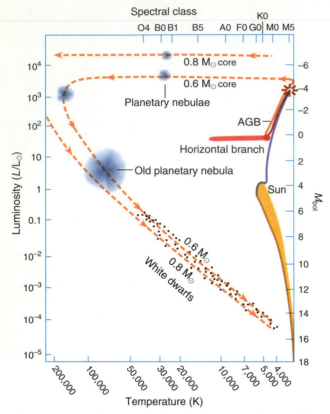

Figure 24.15 *When mass loss has nearly exposed the core of the AGB star, the star tracks move to the left on the HR diagram (dashed orange curve). When the star's effective temperature is high enough, it illuminates the surrounding gas to produce a planetary nebula. The core eventually cools and dims, becoming a white dwarf. Higher-mass cores will be brighter when they leave the asymptotic giant branch and will become hotter, but when they cool as white dwarfs they are smaller and dimmer.*

stellar wind blowing at a speed of a thousand or more km/s begins to shovel the inner edge of the circumstellar cloud into a dense shell (Figure 24.14b). At the same time, the hydrogen-burning shell is eating away at the star's envelope. As the envelope shrinks, the core becomes progressively more exposed, the effective temperature climbs, and the star moves evenly to the left on the HR diagram (Figure 24.15).

When the effective temperature hits about 25,000 K, the star begins to produce enough ultraviolet radiation to ionize and light the compressed shell (Figure 24.14c). The result is one of the most graceful apparitions in the sky, a **planetary nebula** (Figure 24.16), which appears in the telescope as a gaseous ring or disk with a single blue star at its center. (The name, given by William Herschel, refers to the nebula's superficial resemblance to a planetary disk.) Deep-imaging, as in Figure 24.16,

frequently shows the huge surrounding envelope in which the nebula is embedded.

At a maximum effective surface temperature that exceeds 100,000 K, whatever nuclear burning is left begins to shut down, and the star begins to cool and dim (see Figure 24.15). The planetary nebula, with a mass of a few tenths solar, is now expanding into space at around 20 km/s, and when it reaches a radius of about a parsec, it becomes faint and hard to see (Figure 24.17). Finally, some 50,000 years after it was created, the nebula dissipates, leaving behind a white dwarf, a star stabilized forever by the pressure of degenerate electrons. All that is left is for the corpse to cool and dim, a process that will take billions of years.

24.5
White Dwarfs

White dwarfs, the end products of intermediate-mass stellar evolution, begin as hydrogen-burning cores of stars between 0.8 and 8 solar masses; they end as dense stars made largely of carbon and oxygen. As they cool and dim, they string out on the HR diagram in a long line far below the main sequence (see Figure 24.15). The specific evolutionary tracks depend on mass. The higher the original mass of the star, the greater that of the remnant. Most white dwarfs are by-products of near-solar-mass stars and have masses around 0.6 $M_\odot$. Because the number of stars declines upward along the main sequence, more-massive white dwarfs are rarer. Only one (Sirius B) is known that approaches even 1 $M_\odot$.

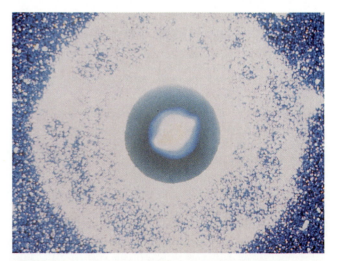

Figure 24.16 *The planetary nebula NGC 6826 is encased in a huge shell, the illuminated remains of much-earlier mass loss.*

Figure 24.17   *The magnificent* Helix Nebula *in Aquarius, NGC 7293, over 0.6 pc across, is dissipating. The central star is now a real, although still very hot (112,000 K), white dwarf.*

On his way by ship to work in England in 1930, a young Indian astrophysicist named Subramanyan Chandrasekhar had the time to think about white dwarfs. As the mass of a white dwarf increases, so does its interior compression. As a result, the electrons are forced by the exclusion principle to move faster. Chandrasekhar realized that at some point their velocities approach that of light, and the theory of relativity must be considered. When he incorporated relativity into white dwarf theory, he found a mass limit beyond which degenerate electrons can no longer provide support and the star must collapse under the force of gravity. The **Chandrasekhar limit** is 1.4 solar masses. No heavier white dwarf can exist, as amply confirmed by observation.

Here is the reason for the distinction between the intermediate and upper main sequences, between giants and supergiants. Between initial main-sequence masses of 0.8 and 8 $M_\odot$, the nuclear-burning cores remain below the Chandrasekhar limit. Winds strip the envelopes away during the two giant stages and leave the cores behind to die as white dwarfs. However, the core of a star with an initial mass greater than about 8 $M_\odot$ ends up above the Chandrasekhar limit and *cannot form a white dwarf.* Something far stranger happens to it, as we will see in the next chapter.

Evolution and the white dwarfs' immense densities explain their division into the DA (hydrogen-rich photosphere) and DB (helium-rich) categories. In the crushing gravity of the DA stars, the heavier atoms of helium have sunk so far down that they cannot produce spectrum lines. The DB stars, however, have lost all their hydrogen skins through winds, leaving the helium behind.

As the stars cool to effective temperatures of roughly 4,000 K, the interiors are expected to turn solid and *crystallize!* The time it takes for a white dwarf to cool to this point is comparable to the age of the Galaxy. Every white dwarf that has ever been created is still shining.

24.6
Binary Evolution

Evolution among binaries can produce a variety of effects and an array of variable stars (see Section 21.5). The two components will usually have different masses and the more massive one will be the first to evolve. We then find a giant paired with a main-sequence star. If sufficiently close, the evolving giant may be tidally distorted and may pass matter onto the dwarf. If the giant's outer layers become enriched with by-products of nuclear reactions, the dwarf will then become enriched as well. This process appears to explain a strangely high abundance of metals in the atmosphere of Sirius A: Sirius B was once a giant that seeded Sirius A with its enriched wind.

If the stars are even closer, the dwarf may find itself within the giant's extended envelope. Friction will dissipate orbital energy, and the two will spiral closer together. When the giant's planetary nebula has been formed and its evolution is complete, the resulting white dwarf has been drawn very close to the main-sequence star. Tides raised by the white dwarf in its unevolved companion may cause mass to flow onto its dense surface, and eventually a nova blossoms into the sky.

There are a variety of other possibilities. The white dwarf and main-sequence star may be fairly widely separated. But then the ordinary dwarf evolves and expands so much as a giant that mass transfer produces a symbiotic star. Dwarf and recurrent novae can be made by differences in age and proximity between the evolving pair. Interacting binary connections may even cause annihilating explosions, as will be seen in the next chapter, to which we now turn to explore the wonders of high-mass evolution.

KEY CONCEPTS

Asymptotic giant branch (AGB): The state of stellar evolution in which intermediate-mass stars grow as giants for the second time, their contracting carbon-oxygen cores surrounded by burning shells of helium and hydrogen.

Chandrasekhar limit: A limit of 1.4 $M_\odot$, above which white dwarfs cannot be supported by the pressure of degenerate electrons.

Degeneracy: A state in which the particles of a gas cannot get any closer because of the **Pauli Exclusion Principle** (which states that no two identical particles such as electrons, protons, and neutrons at a specific velocity can occupy a specified minimum volume of space).

Helium flash: The explosive ignition of helium in the partially degenerate core of a red giant star of about a solar mass.

Intermediate main sequence: The main sequence between 0.8 and 8 $M_\odot$; these stars evolve into white dwarfs.

Lower main sequence: The main sequence between 0.08 and 0.8 $M_\odot$; no lower main-sequence star has ever had time to evolve away from the lower main sequence.

Planetary nebula: A shell of illuminated gas around a hot star; the last part of the ejected wind of an AGB star lit by the old nuclear-burning core.

Red giant branch: The state of stellar evolution in which intermediate-mass stars grow into giants for the first time, their contracting helium cores surrounded by hydrogen-burning shells.

s-process: A set of nuclear reactions in which atoms slowly capture neutrons and then decay into heavier elements.

Upper main sequence: The main sequence above 8 $M_\odot$; these stars are too massive to evolve to white dwarfs because their cores are above the Chandrasekhar limit.

KEY RELATIONSHIPS

Carbon cycle:

The fusion of four atoms of hydrogen into one of helium using carbon as a nuclear catalyst.

$$^{12}C + p \rightarrow {}^{13}N + \gamma$$
$$^{13}N \rightarrow {}^{13}C + e^+, \quad (e^+ + e^- \rightarrow 2\gamma)$$
$$^{13}C + p \rightarrow {}^{14}N + \gamma$$
$$^{14}N + p \rightarrow {}^{15}O + \gamma$$
$$^{15}O \rightarrow {}^{15}N + e^+, \quad (e^+ + e^- \rightarrow 2\gamma)$$
$$^{15}N + p \rightarrow {}^{12}C + {}^4He$$

Triple-alpha (3-α) process:

The fusion of three atoms of helium into one of carbon.

$$^4He + {}^4He \rightleftarrows {}^8Be$$
$$^8Be + {}^4He \rightarrow {}^{12}C + \gamma$$
$$(^{12}C + {}^4He \rightarrow {}^{16}O + \gamma)$$

EXERCISES

Comparisons

1. Compare the three domains of the main sequence. What are they like now? What will happen to the stars within them?
2. What is the difference between clump stars and horizontal-branch stars?
3. What is the essential difference between a giant and a supergiant?
4. Compare the interiors of first and second ascent giant stars.
5. What are the similarities and differences between planetary and diffuse nebulae?

Numerical Problems

6. Using the average mass-luminosity relation, what are the expected lifetimes of stars of 10 $M_\odot$ and 0.1 $M_\odot$? Why would the actual lifetimes differ from your answer?
7. Estimate the main-sequence lifetimes of 61 Cygni A and θ^1 Orionis C.
8. Draw a schematic graph showing how mass-loss rates change over the lifetime of an intermediate main-sequence star.
9. The Sun will someday produce a planetary nebula. At a typical expansion rate of 20 km/s, how long will it take the edge of the planetary nebula to reach α Centauri (assuming the two stars stay the same distance apart—which they will not).

Thought and Discussion

10. What is meant by a main-sequence star?
11. Why do α Centauri A and the Sun, both G2 V stars, differ in absolute magnitude?
12. Why do high-mass stars die sooner than low-mass stars?
13. Why is the carbon cycle important?
14. Give two reasons why the main sequence is a band and not a line.
15. If it is the evolutionary fate of a star to shrink under the force of gravity, why do stars become giants? What part of the star is shrinking?
16. How (in an evolutionary sense) are Cepheid variables created?
17. Why cannot AGB stars be supported by fusing carbon into heavier elements?

18. ^{95}Mo can be produced in two ways by the s-process. What are they?
19. What proofs do we have of the change of one element into another inside stars?
20. How does stellar evolution produce S and carbon stars?
21. What is the evidence for stellar mass loss?
22. What is a degenerate gas? What are the effects of a degenerate gas if it develops inside a star?
23. How are planetary nebulae made?
24. What is meant by the Chandrasekhar limit?
25. In what ways can binary companions affect and be affected by stellar evolution?

Research Problem

26. What were the views on stellar evolution shortly after Russell and Hertzsprung developed the concepts of dwarfs, giants, and supergiants?

Activities

27. It is difficult to describe in words the changes in stellar size wrought by evolution. Demonstrate it outdoors. Start with the Sun as a golf ball and use some means to portray the solar size when it will be at the tops of its red giant and asymptotic giant evolutionary tracks. Then show the size of the Sun as a white dwarf. Compare these to a supergiant.
28. Use a telescope to observe the effects of stellar evolution. Find a giant star. Then find an AGB star (use Table 21.2 or the predictions of the maxima of long-period variables in *Sky and Telescope* magazine). Finally, locate at least one of the three planetary nebulae found in the Messier list in the Appendix. In all cases, describe what you see and reflect on its evolutionary significance.

Scientific Writing

29. Some science-fiction magazines have articles about real science. Write such an article that explains how past and future life on Earth may be affected by the evolution of the Sun.
30. A short essay can be more effective than a long one—as well as much harder to write. In as compact a form as you can, write a summary of the evolution of the Sun. Do not exceed one page of 250 words.

25

Catastrophic Evolution

*The spectacular lives
and violent ends
of high-mass stars
and interacting
binaries*

*The Vela Remnant
reveals the devastating explosion
of a star.*

Intermediate main-sequence evolution is relatively quiet. Upper main-sequence evolution, however, is not. The power of the B and the O stars ensures exciting evolution, with episodes of intense mass loss terminating in catastrophic explosions.

25.1
Supergiant Evolution

That supergiants are the progeny of the upper main sequence and belong to Population I is vividly revealed by spatial relations: supergiants congre-gate with O and B stars—in young clusters, OB asso-ciations, binary systems, and the plane of the Milky Way (Figure 25.1). They are also as rare as O stars. Their luminosities are so great, however, that unless they are obscured by dust, we can see them over the entire Galaxy, no matter how far away. They are even visible in the disks of other galaxies.

The story of upper main-sequence evolution, begun in Section 24.2.3, continues in Figure 25.2. Stellar winds are of paramount importance. Above 40 $M_\odot$, stellar mass-loss rates top 10^{-6} $M_\odot$/year, and as a result, O stars begin to whittle themselves

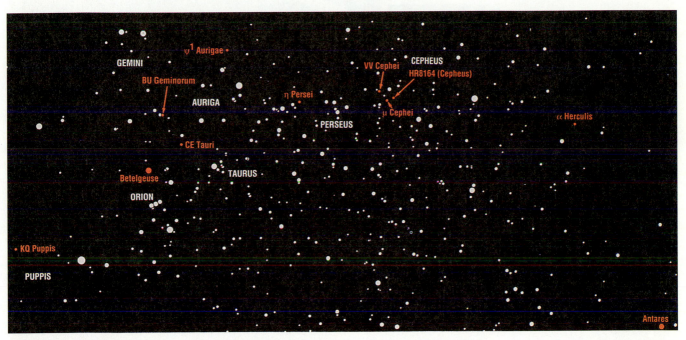

Figure 25.1   *Naked-eye red supergiants follow the plane of the Milky Way (indicated by the named constellations), showing their relation to O and B stars.*

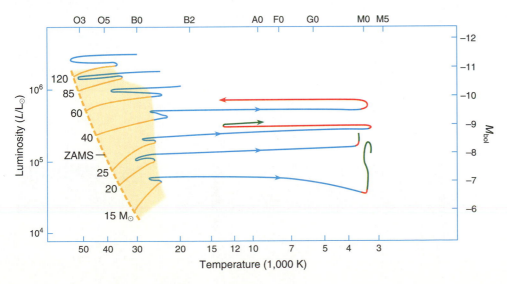

Figure 25.2   *Evolutionary tracks show high-mass stars evolving within the main sequence and then away from it, becoming blue and red supergiants. The various phases of evolution are: hydrogen burning (yellow), helium core contraction (blue), helium burning (red), and carbon burning (green).*

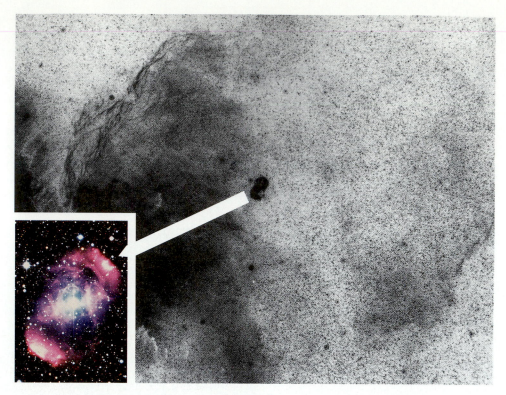

Figure 25.3 *The evolving supergiant HD 148937 in Ara is surrounded by its own ejecta, NGC 6164-5 (inset); it is inside a wind-blown bubble that in turn is in an immense Strömgren sphere.*

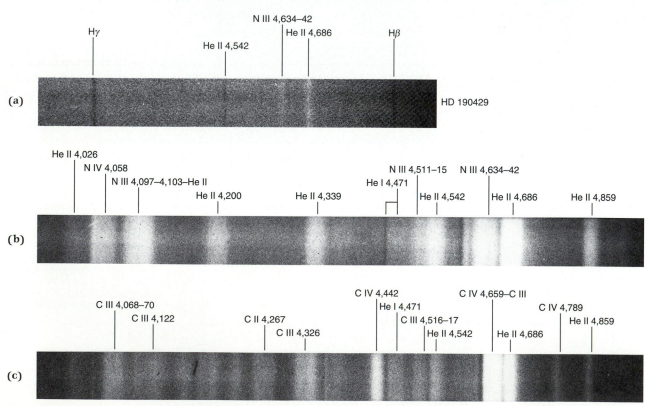

Figure 25.4 **(a)** *An O supergiant is enriched in nitrogen. Below are spectra of Wolf-Rayet stars that have lost their hydrogen envelopes and are rich in nitrogen* **(b)** *and carbon* **(c)**.

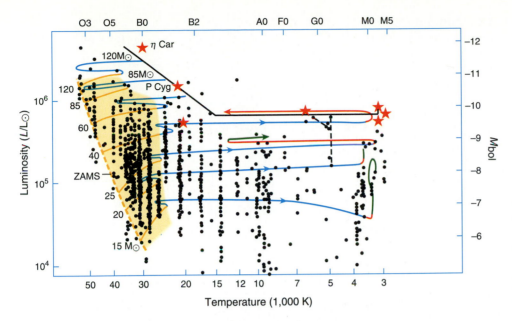

Figure 25.5 *Stars are superimposed on the tracks of Figure 25.2. The solid line at the top is the Humphreys-Davidson limit, which represents the upper envelope to supergiant distribution, the point at which strong winds turn stars into luminous blue variables and other oddities.*

away. As the more luminous stars evolve as blue supergiants, they can become surrounded by thick circumstellar nebulae (Figure 25.3) and can powerfully affect their surroundings by blowing huge bubbles in the interstellar medium. At the same time, the by-products of nuclear burning begin to enter their photospheres, as demonstrated by emission lines of nitrogen (Figure 25.4a).

In Figure 25.5, bright main-sequence stars and supergiants are superimposed on the evolutionary tracks of Figure 25.2. Note two features. First, the number of supergiants drops toward the right and then clumps strongly in class M. This observation supports theoretical evolutionary rates, which predict that the stars should fly across the HR diagram as their cores contract, pausing in class M as they fire their helium.

Second, the maximum luminosity of the O dwarfs and the blue supergiants is greater than that of the red supergiants on the right. The upper envelope to stellar luminosities, the **Humphreys-Davidson limit,** slopes downward to the right and then levels off. Below about 40 $M_\odot$, stars make it all the way across the diagram to become red supergiants. Above 40 $M_\odot$, they run into the limit, bounce off it, and turn around to remain blue supergiants. The limit is imposed by high mass-loss rates that severely affect evolution. Along the limit we find the luminous blue variables (see Section 21.4), as well as some red supergiants with high mass-loss rates. Some of these stars are hiding themselves in their own ejecta and are losing much of their initial masses.

The stars near the Humphreys-Davidson limit may be turning themselves into bizarre **Wolf-**

Rayet (WR) **stars,** named after two nineteenth-century European astronomers. They are hot like class O, but have distinctive spectra that exhibit only emission lines superimposed on a background continuum. The WR stars have stripped themselves of their hydrogen envelopes down to their helium cores. There are two kinds of Wolf-Rayet stars, nitrogen-rich (WN) and carbon-rich (WC); they appear to represent successive stages of evolution, the first showing by-products of hydrogen burning on the carbon cycle, the second by-products of helium burning via the 3-α process.

Wolf-Rayet mass-loss rates are among the highest found, reaching 10^{-4} $M_\odot$ per year. The result is commonly a surrounding *ring nebula* consisting of both stellar ejecta and swept-up interstellar gas (Figure 25.6). From binary systems, we estimate masses between about 10 and 50 $M_\odot$ with an aver-

Figure 25.6 *NGC 2359 surrounds HD 56925, a WN star in Canis Major. The nebula, rich in nitrogen, is sweeping up the interstellar medium.*

age near 20 M$_\odot$. WR stars are commonly paired with higher-mass O dwarfs, showing that they started out near the upper mass limit and have cut their masses about in half, consistent with their odd atmospheric enrichments. We are looking at stars that are transforming themselves almost before our eyes.

25.2
Advanced Evolution: The Road to Disaster

Intermediate main-sequence evolution ends with the creation of a degenerate carbon-oxygen (C/O) core. Upper main-sequence stars, however, have cores larger than the Chandrasekhar limit and cannot become white dwarfs. Once the helium runs out and a C/O core is developed, it contracts and heats (Figure 25.7a), allowing nuclear burning to continue to more advanced stages.

The shrinking C/O core is surrounded by shells of fusing helium and hydrogen. At a temperature near 1 billion K, contraction is halted as carbon begins to fuse to oxygen and neon (through the reactions $^{12}C + \alpha \rightarrow \,^{16}O$ and $^{12}C + \,^{12}C \rightarrow \,^{20}Ne + \alpha$, where α represents ^4He) and further to magnesium (Figure 25.7b), the core still surrounded by two nuclear-burning shells. C burning is indicated on the tracks in Figure 25.2 (where it is not, it occurs off the diagram after the star has looped back to the left). More energy is now being produced in neutrinos than in photons.

When carbon is exhausted at the center, the nuclear fire shuts down and the O/Ne/Mg core contracts under the inexorable force of gravity (Figure 25.7c). At 2 billion K, contraction is halted as the oxygen, neon, and magnesium burn by a complex set of reactions to a mixture of silicon and sulfur (Figure 25.7d). The core is now wrapped in successive shells of burning carbon, helium, and hydrogen, and the star is beginning to look like a nuclear onion. As the oxygen runs out, the dominantly Si/S core starts to contract (Figure 25.7e). The temperature now approaches 3 billion K and the nuclear reactions become ever more complex. When the temperature is sufficiently high, the silicon in the core starts reacting with itself to form iron, ^{56}Fe (Figure 25.7f).

As energy is generated in thermonuclear reactions, the resulting nuclei become more tightly bound together. In fact it is the very *act* of binding protons and neutrons that releases the nuclear energy. As fusion proceeds toward heavier ele-

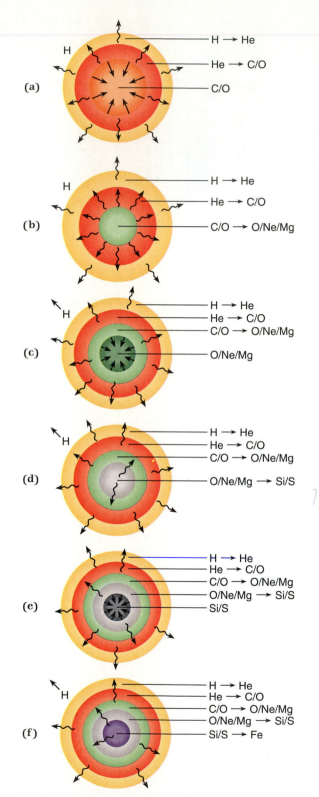

Figure 25.7 *Stages of nuclear burning in supergiant cores are* **(a)** *a contracting C/O core;* **(b)** *C burning to oxygen, neon, and magnesium;* **(c)** *a contracting O/Ne/Mg core;* **(d)** *burning of O, Ne, and Mg to a mixture of silicon and sulfur;* **(e)** *a contracting Si/S core;* **(f)** *Si burning to iron. These zones are surrounded by vast hydrogen and/or helium envelopes. The diagrams are not to scale; each is successively smaller.*

ments, the *increase* in the degree of binding becomes progressively smaller, and as a consequence, less energy per reaction is released: helium burning supplies only 10% of the energy of hydrogen burning. However, no matter what atoms happen to be fusing, they must supply enough energy to support the core. Consequently, the durations of the various burning states decrease as the core becomes layered. An O7 star of 20 solar masses takes 8 million years to use up its initial hydrogen. Helium burning then takes about 10% of the main-sequence lifetime, or only about a million years. Once the carbon starts to fuse, it is gone in about 10% of the He-burning lifetime, or a mere 100,000 years.

What happens next is astonishing. Well over a solar mass of oxygen burns completely in only 20 years, and once silicon starts to fuse to iron, the process is completed in a week! At higher initial masses, the processes are speeded up even more. Clearly, some monumental event is about to take place.

25.3
Supernovae

Supernova explosions are among nature's grandest events. Grander still, we are finding reasonable explanations for them.

25.3.1 *Core Collapse*

We are now left with a star that is growing an iron core of about 1.4 solar masses (near the Chandrasekhar limit) a couple of thousand km across. Finally, the silicon is all gone, and with support removed, the iron core begins its inevitable compression. However, iron is the most tightly bound of all atomic nuclei, and as a result, *no energy can be gained by its fusion.*

The core is briefly supported by degenerate electrons. But under the core's extreme conditions, iron is broken down into manganese by the absorption of electrons and smashed into α particles by encounters with gamma rays. Removal of electrons reduces degeneracy support, and the absorption of gamma rays removes heat. The iron core cannot hold itself up, and it contracts catastrophically at speeds up to a quarter that of light (Figure 25.8a). In *less than a tenth of a second* it collapses, the radius dropping from 1,000 km to less than 50 km, and within a few seconds more down to a mere 10–20 km, releasing a spectacular amount of gravitational energy.

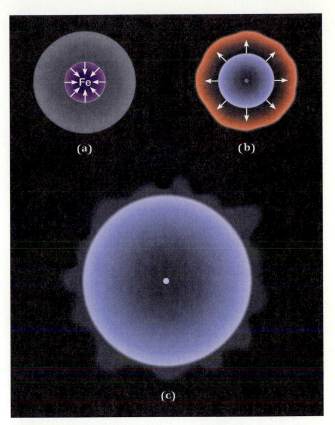

Figure 25.8   *When the iron core of a supergiant is completed, it collapses* **(a)**, *generating a shock wave that rips through the stellar interior* **(b)** *and erupts through the outer envelope to produce a supernova* **(c).**

In less time than it takes to snap your fingers, 10^{46} joules fly outward, 99% of them in the form of neutrinos. (By comparison, our Galaxy of 200 billion suns radiates about 10^{38} joules/s.) At the moment of collapse, the star generates a power output (rate of energy release per second) *comparable to that of all the stars in the observed Universe combined.* The temperature now approaches 200 billion K. The density is over 10^{12} g/cm^3, that of the nucleus itself, a million times greater than that in a white dwarf. Protons and electrons merge into neutrons, which, like the electrons in a white dwarf, are degenerate. Their outward pressure finally halts the collapse.

Part of the imploding core rebounds and produces a shock wave that rips through the layers surrounding the core and then through the vast envelope (Figures 25.8b and c). The density near the core's boundary is so great that the gas is actually opaque to neutrinos, which contribute to the outward pressure tearing the star apart. After an interval of several hours, the shock wave makes it to the stellar surface and a supernova erupts into

Figure 25.9   *A Type II (core collapse) supernova erupted in March of 1993 in the nearby galaxy M 81 in Ursa Major. Although 1.4 million parsecs distant, the supernova, called 1993J, reached apparent magnitude 10, bright enough to be seen with a small amateur telescope. Population types are vividly outlined by color, the blue Population I spiral arms (which contain the supernova) contrasting with reddish Population II in the central regions.*

the sky, its brilliance announcing that another massive star has died (Figure 25.9).

Since these **core-collapse supernovae** occur only in massive stars that lie in the Population I disks of galaxies, we relate them immediately to Type II supernovae (see Section 21.6). Because Type II events have hydrogen in their spectra, they must be produced by stars with hydrogen envelopes—by supergiants like Betelgeuse, Antares, and Mu Cephei and luminous blue variables like Eta Carinae and P Cygni. If Betelgeuse's core were to collapse, the star would brighten to apparent visual magnitude −10, similar to that of a quarter Moon! Wolf-Rayet stars, on the other hand, are good candidates for Type Ib supernovae, those that occur in galactic disks but whose spectra lack hydrogen.

Within the shock wave of a developing supernova, nuclear reactions go berserk. About half a solar mass of the expanding debris is converted into nickel in the form of ^{56}Ni, which decays into cobalt (^{56}Co) and then into iron (^{56}Fe). The myriad reac-

tions create vast numbers of neutrons that are rapidly absorbed by atomic nuclei. This **r-process** (for *r*apid capture) can jump radioactive barriers that the s-process (see Figure 24.13) cannot. Even if an isotope is very unstable, it has time to catch successive neutrons, thereby continually raising its atomic weight. When the neutrons in the nucleus decay by ejecting electrons, thereby turning into protons, the nucleus can jump to high atomic numbers, creating elements beyond lead all the way to plutonium. The by-products of this nuclear burning are then blasted into space, further enriching the interstellar gas.

25.3.2 Type Ia Supernovae and Supernova Significance

The association of Type Ia supernovae with Population II (see Section 21.6) shows that their progenitor stars have masses too low to produce iron cores. There are two current and related hypotheses that can explain them. Ordinary novae are produced when mass from a main-sequence star accumulates on a white dwarf's surface and explodes. If the star were close to the Chandrasekhar limit, the falling matter might push the star *over* the limit before it burns away, causing a catastrophic collapse. Rapidly increasing temperature and density would cause runaway nuclear reactions and the star would go off in a massive thermonuclear explosion, the C/O core fusing to nickel and then back to iron. In support, the light curves are consistent with the half-life of ^{56}Ni as it decays to ^{56}Fe with the production of gamma rays that heat the expanding stellar envelope. The result is a load of enriched debris, including some 0.1 $M_\odot$ of iron, delivered into the interstellar medium.

Another possibility involves double white dwarfs. During normal evolution, each star of a binary may have found itself within the giant envelope of the other. Friction would reduce orbital radii. By the time the second star has developed into a white dwarf, the stars are in close proximity. The theory of relativity predicts that accelerating masses radiate *gravity waves*—disturbances in the gravitational fields—that dissipate energy and bring the stars even closer together. (Such gravity waves are inferred by changes in binary orbits.) Given enough time, the white dwarfs will spiral together. Their combined masses could again exceed the Chandrasekhar limit, and the merged stars would catastrophically collapse.

The observed number of both kinds of supernovae suggest that they occur in the Galaxy at a rate of about one every 200 years. However, a con-

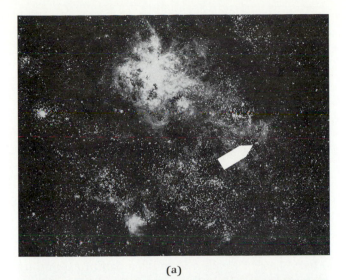

(a)

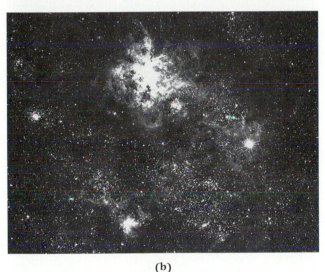

(b)

Figure 25.10 *In "before" (**a**) and "after" (**b**) pictures, we see Supernova 1987A erupting near the Tarantula Nebula from a B3 Ia supergiant in the Large Magellanic Cloud.*

siderable number must have been hidden from view behind thick clouds of interstellar dust. Accounting for dust gives a rate of at least one a century and possibly as high as one every 25 years.

Over the age of the Galaxy (around 15 billion years), perhaps a quarter billion or more supernovae have occurred. If we average Type I and Type II, each blast sends about 10 M$_\odot$ back into space, for a total yield of over a billion M$_\odot$, about 1% of the stellar mass in the Galaxy. More important, this material is highly enriched in heavier elements. We believe that *all* the iron and heavy elements in the Universe—including the metal in your chair and the gold in your ring—came from supernovae.

25.3.3 Supernova 1987A

The most dramatic proofs of supernova theory were provided by Supernova 1987A in the Large Magellanic Cloud (Figure 25.10). It was discovered and announced on February 24, 1987, by Ian Shelton of the University of Toronto during routine photography at Las Campanas Observatory in Chile, by which time it was already visible to the naked eye although 52 kpc away. A sighting had actually been made a day before by the Australian amateur Robert McNaught. An hour before McNaught's photograph, another amateur had seen nothing, so McNaught had caught the exploding star within an hour of its sharp rise from obscurity. Earlier photographs of the region showed the progenitor star to be a 20-solar-mass B3 Ia supergiant called Sk-69° 202. The supernova's spectrum had hydrogen lines, revealing it to be a Type II core-collapse event.

At 7^{h}36^m UT on February 23, 1987, three hours before McNaught photographed the supernova's optical light, 11 neutrinos smashed into electrons in the Kamioka neutrino detector in Japan (see Section 18.5.3). They came from the direction of the Large Magellanic Cloud. At the same time, eight more hit another neutrino detector built in an old salt mine under Lake Erie (Figure 25.11). The interaction rates between electrons and neutrinos indicated a total neutrino flux of 10 billion/cm^2 at the Earth, approximately the number we would expect from a Type II supernova 52 kpc distant. The neutrinos directly revealed the core collapse before the shock wave had reached the stellar surface and anyone saw the star brighten. The delay is about that expected from the size of a B supergiant.

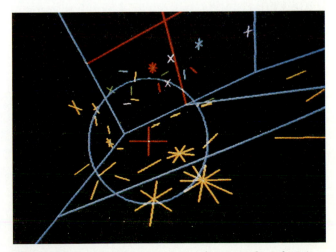

Figure 25.11 *You look into a neutrino detector under Lake Erie, the Large Magellanic Cloud behind you. The blue circle indicates a burst of light from a supernova neutrino hitting an electron. The lines, crosses, and stars show detectors registering the light.*

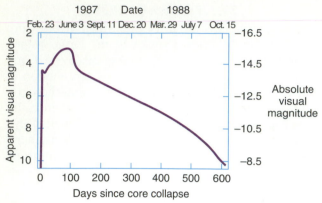

Figure 25.12 *Supernova 1987A took 100 days to reach visual maximum. The decline was controlled by radioactive decay.*

However, supernova experts expected supernova progenitors to be red supergiants, luminous blue variables, or Wolf-Rayet stars, not ordinary blue supergiants. They were not really wrong. In our Galaxy, Sk-69° 202 *would* have been a red supergiant, but the low metal content of the Large Magellanic Cloud (about midway between those of our Galaxy's Population I and II) kept the star smaller and its photosphere hotter, as it does for subdwarfs and horizontal branch stars (see Sections 19.5.4 and 20.5.1). The smaller size of the progenitor also explains a two-month delay between first eruption and maximum brightness (Figure 25.12): it caused such a high surface temperature that most of the light was initially emitted in the ultraviolet where it could not be seen. As the star cooled, its radiation shifted into the optical and the star brightened. After 100 days, the decline of the light curve indicated 0.1 $M_\odot$ of ^{56}Ni decaying into ^{56}Fe. When the gas thinned out, an orbiting observatory even detected the gamma rays released by the reaction.

Shortly after the explosion, we saw *light echoes* (Figure 25.13a) as the light from the supernova was reflected from dust sheets between the star and the

Earth. In addition, the Hubble Space Telescope captured a ring of debris around the star (Figure 25.13b) that was probably ejected during earlier phases of mass loss and heated by the powerful ultraviolet radiation. What we will see next is anticipated by the remains of other great blasts that have occurred within our own Galaxy.

25.4
Supernova Remnants

The places of exploded stars are marked by clouds of expanding debris—*supernova remnants*—of which the Crab Nebula (see Section 21.6 and Figure 21.27) is the most famous. There are many others. East of Epsilon Cygni is a pair of expanding arches (Figure 25.14) that mark the death of a star some 100,000 years ago. Unlike the young Crab, which is filled with gas, the old Cygnus Loop is an empty shell. We look not at exploded debris but at the powerful expanding shock wave of the supernova sweeping up interstellar matter. A photograph of another such object, the Vela Remnant, opens this chapter.

In general, however, supernova remnants are hard to see in the optical spectrum both because they are intrinsically faint and because of absorption by interstellar dust. Radio observations (Figure 25.15) offer a better view, allowing us to see about 150 galactic supernova remnants. Radio spectra show remarkable processes at work. The intensity of radio radiation from diffuse nebulae (free-free radiation: see Section 22.1) is fairly constant with frequency. In stark contrast, the Crab and other supernova remnants have **power-law spectra** that rise steeply according to an inverse power of the frequency (for example, the intensity of radiation may depend on $1/\nu^2$). A power-law spectrum is distinctive of synchrotron radiation (see Section 14.1),

Figure 25.13   **(a)** *Light echoes from Supernova 1987A reflect from intervening dust sheets.* **(b)** *The supernova's radiation heats a cloud of matter ejected by the supergiant prior to detonation.*

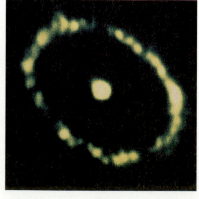

(a)   (b)

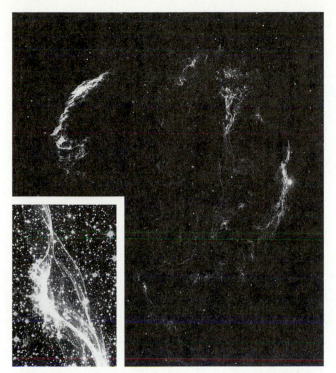

Figure 25.14 *The Cygnus Loop, nearly 3° across, is a blast wave sweeping up and heating interstellar matter. The inset shows a delicate shock wave moving toward upper right and wrapping around an interstellar cloud. The leading edge of the shock has a temperature near 60,000 K.*

created from fast electrons spiraling in a magnetic field at speeds near that of light. The fields have been swept up from the Galaxy and compressed by expanding shock waves. The Crab's field is intrinsic to the object and related to the parent star, and its synchrotron radiation is visible through the optical spectrum and even into the X ray.

Thermal radiation—that produced as a result of heat—also plays an important role. We observe recombination and forbidden lines (see Section 22.1), from which we derive temperatures between 10,000 and 60,000 K and densities in the gaseous filaments near $100/cm^3$. The existence of more highly ionized atoms indicates hot gas within the shock waves with temperatures of 10^6 K. Thermal emission of X rays reveals matter within the hollow shells that reaches 10^7 K and is the origin of the hot coronal gas that threads throughout interstellar space (see Section 22.2 and Figure 22.10).

The compressing blast waves of supernovae are an important factor in inducing star formation, stellar death begetting stellar birth. Meteorites show isotope ratios expected in supernovae, demonstrating that our existence may have been linked to a supernova that began the condensation of the Sun's birth cloud.

25.5
Neutron Stars and Pulsars

A Type II supernova collapses to form a dense ball of self-supporting degenerate neutrons with roughly a Chandrasekhar mass and a radius of about 10 km—a **neutron star.** From the number of supernovae that have taken place over the Galaxy's lifetime, the Galaxy should contain tens of millions of these bizarre objects.

25.5.1 Discovery

In 1934 Walter Baade of the Mount Wilson Observatory suggested that a 13th magnitude star in the Crab Nebula might be a neutron star because its spectrum lacked the normal absorption lines. Unfortunately he could give no proof, and discovery had to wait until 1967. Anthony Hewish, a British radio astronomer, wanted to study fluctuations in the solar wind. So he and his students built a telescope with a rapidly responding detector that could follow the "twinkling" observed in point radio sources when they are refracted by the wind.

He turned the routine monitoring over to a graduate student, Jocelyn Bell, who one morning found a series of small pulses spaced only 1.3 seconds apart (Figure 25.16). She was unable to trace the source, and they disappeared for two months. But then they returned, and most remarkably, right on schedule, for a measured pulse period of 1.337011 seconds. Continued observation established that the source was not on Earth, as pulse visibility correlated with the sidereal day.

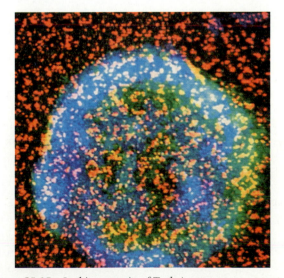

Figure 25.15 *In this composite of Tycho's supernova remnant, blue, red, and green respectively show radio, optical, and X-ray radiation. Most of the radiation is produced by a shock wave interacting with the interstellar medium.*

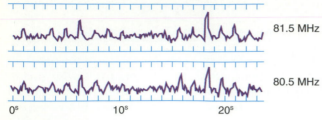

Figure 25.16 *Pulses at two frequencies are shown for the first known pulsar from the discovery paper. Those at 80.5 MHz arrive 0.2 seconds after those at 81.5 MHz, the delay allowing a measure of distance.*

The **pulsar** (for *puls*ating *r*adio source) was traced to the constellation Vulpecula. Radio signals in space are refracted and dispersed by the electrical properties of the interstellar medium, longer waves traveling more slowly. The farther the pulsar and the denser the interstellar medium, the greater the difference between pulse arrival times at two frequencies. By measuring the difference, and by assuming a density for interstellar gas, the distance to the pulsar was estimated to be about 500 pc. For a time, the discoverers entertained the possibility that the pulsar might be an interstellar communications beacon. That idea could not survive the discovery of three more pulsars and the more reasonable view that they were natural phenomena. The signals came too fast to be Cepheid-like pulsations. The only real possibility was rotation. However, for a body to rotate that quickly it would have to be small, no more than a few tens of km across. Bell had finally discovered a neutron star.

The argument was clinched with the discovery of a pulsar in the Crab Nebula with an astonishing period of 0.0316 seconds (Figure 25.17). It pulses not only in the radio spectrum but in the optical and X-ray spectral regions as well, and was the star identified by Baade. We at last could see the stellar end product of a core-collapse supernova, a star of more than a solar mass, some 20 km across, spinning over 30 times a second! Over 400 pulsars are now known, but only three supernova remnants have them. To understand the discrepancy, we look at the radiation mechanism.

25.5.2 Physical Nature

The pulses are complex. Commonly there is an interpulse between the main pulses, and pulse activity may cease for long intervals only to resume right on time. Though pulsars are the most regular natural phenomena known, long-term observations show that pulse periods lengthen as rotation slows. The Crab pulsar lowers its rate by 0.0001% per day. A pulsar may also occasionally display a *glitch*, dur-

ing which it suddenly speeds up, only to resume its slow decay.

The radiation mechanism most likely originates in a magnetic field that has been concentrated along with the mass to a strength about 10^{12} times that of the Earth's field (Figure 25.18). Like almost all magnetic fields driven by a rotational dynamo, a pulsar's field axis is oblique to its rotation axis. By a process still not understood, the magnetic field creates a powerful electric field that accelerates electrons out along the magnetic poles. The result is a pair of oppositely directed beams of tightly focused radiation.

The pulsar acts like the beacon of a lighthouse. If the Earth is in the right position, a beam will sweep by us and we will see a burst of radiation (upper inset, Figure 25.18). If not, we see nothing (lower inset). If the magnetic axis is highly inclined to the rotation axis, we may get hit by part of the opposite beam and see a faint interpulse. For reasons unknown the beams can be temporarily suppressed, but since they are controlled by rotation, they are exactly on schedule when they come back.

Pulsars spin fast because conservation of angular momentum speeds up the supernova core as it collapses. Since the pulses are rotationally driven, radiation must suck energy from the star, which therefore slows with time. The Crab pulsar, only 940 years old, is still highly energetic and is rotating quickly, so it even produces X-ray pulses. As a pulsar slows, it becomes incapable of producing high-energy radiation, and by the time the period is

Figure 25.17 *To the eye, the light from the Crab pulsar (arrow) appears continuous. It actually flashes 30 times per second and is "on" for only 30 milliseconds at a time (inset).*

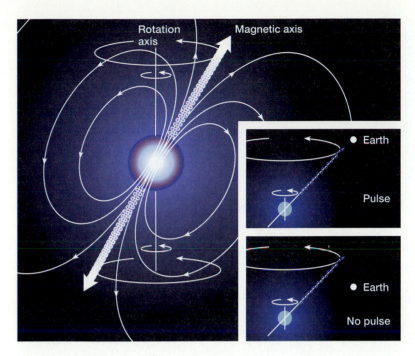

Figure 25.18 *A pulsar's magnetic field spins in space around its rotation axis. Radiation is beamed along the magnetic axis, and if the Earth is in the way (upper inset), our telescopes will pick up pulses; otherwise they will not (lower inset).*

measured in seconds, the little star is bright only in the radio spectrum. After millions of years, when the period lengthens to 10 seconds or so, it fades from sight.

The interiors of white dwarfs are expected to crystallize; a neutron star crystallizes on its surface. Any normal fluid, even water, has some degree of *viscosity*. As it flows, it rubs against itself, producing friction and heat, which dissipates energy. Within the neutron star, matter is in a *superfluid* state in which viscosity disappears. Motions within the superfluid chaotically interact with the surface and can cause the glitches, increases in rotation speed. We know at least something of how these stars work, even if many aspects still mystify.

Now we can understand the poor correlation between pulsars and supernova remnants. Type Ia supernovae may annihilate themselves, so no pulsar is produced. In most Type II (and Type Ib) supernovae, the pulsar is likely to be aligned in the wrong direction, and therefore, we cannot see it. On the other hand, the pulsars we do see long outlive their supernova remnants, so most of them are bare. The number we see in supernova remnants—the Crab, the Vela Remnant, and one other—is about what we expect to find.

Just as there is a maximum mass that can be supported by degenerate electrons (1.4 $M_\odot$), so is there a maximum mass that can be held up by degenerate neutrons. This limit is more difficult to calculate, but it is around 3 $M_\odot$. No neutron star with a greater mass should exist.

25.5.3 Binary Systems

High-mass stars are found in binaries, so ordinary stars can have neutron-star companions. Masses of pulsars derived from the motions of their mates confirm the neutron-star limit: none has been found greater than 3 $M_\odot$. Most pulsars with estimated masses hover near the 1.4 $M_\odot$, the Chandrasekhar limit.

If the binary components are close together, the ordinary star can transfer mass to an accretion disk around the neutron star. Because of the intense gravity of the neutron star, the disk will be hot enough to radiate X rays, accounting for a large number of observed **X-ray binaries.** Hydrogen falls from the accretion disk onto the neutron star, where it burns to helium via the carbon cycle. When around 10^{-7} Earth masses has accumulated, the 3-α process ignites explosively, producing an X-ray burst that lasts for a minute or two (Figure 25.19). This **X-ray burster** will repeat its behavior every few hours or days.

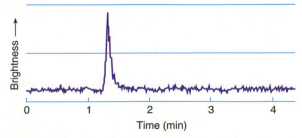

Figure 25.19 *The X-ray burster MXB 1636-53 lets loose a powerful blast of X rays far stronger than the total luminosity of the Sun.*

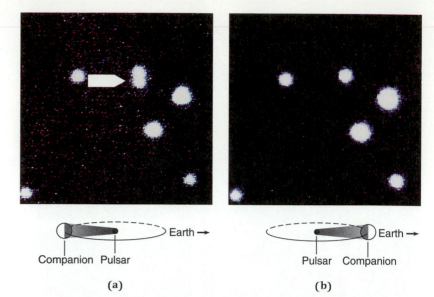

Figure 25.20 (a) *The Black Widow pulsar heats its companion, rendering the companion visible, as indicated by the left-hand drawing (the pulsar is detected in the radio).* (b) *During the eclipse, the pulsar is hidden, the heated side of the companion turns away from us, and the companion disappears as well.*

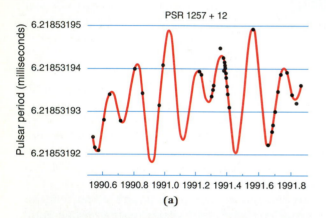

(a)

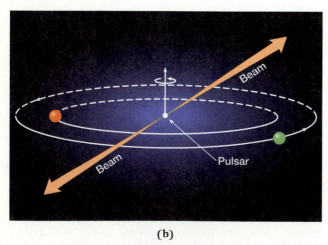

(b)

Figure 25.21 (a) *Changes in the period of the millisecond pulsar PSR 1257 + 12 indicate* (b) *a pair of orbiting planets.*

The Crab pulsar is a fast rotator, but the record is held by a body spinning *885 times per second*, a **millisecond pulsar** that pulses every 0.00113 seconds! Millisecond pulsars are weak radiators, implying both a weak magnetic field and great age. But rapid rotation implies youth. These stars may be the successors to the X-ray bursters. As matter from the accretion disk rains down onto the neutron star, it transfers not only mass but also angular momentum and the star spins faster. As a result, the old pulsar is reborn, now with a fantastically rapid rotation speed. In support, we see millisecond pulsars that are indeed binaries. The companions are inferred from a kind of Doppler shift, not in spectrum lines, but in the apparent frequencies of the pulses. As a pulsar moves away from us, the pulses arrive later and later as the intervals between them are stretched. As it moves toward us, the intervals are shortened.

However, two millisecond pulsars have no companions. The anomaly is explained by observation of a third pulsar that is in an eclipsing binary system. The pulsar's companion has a radius about 0.2 that of the Sun and an extremely low mass of only 0.02 $M_\odot$. This *Black Widow pulsar* (Figure 25.20) is stripping matter from its companion, devouring its mate (which real black widows do not do, demonstrating poor cross-training in the sciences). In time it will join the two companionless pulsars, then slow down and take its place among the other silent remnants of devastating stellar collapse.

This phenomenon has apparently given rise to planets. The pulse times from the millisecond pulsar PSR 1257 + 12 in Virgo exhibit variations with

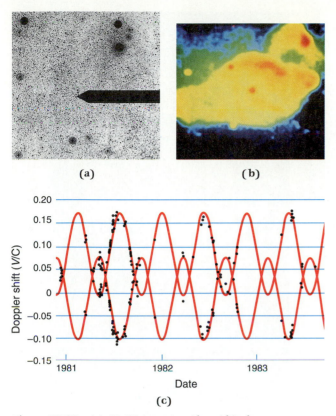

(a)   (b)

(c)

Figure 25.22   **(a)** *SS 433 (arrow) resides within the supernova remnant W 50, represented by a radio image in yellow* **(b).** **(c)** *Doppler shifts of two sets of hydrogen lines are plotted against time. One set is mostly blue-shifted, the other red-shifted, the pairs crossing over once every 164 days.*

periods of 66 and 98 days (Figure 25.21a). The fluctuations are ascribed to the gravitational effects of a pair of orbiting bodies (Figure 25.21b) with semi-major axes of only 0.36 and 0.47 AU and minimum masses of 3.4 and 2.8 Earth masses. These bodies may have been created from the orbiting debris of an evaporated companion star. Apparently, planets will form almost anywhere given a source of matter for their assembly.

Finally, the variety and strangeness of binary interactions is capped by a B supergiant in Aquila called SS 433 (Figure 25.22). Its spectrum displays three sets of emission lines. One consists of hydrogen and helium with a reasonable Doppler shift of 70 km/s. The other sets consist of hydrogen lines that shift back and forth through several hundred Ångstroms during a 164-day period, indicating matter moving at a tenth the speed of light. Mass apparently flows from the B star into a tilted accretion disk around a collapsed companion (Figure 25.23). As it spirals into the central object, some of it is shot through the poles in bipolar jets at a quarter the speed of light. The B star causes the accretion disk to precess over the 164-day interval, which makes the jets wobble. We then see radiation from the jets projected along the line of sight. Orbital analysis indicates that the B star's companion has a mass of about 0.8 $M_\odot$. The system's alliance with a supernova remnant strongly suggests that the companion is a neutron star.

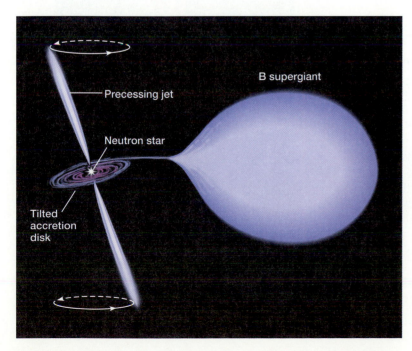

Figure 25.23   *SS 433 is a B supergiant that is feeding mass into a tilted accretion disk around a neutron star. Jets of gas shoot from the poles of the disk at a quarter the speed of light. The disk precesses, causing the jets to wobble and produce shifting emission lines.*

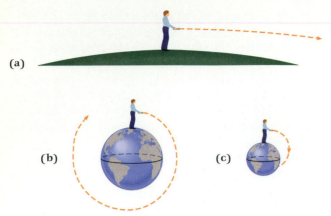

(a)

(b)   (c)

Figure 25.24   *This figure and the next two make a continuous sequence that shows the development of a black hole.* **(a)** *A horizontally directed light ray will bend imperceptibly to Earth because of the curvature of spacetime.* **(b)** *The Earth has been shrunk, and the curvature is great enough to put the light nearly into orbit.* **(c)** *The curvature is so large that the light falls back to the surface and cannot escape. These diagrams are schematic and are not drawn to any scale.*

25.6
Black Holes

If the mass of a supernova core were beyond the neutron degeneracy limit of 3 $M_\odot$, it would no longer have any known means of support. We now examine the possibility—indeed, the *probability*—that stars can go into total collapse and disappear, pulling the fabric of the Universe over their heads.

25.6.1 Theory

Throw a ball upward and gravity brings it down. But if you could throw it fast enough, at the Earth's escape velocity of 11.2 km/s, the ball would not return. Now do an experiment in your mind. Build a machine that shrinks the Earth and increases the acceleration of gravity and escape velocity, v_{esc}, at its surface. From Section 7.4, $v_{esc} = \sqrt{2GM/R}$. When the Earth is a quarter its present size, v_{esc} doubles to 22.4 km/s, and when the Earth is as dense as a neutron star, with a radius of only 50 meters, v_{esc} is 3,997 km/s. At a radius of 9 mm, v_{esc} = 299,793 km/s, the speed of light. Light can no longer escape; to an outside observer, the Earth *disappears from view*. It has become a **black hole.**

When you threw the ball upward, it lost energy and slowed as it worked against the gravity. Shine a flashlight toward the zenith. The light must work against gravity too, but must always move at c.

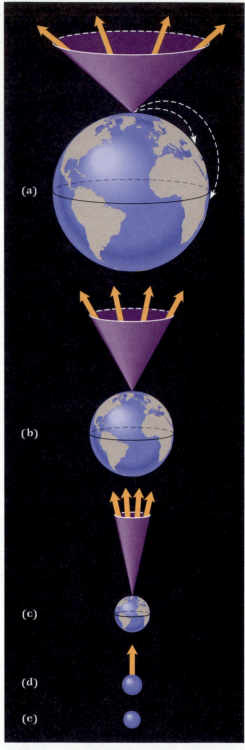

(a)

(b)

(c)

(d)

(e)

Figure 25.25   *The sequence of Figure 25.24 continues.* **(a)** *Only light within the exit cone can escape. Outside the exit cone, all light falls back to the surface.* **(b)** *and* **(c)** *The Earth becomes smaller, and the cone shrinks.* **(d)** *The Earth is just over 9 mm in radius, and only light directed vertically escapes.* **(e)** *The radius is 9 mm, spacetime at the surface is infinitely curved, nothing escapes, and the Earth is a black hole. The diagrams are not to scale.*

Since E_{photon} = hv, light loses energy by reddening. Gravitational redshifts are observed in the laboratory and in the spectra of white dwarfs and the Sun. Assume a friend is orbiting the Earth and observing your flashlight. As the Earth shrinks, the beam reddens, shifting to the infrared and then the radio. When v_{esc} = c, the light redshifts to infinity and disappears from view.

Both these explanations invoke a Newtonian view of gravity. Einstein's theory, however, considers gravity as a curvature of spacetime (see Section 7.8). If you shine your flashlight horizontally, the beam will bend to Earth (Figure 25.24a), as starlight does when it passes near the Sun. Earth's gravity is so weak that such bending is undetectable. But if the Earth were shrunk sufficiently, the bending would be so severe that the light could go into a spiral orbit before escaping (Figure 25.24b); if gravity is further increased, it would fall back to the surface (Figure 25.24c).

As the Earth is made smaller yet, only light within an *exit cone* around the zenith would be able to leave (Figure 25.25a); the rest of it curves back. As the planet shrinks, the cone becomes smaller (Figures 25.25b and c). When the radius is 9 mm, only a vertical ray can escape (Figure 25.25d). If the body shrinks even more, spacetime folds back on itself, *all* light is returned, and the beam and the Earth disappear from view (Figure 25.25e). The light still travels at c, but it will not emerge from an Earth of radius less than 9 mm because of the infinite distortion of spacetime.

The idea of the black hole goes back to the eighteenth century. The German physicist Karl Schwarzschild introduced the modern view in 1916 with the first exact solution of Einstein's equations of gravity. The black hole described above is therefore called a *Schwarzschild black hole.* Every physical body has a radius, R_{bh}, at which it will disappear, which is found by substituting c into the escape velocity equation:

$$R_{bh} = 2\ GM/c^2.$$

R_{bh} of the Sun is 3.0 km, and that of a body at the neutron star limit of 3 $M_\odot$ is 5.9 km.

Under these conditions, a body will contract to a point, a **singularity,** leaving a sphere of radius R_{bh} around it called the **event horizon** (Figure 25.26). Nothing—neither light, mass, nor information—can return from within it. We can know nothing of conditions within the event horizon, and the laws of physics as we know them have no meaning.

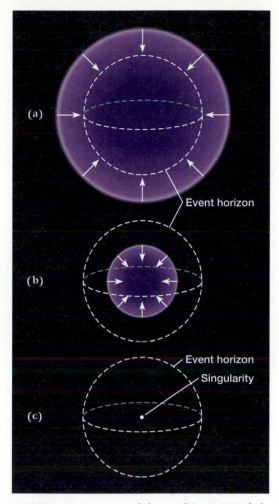

Figure 25.26 **(a)** *In an expanded view of Figure 25.25d, the dashed curves represent the event horizon. The body is not yet a black hole, as part of its contracting mass is outside its event horizon.* **(b)** *The body has contracted within the event horizon and is no longer visible.* **(c)** *The body has contracted to a point, or singularity.*

In spite of its strangeness, a black hole is a simple object. Its description requires only mass, electric charge, and rotation. The Schwarzschild black hole has only mass. Like stars, black holes should be neutral, so we can likely ignore electric fields. However, everything in the Universe rotates, and anything collapsing into a black hole should spin faster and faster. Therefore, all black holes should be rotating. A rotating *Kerr black hole* (after the New Zealand mathematician Roy Kerr) has its singularity in the shape of a ring. A major characteristic of a Kerr black hole is a flattened zone outside the event horizon called the *ergosphere,* within which everything, even spacetime, is dragged around by the rotation.

Black holes are popularly misconceived. Poorly written stories and films give the idea that they are inherently dangerous, that they will suck in matter—and you—if you venture too close. The black hole is no different gravitationally from any star. If the Sun were to become a black hole, it would disappear from sight, but the curvature of spacetime and the gravitational acceleration at the distance of the Earth would remain precisely the same. An approaching or orbiting body can fall in only if some outside force dissipates its orbital energy. Orbiting a black hole is no more dangerous than orbiting the Sun.

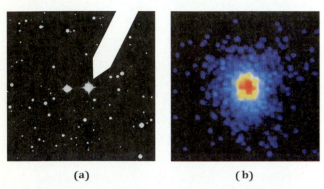

(a)   (b)

Figure 25.27 *Cygnus X-1 is identified with HDE 226868, a B0 supergiant seen in the optical photograph* **(a)** *that uncharacteristically radiates powerful X-ray emission* **(b)**. *The X rays come from matter falling onto a companion, quite probably a black hole.*

25.6.2 Observation

Do such strange bodies exist? As an aid to credibility, the difference between a black hole and a neutron star is not large, only a factor of 2 or 3 in radius. And as odd as black holes may seem, some observed objects are odder yet and require black holes to explain them. If the core of a collapsing supernova is greater than 3 $M_\odot$, a black hole seems inevitable, and there may be large numbers of them in the Galaxy.

Though black holes radiate no energy, we can find them by their effects on their surroundings. Cygnus X-1 (Figure 25.27) is a powerful source of X rays. It has been identified with a B0 supergiant that is a single-lined spectroscopic binary (see Section 20.1.4). The mass of the B0 star is found by comparing it with theoretical evolutionary tracks. From the orbital velocity of the B star (as projected along the line of sight), we find the companion's mass to be greater than 3 $M_\odot$ and possibly as large as 16 $M_\odot$. Any normal star with that high a mass would be visible. The companion is too massive to be a neutron star, so we believe it to be a black hole. Its gravity raises tides that make the nearby B0 star fill its Roche lobe (see Figure 21.22). Mass is then transferred from the supergiant into an accretion disk around the black hole. As the matter loses

Figure 25.28 *Matter falling into a black hole can illuminate the event horizon's surrounding neighborhood.*

BACKGROUND 25.1 Mystery and Reality

Our view of the world and of the Universe is very limited. We live on a quiet planet that belongs to an old, average, settled star off to the side of a large but otherwise ordinary galaxy. We have been pondering the heavens for over 5,000 years. However, counting from the time of the first parallax measurement, we have been viewing the Universe scientifically for a mere century and a half. As sophisticated as the science of astronomy may seem, we must admit that we are still naive. We do not have a clear idea of how the Universe actually works and do not even yet have a census of the kinds of objects that inhabit our environs. Every time a new instrument is constructed, one that allows us to look deeper, with more sensitivity, or in a different wavelength band, new and undreamed-of discoveries are made. The Universe is still a mysterious place.

There are two ways to deal with mystery. We can make up explanations, or we can probe the world and examine the facts, all that we can find. Thousands of years ago people attempted to explain nature by populating the sky with gods and the woods and streams with fairies and nymphs. A pair of modern examples relating directly to astronomy are astrology and the interpretation of *UFOs* as alien craft from other planets. Both seemingly dispel mystery in a comforting way that people might quickly understand. However, they merely displace mystery and keep it deliberately alive. No astrologer knows how astrology "works." No astrologer really *cares* to know. One who did would then be a scientist, and would quickly see that the subject is nothing but a vapor that is quickly dissipated under the light of facts. The same is true for UFO studies. Better to repeat the old saw that the government has a conspiracy to keep the facts from us rather than to look at all the facts available, whence another vapor vanishes.

Why make up fantasies when the truth is so much more fascinating? On first encounter, some modern hypotheses—invisible neutrinos, black holes, and the Big Bang creation of the Universe—may seem, like astrology, to be just made-up ideas. Instead, they are real attempts at explaining what is actually seen, *all* that is seen, which is more amazing than the ideas themselves. Even more intriguing is the knowledge that we know so little. Mystery thus pushes us forward to seek reality.

---

angular momentum by friction and spirals inward, it heats and radiates (Figure 25.28). Other candidates are LMC X-3 in the Large Magellanic Cloud, an object in Monoceros called A0620-00, the X-ray flare star V404 Cygni, and Nova Muscae 1991 (an X-ray nova), which may harbor black holes between 3 and 10 $M_\odot$.

Another black hole candidate is The Great Annihilator, a source of gamma rays in Sagittarius near the galactic center. The object radiates a spectrum line at 0.0243 Å, a characteristic of the mutual annihilation of matter and antimatter in the form of electrons and positrons. Matter thought to be falling from a binary companion into the Great Annihilator is heated to such high temperatures—perhaps 10^9 K—that energetic gamma-ray pairs collide to *produce* positrons and electrons (the reverse process). The positrons emerge in a flow (Figure 25.29), and when they hit the electrons in the sur-

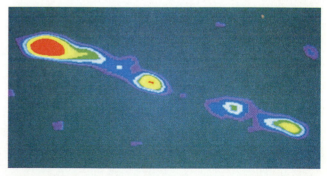

Figure 25.29 *A radio image shows jets of electrons and positrons streaming from the Great Annihilator. The jets are illuminated as the positrons and interstellar electrons mutually annihilate.*

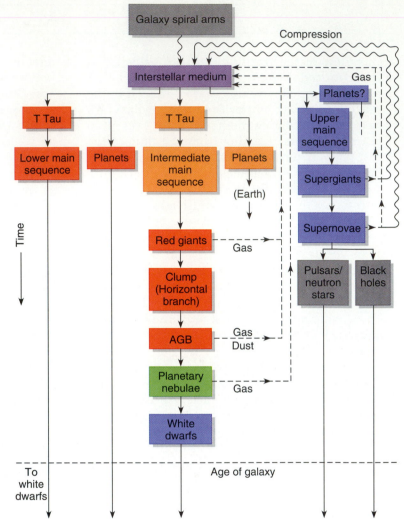

Figure 25.30 *In the cycle of stellar evolution, each generation of stars depends on the those that came before for its heavy elements and for its formation. Time runs downward, the flow of stellar evolution indicated by black arrows. (The relative speeds of evolution are schematic only and are not to scale.) Dashed lines show the return of matter to the interstellar medium and wavy lines show the origin of compressive forces that can initiate star formation.*

rounding interstellar medium, the particles annihilate each other and produce the gamma-ray spectrum line.

Gamma-ray bursters may represent an even more bizarre possibility. The orbiting *Compton Gamma Ray Observatory* shows them all over the sky, and their sources are believed to be extragalactic. It has been suggested that they are produced by collisions between black holes and neutron stars in other galaxies.

Though we have never actually detected a black hole, we are becoming convinced that they exist. If they do, then we have finally found the ultimate in contraction: some stars, as they are born out of the mists of interstellar space, are destined to shrink all the way down to points.

25.7
The Cycle of Evolution

The intermediate-mass and high-mass stars can now be combined into the full cycle of stellar evolution (Figure 25.30). Giants and supergiants manufacture helium, carbon, and nitrogen in their interiors by the cycles of nuclear energy generation, make heavier elements by the s-process (and other nuclear processes), and transfer them to the surface by convection. Mass loss through winds, (producing planetary nebulae, ring nebulae, and the like) strip the hydrogen envelopes from the stars and spread the chemical by-products of evolution into interstellar space. Supernovae create and eject yet more heavy elements, including all the galaxy's iron and the heavy atomic nuclei of the r-process, as stellar evolution continually enriches the interstellar medium.

AGB stars make dust and spew it into the interstellar medium, creating the seeds out of which interstellar grains and molecules—and eventually stars and planets—will grow. O stars and supernovae (as well as the Galaxy's spiral arms, to be discussed in the next chapter) act as triggers that compress the interstellar medium and initiate star formation. The characteristics of stars and planets therefore depend on earlier stellar generations in a tightly interlocked cycle. We now see why young Population I stars are richer in metals than the older ones of Population II. Earth is not isolated from other stars but is a part of them. We are a distillate of stellar evolution.

KEY CONCEPTS

Black hole: A mass inside an **event horizon** where v_{esc} = c, preventing light's escape.

Core-collapse supernovae: Supernovae produced by the collapse of the iron cores that have developed in high-mass stars.

Humphreys-Davidson limit: The upper limit of stellar luminosity, above which stars lose enough mass to alter their evolution.

Millisecond pulsar: A pulsar that spins hundreds of times per second, its rotation speeded up by mass accretion from a companion.

Neutron star: A collapsed star made of neutrons and supported by degenerate neutron pressure; its mass is less than about 3 $M_{\odot}$.

Power-law spectra: Continuous spectra in which the intensity of radiation depends on an inverse power of frequency.

Pulsar: A rotating neutron star that beams energy along a tilted magnetic field; it is visible if a beam hits Earth.

r-process: A set of nuclear reactions that build heavy elements (beyond those created by the s-process) in supernovae by the rapid capture of neutrons.

Singularity: A dimensionless mass; a point-mass.

Wolf-Rayet (WR) **stars:** High-mass, hydrogen-poor stars with emission lines of helium and nitrogen (WN) or carbon (WC).

X-ray binaries: Binaries that radiate X rays from accretion disks around neutron-star components; they may develop into **X-ray bursters** that explosively initiate the 3-α process on their surfaces and emit bursts of X-ray radiation.

KEY RELATIONSHIP

Radius of a black hole:

$$R_{bh} = 2\,GM/c^2.$$

EXERCISES

Comparisons

1. Compare WN and WC Wolf-Rayet stars.
2. What are the differences between the r- and s-processes?
3. Compare the progenitors of Type Ia and Type II supernovae.
4. How do young and old supernova remnants differ?
5. How do ordinary novae and X-ray bursters differ?
6. Compare the limits to electron and neutron degeneracy.
7. Compare the dimensions of white dwarfs, neutron stars, and black holes.
8. How does a black hole's singularity differ from its event horizon?
9. How do Schwarzschild and Kerr black holes differ?

Numerical Problem

10. What is the radius of the event horizon for a black hole with as much mass as the most massive star in the Galaxy?

Thought and Discussion

11. How do we know that supergiants are related to high-mass main-sequence stars?
12. What effects do stellar winds have on high-mass stars and their surrounding environments?
13. How do we know that Wolf-Rayet stars have lost perhaps half their original mass?
14. What is the Humphreys-Davidson limit? What is its effect?
15. What does the binding energy of an atomic nucleus have to do with stellar evolution?
16. Summarize the evidence for the transformation of light elements into heavy elements.
17. What evidence supports the core-collapse hypothesis of supernova explosions?
18. In what form does most of the energy from supernovae emerge?
19. What is the illumination mechanism of a supernova remnant?
20. Why do pulsars and supernova remnants not correlate very well?
21. What effects do supernovae have on the interstellar medium?
22. Why may the pulsar in Supernova 1987A never be seen?

23. What kinds of pulsars radiate in the X-ray spectrum?
24. Why do young pulsars spin so fast? Why do they slow with time?
25. If pulsars slow down, how can you explain old millisecond pulsars?
26. What are the effects of having a neutron star in a binary system?
27. What effect does the discovery of planets orbiting a pulsar have on the Drake equation?
28. How does gravity work to produce a black hole?
29. Summarize the evidence for the existence of black holes.

Research Problem

30. Using photocopies of illustrations in magazines and books, assemble an atlas of supernovae that have taken place in other galaxies. On the basis of their locations within the galaxies, suggest what types they might be.

Activities

31. Construct a picture of the evolution of a 75-solar-mass star. Draw a series of sketches showing what will happen to it over the course of its lifetime.
32. List the different nuclear reactions encountered so far in this book.
33. List the known pause and contraction phases found in stellar evolution and the reasons for the known pauses in contraction.

Scientific Writing

34. Write a feature article for a newspaper describing some known phenomenon that might at first appear so fantastic as to be unbelievable.
35. Write a script for a lecture to a high school chemistry class that explains the origins of the chemical elements.

26

The Galaxy

The local assembly of stars and the structure of the Milky Way

Vast clouds of stars crowd the Milky Way through the galactic center in Sagittarius.

457

Stars, nebulae, molecular clouds, and collapsed objects interact both gravitationally and by way of cyclic evolution. All these objects are interlocked in a grand picture, not only of stellar evolution, but of the evolution of the Galaxy as a whole.

26.1
Basic Structure

The Galaxy (Figure 26.1) consists of a Population I disk made of billions of stars, visible to us as the Milky Way, enclosed in a spherical Population II halo. Disk stars orbit the galactic center on roughly circular paths of low inclination; those in the halo move on elliptical orbits with high inclination (see Figure 19.7). Population I contains stars of all spectral classes, but from a distance appears bluish because its light is dominated by brilliant young blue O and B stars (see Figure 25.9). Population II is reddish because it contains only older stars and their evolved red giants, and it is deficient in metals compared with Population I (see Section 19.5.4 and Table 26.1).

Open clusters and associations belong to Population I, whereas the densely packed globular clusters relate to Population II. The interstellar medium pervades the Milky Way and the Population I galactic disk, and it is therefore within the disk that new stars are born. The halo has little gas and dust and creates no new stars. Population II formed first, giving rise to Population I as dying stars continually enrich the Galaxy, accounting for the composition difference between populations (see Section 25.7). The disk and halo then merge together into a **galactic bulge.**

We must find our way within this structure. To navigate on the Earth we use latitude and longi-

TABLE 26.1
Population Characteristics

Population I	*Population II*
Galactic disk	Galactic halo
Blue O and B stars, interstellar medium, supergiants, and all other classes	Red dwarfs and giants
Open clusters	Globular clusters, subdwarfs
Metal-rich	Metal-deficient
Star formation and young stars	Old stars only
All kinds of supernovae	Type Ia white dwarf supernovae

Figure 26.1 *A schematic model of the main portion of our Galaxy (that contains 90% of the stars) shows a disk (in yellow) studded with blue O and B stars. Surrounding the disk is a sparsely populated halo (in red) with globular clusters. The Sun is located 8.5 kpc, roughly two-thirds of the way out, from the center. The dashed ring around the Sun indicates the zone observed by Herschel. In reality, the Galaxy does not have a sharply defined edge but fades out to a much larger extent than shown here.*

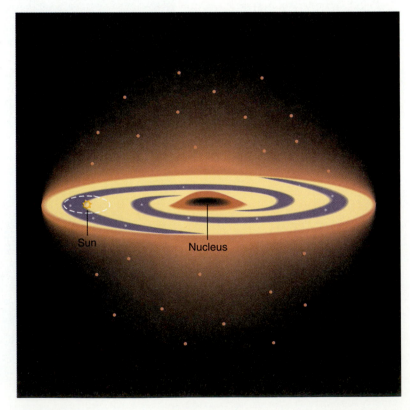

Sun

Nucleus

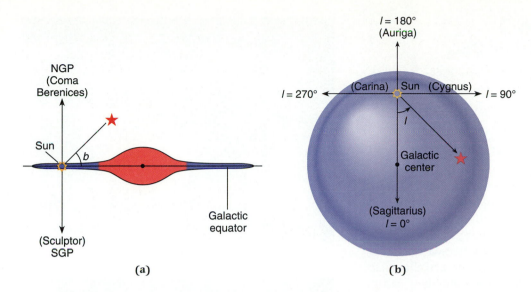

(a)   **(b)**

Figure 26.2   **(a)** *The galactic equator runs through the center of the galactic disk perpendicular to the directions to the north and south galactic poles (NGP and SGP). Galactic latitude (b) is measured perpendicular from the galactic equator.* **(b)** *You look down on the plane of the Galaxy from the NGP. Galactic longitude (l) is measured counterclockwise along the galactic equator from the galactic center.*

tude; on the celestial sphere, hour angle, right ascension, and declination. In the Galaxy we use **galactic coordinates** centered on the Sun. The galactic equator runs down the middle of the Milky Way, dividing the Galaxy into northern and southern galactic hemispheres centered on the *north and south galactic poles* (NGP and SGP, Figure 26.2a). *Galactic latitude* (*b*) is measured north and south of the galactic equator toward the galactic poles. *Galactic longitude* (*l*) is measured along the galactic equator from the galactic center in Sagittarius (defined by a powerful source of radio radiation) counterclockwise as seen looking down from the NGP, in the direction toward Cygnus (Figure 26.2b). The galactic equator, galactic poles, and galactic longitude are marked on the star maps in Appendix 1.

To navigate the Galaxy, we also need to know its size. William Herschel made the first attempt to reach outward by counting the number of stars seen in different directions to successively fainter magnitude limits. By assuming that all stars have the same luminosity (or absolute magnitude), he estimated their relative distances and thereby created a map of the Galaxy (Figure 26.3) that demonstrated the disklike stellar distribution. However, Herschel did not know about interstellar dust, which limited his view in the disk to only a few hundred parsecs and caused him to place the Sun near the center of its own local neighborhood (marked by a circle in Figure 26.1).

The breakthrough in measuring the Galaxy's size was made at Mount Wilson around 1918 by Harlow Shapley when he looked *outside* the galactic plane at the halo's globular clusters and made the logical assumption that the center of the whole system of globulars should coincide with the center of the Galaxy (see Section 20.5.2). From modern HR diagrams of globular clusters, spectroscopic distances of the halo's RR Lyrae stars (whose system must also coincide with the Galaxy's center), and other methods, astronomers now adopt a distance of the Sun to the galactic center (R_0) of 8.5 kpc, with an uncertainty of about 0.5 kpc. From the symmetry of the Milky Way around the sky relative to a great circle, we find that the Sun lies only about 15 pc above the central plane of the galactic disk.

It is difficult to express the Galaxy's radius. There is no clearly defined edge, only a gradual diminution in the number of stars and in the density of the interstellar medium. As a result, we have to rely on rather arbitrary criteria. About 90% of the light of the galactic disk falls within a circle around the center with a radius of 12.5 kpc, which can be taken as the "traditional" radius as expressed by Figure 26.1, and which places the Sun about two-thirds of the way from the center. As we will see, however, the Galaxy really extends much farther.

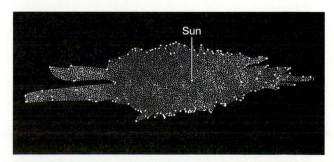

Figure 26.3   *Herschel's system of the stars shows a flattened, fat disk with the Sun near the center. He could observe only stars within a few hundred parsecs and had no idea of the extent of the system. Most of the structure is the result of obscuration by local dust.*

26.2
The Galactic Disk

Galaxies like ours are dominated by disks that contain the vast majority of the visible mass. The disk is the location of most of the action, and where we first focus attention.

26.2.1 Structure, Thickness, and Composition

A closer look at the Galaxy shows a simple division of stars into two populations to be inadequate. Each population has fine structure and subdivisions related to age and chemical composition. The disk can be separated into layers (Figure 26.4). The thickness of any layer is given by its **scale height,** defined as the distance from the center of the galactic plane to the point at which the density of matter has fallen to 0.37 of the central density ($1/e$, where $e = 2.713 \ldots$ is the base of the so-called natural logarithms; ordinary logarithms use 10 as a base).

The thinnest portion of the disk, with a scale height of only about 100 pc, contains the Galaxy's gas and dust. As a result, it also contains the youngest stars, including the blue O and B stars. This thin disk is readily apparent in other galaxies as a dark band of dust, and in our own Galaxy as the central rift in the Milky Way (Figure 26.5). A thicker disk with a scale height of roughly 350 pc contains 98% of the disk's—and the Galaxy's—stars. It encompasses both the thinner disk and the older stars, those born over the past several billion years.

Spreading outward is a thick Population I disk with a scale height of 1,000 pc. This portion contains only about 2% of the disk's stars and is something of a bridge to Population II, since its metal content is about half that of the Sun and the inner disk.

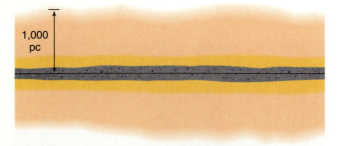

Figure 26.4 *The Population I disk of the Galaxy is made of layers that gradually thin out in the perpendicular direction. The thinnest layer (gray) is made of dust and gas and is studded with young blue O and B stars. A thicker layer (yellow) embraces most of the Galaxy's stars and the sparsely populated thick layer (orange) has a somewhat lower metal content.*

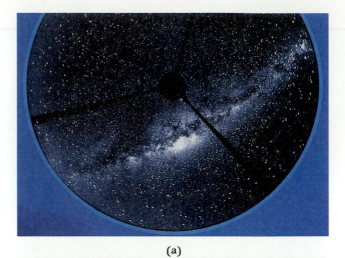

(a)

(b)

Figure 26.5 **(a)** *A wide-angle view of the Milky Way centered on the galactic nucleus bears a remarkable resemblance to* **(b)** *NGC 891, a disk galaxy seen edge-on. Both exhibit the dark dusty band of the thinnest part of the disk.*

26.2.2 Galactic Rotation

The disk of the Galaxy rotates about the galactic center under the force of its own gravity. To find the nature of the rotation, astronomers examine the orbits of the Galaxy's stars and clouds of interstellar matter.

Start with the Sun. One way to find the solar orbital direction and velocity is to observe objects outside the galactic disk that do not partake in galactic rotation, such as globular clusters and RR Lyrae stars (Figure 26.6). These objects have their own motions. But those toward which the Sun is moving directly will have *average* velocities of approach that equal the solar orbital velocity. Those in the opposite direction will have the same *average* velocity of recession. From such studies we find the Sun to be moving in a direction close to 90° galactic longitude at about 240 km/s.

The solar orbit, however, is somewhat elliptical. The average motions of the local disk stars define the local standard of rest, the *LSR* (see Section 19.2), which follows a *circular* path around the galactic center. When we correct for the motion of the Sun relative to the LSR (and average in other methods), we find that the circular orbital velocity of the Galaxy at a distance of 8.5 kpc from the galactic center is about 220 km/s.

The next task is to measure orbital velocities at different distances from the galactic center. Because of obscuring interstellar dust, it is best to use radio observations. At 21 cm, we can see neutral interstellar atomic hydrogen (H I) all the way across the disk. For simplicity, assume the Sun to be on a circular orbit. In Figure 26.7a, a radio telescope is pointed into the disk at a galactic longitude of 35° and receives 21-cm radiation from H I clouds along the line of sight (Figure 26.7b). The observed wavelengths depend on the radial velocities, which in turn depend on the differences between the orbital velocities of the clouds and the orbital velocity of the Sun as projected onto the line of sight. Cloud D, on the solar orbit, moves along the line of sight with the same speed as the Sun, and therefore appears to be at rest and radiates the 21-cm line at the rest wavelength of λ_0. However, greater proportions of the orbital velocities of clouds A and C will be projected along the line of sight, and their spectra will be Doppler-shifted toward longer wavelengths.

Cloud B lies at the tangent point to its orbit, and *all* of its orbital velocity is along the line of sight. Radiation from B then has the greatest Doppler shift

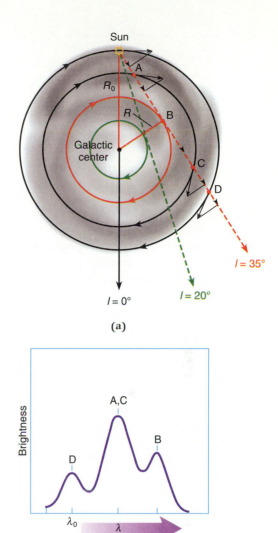

(a)

(b)

Figure 26.7 **(a)** *A radio telescope looks outward at a galactic longitude of 35°. Clouds A, B, C, and D and the Sun are assumed to have circular orbits. The clouds' radial velocities depend on the projections of the orbital velocities along the line of sight.* **(b)** *The telescope records a complex 21-cm line radiated by clouds moving at different radial velocities: Cloud D radiates at the rest wavelength, λ_0; cloud B has the greatest radial velocity, from which we can find the orbital speed at distance R from the galactic center; clouds A and C have equal intermediate velocities. Clouds beyond D have negative velocities. Observation at a longitude of 20° (dashed green arrow) gives the rotation velocity at a different distance from the center.*

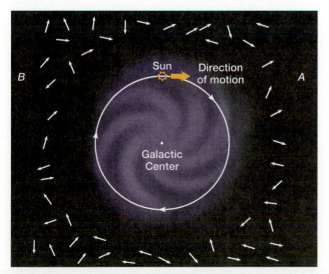

Figure 26.6 *The Sun is moving in the direction of the yellow arrow. Objects not participating in galactic rotation are moving along randomly oriented orbits (white arrows). Those in the A direction will on the average appear to be approaching the Sun at the solar velocity; those in the B direction will on the average appear to be receding.*

and the longest wavelength. Since we know how the solar velocity is directed along the line of sight, we can find the actual orbital velocity of cloud B. The Sun, the galactic center, and cloud B define a right triangle. Since we know the distance of the galactic center, R_0, we can compute the distance of cloud B from the center and the radius of its orbit, R. Observations along different galactic longitudes (such as the dashed green line in Figure 26.7), plus correc-

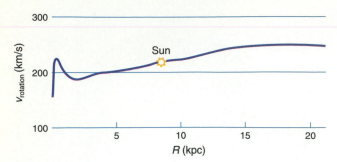

Figure 26.8   *The rotation speed of the Galaxy quickly increases from the galactic center, then levels off.*

tions from the velocity of the Sun to that of the LSR, allow us to determine speeds at different distances from the galactic center and to construct the **galactic rotation curve** (Figure 26.8).

This method does not work in the outer part of the Galaxy because we can no longer look tangent to an orbit. However, the dust there is less opaque, and we can see better in the optical part of the spectrum. Radial velocities of diffuse nebulae and molecular clouds coupled with spectroscopic distances of their embedded O and B stars provide orbital speeds relative to the Sun. We can also judge the distance of the galactic layer of neutral hydrogen moving at a specific velocity by its angular diameter. A combination of all methods allows calculation of the rotation curve farther than 20 kpc from the center, well beyond the "traditional" radius of 12.5 kpc.

26.2.3 *Spiral Structure*

When we look into the Universe, we see that the disks of other galaxies contain matter concentrated into graceful sets of winding **spiral arms** (Figure 26.9). The Milky Way Galaxy ought to have them, too. How do we find them? Examination of other galaxies shows clearly that the arms are made of OB associations, H II regions, and associated giant molecular clouds. We can therefore apply the method of spectroscopic distances to our Galaxy's O and B stars and plot them against galactic longitude to establish local structure and the locations of the arms outside the solar orbit, where the dust is thinner.

Beyond our local neighborhood but inside the solar orbit, we use the galactic rotation curve. As shown in Figure 26.7, the radial velocities of the neutral hydrogen clouds depend upon their distances along the line of sight. Once the rotation curve is known, we can calculate the distances of the clouds from their radial velocities. The result, first obtained over 30 years ago, is a map of neutral

hydrogen that displays clear evidence for spiral structure (Figure 26.10a). We obtain similar evidence by mapping the distribution of giant molecular clouds (Figure 26.10b).

Different tracers, however, chart different arm positions, reflecting the Galaxy's complexity. The 21-cm concentrations are different from those of the giant molecular clouds. In particular, the latter show a distinct enhancement, absent in H I, at a distance of 5 kpc from the center. From all available evidence, astronomers have compiled a simplified map of the major arms, shown in Figure 26.11. However, as prominent as the spiral arms appear to be in our Galaxy and in other galaxies, they constitute a surprisingly minor part of the galactic mass. Take the rare O and B stars away and the arms disappear from optical view. The arms would then appear only in the radio, where they would still be outlined by the interstellar medium.

The origin of the spiral structure has long been debated. The arms cannot be permanent collections of stars: the rotation of the Galaxy would make them wind up and disappear. Instead, they appear to be waves of density. The waves rotate around the Galaxy more slowly than the stars, so that the stars continually pass through the waves in the direction of rotation, slowing down temporarily in response to the increased gravitational field. The effect is not unlike the behavior of cars on a highway slowing for an accident. The vehicles keep moving, but a wave in which they are close together propagates toward the rear of the line. The cars move through the wave, then speed up.

The Galaxy's density waves, however, are not a

Figure 26.9   *NGC 2297 in Antlia displays a beautiful set of spiral arms in its disk, outlined with blue O and B stars and red diffuse nebulae.*

Figure 26.10 (a) *The first detailed map of the distribution of interstellar neutral hydrogen showed the existence of galactic spiral arms. (Velocities could not be defined in the blacked-out wedge.)* (b) *Giant molecular clouds also outline spiral structure.*

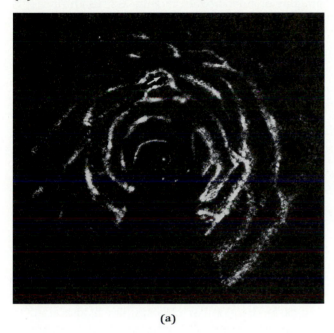

(a)

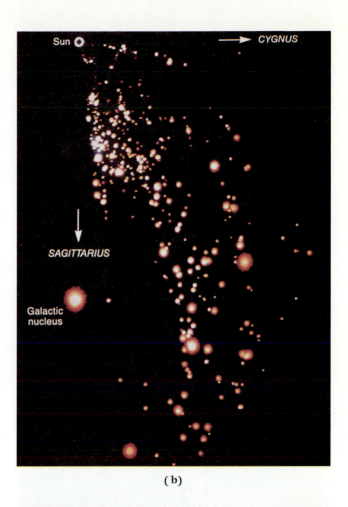

(b)

product of the stars but of the interstellar medium. Some disturbance causes a clumping of interstellar gas. The increased gravity produces further clumping away from the disturbance, which causes even more clumping, and the whole pattern moves outward through the Galaxy in response to the rotation. Although modern theory can explain the propagation of the arms, we still do not know the origins of the initial disturbances. The gravitational effect of the Magellanic Clouds is one possibility.

Density waves may be the initial factor in causing the compression of interstellar material that leads to star formation. Stars develop within the arms, and the arms are subsequently lit by the brilliant, ephemeral blue O and B dwarfs (and blue supergiants) that give the arms their characteristic color. These stars evolve and explode, leading to further compression and more star formation. As time proceeds, the lower-mass stars leave the arms in the forward direction and populate the general disk. This picture explains why different indicators reveal different arm structures, as each—the neu-

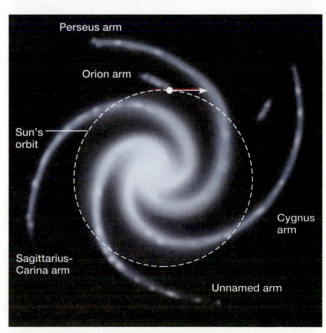

Figure 26.11 *A highly idealized map of the structure of the galactic disk shows a set of major spiral arms and a pair of short arm-segments, one of which, the Orion arm, contains the Sun.*

tral hydrogen clouds, the giant molecular clouds, and the OB associations—relates differently to star formation.

26.2.4 *The Magnetic Field and Cosmic Rays*

The galactic magnetic field is clearly indicated by the polarization of starlight caused by aligned interstellar grains. The field strength is extremely low, only a millionth or so that of the Earth. However, when the field strength is summed over the Galaxy's vast size, the total amount of energy involved is staggering. The origin of the field is still not well understood, though it is undoubtedly related to the rotation and movement of ionized matter in the Galaxy.

The galactic magnetic field controls **cosmic rays,** high-speed atomic nuclei that aid in the ionization of molecular clouds and dense cores. Cosmic rays are observed directly with high-flying balloons and spacecraft, and indirectly when they smash into atoms in the Earth's atmosphere, producing atomic debris that rains to the ground. Cosmic rays are primarily protons and helium nuclei, but a few as heavy as bismuth, element 83, have been observed. With velocities near that of light, they can carry astonishing amounts of energy. At maximum they

have been observed with energies of 10 joules, about that carried by a served tennis ball!

Cosmic rays are almost certainly related to supernovae. Atomic nuclei blasted out in the explosions may be accelerated to high velocities by riding supernova shock waves and then may be given further kicks by stellar winds and bubbles blown by O stars. They may also be accelerated by collapsed progeny of supernovae. Cygnus X-3, 15 kpc away, is a variable X-ray source with a period of 4.8 hours (Figure 26.12), and is thought to consist of an evolving star that dumps mass into an accretion disk around a neutron star. It is also a powerful source of gamma rays that vary with the same period. The gamma rays are thought to be caused by acceleration of atomic particles in a pulsar's powerful magnetic field. Cygnus X-3 and a handful of other similar systems could produce all the Galaxy's cosmic rays. Cosmic rays are charged particles, and once they have been injected into interstellar space, the galactic magnetic field bends their paths into huge irregular orbits around the Galaxy, a tiny fraction crashing into Earth.

26.3
The Halo and the Bulge

Except for its globular clusters, the halo is hard to see. Sparsely populated with red giants, horizontal-branch stars, and red dwarfs, it contains only about 2% of the number of stars in the disk. There is little interstellar gas and dust. It is then no surprise that halos of other spiral galaxies (Figure 26.13) are not especially prominent.

The halo is old and metal-deficient, with a ratio of iron to hydrogen about 1% solar. Though all its components orbit the Galaxy's center, there is little or no rotation of the system as a whole. Like the disk, the halo is structured. There are two populations of globular clusters (see Section 20.5.2): one occupies the full spheroidal halo, the other a much flatter volume that has a scale height perpendicular to the disk of only about 2 kpc (that is, it is 4 kpc thick). The inner globulars have a higher metal content, about a third that of the Sun, implying an evolutionary sequence as the metal content of the gas out of which the clusters formed built up with time.

The bulge of the Galaxy (see Figures 26.1, 26.5, and 26.13), some 3 kpc in radius, is visible through only a few "windows" in the intervening interstellar dust. It consists primarily of old stars, its characteristic color provided by red giants. The region where

Figure 26.12 *Cygnus X-3 is an X-ray and gamma-ray binary with a neutron-star component. Particles are thought to be accelerated to high energies and sent into long, curving orbits around the Galaxy by the galactic magnetic field. A tiny fraction slam into the Earth's atmosphere, where they smash atoms and send showers of subatomic particles to the ground.*

Figure 26.13 *M 104, the Sombrero Galaxy, seen here nearly edge-on, displays its thin Population I layer of interstellar dust. The halo is made visible by the large number of globular clusters seen as fuzzy, almost starlike, images. This galaxy also has a huge and very obvious bulge.*

the halo and disk merge, it has a huge range of metal abundances. From its Population II component it gets metal-deficient stars. As the inner extension of the disk, however, it also contains stars with remarkably high metal contents, up to 3 or so times solar. These are certainly the result of rapid star formation and evolution that have caused great amounts of heavy elements to be deposited into interstellar space.

26.4
The Galactic Nucleus

One of the strongest radio sources in the sky lies 6° west of γ Sagittarii, the star at the tip of Sagittarius's arrow (Figure 26.14). Since this source, *Sagittarius A*, is in the thickest part of the Milky Way and is very compact, it was identified with the true center of the Galaxy, the **galactic nucleus,** and thereby serves as the zero point for galactic longitude. By wonderful coincidence, Sagattarius's arrow points toward it, as if the ancients were whispering a clue down over the ages.

At a distance of 8.5 kpc, the nucleus and its environs are hidden in the visual spectrum by 30 magnitudes of dust absorption. In the infrared and radio spectral regions, however, where the view clears, astronomers find extraordinary complexity. Infrared observations of the box around the nucleus in Figure 26.14 (upper inset) show dense masses of stars within a region only 60 pc across. A radio view (lower inset) shows streamers of gas that emit synchrotron radiation, revealing a magnetic field a

thousand times the strength of that in the outer Galaxy.

Within the inner 2 pc of the Galaxy (Figure 26.15a), we see a tilted spiral whose matter may either be falling into the center or shooting from it. Turbulence in the gas suggests a violent event within the last 100,000 years. Above the spiral's center is a pointlike radio source called Sagittarius A* (Figure 26.15b). Very Long Baseline Interferometry shows Sgr A* to be no more than a thousandth of a second of arc, or 10 AU, across—about the size of Jupiter's orbit! Sgr A* is apparently the actual nucleus of the Galaxy.

Infrared spectra of stars and gas in the Galaxy's central regions show that orbital speeds increase steadily toward the center. Limits on distance from the center and Kepler's generalized third law suggest that the motions are controlled by a mass of perhaps 5 million $M_\odot$. Infrared observations, however, do not show a sufficient number of stars to account for this mass. If it is concentrated inside Sgr A*, we have little recourse but to believe that the galactic nucleus contains a massive black hole and that Sgr A* is its bright accretion disk. The disk is created when stars, perturbed in their orbits by their neighbors, wander too close to the black hole, are tidally disrupted, and fall in. To produce a mass this size would require the infall of a star about every 1,000 years, events that may keep the Galaxy's core in a state of continual agitation. To be sure, the existence of such a "monster in the middle" is uncertain, and we may yet find that a dense cluster of O and B stars is responsible. But we have opened a fascinating possibility, evidence for which will grow in later chapters.

26.5
The Mass of the Galaxy
and Dark Matter

Any body in the Galaxy orbits the galactic center in response to the total mass inside that orbit (including the mass in the halo). The Sun and the inner Galaxy—that portion inside the solar orbit—can therefore be considered a two-body system, and we can apply Kepler's generalized third law to estimate the inner Galaxy's mass. The period of a body in a circular orbit about the galactic center is its orbital circumference ($2\pi a$, where a is the orbital radius) divided by its velocity, or $P = 2\pi a / v$. With an orbital radius of 8.5 kpc = 1.75×10^9 AU = $2.6 \times$

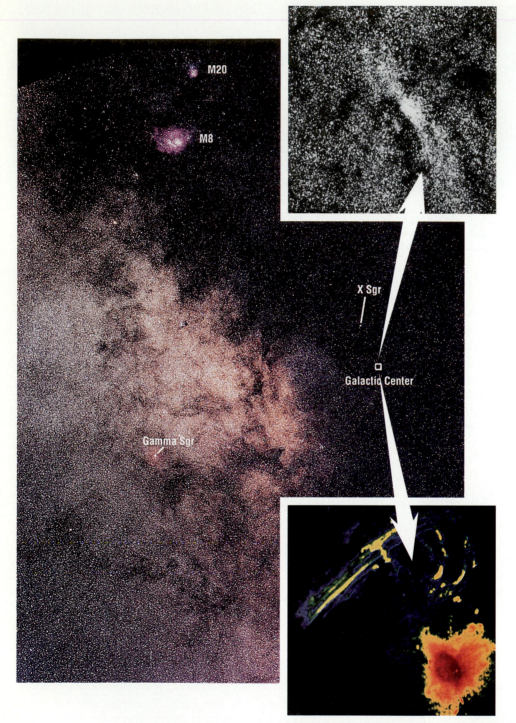

Figure 26.14   *The small box to the west of the dense star clouds of Sagittarius, 25 minutes of arc and 60 pc across, shows the location of the galactic center and Sagittarius A. (Upper inset) An infrared view within the box reveals masses of stars hidden by interstellar dust. (Lower inset) A radio view centered a bit to the northeast shows streamers of gas set within a relatively strong magnetic field. Ionized gas (red) hides the actual nucleus.*

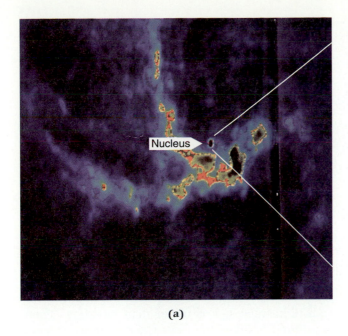

(a)

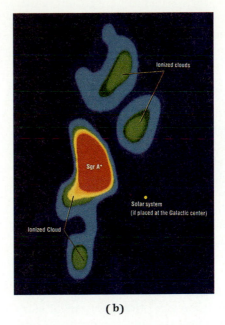

(b)

Figure 26.15
(a) *The center of the Galaxy, Sagittarius A, consists in part of a gaseous spiral 2.5 pc across. A pointlike source buried within Sagittarius A, Sagittarius A*, appears to be the actual nucleus.* **(b)** *The view is only 0.05 pc across. Sgr A*, within the elongated red area, is seen surrounded by ionized clouds of gas.*

10^{17} km and a velocity of 220 km/s, the local standard of rest (and to a good approximation, the Sun) has a period around the galactic center of 7.3×10^{15} seconds, or 240 million years. From the simple formula $M_{\text{inner Galaxy}} + M_{\text{Sun}} = a^3/P^2$, where M, a, and P are respectively in solar masses, AU, and years (see Section 21.1), we find the mass of the inner Galaxy to be 9×10^{10} $M_{\odot}$. Since the Galaxy's disk outside the solar orbit contains only about 20% of the galactic light, and since the halo has a low stellar population, the mass of visible stars and interstellar gas and dust in the whole Galaxy is around 10^{11} $M_{\odot}$. Since the average stellar mass is 0.5 $M_{\odot}$, the Galaxy contains about 200 billion stars.

However, in 1932 Jan Oort (of comet-cloud fame) examined the local density of matter in the galactic plane. Throw a ball in the air: its speed decreases with height. The ball's acceleration depends on Earth's gravity, from which you can find the terrestrial mass. Oort determined the rate at which the velocities of stars decrease as their distances in the direction perpendicular to the galactic plane increase, yielding him the amount of matter in the disk that attracts the stars. The average local density of matter so calculated was about 50% greater than could be accounted for by visible stars and interstellar gas and dust—possible evidence for mass that had not yet been found or could not be seen. The subject is currently controversial. Some stellar data indicate that we still have not found all the matter in the disk, whereas other investigations

demonstrate that the vertical motions can indeed be explained by what we see.

Whatever the outcome, Oort opened the possibility that we have not found all the matter that produces the Galaxy's gravity. Subsequent investigations of galactic motions clearly indicate a severe problem. If all the mass of the Galaxy were effectively inside the solar orbit, the orbital velocities of stars and H I clouds farther from the center would be lower than ours: Mars, for example, moves more slowly around the Sun than does the Earth. Therefore, the galactic rotation curve beyond the solar orbit should decrease in accord with Kepler's laws of motion (Figure 26.16). Since 90% of the light of the Galaxy is inside a radius of 12.5 kpc, we expect some kind of drop in rotation speed at large distances from the galactic center. However, the

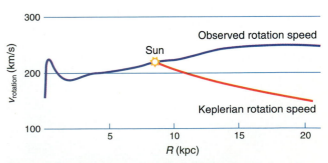

Figure 26.16 *The observed rotation curve of Figure 26.9 is shown by the blue line. If nearly all the mass of the Galaxy were inside the solar orbit, stars outside the orbit would, according to Kepler's laws, move more slowly (red curve).*

speeds do *not* drop. The rotation curve either rises or remains flat as far as we can see, to at least a radius of 20 kpc.

From Kepler's third law, $P^2 \propto a^3/M$, where M is the galactic mass inside any orbit. Therefore, $M \propto a^3/P^2$. Since the orbital period is $2\pi a/v$, $M \propto av^2$. From the observed rotation curve in Figure 26.26, we see that there is about is $2\frac{1}{2}$ times as much mass within 20 kpc as there is inside the solar orbit. Nor does that seem to be the end of it, since the rotation curve shows no signs of dropping off at all!

But only 20% of the galactic light resides outside the solar orbit. We cannot see the matter that causes the high velocities at large galactic radii, and it is therefore called **dark matter.** Do not confuse dark matter with dark nebulae or globules, whose compositions we know. Dark matter can be "seen" only by its gravitational effects.

From investigations of vertical motions of stars, like those made by Oort, we know that at most only about 50% of the dark matter can be in the disk; the majority, or even all of it, resides in a greatly extended halo sometimes called the **galactic corona.** Analyses of the velocities of other galaxies close to ours, those under the influence of our Galaxy's gravity, indicate that the corona may have a radius of 100 kpc or more, and that it and the extended disk may contain 10^{12} solar masses, ten times the mass interior to the solar orbit. Dark matter may constitute 90% of the Galaxy's mass. The "traditional" Galaxy described at the beginning of this chapter is therefore seen to be only a small part of the overall structure (Figure 26.17).

We do not know the nature of dark matter, though several possibilities have been proposed. The Galaxy may be filled with submassive and subluminous brown dwarfs (see Section 23.4). If they are responsible for the dark matter, there must be extraordinary numbers of them, more than there are main-sequence stars. Yet to date no brown dwarf has ever been positively identified. Other researchers have proposed massive interstellar "grains" the size of bowling balls, myriad black holes, and even vast numbers of exotic subatomic particles (to be examined in the Chapter 30). But whether or not we know its form, dark matter clearly exists. Gravity is as good a probe of mass as radiation; indeed, it is better than radiation. There is nothing unique about our Galaxy. As we will see in the following chapters, dark matter is everywhere, in immense amounts.

26.6
The Origin and Evolution of the Galaxy

Although we do not yet have a complete galactic census, we have enough evidence to construct a picture of how the Galaxy may have developed with time, how it came to be in its present state, and how it will continue to evolve.

26.6.1 *Age*

We can tell the age of a cluster by where the main sequence stops (see Section 24.1.2 and Figure 24.3). In practice, we use stellar evolution theory to calculate *isochrones,* positions that evolving subgiants and red giants should have at specific ages (Figure 26.18). The age of the cluster can then be found from the best fit between theory and the data. The oldest open clusters are in the range of 7 to 10 billion years old. Since open clusters belong to Population I, they provide an age for the galactic disk. However, open clusters are subject to disruption, and it is likely none has lived for the full lifetime of the disk. An age derived from open clusters must therefore be a lower limit.

The Milky Way, the great collection of stars in the galactic disk, can be treated in the same way. Unlike an open cluster, however, it has stars of all ages and masses in various states of evolution.

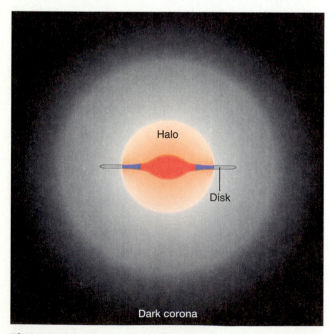

Figure 26.17 *The traditional disk-halo structure of the Galaxy is surrounded by a huge corona filled with dark matter.*

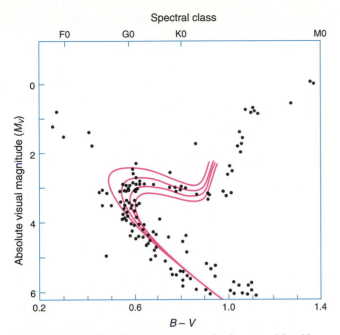

Figure 26.18 *The HR (or color-magnitude) diagram of the old open cluster M 67 is fitted with four isochrones that show where the giant-branch turnoff should fall for different ages (from top to bottom, 3, 4, 5, and 6 billion years). The best fit implies an age of around 4.5 billion years.*

Therefore, the Galaxy's HR diagram has a full main sequence in addition to supergiants, bright giants, and ordinary giants. However, the age of the Galaxy imposes a lower limit to the mass that has evolved, and as we proceed down the main sequence the number of stars rapidly climbs. The result is a thickly populated giant branch that joins the main sequence near class G, the oldest stars that can have evolved. As the Galaxy ages, its pronounced giant branch will slowly move downward. In principle, we could measure the age of the disk by noting where the giant branch joins the main sequence. In practice, we are somewhat confounded not only by the mix of ages but also by the presence of Population II subdwarfs, whose positions on the HR diagram are slightly different from those of Population I stars. At the least, such an age is consistent with that derived from open clusters.

We fare better with the galactic halo and its globular clusters. The globulars have ages between about 13 and 17 billion years, and we will adopt 15 billion as the age of the Galaxy.

26.6.2 Evolution

The Galaxy is a changing, evolving structure. The halo and its ancient globular clusters are low in metals, while the younger disk has a higher metal abundance. Even within the halo and disk, we see evidence for continuous evolution. The flattened disk component of the system of globulars has a higher metal content than the extended halo component, and the metal abundances of disk stars decrease from the bulge outward. The Galaxy appears to have contracted from a huge, somewhat spherical structure into the disk at the same time that stellar evolution pumped heavy elements into the star-forming interstellar medium.

We believe that some 15 billion years ago the gaseous mass that composed the Universe frag-

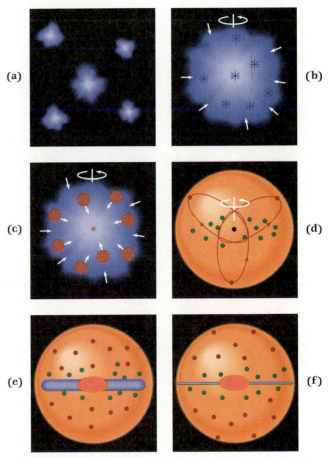

Figure 26.19 **(a)** *The Galaxy was likely created from a blob of hydrogen and helium fragmenting out of the early intergalactic medium.* **(b)** *The initial generation of stars, Population III (shown in blue), seeded the gas with the first heavy atoms.* **(c)** *The contracting cloud then produced the ancient halo component of globular clusters (red), Population II, which were given elliptical orbits toward the galactic center. Conservation of angular momentum caused the interstellar medium to begin to spin into a disk.* **(d)** *The next generation of globulars (green) were then distributed in a thick halo disk and had higher metal abundances, which in turn contracted further to* **(e)** *a thinner Population I disk (light blue).* **(f)** *Five billion years after initial formation, the interstellar medium had collapsed to the layered Population I disk seen in Figure 26.4 (blue), leaving the globulars in elliptical orbits.*

mented into billions of clouds that contracted under the force of their own gravity (Figure 26.19a). One of them was to become our Galaxy. Within a short time, the gas of the cloud started to fragment, producing the first generation of stars (Figure 26.19b). Study of the origin of the Universe (Chapter 30) strongly indicates that the initial contracting cloud was made only of hydrogen, helium, and a tiny amount of lithium; therefore, so was the first stellar generation, called **Population III.** As this population's stars evolved into giants and supergiants, they created the first heavy atoms, which were spread through the interstellar medium by stellar winds and explosions and incorporated into the next stellar generation.

This second generation had a low metal content, and we identify it with the most metal-poor stars of Population II. We thus encounter the weakest link in the chain of reasoning that leads to the present. Where are the remaining low-mass stars of Population III, those with no metals? Even the ancient globular clusters have a metal content about 1% that of the Sun. Among halo field stars (those not in clusters) we can find metal abundances all the way down to about 10^{-4} solar, but we never find zero.

Such stars may simply be very rare and hard to locate. It would not have taken many Population III stars to seed the Galaxy with enough metals to begin making Population II. The higher-mass component of Population III would long ago have burned away, leaving only a few dim dwarfs still on the main sequence. If we look hard enough we may yet find them. Alternatively, early conditions may have allowed the formation of only massive stars that died long ago. In either case, the compositions of Population II stars are consistent with the synthesis of atoms in an earlier generation, so we are fairly confident that some kind of Population III must once have existed.

We are relatively secure in the broad outlines of the rest of the story. As the Galaxy continued to contract, it further divided into large lumps of material that in turn fragmented into stars, creating the globular clusters we see today (Figure 26.19c). The first generation of them had few metals. Since the Galaxy was rapidly contracting, these assemblies had strong components of velocity toward the galactic center. The result is that the halo globulars today have highly elliptical orbits. As they evolved,

some of them partially disrupted and populated the halo with the field stars we now see.

The Galaxy continued to contract, and conservation of angular momentum began to spin it out into a thick Population II disk (Figure 26.19d). The next generation of globular clusters then had a higher metal content, was distributed in a more disklike fashion, and had less elliptical orbits. The metal content of the Galaxy continued to increase as contraction continued. A few billion years after formation, the interstellar medium had spun itself into a flatter disk (Figure 26.19e) in which Population I began to form, and then into the current thin disk (Figure 26.19f) with its O and B stars, leaving behind the globular clusters and the Population II field stars of the galactic halo. The Sun, which came along 10 billion years after the formation of the Galaxy and maybe 5 billion years after the development of the disk, is a relative newcomer.

This theory, as neat as it sounds, has some serious problems. The relations among stellar orbits, ages, and metal abundances are very complex. Galactic evolution was apparently highly chaotic. We do not know how Population III stars could have formed without interstellar dust generated by stars, why the halo took so long to collapse into the disk, why the globular clusters exhibit a fairly large range of ages (about 4 or 5 billion years), or why globular clusters are no longer being made. Most important, we do not know the nature of dark matter, or how it and the galactic corona have influenced star formation and galactic evolution. And what of the future? The Galaxy's metal content will certainly continue to increase, but the effect of chemical enrichment on star formation is unknown. Our Galaxy will provide astronomers with years of exciting research as the layers of mystery continue to be peeled away.

Think, however, not of our ignorance, but of what we have learned. As we stand under the stars, admiring the Milky Way, we can now ponder its meaning and its powerful relation to the Earth and to ourselves. The Galaxy is a vast recyclng engine, its magnetic field instrumental in the formations of stars. The heavy atoms out of which we are made came from aeons of stellar evolution and mass loss. We now know we are not only citizens of our town or even of the world, but are truly residents of the Galaxy. We are no more separate from it than we are from our own planet.

KEY CONCEPTS

Cosmic rays: High-energy atomic nuclei accelerated by the Galaxy's magnetic field; possibly produced in supernovae or in neutron-star binaries.

Dark matter: Unilluminated mass, detectable only through its gravitational effect; its content is unknown.

Galactic bulge: The central, thick region of the Galaxy where the disk and the halo come together.

Galactic coordinates: Coordinates (galactic latitude and longitude) based on the Galaxy's plane and nucleus.

Galactic corona: The volume of space filled with dark matter encompassing the traditional disk and halo.

Galactic nucleus: The energetic center of the Galaxy; it may be a black hole.

Galactic rotation curve: The rotation velocity of the Galaxy graphed against distance from the galactic center.

Population III: The original population of the Galaxy, made only of hydrogen and helium, that seeded the Galaxy with the first heavy elements; no representatives are known.

Scale height: The distance over which density drops to 37% of its original value.

Spiral arms: Arms in the galactic disk that wind outward from near the galactic center and that contain O and B stars and clouds of interstellar matter; they are sites of star formation.

EXERCISES

Comparisons

1. Compare the metal abundances and ages of Population I, Population II, and the galactic bulge.
2. Why are spiral arms bluish and the halo reddish?
3. Compare different-wavelength views of the galactic nucleus.
4. What is the difference between Sagittarius A and Sagittarius A*?

Numerical Problems

5. From the constellation maps in Appendix 1, estimate the galactic latitude and longitude of the stars Deneb and Rigel.
6. What is the mass of the Galaxy within 12.5 kpc of the center?
7. How many times has the Sun orbited the Galaxy since birth?

Thought and Discussion

8. Why did Herschel think he was near the Galaxy's center?
9. Could Shapley have used O stars to find the distance to the galactic center?
10. Why are Population II stars found in the galactic disk?
11. How can observations of the 21-cm radio line be used to find the rotation curve of the Galaxy? Show why the tangent method will not work outside the solar orbit.
12. How would you use Cepheid variables to define the Galaxy's spiral arms? Why?
13. Describe three processes that can compress the interstellar medium to aid in the creation of new stars.
14. What are the effects of the Galaxy's magnetic field?
15. What are cosmic rays? What are their possible origins?
16. Why do we think there is a black hole in Sgr A*?
17. Summarize the evidence for dark matter in the Galaxy.
18. Why do we believe that most dark matter is an extended galactic corona?
19. What are some possible candidates for dark matter?

Research Problem

20. When did astronomers become aware of the galactic nucleus and its bizarre properties? Find properties of the galactic center not described in this book.

Activities

21. Use a pair of binoculars to locate the galactic center. Sketch the stars you see in the region and draw in the nucleus.
22. Make a three-dimensional scale model of the Galaxy that includes the halo and disk, globular clusters, and the galactic nucleus.

Scientific Writing

23. A philanthropist is willing to support galactic research, a subject in which you are an expert. Write a proposal to obtain the funds, explaining the unknown aspects of the Galaxy and the research that must yet be done.
24. You wish to communicate your enthusiasm about the Milky Way to as large a group of people as possible. Write a nontechnical article for a newspaper's Sunday supplement in which you describe the beauty and nature of the Milky Way and its significance to us. Include two illustrations with captions.

27 Galaxies
28 The Expansion and Construction of the Universe
29 Active Galaxies and Quasars
30 The Universe

Galaxies and the Universe

PART VI

27

Galaxies

Normal galaxies and their relation to our own

The magnificent spiral galaxy NGC 6946 is seen nearly edge-on.

One of the profound scientific discoveries of the twentieth century is that stars are organized into separated systems called **galaxies.** We have examined one in detail, our own. Now we look around us at the vast number of others as we begin to assemble the units that make the Universe.

27.1
Revelation

With the development of the telescope, astronomers began to find vast numbers of small fuzzy patches of light. Some were readily resolved into stars, showing them to be clusters, but others defied resolution. A few, the diffuse and planetary nebulae, were found in the nineteenth century to have pure emission-line spectra and thus had to be gaseous. A fraction, however, were neither resolvable *nor* gaseous.

As early as 1750, the English astronomer Thomas Wright suggested that our Galaxy is not limitless but has boundaries, and that the unresolved nebulae are comparable but distant collections of stars. Five years later, the philosopher Immanuel Kant extended this concept in his theory of *island universes,* speculating that our stellar system was shaped like a highly flattened disk and that the nebulae were similar and strewn randomly throughout space, their apparent forms dependent on their orientations. In 1845 William Parsons, the third earl of Rosse, discovered that one, M 51 (Figure 27.1), has a spiral structure. He immediately surmised that it was a spinning disk of stars. Photography eventually disclosed that such spiral structure is common, and spectroscopy revealed absorption lines, supporting evidence that the "spiral nebulae" are indeed made of stars.

Yet the significance of the spiral nebulae remained controversial. The 1885 supernova in M 31, S Andromedae (see Section 21.6), had erroneously been thought to be an ordinary nova; its apparent brightness indicated M 31 to be only 10,000 pc away and a mere 350 pc across, denying the views of Wright and Kant. The puzzle of these mysterious objects deepened in 1917 when V. M. Slipher of Lowell Observatory found that the absorption lines of most nebulae are shifted to the red, suggesting that the objects are moving away from us.

The controversy peaked with the *Great Debate* that took place in 1920 in Washington, D.C., between Heber D. Curtis of Lick Observatory and Harlow Shapley, then of Harvard. Curtis supported Kant's view. By that time several real novae had been observed in M 31, and they indicated a then-amazing distance of 150 kpc. It, and presumably the other spiral nebulae, must therefore be huge. Curtis also noted that the nebulae are moving away from us at speeds that clearly release them from our Galaxy's gravitational grasp, so they could not belong to us. Shapley championed the concept of one "Big Galaxy," believing the spiral nebulae to be nearby. He had measured its size from the globular cluster distribution (see Section 20.5.2), but because he had not known about the obscuring effects of interstellar dust, his calculated distance between the Sun and the galactic center was twice the modern value.

Three years later Edwin Hubble settled the issue forever. With the 100-inch at Mount Wilson, he finally resolved M 31 and its nearby neighbor M 33 in Triangulum into stars (Figure 27.2). If they were anything like the stars of the solar neighborhood, M 31 and M 33 had to be far away and comparable to our Galaxy. The final nail was hammered home in 1924 when Hubble discovered Cepheid variables in M 31. From the period-luminosity relation (see Section 21.1), Hubble measured the distance of M 31 at 300 kpc.

The debate was over. The word nebulae, though still in use, is a misnomer. The mysterious "clouds" are true *galaxies* at great distances. Our Galaxy is not unique, but belongs to a family of billions of others that extends as far as our vision can reach.

Figure 27.1 *The Whirlpool Nebula, M 51, has massive double arms. Below it is NGC 5195, a tidally distorted companion.*

Figure 27.2   *The outer disk of M 31 swarms with millions of stars and a handful of Cepheids, from which its distance can be found. The entire galaxy is shown in the inset.*

27.2
Types of Galaxies

There are several kinds of galaxies. To begin to understand them, they must be classified. The scheme in general use today was initially developed by Edwin Hubble in the 1920s. The original **Hubble classification** contains three major classes and several subclasses, to which other divisions have been added.

27.2.1 *Elliptical Galaxies*

The simplest class is composed of **elliptical galaxies** (E). They have smooth outlines that appear as ellipses (Figure 27.3) and, unlike our Galaxy, have no disks. The most familiar are two companions of the Andromeda Galaxy, M 32 and NGC 205 (see Figures 1.15 and 19.25). Subclassification is indicated by following E with the number $10(a - b)/a$, where a and b are respectively the major and minor axes of the ellipse. For a round system like NGC 4636 (Figure 27.3a), $a = b$, $10(a - b)/a = 0$, and the class

is E0. If a is twice b, $10(a - b)/a = 5$, and the galaxy is E5. NGC 3377 in Figure 27.3b is E6. Any galaxy more elongated than E7 ($a = 3.3b$) has lost its elliptical shape and shows evidence of a disk. The scheme therefore stops at E7.

An elliptical galaxy's apparent shape depends on its actual three-dimensional structure and its orientation. A spherical galaxy will always appear as an E0 system. However, an elongated galaxy, like that in Figure 27.3c, can take a variety of apparent forms. If we view it end-on, we see an E0 galaxy like NGC 4636. However, if we look at it sideways, we see an E6 system like NGC 3377. If we assume that galaxy orientations are random, we find roughly equal numbers of galaxies in each subclass.

Elliptical galaxies are nearly pure Population II, and generally contain no dust, no O or B stars, no active star formation, and only a little interstellar gas. They are similar to the halo of our own Galaxy except that the star density (the number per cubic parsec) is much greater, and—especially in large ellipticals—the metal abundance may be much higher. The distribution of light is very smooth, dropping steadily from the center to an ill-defined boundary. Nearby elliptical systems can be resolved into a grainy structure, into millions of bright red giants that give the ellipticals their char-

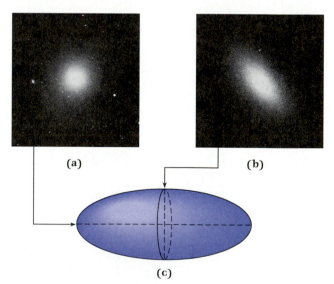

Figure 27.3   *The elliptical galaxies NGC 4636 in Coma Berenices* **(a)** *and NGC 3377 in Leo* **(b)** *have different ellipticities and are respectively classified as E0 and E6. Both could in fact have the shape of the drawing in* **(c)**, *but may be viewed from different angles as indicated by the arrows; NGC 4636 in (a) may also be spherical. In some elliptical galaxies, the three axes of the ellipsoid may all have different lengths.*

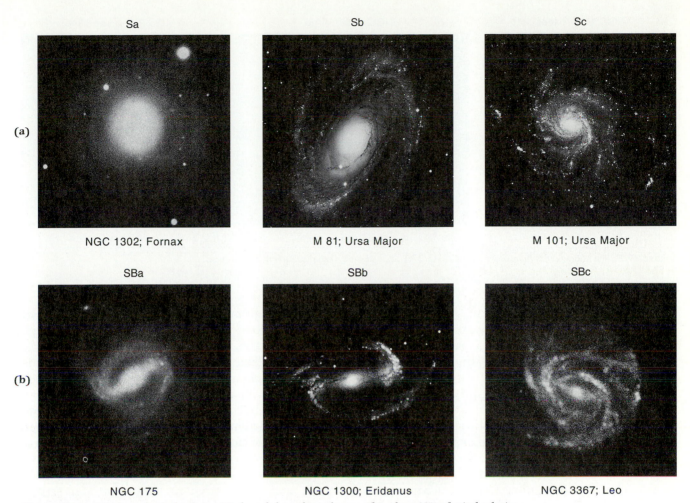

Sa   Sb   Sc

(a)

NGC 1302; Fornax   M 81; Ursa Major   M 101; Ursa Major

SBa   SBb   SBc

(b)

NGC 175   NGC 1300; Eridanus   NGC 3367; Leo

Figure 27.4   **(a)** *Normal spiral galaxies (S) show disks and spiral arms.* **(b)** *About 20% of spiral galaxies are barred spirals (SB). A spiral's subclass (Sa, Sb, Sc or SBa, SBb, SBc) depends on the openness of its arms.*

acteristic reddish color. Stars like the Sun and dimmer dwarfs farther down the main sequence are present, but are too faint to be seen individually.

27.2.2 *Spiral Galaxies*

Ellipticals are not nearly so captivating as the **spiral galaxies,** those with thin disks and complex spiral arms comparable to those in our own Galaxy. Spirals fall into two main categories. **Normal spirals** (S) are like our Galaxy, and have disks and arms that emerge from a spheroidal bulge. They are further subdivided according to how tightly the arms are wound (Figure 27.4a). Sa systems have nearly circular arms that wind outward slowly, the arms of Sc galaxies open out quickly, and those of Sb are in between. The distribution in types is continuous, and intermediate classes like Sab are common. M 31 is classed as Sb, and our Galaxy (see

Figures 26.10 and 26.11) variously as Sab, Sb, or Sbc. About 20% of spirals are **barred spirals** (SB) (Figure 27.4b), in which the arms branch almost perpendicularly from a straight bar (or from a ring that encircles the bar) punched through the bulge like a knitting needle through a ball of yarn. They divide into SBa, SBb, and SBc according to arm structure in parallel with normal spirals.

Both kinds of spiral galaxy are a mixture of Population I, confined to the disk, and Population II, found in a spheroidal halo that condenses into the bulge. As the arms open up, the ratio of young Population I to old Population II increases. The bulges become progressively smaller and the arms break into knots and clumps that are made of O and B associations and complexes of H II regions. The more open spirals have more gas and dust, and consequently star formation proceeds at a

NGC 524; Pisces   NGC 4762; Virgo   NGC 2859; Leo Minor

(a)   (b)   (c)

Figure 27.5   **(a)** *S0 galaxies have disks but no arms.* **(b)** *Seen edge-on, they display little if any dust.* **(c)** *SB0 galaxies have armless bars.*

greater pace. The Hubble system, however, leaves out a number of nuances. Some spirals, like M 81 (Figure 27.4a, center), have a pair of opposed arms, whereas others, like M 101 (Figure 27.4a, right), have multiple arms. M 101's arms are thin and stringy; those of M 51 (Figure 27.1) are thick and fat.

Between the ellipticals and the spirals is a set of galaxies that display obvious disks but no spiral arms called S0 (Figure 27.5a). Seen edge-on, they have no or minimal central dust lanes (Figure 27.5b). The barred version, SB0, has a bar but no projecting arms (Figure 27.5c).

The different types are summarized, and their continuity displayed, by Hubble's **tuning fork diagram** (Figure 27.6). One kind of galaxy merges smoothly into the next. However, the diagram does *not* represent an evolutionary sequence, only a classification scheme.

27.2.3 Irregular Galaxies

The tines of the tuning fork in Figure 27.6 close again at the far right as they merge into chaotic systems of stars and interstellar matter called **irregular galaxies** (I or Irr). Both the Large and the Small Magellanic Clouds, depicted in Figure 21.4, are con-

sidered irregular systems, although inspection of the Large Cloud shows a distinct bar and crude spiral arms.

Most galaxies of this type, like the Magellanic Clouds, are dominated by stars; others, like M 82 (Figure 27.7), have so much dust that no stars can be seen at all. There are no density waves or spiral arms in irregulars, so star birth is sporadic. There is a subset of both kinds, however, the **starburst galaxies,** that produce stars at prodigious rates. An irregular may sit quietly for most of its life— then some event triggers a round of star formation. Supernovae compress the surrounding material, producing more stars, and the whole galaxy comes alive for a relatively brief period. M 82 may be an extreme example, a *nuclear starburst galaxy*, in which supernovae in the core have apparently created a huge bubble of gas around the object.

27.3
Distances

Before we can learn anything about the physical nature of a galaxy, we must know its distance. The best way to find the distance is to resolve the

Figure 27.6   *Hubble's classification sequence is in the form of a closed tuning fork. Ellipticals merge smoothly into either spirals or barred spirals that become more open and then blend into chaotic irregulars. There is no evolution between kinds.*

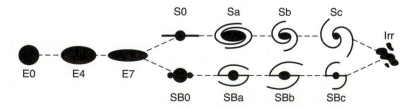

BACKGROUND 27.1 Observe a Galaxy

Galaxies extend to the visible limit of the Universe, and you might be tempted to think of them as exotic entities to which you cannot very well relate. The truth is quite the opposite. Galaxies may be distant, but they are luminous and can easily be seen: you need only modest optical power to bring a great many into view. Start with the naked eye and look for M 31 in Andromeda (see Figures 1.15, 19.25, and 27.2). Its position is marked on the star maps in Appendix 1. A view with binoculars in a dark sky discloses an amazing sight, the disk stretching over a degree. The Large and Small Magellanic Clouds are not visible from the continental United States, but they can be seen with the naked eye from the tropics, and the LMC is visible even from Hawaii.

Although the Triangulum spiral M 33 (Figure 21.6) is not ordinarily classed as a naked-eye object, you can in fact see it if conditions are right (see the star maps). You need an absolutely clear, dust-free sky with no Moon and no artificial lighting. M 33 is a relatively easy object to find with binoculars. Locate it first, then pull the binoculars away. Do not look directly at the object, but use *averted vision,* as the eye is more sensitive just off center. A variety of other galaxies are also visible with binoculars, including M 81 (Figure 27.4a, center). With even a small telescope, a large number of galaxies pop into view. You can roam through the Virgo cloud at random and find a variety of small smudges that belong to the cluster.

It is often disappointing to people that telescopic images do not look like the photographs that grace so many books. A telescopic view of M 31 still shows you only the bulge and the inner disk. However, with a modest instrument of 12 inches or so, you—like the earl of Rosse—can view the spiral arms of the Whirlpool Nebula, M 51 (see Figure 27.1). The excitement of observing galaxies lies not so much in viewing detail but in realizing that you can encompass the light of billions of stars in one sight and that you are looking back into time at light that left the systems before human beings ever walked the Earth.

Galaxy into **distance indicators** or **standard candles,** stars or other recognizable objects whose luminosities (absolute magnitudes) we know. We then measure the distance indicators' apparent magnitudes, apply a correction for absorption of light by dust, and determine distance from the magnitude equation (see Section 19.3).

The most important standard candles are Cepheid variable stars. We can find accurate distances with the period-luminosity relation. As bright giants and supergiants, Cepheids can be seen over large distances, and their obvious variability makes them relatively easy to find. They can be discerned as far as M 101, about 5 million parsecs, or 5 *megaparsecs* (Mpc), away. RR Lyrae stars, whose absolute magnitudes are well known, are also good standard candles. With the Cepheids, they produced the modern distance scale for galaxies (the story told in Background essay 21.1).

O and B stars are so bright that they too should be good distance indicators. Unfortunately, it is hard to distinguish individual stars from OB associations and their attendant diffuse nebulae. Planetary nebu-

lae are more useful, as the bright forbidden oxygen line at 5,007 Å makes them easy targets. The majority of bright planetaries evolve from lower-mass stars, and there is consequently a sharp upper limit to the brightnesses of both the stars and their surrounding nebulae. This limit has been calibrated on the Magel-

Figure 27.7 *M 82 is thought to be a dusty irregular starburst galaxy, within which star formation is proceeding at a rapid pace. (Alternatively, it may be a barred spiral seen edge-on.)*

(a)   **(b)**

Figure 27.8   **(a)** *The giant elliptical galaxy M 87 belongs to the Virgo Cluster and is surrounded by hundreds of globular clusters. It is 50 times the diameter and over 20,000 times more luminous than* **(b)** *the dwarf spheroidal Leo I.*

TABLE 27.1
The Local Group of Galaxies[a]

Name	α h	(2000) m	δ °	,	Type	Distance[b] (kpc)	Diameter (kpc)	Mass[b] ($M_\odot$)	M_V	v_r (km/s)
Milky Way[c]	—		—		Sb	—	24	2×10^{11}	−21.1	—
LMC	05	24	−69	46	Irr	52	6.5	1×10^{10}	−18.7	−270
SMC	00	52	−72	53	Irr	55	2.9	2×10^{9}	−17.7	168
Draco	17	20	+57	54	E3/d[d]	70	0.3	1×10^{5}	−8.5	—
Ursa Minor	15	14	+67	07	E5/d	70	1	1×10^{5}	−9	—
Sculptor	01	00	−33	41	E3/d	90	1	3×10^{6}	−12	—
Fornax	02	40	−34	34	E3/d	170	2	2×10^{7}	−13	40
Leo I	10	08	+12	21	E3/d	230	1	4×10^{6}	−11	—
Leo II	11	13	+22	10	E0/d	230	1	1×10^{6}	−9.5	—
Sag DIG	19	30	−17	41	Irr	460	1	1×10^{7}	−9.3	−58
NGC 6822	19	44	−14	50	Irr	500	2	3×10^{8}	−15.6	−40
IC 1613	01	04	+02	04	Irr	660	2	3×10^{8}	−14.8	−240
M 31	00	43	+41	15	Sb	690	20	3×10^{11}	−21.1	−275
NGC 147	00	33	+48	27	E5	690	1	1×10^{9}	−14.8	−250
NGC 185	00	39	+48	17	E3	690	1	1×10^{9}	−15.2	−300
NGC 205	00	40	+41	38	E5	690	2	8×10^{9}	−16.3	−240
M 32	00	43	+40	51	E2	690	1	3×10^{9}	−16.3	−210
And I	00	46	+38	01	E3/d	690	0.5	—	−11	—
And II	01	16	+33	26	E2/d	690	0.5	—	—	—
And III	00	35	+36	31	E5/d	690	1	—	—	—
M 33	01	34	+30	38	Sc	720	10	1×10^{10}	−18.8	−190
IC 10	00	20	+58	45	Irr	1260	2	1×10^{10}	−15.3	−343
IC 5152	22	06	−64	27	Irr	1600	2	—	−14	78
Leo A	10	00	+30	45	Irr	1600	2	—	−13.4	26
WLM	00	02	−15	31	Irr	1600	3	—	−15.0	−78
Pegasus	23	29	+14	46	Irr	2300	2.5	—	−16.7	−181

[a]Lists from different authorities differ slightly.

[b]Diameters and masses relate to the optically bright parts of the galaxies and do not encompass much dark matter; they are provided only for comparison with our own Galaxy.

[c]Has additional faint companions not listed.

[d]"d" denotes dwarf spheroidal.

lanic Clouds and on M 31. We therefore have well-known absolute magnitudes for the brightest observed objects. Novae are useful, too. Type Ia (Population II) supernovae are potentially among the most important standard candles, since they are all thought to reach a common absolute magnitude and can be seen over enormous distances. We must, however, be careful that we catch an exploding star at its maximum and apply the proper correction for absorption by the galaxy's dust. From all the data, we now find M 31 to be 690 kpc away, about double Hubble's original measurement.

Beyond about 10 Mpc, we can (except for occasional supernovae) no longer see a galaxy's individual stars. However, we can still find its globular clusters, whose absolute magnitudes are inferred from galactic studies. Globulars are so bright they can be observed over 20 Mpc away (Figure 27.8a; see also Figure 26.13). Moreover, they are also seen in elliptical galaxies, which contain none of the Population I standard candles.

At distances greater than about 20 Mpc, the problem of distance determination becomes much more difficult, as even the globulars can no longer be seen. We can, however, determine typical absolute magnitudes of different classes of galaxies. (A galaxy is brought to 10 pc for absolute magnitude determination by assuming that all its light comes from a point.) For example, M_V for the Andromeda Galaxy is –21.1. If we find a faraway Sb spiral similar to M 31, its distance can be estimated from its apparent magnitude and the magnitude equation.

This method is limited because any class of galaxy has a great range in size and absolute magnitude. The ellipticals are the most extreme. At one end we find brilliant (and as we will see, massive) **giant elliptical galaxies** that considerably outshine our own (Figure 27.8a). From these we descend in luminosity through smaller **dwarf ellipticals** like M 31's companions (M 32 and NGC 205; see Figures 1.15 and 19.25) to the dim **dwarf spheroidals** (Figure 27.8b). The dwarf spheroidals are tens of thousands of times fainter than our Galaxy and not much more luminous than big globular clusters. Moreover, at great distances, we cannot even accurately determine the Hubble type. As a result, estimates of large distances by this simple method are subject to considerable error. Additional methods, leading to improved accuracy, will be developed below.

Once distances are known we can easily derive the diameters in kiloparsecs from the angular sizes. Distances and diameters for a variety of galaxies out to about 50 Mpc are given in Tables 27.1 and 27.2.

27.4
Clusters of Galaxies

Galaxies have a powerful tendency to gather together. M 31's prominent companions lie next to it, and all three are at about the same distance. Since they have apparently stayed together over their lifetimes (which are comparable to the age of the Galaxy), they must be gravitationally bound,

TABLE 27.2
Prominent Galaxies Not in the Local Group

Name	Constellation	α (2000) h	m	δ °	′	Type	Apparent Magnitude	Distance (Mpc)	Diameter[a,b] (kpc)	Mass[b] ($M_\odot$)	v_r (km/s)
M 81	Ursa Major	09	55	+69	04	Sb	8.3	1.4	8.3		–43
M 83	Hydra	13	37	–29	52	Sc	8	4.7	16	1.2×10^{11}	+520
NGC 5128	Centaurus	13	25	–42	59	E0pec	8	5	20	2×10^{11}	+526
M 101	Ursa Major	14	03	+54	20	Sc	8	5.4	35	1.6×10^{11}	+231
M 82	Ursa Major	09	55	+69	41	Irr	9.0	5.2	16	—	+210
M 51	Canes Venatici	13	30	+47	15	Sc	7.4	7.7	27	4×10^{10}	+470
M 87	Virgo	12	31	+12	23	E0	10	18[c]	45	4×10^{12}	+1,260
M 104	Virgo	12	40	–11	38	Sa	8	20[c]	49		+1,130

[a]Long axis.
[b]See Table 27.1, footnote b.
[c]Different measurements give 15.6 to 21 Mpc.

Figure 27.9   *Groups like Stephan's Quintet (the fifth is off the picture), 80 Mpc away in Pegasus, are common. It is probably actually a quartet, as the spiral at lower left is not believed to be a true member, lying in the line of sight at a different distance from others.*

(a)

(b)

Figure 27.10   **(a)** *The rich Coma Berenices Cluster is highly concentrated at the center and dominated by a pair of ellipticals, and is thereby classed as a* regular *cluster.* **(b)** *The Hercules Cluster is a spread-out* irregular *system with a large number of spirals.*

just as the Magellanic Clouds are tied together with our Galaxy. Multiples like Stephan's Quintet (Figure 27.9) are also common. The real organization of the Universe, however, involves gravitationally bound **clusters of galaxies** (Figure 27.10). These range from poor assemblies of a dozen or so members to rich clusters that can contain thousands. The rich cluster nearest to us is the great Virgo Cluster, which has more than 1,000 members and is dominated by the giant elliptical galaxies M 87, M 84, and M 86. From various distance methods, we find it to be about 18 Mpc away. With a spatial extent of some 12° and covering a good fraction of the constellation, it is roughly 4 Mpc across. Especially prominent clusters are listed in Table 27.3.

In the simplest classification, galaxy clusters are categorized as *regular* and *irregular*. Regular clusters are rich, consist mostly of elliptical and S0 systems, and are highly concentrated toward their centers. The closest, about 100 Mpc distant, is the magnificent Coma Berenices Cluster (Figure 27.10a), which contains over 2,000 members in a space about 3° (5 Mpc) across. Regular clusters are dominated by one or a pair of giant elliptical galaxies, or by systems known as *supergiant diffuse* (cD) *galaxies,* enormous systems with low surface brightnesses. The Hercules Cluster (Figure 27.10b) is a fine example of an irregular system. It contains mostly spirals and, although rich, it is not centrally condensed. The Virgo Cluster, in spite of its giant ellipticals, is the nearest example of an irregular cluster.

Once we know that such clusters exist, we can look into space to recognize the cluster to which we belong, the sparsely populated irregular **Local Group** (its members are listed in Table 27.1). It is dominated by two great spirals, our Galaxy and

TABLE 27.3
Some Important Clusters of Galaxies

Name	Distance (Mpc)	Class
Local Group	—	Irregular
Virgo	18[a]	Irregular
Coma Berenices	100	Regular
Hercules	145	Irregular
Corona Borealis	300	Regular
Hydra	800	Regular

[a]Different measurements give 15.6 to 21 Mpc; other distances scale accordingly.

M 31 (Figure 27.11). Each of these systems has numerous other small galaxies attached to it, so our system is not triple but multiple. M 33, a prominent but smaller spiral featured in Figure 21.6, lies near M 31, and a few small assemblies hover well outside the boundaries of Figure 27.11.

Clusters give us the opportunity to extend the distance scale. Once a cluster is classified, we have a reasonable idea of the types and absolute magnitudes of its galaxies. We then assume that the cluster's brightest galaxies are similar to those in the Virgo Cluster, whose distances and absolute magnitudes we know reasonably well. Of course, every time a new link in the distance chain is added, the potential for errors is multiplied. Nevertheless, the method allows us at least to estimate the distances of clusters that are very far, over a billion parsecs, away.

27.5
Frequency of Galaxy Types

With the distances of galaxies known, we can assess the numbers of the different kinds. The chief prob-lem in this task is a selection effect similar to that encountered for stars: the galaxies most commonly observed are the most luminous. We therefore find that 77% of catalogued galaxies are spirals, 20% are ellipticals (which, despite the existence of giants, on the average are intrinsically dimmer), and only 3% are irregulars.

To find the true numbers, we must look not at a sample defined by *brightness*, but one defined and limited by *distance*. As we do for stars, we must work statistically, using a large volume for the brighter but rarer galaxies and smaller volumes for the more numerous, but fainter, systems that cannot be seen far away. The results of such studies show the true numbers of different kinds to be the opposite of those observed. Elliptical galaxies actually account for 60%, and only 20% are spirals (in the Local Group there are only three spirals). Of the ellipticals, the vast majority are dwarf or dwarf spheroidals. Clusters of galaxies appear to be dominated by impressive spirals and giant ellipticals, but most of the systems are the faint dwarfs and irregulars that we cannot even see once the distances are sufficiently large.

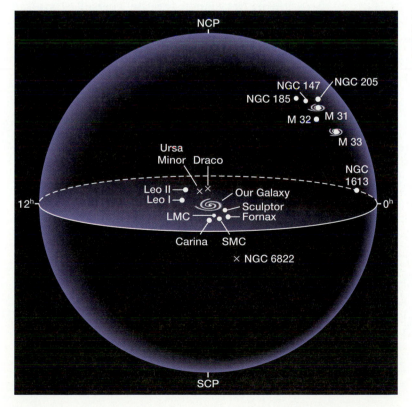

Figure 27.11 *The main portion of the Local Group of galaxies is placed within the celestial sphere, with our Galaxy at center. Galaxies noted by an* **X** *are in the foreground; others are in the background.*

27.6
Dynamics and Masses

Galaxies are gravitational systems that control the motions of the stars within them. Some rotate rapidly, others slowly. Measurements of the speeds of their stars and interstellar clouds help us learn a great deal about them.

27.6.1 Masses

Astronomers have expended considerable effort to find the mass of our Galaxy. To find the mass of an external spiral galaxy is surprisingly simple. Spirals are filled with diffuse nebulae that radiate emission lines. Set the slit of a spectrograph across the long axis of a tilted spiral galaxy, as in Figure

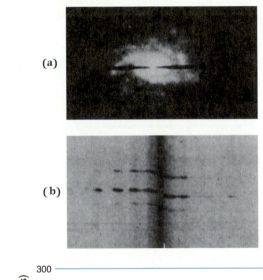

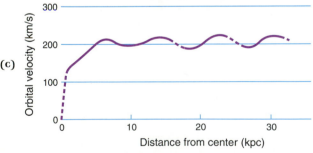

Figure 27.12   **(a)** *The slit of a spectrograph is set across the long axis of NGC 2998, an Sc spiral in Ursa Major 96 Mpc away.* **(b)** *The emission lines from its H II regions are tipped by rotation and the Doppler effect, from which we can find rotation velocity at different angular distances from the center.* **(c)** *If we know the distance of the galaxy, we can convert angular distance into physical distance, average the velocities of the two sides, correct for the tilt to the line of sight, and construct the rotation curve. Except for some wiggles, the rotation curve is flat as distance from the center increases.*

27.12a. Because the galaxy is rotating, one side is coming at the Earth relative to the average speed, the other side moving away. As a result, the emission lines are Doppler-shifted to either the red or the blue by an amount that depends on the angular distance from the galaxy's center (Figure 27.12b). If we know the distance of the galaxy, angular distances from the center can be translated into kiloparsecs. The Doppler shifts from the two sides are then converted to velocities and averaged to produce a rotation curve (Figure 27.12c). Unless the galaxy is seen exactly edge-on, the observed rotation velocities will be less than the true ones. (If the galaxy is face-on, no rotation can be detected at all.) The astronomer then estimates the degree of tilt to make a correction. The corrected rotation velocity at any distance from the center finally allows the calculation of the mass interior to that radius.

Elliptical galaxies are more difficult to evaluate because they exhibit little rotation as a unit. However, the stars of any galaxy orbit the center at different velocities that result in different Doppler shifts. If we examine the spectrum of a galaxy's combined light, we find the absorption lines to be smeared out or broadened. From the degree of broadening we can infer the spread of stellar speeds, from which we can (with the aid of theory) calculate the strength of the gravitational field and the mass. The method can be applied to spiral and irregular galaxies as well.

Tables 27.1 and 27.2 present the results of a variety of studies. The masses given are appropriate to those that dominate the galaxies' bright interior portions, and are presented only for comparison with the stellar mass of our Galaxy. The listed values may include some dark matter, but not any extended dark coronae. We see that our own system, although large, is not the most massive. In the Local Group, it is exceeded by M 31, and the giant elliptical in the Virgo Cluster M 87 has over 10 times the luminous mass of our system. At the other extreme, the irregulars and dwarf spheroidals have luminous masses only a hundred-thousandth of ours. Although there are more small galaxies than large ones, their combined mass is not too significant. The total mass of all the galaxies in the Local Group excepting the three spirals equals a mass less than 20% of our Galaxy's alone. The vast majority of the stellar mass in the Universe belongs to the big spirals and ellipticals.

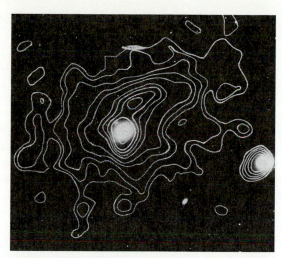

Figure 27.13 *This view of the central region of the Virgo Cluster is only 50 minutes of arc across, yet contains two massive ellipticals (M 84 at right, M 86 near the center) and half a dozen bright spirals. M 87 lies off the picture to the lower left. The cluster has a linear extent some 30 times larger than shown here. M 86 has an extensive X-ray-emitting halo from which the mass of the galaxy can be found.*

27.6.2 More Dark Matter

Other spiral galaxies seem to be structured like ours. Rotation curves derived both from optical spectra and from the 21-cm line of neutral hydrogen do not fall off as expected from Kepler's laws, but instead remain constant even as the luminosity of the galaxy, and the mass of stars and interstellar matter, drop nearly to zero (see Figure 27.12). The total masses within the visible confines of these spirals is about double the illuminated mass. Observations of tidal effects of spirals on their small satellite galaxies suggest dark outer coronae that extend over 100 kpc outward and that contain *ten* times as much dark matter as illuminated. There is some evidence that less massive spirals have a higher proportion of dark matter than the more massive ones.

We find comparable amounts of dark matter in elliptical galaxies. Some ellipticals (like the Virgo Cluster's M 84, seen in Figure 27.13) are enmeshed in large X-ray-emitting halos of hot gas at temperatures of over 10^6 K. The gas may have been blown out of the galaxy by stars, or it may be matter from the surrounding cluster. The gas itself does not have much mass, but its very existence means that it must be trapped in the grip of a powerful gravitational field, from which the mass of the system can be calculated. M 87 (see Figure 27.8a) may have a total mass of 3×10^{13} $M_\odot$, roughly ten times the amount found in stars.

Clusters of galaxies tell the same general story. In clusters, individual members orbit a common center of mass. The actual orbits cannot be determined, but, as for the stars within elliptical galaxies, the spread of individual radial velocities relative to the average will respond to the cluster's total mass. The larger the total mass, the faster the galaxies must move in orbit and the greater the spread of radial velocities. Once again, the clusters typically contain 10 times the combined luminous masses of the individual galaxies. We cannot see the dark matter and we do not know what composes it. The deeper we look the more we realize how little we know about the Universe.

27.6.3 More Distances

The more luminous a spiral, the greater the spread in the radial velocities observed within it as found from the 21-cm line. This **Tully-Fisher relation** (named after its discoverers) makes theoretical sense. If you could increase the number of stars in a galaxy, it would get brighter, its mass would increase, and the maximum velocities of stars and interstellar gas would increase in response to the higher gravitational field. The relation is calibrated with nearby galaxies whose distances are known by other means. The observed spread in 21-cm velocities of a distant galaxy gives its absolute magnitude, and comparison with the apparent magnitude yields its distance.

27.6.4 Galactic Nuclei

The nucleus of our own Galaxy cannot be seen in the optical, or even in the infrared, because of absorption by interstellar dust. We fare better with other spiral systems that are turned more face-on so that the thickness of the dust is less. The nuclei of elliptical galaxies are also accessible, since little dust is present. Figure 27.14 zooms in on the center of the Local Group spiral M 33, whose nucleus appears as a starlike point. However, both the luminosity and the mass of the nucleus (the latter derived as usual from the velocities of the surrounding matter) are comparable to that of a large globular cluster. This galactic core is also a copious producer of X rays. It is not clear whether the nucleus is really a densely packed cluster of luminous stars or whether it harbors a black hole.

The evidence for central black holes in other galaxies is stronger. The Andromeda spiral, M 31, is so close that an inner region only 2 pc across can be resolved. The velocities around this core imply a

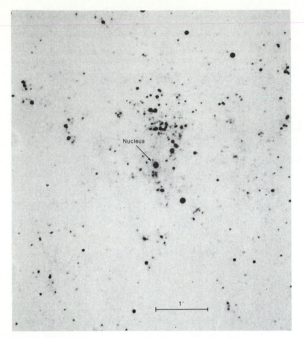

Figure 27.14   *The nucleus of the Triangulum spiral M 33 is a starlike point that may contain only stars or a black hole.*

Figure 27.15   *The extensions at the end of NGC 3115 show it to be an S0 disk galaxy seen edge-on. High velocities at the center suggest a billion-solar-mass black hole.*

Figure 27.16   *A horizontal dust lane in the center of M 51 may be the dark exterior of a 30-pc-wide accretion disk, and the hot spots above and below it may originate in jets of gas emerging from a massive central black hole. The meaning of the tilted dust lane is not known.*

mass 10^7 times that of the Sun, strongly suggesting the existence of a black hole. Several similar systems have been observed. Velocities as high as 300 km/s have been found in the material immediately around the core of the S0 disk galaxy NGC 3115 (Figure 27.15), implying a black hole with a mass a *billion* times solar. M 32, the Andromeda Galaxy's companion, and the Sombrero Galaxy, M 104 (see Figure 26.13), seem to have similar beasts lurking at their centers. It is beginning to appear likely that most—or even all—galaxies above a certain mass have them.

M 51 (see Figure 27.1) provides more direct evidence. A high-resolution image made by the Hubble Space Telescope (Figure 27.16) reveals a dust lane about 30 pc across and bright spots on a line perpendicular to it. The structure is consistent with mass flowing into a black hole from an accretion disk and then blowing outward in a bipolar flow similar to those associated with neutron stars like SS 433 (Figure 25.22) and even T Tauri stars (Figure 23.2), but vastly more powerful. The bright spots are created where the flows hit interstellar matter. The accretion disk is caused by the tidal disruption of whole stars.

Such evidence for the existence of massive black holes is still circumstantial, however. More evidence presented in the next chapter will make the concept much more plausible.

27.7
Peculiar and Interacting Galaxies

Sprinkled across the sky are hundreds of **peculiar galaxies** (Figure 27.17) that do not fit standard categories. In most cases, they are actually pairs of **interacting galaxies** affecting one another gravitationally or even undergoing collisions.

Stars do not collide. Compared to the distances between them, they are essentially points that rarely come close to one another. The same is not true for galaxies, however. The separation between the Milky Way and M 31 is only 30 times the galaxies' diameters (depending on where you establish the vaguely defined "edges"). In rich clusters, galaxies are even closer. The odds of a galactic collision are therefore high. In such an event, the stars pass by one another without any direct hits. In spirals, however, the interstellar clouds are large enough to bump, sweeping gas from the colliding systems as they pass through one another, one reason for gas-poor spirals within galaxy clusters. (Gas can also be

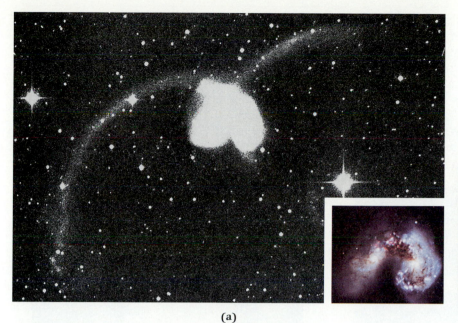

Figure 27.17 **(a)** *NGC 4038 and 4039 in Corvus are caught in the act of collision, the tidal forces spreading gas and stars into great long streamers. The interaction in the center of the system (inset) gives the pair the name, the Ring-Tail Galaxy.* **(b)** *This polar ring type of galaxy has a ring of tidal debris circulating around the poles of an S0 disk.*

(a)   (b)

stripped away by the pressure of intergalactic gas as a galaxy moves through its cluster.)

Figure 27.17a shows that the interaction can raise huge tides in the systems and throw long streamers of matter into intergalactic space. Tidal debris then sometimes lingers in a stable ring that passes around the polar axis of an S0 galaxy (Figure 27.17b). Models of such celestial crashes (Figure 27.18) are quite successful in showing how the interaction takes place. Even a close encounter has a powerful effect. M 51's small companion (see Figure 27.1) produces noticeable tidal distortion and may trigger the formation of the spiral arms.

If the circumstances of the encounter are right, the galaxies can merge. NGC 6240 (Figure 27.19a) is probably the combination of two spirals. Double nuclei (Figure 27.19b) also announce the result of a merger. Collisions and mergers can have great impacts on the resulting systems. Much of the matter can fall to the center to feed a black hole that may have been at the nucleus of one or both of the galaxies. NGC 6240 may harbor a black hole of an astonishing 100 billion solar masses. Collisions can also trigger intense nuclear starburst activity—rapid star formation—in and near the core. Evidence is supplied by ultraluminous infrared galaxies that were discovered with the aid of the Infrared Astronomical Satellite, IRAS (see Section 9.7). The radiation produced by the immense numbers of massive high-luminosity stars created is transformed by the dust in the galaxies into infrared radiation, making these systems enormously brilliant at long wavelengths.

27.8
The Origins of Galactic Forms

Why are there so many different galactic forms? The answer or answers are far from clear, and we can only offer tentative ideas. One possibility is that the initial clouds from which the galaxies developed may have had different rotational characteristics, that is, different values of angular momenta. The clouds that had slow rotations or low angular momenta fell in on themselves to produce ellipticals and irregulars and never had enough spin to make disks and spiral arms. As a result, star formation proceeded quickly, the galaxy exhausted much of its interstellar material in its youth, and winds and explosions resulting from stellar evolution blew the rest away.

The clouds that formed the spirals, on the other hand, were spinning more quickly and followed the sequence that led to the formation of our Galaxy (see Figure 26.19). Bars present a secondary problem. They seem to develop naturally as a result of gravity and the conservation of angular momentum. The real question may be: Why do many galaxies *not* have bars? Or do they? Perhaps most galaxies will be found to have bars upon sufficiently close scrutiny. There is a strong suggestion that our own Galaxy has one.

Mergers may also play a powerful role. The collision of two spiral galaxies has two results: an intense burst of star formation that uses the supply of raw material, and the modification of rotational characteristics. At least some starburst galaxies are produced this way, although only a close encounter and not an actual collision may be necessary. After the merger, the pair may fall into a heap (Figure 27.20), resulting in an elliptical galaxy. Mergers may also be instrumental in making the giant ellipticals and cD galaxies that lie at the hearts of many rich clusters. Much of the gas removed from the spirals during collisions falls to the center of the cluster, where it is consumed by one or more of the massive central galaxies that then grow at the expense of the others.

Another controlling factor is mass. Smaller systems take on the forms of dwarf ellipticals and irregulars. Too much dark matter apparently suppresses spiral arms, so small star-forming galaxies with a high Population I content remain irregular, since they seem to have a higher proportion of dark matter than their larger cousins.

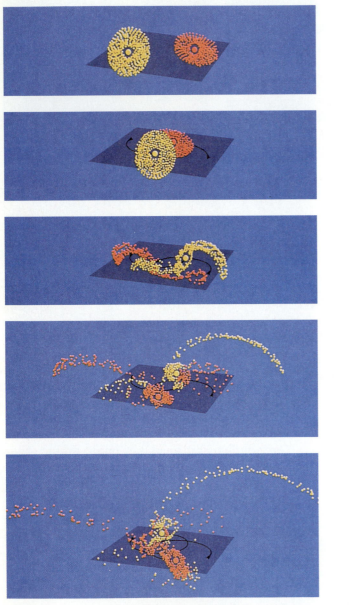

Figure 27.18 *Two galaxies approach and collide in a computer simulation. The tidal filaments are like those surrounding NGC 4038 and NGC 4039 (see Figure 27.17).*

(a)

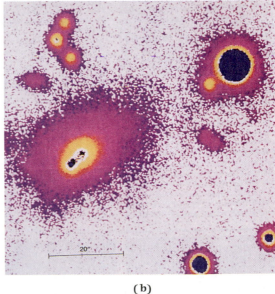

(b)

Figure 27.19 **(a)** *NGC 6240 is the result of the merger of a pair of spirals.* **(b)** *The giant elliptical 3C 348 (Hercules A) has two nuclei.*

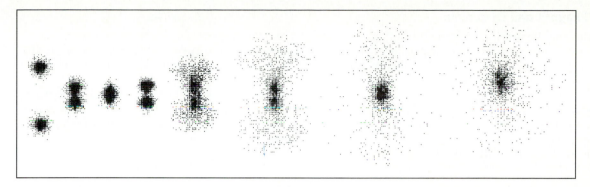

Figure 27.20 *Computer-simulated spiral galaxies collide head-on. As a result of gravitational forces, the two merge into a large elliptical.*

These ideas are necessarily incomplete; problems and contradictions abound. We have yet really to consider the dark matter, which seems to constitute at least 90% of the mass of the Universe, and which must have a powerful controlling effect on the formations of galaxies. We have also not considered the large-scale structure of the Universe, which must also have an influence over its constituents, and to which we now turn.

KEY RELATIONSHIPS

Elliptical galaxy subclassification:

$10(a - b)/a$, where a and b are respectively the galaxy's apparent semimajor and semiminor axes.

Tully-Fisher relation:

Luminosities of spiral galaxies increase with increased spread in internal velocities.

KEY CONCEPTS

Clusters of galaxies: Gravitationally bound groups of galaxies.

Distance indicators/standard candles: Objects with known absolute magnitudes that can be used to find galaxy distances.

Elliptical galaxies: Population II galaxies without spiral arms; they range in mass and luminosity from **giant ellipticals** through **dwarf ellipticals** to **dwarf spheroidals.**

Galaxies: Self-contained collections of mass that include stars, interstellar matter, and dark matter.

Hubble classification: A system of classifying galaxies that branches from ellipticals through the two kinds of spirals to irregulars; the scheme is depicted by the **tuning fork diagram.**

Irregular galaxies: Galaxies without much structure, a subset of which are starburst galaxies, which show rapid star formation.

Local Group: The sparse cluster that contains the Galaxy.

Peculiar galaxies: Galaxies outside the standard classes; most are interacting galaxies in collision.

Spiral galaxies: Galaxies displaying spiral arms and a strong Population I component; they divide into **barred spirals** (with bars through their centers) and **normal spirals** (no bars).

EXERCISES

Comparisons

1. What are the similarities and differences among Sa, Sb, and Sc spirals? Between normal and barred spirals?
2. Compare the stellar populations of spirals and ellipticals.
3. Compare the extreme forms of elliptical galaxies.
4. How can you tell an elongated elliptical from an edge-on S0?
5. Distinguish between regular and irregular galaxy clusters.
6. How do the Local Group, Virgo, and Coma clusters differ?

Numerical Problems

7. Make sketches of E2 and E7 galaxies.
8. What is the diameter of a galaxy 5 minutes of arc across and 10 Mpc away?
9. What do you expect the apparent magnitudes of the brightest stars in M 32 to be? State your assumptions.
10. If the Sun were in M 31, what would be its apparent magnitude?
11. What fraction of the mass of the Local Group is in M 31?

Thought and Discussion

12. What theories and discoveries led to the concept that the Universe contains separate, individual galaxies?
13. Why would an E0 galaxy not necessarily be spherical?
14. How do normal spirals of a subclass differ from one another?
15. What is a starburst galaxy?
16. How can astronomers derive the masses of spiral and elliptical galaxies?
17. List the evidence for dark matter in galaxies.
18. What do we find at the centers of regular galaxy clusters?
19. How do we know that galaxies collide? What are the major results of galaxy collisions?
20. In what two ways might elliptical galaxies form?

Research Problem

21. Examine astronomers' concepts of the distance of M 31 over the twentieth century. Using library materials, find these distances and plot them on a graph. Explain why measurements of the distance kept increasing up to about the middle of the century.

Activities

22. Make a flowchart in which you show how the distance scale of the Universe is established, beginning with the nearest stars. Indicate possible sources of error.
23. If you have access to a telescope, spend an evening trying to find as many galaxies as you can. Sketch each and use your library to find photographs for comparison.

Scientific Writing

24. Black holes are among the most mysterious objects in nature, so strange that nonscientists will not readily believe such things can exist. Write a two-page pamphlet in which you argue for the existence of black holes.
25. Galaxies are among the most beautiful and dramatic of all celestial objects; an introduction to them often spurs an interest in astronomy. Write a one-page description for an astronomy course in which you showcase galaxies to interest potential students. Use two photocopied illustrations.

28

The Expansion and Construction of the Universe

Large-scale motions and the distribution of matter in the Universe

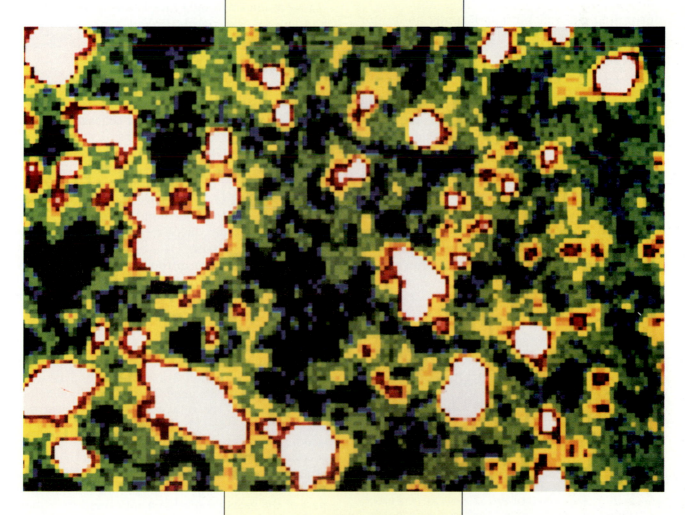

*Galaxies fill this image,
only a minute of arc across,
to magnitude 29.*

Galaxies and their clusters are the units of the Universe. We now look at their large-scale organization, their motions, and their distribution in three-dimensional space in order to probe the construction of the Universe.

28.1
The Velocity-Distance Relation

In 1912 Vesto M. Slipher of the Lowell Observatory measured the first Doppler shift and radial velocity for a spiral nebula. M 31, the Andromeda Galaxy, was seen to be approaching at an unprecedented speed of 300 km/s. By 1925 he had observed the spectra of 41 nebulae (by then known to be galaxies). All but three exhibited **redshifts** (z) in their spectra (Figure 28.1), specified by

$$z = (\lambda_{observed} - \lambda_{true})/\lambda_{true}.$$

Interpreting the redshifts as Doppler shifts, for which

$$v = \frac{(\lambda_{observed} - \lambda_{true})}{\lambda_{true}} c = cz$$

(see Section 19.2), his results implied that the galaxies are moving away from us at speeds approaching 1,800 km/s, far greater than any of our Galaxy's stars. Only M 31, its companion M 32, and M 33 were seen to be approaching.

Full disclosure of the nature of the effect had to wait for distances. By the end of the decade, Edwin Hubble had distance estimates for two dozen galaxies. A graph of the velocities against his distances (Figure 28.2) forever changed our view of the Universe and led to the present theories of its origin,

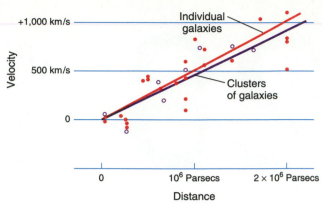

Figure 28.2 *Edwin Hubble's plot of the velocities of galaxies (red symbols) and their clusters (violet symbols) against their distances revealed the expanding Universe.*

structure, and evolution, a study called **cosmology.** The points that represent the galaxies in the graph scatter around a straight line, showing that distance and velocity are tightly linked. Only within the Local Group do orbital motions make a few galaxies approach and have negative velocities.

To see if this **Hubble relation** extends throughout the Universe, we must look outward to much greater distances and redshifts. By 1936 Milton Humason (Hubble's assistant and colleague) pushed the known velocities of clusters of galaxies to a remarkable 40,000 km/s, 13% the speed of light. When Hubble and Humason estimated distances from the magnitudes of the brightest cluster members, they could see the correlation to nearly half a billion parsecs (on the modern distance scale). Examples in Figure 28.3 extend the correlation to nearly a *billion* parsecs.

Greater distances introduce some difficulties. A large redshift will widen a specific segment of spectrum. At $z = 0.225$, light originating between 5,000 and 5,100 Å will be shifted to between 6,125 Å and 6,248 Å and will cover 123 Å. As a result, there is less light per Ångstrom and the galaxy will appear fainter than it would if it were not moving. Moreover, each photon from a receding galaxy has to cover a greater distance than one emitted before it. Photons from the galaxy arrive at a slower rate than they would if the galaxy were stationary, making it fainter yet. We must therefore apply corrections—which can be calculated exactly—to the observed magnitudes. The redshift causes a somewhat more difficult problem by shifting a different part of the galaxy's spectrum into the B or V magnitude band (Figure 28.4). Consequently, the assumed absolute magnitude of the galaxy changes

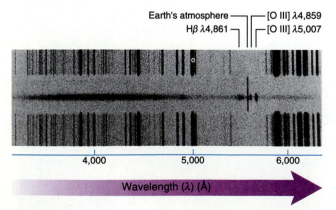

Figure 28.1 *A spectrogram of a galaxy taken in the 1970s runs between two comparison spectra that provide the wavelength scale. Hβ and the [O III] emission lines are indicated by arrows. The galaxy's spectrum is shifted to the red by 13%.*

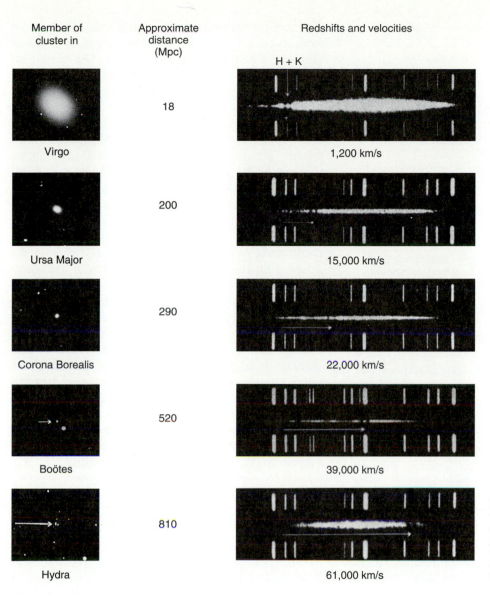

Member of cluster in	Approximate distance (Mpc)	Redshifts and velocities
Virgo	18	1,200 km/s
Ursa Major	200	15,000 km/s
Corona Borealis	290	22,000 km/s
Boötes	520	39,000 km/s
Hydra	810	61,000 km/s

H + K

Figure 28.3 *The recession of galaxies with distance is shown by the spectra of the brightest ellipticals in five clusters. Velocities are found from the observed wavelengths of the H and K lines of ionized calcium. The distances are modern values.*

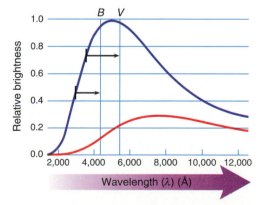

Figure 28.4 *The continuous spectrum of a typical galaxy is similar to that of a blackbody at 6,000 K (blue curve). If the galaxy is redshifted (red curve), the spectrum is stretched and the galaxy dims; at z = 0.5, the redshift shown here, the ultraviolet part of the spectrum is moved into the visual.*

with redshift. To make the corrections, we must know the luminosities of galaxies at shorter wavelengths, a problem solved by observation of nearby (relatively unredshifted) galaxies in the violet and ultraviolet.

Much more difficult is the problem imposed by the travel time of light. We never see anything as it is now, but are always looking into the *past,* because it takes time for light to get from one point to another. Between stars in the Galaxy, or even across the Local Group, light-travel time poses no problem, since a few thousand or million years is small compared with stellar or galactic lifetimes. But when we probe to distant galaxies, we look back to a time when they were in different states of evolution and might have been brighter or fainter than they are

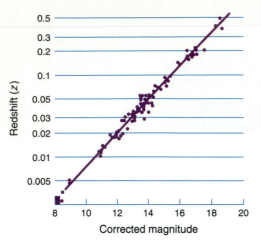

Figure 28.5   *The logarithm of redshift plotted against magnitude (corrected for redshift—but not evolutionary—effects) is a straight line out to z = 0.5. Hubble's original diagram fits into the rectangle at lower left.*

today. All we can do is to make theoretical adjustments based on our uncertain ideas of galaxy evolution. If we make a wrong guess, our distances will be wrong.

With these corrections in mind, astronomers have extended the Hubble relation far into space (Figure 28.5). To avoid problems of interpretation, Figure 28.5 is based almost purely on observational data: the redshifts (z) of 82 clusters are plotted against the apparent magnitudes of their most luminous galaxies, which are assumed to have similar absolute magnitudes (see Section 27.4). The only corrections involve those for absorption by dust and for the effect of the redshift (evolutionary efforts are not included). Since magnitude is a logarithmic scale, z must be plotted similarly. Up to a redshift, z, of 0.5 the observational data points fall along a straight line. Equally important, the observed clusters are found in all directions from the Earth. It does not matter if we look outward toward Virgo or Coma Berenices or in the opposite direction toward Fornax and Eridanus. We now have sufficient evidence to assume with some safety that the Hubble relation is a property of the Universe at large and is **isotropic,** or independent of direction. Everywhere we look, galaxies are flying away from us according to the same law.

Figures 28.2 and 28.5 show that velocity (for the moment our interpretation of the redshift) increases in direct proportion to distance; a galaxy twice as far away recedes at twice the speed. The system of galaxies therefore appears to be *expanding*. At first, our Galaxy appears to be at the center, but that is an illusion. All clusters of galaxies are getting farther away from one *another*. In Figure 28.6, galaxy B is twice as far from our Galaxy as galaxy A, and galaxy C four times as far. A, B, and C are respectively moving away from our Galaxy at speeds of 1,000 km/s, 2,000 km/s, and 4,000 km/s. By imaginary spaceflight, transport yourself to galaxy B. Now our Galaxy appears to be moving away from *you* at 2,000 km/s in the opposite direction. To find the speeds of A and C relative to B, subtract B's arrow from the others. You find that A moves away from B at 1,000 km/s, and C—which is twice as far away—departs at double that value, 2,000 km/s. The speeds seen from galaxy B show the same effect as seen from our Galaxy. You would see exactly the same Hubble relation from any of the galaxies in Figure 28.6 as you see at home. If we allow ourselves to extrapolate the relation beyond the bounds of Figures 28.5 and 28.6, we would see the same expansion *from any point in the Universe.* There is no favored position and, as we will see, no center. Humanity has now received the ultimate displacement from the center of attention.

28.2
The Meaning of the Redshift

The basic observational data are *redshifts* that must be interpreted and explained. Initially it was assumed that they are caused by the Doppler effect, but there are other possibilities. Gravity, for example, also produces redshifts (see Section 25.6.1). However, a noticeable gravitational redshift requires a powerful field like that surrounding a black hole, which is not appropriate to the gravities of the stars providing the galaxies' light. Therefore, gravity can quickly be discounted.

There is a better interpretation. Even before Hubble disclosed the natures of galaxies, mathematical physicists were developing theories of the structure of the Universe using the equations of Einstein's new general theory of relativity. There

Our Galaxy

Figure 28.6   *Our Galaxy appears to be at the center of an expanding system of galaxies (black arrows). An observer in galaxy B also sees the galaxies receding (pink arrows), and would also seem to be at the center of the expansion.*

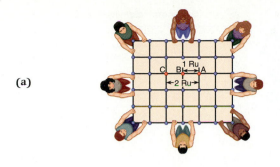

(a)

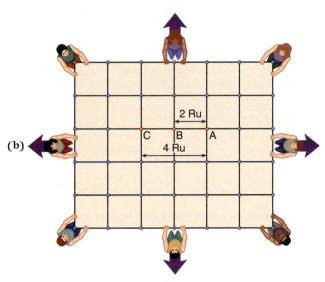

(b)

Figure 28.7   (a) *A rubber sheet with a grid is stuck with colored pins. Pins A, B, and C are each one "rubber unit" (RU) apart. As the sheet is stretched, the pins ride along with it and are separated.* (b) *The sheet has doubled in size, B is 2 RU from A and C is 4 RU from A. An observer on A (and on every other pin) would see an expanding "rubber universe."*

are three simple possibilities. The Universe can expand, contract, or do neither, that is, be *static.* Einstein solved his equations in 1917 to obtain a *static Universe.* To prevent contraction and to counter the attractive force of gravity provided by the mass within it, he postulated a **cosmological constant,** which describes a *repulsive* force whose strength *increases* with distance. (There is no laboratory evidence for such a force. Einstein later referred to it as his most embarrassing error.) However, at the same time the Dutch mathematician Willem de Sitter found a solution for an **expanding Universe.** By 1924 others had obtained more solutions for both expanding and contracting Universes. Hubble's discovery of the actual expansion of the system of galaxies then fit into a theoretical framework already in place.

The solutions of general relativity show that the clusters of galaxies are not expanding *into* space but *with* space, in a smooth motion called the **Hubble flow.** Clusters travel with the flow rather like chips of wood in a stream. The distinction is profound and has vast implications. Space itself is expanding, carrying the clusters along with it. However, neither the galaxies nor their clusters are getting larger because they are held together by gravity.

To see what is happening, begin with a simple two-dimensional analogy. Several people hold a sheet of rubber containing a grid of lines and colored pins representing galaxy clusters (Figure 28.7a). The people step back and the sheet expands (Figure 28.7b), carrying the clusters with it. Pins A, B, and C are each one "rubber unit" (RU) apart in Figure 28.7a. In Figure 28.7b, A and B are pulled to 2 RU apart, but B and C are *also* now 2 RU apart, so C is *4* RU from A. To move to those separations in the same amount of time, the velocity of C relative to A has to be twice that of B. An observer on A, or on *any other pin,* would then see an expanding two-dimensional rubber universe.

What, however, is expanding? The grid lines represent coordinates in the Universe, and they move apart with the pins. Mathematically, the pins are not moving because they stay at the same coordinates. It is the coordinate framework, or *space,* that is expanding, carrying the pins with it. During the expansion, pin B emits a photon in the direction of pin A. While the photon travels, "rubberspace" expands, stretching the wave. When the wave arrives at pin A, the photon's wavelength is longer than it was and the light is redshifted. The amount of redshift (z) depends on how long the photon has been in flight, or on the distance between pins. The observer on A (or at any other point) will see increases in recession velocities in proportion to distance, or a rubber-Hubble relation.

This redshift is *not* a Doppler shift. The Doppler effect occurs when objects change their coordinates relative to one another with time. In the expanding rubber universe, the redshift is caused by the expansion of the coordinates space itself. To calculate the redshift, we use a concept called the **scaling factor of the universe** (R) that represents the relative distance between grid lines in Figure 28.7. If the universe doubles in size, R doubles, and the separations between the lines of the grid all double. The wavelength of the photon observed at pin A depends on the change in R during the time the photon was in flight, or in terms of the expanding rubber universe depicted in parts (a) and (b) of Figure 28.7,

$$\lambda_{observed}/\lambda_{emitted} = R_b/R_a.$$

Therefore, the redshift

$$z = (\lambda_{observed} - \lambda_{emitted})/\lambda_{emitted} = R_b/R_a - 1.$$

There are crucial differences between the rubber universe and our own real Universe. Ours has three dimensions of space, not two. There is no one pulling on it, and as we will see in Chapter 30, it may be infinite and unbounded, and even curved in spacetime. Nevertheless, space is getting larger. If Figure 28.7b represents our Universe *now* and Figure 28.7a the Universe *then* (when the light we see was emitted),

$$z = (\lambda_{observed} - \lambda_{emitted})/\lambda_{emitted} = R_{now}/R_{then} - 1,$$

or

$$R_{now}/R_{then} = z + 1.$$

The redshift, z, therefore gives the fractional increase in size of the Universe between now and when the light was emitted. For example, if z for a cluster of galaxies is 0.5, we know that the Universe has expanded by a factor of 1.5 since the light departed.

At low redshifts (and velocities), below z of a few tenths, the correct equation relating velocity to redshift is about the same as the simple Doppler formula, explaining why redshifts of galaxies are often considered Doppler shifts. However, at higher z, the formulae are *not* the same. As we will see in Chapter 29, z can easily exceed 1. Application of the simple Doppler formula implies that the receding body is going faster than light (the blue line in Figure 28.8), which from Section 7.8 should not be possible. That problem could be reconciled by adopting a more complicated formula from the special theory of relativity, in which as z passes 1 and goes to infinity, the velocity approaches c (the green curve in Figure 28.8). However, in the Universe we deal with *general* relativity (which involves spacetime curvature), not *special* relativity. Any formula relating velocity and redshift *depends on the exact nature of the spacetime curvature of the Universe,* a subject to be dealt with in Chapter 30. An example, from a model of the Universe devised by Einstein and de Sitter in 1932, is shown by the red curve in Figure 28.8, where we see that the expansion of the Universe *can* exceed the speed of light. Such an expansion violates no law, because it involves space itself and not the objects within it.

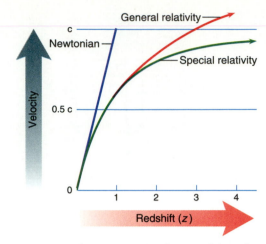

Figure 28.8   *According to Newtonian theory and the ordinary Doppler formula (blue line), a body with $z = 1$ is receding at the speed of light. In special relativity, as the Doppler shift z goes to infinity, v goes to c (green curve). In a model of the Universe developed according to the principles of general relativity by Albert Einstein and Willem de Sitter, as z increases, the expanding Universe can exceed the speed of light (red curve).*

28.3
The Hubble Constant

To find the rate of expansion of the Universe, we need only divide the recession velocity of a cluster of galaxies by its distance from us. The result is a number called the **Hubble constant** (H_0), expressed in kilometers per second per megaparsec, km/s/Mpc. Since the relation between velocity and distance is a straight line, it does not matter which cluster we choose: they all give the same value. For each megaparsec outward, recession velocity increases by H_0. Moreover, residents of any galaxy would derive the same value of H_0. The Hubble constant describes the rate of expansion everywhere in the Universe. Since the redshift-velocity relation within the expanding Universe is close to the Doppler effect for low z (well under 1), astronomers can apply the Doppler formula to the redshifts of nearby galaxies to find the velocities. This locally determined Hubble constant is assumed to be the true H_0 for the Universe at large.

In spite of the apparent simplicity of the procedure, the value of H_0 has been elusive. Hubble's first diagram (see Figure 28.2) gave an expansion rate of 500 km/s/Mpc: for every increase in distance of 1 Mpc, the speed of a galaxy was seen to go up by 500 km/s. His distances, however, were wrong. In 1931 he placed M 31 263 kpc away, less than half the modern value. The distances of farther galaxies, the ones he used to define the Hubble relation, were even more greatly underestimated.

Modern measurements, though much better, are still plagued by severe difficulties. Distances of even relatively nearby galaxies are uncertain. Eighteen megaparsecs is quoted in Table 27.3 for the Virgo Cluster, but different methods give between 15.6 and 21 Mpc. The uncertainty becomes worse with greater distance. Equally confusing, galaxies within a cluster move about as a result of gravitational orbits. And even though clusters of galaxies move apart as a result of the expansion of space, there are still significant gravitational effects causing them to move relative to one another. Both kinds of motion contaminate the smooth Hubble flow with *real* Doppler shifts. If we look so far away that these gravitational motions are unimportant relative to the Hubble flow, then we can no longer find accurate distances; neither can we relate redshift and velocity, since we do not know the nature of spacetime curvature.

These gravitational motions have proven to be quite complex. Galaxies in the direction of the great Virgo Cluster have slightly lower redshifts than they do in the other direction, indicating that our Galaxy and the Local Group are "falling" toward Virgo. The Local Group is apparently part of a cluster of clusters, a **supercluster,** centered on the massive Virgo system. The speed is hard to measure; estimates range from 150 to 300 km/s. However, since the Hubble flow is moving us away from the Virgo Cluster at 1,200 km/s, "falling" simply means that we are not moving away quite as fast as we would if there were no gravitational attraction: we will never arrive. Redshifts of galaxies examined over larger volumes of space show other streaming motions not related to the Hubble flow. We *and* the Virgo cloud may be "falling" in the direction of Hydra and Centaurus at roughly 600 km/s, the result of a somewhat mysterious *Great Attractor* that appears to be a massive supercluster some 50 Mpc away. Other measurements show that within the surrounding 150 Mpc, we are moving toward Lepus at about 550 km/s. We seem to be a long way from sorting out all the motions.

Different distance methods and velocity corrections give estimates for the Hubble constant that range from about 50 km/s/Mpc, about a tenth that derived by Hubble, to as high as 100 km/s/Mpc. Considering all the results, the most likely value falls around 75 or 80 km/s/Mpc. An intermediate value of 75 km/s/Mpc, used in Figure 28.3, is generally adopted here. But be aware that H_0 could be over 30% higher or lower.

28.4
The Big Bang and the Age of the Universe

Los Angeles and San Francisco are 600 km apart. You drive between them at a steady 50 km/hr. You calculate the time it will take (t) by dividing distance (d) by your speed (v),

$$t = d/v,$$

so it will take 600 km/50 km/hr = 12 hr. If you now want to drive at the same speed from Los Angeles to Klamath Falls, Oregon, a distance of 1,200 km, it would take 24 hr. But if you double your speed to 100 km/hr, you can get to Klamath Falls in the same time of 12 hr.

Similarly, you can figure how much time, t, has elapsed since galaxy A in Figure 28.6 and our Galaxy were together by dividing galaxy A's distance from us by its recession velocity. If you calculate the elapsed time since galaxy B and ours were together, you get the same answer: galaxy B has twice the distance and twice the velocity. Since velocity increases in direct proportion to distance, it does not matter which galaxy you use: all give the same answer, suggesting that at one time all the observed matter in the Universe—from which the observed galaxies were made—was concentrated into a small volume and dispersed in a sudden event now called the **Big Bang.** The elapsed time since the Big Bang can then be defined as the **age of the Universe** (t_0), more properly called the **Hubble time.** However, note that this event was *not* an explosion in which matter, and subsequently galaxies, began to fly through space. It was a sudden expansion *of* space that carried the decompressing matter along with it.

The Hubble constant is defined as velocity divided by distance, and age is defined as distance divided by velocity; one is the reciprocal of the other, or

$$t_0 = 1/H_0$$

and

$$H_0 = 1/t_0.$$

Since an observer in every galaxy would see the same Hubble expansion, each would get the same value of t_0.

The distances of galaxies are measured in megaparsecs, and velocities in kilometers per second. To find t_0 we must use consistent units. The Hubble constant H_0 must first be converted to (km/s)/km. Since there are 3.09×10^{19} kilometers in a megaparsec, the Hubble constant can be written as $H_0/3.09 \times 10^{19}$ (km/s)/km $= 3.24 \times 10^{-20} \, H_0$ (km/s)/km (an expansion rate of a hundredth the diameter of an atomic nucleus/s/km). The Hubble time, t_0, is then

$$t_0 = 3.09 \times 10^{19}/H_0 \text{ seconds.}$$

Since there are 3.16×10^7 seconds in a year, the Hubble time can be expressed as

$$t_0 = 978 \times 10^9/H_0 \text{ years.}$$

If H_0 is 50 km/s/Mpc, $t_0 = 19.6$ billion years, and if $H_0 = 100$ km/s/Mpc, $t_0 = 9.8$ billion years. Roughly speaking, the Hubble time falls somewhere between 10 and 20 billion years, with a most likely value of around 13 billion if the value of $H_0 = 75$ km/s/Mpc is adopted. It is not unreasonable to think of the age of the Universe as its most important property. However, current measurements give a factor-of-two range in its possible value, at best making it difficult to arrive at a viable theory for the formation and evolution of the Universe.

This discussion ignores the powerful drag of gravity, which causes the expansion to slow over time and has the effect of shortening the age of the Universe for given values of H_0 and t_0. It also leaves out other age determinations. The measured ages of the oldest globular clusters are greater than some of the values of t_0. The Universe, however, cannot be younger than its oldest stars. These critical matters, as well as arguments both for and against the Big Bang, will be discussed in Chapter 30.

28.5
The Distribution of Galaxies

Exploring the Universe, like exploring a large city, requires a map. We therefore need the locations of the galaxies and their clusters, including directions and distances. We then make plots of them to find how they are organized and distributed.

28.5.1 Redshift Distances

Beyond a few tens of megaparsecs, direct distance measurement becomes very difficult because the galaxies cannot be resolved into recognizable objects. If we look much farther, we cannot even classify the galaxies and must rely on the brightest members of large clusters. That procedure, however, does not serve to locate the myriad faint galaxies and smaller clusters.

However, once the Hubble relation has been established, we can turn the problem around and use the Hubble constant to *find* distance. With the spectrograph, we measure a galaxy's redshift. If the redshift is relatively small, so that there is little or no difference among the redshift-velocity curves in Figure 28.8, we can convert the redshift to velocity with the Doppler formula. The **redshift distance** is then

$$D = v/H_0 = cz/H_0.$$

As an example, if we observe a galaxy moving away with a redshift, z, of 0.2, and if $H_0 = 75$ km/s/Mpc, the distance is about 800 Mpc. As long as a spectrum can be secured, we can locate galaxies in three-dimensional space.

The method has three problems. First is the inevitable deviation of individual galaxies from the smooth Hubble flow. Gravitational influences will make galaxies go either faster or slower than expected, leading to errors in distance. Second is our ignorance of the correct value of the Hubble constant. In the example, the distance could be as low as 600 Mpc, for $H_0 = 100$ km/s/Mpc, or as high as 1.2 billion parsecs (*gigaparsecs, Gpc*) for $H_0 = 50$ km/s/Mpc. Third is the repeated difficul-

Figure 28.9   *A three-dimensional representation shows the local supercluster centered on the Virgo Cluster. The Local Group is at its edge.*

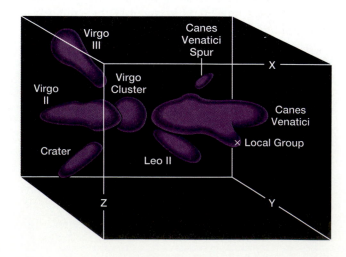

ty that once z becomes sufficiently high, we can no longer translate it into velocity. Perhaps surprisingly, the latter two problems are not terribly serious, as we are more interested in *relative* distances. Differences in our perception of H_0 and in the conversion of z to velocity then simply involve a matter of scale. It is therefore common to avoid the problem altogether and to express distances in terms of z.

28.5.2 The Local Supercluster

The word *local* has successively referred to the Earth, the Solar System, the nearby stars, the Galaxy, and the Local Group. Now we expand our sphere again and use *local* to mean the volume of space under some significant mutual gravitational influence.

The *local supercluster* can be examined in depth better than any other such assembly, because we can see the faint galaxies within it. It is spread all over the sky, allowing us to probe its three-dimensional structure with considerable accuracy. Its effect is found in the remarkable concentration of galaxies in the northern celestial hemisphere centered on the constellation Virgo, which can be noted even in the simple map of Figure 22.13 that shows the zone of avoidance. A three-dimensional map made both from directly determined and redshift distances shows that the Virgo Cluster is indeed a center of attraction (Figure 28.9). Our own Local Group is but a spur in a much larger collection of galaxies related to the massive system in the center. Even though the local supercluster is a region over which gravity has a strong influence, enough to produce significant variations in the Hubble flow, *it is still expanding,* the individual clusters and spurs getting farther apart.

We can get a different view by using the gravitational fields of galaxies and their clusters to punch through the obscuring dust of our own Galaxy. Careful analysis of the deviations from the smooth Hubble flow allows the mapping of mass within a hundred or so megaparsecs of the Galaxy (Figure 28.10), showing that the Virgo supercluster—as big as it is—is only a wrinkle in the Universe's organization and dwarfed by other structures. The gravity map shows the distribution of all matter, both illuminated and dark. The agreement between the distribution of illuminated matter and that derived from gravity suggests that illuminated and dark matter go hand in hand.

28.5.3 Filaments, Sheets, and Voids

An early belief about the Universe is that its galaxies are uniformly (or homogeneously) distributed,

and that if we were to sample different locations we would find the same average number of galaxies per unit volume. The only deviations from this uniformity were believed to be local, caused by clustering or some level of superclustering. The observations show otherwise.

The first hints of the real distribution of galaxies go back to the late 1970s, when astronomers began finding **voids,** huge regions 100 Mpc or more across containing barely any galaxies. A number of groups of astronomers then mounted massive programs to map the Universe on a grand scale. The goal is not only to find how galaxies are grouped and distributed, but also to see if we can actually find a scale of size over which the Universe really *does* become homogeneous. Is there a scale of distance over which the hierarchical groupings (successive groupings within groupings) stop, and over which the number of galaxies per unit volume stays the same?

Various methods have been used. In the simplest, we plot the distribution of galaxies on the celestial sphere, ignoring distance. An enormous number of galaxies is then available for study. The northern hemisphere (Figure 28.11a) was mapped in the 1970s by plotting a million galaxies from a catalogue compiled at Lick Observatory. A recently completed southern survey (Figure 28.11b) undertaken at Oxford University plots 2.5 times as many. Some sense of three-dimensional structure can be obtained by assuming that fainter galaxies are farther away. The Lick map extends to a distance of nearly 500 Mpc, the Oxford map nearly half again as far. From the distribution of galaxies, astronomers find a characteristic scale to the clumping of about 25 Mpc, roughly the size of the local supercluster (see Figure 28.9).

More dramatic results are obtained by mapping galaxies in three dimensions, where distances are

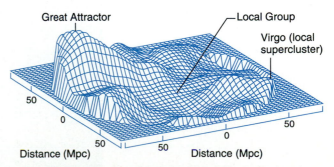

Figure 28.10 *This three-dimensional graph shows the mass in a plane 100 Mpc wide sliced through the nearby Universe and centered on the Galaxy. The density of matter is proportional to the height of the sheet above the plane and is found by determining where mass must be to give the observed deviations from the Hubble flow.*

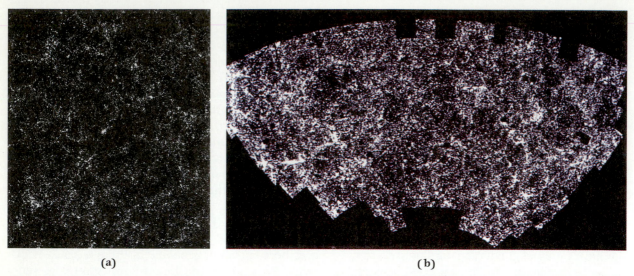

(a)   **(b)**

Figure 28.11   *The 3.5 million galaxies plotted in* **(a)** *the northern and* **(b)** *southern celestial hemispheres show their clumpy distribution.*

Figure 28.12   **(a)** *The distances of galaxies between δ = 27° N and 32° N are plotted against their right ascensions. The Coma Cluster is at the center and the Great Wall (in the blue band) runs from side to side.* **(b)** *Similar plots for four adjacent declinations are stacked together, increasing the numbers and the concentration within the Great Wall.*

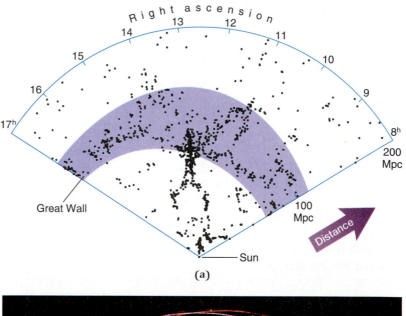

(a)

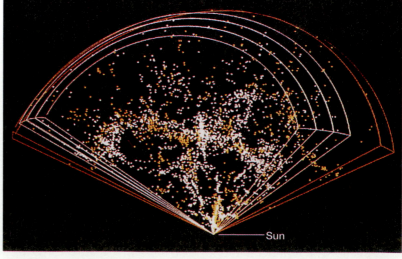

(b)

BACKGROUND 28.1 Cosmology and Technology

The technological revolution in astronomical instrumentation over the past two decades has had its greatest impact on cosmology. The pioneers of the subject labored heroically to obtain the data they needed. It took Slipher *13 years* to assemble radial velocities for a mere 41 bright (by today's standards) galaxies. The only available detectors were photographic plates. Even today, photography is a slow way of collecting photons; the plates in use in the 1920s were a hundred times slower.

With extraordinary patience and dedication, astronomers ran some exposures over two or more nights. The camera shutter would be closed at the end of the first night and the plateholder sealed against stray daylight. It would then be opened on the same object the next night. In attempting a redshift record in 1928, Milton Humason, Hubble's talented observing colleague, exposed the spectrum of NGC 7619 for 45 hours over five consecutive nights, during which time he had to keep guiding the tele-

scope on the object. Flexure in the telescope and changes wrought by temperature fluctuations resulted in a poor spectrum, but one in which the H and K lines of ionized calcium were identifiable, allowing Humason to measure a redshift of 3,780 km/s, close to the modern value.

The same spectrum can be obtained with much better quality with a CCD in under 10 minutes. All of Slipher's work could now be done (provided the objects were above the horizon) in a single night! Such technological ability has allowed astronomers to determine the redshifts of tens of thousands of galaxies and to map the Universe. Even so, the science of observational cosmology is still young. For all our telescopic and detector capability, we have still not been able to nail down the exact value of the expansion rate of the Universe, the Hubble constant, a testimony to the extreme remoteness of the objects and the difficulty of the subject.

found from redshifts. This technique is much more laborious than plotting flat maps, because spectra must be obtained. So far, various groups of astronomers have obtained redshift distances of about 30,000 galaxies, and the goal is many times higher. The results are exemplified by Figure 28.12a, which shows the locations of galaxies with depth within an area of the sky 5° of declination wide (centered at $\delta = 29.5°$N) and 9 hours of right ascension across. The irregular distribution is obvious. We see great voids over 50 Mpc across outlined by filaments of galaxies up to 200 Mpc long.

The clusters are only the most intense concentrations within the filaments. The most prominent filament is the Great Wall, a long string of galaxies that extends from one side of the diagram to the other at a distance of between 50 and 150 Mpc and that contains the massive Coma Cluster. A three-dimensional plot within the same span of right ascension but with four adjacent declination bands stacked together (Figure 28.12b) makes the Great Wall even more impressive.

The concentration produced by the Coma Cluster in the center of the Great Wall points at Earth, a clue that some systematic effect is causing distance errors. In this case, it is the motions of the individ-

ual galaxies within the cluster. In the large scale, such motions are not considered important, and the three-dimensional maps are thought to represent a reasonably accurate picture of galaxy distribution.

The concept of clustering and superclustering does not adequately describe the way galaxies are distributed. What we actually find is a texture more like that of a sponge, holes (voids) surrounded by connected sheets that contain the gravitationally bound clusters of galaxies (Figure 28.13). The Hubble flow then expands the sizes of the sheets and the voids, stretching them, and leaving the clusters together as knots in the connective tissue.

Astronomers would like to extend such maps as far as they can see, and are now even building telescopes and spectrographs dedicated to the task. In the meantime, we rely on sampling, observing galaxies within a narrow but extremely long "pencil beam" only a degree or so across. Such investigations at the north and south galactic poles have probed the distribution of galaxies to roughly 1.3 Gpc (over 4 billion light years) on either side of the Galaxy (Figure 28.14). We see the same kinds of structures, the Great Wall being only the closest knot of galaxies along the line of sight. Pencil beam surveys show clumpings nearly 200 Mpc across.

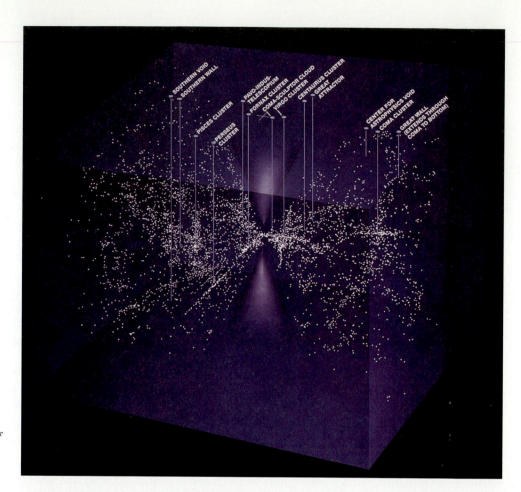

Figure 28.13 *A compilation of redshift surveys provides a map of the local Universe within about 200 Mpc, showing its spongelike texture.*

Figure 28.14 *The Galaxy lies at the center of twin narrow pencil beams projected toward the galactic poles to distances of over 10^9 pc. The sizes of the knots indicate the number of galaxies per unit volume. "Great walls" lie in either direction.*

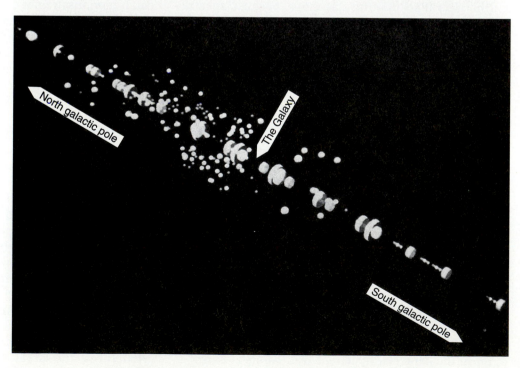

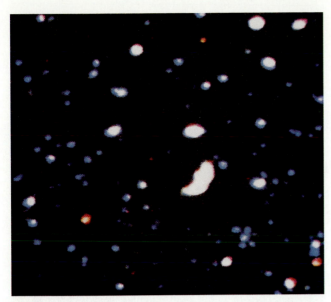

Figure 28.15   *Hundreds of faint blue galaxies lie within a square only 2.2 minutes of arc on a side.*

Beyond scales of about 200 Mpc, the clumps smooth out. Surveys of radio emission from distant galaxies show a remarkably smooth distribution, as do CCD images of faint blue galaxies (Figure 28.15 and the chapter opening image). Spectrograms reveal redshifts up to $z = 0.8$, showing that over distances measured in gigaparsecs the galaxy distribution is uniform.

Distances to these faint blue galaxies are great enough that we now look considerably back into the past. The faintest of them is billions of light-years away, so we see the systems as they were billions of years ago. Their notable blue colors are different from the colors of the nearby galaxies, apparently revealing the effects of galaxy evolution. The redshifts are so high that the ultraviolet is shifted into the blue and visual parts of the spectrum, and the high level of UV radiation attests to a rapid rates of star formation. This evident evolution of galaxies with time clearly complicates our ability to derive distances from average apparent magnitudes (see Section 28.1).

These deep surveys vividly reveal the vast number of galaxies in the Universe. The image in Figure 28.15 contains 4.8 square minutes of arc. A square degree of sky holds 3,600 square minutes of arc and over 100,000 faint blue galaxies. The celestial sphere contains some 41,000 square degrees, so over the entire sky we are capable of detecting *10 billion galaxies*. If we could examine the Universe out twice as far, we could conceivably find some ten times as many, and obviously these are only

the brightest. The visible Universe could contain upward of a *trillion* (10^{12}) galaxies. If each averages 10 billion (10^{10}) stars, the Universe accessible to us could contain $10^{12} \times 10^{10} = 10^{22}$ stars!

KEY CONCEPTS

Big Bang: The concept that 10 to 20 billion years ago, the Universe (or at least the present phase of the Universe) began with the sudden expansion of space from a condition of very high density.

Cosmological constant: A constant that describes Einstein's suggested repulsive force.

Cosmology: The study of the structure, origin, and evolution of the Universe.

Expanding Universe: The steady growth of space, the **Hubble flow,** that increases distances between clusters of galaxies.

Hubble constant (H_0): The rate at which the Universe expands.

Hubble relation: The linear correlation between redshifts and distances of galaxies.

Isotropic: The same in all directions.

Redshift: The general spectral shift to longer wavelengths (redward) observed from receding galaxies.

Scaling factor of the Universe (R): The factor that represents relative distances in the Universe and that increases with time.

Supercluster: A cluster of clusters of galaxies.

Voids: Volumes, megaparsecs across, devoid of galaxies.

KEY RELATIONSHIPS

Hubble time (age of the Universe):

$$t_0 = 1/H_0 \text{ if } H_0 \text{ is in consistent units}$$
$$= 978 \times 10^9/H_0 \text{ years if } H_0 \text{ is in km/s/Mpc.}$$

Redshift:

$$z = (\lambda_{observed} - \lambda_{emitted})/\lambda_{emitted} = R_{now}/R_{then} - 1,$$
where R is the scaling factor of the Universe.

Redshift distance:

$$D = v/H_0 \text{ for small } z.$$

EXERCISES

Comparisons

1. Compare the advantages and disadvantages of the three ways to examine the distribution of galaxies in the Universe.

2. Compare the natures of, and relations among, clusters, superclusters, and walls of galaxies.

Numerical Problems

3. What is the approximate redshift (z) of the Hydra Cluster in Figure 28.3. What is the wavelength of the Ca II K line normally at 3,934 Å?

4. We measure $z = 2.4$ for a distant receding source. What is the recession velocity from **(a)** the Doppler formula; **(b)** the special relativity relation; **(c)** the general relativity relation of Einstein and de Sitter.

5. A distant galaxy has $z = 0.75$. How many times bigger is the Universe than it was at the time the light left the galaxy?

6. What is the Hubble time if $H_0 = 60$ km/s/Mpc?

7. What is the distance to a galaxy receding at 15,000 km/s for $H_0 = 50$, 75, and 100 km/s/Mpc?

Thought and Discussion

8. What complicates the calculation of a galaxy's distance from its magnitude?

9. What is the evidence that the Universe is expanding?

10. Why are we not actually at the center of an expanding Universe?

11. What is meant by the scaling factor of the Universe (R) and by the Hubble constant (H_0)?

12. Why is our estimate of the Hubble constant now so much lower than Hubble's original value?

13. Why do some galaxies exhibit blue spectral shifts? Why are there deviations from the smooth Hubble flow?

14. Why is our Galaxy not expanding with the Universe at large?

15. How does the redshift of a distant galaxy differ from a Doppler shift?

16. Relate the Local Group to the Virgo Cluster and to the Local Supercluster.

17. What is the Great Wall? Why is it important in our understanding of the Universe?

Research Problem

18. Use your library to find a textbook on astronomy written before 1925. Comment on the views of the Universe presented there and contrast them with those held today. What observations might make this textbook appear out of date in another 70 years?

Activity

19. Construct your own Hubble diagram from Tables 27.1 and 27.2. Show how the two tables give different results and explain why Table 27.1 confuses the results from Table 27.2.

Scientific Writing

20. A politician is publicly critical of astronomers, writing that with all the instruments we have and the research money that has been spent we should at least have some firm answers about the nature of the expanding Universe. Answer the criticism, describing how much we have learned in a short time and explaining why the problems are difficult.

21. Edwin Hubble is known for discovering the expansion of the Universe. However, no scientific work lacks a history. Write an essay on work carried on before Hubble that allowed him to make his discovery.

29

Active Galaxies and Quasars

Massive black holes, mighty jets, and a view to the fringe of the Universe

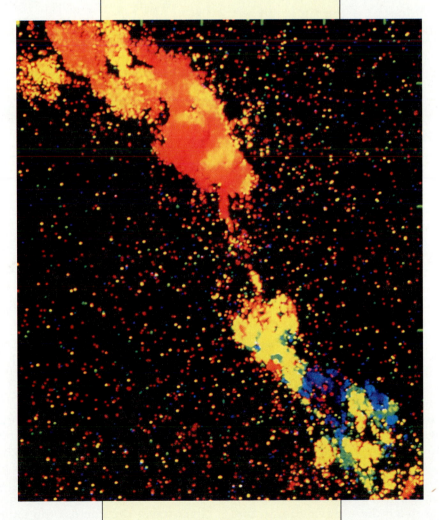

Streamers of gas extend a third of a megaparsec from Hydra A.

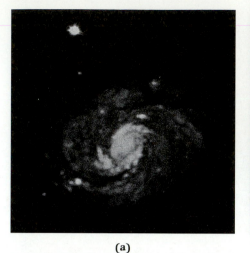

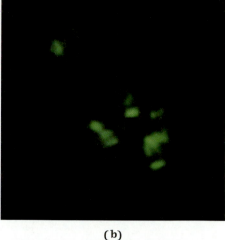

Figure 29.1 **(a)** *Within the central bulge of the Type 2 Seyfert galaxy NGC 1068 is a brilliant nucleus.* **(b)** *The Hubble Space Telescope shows the nucleus surrounded by clouds of ionized gas.*

(a)   **(b)**

Most of the galaxies around us are illuminated by stars, and their nuclei are relatively faint and calm. In some galaxies, however, the relation is reversed, and the nuclei dominate. At the limit, all we observe is a powerful nucleus and its extraordinary effects. Some of these brilliant objects allow us to look to the limits of the observable Universe and back in time toward the Big Bang itself.

29.1
Active Galaxies

Unusually bright galactic nuclei are commonly associated with several kinds of activity. They may be variable, display evidence for high-velocity mass motions, radiate copiously in the ultraviolet and X ray, and hurl jets of matter to great distances. These **active galaxies** seem to distill to a model based on the infall of matter to a massive central black hole.

29.1.1 *Seyfert Galaxies*

In 1943 Carl Seyfert of Mount Wilson Observatory completed a systematic study of spiral galaxies whose cores display strong emission lines. **Seyfert galaxies** (Figure 29.1) constitute about 1% of the spirals, are generally confined to those with more tightly wound arms, and have brilliant, starlike nuclei. There are two principal kinds. The nuclei of *Type 1 Seyferts* have broad hydrogen Balmer lines that reveal gas moving at velocities up to 10,000 km/s (Figure 29.2a) and narrow forbidden lines associated with lower velocities in the hundreds of km/s. Powerful continuous spectra underlying the emissions extend through the ultraviolet into the X ray; like supernova remnants, they behave roughly according to a power law (see Section

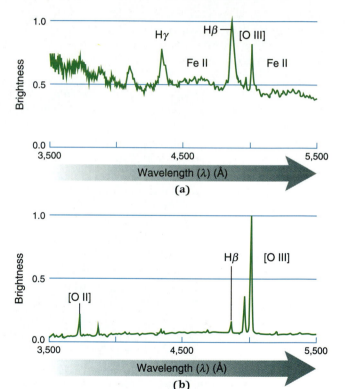

Figure 29.2 **(a)** *The Type 1 Seyfert galaxy Mk 1243 has broad hydrogen and narrow forbidden lines.* **(b)** *The Type 2 Seyfert Mk 1157 displays only narrow lines. Both spectra are presented as you would see them with no redshift.*

25.4). There are no absorption lines, so the continuous spectra are obviously not produced by stars.

The broad lines vary in brightness with periods on the order of weeks. For a body to vary in a coherent fashion, its parts must communicate—that is, exchange information—with one another. The light-travel time across the emitting region must therefore be less than the time it takes for the body to change its brightness. The broad lines come from

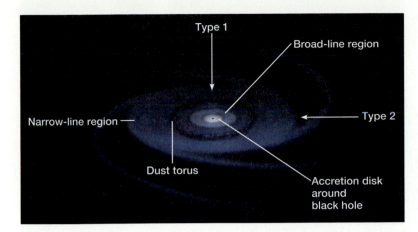

Figure 29.3 *A Seyfert galaxy may be powered by a mass-accreting black hole at its center that is successively surrounded by a rapidly spinning disk that produces the broad spectrum lines, a ring of dust, and a more slowly rotating disk that generates the narrow lines. Observed face-on, the galaxy is a Type 1 Seyfert. From the side, the dust hides the inner ring, and the observer sees a Type 2. The drawing is not to scale.*

regions only a few light-weeks—a few thousand astronomical units—across. The continuous spectra can vary on time scales of minutes! The brightest Type 1 systems radiate up to 10^{38} joules/s, a trillion solar luminosities, all from a region *at most* a few thousand AU wide.

In the spectra of *Type 2 Seyferts,* all the emission lines are narrow (Figure 29.2b) and the continuous spectra are weak or absent. The exception is in the infrared, where blackbody radiation indicates heated dust. The strengths of the emission lines suggest they are excited by a power-law continuum, leading us to believe that Type 2 Seyferts have the same kind of powerful tiny nuclei as Type 1, but are simply not seen.

Seyfert nuclei are too small and bright to be energized by stars and thermonuclear fusion. However, we have already established the likelihood of massive black holes at the centers of normal galaxies (see Sections 26.4 and 27.6.4). The Seyferts can then be explained by having large amounts of matter—disrupted stars and interstellar gas—fall erratically into an energetic accretion disk around a black hole of some 10^7 $M_\odot$ (Figure 29.3). The radiation from the accretion disk produces the bright continuum and ionizes the gas clouds around it. The broad-line region of Type 1 Seyferts is produced by gas rapidly orbiting within a few pc of the black hole. Surrounding the broad-line region is a doughnut, or torus, of dust that radiates in the infrared. The narrow-line region comes from a much larger zone, perhaps hundreds of pc across, in which gas circulates more slowly. Type 2 Seyferts are thought to be the same as Type 1 except that the orientations of the dusty doughnuts hide the nuclei and the broad-line regions (see Figure 29.3).

The Seyferts are only the tip of a big iceberg. One-third of all spiral galaxy nuclei display emission lines from atoms of low ionization, and are called **LINER**s, "Low Ionization Nuclear Emission Regions." The intensities of the emission lines again suggest ionization by a power-law continuum. LINERs are mini Seyferts, and they may be a bridge that connects active and ordinary galaxies. We therefore see a continuum of activity from nearly inactive systems like our own through the LINERs to the true, powerfully active Seyferts.

29.1.2 *Radio Galaxies*

The active versions of elliptical galaxies are **radio galaxies.** Our own Galaxy produces considerable radio radiation both from the gas in its disk and from its bright nucleus. From a distance, however, optical radiation would dominate, as it does in most galaxies. The opposite is true for radio galaxies.

M 87, one of the giant elliptical galaxies in the Virgo Cluster (see Figure 27.8a), is a superb example. As one of the brightest radio sources in the sky, it is also called Virgo A. A straight jet of matter about 2.5 kpc long, roughly 10% of the galaxy's radius, streams from its nucleus (Figure 29.4). The jet radiates powerfully at radio wavelengths and is

Figure 29.4 *M 87's jet pours from the brilliant galactic nucleus in this near-infrared image taken with the Hubble Space Telescope. Knots are regularly spaced from the center outward.*

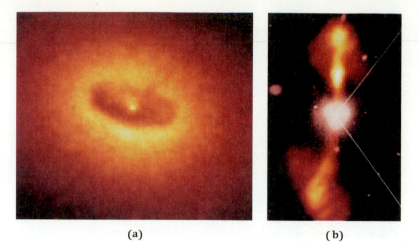

Figure 29.5   **(a)** *The Hubble Space Telescope shows that NGC 4261 has a dusty accretion disk only 100 pc across.* **(b)** *The view expands, combined optical and radio observations showing jets flowing perpendicular to the disk.*

(a)   **(b)**

Figure 29.6   *Powerful jets squirt from the poles of an accretion disk as mass falls into a black hole.*

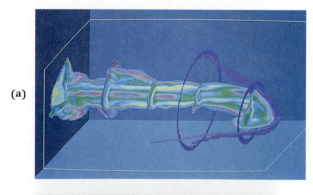

(a)

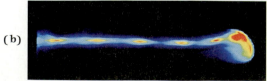

(b)

Figure 29.7   **(a)** *A computer simulation of a galactic jet shows gas blowing turbulently through a magnetically confined tube.* **(b)** *A simulated radio observation matches the observed jet in Figure 29.4.*

seen even in the X ray. Radio spectra show that the jet is producing synchrotron radiation, so it must be related to magnetic fields and energetic electrons moving near the speed of light. The jet is not smooth but broken into knots spaced about 2.5 seconds of arc apart. Proper-motion studies with radio telescopes suggest that the knots are shock waves moving outward through an ionized plasma at nearly $\frac{1}{2}$ c, implying enormous generating energy. There is also a faint counterjet that points in the opposite direction.

The velocities of stars close to the brilliant, unresolved nucleus indicate the existence of a central black hole with an extraordinary mass of 3 billion $M_\odot$ that is likely the source of the jets. Matter falls into the black hole from an accretion disk (Figure 29.5a). The magnetic fields are twisted along the axis of rotation and cause mass to be expelled perpendicular to the disk in a tight bipolar flow (Figures 29.5b and 29.6). Supercomputer simulations (Figure 29.7) indicate that the jets form in turbulent, low-density tunnels in which gas blows at relativistic speeds. Both jets eventually ram into the surrounding medium and are braked to a halt.

The imbalance in the brightnesses of M 87's jets is explained by orientation. Relativity theory shows that radiation from a beam of matter moving toward the observer at nearly the speed of light will be greatly amplified, whereas that from matter rapidly moving away will be suppressed. Therefore, the central black hole is so oriented that the jets lie more or less along the line of sight (Figure 29.8, view B) and are consequently much longer than they appear in Figure 29.4.

This **relativistic beaming** reaches an extreme in the **BL Lacertae objects** (Figure 29.9a). BL Lacertae itself varies between about 14th and 17th magnitudes and flickers on time-scales of minutes. Once thought to be a variable star, it was eventual-

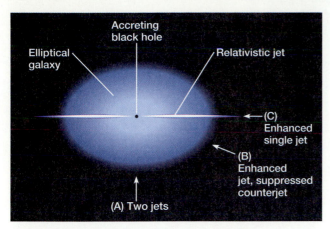

Figure 29.8   *If an active elliptical galaxy is viewed from the side (A), both jets will be visible. If viewed from an angle (B), the jet coming toward the observer will be enhanced, the one going away suppressed. If viewed head on (C), the observer sees the jet as a brilliant point set against the galaxy.*

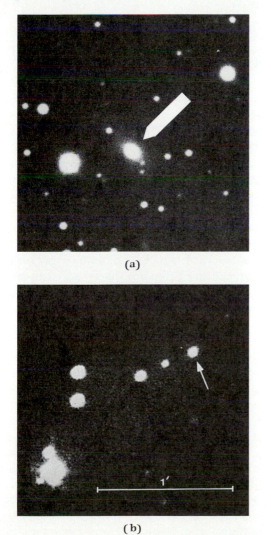

(a)

(b)

Figure 29.9   **(a)** *BL Lacertae consists of a brilliant variable stellar nucleus surrounded by the faint stellar component of a giant elliptical galaxy.* **(b)** *The BL Lac object 0317+186 appears only as a star; the surrounding galaxy is too faint to be seen.*

ly identified with a powerful radio source. These "stars" are the overwhelmingly bright centers of faint, giant elliptical galaxies. In many cases, the surrounding galaxy is too faint to be seen, and the BL Lacertae object is represented by only its bare nucleus (Figure 29.9b). The objects are thought to be oriented so that the relativistic jet is coming straight at the observer (Figure 29.8, view C), resulting in the intense amplification of radiation.

NGC 5128 or Centaurus A (Figure 29.10), the closest of the big radio galaxies, is so oriented that both jets are visible (Figure 29.8, view A). This unusual object is one of the most stunning sights of the sky, and consists of a dusty disk set perpendicular to the jets and surrounded by a huge bright halo. It appears to be the result of a collision between an edge-on spiral galaxy and a giant elliptical. Radio and X-ray images show the jets extending over 100 kpc from the center, five times the galaxy's radius (inset, Figure 29.10). Dust obscures an optical view of the nucleus, but radio observations reveal that the nucleus is similar to M 87's, that it is the source of the jets, and that it is probably a massive, accreting black hole.

In other radio galaxies, the jets extend for vast distances and end in a pair of huge lobes illuminated in the radio spectrum by the synchrontron mechanism. Cygnus A (Figure 29.11) is a brilliant celestial radio source identified with a modestly faint, distant galaxy. Radio observations show a single jet pointing straight from the galaxy into one of

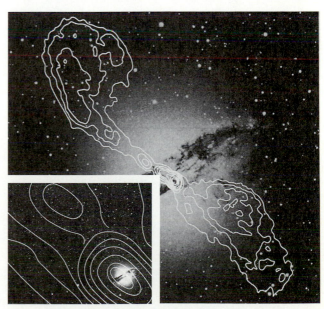

Figure 29.10   *The radio galaxy NGC 5128 is an elliptical that has collided with a spiral. Radio jets, indicated by the contours, pour out of the center. (Inset). Radio-emitting gas extends to huge distances; X-ray emission follows the same pattern.*

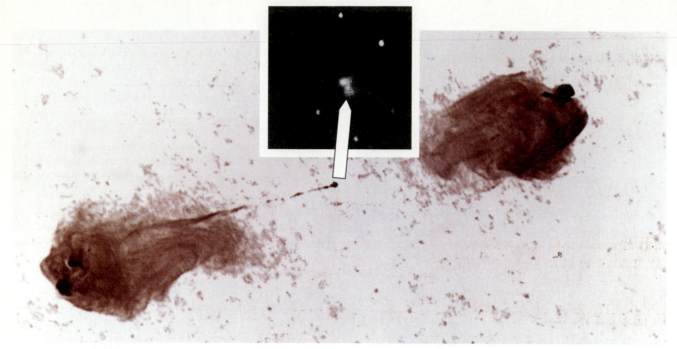

Figure 29.11 *A radio image of Cygnus A shows a jet emerging from the left side of the bright spot at the center and blossoming into a huge radio lobe. The lobe on the other side is fed by an invisible jet. (Inset). An optical close-up of the center shows a peculiar distant elliptical galaxy.*

the lobes, another example of relativistic beaming. Apparently in galaxies like these there is nothing to stop the jets until they shovel enough intergalactic medium in front of them that they brake to a halt and form the huge lobes. Computer simulations suggest speeds within the jets even greater than those in galaxies like M 87 and NGC 5128. These structures are among the largest in the Universe and can reach distances of *megaparsecs* from the galactic nuclei. The jets are the straightest struc-

tures known and can extend on a perfect line from the lobes right into the galactic cores.

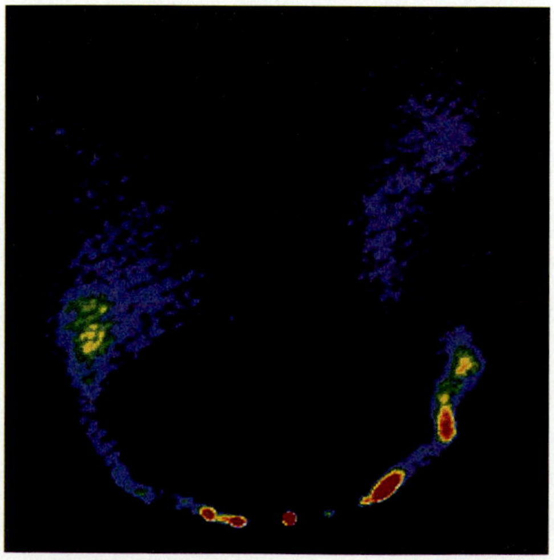

There are numerous variations within this set of double-lobed radio sources. Some beams are twisted, apparently as a result of precession within the central source. More intriguing perhaps are the *head-tail* systems (Figure 29.12), in which the jets and lobes bend away from the nuclei. These are members of clusters in which there is a significant intergalactic medium. As the radio galaxy orbits the common center of mass of the system, the medium rushes past it like a vast wind, pushing the jets backward much as a scarf is blown to the rear on a windy winter day.

29.1.3 *The Origins of Active Galaxies*

What makes a galaxy active, and why do spirals produce one set of phenomena and ellipticals another? An active galaxy requires a significant input of raw material to feed the black hole. One source of matter is the galaxy's spheroidal component, which in spirals is the halo. As a result, Seyferts are confined to spirals with tightly wrapped arms and big bulges and a greater proportion of Population II. Looser spirals—including our own Galaxy—do not have well-fed nuclei. There is a continuum of activity that depends on a continuum of mass inflow, from our Galaxy through the LINERs to the powerful Seyferts. It is not clear whether the Seyfert phenomenon is a permanent

Figure 29.12 *NGC 1265 in Perseus is a fine example of a head-tail radio galaxy, the jets and lobes blown backward by the intergalactic medium as the galaxy moves through it.*

state involving a few galaxies or whether all spirals can be Seyferts for a short time. A large amount of mass that begins to fall into the black hole in the center may suddenly trigger Seyfert activity in a quiet galaxy.

An elliptical galaxy, however, is *all* spheroidal, and the central black holes can be extremely massive. The result is a radio galaxy with powerful jets that can extend megaparsecs into space, the distance depending on how much resistance they meet. Yet not all ellipticals are active: M 32, with a massive black hole inferred from velocities of stars near its center (see Section 27.6.4), is quiet. There must be special sources of food for the black hole to produce this kind of power. In giant ellipticals that reside in clusters, like M 87, the nourishment is probably the mass stripped from other galaxies, the same mass that made the galaxy and the black hole so big to begin with (see Section 27.8). In other systems, like NGC 5128, the collision sends fresh matter into the center. Several Seyferts are apparently the products of collisions as well.

We therefore converge on a single hypothesis that unifies the variety of active galaxies, that of matter falling into a massive black hole in the nucleus. Differences among systems depend on local conditions, the mass-infall rate, and orientation with respect to the observer. Normal galaxies have an insufficient mass flow to be active, starving "the monster in the middle." This idea supplies the basic framework, but it is still under contention, and many aspects, such as the origin of the extreme variability, are far from understood. The simplicity of the hypothesis, however, lends it credence.

29.2
Quasars

Active galaxies include the most luminous objects known and provide a platform from which to leap to the edge of the Universe. We can now see galactic evolution as we look back into the depths of time.

29.2.1 Properties

In the 1950s and 1960s, great effort was spent cataloguing new radio sources and attempting to identify them with optical counterparts. Many were found to be radio galaxies, but one, 3C 48 in Triangulum (Figures 29.13a), was identified in 1960 with a blue star; another, 3C 273 in Virgo (Figure 29.13b), was identified with a short streak of light coming from yet another blue star. Spectra of 3C 48 showed this "star" to be unusual as it displayed emission lines at odd and unidentifiable wavelengths.

Early in 1963 Maarten Schmidt of Palomar Observatory examined his spectrogram of 3C 273 and found four regularly spaced emission lines also at odd wavelengths (Figure 29.14a). He suddenly realized they were just lines of the hydrogen Balmer series redshifted by a then-remarkable 16%. A look back at 3C 48 showed that its lines were hydrogen as well, but with $z = 0.37$, a value so high that no one had thought of a redshift as an explanation. These objects, and others quickly found, were definitely *not* stars, and were dubbed **quasars,** short for "quasi-stellar radio sources."

Shortly after the discovery of quasars, optical surveys (in which astronomers looked for these objects on the basis of their blue colors) revealed that over 90% of the quasars are radio-quiet. The term quasar is then a misnomer, and the sources are now known as quasi-stellar objects, or QSOs, instead. Nevertheless, quasar is still used as a general term that includes the radio-quiet ones.

All quasar redshifts are large. Values over 1 are common, and many quasars have z over 2, the ultraviolet redshifted into the visual spectrum (Figure 29.14b). The current record is $z = 4.897$, in

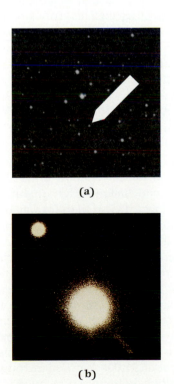

(a)

(b)

Figure 29.13   *An image of the quasar 3C 48 shows it to be stellar.*
(b) *3C 273, the brightest and closest quasar, is the source of a short jet.*

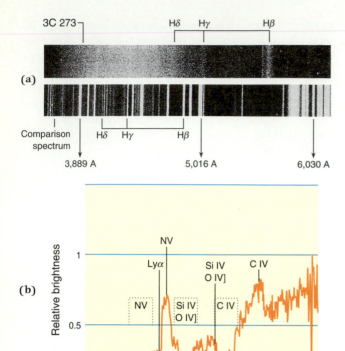

Figure 29.14   (a) *The spectrum of 3C 273 shows four hydrogen Balmer lines, Hα through Hδ, red-shifted by 16%. The rest positions of the lines are shown at the bottom.* **(b)** *Quasar Q0051-279, z = 4.43, recedes so fast that the Lyman α line at 1,216 Å is shifted into the red part of the spectrum.*

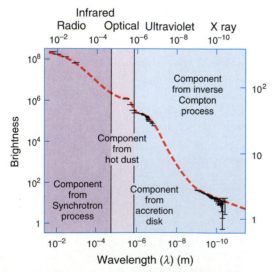

Figure 29.15   *The spectral intensity of the discovery object, 3C 273, drops steadily from the radio roughly according to* ν^{-1} *and is typical (except that most quasars are radio-quiet). The origin of the radiation observed in different wavelength bands is discussed in the text. Adapted from "The Quasar 3C 273," by Thierry J.-L. Courvoisier and E. Ian Robson from* Scientific American, *June 1991, page 53. Copyright © 1991 by Scientific American, Inc. All rights reserved.*

which Lyman α, normally at 1,216 Å, appears red at 7,149 Å!

Some quasars are surrounded by faint material or, like 3C 273, have jets, but the bright central cores all appear stellar. Very long baseline interferometry shows them to be unresolved down to 0.0001 seconds of arc. Quasar spectra all exhibit broad hydrogen emission lines, other broad lines from elements like carbon (see Figure 29.14b), and narrow, forbidden lines superimposed on a continuous spectrum. The emissions must be produced by low-density gas that presumably surrounds the central source. The continuum extends shortward of the visual into the ultraviolet and the X ray and longward into the infrared (Figure 29.15) according to a rough power law. In a few cases, it even extends into the radio, where it produces the classic quasars.

Quasars are also erratically variable over periods of years (Figure 29.16a) and over intervals as short as 10 days (Figure 29.16b). The central power sources must therefore be smaller than 10 light days, or about 2,000 AU, consistent with their stellar appearances. Quasars also occasionally undergo violent outbursts in which their brightnesses can increase by a factor of 100 over only a few months (see Figure 29.16a).

The spectra of quasars commonly exhibit absorption lines as well as emissions. These are also redshifted, but with z always less than or equal to the redshifts displayed by the emission lines. Sometimes there are several sets of absorptions, each with a different redshift.

29.2.2 Interpretation and Controversy

The discovery of quasars launched a controversy over interpretation that rages yet today. What kind of object could have such huge redshifts? The major clue is that *all* quasar spectra are redshifted. The most obvious conclusion is that quasars take part in the Hubble flow and are therefore at **cosmological distances,** distances that span a large portion of the Universe and that relate to cosmology.

If the redshifts do not exceed a few tenths, we can use the ordinary Doppler formula to determine recession velocities, and find distances (d) from $d = v/H_0$ and an adopted value of the Hubble constant (see Section 28.5.1). The nearby quasar 3C 273 ($z = 0.16$) recedes at 40,000 km/s and (for $H_0 = 75$ km/s/Mpc) is 640 Mpc away; 3C 48 recedes at about 91,000 km and is roughly 1.2 Gpc (nearly 4 billion light-years) away. To find velocities and distances for quasars with large z, however, we also need a model for spacetime curvature.

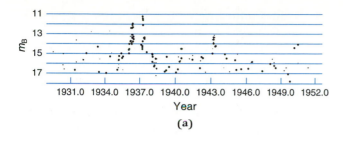

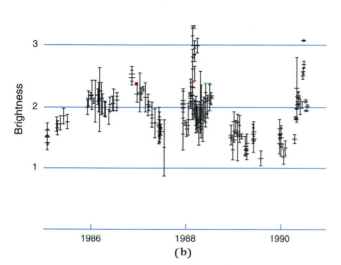

Figure 29.16   (a) *The light curve of 3C 279, reconstructed from old photographic plates, shows considerable variation as well as powerful outbursts over a 20-year interval.* (b) *A short-term examination of 3C 273 reveals variations over a mere ten days.*

Since there are several possible models (to be described in Chapter 30), we are unable to assign unique values. The velocity and distance of Q0051-279 (see Figure 29.14) is therefore indeterminate. However, the velocity must be very high—close to that of light—and its distance must vastly exceed that of any observed galaxy.

The apparent visual magnitude of 3C 273 is 12.9. If the object is indeed at a cosmological distance, then its absolute visual magnitude is –26.1, a phenomenal 2×10^{12} times more luminous than the Sun, and 100 times more luminous than our whole Galaxy! Other quasars are similar, and if we take nonoptical radiation into account (see Figure 29.15), they are more luminous yet. The current record holder, a quasar with a redshift of 4.7, is estimated to radiate at 10^{42} joules/s, over 10,000 times the energy radiated by our Galaxy, all pouring from a volume not much larger than our Solar System.

There are two other possible explanations for quasars. First, the enormous luminosities implied by cosmological distances suggested to some astronomers that quasars could *not* be far away, and had to be "local." Perhaps they are small, high-

speed bodies ejected from the core of our Galaxy in some major calamity. But if that were true, we would expect some quasars to be associated with other galaxies, and we would then also expect to see blueshifts from those ejected *toward* us. The second possibility is that the redshift is not caused by the Doppler effect but by something else, perhaps gravity. Some combination of the two possibilities might also be involved.

For a number of strong reasons, however, most astronomers believe quasars to be at cosmological distances. First, some quasars seem to be allied with other galaxies or clusters of galaxies that have the same redshift (Figure 29.17). Second, the absorption lines can be interpreted as originating in clouds of low-density gas—even galaxies—lying along the line of sight to the distant but brilliant sources of continuous emission. They are best seen as huge numbers of absorptions to the short-wavelength side of the redshifted hydrogen Lyman α emission line, a phenomenon called the **Lyman alpha forest** (Figure 29.18a). Any cloud that lies in front of a quasar must be closer to us and receding more slowly as a result of the Hubble flow. In Figure 29.18b (ignoring the effects of spacetime curvature), cloud 2, half the distance of the quasar, will produce an absorption line with z equal to half that of the quasar; clouds 1 and 3, respectively one-quarter and three-quarters the distance, will have $z = \frac{1}{2}$ and $\frac{3}{4} z$(quasar). A thousand clouds will give a thousand lines.

Third, deep imaging reveals that a number of quasars are surrounded by faint clouds of fuzz (see Figure 29.17) that have the spectra of stars with the

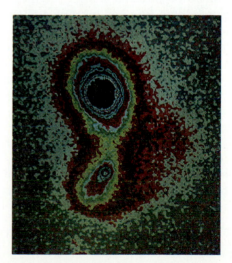

Figure 29.17   *The quasar 0351 + 026 (top) appears to be tidally distorting a galaxy below it. It is surrounded by fuzz that may be a galactic disk. The quasar, the galaxy, and the fuzz have identical redshifts, establishing the quasar at a cosmological distance.*

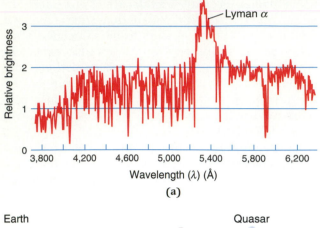

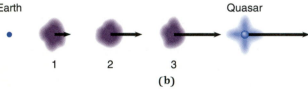

Figure 29.18 **(a)** *Distant quasars such as S5 0014+81, z = 3.4, display a vast number of absorption lines to the blue side of Lyman α.* **(b)** *The Hubble flow forces intervening clouds and galaxies to move more slowly along the line of sight (at velocities given by the arrows), thus producing Lyman α absorptions with smaller Doppler shifts.*

same redshifts as the quasars themselves. In these instances, we interpret the quasars as the brilliant nuclei of galaxies and the fuzzy, resolved outskirts as galactic disks or halos. We then make the assumption that the quasars without the accompanying fuzz are also galactic nuclei, but that the surrounding galaxies are too faint to be seen.

Finally, the similarity between most quasars and Type 1 Seyfert galaxies is striking. If we could make a Seyfert's nucleus a little brighter and the disk a little fainter, it would become a quasar. Similar evidence is provided by some quasars that are associated with double radio sources, showing there is another subset related to elliptical radio galaxies (Figure 29.19). Even the nucleus of Cygnus A (see Figure 29.11), the prototype of radio galaxies, has the characteristics of a quasar buried in the galaxy's dust. Moreover, the first discovered quasar, 3C 273, possesses a jet (see Figure 29.13b) reminiscent of the one emanating from M 87.

These discoveries take some of the mystery out of the quasars and leave us on familiar ground. Quasars are not different kinds of objects but are active galaxies with hyperactive nuclei. The Seyferts and radio galaxies are quasars in which the surrounding galaxy can easily be seen. The power source of the quasar is interpreted as a massive black hole accreting matter from its surroundings.

The radio luminosity comes from the synchrotron mechanism (see Figure 29.15), the infrared from hot dust (as it does for Seyferts), the optical and some of the X ray from the hot accretion disk that surrounds the black hole, and another part of the X ray from the *inverse Compton effect*. In this process, radio photons hit electrons moving near the speed of light, boosting the photons' energies into the X-ray region. The emission lines come from ionized clouds of matter orbiting outside the accretion disk. The models for quasars are still far from perfect, but at least we now have a rational, working explanation for them.

29.2.3 Number and Evolution

There are too many quasars to catalogue, so astronomers must use statistical and sampling methods. To blue magnitude 22.5, we find about 100 per square degree, producing a total number (ignoring obscuring dust in the Milky Way) of around 4 million. At 28th magnitude, near the limit of observation, there should be 30 times as many. Still, this number is only a fraction the count of normal galaxies. The quasars may account for a large portion, perhaps up to half, of an observed diffuse X-ray background (Figure 29.20), the remainder coming from starburst galaxies and perhaps from as yet unrecognized sources.

We actually see significant numbers of quasars only when we look back in time to when the Uni-

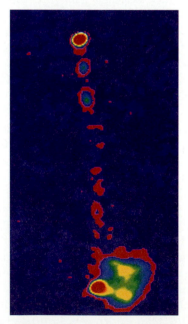

Figure 29.19 *Quasar 1007+417 has the appearance of a radio galaxy with a relativistically beamed jet that terminates in a bright lobe.*

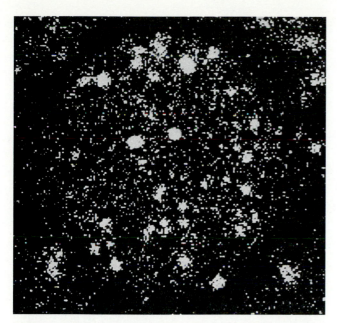

Figure 29.20 *Over 40 quasars are seen in an X-ray view of the sky less than a degree across.*

verse was young. Beyond about $z = 2$, which corresponds to a distance over two-thirds of the way back to the time of the Big Bang (Chapter 30), the count of quasars begins to drop off. They are rare above $z = 3$ and almost unknown above $z = 4$. More-distant quasars, those with the higher redshifts, tend strongly to be the more luminous.

We may be witnessing an evolutionary sequence. The most distant quasars, those with the highest redshifts, are the youngest. As we proceed inward from the edge of the visible Universe, we look less and less far back into time and the objects we see are progressively older. The first things to develop were not ordinary galaxies but brilliant quasars. The act of formation may have sent vast amounts of matter inward to feeds growing black holes. As the quasars aged, the amount of infalling mass dropped and the activity dimmed. At the same time, star formation began to create visible galactic halos and disks, their young stars making them at first appear very blue. Some quasars dimmed into Seyfert and radio galaxies. Others perhaps turned into more ordinary systems, those like our own, with relatively inactive nuclei. There may be a nearly dead quasar at the center of our Galaxy behind the vast dust clouds of the Milky Way.

What other paths of evolution there may have been is unknown. Observations at high redshifts are exceedingly difficult, as apparent brightnesses are low and resolution poor. Further advances in telescopes and detectors should lead us to a better understanding of how our Galaxy of today came to be.

29.3
Gravitational Lenses

There is yet more evidence that quasars are at cosmological distances. In 1979 University of Arizona astronomers discovered two quasars in Ursa Major—Q0957 + 561 A and B—only 6 seconds of arc apart (Figure 29.21a). More surprisingly, their spectra are identical, as are their redshifts of 1.4136. A true binary can quickly be ruled out: the components would be somewhat different and we would expect a velocity difference because of orbital movement.

The theory of relativity provides an explanation. Spacetime is bent, and light deflected, by a gravitational field. The double quasar is produced by an intervening **gravitational lens,** perhaps a

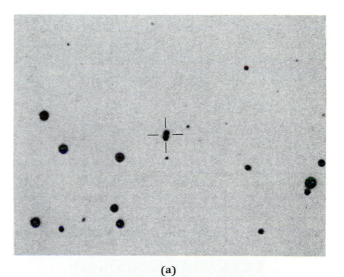

(a)

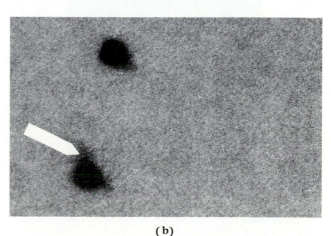

(b)

Figure 29.21 **(a)** *A photograph with the Palomar Schmidt shows a double quasar, Q0957 + 561 A and B, in Ursa Major.* **(b)** *A magnified view with a larger telescope reveals an intervening giant elliptical galaxy (arrow), which has gravitationally focused and split the light of a single distant quasar.*

galaxy, that gravitationally deflects the light and focuses it, causing a distant *single* quasar to *appear* double. Deep imaging of Q0957 + 561 A and B in fact found the lensing object (Figure 29.21b), a giant elliptical situated between the two quasar images. With $z = 0.36$, it is considerably closer than the quasar.

Figure 29.22a shows how an ideal gravitational lens works. A quasar, a point source of radiation, is positioned behind an extended galaxy. The total mass that affects a light ray increases as we proceed from the galaxy's center to its edge. As a result, the rays that pass farther from the center are increasingly deflected, and the light is focused. An observer on Earth would see light coming from all around the galaxy in the form of an *Einstein ring*. If the alignment is shifted slightly, the ring becomes a pair of asymmetrically placed arcs, one considerably brighter than the other (Figure 29.22b). If the alignment is offset a little more, the arcs become more like actual point images of the source and the placement becomes more asymmetrical (see Figure 29.21). Because the intervening galaxy is extended,

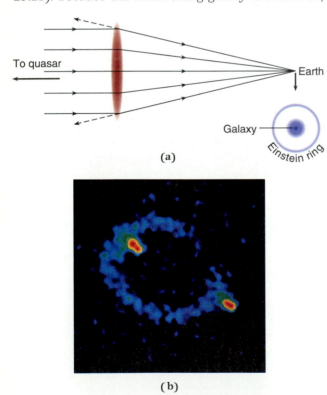

(a)

(b)

Figure 29.22 **(a)** *A distant single quasar lies directly behind an intervening mass. The farther away from the center the light hits, the more mass it "sees" and the more it is deflected. The observer then views the quasar as a lensed Einstein ring around the galaxy.* **(b)** *This radio image of a gravitational lens is a fair approximation to the ideal Einstein ring.*

Figure 29.23 *The Hubble Space Telescope views an Einstein cross, a configuration in which an intervening galaxy at the center of the image splits the light of a distant quasar into four components.*

we might also see multiple images like the *Einstein cross* (Figure 29.23). Depending on alignments, the brightness of the quasar can also be amplified. Such gravitational lenses would not be possible if the lensed quasars were not at great, and probably cosmological, distances.

Gravitational lenses are of considerable importance. First, they can be used to probe the distribution of matter in the Universe, as the intervening mass lenses the light of distant quasars and galaxies into numerous arcs (Figure 29.24). The technique allows an assessment of the amount of dark matter within clusters of galaxies and even allows it to be mapped. Dark matter up to 50 times as abundant as luminous matter has been found in this way.

Second, lensed quasars can be used to derive the Hubble constant. Passage of the light past the lensing galaxy not only deflects and amplifies the rays but also retards them. Since the different images result from different gravitational field strengths, one image will arrive after the other. Quasars are variable, and consequently the images will vary, but out of synchronism with each other: that is, one image will change first and the other will follow. The delay depends on the positioning and distribution of mass of the intervening galaxy and on the distances of the two objects.

The effect has actually been seen and measured for the original lensed quasar, Q0957 + 561: the B variation follows that of A by about 500 days. The placement of the lensing galaxy is known, so the delay gives the quasar's distance, which coupled to its redshift yields H_0. The investigation epitomizes the problem with H_0. The first analysis of the data gave $H_0 = 100$ km/s/Mpc. Another astronomer, with the *same data,* found a slightly dif-

ferent delay time, and using a different mass distribution, derived 50 km/s/Mpc! Improved observations and measurement of the phenomenon for
other lenses might finally settle the issue.

29.4
An Alternative View

Not all astronomers agree that all or even most of
the quasars are galaxies or are at cosmological distances. Some yet hold to some form of "local"
hypothesis, a view still supported by a variety of
data. First, very long baseline interferometer measurements of some quasars show blobs of emission
that move (Figure 29.25). If the quasars are very
far away, the speeds of the blobs are calculated to
be moving faster than light, that is, they are *superluminal*—which by the fundamental standards of
modern physics is impossible. The effect can be
explained as an illusion caused by the ejection of a
blob from the core at a speed *near* that of light in a
direction almost exactly toward the Earth. However, the supporters of the local hypothesis feel that
there are too many of these coincidences to accept,
and therefore the quasars must be close enough to
render the speeds below that of light.

Of considerably more interest are a variety of
observations that relate quasars and galaxies of different redshifts to one another. Figure 29.26a shows
three quasars nestled within the arms of the barred
spiral NGC 1073. The chance of a coincidental

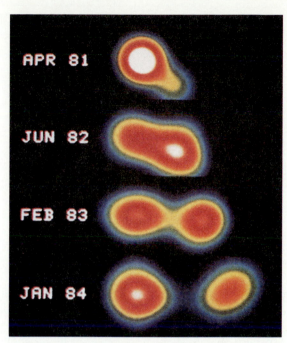

Figure 29.25   *Two blobs of radio-emitting material within 3C 345
separate at a speed that—if the quasar is at a cosmological distance—appears faster than that of light. The blob has actually been
ejected almost toward the Earth at near the speed of light and only
seems to go faster than c.*

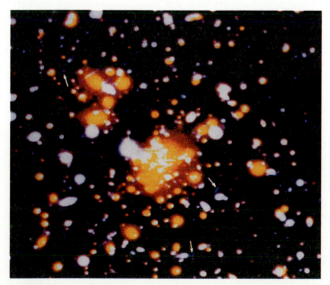

Figure 29.24   *Dark matter in a galaxy cluster lenses more-distant
galaxies into arcs (arrows).*

alignment is argued to be impossibly low. In other
instances (Figure 29.26b), bridges of luminous matter apparently connect quasars and galaxies. In
both cases, the conclusion is that the quasars are
local to the nearby galaxies and are not cosmological. Yet the quasars have redshifts very different
from those of their associated galaxies. Perhaps the
quasars were ejected from the galaxies at high
speeds, or perhaps some redshift mechanism other
than the Doppler effect or the cosmological expansion of space is involved. Even if some quasars *are*
at cosmological distances, the velocities derived
from the Doppler effect and the distances estimated from the Hubble flow cannot then be trusted.

These ideas face formidable challenges. If
quasars have high redshifts because they have been
violently expelled from galaxies, the question
remains: Where are the blueshifts of those that
should have been ejected toward us? Perhaps the
quasars radiate only through their fleeing tails, in
the manner of the flame from a jet engine. If so, we
will see only those going away and none of the
ones that approach. There is no physical foundation for such a phenomenon, however. Furthermore, there are no good alternative explanations
for the redshifts. Gravity does not seem to be signif-

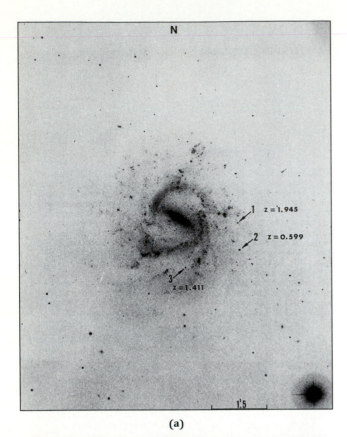

(a)

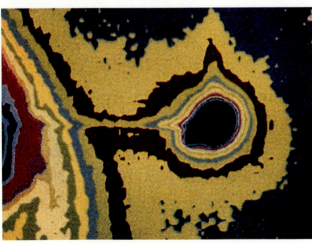

(b)

Figure 29.26 **(a)** *Three quasars with quite different high red-shifts appear to be associated with the nearby spiral NGC 1073, which has a very low redshift.* **(b)** *Mk 205 (center) appears to be connected to the galaxy NGC 4319, part of which is seen at far left. The redshifts of the two are quite different.*

icant. The astronomers who accept cosmological distances dismiss the associations between quasars and galaxies as the coincidences that are expected given the enormous numbers of such bodies in the sky. The statistics of such associations are difficult to treat, and each side has been able to use statistical analyses to support its own theories. The weight of evidence seems to fall heavily toward the cosmological side of the argument, but it must be admitted that all the answers are not in.

This subject and its controversies are not yet finished. As we approach the conclusion of our investigations in the next (and last) chapter, where we discuss the theories of the origin and evolution of the Universe, challenges to conventional ideas will rise again, as they should in any developing science.

KEY CONCEPTS

Active galaxies: Galaxies with bright, variable nuclei sometimes associated with bipolar jets; they are probably caused by matter falling into massive black holes from accretion disks.

BL Lacertae objects: Bright nuclei of active elliptical galaxies from which a jet is coming directly at the observer.

Cosmological distance: A large distance that relates to the expanding Universe and thus to cosmology.

Gravitational lens: Relativistic distortion or amplification of the light from a distant quasar caused by the gravitational field of an intervening galaxy.

LINERs: Low-ionization emission line regions; probably low-activity Seyferts.

Lyman alpha forest: Absorption lines observed in quasar spectra to the short-wavelength side of the Lyman α emission line; caused by clouds of matter between our Galaxy and the quasar.

Quasars: Quasi-stellar radio sources (quasi-stellar objects); bright, starlike objects with high redshifts; some are strong radio emitters. They are probably distant, young versions of active galaxies.

Radio galaxies: Active elliptical galaxies with powerful central radio sources, bipolar jets, and, sometimes, extended double radio sources.

Relativistic beaming: Amplification of the radiation from a jet moving near the speed of light toward the observer.

Seyfert galaxies: Active spirals with abnormally bright nuclei whose centers radiate strong emission lines.

EXERCISES

Comparisons

1. List the similarities and differences between (a) active and nonactive galaxies; (b) Types 1 and 2 Seyfert galaxies; (c) a Type 1 Seyfert and a LINER; (d) radio galaxies and Seyfert galaxies.
2. What does the nucleus of our Galaxy have in common with active galaxies?

Numerical Problems

3. The magnitude of the nucleus of a distant galaxy varies with a period of five days. What is the maximum size of the emitting region in AU?
4. What is the angular diameter of a quasar that has a physical diameter of 1,000 AU and is 1 Gpc away?
5. What would you expect the redshift of the quasar in the previous question to be? Explain your assumptions.

Thought and Discussion

6. Describe the characteristics of BL Lacertae objects.
7. How does orientation affect the jets from active galaxies?
8. What do we think is the driving force behind active galaxies?
9. What are head-tail radio galaxies and what produces them?
10. List the defining characteristics of quasars.
11. Explain the possibilities for the interpretation of quasars.
12. What is the evidence that quasars are at large distances?
13. What is the Lyman α forest? What is its significance?
14. Why might quasars be distant Seyfert galaxies or radio galaxies?
15. What do gravitational lenses prove about quasars?

16. How are quasars useful as probes of the Universe?
17. How can quasars be used to determine the Hubble constant?
18. Under what conditions can an Einstein ring be formed?
19. Why do we believe that quasars evolve into normal galaxies?
20. What is the evidence that quasars are local phenomena?

Research Problem

21. Examine science and astronomy magazines from about 1963 on and summarize the changes that have taken place in astronomers' views of quasars as new data were acquired.

Activities

22. Draw a poster that explains how the kinds of active galaxies and quasars may be related.
23. Create your own "gravitational lens" using optics in place of gravity. Point the stem of a wine glass at a small, bright light. The wine glass is the galaxy and the light the quasar. You will see a ring of light around the stem similar to the Einstein ring. Then displace the center of the glass slightly from the light and you will see a pair of arcs that become progressively more pointlike. Make notes describing what you see.

Scientific Writing

24. Write an essay in which you draw a parallel between the way early astronomers brought order out of the observations of stars and the way modern astronomers are bringing order out of the chaos of the different kinds of active galaxies.
25. Find another topic in astronomy in which there are alternative points of view. Compare the two views and render an opinion as to the correct one.

30

The Universe

How the Universe is constructed, how it began, and what its fate may be

Einstein and Hubble, placed against distant galaxies of the expanding Universe.

We now arrive at the culmination of our voyage through astronomy and space. Gathering all our knowledge, we attempt to create theories about the Universe, the grand concept that incorporates everything, to understand its structure, origin, and evolution. We also enter an arena of uncertainty in which astronomers seek the origins of space and time.

30.1
A Fundamental Observation

You can make a simple cosmological observation from your backyard. The nighttime sky is dark. Why?

Adopt the simplest possible Universe. First, assume it is **infinite** in size. Infinity does not simply imply a big number, but means never-ending: no matter how large the number, you can always add another. Assume also that the Universe is populated uniformly with stars and galaxies and is static in time and space so that nothing moves or evolves. Construct thin concentric shells around the Earth (Figure 30.1). The volume of a shell depends on its surface area ($4\pi r^2$, where r is the shell's radius) times the shell's thickness. The number of galaxies—and stars—within each shell therefore depends on r^2. However, the observed brightness of a star decreases as $1/r^2$. The increased number of stars within a larger shell is exactly compensated by the decrease in an individual shell's apparent brightness. As a result, each shell sends the same amount of light toward the Earth. Since there are an infinite number of shells, the sky should be infinitely bright. Such brilliance is actually reduced by the angular diameters of the stars, as a nearby star will block the light of a more distant one. Any line of sight, however, will eventually encounter some star, and the sky should be as bright as the surface of the Sun.

It is not. In fact, the night sky is so dark that no general optical background from the combined light of distant galaxies has ever been detected. Discussion of this apparent contradiction can be traced to Edmund Halley and maybe even to Kepler. Among the many who examined it was H. Wilhelm Olbers, a nineteenth-century German physician and astronomer, and it has been known as *Olbers' paradox* ever since.

We know by naked-eye observation alone that at least one of the initial assumptions must be wrong. But which one or ones? The expanding Universe plays a small role. The Hubble expansion reddens the radiation from distant stars and reduces

the energies of their photons. As a result, a distant shell filled with stars will provide somewhat less light than a nearby shell. By far the principal reason for nighttime blackness, however, is simply that stars do not live forever. As we look outward, we also look back in time toward the moment when galaxies and their stars were first created. To see a sky filled with stars we would have to look farther in light years—and farther back in time—than the ages of the stars, farther than the age of the Universe itself. Stars therefore cannot fill the sky and it is dark. We still, however, cannot tell if the Universe is infinite in space or time. So we dig deeper.

30.2
Cosmological Principles

A set of three theoretical principles provides frames within which to construct possible pictures of the Universe. Long ago, we learned that we are not at the center of the Solar System, and in our own century discovered we were not even centered within our Galaxy. We are near one edge of a small cluster of galaxies, the Local Group, which is on the edge of a much larger supercluster. The location of the Earth is not special, a concept called the **Copernican Principle** after the man who displaced humanity from the center of attention.

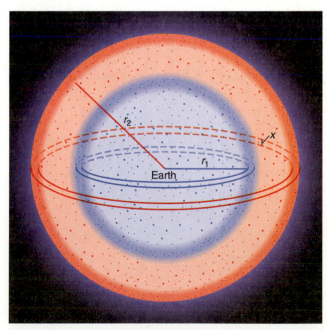

Figure 30.1   *An outer shell of radius r_2 and thickness x (red) has a volume $4\pi r_2^2 x$ and holds $(r_2/r_1)^2$ as many galaxies as a similar inner shell of radius r_1 (blue). Since the apparent brightness of a star at the Earth (center) declines as $1/r^2$, each shell contributes the same amount of light to the nighttime sky.*

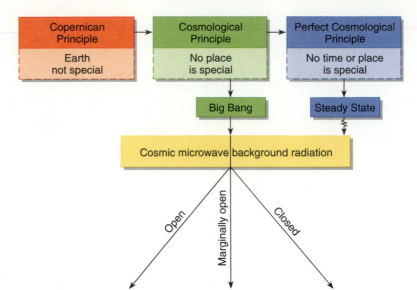

Figure 30.2   *The chart summarizes properties of three fundamental principles. Cosmic background radiation disallows the Perfect Cosmological Principle, but allows the Cosmological Principle and the Big Bang. Three possible structures for the Universe—open, closed, or marginally open—are discussed in the text.*

We appear to live in an expanding Universe filled with galaxies as far as we can see. The linearity of the expansion shows that we would see the same kind of expansion from any galaxy in which we might live. Furthermore, over distance scales of a few hundred megaparsecs, the lumpiness of the Universe smooths out and its structure becomes uniform and isotropic. These observations lead to a broader concept, the **Cosmological Principle,** which states that the Universe must look essentially the same (ignoring local effects) from *any* vantage point, that there are *no* special places. The extension of our observations through the Cosmological Principle means that the Universe can have neither edge, boundary, nor center.

The Cosmological Principle addresses space. What about time? Either the Universe has existed for all time or it has not; it may or may not have an end. A timeless universe is defined by a yet broader concept, the **Perfect Cosmological Principle,** which states that the Universe is the same from all vantage points in space *and* time. If the Perfect Cosmological Principle is correct, the Universe will look the same no matter from *where* you look and no matter *when,* whether vastly long ago or vastly far into the future.

Big Bang theory incorporates a beginning to the Universe as well as its evolution, maintains the Cosmological Principle, but contradicts the Perfect Cosmological Principle. Formulated in response to the redshift-distance relation found by Hubble and from theory derived from general relativity, it was intensively developed by George Gamow of Washington University and his group in the late 1940s. An alternative, which embraces the Perfect Cosmological Principle, is the *Steady State theory,* proposed in 1948 by Hermann Bondi, Thomas Gold, and Fred Hoyle (who scornfully coined the phrase "Big Bang"). The Steady State theory also incorporates the expanding Universe. Clusters of galaxies move farther apart over time, and the spaces between them enlarge. To keep the Universe always looking the same, new matter, which condenses into new galaxies, must continually be created within the spaces.

We do not know how new matter could be so created, but ignorance cannot rule out such a process. Nor can we rule out continuous creation with laboratory tests, since to fill the space left by the expansion of the Universe requires only that three or four new atoms appear per year per cubic

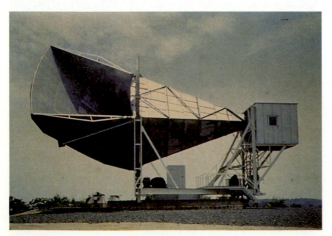

Figure 30.3   *The cosmic background radiation was discovered with this antenna at Holmdell, New Jersey.*

kilometer. The three principles and possible states of the Universe are summarized in Figure 30.2. To determine which are correct, we must expand on the observations.

30.3
Cosmic Background Radiation

In 1965 Arno Penzias and Robert Wilson of Bell Laboratories observed the radio sky at a wavelength of 10 cm with a high signal-to-noise ratio (see Section 19.1) and made a discovery for which they won a Nobel prize. No matter where they pointed their telescope (Figure 30.3), there was always a minimum, uniform level of radiation unassociated with any of the usual galactic or extragalactic radio sources. They found this **cosmic microwave background radiation** everywhere, filling all space. Further measurements showed that this radiation has the characteristic spectrum of a cold blackbody near a chilling 3 K. It is consequently also known as the *3° background radiation*. The *Cosmic Background Explorer, COBE* (Figure 30.4), launched by NASA in 1990 to explore the background radiation free of the deleterious effects produced by the Earth and its atmosphere, found that its average spectrum does not deviate from that of a blackbody at 2.735 K by more than 0.04 percent (Figure 30.5).

The cosmic background radiation is remarkably isotropic. The temperature is a mere 0.003 K higher in the direction of Leo and 0.003 K cooler oppositely, toward Aquarius, as seen in the color-keyed all-sky image in Figure 30.5. This variation is interpreted as a Doppler shift (which changes the peak of the blackbody curve and gives the illusion of temperature change) caused by the combined relative motions of the Sun to the Galaxy, the Galaxy to the Local Group, the Local Group to the

Figure 30.4   *The Cosmic Background Explorer (COBE) orbits the Earth, observing the sky in the radio and infrared.*

local supercluster, and the local supercluster to large-scale homogeneous structure of the Universe, all of which make the Sun move toward Leo at about 620 km/s. When these local motions are accounted for, the background radiation map in Figure 30.5 evens out. There is more structure on yet finer scales that will be explored below.

Cosmic background radiation had actually been predicted in 1948 by Gamow's collaborators Ralph Alpher and Robert Hermann as the cooled remnant of the hot fireball created in the Big Bang. The Universe therefore appears to be changing with time. This conclusion is strongly supported by the apparent evolution of galaxies and quasars as we look to distant, and therefore younger, reaches of the Universe. The 3° radiation is the most important evi-

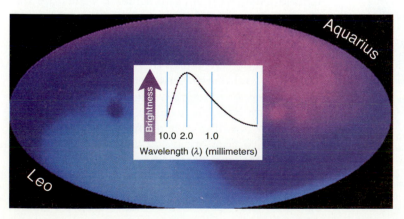

Figure 30.5   *The average spectrum of the cosmic background radiation is a blackbody at 2.735 K. The curve graphed here peaks at 2 mm rather than 1 mm as predicted by the Wien law because units of frequency are used along the lower axis instead of units of wavelength. The colors show a slight change in temperature caused by the Doppler motions of the Sun, Galaxy, and Local Group, blue approaching and red receding.*

dence that the Big Bang actually occurred. We therefore exclude the Perfect Cosmological Principle and the Steady State theory, and accept the Cosmological Principle and the Big Bang as basic premises.

30.4
The Structure of the Universe

Given the Cosmological Principle and the Big Bang, we must deal with the origin of the Universe and with a variety of possible structures. These relate to whether the Universe will ever come to an end.

30.4.1 *Possible Expansion Models*

There are three possibilities for the structure of the Universe. They can be described by the behavior of a ball thrown in the air. If the ball's velocity is less than the escape velocity, it will return. The ball's path or orbit is said to be *closed;* that is, it comes back upon itself. If the velocity is greater than the escape velocity, however, the ball will never return: the path is now *open.* If the two are equal, the ball will coast to a stop, but only after an infinite time when it has reached an infinite distance from Earth. The path could then be said to be *marginally open.*

The expansion of space must similarly be affected by gravity (Figures 30.2 and 30.6), which by its attractive nature must act to slow the expansion over time. Look at a shell like one of those in Figure 30.1. It is large enough that the distribution of matter is smooth—that is, it is larger than the largest clumping scale. Consequently, this shell is

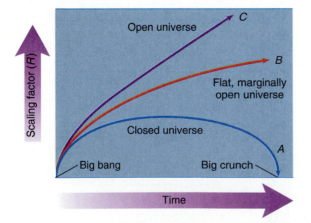

Figure 30.6 *Curves A and C represent the way in which space expands, represented by the scaling factor R, for the two Friedmann-Lemaître universes. In A, gravity will stop the expansion and bring the closed universe back to collapse in the Big Crunch. In C, the open universe expands too fast to be brought back, and will always expand. The Einstein-de Sitter universe (B) is barely open, bringing the rate of expansion to zero only after an infinite amount of time.*

like any other and can represent the Universe at large. As it expands, the shell has an escape velocity that depends on the mass of matter enclosed within it and on its size, or on the average density of its contained matter. If its outbound velocity is less than the escape velocity, it will slow to a halt in a *finite* (less than infinite) time, and we have a **closed universe.** If our Universe is closed, it will someday begin to shrink, its scaling factor, R (see Section 28.2), decreasing rather than increasing. It is fated eventually to contract to the dense structure from which it came in a *Big Crunch.* How long that will take depends on the density of matter.

If, however, the outbound velocity is greater than the escape velocity, the shell and the Universe will expand forever and we have an **open universe.** As it expands it cools, heading for what has been called the *Big Chill.* As in the example of the thrown ball, there is also a dividing case of a **marginally open universe:** it will coast to a stop, but only after an infinite amount of time has elapsed, and it will never contract. We believe we live in one of these kinds of universes. But which one?

30.4.2 *Curved Spacetime*

These arguments are based on a Newtonian view of gravity. Similar concepts are derived from the equations of general relativity. The characteristics of closed and open expanding universes were developed by the Russian mathematician Alexander Friedmann and a Belgian scientist and priest, Abbé Georges Lemaître, in the 1920s. These cases are therefore sometimes called **Friedmann-Lemaître universes.** The relativistic solution for the dividing case, the marginally open universe that comes to a stop only in infinite time, was found in 1931 by Einstein and Willem de Sitter and is called the **Einstein-de Sitter universe.** (There are many other possibilities, but these are considered to be the most realistic.) Gravity, which controls the rate of expansion and the degree of closure, is the result of curved spacetime. Each of the three possible universes described is therefore associated with a unique kind of curvature that actually controls the expansion and determines whether the Universe is open, closed, or marginally open.

It is not possible to draw examples of curved spacetime because it is a four-dimensional construction. To visualize it and to examine its characteristics we must rely on three-dimensional analogues. Look first at the most familiar case. One of the five postulates, or unprovable assumptions, on which the early Greek mathematician Euclid based plane geometry states that "through a point not on a straight line, one and only one straight line may be drawn parallel to the given line." (Parallel lines stay the same distance apart if drawn even to infinity.) This statement is expected to be true only if spacetime is not curved, a condition in which space is called *flat* or *Euclidean* (Figure 30.7a). However, if spacetime is curved over the whole Universe, Euclid's parallel postulate is not true, and we must consider a variety of *non-Euclidean geometries* developed by the great mathematicians (Gauss, Riemann, and others) of the eighteenth and nineteenth centuries. Spacetime might be curved *negatively* or *hyperbolically;* the three-dimensional analogue is a saddle (Figure 30.7b) that curves outward to an infinite distance. In that case, there can be *more* than one line through a point parallel to a given line. It is equally possible that spacetime is curved *positively,* closing back on itself in the four-dimensional equivalent of a sphere (Figure 30.7c). If so, then *no* line can be drawn through a point parallel to a given line, as the lines must inevitably converge.

A test of curvature can be made by summing the angles of a sufficiently large triangle. In Euclidean space, the angles add to 180°, but in hyperbolic space to less than 180°, and in spherical space to more than 180°. In the real Universe, any curvature would be detectable only over vast distances. On Earth, or within the Local Group for that matter, our geometry would look Euclidean even if the Universe as a whole is hyperbolic or spherical.

The three kinds of curvature represented in Figure 30.6 make the terms *open* and *closed* more meaningful. If the Universe is destined to expand forever and is open, it is also hyperbolic, infinite, and geometrically open. If it is closed and destined for the Big Crunch, it is spherical and finite. If the Universe is borderline between the two cases and marginally open, space is flat and Euclidean, but still infinite.

If the Universe is open or flat, it clearly has no center. Such a conclusion about a closed universe is not so obvious. Figure 30.8 again represents a four-dimensional spherical universe in three dimensions. Its inhabitants (its observers) are represented by two-dimensional flat people on the sphere's surface and its star systems by flat galaxies. The observers can move only *around* one another: there is no over and under. They are incapable of visualizing three dimensions except through their mathematics. The world they live in, however, is closed and finite: there is only a certain amount of space,

what we would call the surface area of the sphere. Yet this world is also unbounded and has no edge. A person could travel forever among the galaxies without encountering an impassable barrier. Imagine the surprise of a flat traveler, who after going in a straight line, returns to the starting point! Neither does the surface of a sphere have a center, a term that would have meaning only if three dimensions are considered.

Now scale everything back to our real four-dimensional world. A closed universe is finite and has only a certain volume, but it is bounded and has no center. Everyone and everything only *appear* to be at a center. Moreover, a traveler moving in a straight line (straight in the relativistic sense) will return to the point of origin. As the two-dimensional sphere in Figure 30.8 expands and its scaling factor (R) increases, its galaxies get farther apart, but the two-dimensional world does not exist outside the sphere. Likewise, the real closed Universe is not expanding *into* anything; it *is* everything and is merely getting larger, carrying the galaxies along with it and making them (rather their clusters) recede from one another. The same is true for open or flat universes. Space does not exist outside the Universe.

30.4.3 The Age of the Universe

If gravity slows the Universe, it must affect its age (Figure 30.9). Assume that the cosmological constant is zero, that there is no Einsteinian expansive force. The Hubble time, $t_0 = 1/H_0$ (see Section 28.4), is the age of the Universe only if there is no gravity and consequently no mass. However, in a real universe in which there is mass, gravity slows the expansion, and the Universe was once expanding faster than it is today. The time since the *actual* birth of the Universe, the real age t, must then be smaller than t_0.

The ratio t/t_0 depends on the degree of gravitational drag. For an empty universe, the value is 1 (curve A in Figure 30.9). Even for an open, negatively curved Friedmann-Lemaître universe (curve B), which does not have enough mass to cause the Universe to begin to contract, t/t_0 is reasonably close to 1, and the age of the Universe can be approximated by $1/H_0$. For a closed, spherical, positively curved Friedmann-Lemaître universe (curve D), which has sufficient mass to cause a collapse, t/t_0 is considerably less than 1, and (depending on the mass) could be only 0.5. For the special, marginally open, flat Einstein-de Sitter universe (curve C), t is two-thirds of t_0.

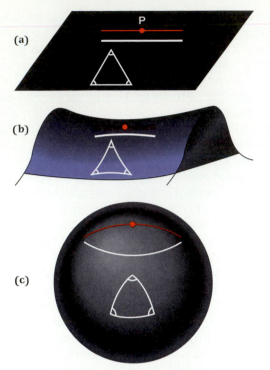

Figure 30.7 *Four-dimensional space is represented here on a three-dimensional surface. In* **(a)**, *it is flat and Euclidean, and the Universe is marginally open. One line (red) can be drawn through a point parallel to another line, and the angles of a triangle add to 180°. In* **(b)**, *space is curved negatively, like a saddle. The Universe is open, more than one line can be drawn parallel to a given line, and the angles of a triangle sum to less than 180°. In* **(c)**, *space is positively curved and folds back on itself, there are no parallel lines, and the triangle's angles sum to more than 180°.*

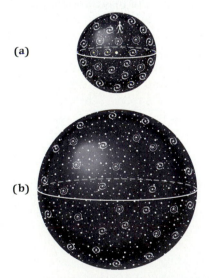

Figure 30.8   **(a)** *The Universe is represented here by the surface of a three-dimensional sphere; its people are two-dimensional and unaware of a third dimension. Their Universe is closed and finite but is unbounded and has no center.* **(b)** *As the sphere expands the scaling factor R increases, and its clusters of galaxies become farther apart.*

In an open universe with little matter, if $H_0 = 50$ km/s/Mpc, t could be close to $1/H_0$, or 19.6 billion years. In the flat case, if $H_0 = 50$ km/s/Mpc, the true age is two-thirds of 19.6 billion years or 13.1 billion. If the Universe is flat and H_0 is 100 km/s/Mpc, t_0 and t respectively equal 9.8 and 6.5 billion years. (For our adopted value of $H_0 = 75$ km/s/Mpc, the two numbers are 13.0 and 8.7 billion years.) Assuming the Universe to be either open or flat, we have restricted the true age to between about 6.5 and 20 billion years, an embarrassingly large range.

30.4.4 Distances, Look-Back Times, and the Cosmic Horizon

Distances to cosmological objects are not uniquely defined. The distance to an object can be measured by comparing its apparent and absolute brightnesses, by comparing its angular diameter with an assumed physical size, or from the time it has taken for its light to get to us. In our local neighborhood, these concepts give the same answer. However, once z exceeds a few tenths, curvature destroys the uniqueness of the concept of distance and each of the three methods gives a different value. It is simplest to choose light-travel time, which has a clear physical and philosophical meaning.

There is a more disconcerting problem. Think of an imaginary observer who can see the whole Universe at once. This observer is unaffected by the finite speed of light, can see where all the galaxies are, and can measure their distances, d_{now}, which come straight from the Hubble relation, $d_{now} =$ velocity $(v)/H_0$. Now assume that you are a real observer looking outward into the Universe. The light from the galaxy has taken time to get to you. You see the galaxy not where it is now, but where it used to be, at a distance d_{then}. Like the relation between velocity and redshift, that between d_{now} and d_{then} and redshift depends on the model of the Universe and its curvature. Only if z is small, under a few tenths, is the distinction between d_{now} and d_{then} unimportant. At high z, they differ greatly and d_{then} can be much less than d_{now}. For an Einstein-de Sitter universe, d_{then} actually decreases with increasing redshift once z is greater than 1.25. Distances for high redshift objects almost always refer to d_{now}, which always increase with redshift in accord with the Hubble relation. With all these qualifications, it is no wonder that cosmologists are reluctant to specify distances of distant galaxies and quasars.

In spite of these complications, examine some of the curiosities of an expanding universe. Galax-

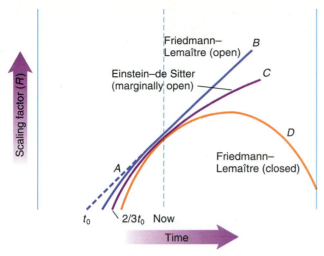

Figure 30.9   *The curves show how the expansions of different models of the Universe are affected by gravity. An empty universe, or one with very little mass (A), has an age of $t_0 = 1/H_0$. The open Friedmann-Lemaître universe (B) is younger than t_0, the flat, marginally open Einstein-de Sitter universe (C) has an age 2/3 of t_0, and a fully closed Universe (D) is even younger.*

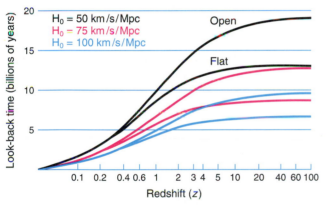

Figure 30.10   *Look-back times in billions of years for three values of H_0 are indicated by different colors. The upper and lower curves for each refer respectively to open and to flat, marginally open universes.*

ies recede faster and faster from us in accordance with the Hubble law. At some distance L (the *Hubble length*), the recession velocity equals that of light, or $L = c/H_0 = c\,t_0$. Beyond the Hubble length, galaxies and quasars are now receding faster than light. Such speeds are possible because the distant objects are not moving *through* space but are being carried along *with* space. In a flat Einstein-de Sitter universe, the redshift of a galaxy now a Hubble length away (as seen by the above imaginary observer) is $z = 3$. But because we see objects as they used to be, that redshift corresponds to a distance d_{then} of only a quarter of a Hubble length, when the galaxy had a recession velocity

that was twice as great (because the universe has decelerated), or 2c.

The **look-back time** to any galaxy is the time it took the light to reach us; it is shown in Figure 30.10 for fully open and flat universes and three values of H_0. For a redshift of $z = 4.9$ (the current quasar record) the look-back time for a flat universe is 93% of the way back to the Big Bang, at which time the Universe was less than a billion years old. We can penetrate time and see the evolution of the Universe.

As we move outward, z continues to increase. At some point z becomes infinite, we can see no farther, and we have reached the **horizon** of the Universe (Figure 30.11). In the Einstein-de Sitter version, it lies at a distance of twice the Hubble length. We have no knowledge of the Universe beyond the horizon, since no light has yet had time to reach us. As time proceeds, the Hubble constant drops under the drag of gravity, and the horizon expands into the Universe, allowing us to see a greater volume of space. At the highest observed values of z we look back in time to see youthful quasars, their number rapidly thinning out. If we could watch for billions of years we would see those quasars age to galaxies, their places continuously taken by additional youthful quasars as the horizon expands outward.

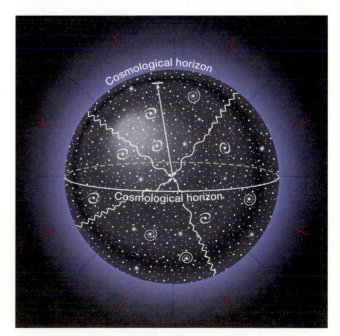

Figure 30.11   *Light coming from the horizon of the Universe is red-shifted to infinity. As the Universe expands (red arrows), the horizon expands even faster (blue arrows), bringing new sights into view.*

30.4.5 Observational Tests

We cannot know distances or the age of the Universe until we know both the Hubble constant and the degree of curvature. H_0 can be derived from observation once we are able to measure accurate distances to galaxies and correct for Doppler motions induced by local gravity (see Section 28.3). But which model of the Universe is correct?

There are several ways to tell. Remember that we see galaxies where they used to be: the farther away they are, the less time gravity has had to drag them back. The degree of the effect depends on how much mass is pulling on them, and consequently on the curvature (and model) of the Universe. Figure 30.12 compares an observational Hubble diagram expressed as a redshift-magnitude relation with different predictions (where magnitude is corrected for redshift effects). Curves *A*, *B*, and *C* respectively describe open, flat, and closed universes with no evolutionary effects, and curve *D* is a flat universe that includes a model for the evolution of the galaxies with time (see Section 28.1). The data lean toward evolutionary effects but are clearly not good enough to discriminate curvature.

A second, more effective, method relies on local measurements. The curvature of the Universe depends on the amount of mass enclosed in a shell such as one in Figure 30.1, and therefore upon average density ρ. If the density is low, the shell will expand forever, and the Universe is open and hyperbolic; if sufficiently high, the expansion will be halted and the Universe is closed and spherical. For any value of H_0 there will be a **criti-**

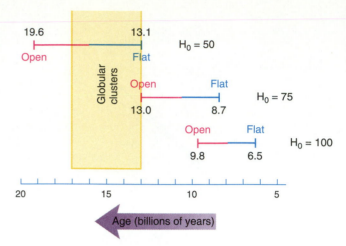

Figure 30.13 *Three horizontal lines illustrate the allowable age ranges for universes from fully open (red) to marginally closed and flat (blue). The yellow box represents the currently estimated range of age of the oldest globular clusters, which allows Hubble constants only from about 75 km/s/Mpc on down.*

cal density ρ_c for which the shell expands forever but comes to a halt in an infinite time, giving us an Einstein-de Sitter, flat, marginally open universe. For H_0 equal to 50, 75, and 100 km/s/Mpc, ρ_c is respectively 5×10^{-30} g/cm^3, 1.1×10^{-29} g/cm^3, and 2×10^{-29} g/cm^3 (ρ_c is proportional to H_0^2). The curvature of the Universe can then be measured by a quantity called *Omega* (Ω), the ratio of the true density to the critical density, or $\Omega = \rho/\rho_c$. If Omega is less than 1, the Universe is open; if greater, it is closed; if exactly 1, it is flat. Omega is probably the most sought-after number in modern astronomy.

To evaluate Ω, sum the masses of the constituents of a large and representative region of space and divide by the volume to measure the average density. From the luminous masses of galaxies, we find a small value for Ω (the exact value of H_0 is not critical), only about 0.01. However, when we include dark matter, whose mass exceeds that of luminous matter by a factor of at least 10, Ω climbs to over 0.1. Favorable selection of data, evaluations of dark matter from observations of gravitational lensing (see Section 29.3), and the mass required to produce the gravitational Doppler shifts seen in the *COBE* data (Figure 30.3) suggests an even larger value, one that could approach $\Omega = 1$.

A third and crucial test compares the age of the Universe with the ages of the oldest globular clusters. For any given value of H_0 there is a range in ages that depends on the Universe's curvature (see Figure 30.9). If H_0 is known, the ages of the globulars (ignoring any time it took to make stars after

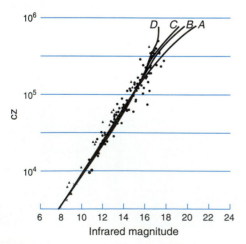

Figure 30.12 *Curves A, B, and C respectively show theoretical Hubble relations for open, flat, and closed universes, and curve D a relation for a flat universe that includes the evolution of galaxies. The observational points cannot discriminate curvature.*

the Big Bang) will then give the true age of the Universe and consequently t/t_0 and the degree of curvature. From stellar evolutionary theory, the ages of the oldest globulars appear to lie between about 13 and 17 billion years (see Section 26.6.1). The lower limit excludes any value of H_0 over about 75 km/s/Mpc (Figure 30.13). For H_0 = 75 km/s/Mpc, there is overlap between the age of the Galaxy and the age of the Universe only if the Universe is both fully open and we can allow the oldest globulars to be only 13 billion years old. To obtain a flat, marginally open universe, we must accept a Hubble constant as low as 50 km/s/Mpc. If H_0 is eventually confirmed to be significantly above about 75 km/s/Mpc, we will somehow have to revise stellar evolution theory or revive Einstein's old idea of the cosmological constant (see Section 28.2). An expansive force pushing on spacetime would make the Hubble constant higher than expected for a given age; with such an expansive force the Universe could be old enough to fit the globulars and H_0 could still be high.

None of the tests provides a firm answer. There are too many uncertainties about the value of H_0, the ages of the globular clusters, and the value of the cosmological constant. Density measurements suggest that the Universe is either open or flat and marginally open, the latter only if we adopt an extreme value for the average density. There is one other avenue, but it involves further theory.

30.5
The Origin and Evolution of the Universe

We now have an idea of the structure of the Universe, even if we still do not know its fate. The next question asks how the Universe might have come to be. The theories take us back to the beginning of time.

30.5.1 *Back to the Beginning*

The photons of the cosmic background radiation were created in the Big Bang. As the Universe aged and expanded, they stretched along with space, their wavelengths steadily increasing and their energies steadily decreasing: as the Universe enlarged, it cooled. The measure of the Universe's expansion involves its scaling factor, R, where redshift $z = R_{now}/R_{then} - 1$. The temperature of the background radiation—the temperature of the Universe at large—is therefore proportional to $1/R$. Figure 30.14 shows the temperature of the Universe plotted against time in seconds. Because the graph is plotted logarithmically, according to powers of 10, it is highly distorted, stretching out the earliest phases of the Universe in order to show its development. The graph is also meant to be highly schematic, since we do not yet know the age of the Universe or the degree of curvature, that is, the correct model to apply.

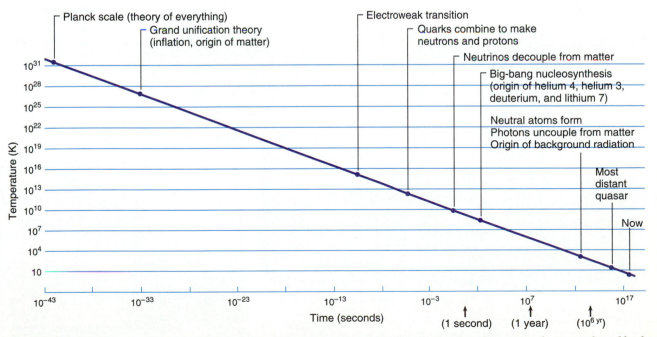

Figure 30.14   *The history of the Universe is shown by plotting temperature against time in seconds with various milestones indicated by dots.*

As we proceed backward in time and the scaling factor shrinks, the Universe becomes hotter and denser (density depending on $1/R^3$). The background temperature is currently 2.7 K; for most of the way back to the beginning the changes are slow. Even looking out to the most distant quasar, with z nearly 5, the cosmic background temperature is still only a little over 10 K. But as we regress to earlier times, the temperature and density mount. When the Universe was only a million years old, T was roughly 1,000 K. There could then have been no galaxies, only radiation and matter. At 100 seconds, the conditions of the whole Universe were like those inside a star. We can in principle watch the Universe shrink backward to a *singularity* (see Section 26.5.1) where $R = 0$ and density and temperature both become infinite.

Whether a true singularity was involved is contended, but it seems very likely that at the time of its origin—the beginning of spacetime itself—the Universe was exceedingly hot and dense. What we see today is the result of steady expansion and cooling.

30.5.2 Particles and the Unification of Forces

Such high temperatures and densities produce curious effects that are crucial to a theory of the development of the Universe. We have identified four forces of nature: gravity, electromagnetism, the weak force (involved in particle decay), and the strong or nuclear force (see Section 8.1). Why should there be four? In the nineteenth century James Clerk Maxwell unified the apparently different forces of electricity and magnetism into one entity. Is it possible that the four forces are really but four manifestations of a single force?

To see how such unification can work, we have to look deeper into the atom. In the modern view, nature is symmetric: waves act like particles (the photon is the particle of the electromagnetic wave) and particles act like waves (see Sections 8.2.3 and 8.3.3). The four forces are also *carried* by particles.

Figure 30.15 *The Tevatron is a huge, high-speed atomic accelerator at Fermilab near Chicago. Particles are accelerated by magnets to almost the speed of light in an underground tube (lower inset) 2 km in diameter, which is outlined in the main photograph by the lights of a moving car. In the upper inset, two protons are seen to collide (arrow) in a spray of particles.*

The electromagnetic force between two bodies is produced by invisible *virtual photons* thrown back and forth between them. Under some conditions, the virtual photons can escape and become real photons that can be seen, and the body is said to *radiate*.

The weak force is carried by exotic particles called *W* and *Z*. The photon has no mass: therefore, it can fly at the speed of light to infinity, rendering the electromagnetic force an inverse-square law. W and Z, however, are heavy particles and cannot fly far—their range is limited roughly to the atomic nucleus. They are real and have been found by smashing atomic particles in high-speed atomic accelerators (Figure 30.15) that can throw them to speeds near that of light. The strong force is theoretically carried by yet more massive particles called *gluons,* which also act over only ultrashort ranges. Gravity has defied this concept of mediating particles, but may be carried by *gravitons* that move at the speed of light, producing the familiar inverse-square law.

If temperature and particle energies are sufficiently high, W and Z lose their mass and behave like photons. Electricity, magnetism, and the weak force then become one, the *electroweak force,* reducing the number of forces to three. They appear separate only at lower energies. The required temperature is staggering, about 10^{16} K, 100,000 times higher than even the core of a supernova. Such energies can actually be achieved in the biggest accelerators, where the atomic collisions raise temperature to the required heights. They were achieved naturally in the early moments of the Big Bang when the Universe was a mere 10^{-12} s old. At even higher energies, we expect the strong force to join in. Such **Grand Unified Theories** (GUTS) cannot be experimentally verified: the required temperature is an unreachable 10^{27} K and would require an accelerator light-years across. However, such conditions were present in the early Big Bang, 10^{-33} s after birth (see Figure 30.14).

The end of the road in Figure 30.14 is actually not zero but 10^{-43} s, the *Planck time,* when temperature and density respectively approached 10^{33} K and 10^{66} g/cm^3. At the Planck time, the Universe was so small that it was effectively a black hole, and general relativity, our standard theory of gravity, no longer applied. Near that time, gravity is thought to have been united with the other forces in a **Theory of Everything** (TOE). The four forces would then be in perfect **symmetry.**

The interactions of gravity in a TOE require a further symmetry called **supersymmetry** (SUSY). For each force particle there should be a corresponding mass particle and for each mass particle there should be a force-type particle. For example, the photon should have a massive counterpart called the *photino*. None of these particles has ever been detected. But supersymmetry demands that since the photon interacts strongly with matter, the photino does *not,* and thus would be exceedingly difficult to find. The collection of the massive particles that include the photino are therefore called **weakly interacting massive particles** (WIMPS). They represent one possible component for the missing matter needed to bring Ω up to 1.

In one theory of quantum gravity (in which gravity is carried by gravitons), the early Universe contained **superstrings,** "particles" that exist in 11 dimensions. As the Universe expanded, the dimensions collapsed and created forces until we are left with only four and our familiar spacetime.

30.5.3 Challenges to Theory

Standard Big Bang theory, as developed by Gamow's group and its successors, assumes steadily decelerating expansion since the Big Bang. The theory runs into several serious problems, the first of which is the *horizon problem.* Within very small limits, the temperature of the cosmic background radiation is the same in all directions. To establish such constancy, all parts of the Universe must have been in some kind of contact with one another; that is, they must have been able to communicate and exchange information. Yet the radiation we observe from one direction is beyond the horizon of the radiation coming from the other. As we progress backward in time and the size of the Universe shrinks, the horizon shrinks even faster, and the problem gets worse. How did communication take place?

Second is the *flatness problem.* The value of Ω is now not that far from 1. If in the early days of the Universe Ω was even slightly greater than 1, the Universe would have collapsed almost instantly. If it had been slightly below, it would have expanded so quickly that no galaxies could have formed. To be even close to 1 now, it needed to be within 10^{-60} of 1 within the first moments—a remarkable and not very likely accidental balancing act.

A third difficulty that a complete and successful theory must confront is the *matter-antimatter problem.* Antimatter (negative protons and positive electrons) is identical to normal matter except for its electric charge. Why is the Universe made of only normal matter?

Last, we face the *lumpiness problem*. The expanding matter of the Big Bang must have had fluctuations that produced galaxies. How and when did they develop?

30.5.4 An Evolutionary Model of the Universe

Return now to the very beginning for a tour of modern thought and theory. On the quantum level there is no such thing as an empty vacuum, only minimum energy that is always randomly fluctuating. The Universe may have erupted in a random event before the Planck time from such a *quantum foam*. In the first 10^{-43} s, the Universe may have been controlled by an 11-dimensional supersymmetry and all the forces were combined. Immediately after that instant, the symmetry that combined all the forces broke for the first time and gravity separated from the others (Figures 30.14 and 30.16).

At an age of about 10^{-35} s, the strong force began to separate from the electroweak. However, the separation did not take place instantly. The Universe actually cooled below the breaking point; that is, it supercooled. If you keep water very still, you can supercool it, keeping it liquid well below the freezing point. Any disturbance, however, and the water will suddenly convert to ice. Supercooling of the Universe created a *false vacuum* that acquired energy. Then symmetry started to break, the energy of the false vacuum was released, and the false vacuum became a true vacuum. The density of a false vacuum is negative. We know that the gravity of a body is related to its density. If the density is negative, gravity becomes *antigravity* and is repulsive rather than attractive. The collapse of the false vacuum suddenly inflated the Universe to comparatively vast proportions (Figure 30.17). Between 10^{-35} and 10^{-32} s, it expanded by a factor of up to 10^{50}, growing from smaller than a proton to the size of a softball. Such strange behavior is not mere speculation expressed in arcane vocabulary but the result of solid mathematics.

This theory of **inflation,** initially developed in 1980 by Alan Guth of Harvard and since considerably embellished, provides a natural solution to the horizon and flatness problems. Before the inflationary period, the Universe was small enough to allow all its parts to communicate. Rapid inflation separated the parts, but they had already established the smoothness we now see. Inflation also mathematically shows why the Universe is so close to being flat and marginally open. It *is* flat. The inflation was so great that local spacetime was flattened to the point that Ω effectively equals 1. (If you live on a large curved surface and cannot see very far, your world will look flat.) The inflationary Big Bang is thus consistent with the most extreme conclusion found from the study of dark matter. Luminous matter then accounts for only 1% of the total.

Energy and mass can be converted back and forth into each other. As the temperature dropped at the end of the GUTS era, the energy decayed into the fundamental mass units of nature, quarks, and antiquarks (see Section 8.1.1), which then combined to make **baryons,** the collective name for protons, antiprotons, and neutrons. However, the temperature was still so high that collisions among the baryons immediately broke them back into quarks, establishing a temporary equilibrium. About 10^{-11} s into the Big Bang, symmetry broke again, separating the weak and electromagnetic forces (see Figures 30.14 and 30.16). A millionth of a second later, when the temperature was down to about 10^{13} K, collisions were no longer sufficiently energetic to smash the baryons apart; they stabilized, their creation absorbing all the quarks and antiquarks.

In 1956 Tsung Dao Lee and Chen-Ning Yang discovered that nature is not perfectly mirror-symmetric, and thereby earned a Nobel prize. There is a slight difference between matter and antimatter.

Figure 30.16   *During the expansion of the Universe, symmetry broke three times to separate out the four forces.*

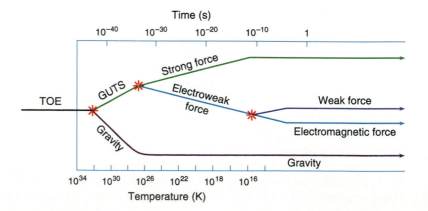

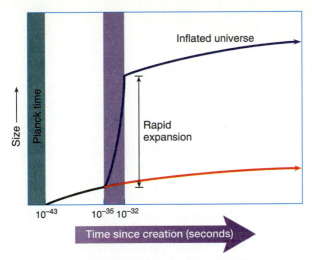

Figure 30.17 *Inflation that began at an age of 10^{-35} s produced a rapid increase in the size of the Universe (blue curve) compared to that expected in a Big Bang theory with no inflation (red curve).*

As a result, the new Universe produced slightly more quarks than antiquarks (by about 1 part in 10^{10}) and consequently more protons than antiprotons. The protons and antiprotons collided and annihilated each other, creating huge numbers of gamma rays. This reaction can also go both ways. Two energetic gamma-ray photons can collide to make a pair of protons and antiprotons. As the temperature dropped, gamma-ray collisions were no longer sufficiently energetic to make protons and antiprotons. Matter and antimatter then continued to annihilate each other until only the excess protons were left. The theory accurately accounts for the lack of antimatter and the observed ratio of protons to photons.

Collisions between the gamma rays could still produce electrons and positrons, but eventually, by about one second, the lowered temperature caused this reaction to stop as well, leaving a slight excess of normal negative electrons. At about this time, the density of the Universe had dropped low enough to make it transparent to neutrinos (produced in abundance by reactions involving the weak force) that then were free to fly through space unimpeded. Solar experiments indicate that neutrinos may have a tiny amount of mass; there were so many created in the Big Bang that if they do they could provide some of the dark matter needed to flatten the Universe.

30.5.5 Transition

At an age of about three minutes, the Universe began to take on more of its present character. The temperature was then about 10^9 K. Deuterium was created by colliding protons and neutrons; as the temperature dropped it could no longer be broken apart by collisions and could survive. Deuterium atoms reacted with other deuterium atoms and with protons and neutrons to make helium, both ^{3}He and ^{4}He. The chain built to lithium (^{7}Li), but then the temperature dropped too low to allow anything else. We were then left with the basic building blocks of Population III stars (see Section 26.6.2). Theory predicts an initial He/H ratio of about 0.08 (by number of atoms), exactly what we see in extreme Population II. Moreover, the abundance of ^{4}He is sensitive to the number of families of matter, of which the electron and the proton constitute the lightest (see Section 18.5.3). The observations restrict the number of families of matter to four. Experiments made with particle accelerators indicate the existence of two heavier (and much rarer) families. Modern Big Bang theory works!

The abundances of deuterium, ^{3}He, and ^{7}Li are very sensitive to the density of baryons. From Big Bang theory we can calculate these abundances in terms of what the baryonic density of the Universe ought to be *now* (Figure 30.18). From the observed abundances we find a maximum current mean density of about 4×10^{-31} g/cm^3. This value is several

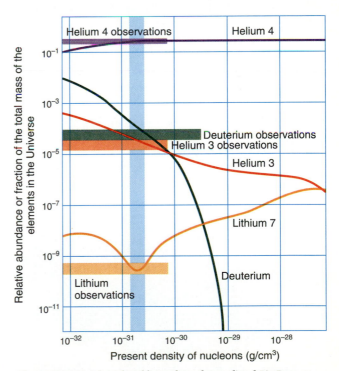

Figure 30.18 *The colored lines show the predicted Big Bang (primordial) abundances of ^{2}H, ^{3}He, ^{4}He, and ^{7}Li (by mass) plotted against the density the Universe should now have (calculated for H_0 = 50 km/s/Mpc). The thick horizontal bars are the observations, which cross the predictions within the vertical bar.*

BACKGROUND 30.1 Interactions Within Science

One branch of science or, more parochially, one branch of astronomy, cannot be isolated from other branches. Fundamental input that increases knowledge in one area can come from unlikely sources. Look at some examples. The number of families of matter is predicted both by atomic experiments with accelerators and by observation of cosmic helium abundances with telescopes: the findings are consistent and provide evidence for the Big Bang. The neutrino detectors in Japan and under Lake Erie were originally built to detect proton decay and to measure its rate, but they were most successful in detecting neutrinos from the Sun and Supernova 1987A. Solar neutrino experiments demonstrated that neutrinos may have mass, mass possibly needed to close the Universe to make $\Omega = 1$, as suggest-

ed by the inflationary Big Bang. However, observation of Supernova 1987A limited the masses of neutrinos, bringing together three aspects of astronomy and three very different kinds of astronomers. The breaking of Grand Unified symmetry has a counterpart in, remarkably, liquid crystals in the physics laboratory.

Scientists, as well as government and scientific funding agencies, must be very careful in trying to direct research toward particular areas. All things connect to all other things, and we never know from which direction a fundamental breakthrough will come that will affect another apparently unrelated aspect of science. Science is a single entity, not a collection of separate subjects.

---

times that found from luminous matter, confirming the existence of dark matter. However, it still yields an Ω of under 0.1 (depending on the value of H_0). If Ω is actually 1, baryons—what we think of as normal matter—constitute only a small percentage of the dark matter. The additional mass must then be nonbaryonic, in the form of other kinds of particles—perhaps WIMPS or neutrinos.

The Universe then coasted. The density was still so high that photons interacted with protons and electrons to keep the matter ionized. About 100,000 years after helium was formed, when T was a few thousand Kelvins, the density dropped to

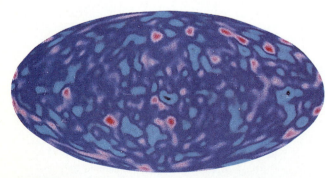

Figure 30.19 *This all-sky map of background radiation from* COBE *has Doppler effects removed, and shows temperature fluctuations of a mere 10^{-5} K that could have given rise to the lumpy structure of the Universe.*

the point where electrons could combine with protons and other nuclei to form permanent neutral atoms. Photons no longer coupled with matter to keep it ionized, and the Universe became transparent to light. This was the critical moment of the creation of the cosmic background radiation. Since that time, the scaling factor of the Universe (R) has increased by over 1,000 and the background temperature (which is proportional to $1/R$) has dropped by a like factor to 2.7 K. When we look into space with our radio telescopes, we see back only to the moment of the onset of transparency, to a redshift of about 1,000. No matter how good our equipment, we can never see beyond or farther into the past, except indirectly, by application of theory and its test by observation. As the horizon of the Universe expands, could we live long enough, we would see new quasars and then galaxies emerge from the murk of this opaque background.

30.5.6 The Formation of Galaxies

One of the challenges to theory remains. The first galaxies developed as quasars within about a billion years of the Big Bang. How did they form out of the expanding medium? Continued observations of the cosmic background radiation with COBE have produced extraordinary signal-to-noise ratios. After correction for various confusing sources of radiation (such as the Galaxy) and removal of the Doppler effect, seen in Figure 30.5, the background radiation is seen to break into a pattern of fluctuations of no more than 10^{-5} K (Figure 30.19). They are interpreted as density ripples that developed immediately after the Big Bang, as mass and gravity variations that ultimately produced galaxies and their clusters and superclusters (Figure 30.20).

However, the amount of ordinary baryonic matter in the Universe (at least the amount inferred from most observations) cannot by itself produce sufficient gravitational attraction to allow the ripples to grow. They may instead be markers for much larger fluctuations in the distribution of nonbaryonic dark matter that acts as *seeds* that can cause the baryonic matter (both currently luminous and dark) to accumulate (Figure 30.21).

There are two candidates for such seeds. The breaking of GUTS symmetry, which has been likened to the freezing of water, introduces **defects** into spacetime that are conceptually akin to the cracks in an ice cube. The defects are regions

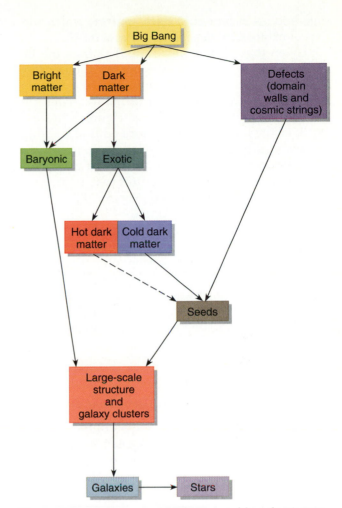

Figure 30.21   *Galaxies may have been created from the Big Bang, through seeds in the form of cold dark matter and defects that accumulate baryonic matter into large-scale structures (like the Great Wall), galaxies, and stars.*

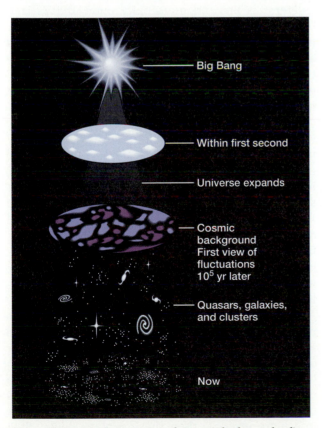

Figure 30.20   *The fluctuations in the cosmic background radiation had their start in the first second following the Big Bang. They grew into the structures we see today.*

where GUTS is still in effect. They come in one-, two-, and three-dimensional varieties, respectively, called *magnetic monopoles* (units of single magnetic charge), *cosmic strings*, and *domain walls* (Figure 30.22). Magnetic monopoles were removed from the Universe by inflation. They are so massive that if there were as many as the original (no-inflation) Big Bang theory predicts, they would have caused the Universe to collapse. The strings (which have never been observed, but which have mass and gravity) and walls may make the necessary seeds that can quickly attract baryonic matter.

The other candidate is nonbaryonic dark matter. Neutrinos, which move at or near the speed of light, are considered as **hot dark matter.** Though they may act to close the Universe, they cannot easily accumulate to act as seeds. An alternative suggestion is slower **cold dark matter,** which consists of exotic particles, perhaps the WIMPS.

Given that matter can accumulate from the Big Bang, there are two more possibilities. Either galaxies formed first and then accumulated into clusters, superclusters, and the larger structures (like the Great Wall) seen in such plots as Figures 28.12 and 28.13, or the larger structures formed first and then collapsed into galaxies. The latter theory is the more accepted. Computer simulation of galaxy formation by gravity shows structures that are remarkably similar to those observed (Figure 30.23).

As of today, however, none of the theories is really successful in showing how galaxies can form. Is this problem a flaw in the Big Bang? It is important to realize that the Big Bang is still a partial theory. Parts of it are successful and other parts are not; a decade or two ago, it was even less successful. The Big Bang provides a needed theoretical framework that seems to work, one on which astronomers and cosmologists continue to build.

30.6
Alternative Cosmologies

As good as Big Bang theory seems to be, it is not unanimously accepted. And until it is proven beyond significant doubt, there should be opponents nibbling at its edges, trying to bring it down. Only a

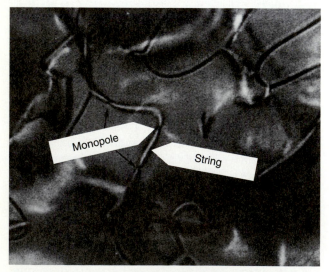

Figure 30.22 *Symmetry breaking, like the freezing of water into ice, is a normal part of nature. Here we see symmetry breaking in a liquid crystal; the defects may be similar to those created when GUTS symmetry broke. The defects, reminiscent of large-scale structure in the Universe, may have been seeds around which the matter that created the galaxies accumulated.*

theory that can withstand such attempts is acceptable. Remember that there are still some questions, small though they may seem, about the interpretation of redshifts and about whether all quasars and galaxies take part in the Hubble flow.

An entire field called *plasma cosmology* suggests that there was no Big Bang and that the observations can be explained by plasmas and magnetic fields. Even the Steady State theory is still alive, though in vastly modified form, in an attempt to account for the observations. In the New Steady State, the Big Bang is replaced by lots of "little bangs" that drive the inflation. The cosmic microwave background comes not from the Big Bang but from accumulations of distant stars and galaxies. In the opinion of the large majority of scientists, however, the inflationary Big Bang has by far the fewest problems and best fits the observations.

30.7
The Future of the Universe

What happens next? The prevailing view is that the Universe is flat and marginally open, that Ω is 1. The Universe will continue to expand and the temperature of the cosmic background will drop forever and approach zero. Within galaxies, interstellar matter will continue to form stars, and stars will continue to pour some of their mass back into space. The metal content will continue to rise. How star formation and the formation of future life in the Universe will be affected is impossible to guess. Eventually— perhaps in 10^{15} years—nearly all the matter will be tied up in the condensed states of white dwarfs, neutron stars, and black holes, the remaining interstellar matter too tenuous to form stars. As the dead stars whirl about their galactic centers they will encounter one another. Some will be ejected into intergalactic space, others sent to massive black holes accumulating as galactic nuclei. The Universe will be made of nothing but massive black holes, isolated dead stars, and tenuous gas.

But that is not the end. Grand Unified Theories predict that protons are not completely stable and will decay into positrons, radiation, and other debris with a half-life of roughly 10^{32} years. So far, however, efforts to detect such decays have been unsuccessful. If protons decay, they release neutrons from atomic nuclei. The free neutrons decay into more protons; then these protons decay. Stellar relics will therefore evaporate. In 1974 Stephen Hawking of Cambridge discovered theoretically

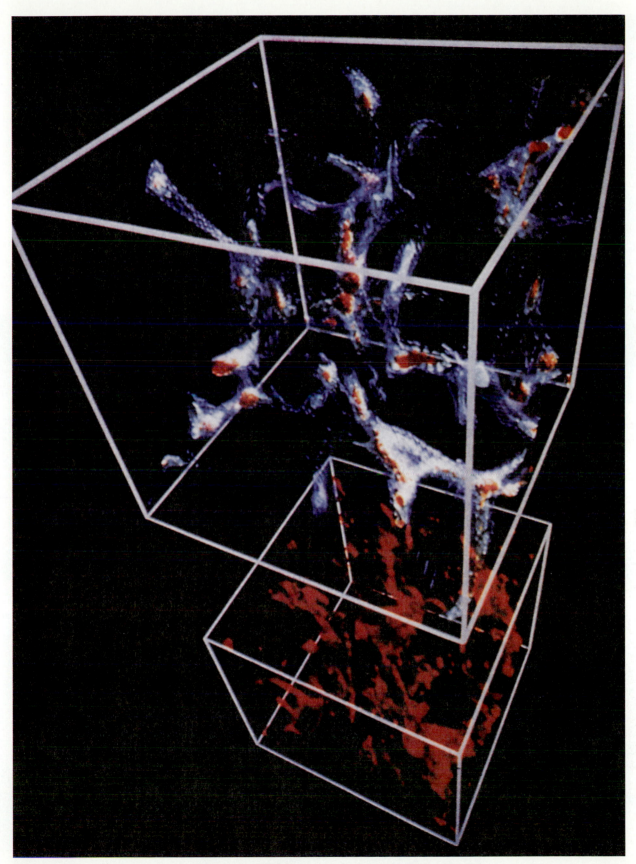

Figure 30.23   *Computer simulation of galaxy formation shows structures similar to those observed.*

Figure 30.24   *We journey from the edge of the Universe to home.*

that even black holes are not entirely stable. Their mass is subject to quantum mechanical tunneling (see Section 18.5.1), which allows matter to escape. The black holes will thus evaporate as well. In 10^{100} years, the Universe may be an expanding entity filled only with radiation and particles; all the stars that had once formed and lit the night will have returned to their origins.

Yet for all our apparent certainty in our knowledge we must bear in mind that every generation has thought it understood the basic aspects of the Universe, and every generation has been wrong in some crucial aspect. Even if we are on the right track, our ignorance is still deep: we do not know whether the Universe is closed or even the nature of the majority of its matter; if Ω is truly 1 then some 99% is shrouded in mystery. Only time and future efforts will tell us what the Universe is really like.

30.8
The Earth and the Universe

The quantum view of the Universe and cosmic inflation opens the possibility that many universes were created at once, or even that one universe can spawn another. Our own vast Universe, of which we probably see only a tiny portion, may be but one of an infinite number of universes that will remain forever out of reach.

There is no question, however, of our connection with our own Universe. Galaxies developed only because of the way in which the Big Bang (or alternative cosmologies) proceeded. We are here because of the way in which the stars evolved, building heavy elements out of light and throwing them to the stellar winds out of which the Sun and Earth and humanity were ultimately created. We are not isolated in space. All the Universe is our birthplace and our rightful domain.

When we first approach astronomy, and especially cosmology, we tend to shrink within ourselves and feel insignificant against the magnificent, perhaps infinite, panorama of the sky. Who are we, two meters tall, compared with megaparsecs, billions of solar masses, ages of billions of years? But the Universe, however grand, is inanimate. All of it can exist—can be comprehended—within our own minds, and all of it was necessary to create us. Think about that when you next stand outdoors amidst the sparkling stars.

KEY CONCEPTS

Baryons: Collectively, protons, antiprotons, and neutrons.

Closed, open, and **marginally open universes:** Respectively, a four-dimensional, positively curved, spherical universe that folds back on itself, a negatively curved, hyperbolic universe that does not close back (both of these are **Friedmann-Lemaître universes**), and a flat, Euclidean **Einstein-de Sitter** universe.

Cosmic microwave background radiation: All-pervading blackbody radiation near 2.7 K, cooled from the Big Bang fireball.

Copernican Principle: The Earth is not special; extended to the **Cosmological Principle** (no place in the Universe is special); extended further to the **Perfect Cosmological Principle** (there are neither special places nor times in the Universe).

Critical density: The average density of matter needed to close the Universe.

Defects: Places in the Universe where symmetry did not break: magnetic monopoles, strings, and domain walls.

Grand Unified Theories (GUTS): Theories combining the electromagnetic, weak, and strong forces.

Horizon: The maximum distance we can look, which is determined by the age and structure of the Universe.

Hot and cold dark matter: Respectively, fast-moving dark matter that cannot cause the formation of galaxies (for example, neutrinos), and slow-moving dark matter that can (possibly WIMPS).

Infinite: Never-ending.

Inflation: A super-rapid expansion, controlled by symmetry breaking, that occurred just after the Big Bang.

Look-back time: The time it has taken for light to reach us from a distant source.

Superstrings: Eleven-dimensional particles that may have existed at the beginning of the Universe and that created the four forces.

Supersymmetry (SUSY): A relationship in which for each force particle there should be a corresponding mass particle and for each mass particle there should be a force particle; supersymmetry includes **weakly interacting massive particles** (WIMPS) that may contribute to dark matter.

Symmetry: The unification of the forces of nature at high temperature.

Theory of Everything (TOE): A theory that combines all the forces of nature.

KEY RELATIONSHIPS

Age of an empty universe:
$$t = t_0 = 1/H_0$$

Age of a flat universe:
$$t = \tfrac{2}{3}t_0 = \tfrac{2}{3}(1/H_0)$$

EXERCISES

Comparisons

1. Compare the Copernican Principle, the Cosmological Principle, and the Perfect Cosmological Principle.
2. Describe the differences in fates of open and closed universes.
3. Compare the forces in GUTS and TOE.
4. Distinguish between hot and cold dark matter.

Numerical Problems

5. What is the critical density if $H_0 = 60$ km/s/Mpc?
6. What should be the actual age of a marginally open universe with a Hubble constant of 60 km/s/Mpc?
7. When the Universe doubles in size 13 billion years from now, what will be the temperature of the cosmic microwave background?

Thought and Discussion

8. What does Olbers' paradox tell us about the Universe?
9. What evidence suggests that the cosmological constant might be greater than zero?
10. Which theory of the Universe derives from the Perfect Cosmological Principle? What observation negates it? What does this theory require?
11. What is the most likely origin of the cosmic microwave background radiation?
12. What deviations from perfect uniformity are seen in the cosmic background radiation? What causes them?
13. Which property of quasars negates the Steady State theory?
14. How do parallel lines behave in Euclidean space?
15. How does the curvature of the Universe relate to whether it is open or closed?
16. Why can the Universe have no center?
17. Explain why the actual age of the Universe is related to the curvature of spacetime.
18. Summarize the evidence for the existence of the Big Bang.
19. What is Omega? How does it relate to the structure of the Universe? How can we find its value?
20. Why is it so difficult to define distance at cosmological redshifts?
21. Define the horizon of the Universe.
22. Which problems presented by the Big Bang are solved by inflation?
23. What is meant by broken symmetry with regard to the forces of nature?
24. What particles are associated with each force of nature? Which have mass under current conditions?
25. What possible roles do weakly interacting massive particles play in the Universe?
26. Name the isotopes made in the Big Bang. Why were heavier isotopes not made? What do they tell us about the early conditions in the Universe?
27. How do the different families of matter relate to the Sun and to the Universe?
28. Identify the candidates for the seeds that can accumulate mass for galaxy formation.

Research Problem

29. Using your library, examine and list opinions that have appeared in the press on the reality of the Big Bang.

Activities

30. From the figures in the text, construct a detailed timetable of the sequence of events in the inflationary Big Bang. Include the approximate temperature at each point. Comment on the importance of each event along the way.
31. List the problems and uncertainties of modern cosmology; then list some observations that might resolve them.

Scientific Writing

32. The theory of the Big Bang is commonly ridiculed in the press. Write a rebuttal letter to the editor of a newspaper demonstrating that the Big Bang is a valid theory that predicts and describes many of the properties of the observed Universe.
33. Write a five-page history of our view of the Universe; discuss how it has changed since Aristotle's times, and speculate on how it might change in the future.

The following six maps locate the constellations and brighter stars. The first shows the north polar region down to about 50°N declination. The next four are seasonal equatorial maps that display stars between 60°N and 60°S declination, and the last shows the south polar region. Declinations are indicated along a central hour circle. Right ascensions are noted around the peripheries of the polar maps and along the celestial equator for the equatorial maps.

Stars are generally selected to indicate constellation positions and outlines. They are shown through fourth magnitude, although their census is not complete. A few fainter stars are included to indicate the locations of obscure constellations. All 88 constellations are represented, although several are indicated by only their brightest stars. A small number of nonstellar objects, clusters, nebulae, and galaxies, are also included.

Colors are indicated for the 48 brightest stars, those through magnitude 2.0, down to the brightness of Polaris. The colors are greatly exaggerated and are assigned according to spectral class as follows:

Blue: Class O through B5
Light blue: B6 through B9
White: Class A (all stars fainter than magnitude 2.0 are also assigned white)
Pale yellow: Class F
Yellow: Class G
Orange: Class K
Red: Class M

The maps show the broad outline of the Milky Way. However, much of its intricate detail is omitted. The galactic equator, the mid-line of the Galaxy, is marked with galactic longitude starting at the galactic center in Sagittarius.

Complete paths of precession of the North and South Celestial Poles are shown on Maps 1 and 6. Partial paths are shown for the NCP on Maps 2 and 5 and for the SCP on Maps 3 and 4. These paths are distorted from circles as a result of distortions inherent in mapping a sphere onto a plane.

The times of year around the edges of the polar maps and along the tops and bottoms of the equatorial maps indicate the appearance of the sky at approximately 8:30 PM (20^h30^m) local time. To use the north polar map, face north and rotate the map so that the current month appears at the top. In the southern hemisphere use the south polar map similarly. The celestial pole will have an elevation in degrees equal to your latitude. To use the equatorial maps in the northern hemisphere, face south and line up the current month with the celestial meridian. To use them in the southern hemisphere, face north and turn them upside down. The equator point (the intersection between the celestial equator and the meridian) will have an elevation in degrees equal to 90° minus the latitude.

For each hour past 20^h30^m shift or rotate the map one hour to the west; that is, align an hour circle that is one additional hour to the east. For every 2 hours past 20^h30^m add one month to your current month. For example, if it is March 15 at 20^h30^m, you would align "March" on Map 4 with the celestial meridian. If it is 2^h30^m, set "June" (Map 5) on the meridian.

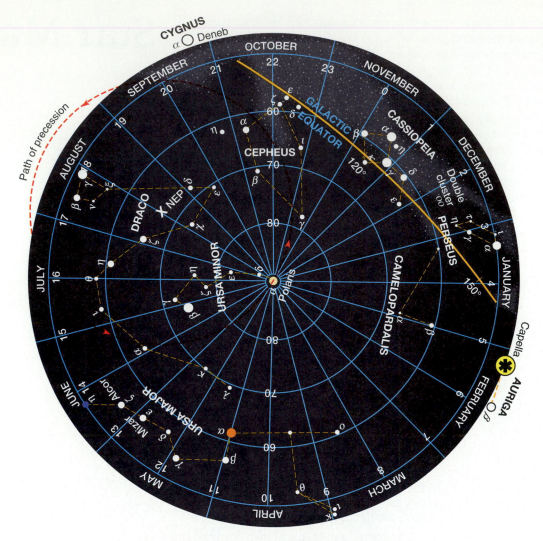

Map 1. *The North Polar Constellations*

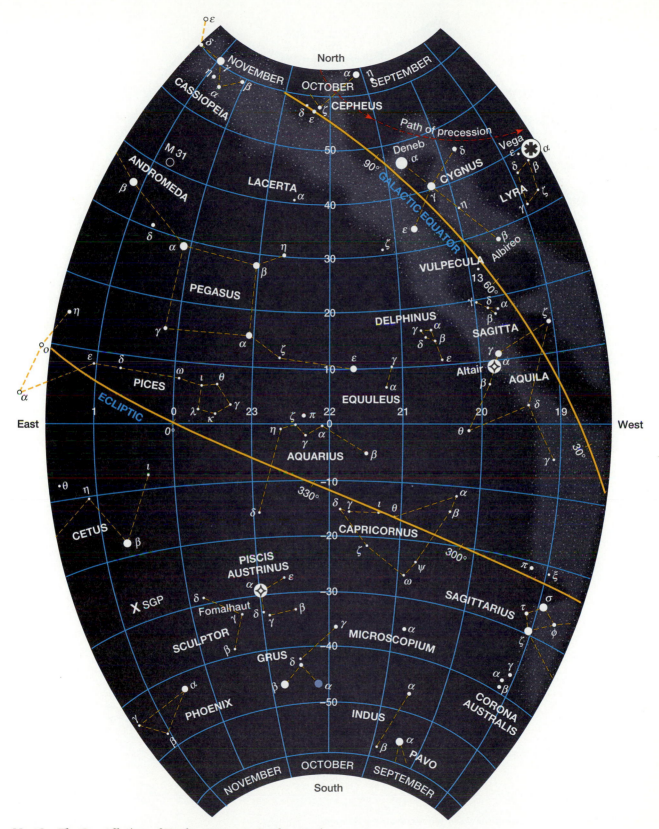

Map 2.   *The Constellations of Northern Autumn, Southern Spring*

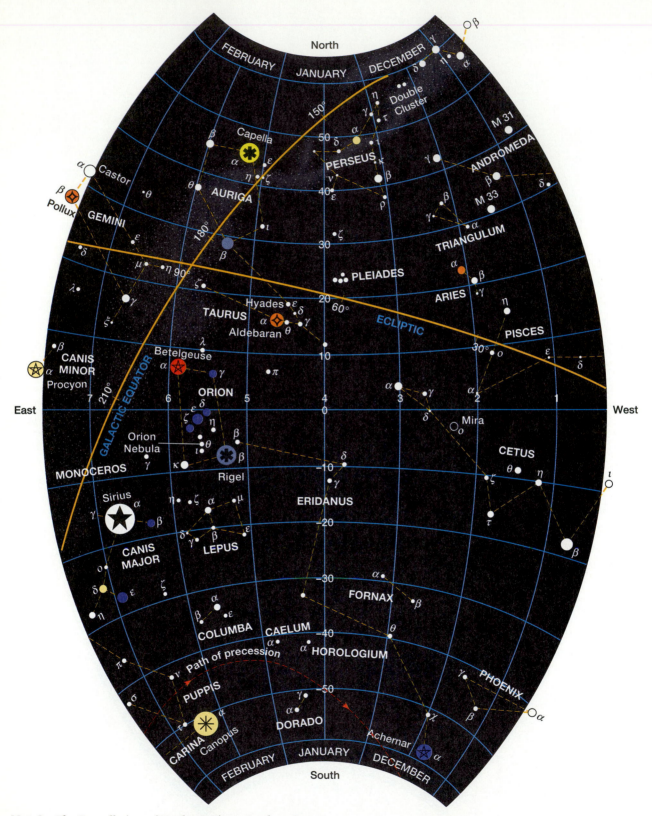

Map 3. *The Constellations of Northern Winter, Southern Summer*

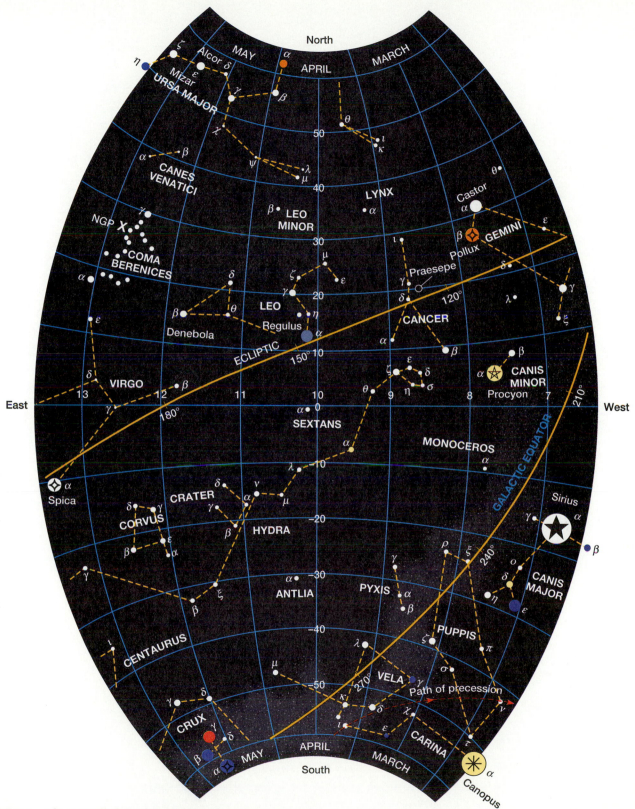

Map 4. *The Constellations of Northern Spring, Southern Autumn*

Map 5. *The Constellations of Northern Summer, Southern Winter*

Map 6.   *The South Polar Constellations*

The Messier Catalogue is a list of 103 clusters, galaxies, and nebulae compiled by Charles Messier between 1781 and 1784 and extended to 109 objects in 1786 by Pierre Méchain. The Catalogue is the fundamental list of such objects for beginning observers and contains some of the most famous and lovely of celestial sights.

The Messier Catalogue

The Messier Catalogue

M	NGC	Constellation	α (2000)		δ		Size	Description
1	1952	Taurus	05	35	+22	01	5'	*Crab Nebula,* remnant of supernova of 1054.
2	7089	Aquarius	21	33	–00	50	12'	Globular cluster.
3	5272	Canes Venatici	13	42	+28	25	19'	Bright globular cluster; binocular object.
4	6121	Scorpius	16	24	–26	31	23'	Globular cluster; binocular object.
5	5904	Serpens	15	19	+02	05	20'	Globular cluster.
6	6405	Scorpius	17	40	–32	12	26'	Open cluster; easy binocular object.
7	6475	Scorpius	17	54	–34	49	50'	Magnificent open cluster; naked eye object.
8	6523	Sagittarius	18	04	–23	23	1°	*Lagoon Nebula;* bright diffuse nebula; naked eye object.
9	6333	Ophiuchus	17	19	–18	31	6'	Globular cluster.
10	6254	Ophiuchus	16	57	–04	06	12'	Globular cluster; binocular object.
11	6705	Scutum	18	51	–06	18	12'	Open cluster; striking in telescope.
12	6218	Ophiuchus	16	47	–01	57	12'	Globular cluster; binocular object.
13	6205	Hercules	16	42	+36	30	23'	*Great cluster in Hercules;* magnificent globular cluster; barely naked eye; easy in binoculars.
14	6402	Ophiuchus	17	38	–03	15	7'	Globular cluster.
15	7078	Pegasus	21	30	+12	09	12'	Globular cluster.
16	6611	Serpens	18	19	–13	46	8'	Open cluster.
17	6618	Sagittarius	18	21	–16	12	40'	*Omega Nebula; Horseshoe Nebula;* diffuse nebula.
18	6613	Sagittarius	18	21	–17	08	7'	Open cluster.
19	6273	Ophiuchus	17	03	–26	15	5'	Globular cluster.
20	6514	Sagittarius	18	03	–23	02	30'	*Trifid Nebula;* diffuse nebula.
21	6531	Sagittarius	18	05	–22	30	12'	Open cluster.
22	6656	Sagittarius	18	36	–23	55	17'	Bright globular cluster; binocular object.
23	6494	Sagittarius	17	57	–19	01	27'	Open cluster.
24	—	Sagittarius	18	18	–18	29	1.5°	Star cloud in Milky Way; naked eye.
25	IC4725	Sagittarius	18	32	–19	15	35'	Open cluster; binocular object.
26	6694	Scutum	18	45	–09	25	9'	Open cluster.
27	6853	Vulpecula	20	00	+22	43	6'	*Dumbbell Nebula;* planetary nebula.
28	6626	Sagittarius	18	25	–24	52	15'	Globular cluster.
29	6913	Cygnus	20	24	+38	32	7'	Open cluster.
30	7099	Capricornus	21	40	–23	11	9'	Globular cluster.
31	224	Andromeda	00	43	+41	15	1 × 2°	*Great Nebula in Andromeda; Andromeda Galaxy;* spiral galaxy; naked eye object.
32	221	Andromeda	00	43	+40	51	3'	Elliptical galaxy; companion to M 31.
33	598	Triangulum	01	34	+30	38	1°	*Triangulum Spiral;* spiral galaxy; binocular object and just visible to naked eye.
34	1039	Perseus	02	42	+42	47	30'	Open cluster.
35	2168	Gemini	06	09	+24	21	30'	Open cluster; easy binocular object.
36	1960	Auriga	05	36	+34	08	16'	Open cluster.
37	2099	Auriga	05	52	+32	33	24'	Open cluster; binocular object.
38	1912	Auriga	05	29	+35	51	18'	Open cluster.
39	7092	Cygnus	21	32	+48	26	32'	Open cluster.
40	—							Does not exist; star.
41	2287	Canis Major	06	47	–20	44	32'	Open cluster; binocular object.
42	1976	Orion	05	35	–05	25	1°	*Orion Nebula;* bright diffuse nebula; easy binocular object.
43	1982	Orion	05	35	–05	16	10'	Diffuse nebula at northern edge of Orion Nebula.

The Messier Catalogue   *(continued)*

M	NGC	Constellation	α (2000)		δ		Size	Description
44	2632	Cancer	08	40	+20	00	1.5°	*Beehive* or *Praesepe Cluster;* open cluster; naked eye object.
45	—	Taurus	03	48	+24	06	2°	*Pleiades; Seven Sisters;* open cluster; obvious naked eye object.
46	2437	Puppis	07	42	−14	49	27'	Open cluster.
47	2422	Puppis	07	37	−14	29	25'	Open cluster; naked eye object.
48	2548	Hydra	08	14	−05	47	30'	Open cluster; binocular object.
49	4472	Virgo	12	30	+07	59	4'	Elliptical galaxy.
50	2323	Monoceros	07	03	−08	20	16'	Open cluster.
51	5194	Canes Venatici	13	30	+47	11	10'	*Whirlpool Nebula;* spiral galaxy.
52	7654	Cassiopeia	23	26	+61	35	13'	Open cluster.
53	5024	Coma Berenices	13	13	+18	10	14'	Globular cluster.
54	6715	Sagittarius	18	55	−30	28	6'	Globular cluster.
55	6809	Sagittarius	19	40	−30	56	15'	Globular cluster.
56	6779	Lyra	19	17	+30	02	5'	Globular cluster.
57	6720	Lyra	18	54	+33	01	1.2'	*Ring Nebula in Lyra;* planetary nebula.
58	4579	Virgo	12	38	+11	48	4'	Spiral galaxy.
59	4621	Virgo	12	42	+11	39	3'	Elliptical galaxy.
60	4649	Virgo	12	44	+11	33	4'	Elliptical galaxy.
61	4303	Virgo	12	22	+04	28	6'	Spiral galaxy.
62	6266	Ophiuchus	17	01	−30	07	6'	Globular cluster.
63	5055	Canes Venatici	13	16	+42	01	6'	Spiral galaxy.
64	4826	Coma Berenices	12	58	+21	41	6'	Spiral galaxy.
65	3623	Leo	11	19	+13	07	6'	Spiral galaxy.
66	3527	Leo	11	20	+13	01	6'	Spiral galaxy.
67	2682	Cancer	08	51	+11	51	18'	Open cluster; one of oldest known.
68	4590	Hydra	12	39	−26	45	9'	Globular cluster.
69	6637	Sagittarius	18	31	−32	21	4'	Globular cluster.
70	6681	Sagittarius	18	43	−32	18	4'	Globular cluster.
71	6838	Sagitta	19	54	+18	47	6'	Globular cluster.
72	6981	Aquarius	20	53	−12	33	5'	Globular cluster.
73	6994	Aquarius	20	59	−12	38	—	Four stars.
74	628	Pisces	01	37	+15	47	8'	Spiral galaxy.
75	6864	Sagittarius	20	06	−21	56	5'	Globular cluster.
76	650-1	Perseus	01	42	+51	34	1'	Planetary nebula.
77	1068	Cetus	05	47	−00	02	2'	Spiral galaxy.
78	2068	Orion	05	47	+00	03	7'	Diffuse nebula.
79	1904	Lepus	05	24	−24	31	8'	Globular cluster.
80	6093	Scorpius	16	17	−22	59	5'	Globular cluster.
81	3031	Ursa Major	09	56	+69	04	13'	*Great Spiral in Ursa Major;* spiral galaxy; binocular object.
82	3034	Ursa Major	09	56	+69	42	7 × 2'	Irregular starburst galaxy.
83	5236	Hydra	13	37	−29	52	9'	Spiral galaxy.
84	4374	Virgo	12	25	+12	53	3'	Elliptical galaxy.
85	4382	Coma Berenices	12	25	+18	11	3'	Elliptical galaxy.
86	4406	Virgo	12	26	+12	56	4'	Elliptical galaxy.
87	4486	Virgo	12	31	+12	23	3'	*Virgo A;* active elliptical galaxy.

The Messier Catalogue   (continued)

M	NGC	Constellation	α (2000)		δ		Size	Description
88	4501	Coma Berenices	12	32	+14	25	6 × 3'	Spiral galaxy.
89	4552	Virgo	12	36	+12	33	2'	Elliptical galaxy.
90	4569	Virgo	12	37	+13	09	6 × 3'	Spiral galaxy.
91	—	—	—		—		—	Erroneous discovery; unexplained.
92	6341	Hercules	17	17	+43	08	12'	Globular cluster.
93	2447	Puppis	07	45	−23	52	18'	Open cluster.
94	4736	Canes Venatici	12	51	+41	07	5'	Spiral galaxy.
95	3351	Leo	10	44	+11	42	3'	Spiral galaxy.
96	3368	Leo	10	47	+11	49	7 × 4'	Spiral galaxy.
97	3587	Ursa Major	11	15	+55	02	3'	*Owl Nebula;* planetary nebula.
98	4192	Coma Berenices	12	14	+14	54	8 × 2'	Spiral galaxy.
99	4254	Coma Berenices	12	19	+14	25	4'	Spiral galaxy.
100	4321	Coma Berenices	12	23	+15	49	5'	Spiral galaxy.
101	5457	Ursa Major	14	03	+54	21	22'	*Pinwheel Galaxy;* spiral galaxy.
102	—	—	—		—		—	Same as M 101.
103	581	Cassiopeia	01	33	+60	41	6'	Open cluster.
104	4594	Virgo	12	40	−11	37	7 × 2'	*Sombrero Galaxy;* spiral galaxy.
105	3379	Leo	10	48	+12	35	2'	Elliptical galaxy.
106	4258	Canes Venatici	12	19	+47	18	20 × 6'	Spiral galaxy.
107	6171	Ophiuchus	16	32	−13	03	8'	Globular cluster.
108	3556	Ursa Major	11	12	+55	41	8 × 2'	Spiral galaxy.
109	3992	Ursa Major	11	58	+53	22	7'	Spiral galaxy.

Bibliography

ASTRONOMY MAGAZINES

Astronomy, Kalmbach Publishing Co., Waukesha, WI.

Griffith Observer, Griffith Observatory, Los Angeles, CA.

Mercury, Astronomical Society of the Pacific, San Francisco, CA.

Planetary Report, Planetary Society, Pasadena, CA.

Sky and Telescope, Sky Publishing Corp., Cambridge, MA.

StarDate, McDonald Observatory, Austin, TX.

GENERAL MAGAZINES WITH ASTRONOMICAL CONTENT

American Scientist, Sigma Xi, New York, NY.

Science News, Science News, New York, NY.

Scientific American, Freeman, New York, NY.

MAPS AND ATLASES

Ridpath, I., ed. 1989. *Norton's Star Atlas.* 19th ed. Cambridge, Mass.: Sky Publishing Corp. (All naked eye stars and many telescopic objects.)

SC1, SC2, SC3 Star Charts. Cambridge, Mass.: Sky Publishing Corp. (Simple charts for beginning constellation study.)

Tirion, W. 1981. *Atlas 2000.* Cambridge, Mass.: Sky Publishing Corp. (Deep atlas to 8th magnitude includes a large number of nonstellar objects.)

Tirion, W., Rappaport, B., and Lovi, G. *Uranometria 2000.* Richmond, Va.: Willmann-Bell. (Includes stars to 10th magnitude and nonstellar objects.)

SLIDES, VIDEOS, AND POSTERS

Contact the Astronomical Society of the Pacific, 390 Ashton Avenue, San Francisco, CA 94112.

FURTHER READING

General

Abetti, G. 1952. *The History of Astronomy.* New York: Henry Schuman.

Berry, A. 1961. *A Short History of Astronomy.* Reprint. New York: Dover.

Goldsmith, D. 1990. *The Astronomers.* New York: St. Martin's Press.

Hermann, D. B., and Krisciunas, K. 1984. *The History of Astronomy from Herschel to Hertzsprung.* Cambridge, England: Cambridge University Press.

Illingworth, V. 1985. *The Facts on File Dictionary of Astronomy.* 2nd ed. New York: Facts on File Publications.

Lang, K. R. 1992. *Astrophysical Data: Planets and Stars.* New York: Springer.

Mallas, J. H., and Kreimer, E. 1978. *The Messier Album.* Cambridge Mass.: Sky Publishing Corp.

Maran, S., ed. 1992. *The Astronomy and Astrophysics Encyclopedia.* New York: Van Nostrand Reinhold.

Pannekoek, A. 1989. *A History of Astronomy.* Reprint. New York: Dover.

Parker, B. 1988. *Creation: The Story of the Origin and Evolution of the Universe.* New York: Plenum.

Sagan, C. 1980. *Cosmos.* New York: Ballantine.

Stott, C., ed. 1991. *Images of the Universe.* Cambridge, England: Cambridge University Press.

Tauber, G. E. 1979. *Man's View of the Universe.* New York: Crown.

Part I. Classical Astronomy

Chapter 1. From Earth to Universe

Morrison, P., Morrison, P., and the Office of C. and R. Eames. 1982. *Powers of Ten.* Scientific American Library. New York: Freeman.

Chapters 2 and 3. The Earth and the Sky; The Earth and the Sun

Aveni, A. F. 1986. Archaeoastronomy: Past, Present, and Future. *Sky and Telescope.* 72:456 (November).

Aveni, A. F. 1989. *Empires of Time: Calendars, Clocks, and Cultures.* New York: Basic Books.

Daniel, G. 1980. Megalithic Monuments. *Scientific American.* 243:80 (July).

Gingerich, O. 1986. Islamic Astronomy. *Scientific American.* 254:74 (April).

Hawkins, G. S. 1965. *Stonehenge Decoded.* Garden City, N.Y.: Doubleday.

Moyer, G. 1982. The Gregorian Calendar. *Scientific American*. 246:144 (May).

Piini, E. W. 1986. Ulugh Beg's Forgotten Observatory. *Sky and Telescope*. 71:542 (June).

Reese, R. L. 1992. Midwinter Sunrise at El Karnak. *Sky and Telescope*. 83:276 (March).

Smart, W. M. 1977. *Spherical Astronomy*. 6th ed. Cambridge, England: Cambridge University Press.

Stott, C. 1984. Greenwich: Where East Meets West. *Sky and Telescope*. 68:300 (October).

Chapter 4. The Face of the Sky

Allen, R. H. 1963. *Star Names, Their Lore and Meaning*. Reprint. New York: Dover.

Cozens, G. 1986. Discover the Southern Skies–I. *Sky and Telescope*. 71:126 (February).

Cozens, G. 1986. Discover the Southern Skies–II. *Sky and Telescope*. 71:237 (March).

Fraknoi, A. 1989. Your Astrology Defense Kit. *Sky and Telescope*. 78:146 (August).

Gingerich, O. 1984. The Origin of the Zodiac. *Sky and Telescope*. 67:218 (March).

Kelly, I. 1980. The Scientific Case Against Astrology. *Mercury*. 9:135 (November/December).

Krupp, E. C. 1991. *Beyond the Blue Horizon*. New York: HarperCollins.

Kunitzsch, P. 1983. How We Got Our "Arabic" Star Names. *Sky and Telescope*. 65:20 (January).

Kunitzsch, P., and Smart, T. 1986. *Short Guide to Modern Star Names and their Derivations*. Wiesbaden, Germany: Otto Harrassowitz.

Menzel, D. H., and Pasachoff, J. 1982. *A Field Guide to the Stars and Planets*. 3rd. ed. New York: Houghton-Mifflin.

Ridpath, I. 1988. *Star Tales*. New York: Universe Books.

Ulansey, D. 1989. *The Mithraic Mysteries*. Scientific American. 261:130 (December).

Urton, G. 1981. *At the Crossroads of the Earth and the Sky: An Andean Cosmology*. Austin: University of Texas Press.

Villard, R. 1989. The World's Biggest Star Catalogue. *Sky and Telescope*. 78:583 (December).

Williamson, R. A. 1984. *Living the Sky: The Cosmos of the American Indian*. Boston: Houghton-Mifflin.

See also George Lovi's "Ramblings" column that ran in *Sky and Telescope* until 1993.

Chapter 5. The Earth and the Moon

Meeus, J., Grosjean, C. C., and Vanderleen, W. 1966. *Canon of Solar Eclipses*. New York: Pergamon.

Newton, R. R. 1970. *Ancient Astronomical Observations of the Accelerations of the Earth and Moon*. Baltimore: Johns Hopkins Press.

Oppolzer, T. 1962. *Canon of Eclipses*. Reprint. New York: Dover.

Schaeffer, B. E. 1992. Lunar Eclipses that Changed the World. *Sky and Telescope*. 84:639 (December).

Stephenson, F. R. 1982. Historical Eclipses. *Scientific American*. 247:170 (October).

Zirker, J. B. 1984. *Total Eclipses of the Sun*. New York: Prentice-Hall.

Chapter 6. The Planets

Gingerich, O. 1982. The Galileo Affair. *Scientific American*. 247:131 (August).

Gingerich, O. 1993. *The Eye of Heaven: Ptolemy, Copernicus, Kepler*. New York: American Institute of Physics.

Gingerich, O. 1993. How Galileo Changed the Rules of Science. *Sky and Telescope* 85:32 (March).

Hodson, D. G., ed. 1974. *The Place of Astronomy in the Ancient World*. Oxford: Oxford University Press.

McPeak, W. J. 1990. Tycho Brahe Lights Up the Universe. *Astronomy*. 18:29 (December).

Pedersen, O. 1974. *A Survey of the Almagest*. Odense, Denmark: Odense University Press.

Part II. Physical Astronomy

Chapter 7. Newton, Einstein, and Gravity

Christianson, G. E. 1987. Newton's Principia: A Retrospective. *Sky and Telescope*. 74:18 (July).

Cohen, I. B. 1981. Newton's Discovery of Gravity. *Scientific American*. 244:165 (March).

Drake, S. 1980. Newton's Apple and Galileo's Dialogue. *Scientific American*. 243:151 (August).

Einstein, A. 1961. *Relativity: The Special and General Theory*. Reprint. New York: Crown.

Gillies, G. T., and Sanders, A. J. 1993. Getting the Measure of Gravity. Sky and Telescope. 85:28 (April).

Gleick, J. 1988. *Chaos*. New York: Penguin.

Killian, A. M. 1989. Playing Dice with the Solar System. *Sky and Telescope*. 78:136 (August).

Stipe, J. G., Jr. 1967. *The Development of Physical Theories*. New York: McGraw-Hill.

Wheeler, J. A. 1990. *A Journey into Gravity and Spacetime*. Scientific American Library. New York: Freeman.

Will, C. F. 1983. Testing General Relativity. *Sky and Telescope*. 66:294 (October).

Will, C. F. 1986. *Was Einstein Right? Putting General Relativity to the Test*. New York: Basic Books.

Chapter 8. Atoms and Light

Atkins, P. W. 1991. *Atoms, Electrons, and Change*. Scientific American Library. New York: Freeman.

Close, F., Marten, M., and Sutton, C. 1987. *The Particle Explosion*. New York: Oxford University Press.

Davies, P. 1987. Particle Physics for Everybody. *Sky and Telescope*. 74:582 (December).

Overheim, R. D., and Wagner, E. L. 1982. *Light and Color*. New York: Wiley.

Sobel, M. I. 1987. *Light*. Chicago: University of Chicago Press.

Weinberg, S. 1983. *The Discovery of Subatomic Particles*. New York: Freeman.

Williamson, S. J., and Cummins, H. Z. *Light and Color in Nature and Art*. New York: Wiley.

Chapter 9. The Tools of Astronomy

Cole, S. 1992. Astronomy on the Edge: Using the Hubble Space Telescope. *Sky and Telescope*. 84:386 (October).

Davis, J. 1991. Measuring the Stars. *Sky and Telescope*. 82:361 (October).

Davis, J. 1992. The Quest for High Resolution. *Sky and Telescope*. 83:29 (January).

Janesick, J., and Blouke, M. 1987. Sky on a Chip: The Fabulous CCD. *Sky and Telescope*. 74:238 (September).

Kellerman, K. I. 1991. Radio Astronomy: The Next Decade. *Sky and Telescope*. 82:247 (September).

King, H. C. 1979. *The History of the Telescope*. Reprint. New York: Dover.

Kondo, Y., Wamsteker, W., and Stickland, D. 1993. IUE: 15 Years and Counting. *Sky and Telescope*. 86:30 (September).

Lerner, R. 1986. The Legacy of the 200-inch. *Sky and Telescope*. 71:349 (April).

Maran, S. P. 1992. Hubble Illuminates the Universe. *Sky and Telescope*. 83:619 (June).

O'Dell, C. R. 1989. Building the Hubble Space Telescope. *Sky and Telescope*. 78:31 (July).

Powell, C. S. 1991. Mirroring the Cosmos. *Scientific American*. 265:113 (November).

Readhead, A. C. S. 1982. Radio Astronomy by Very Long Baseline Interferometry. *Scientific American*. 246:52 (June).

Ressmeyer, R. H. 1992. Keck's Giant Eye. *Sky and Telescope*. 84:623 (December).

Sinnott, R. W., and Nyren, K. 1993. The World's Largest Telescopes. *Sky and Telescope*. 86:27 (July).

Strom, S. E. 1991. New Frontiers in Ground-based Optical Astronomy. *Sky and Telescope*. 82:18 (July).

Sullivan, W. T., III. 1982. Radio Astronomy's Golden Anniversary. *Sky and Telescope*. 64:544 (December).

Svec, M. T. 1992. The Birth of Electronic Astronomy. *Sky and Telescope*. 83:496 (May).

Tucker, W., and Tucker, K. 1986. *The Cosmic Enquirers*. Cambridge, Mass.: Harvard University Press.

Verschuur, G. 1987. *The Invisible Universe Revealed: The Story of Radio Astronomy*. New York: Springer.

Wearner, R. 1992. The Birth of Radio Astronomy. *Astronomy*. 20:49 (June).

Part III. Planetary Astronomy

General

Beatty, J. K., and Chaikin, A., eds. 1990. *The New Solar System*. 3rd ed. Cambridge, England: Cambridge University Press.

Kaufmann, W. J. 1979. *Planets and Moons*. New York: Freeman.

McLaughlin, W. I. 1989. Voyager's Decade of Wonder. *Sky and Telescope*. 78:16 (July).

Miner, E. D. 1990. Voyager 2's Encounter with the Gas Giants. *Physics Today*. 43(no. 7):40 (July).

Morrison, D. 1993. *Exploring Planetary Worlds*. Scientific American Library. New York: Freeman.

Morrison, D., and Owen, T. 1988. *The Planetary System*. New York: Addison-Wesley.

Rothery, D. A. 1992. *The Satellites of the Outer Planets*. Oxford: Oxford University Press.

Chapter 10. The Earth

Akasofu, S.-I. 1989. The Dynamic Aurora. *Scientific American*. 260:90 (May).

Anderson, D. L., and Dziewonski, A. D. 1984. Seismic Tomography. *Scientific American*. 251:60 (October).

Badash, L. 1989. The Age-of-the-Earth Debate. *Scientific American*. 261:90 (August).

Bonatti, E. 1987. The Rifting of Continents. *Scientific American*. 256:97 (March).

Broecker, W. S. 1983. The Ocean. *Scientific American*. 249:146 (September).

Burchfiel, B. C. 1983. *Scientific American*. 249:130 (September).

Davis, N. 1992. *The Aurora Watcher's Handbook*. College, Alaska: University of Alaska Press.

Foxworthy, B. L., and Hill, M. 1982. *Volcanic Eruptions of 1980 at Mount St. Helens*. Geological Survey Professional Paper 1249. Washington, D.C.: U.S. Government Printing Office.

Frencheteau, J. 1983. The Oceanic Crust. *Scientific American*. 249:114 (September).

Greenler, R. 1980. *Rainbows, Halos, and Glories*. Cambridge, England: Cambridge University Press.

Grove, N. 1992. Crucibles of Creation. *National Geographic*. 182:5 (December).

Ingersoll, A. P. 1983. The Atmosphere. *Scientific American*. 249:162 (September).

McKenzie, D. P. 1983. *Scientific American*. 249:67 (September).

Meinel, A., and Meinel, M. 1983. *Sunsets, Twilights, and Evening Skies*. Cambridge, England: Cambridge University Press.

Minnaert, M. *The Nature of Light and Color in the Open Air*. Reprint. New York: Dover.

Siever, R. 1983. The Dynamic Earth. *Scientific American*. 249:46 (September).

White, R. M. 1990. The Great Climate Debate. *Scientific American*. 263:36 (July).

White, R. S., and McKenzie, D. P. 1989. Volcanism at Rifts. *Scientific American*. 261:62 (July).

Chapter 11. The Moon

Alvarez, W., and Asaro, F. 1990. An Extraterrestrial Impact. *Scientific American*. 263:78 (October).

Benningfeld, D. 1991. Mysteries of the Moon. *Astronomy*. 19:51 (December).

Burnham, R. 1993. Galileo Returns to the Earth and Moon. *Astronomy*. 21:34 (March).

Grieve, R. A. F. 1990. Impact Cratering on the Earth. *Scientific American*. 262:66 (April).

Kosovsky, L. J., and Farouk, E.-B. 1970. *The Moon as Viewed by Lunar Orbiter*. Washington, D.C.: NASA.

Masursky, H., Colton, G. W., and Farouk, E.-B. 1978. *Apollo over the Moon: A View from Orbit*. Washington, D.C.: NASA.

Moore, P. 1981. *The Moon*. New York: Rand McNally.

Musgrove, R. G. 1971. *Lunar Photographs from Apollos 8, 9, and 10*. Washington, D.C.: NASA.

Wilhelms, D. E. 1987. *The Geologic History of the Moon*. Geological Survey Professional Paper 1348. Washington, D.C.: U.S. Government Printing Office.

Chapter 12. Hot Worlds

Beatty, J. K. 1993. Working Magellan's Magic. *Sky and Telescope*. 86:16 (August).

Burnham, R. 1993. What Makes Venus Go. *Astronomy*. 21:40 (January).

Davies, M. E. et al. 1978. *Atlas of Mercury*. Washington, D.C.: NASA.

Plaut, J. J. 1993. Venus in 3-D. *Sky and Telescope*. 86:32 (August).

Saunders, R. S. et al. 1991. Magellan at Venus. *Science*. 252:185–312.

Saunders, R. S. 1991. Magellan at Venus: A Magellan Progress Report. *Mercury*. 20:131 (September/October).

Solomon, S. C. 1993. The Geophysics of Venus. *Physics Today*. 46(no. 7):48–55 (July).

Solomon, S. C. et al. 1992. Venus Tectonics: An Overview of Magellan Observations. *Journal of Geophysical Research*. 97:13, 199.

Stofan, E. R. 1993. The New Face of Venus. *Sky and Telescope*. 86:22 (August).

Strom, R. G. 1990. Mercury: The Forgotten Planet. *Sky and Telescope*. 80:256 (September).

Venus Unveiled. 1992. *Sky and Telescope*. 83:258 (March).

Chapter 13. Intriguing Mars

Baker, V. R. 1982. *The Channels of Mars*. Austin: University of Texas Press.

Edgett, K., Geissler, P., and Herkenhoff, K. 1993. The Sands of Mars. *Astronomy*. 21:26 (June).

French, B. M. 1977. *Mars: The Viking Discoveries*. Washington, D.C.: NASA.

Kieffer, H. H. et al., eds. *Mars*. Tucson: University of Arizona Press.

McKay, C. P. 1993. Did Mars Once Have Martians? *Astronomy*. 21:26 (September).

McKay, C. P., Toon, O. B., and Kasting, J. F. 1991. Making Mars Habitable. *Nature*. 352:489.

Sheehan, W. 1988. Mars 1909: Lessons Learned. *Sky and Telescope*. 76:247 (September).

Chapter 14. Magnificent Jupiter

Beebe, R. F. 1990. Queen of the Giant Storms. *Sky and Telescope*. 80:339 (October).

Hunt, G., and Moore, P. 1981. *Jupiter*. New York: Rand McNally.

Ingersoll, A. P. 1981. Jupiter and Saturn. *Scientific American*. 245:90 (December).

Morrison, D. 1980. *Voyage to Jupiter*. Washington, D.C.: NASA.

Morrison, D. 1985. The Enigma Called Io. *Sky and Telescope*. 69:198 (March).

Soderblom, L. A. 1980. *Scientific American.* 243:88 (November).

Chapter 15. Beautiful Saturn

Beatty, J. K. 1981. Voyager at Saturn. *Sky and Telescope.* 62:430 (November).

Hunt, G., and Moore, P. 1982. *Saturn.* New York: Rand McNally.

Eliott, J., and Kerr, R. 1987. *Rings.* Cambridge, Mass.: MIT Press.

Morrison, D. 1982. *Voyages to Saturn.* Washington, D.C.: NASA.

O'Meara, S. J. 1991. Saturn's Great White Spot Spectacular. *Sky and Telescope.* 81:144 (February).

Soderblom, L. A., and Johnson, T. V. 1982. The Moons of Saturn. *Scientific American.* 246:101 (January).

Chapter 16. Outer Worlds

Beatty, J. K. 1986. A Place Called Uranus. *Sky and Telescope.* 71:333 (April).

Beatty, J. K. 1990. Getting to Know Neptune. *Sky and Telescope.* 79:146 (February).

Berry, R. 1989. Neptune Revealed. *Astronomy.* 17:22 (December).

Binzel, R. P. 1990. Pluto. *Scientific American.* 262:50 (June).

Cuzzi, J. N., and Esposito, L. W. 1987. The Rings of Uranus. *Scientific American.* 257:52 (July).

Dowling, T. 1990. Big, Blue: The Twin Worlds of Uranus and Neptune. *Astronomy.* 18:42 (October).

Hunt, G., and Moore, P. *Atlas of Uranus.* Cambridge, England: Cambridge University Press.

Ingersoll, A. P. 1987. Uranus. *Scientific American.* 256:38 (January).

Johnson, T. V., Brown, R. H., and Soderblom, L. A. 1987. The Moons of Uranus. *Scientific American.* 256:48 (April).

Laeser, R. P., McLaughlin, W. I, and Wolff, D. M. 1986. Engineering Voyager 2's Encounter with Uranus. *Scientific American.* 255:36 (November).

Miner, E. D. 1989. Voyager's Last Encounter. *Sky and Telescope.* 78:26 (July).

Moore, P. 1989. The Discovery of Neptune. *Mercury.* 18:98 (July/August).

Morrison, N. D. 1989. A Refined View of Miranda. *Mercury.* 18:55 (March/April).

Neptune and Triton: Worlds Apart. 1990. *Sky and Telescope.* 79:136 (February).

Tombaugh, C. 1991. Plates, Pluto, and Planet X. *Sky and Telescope.* 81:360 (April).

Chapter 17. Planetary Creation and its Debris

Balsiger, H., Fechtig, H., and Geiss, J. 1988. A Close Look at Halley's Comet. *Scientific American.* 259:96 (September).

Binzel, R. P., Barucci, M. A., and Fulchignoni, M. 1991. *Scientific American.* 265:88 (October).

Binzel, R. P., Gehrels, T., and Matthews, M. S. 1989. *Asteroids II.* Tucson: University of Arizona Press.

Brandt, J. C. 1992. Rendezvous in Space: *The Science of Comets.* New York: Freeman.

Brandt, J. C., and Niedner, M. B., Jr. 1986. The Structure of Comet Tails. *Scientific American.* 254:49 (January).

Brophy, T. 1993. Motes in the Solar System's Eye. *Astronomy.* 21:34 (May).

Delsemme, A. H. 1989. Whence Come Comets? *Sky and Telescope.* 77:260 (March).

Dodd, R. T. 1981. *Meteorites.* Cambridge, England: Cambridge University Press.

Durda, D. 1993. All in the Family. *Astronomy.* 21:36 (February).

Gingrich, O. 1986. Newton, Halley, and the Comet. *Sky and Telescope.* 71:230 (March).

Gropman, D. 1985. *Comet Fever: A Popular History of Halley's Comet.* New York: Simon & Schuster.

Kronk, G. W. 1988. Meteor Showers. *Mercury.* 17:162 (November/December).

Kronk, G. W. 1988. *Meteor Showers: A Descriptive Catalogue.* Hillside, N.J.: Enslow.

Lewis, R. S., and Anders, E. 1983. Interstellar Matter in Meteorites. *Scientific American.* 249:66 (August).

McFadden, L. A., and Chapman, C. R. 1992. Interplanetary Fugitives. *Astronomy.* 20:30 (August).

McSween, H. Y., Jr., and Stolper, E. M. 1980. Basaltic Meteorites. *Scientific American.* 242:54 (June).

Spratt, C., and Stephens, S. 1992. Against All Odds: Meteorites That Have Struck Home. *Mercury.* 21:50 (March/April).

Stern, A. 1992. Where Has Pluto's Family Gone? *Astronomy.* 20:41 (September).

Wetherill, G. W. 1981. The Formation of the Earth from Planetesimals. *Scientific American.* 244:162 (June).

Whipple, F. L. 1985. *The Mystery of Comets.* Washington, D.C.: Smithsonian Institution Press.

Whipple, F. L. 1987. The Black Heart of Comet Halley. *Sky and Telescope.* 73:242 (March).

Part IV. Stellar Astronomy

General

Aller, L. H. 1971. *Atoms, Stars, and Nebulae.* Cambridge, Mass.: Harvard University Press.

Burnham, R. R. 1978. *Burnham's Celestial Handbook.* 3 vols. New York: Dover.

DeVorkin, D. H. 1989. Henry Norris Russell. *Scientific American.* 260:127 (May).

Hoskin, M. 1986. William Herschel and the Making of Modern Astronomy. *Scientific American.* 254:106 (April).

Kaler, J. B. 1992. *Stars.* Scientific American Library. New York: Freeman.

Chapter 18. The Sun

Bahcall, J. N. 1990. The Solar Neutrino Problem. *Scientific American.* 262:54 (May).

Fischer, D. 1992. Closing in on the Solar-Neutrino Problem. *Sky and Telescope.* 84:378 (October).

Foukal, P. V. 1990. The Variable Sun. *Scientific American.* 262:34 (February).

Friedman, H. 1986. *Sun and Earth.* Scientific American Library. New York: Freeman.

Golub, L. 1993. Heating the Sun's Million Degree Corona. *Astronomy.* 21:27 (May).

Harvey, J. W., and Leibacher, J. W. 1987. GONG: To See Inside Our Sun. *Sky and Telescope.* 74:470 (November).

Leibacher, J. W. et al. 1985. Helioseismology. *Scientific American.* 253:48 (September).

McCrea, W. 1991. Arthur Stanley Eddington. *Scientific American.* 264:92 (June).

Petersen, C. C. et al. 1993. Yohkoh and the Mysterious Solar Flares. 1993. *Sky and Telescope.* 86:20 (September).

Nicolson, I. 1982. *The Sun.* New York: Rand McNally.

Phillips, K. J. H. 1992. *Guide to the Sun.* Cambridge, England: Cambridge University Press.

Schwarzschild, B. 1990. Solar Neutrino Update: Three Detectors Tell Three Stories. *Physics Today.* 43(no. 10):17 (October).

Wentzel, D. G. 1989. *The Restless Sun.* Washington, D.C.: Smithsonian Institution Press.

Chapter 19. The Stars

Baliunas, S., and Sarr, S. 1992. Unfolding the Mysteries of Stellar Cycles. *Astronomy.* 20:42 (May).

Giampapa, M. S. 1987. The Solar-Stellar Connection. *Sky and Telescope.* 74:142 (August).

Kaler, J. B. 1989. *Stars and their Spectra.* Cambridge, England: Cambridge University Press.

Kaler, J. B. 1991. The Faintest Stars in the Galaxy. *Astronomy.* 19:26 (August).

Kovalevsky, J. 1990. Astrometry from Earth and Space. *Sky and Telescope.* 79:493 (May).

Spradley, J. L. 1990. The Industrious Mrs. Fleming. *Astronomy.* 18:48 (July).

Steffey, P. C. 1992. The Truth About Star Colors. *Sky and Telescope.* 84:266 (September).

Wilson, O. C., Vaughan, A. H., and Mihalas, D. 1981. The Activity Cycles of Stars. *Scientific American.* 244:104 (February).

Chapter 20. Stellar Groupings: Doubles, Multiples, and Clusters

Cannizzo, J. K., and Kaitchuck, R. H. 1992. Accretion Disks in Interacting Binary Stars. *Scientific American.* 266:92 (January).

Harris, W. E. 1991. Globular Clusters in Distant Galaxies. *Sky and Telescope.* 81:148 (February).

Hodge, P. 1988. How Far Are the Hyades? *Sky and Telescope.* 75:138 (February).

King, I. R. 1985. Globular Clusters. *Scientific American.* 252:79 (June).

Kopal, Zdenek. 1990. Eclipsing Binary Stars. *Mercury.* 19:88 (May/June).

Terrell, D. 1992. Close Binary Stars. *Astronomy.* 20:34 (October).

Tomkin, J., and Lambert, D. 1987. The Strange Case of Beta Lyrae. *Sky and Telescope.* 74:354 (October).

Trimble, V. 1984. A Field Guide to Close Binary Stars. *Sky and Telescope.* 68:306 (October).

van den Bergh, S. 1992. Star Clusters: Enigmas in Our Backyard. *Sky and Telescope* 83:508 (May).

White, R. E. 1991. Globular Clusters: Fads and Fallacies. *Sky and Telescope.* 81:24 (January).

Chapter 21. Unstable Stars

Gingerich, O., and Welther, B. 1985. Harlow Shapley and the Cepheids. *Sky and Telescope.* 70:540 (December).

Hoffmeister, C., Richter, G., and Wenzel, W. 1985. *Variable Stars.* New York: Springer.

Petit, M. 1987. *Variable Stars.* New York: Wiley.

Merrill, P. 1940. *The Spectra of Long Period Variable Stars.* Chicago: University of Chicago Press.

Norris, R. 1986. Cosmic Masers. *Sky and Telescope.* 71:284 (March).

Percy, J. R. 1984. Cepheids: Cosmic Yardsticks, Celestial Mysteries? *Sky and Telescope.* 68:517 (December).

Part V. Birth and Death in the Galaxy

Chapter 22. The Interstellar Medium

Blitz, L. 1982. Giant Molecular-Cloud Complexes in the Galaxy. *Scientific American.* 246:84 (April).

Dame, T. M. 1988. The Molecular Milky Way. *Sky and Telescope.* 76:22 (July).

Goldstein, A. 1990. Magnificent Orion. *Astronomy.* 18:79 (November).

Greenberg, J. M. 1984. The Structure and Evolution of Interstellar Grains. *Scientific American.* 250:124 (June).

Kancke, R. 1984. Cosmic Dust and the Comet Connection. *Sky and Telescope.* 68:206 (September).

Malin, D. 1982. The Dust Clouds of Sagittarius. *Sky and Telescope.* 63:254 (March).

Malin, D. 1987. In the Shadow of the Horsehead. *Sky and Telescope.* 74:253 (September).

Smith, D. H. 1985. Reflection Nebulae: Celestial Veils. *Astronomy.* 13:207 (September).

Verschuur, G. 1989. *Interstellar Matters.* New York: Springer.

Verschuur, G. 1992. Interstellar Molecules. *Sky and Telescope.* 83:379 (April).

Verschuur, G. 1992. Star Dust. *Astronomy.* 20:46 (March).

Chapter 23. Star Formation

Beckwith, S., and Sargent, A. 1987. HL Tauri: A Site for Planet Formation? *Mercury.* 16:178 (November/December).

Boss, A. P. 1991. The Genesis of Binary Stars. *Astronomy.* 19:34 (June).

Bruning, D. 1992. Desperately Seeking Jupiter. *Astronomy.* 20:37 (July).

Chyba, C. 1992. The Cosmic Origins of Life on Earth. *Astronomy.* 20:29 (November).

Cohen, M. 1988. *In Darkness Born.* Cambridge, England: Cambridge University Press.

Drake, F., and Sobel, D. 1992. *Is Anyone Out There? The Scientific Search for Extraterrestrial Intelligence.* New York: Delacorte.

Feinberg, R. T. 1990. Bad News for Brown Dwarfs. *Sky and Telescope.* 80:370 (October).

Goldsmith, D., and Owen, T. 1992. *The Search for Life in the Universe.* 2nd ed. New York: Addison-Wesley.

Horgan, J. 1991. In the Beginning. *Scientific American.* 264:116 (February).

Lada, C. J. 1986. Star in the Making. *Sky and Telescope.* 72:334 (October).

Lada, C. J. 1993. Deciphering the Mysteries of Stellar Origins. *Sky and Telescope.* 85:18 (May).

Lada, C. J., and Shu, F. H. 1990. The Formation of Sunlike Stars. *Science.* 248:564.

Naeye, R. 1992. SETI at the Crossroads. *Sky and Telescope.* 84:507 (November).

Sargent, A. I., and Beckwith, S. V. W. 1993. The Search for Forming Planetary Systems. *Physics Today.* 46(no. 4):22 (April).

Schild, R. E. 1990. A Star Is Born. *Sky and Telescope.* 80:600 (December).

Schorn, R. A. 1981. Extraterrestrial Beings Don't Exist. *Sky and Telescope.* 62:207 (September).

Stahler, S. W. 1991. The Early Life of Stars. *Scientific American.* 265:48 (July).

Tipler, F. J. 1982. The Most Advanced Civilization in the Galaxy Is Ours. *Mercury.* 11:5 (January/February).

Chapter 24. The Life and Death of Stars

Balick, B. 1987. The Shaping of Planetary Nebulae. *Sky and Telescope.* 73:125 (February).

Bruning, D. 1993. Neon Nova. *Astronomy.* 21:36 (July).

Kaler, J. B. 1982. Bubbles from Dying Stars. *Sky and Telescope.* 63:129 (February).

Kaler, J. B. 1988. Journeys on the HR Diagram. *Sky and Telescope.* 75:483 (May).

Kaler, J. B. 1990. The Coolest Stars. *Astronomy.* 18:20 (May).

Kaler, J. B. 1990. Realm of the Hottest Stars. *Astronomy.* 18:32 (February).

Kaler, J. B. 1991. The Smallest Stars in the Universe. *Astronomy.* 19:50 (November).

Kawaler, S. D., and Winget, D. E. 1987. White Dwarfs: Fossil Stars. *Sky and Telescope.* 74:132 (August).

Nather, R. E., and Winget, D. E. 1992. Taking the Pulse of White Dwarfs. *Sky and Telescope.* 83:374 (April).

Soker, N. 1992. Planetary Nebulae. *Scientific American.* 266:78 (May).

Trimble, V. 1986. White Dwarfs: The Once and Future Suns. *Sky and Telescope.* 72:348 (October).

Williams, R. E. 1981. The Shells of Novas. *Scientific American.* 244:120 (April).

Chapter 25. Catastrophic Evolution

Bailyn, C. 1991. Problems with Pulsars. *Mercury.* 20:55 (March/April).

Bethe, H. A. 1990. Supernovae. *Physics Today.* 43(no. 9):24 (September).

Bethe, H. A., and Brown, G. 1985. How a Supernova Explodes. *Scientific American.* 252:60 (May).

Croswell, K. 1992. The Best Black Hole in the Galaxy. *Astronomy.* 20:30 (March).

Graham-Smith, F. 1990. Pulsars Today. *Sky and Telescope.* 80:240 (September).

Greenstein, G. 1983. *Frozen Star.* New York: Freundlich.

Hewish, A. 1989. Pulsars after 20 Years. *Mercury.* 18:12 (January/February).

Kaler, J. B. 1990. The Largest Stars in the Galaxy. *Astronomy.* 18:30 (October).

Kaler, J. B. 1991. The Brightest Stars. *Astronomy.* 19:30 (May).

Kamper, K. W., and van den Bergh, S. 1991. Capturing a Stellar Explosion: A 31-Year Time Exposure of Cassiopeia A. *Mercury.* 20:176 (November/December).

Kaufmann, W. 1979. *Black Holes and Warped Spacetime.* New York: Freeman.

Lattimer, J. M., and Burrows, A. S. 1988. Neutrinos from Supernova 1987A. *Sky and Telescope.* 76:348 (October).

Luminet, J.-P. 1987. *Black Holes.* Cambridge, England: Cambridge University Press.

Malin, D. 1987. The Splendor of Eta Carinae. *Sky and Telescope.* 73:14 (January).

Malin, D., and Allen, D. 1990. Echoes of the Supernova. *Sky and Telescope.* 79:22 (January).

Marshall, L. A. 1988. *The Supernova Story.* New York: Plenum.

McClintock, J. 1987. Stalking the Black Hole in the Garden of the Unicorn. *Mercury.* 16:108 (July/August).

McClintock, J. 1988. Do Black Holes Exist? *Sky and Telescope.* 75:28 (January).

Naeye, R. 1993. Supernova 1987A Revisited. *Sky and Telescope.* 86:39 (February).

Schorn, R. A. 1986. Binary Pulsars: Back from the Grave. *Sky and Telescope.* 72:588 (December).

Seward, F. 1986. Neutron Stars in Supernova Remnants. *Sky and Telescope.* 71:6 (January).

Seward, F. D., Gorenstein, P., and Tucker, W. H. 1985. Young Supernova Remnants. *Scientific American.* 253:88 (August).

Shaham, J. 1987. The Oldest Pulsars in the Universe. *Scientific American.* 256:50 (February).

Shipman, H. 1980. *Black Holes, Quasars, and the Universe.* New York: Houghton Mifflin.

Straka, W. 1987. The Cygnus Loop: An Older Supernova Remnant. *Mercury.* 16:150 (September/October).

van Buren, D. 1993. Bubbles in the Sky. *Astronomy.* 21:47 (January).

Wheeler, J. C., and Harkness, R. P. 1987. Helium-rich Supernovas. *Scientific American.* 257:50 (November).

Will, C. 1987. The Binary Pulsar: Gravity Waves Exist. *Mercury.* 16:162 (November/December).

Woosley, S., and Weaver, T. 1989. The Great Supernova of 1987. *Scientific American.* 261:32 (August).

Chapter 26. The Galaxy

Bok, B. J. 1981. The Milky Way Galaxy. *Scientific American.* 244:93 (March).

Bok, B., and Bok, P. 1981. *The Milky Way.* Cambridge, Mass.: Harvard University Press.

Burnham, R. 1990. Strange Doings at the Milky Way's Core. *Astronomy.* 18:39 (October).

Crosswell, K. 1992. Galactic Archaeology. *Astronomy.* 20:29 (July).

Friedlander, M. 1990. Cosmic Rays: "A Thin Rain of Charged Particles." *Mercury.* 19:130 (September/October).

Helfrand, D. J. 1988. Fleet Messengers from the Cosmos. *Sky and Telescope.* 75:263 (March).

Oort, J. 1992. Exploring the Nuclei of Galaxies (Including Our Own). *Mercury.* 21:57 (March/April).

Robinson, L. J. 1982. The Black Heart of the Milky Way. *Sky and Telescope.* 64:133 (August).

Smith, D. H. 1990. Seeking the Origin of Cosmic Rays. *Sky and Telescope.* 79:479 (May).

Townes, C. H., and Genzel, R. 1990. What Is Happening at the Center of Our Galaxy? *Scientific American.* 262:46 (April).

Twarog, B. A. 1985. Chemical Evolution of the Galaxy. *Mercury.* 14:107 (July/August).

Verschuur, G. L. 1990. The Magnetic Milky Way. *Astronomy.* 18:32 (June).

Verschuur, G. L. 1993. Journey into the Galaxy. *Astronomy.* 21:33 (January).

Part VI. Galaxies and the Universe

General

Jones, B. 1989. The Legacy of Edwin Hubble. *Astronomy.* 17:38 (December).

Osterbrock, D., and Brashear, R. 1990. Young Edwin Hubble. *Mercury.* 19:2 (Jan./Feb.)

Sadoulet, B., and Cronin, J. W. 1992. Subatomic Astronomy. *Sky and Telescope.* 83:25 (January).

Sandage, A., Sandage, M., and Kristian, J., eds. 1975. *Galaxies and the Universe.* Chicago: University of Chicago Press.

Tucker, W., and Tucker, K. 1988. *Dark Matter*. New York: Morrow.

Chapter 27. Galaxies

Allen, D. A. 1987. Star Formation and IRAS Galaxies. *Sky and Telescope*. 73:372 (April).

Barnes, J., Hernquist, L., and Schweizer, F. 1991. Colliding Galaxies. *Scientific American*. 265:40 (August).

de Vaucoleurs, G. 1983. The Distance Scale of the Universe. *Sky and Telescope*. 66:511 (December).

Dressler, A. 1993. Galaxies Far Away and Long Ago. *Sky and Telescope*. 85:22 (April).

Elmegreen, E. M., and Elmegreen, B. 1993. What Puts the Spiral in Spiral Galaxies? *Astronomy*. 21:34 (September).

Ferris, T. 1980. *Galaxies*. New York: Stewart, Tabori, and Chang.

Hodge, P. 1986. *Galaxies*. Cambridge, Mass.: Harvard University Press.

Hodge, P. 1987. The Local Group: Our Galactic Neighborhood. *Mercury*. 16:2 (January/February).

Keel, W. W. 1989. Crashing Galaxies, Cosmic Fireworks. *Sky and Telescope*. 77:18 (January).

Lake, G. 1992. Cosmology of the Local Group. *Sky and Telescope*. 84:613 (December).

Lake, G. 1992. Understanding the Hubble Sequence. *Sky and Telescope*. 83:513 (May).

Miley, G. K., and Chambers, K. C. 1993. The Most Distant Radio Galaxies. *Scientific American*. 268:54 (June).

Morrison, N. D. 1988. The Extragalactic Distance Scale. *Mercury*. 17:171 (November/December).

Parker, B. 1990. *Colliding Galaxies: The Universe in Turmoil*. New York: Plenum.

Rubin, V. C. 1983. Dark Matter in Spiral Galaxies. *Scientific American*. 248:96 (June).

Sandage, A. 1961. *The Hubble Atlas of Galaxies*. Washington, D.C.: Carnegie Institution of Washington.

Silk, J. 1986. Formation of the Galaxies. *Sky and Telescope*. 72:582 (December).

Smith, D. H. 1987. Secrets of Galaxy Clusters. *Sky and Telescope*. 73:377 (April).

Smith, R. W. 1983. The Great Debate Revisited. *Sky and Telescope*. 65:28 (January).

Struble, M. F. 1988. Diversity among Galaxy Clusters. *Sky and Telescope*. 75:16 (January).

Sulentic, J. W. 1992. Odd Couples. *Astronomy*. 20:36 (November).

Tremain, S. 1992. The Dynamical Evidence for Dark Matter. *Physics Today*. 45(no. 2)82:28 (February).

Chapter 28. The Expansion and Construction of the Universe

Burns, J. O. 1986. Very Large Structures in the Universe. *Scientific American*. 255:38 (July).

Burstein, D., and Manly, P. L. 1993. Cosmic Tug of War. *Astronomy*. 21:40 (July).

Dressler, A. 1987. The Large-Scale Streaming of Galaxies. *Scientific American*. 257:46 (September).

Freedman, W. 1992. The Expansion Rate and Size of the Universe. *Scientific American*. 267:54 (November).

Geller, M. J., 1990. Mapping the Universe: Slices and Bubbles. *Mercury*. 19:66 (May/June).

Geller, M. J. and Huchra, J. P. 1991. Mapping the Universe. *Sky and Telescope*. 82:134 (August).

Hodge, P. 1993. The Extragalactic Distance Scale: Agreement at Last? *Sky and Telescope*. 86:24 (August).

Jacoby, G. H. et al. 1992. *A Critical Review of Selected Techniques for Measuring Extragalactic Distances*. Astronomical Society of the Pacific. 104:599.

Osterbrock, D. E., and Brashear, R. S. 1993. Edwin Hubble and the Expanding Universe. *Scientific American*. 269:84 (July).

Spergel, D. N., and Turok, N. G. 1992. Textures and Cosmic Structures. *Scientific American*. 266:52 (March).

Tully, R. B. 1982. Unscrambling the Local Supercluster. *Sky and Telescope*. 63:550 (June).

Chapter 29. Active Galaxies and Quasars

Arp, H. 1987. *Quasars, Redshifts, and Controversies*. Berkeley, Calif.: Interstellar Media.

Blandford, R., and Königl, A. 1993. The Disk-Jet Connection. *Sky and Telescope*. 85:40 (March).

Burbidge, G. 1988. Quasars in the Balance. *Mercury*. 17:136 (September/October).

Burns, J. O. 1990. Chasing the Monster's Tail: New Views of Cosmic Jets. *Astronomy*. 18:29 (August).

Courvoisier, T. J.-L., and Robson, E. I. 1991. The Quasar 3C 273. *Scientific American*. 264:50 (June).

Croswell, K. 1993. Have Astronomers Solved the Quasar Enigma? *Astronomy*. 21:29 (February).

Finkbeiner, A. 1992. Active Galactic Nuclei: Sorting Out the Mess. *Sky and Telescope*. 84:138 (August).

Gregory, S. A. 1988. Active Galaxies and Quasars: A Unified View. *Mercury*. 17:111 (July/August).

Hurley, K. 1990. What Are Gamma-Ray Bursters? *Sky and Telescope*. 80:143 (August).

Kaufmann, W. 1979. *Galaxies and Quasars.* New York: Freeman.

Powell, C. S. 1991. Star Bursts: The Deepening Mystery of the Gamma-Ray Sky. *Scientific American.* 265:32 (December).

Preston, R. 1988. Beacons in Time: Maarten Schmidt and the Discovery of Quasars. *Mercury.* 17:2 (January/February).

Rees, M. J. 1990. Black Holes in Galactic Centers. *Scientific American.* 263:56 (November).

Schild, R. E. 1991. Gravity Is My Telescope. *Sky and Telescope.* 81:375 (April).

Sheldon, E. 1990. Faster than Light? *Sky and Telescope.* 79:26 (January).

Smith, D. H. 1985. Mysteries of Galactic Jets. *Sky and Telescope.* 69:213 (March).

Talcott, R. 1991. A Burst of Gamma Rays. *Astronomy.* 19:46 (October).

Turner, E. L. 1988. Gravitational Lenses. *Scientific American.* 259:54 (July).

Tyson, A. 1992. Mapping Dark Matter with Gravitational Lenses. *Physics Today.* 82(no. 6):24 (June).

Weedman, D. W. 1986. *Quasar Astronomy.* Cambridge, England: Cambridge University Press.

Wilkes, B. J. 1991. The Emerging Picture of Quasars. *Astronomy.* 19:35 (December).

Chapter 30. The Universe

Bartusiak, M. 1987. The Cosmic Burp: The Genesis of Inflation. *Astronomy.* 16:34 (March/April).

Brush, S. G. 1992. How Cosmology Became a Science. *Scientific American.* 267:62 (August).

Collins, G. P. 1992. COBE Measures Anisotropy in Cosmic Microwave Background Radiation. *Physics Today.* 82(no. 6):17 (June).

Davies, P. 1985. New Physics and the Big Bang. *Sky and Telescope.* 70:406 (November).

Davies, P. 1985. Relics of Creation. *Sky and Telescope.* 69:112 (February).

Davies, P. 1990. Matter–Antimatter. *Sky and Telescope.* 79:257 (March).

Davies, P. 1991. Everyone's Guide to Cosmology. *Sky and Telescope.* 81:250 (March).

Davies, P. 1992. The First One Second of the Universe. *Mercury.* 21:82 (May/June).

Dressler, A. 1991. Observing Galaxies Through Time. *Sky and Telescope.* 82:126 (August).

Green, M. B. 1986. Superstrings. *Scientific American.* 255:48 (September).

Guth, A. H., and Steinhardt, P. J. 1984. The Inflationary Universe. *Scientific American.* 250:116 (May).

Haber, H. E., and Kane, G. L. 1986. Is Nature Supersymmetric? *Scientific American.* 254:52 (June).

Harrison, E. R. 1981. *Cosmology.* Cambridge, England: Cambridge University Press.

Hawking, S. 1988. *A Brief History of Time.* New York: Bantam.

Horgan, J. 1990. Universal Truths. *Scientific American.* 263:108 (October).

Jayawardhana, R. 1993. The Age Paradox. *Astronomy.* 21:39 (June).

Kippenhahn, R. 1987. Light from the Depths of Time. *Sky and Telescope.* 73:140 (February).

Krauss, L. M. 1986. Dark Matter in the Universe. *Scientific American.* 255:58 (December).

Lederman, L. M., and Schramm, D. N. 1989. *From Quarks to the Cosmos.* Scientific American Library. New York: Freeman.

Mallove, E. F. 1988. The Self-Reproducing Universe. *Sky and Telescope.* 76:253 (September).

McCarthy, P. J. 1988. Measuring the Distances to Remote Galaxies and Quasars. *Mercury.* 17:19 (January/February).

Odenwalf, S., and Fienberg, R. T. 1993. Galaxy Redshifts Reconsidered. *Sky and Telescope.* 85:31 (February).

Overbye, D. 1991. *Lonely Hearts of the Cosmos.* New York: HarperCollins.

Parker, B. 1986. Discovery of the Expanding Universe. *Sky and Telescope.* 72:227 (September).

Phillips, S. 1993. Counting to the Edge of the Universe. *Astronomy.* 21:38 (April).

Riordan, M., and Schramm, D. N. 1991. *The Shadows of Creation.* New York: Freeman.

Schramm, D. N. 1991. The Origin of Cosmic Structure. *Sky and Telescope.* 82:140 (August).

Schramm, D. N., and Steigman, G. 1988. Particle Accelerators Test Cosmological Theory. *Scientific American.* 258:66 (June).

Silk, J. 1989. *The Big Bang.* 2nd ed. New York: Freeman.

Weinberg, S. 1988. *The First Three Minutes.* Updated ed. New York: HarperCollins.

Wesson, P. S. 1989. Olbers' Paradox Solved at Last. *Sky and Telescope.* 77:594 (June).

Glossary

A stars: White stars with strong hydrogen lines and with temperatures of 7,400 to 9,900 K.

AAAOs: Asteroids in the Amor, Apollo, or Aten families, all of which come close to the Earth's orbit.

Aberration of starlight: The shift in the position of a star caused by the motion of the Earth relative to the velocity of light.

Absolute magnitudes (M): Apparent magnitudes that celestial bodies would have at a distance of 10 pc; can be visual (M_V), blue (M_B), bolometric (M_{bol}), or other.

Absolute zero: The lowest possible temperature, achieved when all possible heat has been removed from a body.

Absorption lines: Lack of radiation, or gaps, at specific wavelengths in the electromagnetic spectrum.

Acceleration (A): A change in velocity (speed or direction) with time.

Acceleration of gravity: The acceleration of a falling body in a gravitational field.

Accretion disk: A disk of gas revolving around a star into which matter flows from a companion.

Achondrites: Chondrite meteorites that have no chondrules.

Achromatic lens: A compound lens designed to reduce chromatic aberration.

Active galaxies: Galaxies with bright, variable nuclei sometimes associated with bipolar jets, probably caused by matter falling into massive black holes from accretion disks.

Active Sun: The collective phenomena of the solar magnetic cycle.

Albedo: The percentage reflectivity of a surface.

Alpha particles (rays): Helium nuclei.

Alt-azimuth mounting: A mounting for a telescope in which one axis points to the zenith.

Altimeter: A device (here aboard a spacecraft) that measures the altitude of the ground directly below it.

Altitude (h): the angular elevation of a body above the horizon; $h = 90° - z$.

Amino acids: Building blocks of proteins and of life, found in meteorites.

Amorphotoi: The "unformed" regions between ancient constellations.

Amplitude: The range of magnitudes exhibited by a variable star.

Analemma: A figure-eight graph of the equation of time plotted against solar declination.

Ancient constellations: The original 48 constellations handed down by the ancient Greeks.

Angle of incidence: The angle between a light ray and the perpendicular to the surface it will strike.

Angle of reflection: The angle between a light ray and the perpendicular to the surface from which it has reflected.

Angle of refraction: The angle between a light ray and the perpendicular to the surface from which it has been refracted.

Angular diameter: The angle formed by lines projecting to opposite sides of a body.

Angular momentum: The product of mass, speed, and distance of a revolving body.

Annular eclipse: An eclipse of the Sun wherein the Moon is too far from the Earth to cover it completely, leaving a ring.

Antarctic circle: The parallel of latitude, 66.5° S, below which it is possible to have a midnight Sun between September 23 and March 21.

Ante meridiem (A.M.): Before noon.

Antenna: That part of a radio receiver electrically excited by radiation.

Anticyclone: Motion of an atmospheric gas around a high-pressure zone; clockwise in the northern hemisphere, counterclockwise in the southern.

Antimatter: Particles with reversed charges.

Aperture: The diameter of a telescope objective.

Aphelion: The point of greatest distance between a planet and the Sun.

Aphrodite Terra: The great equatorial rise on Venus.

Apogee: The point in the lunar orbit farthest from Earth.

Apollo program: The program of spaceflight that landed humans on the Moon.

Apparent magnitudes (m): Magnitudes as measured in the sky.

Apparent sun: The real observed Sun.

Arctic circle: The parallel of latitude, 66.5°N, above which it is possible to have a midnight Sun between March 21 and September 23.

Asterisms: Small named portions of constellations, or stellar groupings that extend over constellation boundaries.

Asteroid belt: A zone of debris that lies mostly between the orbits of Mars and Jupiter.

Asteroids: Rocky or metallic bodies smaller than planets that orbit mostly between Mars and Jupiter; C, S, and M types relate to meteorite classes.

Astrology: A system of magic and divining that uses the stars to tell the future.

Astrometric binary: A binary star with an unseen component that orbits the visible component, resulting in a wobbly proper motion.

Astronomical horizon: The great circle defined by the intersection between the celestial sphere and a plane at the observer's feet perpendicular to the line to the zenith.

Astronomical unit (AU)**:** The semimajor axis of the Earth's orbit around the Sun and the average distance between the Earth and the Sun.

Asymptotic giant branch (AGB)**:** The state of stellar evolution in which intermediate-mass stars grow as giants for the second time and in which they have contracting carbon-oxygen cores surrounded by helium- and hydrogen-burning shells.

Atmosphere: The gases that surround a planet or satellite.

Atom: The basic unit that forms the chemical elements.

Atomic number: The number of protons in an atomic nucleus; the atomic number defines the chemical element.

Atomic weight: The number of protons plus neutrons in an atomic nucleus.

Aurora: Lights in the upper atmosphere caused by the interaction between the terrestrial and solar magnetic fields and the solar wind; borealis in the north, australis in the south.

Autumn: The period in the northern hemisphere between September 23 and December 22 when the Sun is moving southward in the southern celestial hemisphere; spring in the southern hemisphere.

Autumnal equinox: The point on the celestial sphere where the ecliptic crosses the celestial equator with the Sun moving south, the crossing taking place about September 23.

Averted vision: A visual observing technique in which the observer does not look directly at an object but uses the more light-sensitive periphery of the eye.

Axis: A line about which a body rotates or moves.

Azimuth: The angular direction along the horizon starting from north at 0°, measuring through east.

B stars: Blue-white stars with strong hydrogen lines and neutral helium lines and with temperatures of 9,900 to 28,000 K.

Baily's beads: Portions of the Sun seen at the lunar edge at the start or end of a total solar eclipse.

Balmer series: A series of hydrogen absorption or emission lines that have the second energy level as their common bottom level.

Barred spirals: Spiral galaxies in which the arms come from the ends of a bar through the galaxies' centers.

Baryons: Collectively, protons, antiprotons, and neutrons.

Basalts: Fine-grained, metal-bearing silicates.

Basin: A depression in a planetary crust.

Bayer Greek letters: A system of assigning Greek letters to stars more or less in order of brightness within a constellation.

BD (Bonner Durchmusterung) or CD numbers: Numbers that catalogue stars by right ascension within 1° declination strips ignoring constellations.

Belts: Strips of dark, lower-altitude, higher-temperature clouds in Jupiter's atmosphere.

Beta decay: The ejection of an electron from a neutron in an atomic nucleus that turns the neutron into a proton.

Beta particles (rays): Electrons.

Big Bang: The concept that 10 to 20 billion years ago, the Universe (or at least the present phase of the Universe) began with the sudden expansion of space from a condition of very high density.

Big Chill: A state of the Universe in which it expands forever and cools to absolute zero.

Big Crunch: The final state of a closed universe after it has contracted back to ultra-high density or into a singularity.

Binary stars: Double stars with gravitationally bound components.

Bipolar flow: Gaseous jets emitted from an object's rotation poles.

Birthline: The line on the HR diagram above and to the right of the main sequence where protostars first become visible.

BL Lacertae objects: Bright nuclei of active elliptical galaxies from which a jet is coming directly at the observer.

Black hole: A mass inside an event horizon where the escape velocity equals the velocity of light, preventing light's escape.

Blackbody: A body that absorbs all the radiation that falls upon it; if temperature is constant, a blackbody emits as much radiation as it absorbs.

Blackbody (Planck) curves: Graphs of the brightnesses of blackbodies plotted against wavelength.

Bode's law: A numerical progression (not a true law) that matches the distances of most planets from the Sun.

Bok globules: Small, opaque, relatively dense, dusty interstellar clouds, typically a few parsecs across containing 10 to 100 solar masses.

Bolometric correction: The difference between the visual and bolometric magnitude $V - m_{bol}$.

Bolometric magnitudes (m_{bol}): Magnitudes that represent all the energy of a star.

Bow shock: A buildup of pressure in front of a body that is moving against a fluid.

Breccia: A rock that consists of fused pieces of smaller rocks.

Bright companion: Bright clouds near Neptune's Great Dark Spot.

Brown dwarfs: Bodies that contract directly from the interstellar medium, but with masses below the main sequence cutoff of 0.08 $M_\odot$, too low to allow the operation of the proton-proton chain.

Buckminsterfullerene: The C_{60} molecule, a possible constituent of the interstellar medium.

Bulge: The central, thick region of a spiral galaxy where the disk and the halo come together.

Butterfly diagram: A graph of the latitudes of sunspots plotted against time.

Caldera: A depression caused by the collapse of a portion of a volcano into an underground chamber emptied of magma.

Callisto: The outermost Galilean satellite of Jupiter.

Caloris: The major basin on Mercury.

Camera mirror: The mirror in a spectrograph that focuses dispersed light onto a detector.

Canals: Illusory linear features on Mars caused by the eye linking real features together.

Cantaloupe terrain: A landform toward Triton's equator filled with grabens and basins.

Carbon cycle: The fusion of four atoms of hydrogen into one of helium using carbon as a nuclear catalyst.

Carbon stars: Stars that have at least as much carbon as oxygen; classes R, N, and S.

Carbonaceous chondrites: Primitive chondrite meteorites with a high carbon content.

Carbonate rock: Rock made from carbon and oxygen in combination with metals.

Cassegrain focus: A configuration of a primary mirror and a curved secondary mirror that extends the focal length and sends the light back through a hole in the primary of a reflecting telescope.

Cassini division: A resonance gap that divides Saturn's A and B rings.

Celestial: Pertaining to the sky.

Celestial equator: The great circle on the celestial sphere above the Earth's equator equidistant from the celestial poles.

Celestial meridian: The great circle through the celestial poles, the zenith, and the nadir.

Celestial navigation: The means by which we find our way on Earth by the stars.

Celestial poles: The points of rotation of the celestial sphere that lie above the north and south poles of the Earth.

Celestial sphere: The apparent sphere of the sky.

Celsius (centigrade) temperature scale (°C): A scale in which water freezes at 0°C and boils at 100°C.

Center of mass: The point at the mutual focus of two orbiting bodies.

Centers of activity: Magnetic regions of the solar surface that contain sunspots and associated phenomena.

Cepheid instability strip: A nearly vertical strip among class F and G giants and supergiants in the HR diagram in which Cepheid variable stars are found.

Cepheid variables: F and G giant and supergiant variables with regular periods.

Ceres: The largest and first-discovered asteroid.

Chandrasekhar limit: A limit of 1.4 M, above which white dwarf stars cannot be supported by the pressure of degenerate electrons.

Chaos: A branch of mechanics that deals with unstable systems.

Chaotic terrain: A landform on Mars with an irregular assembly of large blocks of rock and fractures.

Charge-coupled device (CCD): An electronic imaging device.

Charon: Pluto's satellite.

Chondrites: Stony meteorites with small round inclusions known as chondrules.

Chondrules: Small round inclusions found in stony meteorites.

Chromatic aberration: A smearing of the focal point of a lens caused by the refractive dispersion of light.

Chromosphere: The thin reddish emission-line layer of the Sun that lies between the photosphere and the corona.

Chryse Planitia: The "Plains of Gold," a plain in the Martian northern hemisphere.

Cinder-cone volcano: A steep-sided volcano produced by thick, viscous, explosive lava.

Circle: A conic section derived by slicing a cone with a plane parallel to the cone's base; an ellipse of zero eccentricity in which the foci are together at the center.

Circumpolar stars: Stars that do not set on their daily paths; stars always visible that are north of $(90° - \phi)$ in the northern hemisphere and south of $(\phi - 90°)$ in the southern hemisphere.

Classical Cepheids: Population I Cepheids.

Climate: Long-term conditions and changes in a planet's atmosphere.

Closed universe: A four-dimensional, positively curved, spherical universe that folds back in on itself; parallel lines meet and the angles of a triangle sum to more than 180°.

Clump: A region on the HR diagram about halfway down the giant branch where Population I stars fuse helium to carbon.

Clusters: Gravitationally bound groups of stars.

Clusters of galaxies: Gravitationally bound groups of galaxies.

Coalsack: A famous Bok globule located in Crux.

Cold dark matter: Slow-moving dark matter that can cause the formation of galaxies (possibly WIMPS).

Collimating mirror: The mirror in a spectrograph that makes converging light rays from the telescope parallel to one another.

Color index: Visual minus photographic magnitude, or $B - V$.

Color-magnitude diagram: A plot of color index against magnitude.

Coma: The bright cloud of ions and dust that surround a comet's nucleus.

Coma cluster: The nearest rich, regular cluster of galaxies, in Coma Berenices, about 100 Mpc away.

Comet: A fragile interplanetary body with a nucleus of dirty ice that is heated by the Sun to produce a surrounding coma and tails of gas and dust.

Common proper motion (CPM) **binaries:** Binaries whose components are too far apart to have shown orbital motion, but whose components move together.

Comparative planetology: The science of comparing planets and their varied conditions, using one to learn about another.

Comparison spectrum: A spectrum with lines of known wavelength produced in a spectrograph to calibrate a spectrogram of a celestial object.

Composite volcano: A volcano intermediate between a cinder-cone and a shield volcano.

Conduction: Heat transfer from atom to atom.

Conic sections: The curves (circle, ellipse, parabola, and hyperbola) defined by the intersections of a cone and a plane.

Conjunction: The position in which a planet (as viewed from Earth) is in the same direction as the Sun (or in which two celestial bodies are aligned with each other).

Conservation of angular momentum: The concept that total angular momentum is always constant in a closed system.

Conservation of energy: The concept that energy can neither be created nor destroyed, only changed in form; in relativity, the conservation of mass-energy.

Conservation of mass-energy: The concept that energy and its equivalent in mass (through $E = Mc^2$) can neither be created nor destroyed.

Constellations: Named patterns of naked-eye stars; 88 are now officially recognized and used.

Constructive interference: Reinforcement of waves that fall on top of one another.

Continental drift: The motion of continents across the mantle.

Continents: Raised portions of the Earth's crust made out of light rock.

Continuous radiation: Radiation with a spectrum with no breaks, gaps, or sudden changes in brightness.

Convection: The up-and-down circulation of a heated fluid.

Coordinate system: A grid for the measurement of position.

Coordinated Universal Time (UTC): A hybrid time formulated with a constant-length second adjusted to keep up with the changing rotation period of the Earth.

Copernican Principle: The Earth is not special.

Copernican system: A system of circular heliocentric planetary orbits.

Copernicus: A recent large crater on the Moon.

Core: The metallic (nickel-iron) core of a terrestrial planet, the ice and rock core of a Jovian planet, or the inner portion of a star within which nuclear reactions are produced.

Core-collapse supernovae: Type II or Type Ib supernovae produced by the collapse of the iron cores that have developed in high-mass stars.

Coriolis effect: Motion of a body caused by the Earth's spin.

Corona (atmospheric): A halo of colored light around the Sun or Moon produced by diffraction by water droplets or ice.

Corona (solar): The hot (2 million K) outer envelope enclosing the Sun; seen during a total solar eclipse.

Corona (galactic): The volume of dark matter that surrounds the the disk and the halo of the Galaxy.

Coronae (Venus): Large volcanically uplifted areas with fractured centers, surrounded by concentric fractures.

Coronal gas: A very hot gas in the interstellar medium created and heated by supernova blasts.

Coronal holes: Gaps in the solar corona where there is little hot gas and from which the solar wind blows most strongly.

Coronal loops: Gas in the inner solar corona confined to magnetic loops above centers of activity.

Coronal streamers: Streamers of solar coronal gas above coronal loops that extend far into space.

Coronal transients: Ejections of solar coronal gas.

Cosmic: Pertaining to the cosmos, to the sky, or to the Universe.

Cosmic Background Explorer: An Earth-orbiting satellite designed to observe the cosmic microwave background radiation.

Cosmic microwave background radiation: All-pervading blackbody radiation near 2.7 K, cooled from the Big Bang fireball.

Cosmic rays: High-energy atomic nuclei accelerated by the Galaxy's magnetic field; probably produced in supernovae and neutron-star binaries.

Cosmic strings: Two-dimensional defects that may have been left over from the time of the breaking of GUTS symmetry where GUTS are still in effect.

Cosmological constant: A constant that describes Einstein's suggested repulsive force.

Cosmological distance: A large distance that relates to the expanding Universe and thus to cosmology.

Cosmological Principle: There is no special place in the Universe.

Cosmology: The study of the structure, origin, and evolution of the Universe.

Cosmos: The sky or Universe.

Coudé focus: An arrangement of several mirrors that sends the light from the primary of a reflecting telescope to a fixed position.

Crab Nebula: The remnant of the supernova of 1054 in Taurus.

Craters: Pits caused by impacts.

Crepe ring: Saturn's innermost C ring.

Crescent moon: A phase in which we see less than half the illuminated side of the Moon, at a time when the Moon makes less than a right angle with the Sun.

Critical density: The density of matter needed to close the Universe.

Cross-quarter days: The days that fall right between the four major ecliptic points; they include Groundhog Day, May Day, and Halloween.

Crust: The top, thin, light layer of a terrestrial planet or satellite.

Cyclone: The motion of an atmospheric gas around a low-pressure zone, counterclockwise in the northern hemisphere, clockwise in the southern.

Cygnus Loop: A large, ancient supernova remnant in Cygnus.

Cygnus X-1: The best-known black hole candidate.

D2: A small storm oval on Neptune.

DA white dwarf: A white dwarf with a nearly pure hydrogen photosphere.

Daily paths: The apparent paths taken by celestial bodies as the Earth rotates.

Dark matter: Unilluminated mass, detectable only through its gravitational effect; its content is unknown.

DB white dwarf: A white dwarf with a nearly pure helium photosphere.

Decameter bursts: Bursts of radio radiation from Jupiter caused by the action of a current ring that connects the planet to Io.

Declination (δ): Arc measurement along an hour circle north or south of the celestial equator.

Declination axis: The axis of an equatorial telescope mounting that is perpendicular to the polar axis and that moves the telescope in declination.

Defects: Places in the Universe where symmetry did not break: magnetic monopoles, strings, and domain walls.

Deferent: In the Ptolemaic system, a circular orbit around the Earth or, more properly, around the equant.

Degeneracy: A state of high density in which the particles of a gas at a particular velocity cannot get any closer and in which the perfect gas law no longer works because the particles would violate the Pauli exclusion principle.

Dense cores: Knots of matter in the interstellar medium, observable by ammonia radiation; the first observable step in star formation.

Density: Mass per unit volume, measured in kg/m^3 or g/cm^3.

Density waves: Gravitational accumulations of matter that spread into a spiral pattern as a result of orbital revolution.

Destructive interference: Destruction or cancellation of waves that fall on top of one another.

Deuterium: Hydrogen with a neutron in its nucleus; ^2H.

Diamond ring: The moment of the first appearance of the Sun after a total solar eclipse when the corona is still visible.

Differential rotation: Rotation in which different parts of a body rotate with different speeds, resulting in different periods.

Differentiation: The separation of a planet's interior into layers of different composition.

Diffraction: A phenomenon that spreads light around barriers into areas that otherwise would be dark.

Diffraction disk: The central fringe of a circular diffraction pattern.

Diffraction grating: A plate with narrowly ruled lines that produces diffraction spectra.

Diffraction rings: Circular diffraction fringes formed by a telescope aperture.

Diffuse nebulae: Bright clouds of interstellar gas ionized by stars at least as hot as type B1.

Dipole: A field with two poles.

Disk: The flat, Population I component of a spiral galaxy that contains the majority of the galaxy's stars.

Dispersion: The spreading of radiation into a spectrum by refraction or diffraction.

Distance indicators: Objects with known absolute magnitudes used to find distances, the term most commonly applied to those in other galaxies.

Domain walls: Three-dimensional defects that may have been left over from the time of the breaking of GUTS symmetry where GUTS are still in effect.

Doppler effect: The observed shift in wavelength or frequency caused by relative radial motion.

Double-double star: Two double stars in orbit about each other.

Drake equation: An equation consisting of several factors multiplied together that gives the probability of intelligent life in the Universe.

Dry ice: Frozen carbon dioxide.

Dust: Small, solid grains with a variety of chemical compositions.

Dust tail: The diffuse fan-shaped tail of a comet caused by sunlight scattered from dust released by the nucleus.

Dwarf Cepheids: Short-period Cepheids just above the main sequence.

Dwarf elliptical galaxies: Small elliptical galaxies.

Dwarf novae: Binaries with irregular, low amplitude outbursts caused by instabilities in accretion disks.

Dwarf spheroidal galaxies: The smallest kind of elliptical galaxies.

Dwarfs: Stars of the main sequence on the HR diagram.

E0 galaxy: An elliptical galaxy with a circular outline.

E7 galaxy: The most elliptical form of elliptical galaxy.

Earth: Our world, the third planet from the Sun.

Earthlight: Light from the Earth that illuminates the nighttime side of the Moon.

Earthquake waves: Compressional (P) and transverse (S) waves sent through the Earth by the shocks of earthquakes.

Earthquakes: Vibrations caused by the slippage of the Earth's crust along faults.

Eccentricity: The degree of flattening of an ellipse (center-to-focus divided by the semimajor axis), ranging from 0 for a circle to 1.

Eclipse of the Moon: The passage of all or part of the Moon through the Earth's shadow.

Eclipse of the Sun: The passage of the Moon across the Sun or the lunar shadow across the Earth.

Eclipse season: The interval during which the new or full Moon is near a node and an eclipse is possible.

Eclipse year: The interval of 346 days between successive passages of the Moon across a specific node.

Eclipsing binaries: Binary stars whose components eclipse each other.

Ecliptic: The apparent path of the Sun through the stars.

Ecliptic poles: The points on the celestial sphere defined by the directions of the perpendicular to the Earth's orbit and to the ecliptic.

Effective temperature: The temperature of a blackbody that has the surface area and luminosity of the radiant source.

Egress: The exiting state of an eclipse.

Einstein cross: A multiple gravitational lens produced when a quasar is closely aligned behind an extended mass.

Einstein-de Sitter universe: A flat, Euclidean universe.

Einstein ring: A gravitational lens produced when a quasar is exactly lined up behind a mass of small angular diameter, producing a ring of light.

Ejecta: The debris thrown out of a crater by the impact that made it.

Ejecta blanket: The blanket that ejecta from an impact crater form around the crater.

Electric charge: A property of electrons and protons by which they attract or repel one another; a manifestation of the electromagnetic force.

Electric field: The range of the electric charge as it acts over a distance.

Electromagnetic force: The force of nature that combines electricity and magnetism; manifested by the electric charge and electric and magnetic fields that act over a distance.

Electromagnetic radiation: Alternating electric and magnetic waves, or photons, that transport energy at the speed of light, $c = 299{,}792$ km/s.

Electromagnetic spectrum: The array of the kinds (wavelengths) of electromagnetic waves, including gamma rays, X rays, ultraviolet, optical, infrared, and radio waves.

Electrons: Negatively charged atomic particles.

Electroweak force: The unified electromagnetic and weak forces.

Elements: The basic constituents of common matter; each element is defined by a different number of nuclear protons.

Ellipse: A curve defined by two foci such that the sum of the distance from any point on the curve to each of the foci is constant; a conic section derived by slicing a cone with a plane at an angle less than parallel to the cone's side.

Elliptical galaxies: Population II galaxies without spiral arms.

Elongation: The angle between the direction to a planet and the direction to the Sun as viewed from Earth.

Emission lines: Radiation at specific wavelengths.

Encke division: A narrow gap in the outer part of the A ring produced by a satellite that sweeps up ring material.

Energy: The capacity of a body to do work on (or accelerate) or heat another body.

Energy-level diagram: A graph of energies of atomic or molecular electrons.

Envelope: The thick layer of the Sun that blankets the core and transmits solar energy.

Epicycle: In the Ptolemaic system, a secondary planet-carrying orbit centered on a deferent.

Epoch: The moment for which right ascensions and declinations are valid.

Equal areas law: Kepler's second law: the radius vector of a planet sweeps out equal areas in equal times.

Equant: In the Ptolemaic system, a point set off from the center about which the planets move at constant angular speed.

Equation of time: The difference between local apparent solar time and local mean solar time, LAST − LMST.

Equator (terrestrial): The great circle on the Earth that is equidistant from the poles.

Equator point: The intersection of the celestial equator and the celestial meridian.

Equatorial coordinate system: A system of astronomical coordinates, hour angle, right ascension, and declination, based on the celestial equator.

Equatorial mount: A telescope mounting in which one axis is aligned to the visible celestial pole.

Equinoxes: The intersections between the ecliptic and the celestial equator.

Ergosphere: A surface surrounding a rotating black hole within which spacetime rotates.

Error: A formal limit within which a measured value probably falls.

Escape velocity: The velocity needed to achieve a parabolic orbit such that a body will not return.

Euclidean: Pertaining to Euclidean geometry; space in which one and only one line can be drawn through a point parallel to another line.

Eucrites: Stony meteorites from the asteroid 4 Vesta.

Europa: The second-out and smallest Galilean satellite of Jupiter.

Event horizon: The surface surrounding a black hole where the escape velocity equals the velocity of light.

Evolutionary tracks: Graphs of luminosity against effective temperature (or M_v vs. spectral class) on the HR diagram.

Expanding Universe: The steady growth of space that increases distances between clusters of galaxies.

Exponential notation: A means of expressing numbers by using powers of ten.

Eyepiece: A lens attached to a telescope that makes light rays parallel, thus allowing an object to be imaged by the eye.

F stars: Yellow-white stars with strong hydrogen lines and metal lines and with temperatures of 6,000 to 7,400 K.

Faculae: Bright regions around sunspots.

Fahrenheit temperature scale (°F): A scale in which water freezes at 32° F and boils at 212° F.

False vacuum: A vacuum with positive energy acquired when symmetry broke in the early Universe; responsible for inflation through anti-gravity resulting from the false vacuum's negative density.

Families: Groups of asteroids that result from breakup of a parent body, from assembly by resonances, or grouped by orbital properties.

Families of matter: Different groups of atomic particles analogous to the electron-proton pair, each with different neutrinos.

Farside (of the Moon): The invisible side of the Moon, that facing away from the Earth.

Fast novae: Novae that take about a month to decay three magnitudes from maximum.

Fault: A separation in a planet's crust.

Favorable opposition: An opposition of Mars near its perihelion.

Feldspar: A complex class of silicate that incorporates light metals like potassium, aluminum, and calcium.

Field of view: The area of the sky over which a telescope can form an image.

Field stars: Stars not in clusters.

Filaments: Dark ribbons of cool gas in the solar corona seen against the photosphere and confined by magnetic fields; prominences seen against the Sun.

Finder telescope: A small telescope with a wide field of view attached to the side of the main telescope.

Fireball: A brilliant meteor that can produce a meteorite.

First quarter Moon: A phase in which we see half the illuminated side of the Moon, at a time when the Moon makes a right angle with the Sun following new.

Fission: The breaking of atoms with higher atomic numbers or isotopes with higher atomic weights into those with lower atomic numbers or atomic weights.

Flamsteed numbers: A system of assigning numbers to stars in order of increasing right ascension within a constellation.

Flare stars: Main sequence M stars with sudden eruptions.

Flares: Sudden releases of magnetic energy in the solar corona that also brighten the chromosphere.

Flatness problem: The question of why Omega should be so close to one; resolved by inflation.

Flux (of radiation) (F): The amount of radiation flowing per second through a surface with an area of one square meter (joules/m^2/s).

Focal length: The distance from a focusing lens or mirror to the focal point.

Focal point/plane: The point on the optical axis, and the plane perpendicular to the axis, where images are focused.

Focus (of an ellipse): One of two points that define an ellipse.

Follower spot: A sunspot or a pair of spots that is behind in the direction of solar rotation.

Forbidden lines: Emission lines common in diffuse nebulae that build up great strength because of high cloud mass, even though the electron jumps that make them are improbable; produced by collisional excitation.

Force (F): That which produces an acceleration of a mass.

Foucault pendulum: A swinging weight that demonstrates the Earth's rotation.

Free-free radiation: Continuous radiation caused by the close passages of electrons and protons.

Frequency (ν): The number of waves passing a location per second.

Friedmann-Lemaître universes: A set of models for the Universe that consists of open and hyperbolic, and closed and spherical, universes.

Full Moon: A phase in which we see the full illuminated side of the Moon, when the Moon is opposite the Sun.

Fusion: A short name for thermonuclear fusion.

G stars: Yellow-white stars with metal lines and particularly strong ionized calcium lines and with temperatures of 4,900 to 6,000 K.

Galactic bulge: The central, thick region of the Galaxy where the disk and the halo come together.

Galactic coordinates: Coordinates (galactic latitude and longitude) based on the Galaxy's plane and nucleus.

Galactic corona: The volume of space filled with dark matter encompassing the traditional disk and halo.

Galactic disk: The flat, Population I component of the Galaxy that contains the majority of its stars.

Galactic equator: The center line of the Milky Way, on which galactic longitude is based.

Galactic halo: The sparsely populated, spherical, Population II component of the Galaxy that surrounds the Galaxy's disk.

Galactic latitude: Arc measurement along a meridian of galactic longitude north or south of the galactic equator.

Galactic longitude: A coordinate measured along the galactic equator from the center of the Galaxy, Sagittarius A.

Galactic nucleus: The energetic center of the Galaxy; may be a black hole.

Galactic poles: The points on the celestial sphere perpendicular to the galactic equator.

Galactic rotation curve: The rotation velocity of the Galaxy graphed against distance from the galactic center.

Galaxies: Self-contained collections of mass that include stars, interstellar matter, and dark matter.

Galaxy, the: The collection of 200 billion stars, interstellar matter, and dark matter in which we live; shaped like a disk surrounded by a sparsely populated halo, surrounded by a dark corona.

Galilean satellites: Jupiter's four largest satellites, discovered by Galileo.

Gamma ray bursters: Objects of unknown character that emit gamma ray bursts.

Gamma rays: The shortest-wave electromagnetic radiation with wavelengths under about 1 Å.

Ganymede: The third-out and largest Galilean satellite of Jupiter.

Gas: A state of matter with neither a fixed shape or volume.

Gegenschein: The counterglow, or the brightening of the zodiacal light, in the direction opposite the Sun.

General relativity: That part of relativity that involves accelerations.

Geocentric theory: A theory of the solar system in which the planets orbit the Earth.

Geosynchronous orbit: An orbit around the Earth in which the orbiting body has a revolution period equal to the Earth's rotation period.

Giant elliptical galaxies: The largest kind of elliptical galaxy, more massive than our Galaxy, and commonly found near the centers of clusters of galaxies.

Giant H II region: A massive H II region over 100 or so pc across illuminated by associations of O and B stars.

Giant molecular clouds (GMCs): Massive dusty clouds (typically 200,000 $M_\odot$ and 100 pc across) made mostly of hydrogen molecules.

Giants: Large stars that are brighter than the main sequence for a given temperature; MKK classes II and III.

Gibbous Moon: A phase in which we see more than half the illuminated side of the Moon, at a time when the Moon makes more than a right angle with the Sun.

Giotto: The painter who painted a picture that included Halley's Comet and the European spacecraft that passed close to and imaged Halley's Comet.

Glitch: An event in which the rotation speed of a pulsar suddenly increases.

Globular clusters: Rich Population II clusters of the galactic halo; found in the halos of other spiral galaxies and in elliptical galaxies.

Gluons: The particles that carry the strong force.

Grabens: Cracks caused by the pulling of surface rock.

Grand Unified Theories (GUTS): Theories combining the electromagnetic, weak, and strong forces.

Granules: Bright cells in the solar photosphere about a second of arc across caused by convection.

Gravitational constant (G): The constant of nature used in the law of gravity; the force between two 1-kg bodies 1 m apart.

Gravitational lens: Relativistic distortion, or amplification, of the light from a distant quasar caused by the gravitational field of an intervening mass.

Gravitational redshift: The shift of a photon toward longer wavelengths as a result of its escaping a gravitational field.

Gravitons: The particles that are presumed to carry gravity.

Gravity: The attractive force between masses; the curvature of spacetime caused by the presence of mass.

Gravity waves: Energy-dissipating waves in an accelerating body's gravitational field.

Great Attractor: A nearby massive supercluster of galaxies in the direction of Hydra-Centaurus that changes motions of galaxies away from the smooth Hubble flow.

Great circle: A circle on a sphere whose center is coincident with the center of the sphere.

Great Dark Spot (GDS): A dark, oval, high-pressure storm in Neptune's southern hemisphere probably caused by a convective plume.

Great Debate: The 1920 debate between Curtis and Shapley over the nature of spiral nebulae and the existence of external galaxies.

Great Red Spot (GRS)**:** A huge, reddish zone near Jupiter's south equatorial belt; physically, a high-pressure anticyclone.

Great Wall: A dense line of galaxies some 50 to 150 Mpc away stretching around a third of the sky and containing the Coma cluster.

Great White Spot (GWS)**:** A seasonal northern-hemisphere storm on Saturn that extends around the entire planet.

Greatest brilliancy: The maximum apparent brightness of Venus or Mercury, for Venus occurring in its crescent phases.

Greatest elongation: The maximum elongation possible for an inferior planet.

Green flash: A moment of green sunlight seen at the last moment of sunset caused by refraction in the Earth's atmosphere.

Greenhouse effect: The process by which carbon dioxide and water vapor in the Earth's atmosphere trap radiated heat.

Gregorian calendar: Our modern calendar of 365 days that has an extra day every 4 years except in century years not divisible by 400; replaced the Julian calendar.

Ground state: The lowest energy level of an atom or ion.

Guest star: The supernova of 1054 in Taurus, source of the Crab Nebula, and observed by the Chinese.

Guide star catalogue: A catalogue of positions of 15 million stars made for the Hubble Space Telescope.

H II regions: Ionized clouds of interstellar gas.

Hadley cell: An atmospheric circulation cell that brings warm air from the equator to higher latitudes and cool air back.

Hairy ball: The appearance of the tangled solar magnetic field lines.

Half-life: The time it takes for a specific amount of a radioactive element to decay to half that amount.

Halo (atmospheric): A circle of colored light, usually of 22° radius, around the Sun caused by refraction of sunlight by ice crystals.

Halo (galactic): The sparsely populated, spherical, Population II component of a spiral galaxy that surrounds the galaxy's disk.

Harmonic law: Kepler's third law: the sidereal period of a planet in years squared equals the semimajor axis of the planet in astronomical units cubed.

Harvest Moon: The full Moon nearest the time of the autumnal equinox, following which there is a great deal of early evening moonlight.

Head-tail radio galaxies: Radio galaxies in which the jets bend away from the central source, the result of the pressure of intergalactic gas streaming by.

Heavy bombardment: The high rate of impacting during the early Solar System.

Heliocentric theory: A theory of the Solar System in which the planets orbit the Sun.

Helium flash: The explosive ignition of helium in the partially degenerate core of a red giant star about a solar mass.

Henry Draper (HD) **Catalogue:** A catalogue of stellar spectral classes.

Herbig-Haro (HH) **objects:** Knots of gas in the interstellar medium lit by shock waves from bipolar flows.

Hertzsprung-Russell (HR) **diagram:** A plot of stellar magnitudes against spectral types.

High resolution microwave survey: NASA's modern search for extraterrestrial intelligence.

High tide: The maximum water line at the shore produced by lunar and solar tides.

Hipparcos: An Earth-orbiting satellite that measures stellar parallaxes.

Horizon: The line where the land seems to meet the sky; in cosmology, the maximum distance we can look in the Universe, which is determined by the Universe's age and structure.

Horizon problem: The problem that the cosmic microwave background radiation is the same in all directions (except for a Doppler shift) even though all parts of the Universe could not communicate; resolved by inflation.

Horizontal branch: A branch of roughly constant magnitude extending left from the giant branch in globular cluster HR diagrams in which stars fuse helium into carbon in their cores.

Horsehead Nebula: A famous Bok globule located in Orion.

Hot dark matter: Fast-moving dark matter, for example, neutrinos.

Hot poles: Hot regions on Mercury that alternately face the Sun at perihelion.

Hour: Usually a measure of time in which the day is divided into 24 parts, but also a measure of angle in which the circle is divided into 24 parts, where 1 hour = 15°.

Hour angle (HA)**:** The arc along the celestial equator measured from the equator point to an hour circle; $HA = LST - \alpha$.

Hour circle: A great semicircle that runs between the celestial poles through a celestial body.

HR numbers: Naked-eye stars in the *Bright Star Catalogue* numbered according to increasing right ascension.

Hubble classification: A system of classifying galaxies that branches from ellipticals through the two kinds of spirals to irregulars.

Hubble constant (H_0): The rate at which the Universe expands.

Hubble flow: The smooth expansion of space, within which clusters of galaxies flow apart.

Hubble length: The distance at which the speeds of receding cosmological bodies (galaxies or quasars) reach the speed of light.

Hubble relation: The linear correlation between redshifts and distances (or apparent magnitudes) of galaxies.

Hubble time: The reciprocal of the Hubble constant and the age of an empty universe.

Humphreys-Davidson limit: The upper limit to stellar luminosity, above which stars lose enough mass to alter their evolution severely.

Hydrocarbons: Chemical compounds based on hydrogen and carbon.

Hydrostatic equilibrium: A condition in which the upward push of pressure is balanced by the downward pull of gravity.

Hyperbola: A conic section derived by slicing a cone with a plane at an angle greater than parallel to the cone's side; the hyperbola does not close on itself.

Hypothesis: An idea tested by experiment or observation.

Ice ages: Periods of heavy glaciation.

Igneous rock: Rock that forms directly from the molten state.

Image: The depiction of an object in the focal plane.

Impact basins: Large impact craters, commonly filled with dark, solidified lava.

Impact melts: Stones solidified directly from earlier rock rendered molten in impacts.

Inclination (i): The angle that an orbital plane makes with the ecliptic plane, or in the case of double stars, with the plane of the sky.

Index Catalogue (*IC*): A catalogue of over 5,000 nonstellar celestial objects that supplements the *New General Catalogue*.

Index of refraction: The degree to which a substance is capable of bending light in refraction.

Inequality of the seasons: The fact that northern spring and summer are longer than autumn and winter as a result of the eccentricity of the Earth's orbit.

Inferior conjunction: A conjunction between Venus or Mercury and the Sun in which the planet lies between the Earth and the Sun.

Inferior planets: Mercury and Venus, the two inside the Earth's orbit.

Infinite: Never-ending.

Inflation: A super-rapid expansion controlled by symmetry breaking (separation of the strong force from the GUTS unified force) that occurred just after the Big Bang.

Infrared radiation: Radiation with wavelengths longer than can be seen with the human eye (about 7,000 Å), but less than about 0.1 mm.

Ingress: The entering stage of an eclipse.

Interacting galaxies: Galaxies in some state of tidal interaction or collision.

Intercloud medium: A thin, warm, partially ionized gas that lies in the spaces between interstellar clouds.

Intercrater plains: Volcanic flows on Mercury that formed during the late heavy bombardment.

Interferometer: Two or more linked radio telescopes that improve resolution by detecting interference of radio waves.

Intermediate main sequence: The main sequence between 0.8 and 8 $M_\odot$, whose stars evolve into white dwarfs.

International date line: A line near 180° longitude where you drop a day when going from west to east.

Interpulse: A pulse from a pulsar between two main pulses.

Interstellar absorption: The dimming of starlight by interstellar dust.

Interstellar absorption lines: Narrow absorption lines superimposed on the spectra of stars, produced by atoms and ions in the interstellar medium.

Interstellar dust: Small solid grains made of silicates or carbon coated with ices and embedded with heavier atoms.

Interstellar medium: The lumpy mixture of gas and dust that pervades the plane of the Galaxy.

Interstellar reddening: The reddening of starlight by interstellar dust; correlated with interstellar absorption.

Intrinsic variables: Stars whose brightnesses vary because of internal processes (as opposed to those that vary by eclipsing).

Inverse Compton effect: An effect in which radio photons hit electrons moving near the speed of light, which boosts the photons' energies into the X-ray region of the spectrum.

Inverse P Cygni lines: Stellar emission lines flanked by red-shifted absorption lines; caused by mass flowing onto a star.

Inverse square law: A law in which a quantity varies according to 1 divided by the square of another quantity.

Io: The innermost and most active Galilean satellite of Jupiter.

Io plasma torus: A ring of ions and electrons around Jupiter in Io's orbit.

Ionosphere: A layer at the top of a planet's atmosphere that includes the thermosphere and in which the gases are ionized.

Ions: Atoms or molecules that are electrically charged because electrons are missing or added.

IRAS: The infrared astronomical satellite, which orbited Earth and surveyed the sky at several infrared wavelengths.

Iron peak: The high abundances of the elements around iron.

Irregular clusters of galaxies: Clusters of galaxies not concentrated toward their centers and rich in spirals.

Irregular galaxies: Small galaxies without much structure.

Irregular variables: Giant or supergiant pulsating stars with erratic periods.

Ishtar Terra: The northern great rise on Venus.

Island universe: Kant's original name for an external galaxy, one outside our own.

Isochrones: Curves that show stars of the same age on HR diagrams.

Isotopes: Variations of a chemical element caused by differences in neutron number.

Isotropic: The same in all directions.

Jovian planets: Those planets like Jupiter: Jupiter, Saturn, Uranus, and Neptune.

Julian calendar: The old calendar dating from Julius Caesar, which had an extra day added every four years; replaced by the Gregorian calendar.

K stars: Orange stars with strong neutral and ionized calcium lines and molecular lines and with temperatures of 3,500 to 4,900 K.

Kelvin (absolute) temperature scale (K): A scale that uses Celsius degrees with zero set at absolute zero; water freezes at 273 K and boils at 373 K.

Kepler's first generalized law: The path of an orbiting body is a conic section with the other body at one focus of the curve.

Kepler's generalized laws of planetary motion: Kepler's laws derived from, and generalized by, Newton's laws of motion and the law of gravity.

Kepler's laws of planetary motion: The law of ellipses, the equal areas law, and the harmonic law, which together describe planetary orbits.

Kepler's second generalized law: In any closed system, angular momentum is conserved.

Kepler's third generalized law: The period of an orbiting body in seconds squared is equal to a constant times the semimajor axis in km cubed divided by the sum of the masses in kg.

Kerr black hole: A rotating black hole with a ring-shaped singularity.

Kilogram: The basic unit of mass, equal to 1,000 grams.

Kiloparsec (kpc): A thousand parsecs.

Kinetic energy: Energy of motion.

Kinetic temperature: Temperature defined by atomic or particle velocities.

Kirchhoff's first law: An incandescent solid or a hot gas under high pressure produce a continuous spectrum.

Kirchhoff's laws of spectral analysis: A set of three laws that define the conditions for the appearance of the spectrum.

Kirchhoff's second law: A hot low-density gas will produce an emission-line spectrum.

Kirchhoff's third law: A source of continuous radiation viewed through a cooler low-density gas will produce an absorption-line spectrum.

Kirkwood gaps: Gaps in the distribution of the semimajor axes of asteroid orbits caused by resonances with Jupiter's orbit.

KT boundary: The geologic time that divides the Cretaceous period from the Tertiary, when the dinosaurs disappeared, possibly as the result of a meteorite impact.

Kuiper belt: A disk-shaped reservoir of comets outside the orbit of Neptune that produces the short-period comets.

Lagrangian points: Points of orbital stability 60° ahead of and behind a planet or satellite.

Laser: Light Amplification by the Stimulated Emission of Radiation, resulting in a powerful narrow beam of radiation.

Late heavy bombardment: The last period of meteorite fall before 3.8 billion years ago that produced heavy cratering on planetary and satellite surfaces.

Latitude (ϕ): Arc measurement north and south of the Earth's equator; $\phi = z$(equator point) = h(NCP) = δ(zenith).

Lava: Liquid rock.

Law of ellipses: Kepler's first law: planets orbit the Sun in elliptical paths with the Sun at one focus.

Layered deposits: Layers of ice and dust in the Martian polar caps.

Leader spot: A sunspot of a pair of spots that is ahead in the direction of solar rotation.

Leap second: A second added to or subtracted from Coordinated Universal Time to keep it in synchronism with the Earth's rotation.

Leap year: A year in which a day is added, in the Gregorian calendar as February 29.

Lens: A curved piece of glass that can focus radiation.

Leonids: A famous meteor storm that occurs every 33 years.

Light: Electromagnetic radiation; commonly used for radiation visible with the human eye or near the optical domain.

Light curve: A plot of magnitude against time.

Light echoes: Light reflections off interstellar dust from exploding stars.

Light-gathering power: The degree to which a telescope collects electromagnetic radiation; proportional to the square of the telescope's aperture.

Light-year (ly): The distance a ray of light travels in a year at 299,792 km/s.

Limb: The edge of the solar or lunar disk.

Limb darkening: The darkening of the surface of the photosphere of the Sun or star toward its edge, caused by the transparency of the photosphere and inwardly increasing temperature.

LINERs: Low ionization emission line regions; low-activity Seyfert galaxies.

Liquid: A state of matter that has a fixed volume but not a fixed shape.

Lithosphere: A moving rock layer that consists of the crust and the top part of the mantle.

Local apparent solar time (LAST): Time told by the apparent, or real, sun; the hour angle of the apparent sun plus 12^h.

Local Group: The sparse cluster that contains the Galaxy.

Local mean solar time (LMST): Time told by the average, or mean, sun; the hour angle of the mean sun plus 12^h.

Local sidereal time (LST): Time told according to the stars or the vernal equinox; the hour angle of the vernal equinox; LST = HA + α.

Local standard of rest (LSR): The average motion of the stars near the Sun.

Local supercluster: The supercluster of galaxies that contains, among others, the Local Group and the Virgo cluster.

Logarithms: A system in which numbers are represented by powers of ten.

Long-period comets: Comets that have random orbits with periods over 200 years and that come from the Oort cloud.

Long period (Mira) variables: Giant pulsating stars with periods over about 100 days and large visual magnitude ranges.

Longitude (λ): Arc measurement on the equator east and west of Greenwich; λ(west) = UT − LMST; λ(east) = LMST − UT.

Look-back time: The time it has taken light to reach us from a distant source.

Lorentz contraction: An effect of relativity in which the observer measures a body to be shorter than an observer riding with the body.

Low tide: The minimum water line at the shore produced by lunar and solar tides.

Lower main sequence: The main sequence between 0.08 and 0.8 $M_\odot$, whose stars have never had time to evolve away from it.

Luminosity (L): The amount of energy radiated by a body per second (joules/s).

Luminosity (MKK) classes: The categorization of stars by their luminosities and sizes: I, II, III, IV, V for supergiants, bright giants, giants, subgiants, and dwarfs respectively.

Luminous blue variables (LBVs): Extremely luminous supergiants that sporadically bury themselves in dusty ejecta.

Lumpiness problem: The problem of why the Universe is so lumpy with clusters of galaxies while the cosmic microwave background radiation is so smooth.

Lunae Planum: The "Plateau of the Moon," an ancient cratered plateau in the Martian northern hemisphere.

Lunar calendar: A calendar in which the months correlate with the phases of the Moon.

Lunar highlands: Original lunar crust crushed by heavy cratering during the heavy bombardment.

Lunar orbiters: A series of spacecraft that orbited and photographed the Moon's surface with great detail.

Lyman alpha forest: Absorption lines observed in quasar spectra to the short-wavelength side of the Lyman α emission line; caused by clouds of matter between our Galaxy and the quasar.

Lyman series: A series of hydrogen absorption or emission lines that have the ground state as their common bottom level.

M stars: Reddish stars with a strong neutral calcium line and molecular lines, particularly titanium oxide, and with temperatures of 2,000 to 3,500 K.

Magellanic Clouds: Two small companion galaxies to our Galaxy, the Large Magellanic Cloud (the LMC) and the Small Magellanic Cloud (the SMC).

Magma: Liquid rock.

Magnetic activity cycle: The 22-year cycle in the amount of solar magnetic activity that produces the 11-year sunspot cycle.

Magnetic brake: The process by which a magnetic field tied to a stellar wind slows a star.

Magnetic field: The range of magnetism as it acts over a distance.

Magnetic monopoles: Single poles of magnetism, one-dimensional defects that may have been left over from the breaking of GUTS symmetry.

Magnetodisk: An extended disk of plasma in Jupiter's magnetosphere.

Magnetogram: A map of magnetic field strengths on the Sun.

Magnetosphere: The structure around the Earth filled with its magnetic field and particles trapped from the solar wind.

Magnetotail: A long streamer of magnetosphere that stretches from a planet away from the Sun.

Magnifying power: The amount by which a telescope eyepiece multiplies the apparent angular diameter of an object; equal to the ratio of the focal length of the objective to that of the eyepiece.

Magnitude equation: A relation among absolute magnitude, apparent magnitude, and distance: $M = m + 5 - 5 \log d$ (d is distance in pc).

Magnitudes: A system for gauging brightnesses of celestial objects in which 5 magnitudes correspond to a brightness factor of 100.

Magnitudes: Classes of star brightnesses.

Main belt: The location of the greatest concentration of asteroids, between 2.1 and 3.2 AU from the Sun.

Main (dwarf) sequence: The main band of stars that falls from lower right to upper left (luminosity increasing with temperature) on the HR diagram; luminosity class V.

Main-sequence fitting: The means of deriving cluster distances by comparing their main sequences.

Mantle: The thick layer of rock that surrounds a terrestrial planet's core.

Mare Imbrium: A major lava-filled basin in the Moon's northern hemisphere.

Mare Orientale: A major multiringed basin on the Moon.

Marginally open universe: A four-dimensional flat universe that does not fold back on itself.

Maria: Dark areas made of lava flows that fill many impact basins.

Mariner: Planetary spacecraft; Mariners 2 and 10 went to Venus, Mariner 10 to Mercury, and Mariners 4, 6, and 7 to Mars.

Maser: The microwave version of the laser.

Mass (M)**:** The amount of matter in a body or the degree to which a force is resisted.

Mass-luminosity relation: $L \propto M^{3.5}$ (average) for main-sequence stars.

Matter-antimatter problem: The problem of why the Universe is constructed of matter and why matter and antimatter did not completely cancel each other out.

Maunder minimum: The near-disappearance of sunspots between 1645 and 1715.

Maxwell Montes: The major mountains on Venus, located in Ishtar Terra.

Mean sun: A point that travels the celestial equator keeping pace with the average position of the hour circle through the apparent sun.

Megaparsec (Mpc)**:** A million parsecs.

Meridian transit telescope: A telescope that points only to the celestial meridian, used for telling time or for measuring right ascensions.

Meridians (of longitude): Great semicircles that run between the poles of the Earth perpendicular to the equator.

Mesosphere: A layer above the Earth's stratosphere in which temperature falls with altitude.

Messier Catalogue (M)**:** A catalogue of (now) 109 nonstellar celestial objects compiled by Charles Messier in the eighteenth century.

Metamorphic rock: Rock that is chemically transformed under high heat and pressure.

Meteor: A bright tube of ionized atmosphere caused by the passage of a frictionally heated meteoroid.

Meteor shower: A shower of meteors that emanates from a specific point in the sky (a radiant) and caused when the Earth passes near a comet's orbit.

Meteor storm: An intense meteor shower caused by a concentration of cometary debris.

Meteorites: Meteoroids that land on Earth; they are stones, irons, or stony irons.

Meteoroids: Pieces of interplanetary debris that produce meteors when they enter the Earth's atmosphere and meteorites if they hit the ground.

Metonic cycle: A period of 19 years in which there are an almost-even number of 235 lunar phase cycles.

Metric system: A system of measures in which the units differ by multiples of ten.

Midnight: 0^h solar time.

Milky Way: The band of light around the sky caused by stars in the disk of our Galaxy.

Millisecond pulsar: A pulsar that spins hundreds of times per second, its rotation speeded up by mass accretion from a companion.

Mira variables (LPVs)**:** Giant pulsators with periods over about 100 days and large visual magnitude ranges.

MKK classes: Luminosity classes of stars.

Mock suns (sundogs): Bright spots in the solar halo, an atmospheric effect produced by refraction of sunlight in ice crystals.

Models: Mathematical descriptions of physical systems (how temperature, density, and pressure change with depth, for example) that allow predictions of observations.

Modern constellations: Constellations generally invented since 1600.

Molecules: Combinations of atoms.

Monochromatic radiation: Radiation of one wavelength or with severely restricted wavelengths.

Moon: The Earth's satellite, which orbits Earth under the force of gravity.

Moving cluster method: A method that allows the measurement of the distance to a cluster from the motions of its stars.

Multiringed impact basins: Impact basins with multiple concentric walls.

N stars: Cool red carbon stars with strong lines of C_2 and with temperatures of 1,900 to 3,500 K.

Nadir: The point on the celestial sphere directly beneath the observer's feet.

Nasmyth focus: A configuration of a primary and two secondary mirrors in a reflecting telescope that sends the light to a fixed position on a rotation axis.

Neap tides: Ocean tides with minimum high tide produced at the lunar quarters.

Nearside (of the Moon): The side of the Moon that always faces Earth.

Neutrino: A massless (or nearly massless) particle that carries energy at or near the speed of light.

Neutron star: A collapsed star made of neutrons and supported by degenerate neutron pressure.

Neutron star limit: The upper limit of about 3 $M_\odot$ above which neutron stars cannot be supported by degenerate neutrons.

Neutrons: Neutral atomic particles.

New General Catalogue (*NGC*)**:** A catalogue of over 7,000 nonstellar celestial objects compiled by J. E. L. Dreyer in the nineteenth century.

New Moon: A phase in which we can see none of the illuminated side of the Moon, when the Moon is aligned with the Sun.

Newtonian focus: A configuration of a primary mirror and a flat secondary mirror that sends the light to the side of a reflecting telescope.

Newton's first law of motion: Left undisturbed, a body will continue in a state of rest or uniform straight-line motion.

Newton's laws of motion: The three laws stated by Isaac Newton that describe how things move.

Newton's second law of motion: The degree of an acceleration is directly proportional to the force applied, is inversely proportional to the mass of the body being accelerated, and the direction of the acceleration is the same as the direction of the force.

Newton's third law of motion: For every force applied to a body, there is an equal force exerted in the opposite direction (action equals reaction).

Nodes: The intersections of an orbital path and the ecliptic as viewed from Earth.

Non-Euclidean geometry: A geometry that does not accept Euclid's parallel postulate.

Nonradial oscillation: An oscillation in which some parts of a body expand while others contract.

Nonthermal radiation: Radiation produced by processes that are not the result of heat and that cannot be related to temperature.

Noon: 12^h solar time, when either the apparent sun (for noon LAST) or the mean sun (for noon LMST) crosses the upper part of the celestial meridian.

Normal spirals: Spiral galaxies in which the arms come directly from the galaxy's center.

North celestial pole (NCP)**:** The point of rotation of the celestial sphere that lies above the north pole of the Earth.

North ecliptic pole (NEP)**:** The point in the northern hemisphere of the celestial sphere perpendicular to the Earth's orbital plane.

North galactic pole (NGP)**:** The point in the northern hemisphere of the celestial sphere perpendicular to the galactic equator.

North Star: Polaris, the second magnitude star near the north celestial pole.

Nova: A stellar brightening to absolute visual magnitude −8 or −10 caused by the explosion on the surface of a white dwarf of matter flowing from a main-sequence star.

Nova remnant: The cloud of debris expanding around a nova.

Nuclear burning: A common term for thermonuclear fusion.

Nuclear starburst galaxy: A galaxy with extreme star formation and supernova production near its nucleus.

Nucleons: Protons and neutrons.

Nucleus: The combined protons and neutrons at the atomic center.

O stars: Bluish stars with neutral and ionized helium lines and with temperatures of 28,000 to 50,000 K.

OB associations: Groups of O and B stars found in the disks of spiral galaxies.

Objective: A telescope lens or mirror.

Oblate spheroid: A sphere flattened at its poles.

Obliquity of the ecliptic (ε): The tilt of the Earth's axis relative to the direction to the ecliptic poles, or the angle between the celestial equator and the ecliptic, equal to 23° 27' (23.5°).

Observational selection: An effect in which scientific data preferentially selected by some often-unidentified means distorts statistical findings.

Occultation: An event in which one body hides, or crosses in front of, another.

Oceans: Basins filled with water.

Oceanus Procellorum: The largest recognized impact feature on the Moon.

OH/IR stars: Class M Miras buried in dusty shells, the shells caused by strong stellar winds.

Olbers' paradox: That the night sky is dark.

Olympus Mons: The major shield volcano on Mars, located on the slope of the Tharsis bulge.

Omega (Ω): The ratio of the average density of the Universe to the critical density; the Universe is open if Ω is less than 1, marginally open and flat if equal to 1, and closed if greater than 1.

Oort comet cloud: A reservoir over 100,000 AU in radius that contains a trillion comet nuclei; the origin of the long-period comets.

Opacity: The degree to which a substance blocks electromagnetic radiation; the opposite of transparency.

Open clusters: Loose Population I clusters in the Milky Way or in the disks of spiral galaxies.

Open universe: A negatively curved hyperbolic universe that does not close back on itself; there can be more than one line drawn through a point parallel to a given line and the angles of a triangle sum to less than 180°.

Opposition: The position in which a planet is in the opposite direction from the Sun (or in which two celestial bodies are opposite each other).

Optical axis: The line perpendicular to a lens's or mirror's surface at its center.

Optical double: An apparent binary produced by a chance alignment of stars at different distances.

Optical radiation: Electromagnetic radiation seen with the human eye, with wavelengths between about 4,000 and 7,000 Å.

Orbit: The path that one body takes about another as a result of their mutual gravitational attraction.

Organic molecules: Molecules that contain carbon.

Orion Nebula: The archetype of diffuse nebulae.

Outflow channels: Large channels on Mars created by eruptions or releases of vast amounts of water.

Ozone: The O_3 molecule.

P Cygni lines: Stellar emission lines flanked by blue-shifted absorption lines; caused by mass flowing from a star.

P wave: A compressional earthquake wave.

Palomar Sky Survey: A photographic survey of the northern hemisphere and part of the southern made in two colors with the Palomar Schmidt telescope.

Pangaea: The supercontinent that held most of the world's land masses about 200 million years ago.

Parabola: A conic section derived by slicing a cone with a plane parallel to the cone's side; the curve that separates the ellipse and the hyperbola.

Parallax: The apparent shift in the position of a body when viewed from different directions; in particular, half the angular shift of a star over the course of the year as a result of the orbital motion of the Earth.

Parallel postulate (of Euclid): One and only one line can be drawn through a point parallel to another line.

Parallels of latitude: Lines of constant latitude that run east and west.

Parsec (pc): A distance unit at which the Earth's semimajor axis subtends one second of arc; 3.26 light-years.

Partial eclipse: The passage of the Moon through only part of the Earth's umbra, or the partial passage of the Moon across the Sun.

Particle accelerator: A machine that accelerates subatomic particles to speeds near that of light and breaks them in collisions.

Paschen series: A series of hydrogen absorption or emission lines that have the third level as their common bottom level.

Pateras: Ancient, collapsed shield volcanoes.

Pauli Exclusion Principle: No two identical atomic particles at a specific velocity can occupy a specified minimum volume of space.

Peculiar galaxies: Galaxies outside the standard classes.

Penumbra: A region of partial shadow, or the outer grayish region of a sunspot.

Perfect Cosmological Principle: There are neither special places nor times in the Universe.

Perfect gas law: A law in which the pressure of an ordinary gas is directly proportional to the number of particles in the gas times the temperature.

Perigee: The point in the lunar orbit closest to the Earth.

Perihelion: The point of closest approach of a planet to the Sun.

Period: The time required for a body to go from its starting point or position and return to that point or position; the time required for any sort of oscillation to complete a cycle.

Period-luminosity relation: The correlation between the periods of Cepheids and their absolute magnitudes, which allows derivation of distances.

Periodic table: A tabular arrangement of the chemical elements by atomic number according to their chemical properties.

Permanent polar caps (Mars): Polar caps that never dissipate, consisting of dry ice (and water ice) in the south, water ice in the north.

Perseids: The most famous meteor shower; takes place on August 12.

Perturbations: Orbital changes induced by outside gravitational forces.

Phases (of the Moon): The different apparent shapes of the Moon caused by viewing a segment of the lighted side: crescent, quarter, gibbous, full; also seen for inferior planets.

Photochemical reactions: Chemical reactions that take place under the action of sunlight.

Photoelectric effect: A phenomenon in which photons knock electrons from matter.

Photoelectric photometer: A device that uses the photoelectric effect to measure the brightnesses of celestial objects.

Photographic magnitudes (B)**:** Magnitudes as measured with an untreated photographic plate or with a blue-sensitive detector.

Photoionization: The ionization of gas by energetic radiation.

Photons: Particles of electromagnetic radiation that also incorporate wave motion.

Photosphere: The bright apparent surface of the Sun or a star.

Pioneer: A series of planetary spacecraft; Pioneer Venus Probe Carrier and Pioneer Venus orbiter went to Venus, Pioneers 10 and 11 went to Jupiter and Saturn.

Pixel: Picture element; the individual recording element of a CCD.

Planck time: The smallest age to the known Universe, when the Universe was so small that it was effectively a black hole.

Planck's constant: A constant of nature, the energy carried by a photon with a frequency of 1 Hz.

Planetary nebula: A shell of illuminated gas around a hot star; the last part of the ejected wind of an asymptotic giant branch star lit by the old nuclear-burning core.

Planetesimals: Primitive bodies of the Solar System that preceded the formation of the planets.

Planets: The Sun's family of major orbiting bodies; small bodies orbiting other stars analogous to the Sun's planets.

Plasma: A gas that consists of ions and electrons.

Plasma cosmology: An alternative cosmology to the Big Bang that purports that there was no Big Bang and that the observations can be explained by plasmas and magnetic fields.

Plasma tail: A comet's ionized gas tail; points away from the Sun and is structured by the Sun's wind and magnetic field.

Plate tectonics: The process by which crustal plates move on the Earth's surface.

Plates: The divisions of the Earth's crust.

Plumes: Rising columns of hot mantle material that can break through the crust to produce shield volcanoes and volcanic floods or can cause volcanic rises.

Polar axis: The axis of a telescope mounting that points to the visible celestial pole and moves the telescope in hour angle.

Polar ring galaxy: An SO galaxy with a ring of tidal debris circulating around its disk.

Polarization: An optical process that makes light waves oscillate in a preferential direction.

Poles (celestial): The points of rotation of the celestial sphere that lie above the north and south poles of the Earth.

Poles (terrestrial): The points where the Earth's rotation axis emerges.

Polycyclic aromatic hydrocarbons (PAHs)**:** Organic interstellar molecules built of benzene rings.

Population I: The galactic disk and its components; stars of high metal content, luminous blue stars, young stars, and interstellar matter.

Population II: The galactic halo and its components; red stars of the lower main sequence, evolved red giants, horizontal branch stars, and stars of lower metal content.

Population III: The original population of the Galaxy, made only of hydrogen and helium, that seeded the Galaxy with the first heavy elements; no representatives are known.

Position angle: The angle that the line between two bodies makes with the local hour angle through one of them, measuring from north through east.

Positron: A positive electron.

Post meridiem (P.M.)**:** After noon.

Postulate: An unprovable assumption.

Potential energy: Energy held by virtue of position or configuration.

Power-law spectra: Continuous spectra in which the intensity of radiation depends on an inverse power of frequency.

Pre–main-sequence stars: Stars that have developed from the interstellar medium but have not yet arrived on the main sequence.

Precession: The motion of the celestial poles, celestial equator, and equinoxes caused by a 26,000-year wobble of the Earth's axis.

Pressure: The outward force of compressed matter per unit area.

Primary eclipse: The deeper of the two eclipses in the light curve of an eclipsing binary star.

Primary mirror: The large light-collecting mirror of a reflecting telescope.

Prime focus: The focal position of the primary mirror of a reflecting telescope.

Prime meridian: The meridian of longitude that passes through a specific point in Greenwich England and that defines 0° longitude.

Principle of equivalence: In relativity, that an observer cannot tell the difference between acceleration produced by motion and the acceleration of a gravitational field.

Prograde: In the normal direction of revolution of the Solar System, to the east among the stars, on in the context of a planet, in the direction of rotation.

Project Ozma: The first search for extraterrestrial intelligence.

Prominences: Bright arches of cool gas in the solar corona seen against the sky and confined by magnetic fields; filaments seen against the sky.

Proper motion (μ)**:** The angular motion of a star across the line of sight.

Proper (star) names: Individual names assigned to brighter stars, usually reflecting the stars' properties or positions; most are of Arabic origin.

Proton-proton (p-p) **chain:** A fusion reaction that turns four atoms of hydrogen into one atom of helium.

Protons: Positively charged atomic particles.

Protostars: Stars in the process of formation.

Proxima Centauri: The nearest star and a companion to Alpha Centauri.

Ptolemaic system: A system of geocentric orbits that carry epicycles that carry planets.

Pulsar: A rotating neutron star that beams energy along a tilted magnetic field and is only visible if a beam hits the Earth.

Pulsation: A change in stellar radius.

Pyroclastics: Volcanic ejecta consisting of rocks and finely divided ash.

Quantum: A packet or quantity of energy.

Quantum efficiency: The ratio of the number of photons recorded by a detector to the number striking it.

Quantum foam: The fluctuating vacuum from which the Universe may have erupted.

Quantum mechanics: The science that deals with the atom and its constituents, with photons, and with the quantization of their energies and other properties.

Quarks: Particles that make protons and neutrons.

Quasars: Quasi-stellar radio sources (quasi-stellar objects); bright, starlike objects with high redshifts; some are strong radio emitters; they are probably distant, young versions of active galaxies.

Quiet Sun: The constant phenomena of the Sun, which do not take large part in the solar magnetic cycle.

R Coronae Borealis stars: Stars that suddenly dim as a result of dust formation.

R stars: Warmer, yellow-orange carbon stars with lines of CN and with temperatures of 3,500 to 5,400 K.

r-process: A set of nuclear reactions that build heavy elements (beyond those created by the s-process) in supernovae by the rapid capture of neutrons.

Radar: An active observational technique in which radio waves are reflected from a body to determine its distance, speed, and surface features.

Radial velocity (v_r)**:** The relative speed of a body along the line of sight.

Radian: A unit of angular measure equal to 206,265 seconds of arc; the angle in radians is equal to the arc of the circle cut off by the lines that make an angle divided by the radius of the circle.

Radiant: The point in the sky from which the meteors of a meteor shower seem to come, or from which the stars of a cluster seem to come or to which they seem to go.

Radical: An incomplete molecule, one with an unshared electron that needs an atomic partner.

Radio galaxies: Active elliptical galaxies with powerful central radio sources, bipolar jets, and, sometimes, extended double radio sources.

Radio radiation: The longest wave radiation, with wavelengths longer than about 0.1 mm.

Radio telescope: A telescope that collects and detects radio radiation.

Radioactive dating: Dating of rocks and other objects by establishing the ratio of the amount of the daughter product of radioactive decay to that of the parent.

Radioactive isotopes: Isotopes whose nuclei decay into other nuclei with the release of particles and radiation.

Rainbow: A circle of colored light produced by refraction and reflection of sunlight by raindrops.

Rampart craters: Craters unique to Mars, whose ejecta have flow patterns that end in cliffs.

Rays (lunar): Bright lines that emanate from young craters; caused by secondary impacts that expose light-colored rock.

Recombination: The capture of an electron by an ion.

Recombination lines: Emission lines produced by recombination as the recaptured electron cascades to the ground state.

Recurrent novae: Binaries with repeating nova-like outbursts.

Red giant branch: The state of stellar evolution in which intermediate-mass stars grow into giants for the first time and in which they have contracting helium cores surrounded by hydrogen-burning shells.

Redshift: The general spectral shift to longer (redder) wavelengths exhibited by galaxies.

Redshift distance: A distance to a galaxy derived from its redshift and the Hubble constant.

Reduction of data: The process whereby the raw numbers from observations are converted into useful information.

Reflecting telescope: A telescope that focuses radiation with a mirror (usually a paraboloid) and that has a variety of focal positions.

Reflection: The return of radiation from a surface at an angle equal to the angle of incidence.

Reflection nebula: A bright cloud created by light scattered by dust from a star too cool to cause ionization (cooler than type B1).

Refracting telescope: A telescope that uses a lens to refract light to a focus.

Refraction: The bending of the path of radiation as it goes from one substance to another.

Refractory elements or compounds: Those that melt or boil only at high temperatures.

Regolith: Lunar or planetary soil, devoid of organic compounds, produced by constant pulverization by meteorites.

Regular cluster of galaxies: A clusters of galaxies concentrated toward its center and dominated by a giant elliptical galaxy (or by a pair of them) or by a supergiant diffuse galaxy.

Relativistic beaming: Amplification of the radiation from a jet moving near the speed of light toward the observer.

Relativity: The branch of mechanics developed by Albert Einstein that lets the speed of light be independent of the speeds of the source or the observer and that describes the action of gravity as a curvature of spacetime.

Resolving power: The ability of a telescope to separate objects; proportional to wavelength divided by aperture.

Resonance: A gravitational mechanism in which one orbiting body produces a large gravitational perturbation in another because the two have periods that are simple multiples of each other.

Retrograde motion: The motion of a planet to the west relative to the stars as a result of the Earth passing a superior planet in orbit or of the Earth being passed by an inferior planet; counter to the general revolution of the Solar System; in the context of the planets, opposite the direction of rotation.

Revolution: The movement of one body about another.

Right ascension (α): The arc on the celestial equator between the vernal equinox and the hour circle through a celestial body, measured eastward.

Rilles: Channels caused by running lava.

Ring: In the context of the Jovian planets, an encompassing belt of dust and debris.

Ring arcs: Segments of a planetary ring enhanced by the concentration of matter.

Ring nebula: A nebula, commonly around a Wolf-Rayet star, that consists of matter from a stellar wind sweeping up matter from the interstellar medium.

Ringlets: Small rings only a few hundred km wide that make up Saturn's big rings.

Rise: The act of a body appearing from below the horizon.

Roche limit: A limit surrounding a planet within which a fluid body would be torn apart by tides.

Roche lobe: An equilibrium surface around orbiting stars upon which the acceleration is zero.

Rock and ice: In the context of the Jovian planets, respectively, a mixture of heavier materials like silicates and iron and a mixture of volatiles like water, methane, and ammonia.

Rock cycle: The cycle by which one kind of rock is transformed into another.

Rotation: The spinning of a body on its axis.

RR Lyrae stars: Regular short-period pulsators of classes A and F on the horizontal branch of the HR diagram.

Runaway greenhouse effect: A process in which high temperature produces more atmospheric carbon dioxide, which produces higher temperature, and so on.

Runoff channels: Ancient dry riverbeds of low-volume water flows.

S stars: Reddish, cool, intermediate carbon giant stars with molecular lines and particularly strong zirconium oxide lines and with temperatures of 2,000 to 3,500 K.

s-process: A set of nuclear reactions in which atoms slowly capture neutrons and then decay, building heavier elements up through bismuth.

S-wave: A transverse earthquake wave, one that will not go through a liquid.

Sa galaxy: A normal spiral galaxy with tightly wound arms and a large bulge.

S0 (S zero) galaxy: A disk galaxy with no spiral arms.

Sagittarius A: The bright radio source at the center of the Galaxy as viewed with a single-dish radio telescope.

Sagittarius A*: The unresolved bright radio source within Sagittarius A; the true nucleus of the Galaxy.

Saros: An interval of 18 years 11 1/3 days, after which the circumstances of an eclipse repeat themselves.

Saros cycle: An interval of about 600 years in which similar eclipses separated by a saros work their way from one terrestrial pole to the other.

Satellite: A body that orbits a planet.

Sb galaxy: A normal spiral galaxy with intermediately wound spiral arms and an intermediate-sized bulge.

SBa galaxy: A barred spiral galaxy with tightly wound arms and a large bulge.

SBb galaxy: A barred spiral galaxy with intermediately wound spiral arms and an intermediate-sized bulge.

SBc galaxy: A barred spiral galaxy with open spiral arms and a small bulge.

Sc galaxy: A normal spiral galaxy with open spiral arms and a small bulge.

Scale height: A distance over which density drops to 37% of its original value.

Scaling factor of the Universe (R)**:** The factor that represents relative distances in the Universe and that increases with time in an expanding universe.

Scarp: A steep cliff or fault.

Schmidt telescope: A reflecting telescope with a refracting correcting lens that provides a wide field of view.

Schwarzschild black hole: A nonrotating black hole with a pointlike singularity.

Scientific method: A method of inquiry whereby observations or experiments are used to establish a theory that, in turn, predicts new experiments or observations for verification.

Search for Extraterrestrial Intelligence (SETI)**:** The search for extraterrestrial civilizations by radio.

Seasonal polar caps (Mars): Dry-ice polar caps that extend well toward the equator in winter.

Secondary craters: Craters caused by rocks ejected from impact sites.

Secondary eclipse: The shallower of the two eclipses in the light curve of an eclipsing binary star.

Secondary mirror: A small mirror that intercepts the light from the primary of a reflecting telescope and sends it elsewhere.

Sector boundaries: Curved boundaries in interplanetary space that separate opposite directions in the solar magnetic field.

Sedimentary rock: Rock that is formed from sediments under pressure.

Seeing disk: The apparent disk of a star produced by variable refraction in the atmosphere.

Seismograph: A device for recording earthquake waves.

Semimajor axis (a)**:** Half the major axis of an ellipse, which characterizes the ellipse's size.

Semiregular variables: Giant or supergiant pulsating stars with somewhat erratic periods.

Set: The act of a body disappearing below the horizon.

Sextant: A device for measuring the altitudes of celestial bodies.

Seyfert galaxies: Active spirals with abnormally bright nuclei whose centers radiate strong emission lines.

Shearing: The act of one part of a substance moving past another at a different speed.

Shepherd satellites: Small satellites that organize and preserve narrow rings.

Shield volcano: A volcano with runny lava that creates a broad mountain with a low slope.

Shock wave: A wave of sudden increase in pressure.

Short-period comets: Comets with periods under 200 years that come from the Kuiper belt.

Sidereal: Pertaining to the stars.

Sidereal day: The day defined according to the stars; Earth's sidereal day is equal to 23^h56^m of solar time.

Sidereal period: The orbital period of a celestial body relative to the stars.

Signal-to-noise (S/N) **ratio:** The ratio of the values of observational data to random background "noise" not associated with the source.

Silicate rocks: Rocks (including granites and basalts) made of compounds that contain silicon, oxygen, and a variety of metals, and that include granites and basalts.

Single-lined spectroscopic binary: A binary in which only one set of Doppler-shifted absorption lines can be seen.

Singularity: A dimensionless mass; a point-mass.

Size distribution: The way in which the number of objects varies with their sizes.

Slitless spectrogram: An image made with a slitless spectrograph.

Slitless spectrograph: A telescope with a prism over the objective that takes spectra of all the stars in the field of view.

Slow novae: Novae that take some six months to decay three magnitudes from maximum.

Solar constant: The flux of solar radiation at the Earth.

Solar day: The day defined according to the Sun; the interval between successive passages of the Sun across the upper part of the celestial meridian.

Solar dynamo: The combination of rotation and convection that produces the solar magnetic field.

Solar Maximum Mission: An Earth-orbiting satellite that carried spectrographs and imaging devices for observation of the Sun.

Solar model: A mathematical description of the Sun that gives its temperature, density, and composition at all points.

Solar nebula: The disk of gas and dust around the forming Sun, out of which grew the planetesimals and the planets.

Solar oscillations: Multiple movements or vibrations of the solar surface.

Solar System: The Sun and its collection of orbiting bodies.

Solar wind: A thin, ionized gas blowing from the Sun.

Solid: A state of matter that has a fixed shape and volume.

Solstices: The points on the ecliptic farthest from the celestial equator.

South celestial pole (SCP): The point of rotation of the celestial sphere that lies above the south pole of the Earth.

South ecliptic pole (SEP): The point in the southern hemisphere of the celestial sphere perpendicular to the Earth's orbital plane.

South galactic pole (SGP): The point in the southern hemisphere of the celestial sphere perpendicular to the galactic equator.

Space velocity (v_s): The velocity of a star relative to the Sun.

Spacetime: A four-dimensional construction that consists of the three dimensions of space and the one of time.

Seasons: The four periods of the year—spring, summer, autumn, and winter—defined by the position of the Sun relative to the equinoxes and solstices.

Special relativity: That part of relativity that involves constant speed.

Speckle interferometry: An observing technique designed to produce high resolution; it uses computer-combined images of very short exposure to overcome the effects of turbulence in the Earth's atmosphere.

Spectral sequence: The basic categories of stars organized by temperature, OBAFGKM, and by carbon composition, RNS.

Spectrogram: A photographic or graphical rendering of a spectrum.

Spectrograph: A device that separates electromagnetic radiation by wavelength.

Spectroheliogram: A picture of the solar chromosphere made by taking a photograph in the light of a chromospheric emission line.

Spectroscopic binaries: Binaries with components detected by Doppler shifts.

Spectroscopic distance: Distance estimated from absolute magnitude as determined by spectral class.

Spectrum: An array of properties; here, the array of electromagnetic waves.

Spectrum lines: Radiation, or lack thereof, at specific wavelengths.

Speed: The rate at which a body changes its distance with time.

Spicules: Needlelike projections at the top of the solar chromosphere.

Spin-flip: The reversal of the direction of the spin of an electron in the ground state of hydrogen from the same as the proton to the opposite, the act producing the 21-cm line.

Spiral arms: Arms in a galaxy's disk that wind outward from near its center (or from a bar through its center), contain O and B stars and clouds of interstellar matter, are sites of star formation, and are produced by density waves.

Spiral galaxies: Galaxies displaying spiral arms and a strong Population I component.

Spokes: Dark structures in Saturn's rings possibly caused by electrical effects.

Spring: The period in the northern hemisphere between March 20 and June 21 when the Sun is moving northward in the northern celestial hemisphere; autumn in the southern hemisphere.

Spring tides: Ocean tides with maximum high tide produced at new and full Moon.

Sputnik 1: The first Earth satellite, launched by the Soviet Union.

Sputter: To remove atoms from a material by impact with high-speed atoms or ions.

Standard candles: Objects with known absolute magnitudes used to find distances to other galaxies.

Standard time: Local mean solar time at standard meridians spaced 15° apart starting at Greenwich.

Starburst galaxies: Irregular galaxies that exhibit or have exhibited rapid star formation.

Stars: Gaseous, self-luminous bodies similar in nature to the Sun but with diverse properties.

States of matter: The forms matter may take that include gases, solids, and liquids.

Static universe: A universe that neither expands nor contracts.

Steady State theory: A theory in which the Universe appears the same from all locations and at all times; it satisfies the Perfect Cosmological Principle.

Stefan-Boltzmann law: The flux of radiation from a blackbody is equal to a constant times its temperature to the fourth power.

Stellar evolution: Changes in stars brought about by aging.

Stellar wind: A flow of mass from the surface of a star.

Stonehenge: A stone monument in southern England aligned with the summer solstice sunrise.

Stratigraphy: A means of relative dating by observing how one feature lies on top of another.

Stratosphere: The layer above the Earth's troposphere in which temperature rises with altitude.

Strömgren sphere: A spherical bubble of ionized gas—a diffuse nebula—set within a neutral medium.

Strong (nuclear) force: The force that binds atomic nuclei together.

Subduction: One crustal plate diving beneath another.

Subdwarfs: Low-metal Population II stars to the left of the lower main sequence.

Subgiants: Stars classified between the giants and the main sequence; MKK class IV.

Sublime: To pass directly from the solid to the gaseous state.

Summer: The period in the northern hemisphere between June 21 and September 23 when the Sun is moving southward in the northern celestial hemisphere; winter in the southern hemisphere.

Summer solstice: The most northerly point on the ecliptic, at $\delta = 23.5°$ N, where the Sun is found about June 21.

Summer Triangle: A triangle made of the stars Vega, Altair, and Deneb.

Sun: The gaseous self-luminous body that dominates the Solar System; the nearest star.

Sundial: A device with a stick that casts a shadow on a graduated plate, used for telling local apparent solar time.

Sungrazer: A comet that comes very close to the solar surface at perihelion.

Sunspot cycle: The 11-year periodic increase and decrease in the number of sunspots.

Sunspots: Cool, dark regions on the Sun caused by intense magnetic fields that inhibit the flow of energy.

Supercluster: A cluster of clusters of galaxies.

Supercooling: The cooling of a fluid below its freezing point; in the context of the Universe, a lowering of the temperature below that required to break symmetry.

Superfluid state: A state in which a fluid has no viscosity.

Supergiant diffuse (cD) **galaxy:** An enormous galaxy with low surface brightness.

Supergiants: Huge, luminous stars plotted across the top of the HR diagram; MKK class I.

Supergranules: Large-scale mass-motions in the Sun that enclose sets of granules.

Superior conjunction: A conjunction between Venus or Mercury and the Sun in which the planet lies beyond the Sun.

Superior planets: The planets outside the Earth's orbit, from Mars through Pluto.

Superluminal motion: Motion that appears to be faster than light.

Supernova remnant: The expanding cloud of debris caused by the explosion of a supernova.

Supernovae: Outbursts reaching absolute visual magnitude –17 to –19 that involve the destruction or near-destruction of a star.

Superstrings: Eleven-dimensional particles that may have existed at the beginning of the Universe and that created the four forces.

Supersymmetry (SUSY)**:** A relationship in which for each force particle there should be a corresponding mass particle and for each mass particle there should be a force particle.

Symbiotic stars: Interacting binaries with both hot and cool spectral characteristics.

Symmetry: The unification of the forces of nature at high temperatures.

Synchronous rotation: Identical rotation and revolution periods.

Synchrotron radiation: Radiation produced by fast electrons spiraling in a magnetic field.

Synodic period: The orbital period of a celestial body relative to the Sun; the interval between successive conjunctions or oppositions of a planet with the Sun; for the Moon, the 29.5 day period of the phases.

Systems I, II, and III: The three major rotation periods of Jupiter, respectively equatorial, higher-latitude, and internal.

T associations: Gravitationally unbound groups of T Tauri stars.

T Tauri stars: Active young variable pre–main-sequence stars that group into T associations.

Tangential velocity (v_t)**:** The relative speed of a body perpendicular to the line of sight.

Telescope: A device for gathering and focusing electromagnetic radiation.

Temperature: A measure of the average velocities of the particles in a gas and a measure of heat energy in a substance.

Temperature gradient: The rate at which temperature varies with distance.

Terminator: The line on a celestial body that separates night from day.

Terrestrial: Pertaining to Earth or things like Earth.

Terrestrial planets: Earth and those planets like it: Mercury, Venus, and Mars (and perhaps the Moon).

Tesserae (on Venus): Jumbled disrupted volcanic terrain.

Tharsis bulge: The major volcanic rise on Mars; located in the northern hemisphere, it contains four major volcanoes, including Olympus Mons.

Theory: A model that embraces and explains observational or experimental data.

Theory of Everything (TOE)**:** A theory that combines all the forces of nature.

Thermal radiation: Radiation produced as a result of heat, for example, blackbody radiation.

Thermonuclear fusion: The process of energy generation by the combination of lighter atoms into heavier atoms.

Thermosphere: A layer at the top of a planet's atmosphere in which temperature generally rises with altitude.

Thin disk: The portion of a galaxy that contains its youngest components, the O and B stars, and the bulk of the interstellar medium.

Third quarter Moon: A phase in which we see half the illuminated side of the Moon at the time when the Moon makes a 270° angle with the Sun, the right angle before new.

Three-alpha (3-α) **reaction:** The fusion of three atoms of helium into one of carbon.

Three-degree background radiation: Another name for the cosmic microwave background radiation.

Tide: A distortion in a body caused by differential gravity; on Earth, a periodic flow of water.

Time dilation: An effect of relativity in which time slows down for a moving body.

Time zones: Zones within which everyone is on the same time.

Titan: Saturn's large satellite.

Torus: A doughnut-shaped ring.

Total eclipse: An eclipse in which the Moon is entirely immersed in the Earth's umbral shadow, or in which the observer is immersed in the Moon's umbral shadow.

Transit: Passage of a body across the celestial meridian, or the act of one body crossing in front of another.

Tritium: Hydrogen with two neutrons in its nucleus; ^{3}H.

Triton: The large satellite of Neptune.

Trojan asteroids: Asteroids trapped in the Lagrangian points of Jupiter's orbit.

Tropic of Cancer: The northerly limit, at latitude 23.5°N, at which the Sun can be found overhead; the Sun is overhead there about June 21.

Tropic of Capricorn: The southerly limit, at latitude 23.5°S, at which the Sun can be found overhead; the Sun is overhead there about December 21.

Tropics: Limits of latitude (23.5°N to 23.5°S) between which the Sun appears overhead sometime during the year.

Troposphere: The layer of atmosphere closest to a solid planet's surface.

Tully-Fisher relation: Luminosities of spiral galaxies increase with increased spread in internal velocities.

Tuning fork diagram: Hubble's classification scheme for galaxies that shows ellipticals branching to spirals and barred spirals.

21-cm line: A powerful radio line of neutral hydrogen produced in the interstellar medium by an electron in the ground state reversing the direction of its spin from the same as that of the proton to opposite that of the proton.

Twilight: The period of time before sunrise or after sunset when the Earth's atmosphere, bright from sunlight, illuminates the ground.

Tycho: One of the most recent large craters on the Moon.

Tycho's star: The Supernova of 1572, observed by Tycho.

Type I Seyfert galaxies: Seyfert galaxies with broad hydrogen lines, narrow forbidden lines, and powerful continua.

Type Ia supernovae: Supernovae of either population with no spectral hydrogen lines, reaching absolute visual magnitude –19, produced by the collapsing of white dwarfs that have exceeded the Chandrasekhar limit.

Type Ib supernovae: Supernovae of Population I with no spectral hydrogen lines powered by the collapsing iron cores of Wolf-Rayet stars.

Type II Seyfert galaxies: Seyfert galaxies with weak hydrogen and forbidden lines and weak or absent continua; Seyferts with hidden nuclei.

Type II supernovae: Supernovae of Population I with spectral hydrogen lines, reaching absolute visual magnitude –17.5, powered by collapsing iron cores of supergiant stars.

Ultraviolet radiation: Radiation with wavelengths shorter than the human eye can see but longer than about 100 Å.

Ulysses: A spacecraft orbiting the Sun designed to go over the solar poles.

Umbra: A region of full shadow or the central dark region of a sunspot.

Uncompressed density: The average density of a planet corrected for the compressing effects of gravity.

Unidentified flying objects (UFOs): Objects in the sky for which an observer has no ready explanation, including misunderstandings of real celestial and terrestrial objects, atmospheric effects, illusions, and hoaxes.

Unit circle: A circle with a radius of one unit.

Universal time (UT): Local mean solar time at Greenwich, longitude 0°.

Universe: The all-encompassing structure that contains everything.

Upper main sequence: The main sequence above 8 M$_\odot$ whose stars are too massive to evolve to white dwarfs because their cores are above the Chandrasekhar limit.

Uraniborg: Tycho's observatory in Denmark.

Valles Marineris: The great fault canyon on Mars, associated with the Tharsis Bulge.

Van Allen belts: Doughnut-shaped zones around the Earth filled with high-energy particles.

Variable stars: Stars whose magnitudes change with time.

Velocity (v): The combination of speed and direction.

Velocity curve: A plot of radial velocities against time for the components of a spectroscopic binary.

Velocity equation: An equation that relates stellar space velocity, tangential velocity, and radial velocity: $v_s^2 = v_t^2 + v_r^2$.

Venera: Russian spacecraft that landed on Venus.

Vernal equinox (Υ): The point on the celestial sphere where the ecliptic crosses the celestial equator with the Sun moving north, the crossing taking place on March 20 or 21.

Very long baseline interferometer (VLBI): An interferometer with separate telescopes synchronized by clocks.

Viking: Spacecraft that imaged, then landed on, Mars.

Virgo cluster: The nearest large cluster of galaxies, in Virgo, about 18 Mpc away.

Virtual photons: The particles that carry the electromagnetic force.

Viscosity: Energy-dissipating internal friction in a fluid.

Visual binaries: Binary stars with components that can be separated by eye at the telescope.

Visual magnitudes (*V*): Magnitudes as seen by the human eye or measured with a yellow-green detector.

Visual radiation: Electromagnetic radiation seen with the human eye, with wavelengths between about 4,000 and 7,000 Å.

Voids: Volumes, megaparsecs across, devoid of galaxies.

Volatile elements or compounds: Those that melt or boil at low temperatures.

Volcanoes: Vents in the Earth's crust through which magma, ash, and gas can escape.

Voyagers: The two spacecraft that probed Jupiter and Saturn, *Voyager 2* going also to Uranus and Neptune.

W and Z particles: The particles that carry the weak force.

W Virginis stars: Population II Cepheids.

Waning phases: The diminishing phases between full and new Moon.

Water cycle: The cycle by which water evaporates from the oceans and then condenses to fall back into the oceans or condenses on land to run back to the oceans.

Wavelength (λ): The distance between crests of a wave.

Waxing phases: The growing phases between new and full Moon.

Weak force: The second strongest force of nature; involved with radioactive decay.

Weakly interacting massive particles (WIMPs): Massive counterparts to massless (or nearly massless) particles in supersymmetry theory; candidates for cold dark matter.

Weather: Short-term variations in a planet's atmosphere.

Weight: The force with which a body is pressed to the surface of another body as a result of gravity.

White dwarfs: Small dim stars about the size of Earth plotted across the bottom of the HR diagram; class D; hydrogen-rich DA and helium-rich DB.

Wien law: The wavelength of maximum radiation from a blackbody is a constant divided by the temperature.

Winter: The period in the northern hemisphere between December 22 and March 21 when the Sun is moving northward in the southern celestial hemisphere; summer in the southern hemisphere.

Winter solstice: The most southerly point on the ecliptic, at $\delta = 23.5°S$, where the Sun is found on December 22.

Winter Triangle: A triangle made of the stars Sirius, Betelgeuse, and Rigel.

Wolf-Rayet (WR) **stars**: High-mass, hydrogen-poor stars with emission lines of helium and nitrogen (WN) or carbon (WC).

Work: Force applied over a distance.

Wrinkle ridges: Ridges caused by the compression of surface rocks.

X-ray binaries: Binaries that radiate X rays from accretion disks around neutron star components.

X-ray bursters: Binaries that emit bursts of X-ray radiation from helium accumulating on neutron star surfaces; the progeny of X-ray binaries.

X rays: Radiation with wavelengths between the gamma ray and ultraviolet regions of the spectrum, with wavelengths between about 1 Å and 100 Å.

Zeeman effect: The magnetic splitting of spectrum lines.

Zenith: The point on the celestial sphere over the observer's head.

Zenith distance (*z*): The vertical arc between the zenith and a point in the sky; $z = 90° - h$.

Zero-age main sequence (ZAMS): The line on the HR diagram outlined by new stars that are just beginning to fuse hydrogen into helium.

Zodiac: The band of constellations that contains the ecliptic.

Zodiacal light: A band of light in the Zodiac caused by sunlight scattered from cometary and asteroidal dust.

Zone of avoidance: The region around the plane of the Milky Way where galaxies are not seen because of dust absorption.

Zones: Strips of higher-altitude, bright (reflective), lower-temperature clouds in Jupiter's atmosphere.

ZZ Ceti stars: DA white dwarf nonradial oscillators.

Acknowledgments

LITERARY PERMISSIONS

Figure 3.8(b) Illustration by Snowden Hodges from *Living the Sky* by Ray A. Williamson. Line illustrations copyright © 1984 by Snowden Hodges. Reprinted by permission of Houghton Mifflin Company. All rights reserved.

Figure 3.21 From *Stars* by James B. Kaler, page 9. Copyright © 1992 by Scientific American Library. Reprinted by permission of W.H. Freeman and Company.

Figure 3.24 Adapted from *Stars* by James B. Kaler, page 30. Copyright © 1992 by Scientific American Library. Reprinted by permission of W.H. Freeman and Company.

Figure 5.15 U.S. Naval Observatory.

Figure 10.11 From figure by T. Aherens from *Cambridge Encyclopedia of Earth Sciences*. Reprinted with the permission of Cambridge University Press.

Figure 10.21 From figure by J.B. Pollack from *The New Solar System* by J. Kelly Beatty and Andrew Chaikin, page 95. Reprinted by permission.

Figure 12.4 Right Adaptation of Figure 3 from *Science,* Volume 252, Number 5003, April 12, 1991, page 262. Copyright © 1991 by the American Association for the Advancement of Science. Reprinted by permission.

Figure 13.9 Bottom From *Sky and Telescope,* November 1982, page 421. Copyright © 1982 Sky Publishing Corporation. Courtesy Sky Publishing Corporation.

Figure 13.21 Reprinted with permission from "Ancient oceans, ice sheets and the hydrological cycle on Mars" by V.R. Baker et al., from *Nature,* August 15, 1991, Volume 352, page 589. Copyright © 1991 Macmillan Magazines Ltd. Reprinted by permission.

Figure 14.6 Adapted from *The New Solar System,* Third Edition, edited by J. Kelly Beatty and Andrew Chaikin, page 143. Copyright © 1990 Sky Publishing Corporation. Reprinted by permission.

Figure 14.8(a–b) From *The New Solar System,* Third Edition, edited by J. Kelly Beatty and Andrew Chaikin, page 148. Copyright © 1990 Sky Publishing Corporation. Courtesy Sky Publishing Corporation.

Figure 14.9 From *Jupiter* by Garry Hunt and Patrick Moore, page 7. Copyright © 1981 by Mitchell Beazley Publishers. Reprinted by permission of Reed Consumer Books Ltd.

Figure 14.17 Illustration of model of Callisto reprinted by permission of the Jet Propulsion Laboratory.

Figure 14.19 Illustration of model of Europa reprinted by permission of the Jet Propulsion Laboratory.

Figure 14.20 Illustration of model of Io reprinted by permission of the Jet Propulsion Laboratory.

Figure 15.5 Adapted from *The New Solar System,* Third Edition, edited by J. Kelly Beatty and Andrew Chaikin, page 143. Copyright © 1990 Sky Publishing Corporation. Reprinted by permission.

Figure 15.10 From figure, "Opacity of Saturn's rings" from *The New Solar System,* Third Edition, edited by J. Kelly Beatty and Andrew Chaikin, pages 164–165. Reprinted by permission.

Background 15.1 From "The Voyager Encounters" by Bradford C. Smith from *The New Solar System,* Third Edition, edited by J. Kelly Beatty and Andrew Chaikin, pages 115, 116 and 117. Reprinted by permission.

Figure 16.3 Adapted from *The New Solar System,* Third Edition, edited by J. Kelly Beatty and Andrew Chaikin, page 132. Copyright © 1990 Sky Publishing Corporation. Reprinted by permission.

Figure 16.8 Adapted from *Sky and Telescope,* February 1990, page 148. Copyright © 1990 Sky Publishing Corporation. Reprinted by permission.

Figure 17.5 From *The New Solar System,* Third Edition, edited by J. Kelly Beatty and Andrew Chaikin, page 248. Copyright © 1990 Sky Publishing Corporation. Courtesy Sky Publishing Corporation.

Figure 17.9 Adapted from *Sky and Telescope,* June 1992, p. 609. Reprinted by permission of Lucy McFadden.

Figure 17.13 Reprinted from *Asteroids II, Space Science Series,* edited by Richard Binzel, Tom Gehrels and Mildred Shapley Matthews, page 330. Copyright © 1990 by the University of Arizona Press. Reprinted by permission.

Figure 17.14 Reprinted from *Asteroids II, Space Science Series,* edited by Richard Binzel, Tom Gehrels and Mildred Shapley Matthews, page 323. Copyright © 1990 by the University of Arizona Press. Reprinted by permission.

Figure 17.22 From Figure 19 by J.C. Brandt from *The New Solar System* by J. Kelly Beatty and Andrew Chaikin. Reprinted by permission.

Figure 17.24 Right From *The New Solar System,* Third Edition, edited by J. Kelly Beatty and Andrew Chaikin, page 225. Copyright © 1990 Sky Publishing Corporation. Courtesy Sky Publishing Corporation.

Figure 17.25 From *The New Solar System,* Third Edition, edited by J. Kelly Beatty and Andrew Chaikin, page 227. Copyright © 1990 Sky Publishing Corporation. Courtesy Sky Publishing Corporation.

Figure 17.29(a) From *Sky and Telescope,* October 1985, page 317. Copyright © 1985 Sky Publishing Corporation. Courtesy Sky Publishing Corporation.

Figure 17.33 From *Sky and Telescope,* January 1993, page 26. Artwork by Steven Simpson. Copyright © 1993 Sky Publishing Corporation. Courtesy Sky Publishing Corporation.

Figure 17.35 Adapted from *The New Solar System,* Third Edition, edited by J. Kelly Beatty and Andrew Chaikin, page 287. Copyright © 1990 Sky Publishing Corporation. Reprinted by permission.

Figure 18.17(a) From *The Sun* by Iain Nicolson, page 41. Copyright © 1982 by Michell Beazley Publishers. Reprinted by permission of Reed Consumer Books Ltd.

Figure 18.7(b) From "A Search for Rhenium Lines in the Fraunhofer Spectrum" by J.W. Swensson from *Solar Physics,* Volume 13, No. 1, July 1970. Copyright © 1970 by D. Reidel Publishing Company. Reprinted by permission of Kluwer Academic Publishers.

Figure 18.11 Adapted from *The Restless Sun* by Donat G. Wentzel, page 76. Copyright © 1989 by the Smithsonian Institution. Reprinted by permission.

Figure 18.20(b) From *Sky and Telescope,* June 1987, page 591. Artwork by Steven Simpson. Copyright © 1987 Sky Publishing Corporation. Courtesy Sky Publishing Corporation.

Figure 18.23 Adapted from *Stars* by James B. Kaler, page 114. Copyright © 1992 by Scientific American Library. Reprinted by permission of W.H. Freeman and Company

Figure 18.25 From *The Restless Sun* by Donat G. Wentzel, page 97. Copyright © 1989 by the Smithsonian Institution. Reprinted by permission.

Figure 18.28 Adapted from *The Sun, Our Star* by Robert Noyes, page 68. Reprinted by permission of Richard Noyes.

Figure 18.29 From *Stars* by James B. Kaler, page 125. Copyright © 1992 by Scientific American Library. Reprinted by permission of W.H. Freeman and Company.

Figure 19.16 From *Landolt-Börnstein, Group VI: Astronomy Astrophysics and Space Research, Volume I: Astronomy and Astrophysics* edited by H.H. Voigt, page 284. Copyright © 1965 by Springer-Verlag. Reprinted by permission.

Figure 19.27(b) From figure, "Starspot cycle of 40 Eri" by Sallie Baliunas from *Sky and Telescope.* Reprinted by permission.

Figure 20.5 From figure, "ORBIT of 70 Ophiuchi, 88-year period" by Strand from *Catalog of Photographs from the Yerkes Observatory.* Reprinted by permission of The Yerkes Observatory.

Figure 20.18(a–d) From figures on pages 51 (Hyades), 26 (Pleiades), 14 & 15 (Double Cluster), and 91 (M67) from *An Atlas of Open Cluster Colour-Magnitude Diagrams* by Gretchen L. Hagen, Volume 4, 1970. Reprinted by permission.

Figure 20.18(c) Data from figure from *The Astrophysical Journal,* 105, 492, 1947. Reprinted by permission of the author.

Figure 20.19 From figures on pages 51 (Hyades) and 26 (Pleiades) from *An Atlas of Open Cluster Colour-Magnitude Diagrams* by Gretchen L. Hagen, Volume 4, 1970. Reprinted by permission.

Figure 21.1 Graph Reprinted by permission of the publishers from *The Story of Variable Stars* by Leon Campbell and Luigi Jacchia, Cambridge, Mass.: Harvard University Press. Copyright © 1941 by the Blakiston Company.

Figure 21.10 From *Variable Stars* by Michel Petit, page 54. Copyright © 1987 by John Wiley & Sons, Ltd. Reprinted by permission of John Wiley & Sons, Ltd.

Figure 22.9 From "The large-scale distribution of neutral hydrogen in the galaxy" by W. B. Burton from *Galactic and Extragalactic Radio Astronomy,* edited by Gerrit L. Verschuur and Kenneth I. Kellermann, page 92. Copyright © 1974 by Springer-Verlag. Reprinted by permission.

Figure 22.20 From *Sky and Telescope,* December 1983, page 495. Reprinted by permission of the author.

Figure 23.16 From *The Astrophysical Journal,* Volume 331, 1988, page 902. Reprinted by permission.

Figure 24.3 Allan Sandage.

Figure 24.4 From "The Cl37 Solar Neutrino Experiment and The Solar Abundance" by Icko Iben, Jr. from *Annals of Physics,* Vol. 54, No. 1, August 1969, page 164. Copyright © 1969 by Academic Press, Inc. Reprinted by permission.

Figures 24.5, 24.10, 24.12, and 24.15 Adapted from Icko Iben.

Figure 25.2 From article by A. Maeder and G. Meynet from *Astronomy and Astrophysics Supplements,* Volume 76, page 411. Reprinted by permission.

Figure 25.5 From article by A. Maeder and G. Meynet from *Astronomy and Astrophysics Supplements,* Volume 76, page 411. Reprinted by permission. Observed supergiants on HR diagram from *The Astrophysical Journal,* 284, 565, 1984; 232, 409, 1979. Reprinted by permission of the author.

Figure 25.7 Adapted from *Stars* by James B. Kaler, pages 179 and 189. Copyright © 1992 by Scientific American Library. Reprinted by permission of W.H. Freeman and Company.

Figure 25.12 Adapted from *Sky and Telescope,* February 1989. Copyright © 1989 Sky Publishing Corporation. Reprinted by permission.

Figure 25.18 Adapted from *Stars* by James B. Kaler, page 199. Reprinted by permission of George Greenstein.
Figure 25.19 From *The Astronomy and Astrophysics Encyclopedia,* edited by Stephen P. Maran, page 961. Copyright © 1992 by Van Nostrand Reinhold. Reprinted by permission of Van Nostrand Reinhold.
Figure 25.21(a) Reprinted with permission from "A planetary system around the millisecond pulsar PRS1257 + 12" by A. Wolszczan and D.A. Frail from *Nature,* January 9, 1992, Volume 355, page 147. Copyright © 1992 Macmillan Magazines Ltd. Reprinted by permission.
Figure 25.23 Adapted from *Stars* by James B. Kaler, page 203. Copyright © 1992 by Scientific American Library. Reprinted by permission of W.H. Freeman and Company.

Figure 26.11 From *Sky and Telescope,* May 1990. Copyright © 1990 Sky Publishing Corporation. Courtesy Sky Publishing Corporation.
Figure 26.18 Adapted from "Solar Calibration and the Ages of Old Disk Clusters M 67, NGC 188, and NGC 791" by P. Demarque, E. Green, and D. Guenther. *Astronomical Journal,* vol. 103, p. 151, 1992.

Figure 28.5 Adapted from *Galaxies and the Universe,* edited by Allan Sandage, Mary Sandage and Jerome Kristian, page 782. Copyright © 1975 by The University of Chicago. Reprinted by permission of The University of Chicago Press.
Figure 28.9 R. Hess

Figure 30.6 Adapted from *Sky and Telescope,* March 1991, page 253. Copyright © 1991 Sky Publishing Corporation. Reprinted by permission.
Figure 30.9 Adapted from *The Big Bang: The Creation and Evolution of the Universe* by Joseph Silk, page 95. Copyright © 1980 by W.H. Freeman and Company. Reprinted by permission.
Figure 30.10 Adapted from *Sky and Telescope,* March 1991, page 255. Copyright © 1991 Sky Publishing Corporation. Reprinted by permission.
Figure 30.14 Adapted from *From Quarks to the Cosmos: Tools of Discovery* by Leon M. Lederman and David N. Schramm, page 152. Copyright © 1989 by Scientific American Library. Reprinted by permission of W.H. Freeman and Company.
Figure 30.17 From *Sky and Telescope,* September 1988, page 253. Copyright © 1988 Sky Publishing Corporation. Courtesy Sky Publishing Corporation.
Figure 30.18 From *From Quarks to the Cosmos: Tools of Discovery* by Leon M. Lederman and David N. Schramm, page 144. Copyright © 1989 by Scientific American Library. Reprinted by permission of W.H. Freeman and Company.
Figure 30.20 From the illustration "How the Universe Began" from "Astronomers Detect Proof of 'Big Bang'," *The New York Times,* Friday, April 24, 1992, page A16. Copyright © 1992 by The New York Times Company. Reprinted by permission.

PHOTO CREDITS

Page i Anglo-Australian Telescope Board
Page iii Anglo-Australian Telescope Board
Page ix Mario Grassi
Page x Lick Observatory
Page xi NASA
Page xii NASA/JPL
Page xiii Top NASA/JPL
Page xiii Bottom W. P. Sterne, Jr.
Page xiv Max Planck Society for the Advancement of Science
Page xv Bill Fletcher
Page xvi H. R. Bridle, NR
Page xx Department of Environment, Crown Copyright
Page xxiv Griffith Institute, Ashmolean Museum, Oxford

Page 1 National Optical Astronomy Observatories
Page 1 Inset From Jan Blaeu, *Grand Atlas au Cosmographie Blaviane,* vol. I, Amsterdam, 1667. Bibliotheque Publique et Universitaire, Geneva
Page 2 James B. Kaler
Figure 1.1 Special Collections, San Diego State University Library
Figure 1.3 Camerique/H. Armstrong Roberts
Figure 1.4 NASA
Figure 1.5 NASA
Figure 1.6 Mt. Wilson and Las Campanas Observatories
Figure 1.8 NASA
Figure 1.9 University of Illinois Prairie Observatory
Figure 1.10 James B. Kaler
Figure 1.11(a) Rudolph Schild, Smithsonian Astrophysical Observatory
Figure 1.11(b) Laird Thompson, University of Hawaii
Figure 1.12 Dennis DiCicco
Figure 1.14(a–b) National Optical Astronomy Observatories
Figure 1.15 California Institute of Technology
Figure 1.16 ROE/Anglo-Australian Telescope Board, photo by David F. Malin

Page 13 Adler Planetarium Museum
Figure 2.1(a) James B. Kaler
Figure 2.4 Yerkes Observatory
Figure 2.8 National Maritime Museum, London
Figure 2.13 National Optical Astronomy Observatories
Figure 2.15 Anglo-Australian Telescope Board, photo by David F. Malin
Figure 2.18 The Bettmann Archive

Page 29 James B. Kaler
Figure 3.8 Department of Environment, Crown Copyright
Figure 3.11(a–b) Mario Grassi
Figure 3.14 Dennis DiCicco
Figure 3.19 U. S. Naval Observatory
Figure 3.20 Tom Stimson/FPG International Corp.
Figure 3.25 James B. Kaler

Page 49   University of Illinois Library
Figure 4.1   The Bettmann Archive
Figure 4.2   Akira Fujii
Figure 4.3   University of Illinois Library
Figure 4.4(b)   Rick Olson
Figure 4.5(a)   Akira Fujii
Figure 4.6   University of Illinois Library
Figure 4.7   Akira Fujii
Figure 4.8   Akira Fujii
Figure 4.9   University of Illinois Library
Figure 4.10   University of Illinois Library
Figure 4.11   Istanbul University Library
Figure 4.12   Dennis DiCicco
Figure 4.13(a)   The Metropolitan Museum of Art, The Brisbane Dick Fund 1951
Figure 4.13(b)   The National Gallery
Figure 4.14   Palomar Sky Survey, *Smithsonian Astrophysical Observatory Atlas*

Page 67   Jean-François Millet, "Sheepfold by Moonlight," Walters Art Gallery, Baltimore
Figure 5.1   Zdenek Kopal, *The Moon*
Figure 5.3 (8 photos)   Lick Observatory
Figure 5.4   Rick Olson
Figure 5.5   NASA
Figure 5.12   Gerald Newsom
Figure 5.14   F. Parker
Figure 5.16(b)   Yerkes Observatory
Figure 5.16(c)   Richard Berry
Figure 5.16(d)   Tersch Enterprises
Figure 5.16(e)   National Optical Astronomy Observatories

Page 80   Scala/Art Resource, NY
Figure 6.1(a–b)   James B. Kaler
Figure 6.10 (4 photos)   E. C. Slipher, Lowell Observatory
Figure 6.13   The Bettmann Archive
Figure 6.14   "Copernicus Concerning the Revolutions of the Heavenly Bodies," 1543
Figure 6.17(a)   From Jan Blaeu, *Grand Atlas au Cosmographie Blaviane*, vol. I, Amsterdam, 1667. Bibliotheque Publique et Universitaire, Geneva
Figure 6.17(b)   New York Public Library; Astor, Lenox and Tilden Foundations
Figure 6.18   Scripta Mathematica, Yeshiva University
Figure 6.21   Museo di Storia della Scienza, Florence

Page 95   Bausch & Lomb
Page 95 Inset   California Institute of Technology
Page 96   F. Rickard-Artdia/Vandystadt/Photo Researchers, Inc.
Figure 7.1   Brown Brothers
Figure 7.3   Joe Towers/The Stock Market
Figure 7.9   Rick Steart/ALLSPORT USA
Figure 7.11   James B. Kaler
Figure 7.12(a–b)   Lowell Observatory

Figure 7.13(a–b)   NASA
Figure 7.15   NASA
Figure 7.16   The Bettmann Archive
Figure 7.21   GEOPIC™, Earth Satellite Corporation

Page 116   R. T. (Jack) Gladin
Figure 8.4   The Bettmann Archive
Figure 8.6   The Exploratorium
Figure 8.14   Akira Fujii
Figure 8.15   Deutsches Museum

Page 136   National Optical Astronomy Observatories
Figure 9.5   Bausch & Lomb
Figure 9.7   Yerkes Observatory
Figure 9.10   California Institute of Technology
Figure 9.13   California Institute of Technology
Figure 9.14   Kitt Peak/National Optical Astronomy Observatories
Figure 9.15   James B. Kaler
Figure 9.17   James B. Kaler
Figure 9.18(a) Top   James B. Kaler
Figure 9.18(a) Bottom   Pat Sweitzer
Figure 9.18(b) Top   National Optical Astronomy Observatories
Figure 9.18(b) Bottom   Pat Sweitzer
Figure 9.19(a–b)   European Southern Observatory
Figure 9.20   James B. Kaler
Figure 9.21   Palomar Observatory
Figure 9.23   NASA/Ames Research Center
Figure 9.24   NRAO/AUI, photo by Bell Telephone Laboratories
Figure 9.25   NRAO/AUI
Figure 9.26   Roger Ressmeyer/Starlight
Figure 9.27   California Institute of Technology
Figure 9.28   Roger Ressmeyer/Starlight
Figure 9.29(a–b)   NASA
Figure 9.30   NASA

Page 155   NASA/JPL
Page 155 Inset   NASA/JPL
Page 156   James B. Kaler
Figure 10.2   Mark Burnett/Stock Boston
Figure 10.3   W. Haxby, NOAA/NESDIS/NGDC
Figure 10.4   David Weintraub/Photo Researchers, Inc.
Figure 10.5   Phil Degginger/H. Armstrong Roberts
Figure 10.6(a)   David Weintraub/Photo Researchers, Inc.
Figure 10.6(b)   Gregory C. Dimijian
Figure 10.6(c)   Rich Buzzelli/Tom Stack & Associates
Figure 10.9(a–b)   Lamont-Doherty Geological Observatory
Figure 10.15   Woods Hole Oceanographic Institute
Figure 10.20(b)   James B. Kaler
Figure 10.22   Michael Giannechini/Photo Researchers, Inc.
Figure 10.23   Bruce Morley
Figure 10.25   Inset Louis Frank, University of Iowa
Figure 10.26   Dyballa/Zefa/H. Armstrong Roberts

ACKNOWLEDGMENTS   **593**

Page 177   NASA
Figure 11.1   McDonald Observatory, Texas
Figure 11.2(a–b)   Maine Department of Sea and Shore Fisheries
Figure 11.4(a) Left   Lick Observatory
Figure 11.4(b) Right   Lick Observatory
Figure 11.5(a–c)   NASA
Figure 11.6(a)   NASA
Figure 11.8(a–c)   NASA
Figure 11.9   Lick Observatory
Figure 11.11   Lick Observatory
Figure 11.12   NASA
Figure 11.13   NASA
Figure 11.14   Yerkes Observatory
Figure 11.14 Inset   NASA
Figure 11.15   NASA
Figure 11.16   Lick Observatory
Figure 11.17(a–c)   NASA
Figure 11.18   NASA
Figure 11.20   NASA
Figure 11.21   NASA
Figure 11.22   NASA
Figure 11.23   NASA
Figure 11.25(a)   Yerkes Observatory
Figure 11.25(b)   NASA
Figure 11.26   NASA/Ames Research Center
Figure 11.27(a–d)   Center for Astrophysics

Page 197   NASA
Figure 12.1(a)   Lick Observatory
Figure 12.1(b)   Lowell Observatory
Figure 12.4(a)   NASA
Figure 12.7(a–b)   NASA
Figure 12.10 (a–d)   A.G.W. Cameron
Figure 12.11   NASA
Figure 12.13   Goddard Space Flight Center
Figure 12.14   NASA/JPL
Figure 12.15   NASA/JPL
Figure 12.16   NASA/JPL
Figure 12.17   NASA/JPL
Figure 12.18   NASA/JPL
Figure 12.19   NASA/JPL
Figure 12.20(a–b)   Vernadsky Institute, USSR Academy of Sciences
Figure 12.21   NASA/JPL
Figure 12.22(a–b)   NASA/JPL
Figure 12.23   NASA/JPL
Figure 12.24   NASA/JPL
Figure 12.25   NASA/JPL

Page 215   NASA/JPL
Figure 13.1   Dr. P. James, University of Toledo/NASA
Figure 13.2(a–d)   Lowell Observatory
Figure 13.3(a)   Lowell Observatory
Figure 13.3(b)   Meudon Observatory
Figure 13.4(a)   NASA/JPL
Figure 13.4(b)   NASA
Figure 13.7   U.S. Geological Survey, Flagstaff, Arizona

Figure 13.8   NASA/JPL
Figure 13.9   NASA
Figure 13.10   U.S. Geological Survey, Flagstaff, Arizona
Figure 13.11   NASA/JPL
Figure 13.12(a–b)   NASA/JPL
Figure 13.13   NASA/JPL
Figure 13.14   NASA/JPL
Figure 13.15   NASA
Figure 13.16   NASA/JPL
Figure 13.17(a–b)   NASA
Figure 13.18(a)   NASA
Figure 13.18(b)   A. Treiman
Figure 13.19(a–b)   NASA/JPL
Figure 13.20   NASA/JPL
Figure 13.20 Inset   NASA/JPL
Figure 13.22   NASA/JPL
Figure 13.23   NASA/JPL
Figure 13.24   NASA/JPL
Figure 13.25(a–b)   NASA/JPL
Figure 13.25 Inset   NASA/JPL
Figure 13.26   NASA
Figure 13.27(a–b)   NASA

Page 233   NASA
Figure 14.1   National Optical Astronomy Observatories
Figure 14.3(b)   NASA/JPL
Figure 14.7   NASA/JPL
Figure 14.10   NASA/JPL
Figure 14.12 Inset   M. Mendillo, Boston University
Figure 14.14   NASA
Figure 14.17   NASA/JPL
Figure 14.18   NASA/JPL
Figure 14.18   Inset NASA/JPL
Figure 14.19   NASA/JPL
Figure 14.20   NASA/JPL
Figure 14.21(a–b)   NASA/JPL
Figure 14.22   NASA/JPL
Figure 14.23   NASA/JPL
Figure 14.24   NASA

Page 251   NASA/JPL
Figure 15.1   National Optical Astronomy Observatories
Figure 15.4   NASA/JPL
Figure 15.6   NASA/JPL
Figure 15.7   NASA
Figure 15.8   NASA/JPL
Figure 15.9   NASA/JPL
Figure 15.11   NASA/JPL
Figure 15.12   NASA/JPL
Figure 15.13   NASA/JPL
Figure 15.14   NASA/JPL
Figure 15.16(a–b)   NASA/JPL
Figure 15.17(b–c)   NASA/JPL
Figure 15.18(b)   NASA
Figure 15.19(a)   NASA/JPL
Figure 15.19(b)   J. Lunine, University of Arizona
Figure 15.19(c)   Don Davis

Page 267   NASA/JPL
Figure 16.1(a)   Mauna Kea Observatory
Figure 16.1(b)   Lick Observatory
Figure 16.2   NASA
Figure 16.6(a–c)   NASA/JPL
Figure 16.7   NASA/JPL
Figure 16.9   NASA/JPL
Figure 16.10   NASA/JPL
Figure 16.12   NASA/JPL
Figure 16.12 Inset   NASA/JPL
Figure 16.13   NASA/JPL
Figure 16.14(a–b)   NASA/JPL
Figure 16.16(a–c)   NASA/JPL
Figure 16.17(a–b)   NASA/JPL
Figure 16.18(a–b)   NASA/JPL
Figure 16.19   NASA/JPL
Figure 16.20   NASA/JPL
Figure 16.21(a)   U.S. Naval Observatory
Figure 16.21(b)   Space Telescope Science Institute
Figure 16.23(a–c)   Lowell Observatory
Figure 16.24   NASA/JPL

Figure 17.1   Astrophysical Station, Haute-Provenc
Figure 17.2   W. Menke
Figure 17.3(a)   Smithsonian Institution
Figure 17.3(b)   John A. Wood
Figure 17.4(a–b)   John A. Wood
Figure 17.6   Keith Swinden
Figure 17.7   Yerkes Observatory
Figure 17.8   Richard Binzel, Massachusetts Institute of Technology
Figure 17.10 Top and Bottom   P. Jewitt, University of Hawaii
Figure 17.11(a)   NASA/JPL
Figure 17.12(a–b)   Andrew Chaikin and Clark Chapman
Figure 17.15   Akira Fujii
Figure 17.17   Lowell Observatory
Figure 17.18   NASA/JPL
Figure 17.20(a)   Giraudon/Art Resource, NY
Figure 17.20(b)   Scala/Art Resource, NY
Figure 17.23 (3 photos)   Yerkes Observatory
Figure 17.24 Left and Right   Halley Multicolor Camera Team
Figure 17.27   J. Luu and P. Jewitt, University of Hawaii
Figure 17.28   Maroshi Hayashi
Figure 17.29(b)   Union Pacific Railroad
Figure 17.30   NASA
Figure 17.32   A. and M. Meinel, *Sunsets, Twilights, and Evening Skies,* Cambridge University Press, 1983

Page 309   Anglo-Australian Telescope Board
Page 309 Inset   David F. Malin, Royal Observatory, Edinburgh
Figure 18.2(a–b)   Roger Ressmeyer/Starlight
Figure 18.3(a)   Mt. Wilson and Las Campanas Observatory
Figure 18.4(a)   Royal Swedish Academy of Sciences
Figure 18.4(b)   NASA
Figure 18.5   E. C. Olson, Mt. Wilson and Las Campanas Observatory

Figure 18.6   National Optical Astronomy Observatories
Figure 18.10(a)   Gary McDonald
Figure 18.10(b)   Richard Berry
Figure 18.12   National Optical Astronomy Observatories
Figure 18.13(a)   Perkin-Elmer Corporation
Figure 18.14   Mt. Wilson and Las Campanas Observatory
Figure 18.15(a–b)   Kitt Peak/National Optical Astronomy Observatories
Figure 18.16   Kitt Peak/National Optical Astronomy Observatories
Figure 18.17(a)   John Eddy, UCAR
Figure 18.17(b)   Royal Greenwich Observatory
Figure 18.19   R. Howard
Figure 18.21(a)   NASA
Figure 18.21(b)   Mees Solar Observatory
Figure 18.21(c)   CSIRO Division of Radiophysics
Figure 18.22(a)   Big Bear Solar Observatory
Figure 18.22(b)   NASA
Figure 18.24(a)   W. P. Sterne, Jr.
Figure 18.24(b)   NASA
Figure 18.26   NASA
Figure 18.30   Brookhaven National Laboratory

Page 333   University of Chicago
Figure 19.5(b)   Palomar Observatory
Figure 19.9(a)   James B. Kaler
Figure 19.9(b)   University of Chicago
Figure 19.11(a–c)   *Atlas of Representative Spectra*
Figure 19.12   University of Michigan
Figure 19.13   University of Michigan
Figure 19.14(a–b)   *Atlas of Spectra of Cooler Stars*
Figure 19.19   Sproul Observatory
Figure 19.20(a–b)   Jesse Greenstein, Palomar Observatory
Figure 19.21   Mt. Wilson Observatory
Figure 19.22(a–c)   *Atlas of Representative Spectra*
Figure 19.23   *Atlas of Stellar Spectra*
Figure 19.25   California Institute of Technology
Figure 19.26(b)   E. C. Olson, National Optical Astronomy Observatories
Figure 19.27(a)   Mt. Wilson Observatory
Figure 19.27(b)   S. Baliunas et al.

Page 356   National Optical Astronomy Observatories
Figure 20.1   James B. Kaler
Figure 20.1 Inset   Lowell Observatory
Figure 20.2(a–c)   Yerkes Observatory
Figure 20.8(b)   Yerkes Observatory
Figure 20.14   T. Schmidt-Kaler
Figure 20.15   David F. Malin/Royal Observatory, Edinburgh
Figure 20.16   NASA/JPL
Figure 20.20   Palomar Sky Survey
Figure 20.21   National Optical Astronomy Observatories
Figure 20.22(a–b)   National Optical Astronomy Observatories
Figure 20.23   Space Telescope Science Institute
Figure 20.24   *Dudley Observatory Report No. 11,* A.G.D. Philip, M. F. Cullen, and R. E. White

Page 373 University of Illinois Library

Figure 21.4(a–b) ROE/Anglo-Australian Telescope Board

Figure 21.6 Palomar Observatory

Figure 21.8(a) *Dudley Observatory Report No. 11*, A.G.D. Philip, M. F. Cullen, and R. E. White

Figure 21.9(a–b) Lowell Observatory

Figure 21.10 M. Petit, *Variable Stars*, Masson, Paris, 1982

Figure 21.11 Mt. Wilson Observatory

Figure 21.13 S. Ridgway, National Optical Astronomy Observatories

Figure 21.14(a) American Association of Variable Star Observers

Figure 21.14(b) NASA/JPL

Figure 21.15(a) Anglo-Australian Telescope Board

Figure 21.15(b) Space Telescope Science Institute

Figure 21.16(a) *Atlas of Representative Spectra*

Figure 21.17 Sproul Observatory

Figure 21.18 A. E. Morton

Figure 21.20 University of Michigan

Figure 21.21 Steward Observatory, University of Arizona

Figure 21.24 Mt. Wilson and Las Campanas Observatories

Figure 21.25 University of Illinois Prairie Observatory

Figure 21.27 Smithsonian Institution

Page 391 Smithsonian Institution

Page 391 Inset National Optical Astronomy Observatories

Page 392 Royal Observatory, Edinburgh/AATB/SPL/Photo Researchers, Inc.

Figure 22.1 Anglo-Australian Telescope Board

Figure 22.2 Smithsonian Institution

Figure 22.3 Lick Observatory

Figure 22.5(b) California Institute of Technology

Figure 22.7(a) Palomar Observatory

Figure 22.11(a) Anglo-Australian Telescope Board

Figure 22.12 National Optical Astronomy Observatories

Figure 22.13 James Wehmer

Figure 22.14 R. Schild

Figure 22.15 NASA/JPL

Figure 22.17(a–b) National Optical Astronomy Observatories

Figure 22.22 Columbia and Palomar Observatory

Page 407 National Optical Astronomy Observatories

Figure 23.1(a) Mt. Wilson Observatory

Figure 23.1(b) Palomar Observatory

Figure 23.3 Mt. Wilson Observatory

Figure 23.4 Max Planck Society for the Advancement of Sciences

Figure 23.5 European Southern Observatory

Figure 23.6 Palomar Observatory

Figure 23.7 R. Schild

Figure 23.8 Palomar Sky Survey

Figure 23.10 Royal Observatory, Edinburgh/AATB/SPL/Photo Researchers, Inc.

Figure 23.10 Inset NASA/IPAC

Figure 23.12(a) National Optical Astronomy Observatories

Figure 23.12(b) Space Telescope Science Institute/NASA

Figure 23.13 U. K. Schmidt, M Dopita

Figure 23.14(a–b) Palomar Observatory

Figure 23.14(c) W. Forrest/NASA

Figure 23.15 Top Backman, CTIO

Figure 23.15 Bottom NASA/JPL

Figure 23.17(a–b) Robert Schaeffer

Page 422 California Institute of Technology

Figure 24.13 Inset Mt. Wilson Observatory

Figure 24.16 University of Virginia

Figure 24.17 Anglo-Australian Telescope Board

Page 436 Royal Observatory, Edinburgh/AATB/SPL/Photo Researchers, Inc.

Figure 25.1 *Astronomy Magazine*

Figure 25.3 Royal Observatory, Edinburgh

Figure 25.3 Inset Anglo-Australian Telescope Board

Figure 25.4(a–c) *Atlas of Representative Spectra*

Figure 25.6 Anglo-Australian Telescope Board

Figure 25.9 Bill Fletcher

Figure 25.10(a–b) National Optical Astronomy Observatories

Figure 25.11 L. R. Sulak, Boston University

Figure 25.13(a) Anglo-Australian Telescope Board

Figure 25.13(b) Space Telescope Science Institute/NASA

Figure 25.14 Palomar Observatory

Figure 25.14 Inset University of Michigan

Figure 25.15 NRAO

Figure 25.17(a) California Institute of Technology

Figure 25.17(b) National Optical Astronomy Observatories

Figure 25.20(a–b) Jan van Paradijs

Figure 25.22(a) Palomar Schmidt photo, S. van den Bergh

Figure 25.22(b) VLA/NRAO

Figure 25.27(a) Mt. Wilson Observatory

Figure 25.27(b) R. Giacconi, Einstein Observatory

Figure 25.28 J. Luminet

Figure 25.29 I. F. Mirabel/NRAO

Page 457 Akira Fujii

Figure 26.5(b) Mt. Wilson and Palomar Observatories

Figure 26.9 David F. Malin, Anglo-Australian Telescope Board

Figure 26.10(a) G. Westerhout

Figure 26.10(b) P. Solomon

Figure 26.13 National Optical Astronomy Observatories

Page 473 National Optical Astronomy Observatories

Page 473 Inset Space Telescope Science Institute

Page 474 National Optical Astronomy Observatories

Figure 27.1 National Optical Astronomy Observatories

Figure 27.2 Mt. Wilson and Las Campanas Observatories

Figure 27.2 Inset   Mt. Wilson and Las Campanas Observatories

Figure 27.3(a–b)   Palomar Observatory

Figure 27.4(a) Left and Center   Palomar Observatory

Figure 27.4(a) Right   Mt. Wilson and Las Campanas Observatories

Figure 27.4(b) Left, Center, and Right   Palomar Observatory

Figure 27.5(a   Palomar Observatory

Figure 27.5(b)   Mt. Wilson and Las Campanas Observatories

Figure 27.5(c)   Palomar Observatory

Figure 27.7   U.S. Naval Observatory

Figure 27.8(a)   Mt. Wilson and Palomar Observatories

Figure 27.8(b)   Anglo-Australian Telescope Board

Figure 27.9   National Optical Astronomy Observatories

Figure 27.10(a)   National Optical Astronomy Observatories

Figure 27.10(b)   California Institute of Technology

Figure 27.12(a–c)   Vera Rubin

Figure 27.13   Smithsonian Institution

Figure 27.13 Overlay   J. Cornell

Figure 27.14   Asiago Observatory

Figure 27.15   Palomar Observatory

Figure 27.16   Space Telescope Science Institute

Figure 27.17(a)   European Southern Observatory

Figure 27.17(b)   Palomar Observatory

Figure 27.18 (5 images)   A. and J. Toomre

Figure 27.19(a)   European Southern Observatory

Figure 27.19(b)   J. Hayes, Lowell Observatory

Page 491   European Southern Observatory

Figure 28.1   A. Sandage, Palomar Observatory

Figure 28.3   Mt. Wilson and Las Campanas Observatories

Figure 28.10   Bertschinger and Dekel

Figure 28.11(a)   R. M. Soniera and P. J. E. Peebles

Figure 28.11(b)   Oxford University

Figure 28.12(a–b)   Center for Astrophysics

Figure 28.13   C. Park, Princeton University

Figure 28.14   National Science Foundation

Figure 28.15   T. Tyson, National Optical Astronomy Observatories

Page 505   G. B. Taylor and R. A. Perley, NRAO

Figure 29.1(a)   R. Schild, Smithsonian Center for Astronomy

Figure 29.1(b)   NASA

Figure 29.4   Space Telescope Science Institute

Figure 29.5(a)   Walter Jaffe/Leiden Observatory, Holland Ford/Space Telescope Science Institute, and NASA

Figure 29.5(b)   NRAO

Figure 29.9(a)   Tom Kinman, Kitt Peak National Observatory

Figure 29.9(b)   Smithsonian Institution

Figure 29.10   Smithsonian Institution

Figure 29.11   NRAO/AUI

Figure 29.11 Inset   Laird Thompson, University of Hawaii

Figure 29.12   NRAO/AUI

Figure 29.13(a–b)   Palomar Observatory

Figure 29.14(a)   Palomar Observatory

Figure 29.16(a)   Based on L. Eachus and William Liller and *The Astrophysical Journal,* published by the University of Chicago Press; copyright 1975 The American Astronomical Society

Figure 29.17   National Optical Astronomy Observatories

Figure 29.18(a)   H. Kuhr, University of Arizona

Figure 29.19   H. R. Bridle

Figure 29.20   ROSAT

Figure 29.21(a)   Palomar Observatory

Figure 29.21(b)   A. N. Stockton, University of Hawaii

Figure 29.22(b)   Hewitt and Turner, NRAO

Figure 29.23   Space Telescope Science Institute

Figure 29.24   A. Tyson, National Optical Astronomy Observatories

Figure 29.26(a)   G. Burbidge, Palomar Observatory

Figure 29.26(b)   H. Arp and J. Sulenti, Palomar Observatory

Page 520   Space Telescope Science Institute

Page 520 Inset Left   Wide World Photos

Page 520 Inset Right   *Life* photo by J. R. Eyerman

Figure 30.3   AT&T Bell Laboratories

Figure 30.4   NASA

Figure 30.5   NASA

Figure 30.15   Fermilab

Figure 30.15 Left Inset and Right Inset   Fermilab

Figure 30.19   NASA

Figure 30.22   Yurke and Chuang, AT&T Bell Laboratories

Figure 30.23   J. Centrella

Figure 30.24 Clockwise from Top   California Institute of Technology; David F. Malin, Anglo-Australian Telescope Board; Akira Fujii; Mt. Wilson and Las Campanas Observatory, Carnegie Institution; NASA; James B. Kaler

Index*

A0620-00, 453
AAAOs, 290–291
  origin, 294
Aberration of starlight, 43–44
Absolute magnitudes, 341–342
Absorption lines, 127–128
  strengths of, 132
Acceleration, 97–98
  of a falling body, 101
  of gravity, 98, 101
Accretion disks (stars), 385, 408
  active galaxies, 508
Achernar, 56
Achondrites, 287
Achromatic lens, 140
Active galaxies, 506–511
  BL Lacertae objects, 508–509
  LINERS, 507
  origins, 510–511
  quasars, 511–517
  radio galaxies, 507–510
  Seyfert galaxies, 506–507
Active sun, 312, 319–326
  cycle of, 319–323
Adams, John Couch, 105
Adaptive optics, 146
Adoration of the Magi, **297**, 298
Adrastea, 248
Advance of Mercury's perihelion,
  109, 111
Age of Earth, 157
Age of the Universe, 497–498
Ages of stars, 366
  globular clusters, 424, 469,
    528–529
  open clusters, 366, 424
Aglionice (Venus), **211**
Al Jauza, 54
Alba Patera (Mars), 222
Albedo, 71
Alcor, 357
Aldebaran, 428
Aldrin, Edward, **191**
Aleutian Islands, 166
Alexandria, 15

Algol, 361, 362
All-Saints' Day Eve, 40
Allende meteorite, 286
Alloys, 286
Almagest of Ptolomy, 87
Alpha Centauri, 9, 334, 357, 360
Alpha Herculis, 381
Alpha particles, 120
Alpha Regio, 207
Alpha Ursae Majoris, 428
Alpher, Ralph, 523
Altair, 56, **56**
Altitude, 20
Amalthea, 248, **248**
Amino acids (meteorites), 288
Amorphotoi, 57
Amors (asteroids), 291
Analemma, 38, **38**
Ancient constellations, 50
  table of, 51
Ancient Greece, 15
Ancient planets, 81
Andes, 159
Andromeda, 54
Andromeda Galaxy, *see* M 31
Andromedids, 303
Angles, 19, 44
  measured in time units, 21–22
Ångstrom, 124
Angular diameter, 19
Angular momentum, 103
Annular, 76, **76**
Antarctic, 35
  meteorites, 287
Antares, **52**, 429, 442
Ante meridiem, 37
Antennas, 148–149
Anticyclone, 171, 172
Antigravity, 532
Antimatter, 327
Antinoüs, 57, **57**
Antiquarks, 532
Antlia, 57
Antoniadi, E. M., 217, 218
Aphelion, 30

Aphrodite Terra, 207
Apogee of Moon, 68
Apollinaris Patera (Mars), **223**
Apollo program, 181, 183
Apollos (asteroids), 291
Appalachian mountains, 165
Apparent bolometric magnitudes,
  341
Apparent magnitudes, 339–341
Apparent photographic magnitudes,
  341
Apparent sun, 37
Apparent visual magnitudes, 339
Apus, 57
Aquarius, 52, 53
Aquila, 56, **56**, **57**
Ara, 55
Arabian astronomy, 60–61
Aratos, 50
Archduke Ferdinand, 90
Arctic, 35–37
Arcturus, **53**, 54, 349, 428
  spectrum of, **337**
Arecibo radio telescope, 149, **150**
Argo, 55
Argyre basin (Mars), 220, **221**
Ariel, 278, **278**
Aries, 52, **52**, 53
Aristarchus, 15
  distance to the Sun, 74
  heliocentric theory of Solar
    System, 86
Aristarchus (crater), 189
Aristotle, 15
Armillary sphere, **13**
Asclepius, 55
Associations, *see* OB or T associations
Asterisms, 59
  table of, 60
Asteroid belt, 231, 290
Asteroids, 8, 288–294
  compositions, 292
  discovery, 288–289
  families, 290–291
  heating, 294

* Photographs and other images are indicated by boldfaced numbers.

names, 289
orbits, 290–291
origins, 292–294
physical properties, 291–292
relations to meteorites, 292
size distribution, 292–294
table of, 289
Astrological signs, 52
Astrology, 52, 453
Astronomical horizon, 20
Astronomical Unit, 6, 30
calibration of, 198, 200
Asymptotic giant branch (AGB), 429–431
Atens (asteroids), 291
Athens, 15
Atla Regio, 207
Atlantic Ocean, 159
Atlas, 59
Atmosphere of Earth, 167–173
changes in, 168
composition, 167–168
optical properties, 170–171
origin, 172–173
structure, 169
Atmosphere of Mars, 217, 219–220
Atmosphere of Mercury, 202–203
Atmosphere of Jupiter, 238–241
Atmosphere of Neptune, 269, 273–274
Atmosphere of Pluto, 282
Atmosphere of Saturn, 255–256
Atmosphere of Uranus, 269, 271–273
Atmosphere of Venus, 206–207
Atmospheric absorption of radiation, 124, 147
Atmospheric dispersion, 171
Atmospheric refraction, 171
Atmospheric seeing, 140, **140**, 146, **146**
Atomic number, 117
Atomic weight, 119
Atoms, 117–121
Auriga, 56
Aurora, 173–174, **174**
cause of, 326
Autumn, 33
Autumnal equinox, 33
Azimuth, 21

B ring (Saturn), **257**
Baily's beads, 78, **78**
Balmer, J. J., 128
Balmer limit, 128
Balmer series, 128, **128**
Bar, 163
Barnard, E. E., 248
Barnard 86 (Bok globule), **399**

Barnard's Star, **336**, 337
Barred spirals, 477–478
Baryons, 532
Basalt, 158
Bayer, Johannes, 57
Greek letter star names, 60
Bayeux Tapestry, 297
BD catalogue and numbers, 61
Becquerel, Henri, 120
Bell, Jocelyn, 445
Beltane, 40
Belts (Jupiter), 234
Benzene ring (structure), 404
Bessel, Wilhelm Friedrich, 334, 363
Beta Lyrae, 362
Beta particles, 120
Beta Pictoris, 416, **416**
Beta Regio (Venus), 207, **211**
Beta Scorpii, 368
Beta Ursae Minoris, 428
Betelgeuse, 54, **54**, 60, 341, 348, 349, 429, 442
Bethe, Hans, 327, 423
B^2FH, 430
Biela's Comet, 300, 303
Big Bang, 11, 497–498, 522, 523, 528–536
challenges to theory, 531–533
creation of galaxies, 534–536
defects in, 535, 536
density ripples in, 535
evolution of, 532–536
inflation, 532, 533
singularity, 530
theory, 532–536
Big Chill, 524
Big Crunch, 524
Big Dipper, 53, **53**
proper motions in, 334
Big Horn Medicine Wheel, 34, **34**
Binary stars, 357–364
astrometric, 363
eclipsing, 361–363
evolution of, 433
formation, 412
proper motions, 360
spectroscopic, 360–361
table of, 358
unseen companions, 363
visual, 357–360
Binding energy, 440
Bipolar flows (stars), 409–410, **410**, **411**, 412
galaxies, 486
Birthline, 413
BL Lacertae, 508–509, **509**
BL Lacertae object 0317+186, **509**
BL Lacertae objects, 508–509

Black holes, 450–454
in galaxies, 486
in the Galaxy, 465
observation, 452–454
in radio galaxies, 508, 511
Seyfert galaxies, 507
theory, 450–452
Black smoker, **166**
Black Widow pulsar, 448, **448**
Blackbody, 125–127
curves, 126
Blue sky, 168
Bode, Johann, 288
Bode's law, 288
Bohr, Niels, 129
Bohr atom, 129–130
Bok, Bart, 398
Bok globules, **56**, **392**, 398–399, **399**
Bolometric magnitudes, 341, 342
Boltzmann, 126
Bondi, Hermann, 522
Bonner Durchmusterung, 61
atlas, 63
Boötes, **53**, 54
Boötes Cluster of galaxies, **493**
Bow shock, 173
Bowen, Ira S., 395
Breccia, 186, **186**
Bright Companion (Neptune), 274, **274**
Bright Star Catalogue, 61
Brightest stars (table), 340
Broglie, Prince Louis de, 130
Broken symmetry, 530, 532
Brown dwarfs, 414–415
candidates, **415**
Buckminsterfullerene, 404, 405
Burbidge, E. Margaret, 430
Burbidge, Geoffrey R., 430

C_2 spectrum, 131
C_{60}, 404, 405
Caduceus, the, 55
Calculus, 97
Caldera, 161
Calendar, 39–40
Callisto, 245, **245**
Caloris basin, 203
Calypso, 261
Camera mirror of spectrograph, 146
Canals of Mars, 217–218
Cancer, 52
Candor Chasma (Mars), 224, **224**
Canes Venatici, **53**, 57
Canis Major, 54, **54**
Canis Minor, 54, **54**
Cannon, Annie Jump, 342
Canopus, 55

Cantaloupe terrain (Triton), 280, **280**
Capricornus, 53
Carbon, 119
Carbon cycle, 423
Carbon fusion, 440
Carbon monoxide (interstellar), 404
Carbon stars, 344
   creation of, 431
   Mira variables, 381
Carbonaceous chondrites, 287, **287**
Carbonate rocks, 158
Carina, 55
Carina Nebula, 393, **394**
Carpathian mountains (lunar), **185**
Cassegrain focus, 141
Cassini, Giovanni, 241, 253
Cassini division, **252**, 253, **257**, 258,
   259
Cassini spacecraft, 265
Cassiopeia, 54
Castor, 358
Cataclysmic variables, 385–386
Catholic Church, 89
cD galaxies, 482
   origin, 487
Celestial equator, 20
Celestial meridian, 20
Celestial motions, 21–25
Celestial navigation, 20, 39, 42–43
Celestial poles, 20
Celestial sphere, 18–25
Celsius, Anders, 100
Celsius temperature scale, 100
Centaurus, 55, **56**
Centaurus A, *see* NGC 5128
Center of activity, 319, **320**
Center of mass, 104
   binary stars, 360
   Earth-Moon system, 178
Centigrade temperature scale, 100
Cepheid instability strip, 374
Cepheid variables, 374–378
   cause of, 377
   pulsation, 375
   radius changes, 375
   table of, 375
Cepheus, 54
Cepheus OB 2, 368
Ceres, 8, 289
Cetus, 55, 373
CH Cygni, 386
Chamaeleon, 57
Chandrasekhar, Subramanyan, 433
Chandrasekhar limit, 433
Chaos, 112–113
Chaotic terrain, 224, **224**
Charge-coupled device (CCD),
   145–146

Charles II, 57
Charon, **280**, 281, 282
   data table, 281
Chemical elements, 117–119
   creation in Big Bang, 533
   creation in stars, 430, 431, 442
   table of, 118
Chi Cygni, 380
Chi Persei, *see* Double cluster in
   Perseus
Chiron, 291, 300–301, **301**
Chondrites, 287
Chondrules, 287
Christian IV of Denmark, 90
Chromatic aberration, 140
Chromosphere of Sun, **78**, **316**,
   316–318
   eclipses, 78
   heating of, 317–318
   spectrum, **316**
   temperature, 317–318
Chryse Planitia (Mars), 223, **223**,
   225, **225**, 227, **227**
CHU, 39
Cinder cone volcanoes, 161
Circle, 102
Circumpolar stars, 22–25, **23**
Circumpolar Sun, 35–36, **36**
Clarke, Alvan, 363
Class 0 (zero) stars, 349
Classical Cepheids, 376
Clavius (crater), **186**
Clocks, 40
   relativity, 112
Closed universe, 524–527
Closest stars, table, 335
Clouds (Venus), 206
   Jupiter, 234, 238–240
   Mars, 220
   Neptune, 273–274, **274**
   Saturn, 253
   Uranus, 271, **272**
Clump (in stellar evolution), 428
Clusters of galaxies, 11, **11**, 481–483,
   482
   classes, 482
   dark matter, 485
   distant, **520**
   distances, 483
   Local Group, 480, 483
   table, 482
Clusters of stars, 10, 364–370
Coalsack, **56**, 399
Cold fusion, 218
Cold dark matter, 536
Colette (Venus), **213**
Colliding galaxies, 486–489
   as origin of active galaxies, 511

Collimating mirror, 146
Collision hypothesis (Moon), 194
Color index, 341
Color-magnitude diagrams, 366
Coma Berenices, **52**, 57
Coma Berenices Cluster of galaxies,
   482, **482**, 500, 501
Comets, 8, 295–301
   comas, 295
   Kuiper belt, 303, 306
   names, 296–297
   nuclei, 295, 299–301
   Oort comet cloud, 303, 306
   physical natures, 298–301
   spectra, 295, **295**, 299
   tails, 8, 295, 299
   water return to Earth, 417
Comet Bennett, **8**
Comet Giacobonni-Zinner, **296**, 300,
   303
Comet Ikeya-Seki, **296**, 297
Comet Shoemaker-Levy, **301**
Comet Swift-Tuttle, 303
Comet West, **285**, 297, 300
Command module, 183, **183**
Common proper motion stars, 357
Compact H II regions, 414
Comparative planetology, 198
Comparison spectrum, 147, **147**
Composite volcanoes, 161
Compressional wave, 162
Compton Gamma Ray Observatory,
   454
Conic sections, 102
Condensation in the Solar System,
   305–306
Conduction, 213
Conjunction, 82
   inferior and superior, 85
Conservation of angular
   momentum, 101
   and distance of the Moon, 180
Conservation of energy, 100
Conservation of mass-energy, 111
Constellations, 3, 50–59
   ancient, 50–56
   modern, 57–59
   seasons, 53
Constructive interference, 123
Continental drift, 164–167
Continents, 158
Continuous creation of matter, 522
Continuous radiation or spectrum,
   125, 315
Convection, 163, 164
   in the Sun, 313–314
Coordinate systems, 18
   altitude and azimuth, 20–21

equatorial, 20–21, 42–43
galactic, 458
latitude and longitude, 17–18
Copernican Principle, 521
Copernican system, 88–89
Copernicus, Nicolaus, 88–89, **88**
Copernicus (crater), **185**, 189
Cordelia, 275
Cordoba Durchmusterung, 61
Core collapse supernovae, 441–442
Core of Earth, 162–164
Core of Mercury, 205
Core of the Moon, 192
Core of Sun, 328
Core of Venus, 213
Coriolis effect, 25–26
Corona (atmospheric effect), 170
Corona Borealis, **53**, **382**
Corona Borealis Cluster of galaxies, **493**
Corona of the Galaxy, 468
Corona of the Sun, **78**, **310**, 318–319
eclipses, 78
heating, 319
temperature, 318
Coronae, volcanic, on Venus, 208, **209**
Coronagraph, 319
Coronal gas (interstellar), 398
formation of, 445
Coronal holes, 325, **325**
Coronal loops, 325, **325**
Coronal streamers, 325, **325**
Coronal transients, 326, **326**
Cosmic Background Explorer, 523
Cosmic microwave background radiation, 522, 523–524
creation of, 534
Doppler shift in, 523
evolution of temperature, 529–530
fluctuations in, 534
prediction of, 523
temperature, 523
Cosmic rays, 322, 412, 464
origin, 464
Cosmic strings, 535
Cosmological constant, 495
Cosmological distances, 512
Cosmological Principle, 522
Cosmological principles, 521–533
Cosmology, 492, 521–539
alternative cosmologies, 536
Big Bang cosmology, 11, 497–498, 522, 523, 528–536
expansion models, 524–527
plasma cosmology, 536
Coudé focus, 141
Council of Nicaea, 40
Counterglow, *see* Gegenschein

Crab Nebula, 387, **387**, 444–445, **446**
pulsar in, 446, **446**
Crater Lake, **160**
Craters, impact, *see* Impact craters
Craters, volcanic, 161
Crepe ring, 253
Crescent Moon, 71
Critchfield, Charles, 327
Critical density, 528
Cross-quarter days, 40
Crust of the Earth, 159, 164
Crust of Mars, 229
Crust of the Moon, 186
Crux, **56**
Curie, Marie, 120, **120**
Curie, Pierre, 120
Curtis, Heber D., 475
Curtis-Shapley debate, 475
Cuzco, 34
Cyclones, 171, 172
Cygnus, 56, **56**
Cygnus A, 509, **510**, 514
Cygnus Loop, 444, **445**
Cygnus OB 2, 368
Cygnus X-1, 452

D2 (Neptune), 274, **274**
DA and DB white dwarfs, 347
creation of, 433
Daily paths, 21–25
of Moon, 72
Dark matter (the Galaxy), 467–468
from Big Bang theory, 533–534
clusters of galaxies, 485
cold, 536
galaxies, 485
gravitational lenses, 516, **517**
hot, 535
Davis, Ray, 329
De Chéseaux's Comet, 297, **297**
de revolutionibus orbium coelestium of Copernicus, 89
de Sitter, Willem, 495
Death Star, 263
Decameter bursts, 235, 242–243
Deccan plateau, 161
Declination, 21
Declination axis, 143
Defects in the Big Bang, 535, 536
Deferent, 87
Degeneracy, 427
Deimos, 230–231, **230**
data, 231
Delphinus, **56**
Delta Cephei, 374
Delta Scorpii, 368
Democritus, 117
Deneb, 56, **56**, 60

Deneb Algedi, 60
Deneb Kaitos, 60
Denebola, 60
Dense cores, 410–411
Density, 127
critical, 528
Earth, 162
Galaxy, 462
Galilean satellites, 245
Jupiter, 235, 237
Mars, 216
Mercury, 198
Moon, 191
Neptune, 269
Pluto, 282
ripples in Big Bang, 535
Saturn, 254
Saturn's satellites, 261
solar core, 328
Sun, 311
uncompressed, 162
Universe, 528, 533
Uranus, 269
Uranus's satellites, 277
Venus, 198
waves (Saturn's rings), 259
white dwarfs, 364
Density waves (Saturn's rings), 259
Galaxy, 462
Despina, 276
Destructive interference, 123
Deuterium, 119
Devana Chasma, **211**
Dialogue of Two Chief Systems of the World of Galileo, 93
Diamond ring effect, 78, **78**
Diamonds in the interstellar medium, 401
Diamonds in meteorites, 287
Diana, 54
Differential rotation, 235
Differentiation (planetary), 162
asteroids, 294
Diffraction, 122–123
dispersion by, 138
disk, 140
Diffuse nebulae, 393–396, 398
compositions, 396
illumination, 394–396
spectra, 394–396
table, 393
Dinosaurs, 192–193
Dione, 261–263, **262**
Dipole, 163
Direct relationships, 98
Dirty snowball model, 298
Disconnections of comet tails, 299, **299**

Dispersion, 138
Distance indicators for galaxies, 478–481, 485
  Cepheid variables, 479
  globular clusters, 481
  novae, 479
  O and B stars, 479
  planetary nebulae, 479
  RR Lyrae stars, 479
  supernovae, 481
Distance measures in Universe, 526
Distances (units), 5
  ancient, 74
  clusters of galaxies, 483
  cosmological, 512
  galactic center, 470, 459
  galaxies, 475, 478–481, 485, 498
  inferior planets, 89
  Mercury (Copernicus), 198
  Moon, 68, 74, 178, 180
  open clusters, 365–366
  planets, 89
  pulsars, 446
  quasars, 512–514
  radar, 200
  redshift, 498
  spectroscopic, 349–350
  stars, 334–337, 349–350, 365–366, 379, 399–401
  Sun, 198, 200
  superior planets (Copernicus), 89
  Venus (Copernicus), 89
  Virgo Cluster of galaxies, 496
Domain walls, 535
Domes, volcanic, on Venus, 208
Doppler, Christian, 133
Doppler effect, 132–133
Doppler shift, 133, 337
Double cluster in Perseus, 365–366, **365**
Double stars, *see* Binary stars
DQ Herculis, 383–385
Draco, **53**
Drake equation, 419
Dreyer, J. E. L., 365
Dry ice, 220
Dumbbell Nebula, 422
Dürer, Albrecht, 63
  planispheres, 64
Dust (interstellar), 398–402
  composition, 401
  in galaxies, **401**
  growth of, 402, 403
  interstellar absorption, 399–400
  interstellar reddening, 400
  relation to Mira dust, 401
Dust from Mira variables, 381
  relation to interstellar dust, 401

Dust tails, 295, 299
Dwarf Cepheids, 379
Dwarf elliptical galaxies, 482
Dwarf novae, 385
Dwarf spheroidal galaxies, 482
Dwarf stars, 346

Earl of Rosse, 475
Earliest sunset, 39
Earth, 4, **6**, **113**, 157–174, **177**
  age, 157
  atmosphere, 167–173
  axis, 17
  crust, 159
  data table, 157
  density, 162
  diameter, 17
  equator, 17
  equatorial bulge, 45
  heat, 164
  impact craters, 192–193
  lithosphere, 164
  magnetic field, 163, 165, 173–174
  magnetosphere, 173–174
  mantle, 163–164
  map of geologic activity, **159**
  mass, 104
  oblateness, 16, 18
  orbit, 30–31
  perihelion and aphelion, 30
  plate tectonics, 166–167
  poles, 17
  proof of revolution, 43–44
  proof of rotation, 25–26
  revolution, 30–33
  rotation, 17
  shape, 14–15
  size, 15–16
  structure, 162–164
  surface features, 158–161
  tides on, 178–180
  weather, 113, 170–172
Earth-Moon system, **6**
Earthlight, 71
Earthquakes, 159–161
Earthquake waves, 162–163
East, 20
Eclipse season, 74
Eclipse year, 73
Eclipses of the Moon, 71–75, **75**
  brightness of, 75
  number in a year, 74
  partial, 73
  shape of the Earth, 15–16
  table of, 75
  total, 74
Eclipses of the Sun, 75–78, **78**, **310**, annular, 76, **76**

  how to observe, 77
  number in year, 77
  partial, 76
  phenomena of, 77–78
  saros, 77
  saros cycle, 77
  table of, 76
  total, 76
Eclipsing binaries, 361–363
Ecliptic poles, 31
Eddington, Sir Arthur, 112, 327
Effective temperature, 127
Einstein, Albert, 109, **109**, 494
Einstein-de Sitter universe, 496, 524–525
  age of Universe in, 525
  redshift relation in, 496
Eistla Regio, 207
Ejecta blanket, 184
Electric charge, 117
Electric field, 117
Electromagnetic force, 117
  particles of, 531
Electromagnetic radiation, 120, 122–125
  effects of, 125
  flux, 125
  luminosity, 125
  and matter, 125–132
Electromagnetic spectrum, 122–124
Electron degeneracy, 427
Electrons, 117
  dual nature of, 130
Electroweak force, 531
Elements, *see* Chemical elements
Ellipse, 30, 102
Elliptical galaxies, 11, 476
  masses, 484
  origin, 488
Elongation of planet, 82
  of inferior planets, 85
Elysium Planitia (Mars), 221, **222**, 227, **227**
Emission lines, 127–128
Enceladus, 261, **262**
Encke, J. F., 156
Encke division, 256, **257**, 258, **258**, 259
Encke's Comet, 296
Energy, 99
  conservation of, 100
  of radiation, 124–125
Energy level diagram, 129–131
  of doubly ionized oxygen, 131
  of hydrogen, 130
English Muffins, **209**
Epicycle, 87
Epimetheus, 261
Epoch, 46

Epsilon Aurigae, 362–363
Epsilon Eridani, 416
Epsilon Lyrae, 358, 359
Epsilon Orionis (spectrum), **396**
Equal areas law, 91
Equant, 88
Equation of time, 37–39, **38**
Equator
    celestial, 20
    Earth's, 17
Equator point, 20
Equatorial coordinate system
    celestial, 42–43
    Earth, 21
Equatorial telescope mounting, 141
Equinoxes, 32–33
Eratosthenes, 15, 16, 57
Ergosphere, 451
Eridanus, 56
Eros, 289
Errors of measurement, 334
Eruptive variables, 383–387
Escape velocity, 102
Eta Carinae, 382, **382**, 393, **393**, 442
Ethyl alcohol (structure), 404
Euclid, 111, 524
Euclid's postulates, 524
Euclidean space, 111, 525
Eudoxus of Cnidos, 50
    theory of Solar System, 86
Euler (crater), **184**
Europa, **242**, 245, **246**
European Southern Observatory, 142
European Space Agency, 300
Event horizon, 451
Evolution of the Galaxy, *see* Galactic
    evolution
Evolution of galaxies, 492
Evolution of stars, *see* Stellar evolution
Evolutionary tracks, 413
Ewan, Harold, 397
Eyepiece, 138, 139
Exit cone, 451
Expanding Universe, 11, 494–497
    effect of gravity, 498
Expansion models of the Universe,
    524–525
    tests of, 528–529
Exponential notation, 5

F ring, 259, **259**, 263
Fabricius, David, 374
Fahrenheit, Gabriel, 100
Fahrenheit temperature scale,
    100
False vacuum, 532
Families of asteroids, 290–291
Families of matter, 330
Farside of the Moon, 71, **191**

Faults (Earth), 159, 166
    on Mars, 223
    on Venus, 211
Favorable opposition of Mars, 84
Feldspar, 158
Field stars, 366
Filaments, 323
Fireballs, 286
First quarter Moon, 69
Flamsteed, John, 61
Flamsteed numbers, 61
Flare stars, 383
Flares (solar), 323–325, **324**
Flat spacetime, 525
Flatness problem, 531
    solution by inflation, 532
Fleming, Williamina P., 342
Floras (asteroids), 290
Focal length, 138
Focal plane, 139
Focal point, 138
Fomalhaut (disk around), 415
Forbidden lines, 395–396
Force, 97, 99
Forces of nature, 117, 119
    particle carriers, 530
    symmetry of, 531
    unification of, 530–531
Fornax, 57
40 Eridani B, 347
    spectrum, **348**
Foucault, Léon, 25, **25**
Foucault pendulum, 25, **25**
14 Cephei, 368
Fowler, William A., 430
Fra Mauro formation, 187
Fraunhofer, Joseph von, 127, 317
Fraunhofer lines, 317
Free-free radiation, 395
Frederick II of Denmark, 89
Frequency of radiation, 122
Friction, 97
Friedmann, Alexander, 524
Friedmann-Lemaître universes, 524–525
    age of Universe in, 525–527
Fujiyama, 161
Full Moon, 71
Fundamental particles, 117

Galactic bulge, 338, 458, 464–465
Galactic corona, 468
Galactic disk, 460–464
    Population I structure, 460
    rotation, 460–462
Galactic evolution, 469–470
Galactic halo, 464–465
Galactic longitude and latitude, 459
Galactic equator, 459
Galactic poles, 459

Galatea, 276
Galaxies, 475–481
    classification, 476–478
    clusters, 482–483
    colliding, 486–489
    dark matter, 485
    distances, 475, 478–481, 485, 498
    distant, **491**, **503**
    distribution in Universe, 498–503
    dwarf elliptical, 481
    dwarf spheroidal, 481
    elliptical, 11, 476
    evolution of, 492
    formation of, 534–537
    frequency of types, 483–484
    giant elliptical, 481
    interacting, 486–487
    irregular, 478
    Local group, table, 480
    maps of, 499–503
    masses, 484–485
    mergers, 487–489
    nuclear starburst, 478
    nuclei, 485–486
    number in visible Universe, 503
    origin of forms, 487–489
    peculiar, 486–487
    polar ring, 487
    redshifts, 475, 492–497
    rotation curves, 484
    spiral, 477–478
    superclusters, 497
    starburst, 478
    table, 481
    tides between, 487–488
    X-rays from, 485, **485**
Galaxy, the, 9–10, 338–339, 458–470
    age, 468–469
    black hole in nucleus, 465
    corona, 468
    bulge of, 338, 458, 464–465
    dark matter in, 467–468
    disk of, 9, 460–464
    distance of Sun to center, 370, 459
    evolution, 428, 469–470
    globular clusters, 368–370
    halo of, 10
    magnetic field, 402, 410–411, 464
    maps of, 463
    mass, 465–468
    nucleus, 465
    open clusters, 364
    orbits of stars, 338–339
    radius, 459
    rotation, 460–462
    spiral arms of, 9, 462–464
    structure, 458–465
Galilean satellites, 235, 243–248
    densities, 245

Galileo, **80**, 92–93
  observations, 93
  telescopes, **92**
Galileo spacecraft, 248, **249**
Galle, John, 106
Gallex neutrino detector, 330
Gamma Leonis, 357, 359
Gamma ray bursters, 454
Gamma rays, 120, 124
Gamow, George, 522
Ganymede, 245, **245**, 265
Gaseous state, 122
Gaspra, 291, **292**
Gauss, Karl Friedrich, 289
Gegenschein, 303
Gemini, 52, 53, 54, **127**
General Catalogue, 365
General theory of relativity, 111–112
  expanding universe, 494–497
Geocentric theories of Solar System,
  86–88
Geosynchronous orbit, 106
Gheyn, Jacobo de, 63
Giant branch, 346, 426
Giant elliptical galaxies, 482
Giant H II regions, 394
Giant molecular clouds, 404–405
Giant stars, 346, 426–428
  evolution to, 426
Gibbous Moon, 71
Gigaparsec, 498
Giotto (painter), 298
Giotto spacecraft, 300
Glaciers (Earth), 172
  on Mars, 225, 229
Glitches, 447
Globular clusters, 10, 368–370
  age of Universe from, 528–529
  ages, 424, 469
Globule, *see* Bok globule
Gluons, 531
Goddard, Robert, 108, **108**
Gold, Thomas, 522
Grabens, 187, **188**
  on Venus, 212
Gram, 99
Grand Unified Theories, 531
Granite, 158
Granulation (solar), 313–314
Grating, 138
Gravitational compression of a
  planet, 162, 254
Gravitational constant, 101
Gravitational lenses, 515–517
  detection of dark matter, 516, **517**
  Einstein cross, 516, **516**
  Einstein ring, 516, **516**
  Hubble constant from, 516
Gravitational reddening, 451

Gravitons, 531
Gravity, 6, 97, 100–102
  particles of, 531
  relativistic explanation, 111
Gravity assist for spacecraft, 109, 237
Gravity waves, 442
Great Annihilator, 453, **453**
Great Attractor, 497
Great circle, 17
Great Comet of 1843, 297
Great Comet of 1881, 297, **297**
Great Dark Spot (Neptune), 273–274,
  273, **274**
Great Debate, 475
Great Red Spot (Jupiter), 235, 241, **241**
Great Rift, 212
Great Square of Pegasus, 59
Great Wall, 500, 501
Great White Spot (Saturn), 256, **256**
Greatest elongation, 85
Greek alphabet, 18
Greek constellations, 50
Greek letter star names, 60
Greek measures of distance, 74
Green flash, 171
Greenhouse effect, 168
  on Venus, 206
Greenwich, 17, 18, **18**
Gregorian calendar, 40
Groundhog day, 40
Guest star, 386
Guide Star Catalogue, 65
Gula Mons, 208, **209**
Gulliver's Travels, 231
Guth, Alan, 532

h Persei, *see* Double Cluster in
  Perseus
H II regions, *see* Diffuse nebulae
Hadley cell, 170, 172
Hadley rille, **188**
Hadrian, Emperor, 57
Hairy ball model, 325–326
Half-life, 121–122
Hall, Asaph, 231
Halley, Edmund, 57, 100, 104, 521
Halley's Comet, 104, **295**, 297–298,
  299
  nucleus of, 299–290, **300**
  spectrum of, **295**
Halloween, 40
Haloes around Sun and Moon, 170,
  171
Harmonic law, 92
Harvest Moon, 72
Hawaii, 161
$HC_{11}N$ (structure), 404
HD 93129A, **394**
Head-tail radio galaxies, 510

Heat energy, 99–100
Heavy bombardment, 187
Helene, 261
Helium, 119
Helium condensation in Saturn, 255
Helium flash, 428
Helium fusion, 427
Helix Nebula, **433**
Hellas, 220
Helmholtz, Heinrich von, 327
Henry Draper Catalogue, 334
Heraclides Ponticus, 17
Herbig, George, 409
Herbig-Haro objects, 409, **409**, **410**,
  412
Hercules, 55
Hercules A, **488**
Hercules Cluster of Galaxies, **482**
Hermann, Robert, 523
Herschel, John, 365
Herschel, William, 105, **147**, 365
  infrared observations, 147
  structure of Galaxy, 459
  telescopes, 141
Hertz, Heinrich, 122
Hertzprung, Ejnar, 346
Hertzsprung-Russell diagram,
  346–351, **346**, **347**, **350**
  distribution of stars on, 350–351
  variable stars on, **374**
Hevelius, 57
Hewish, Anthony, 445
HH 34, **409**, **410**
Hidalgo, 291
High resolution microwave survey,
  420
High tide, 178
Highlands, lunar, 181
Hildas (asteroids), 290
Hipparchus, 15
  constellations, 50
  discovery of precession, 45
  distance to Moon, 74
  magnitudes, 60
Hipparcos satellite, 336
Holmdell antenna, **522**
Homestake neutrino detector, 329, **329**
Horizon, 14
  astronomical, 20
  Universe, 527
Horizon problem, 531
  solution by inflation, 532
Horizontal branch, 369
  evolution to, 428
Horsehead Nebula, **392**, 399
Hot dark matter, 535
Hot poles (Mercury), 203
Hour angle, 21
Hour circle, 21

Houtman, Frederik de, 57
Howe (Venus), **211**
Hoyle, Fred, 430, 522
HR numbers, 61
HR 1040 (spectrum), **348**
Hubble, Edwin, 386
 galaxy classification, 476
 galaxy distances, 475
Hubble constant, 496–497
 from gravitational lenses, 516
 uncertainty in, 497
Hubble diagram, *see* Hubble relation
Hubble flow, 495
Hubble relation, 492–494
 curvature of, 528
Hubble Space Telescope, 151–152, **152**
 Guide Star Catalogue, 65
Hubble time, 497–498, 525
Humason, Milton, 492, 501
Humphreys-Davidson limit, 439
Hunter's Moon, 72
Huygens, Christiaan, 253
Hyades, 59, 365–367, **365**
 proper motion, 365
 slitless spectrogram, **343**
Hydra, 55
Hydra A, **505**
Hydra Cluster of galaxies, **493**
Hydrocarbons, 240
Hydrogen, 117, 119
 fusion, 327–328
 interstellar molecular, 404
 liquid molecular, 237
 metallic, 238
 spectrum, 128, **128**, **129**, 130
Hydrostatic equilibrium, 237
Hyperbola, 102
Hyperbolic spacetime, 525
Hyperion, 261, 263, **264**
Hypothesis, 193

Iapetus, 263, **264**
IC 708, **510**
Igneous rocks, 158
Image of a telescope, 138
Imbrium basin, **187**, 189
Impact craters (Moon), 181–185
 Callisto, 245
 central mountain peaks, 181, 184
 Dione, 261–263
 Earth, 192–193
 Enceladus, 261–262
 formation, 184
 Ganymede, 245
 Mars, 220–221
 Mercury, 204
 Miranda, 278

Mimas, 261, 262
 Oberon, 278
 Tethys, 261, 262
 Umbriel, 278
 Venus, 210–211, **211**
Impact basins (Moon), 187
 Mars, 221
 Mercury, 203
Impact melts, 186
Inclination of orbit, 68
Index Catalogue, 365
Index of Prohibited Books, 89
Index of refraction, 137
Inequality of the seasons, 92
Inferior conjunction, 85
Inferior planets, 85
 conjunctions, 85
 distances (Copernicus), 198
 elongations, 85
 motions, 85–86
 phases, 85
Infinity, 521
Inflationary Big Bang, 532, 533
Infrared Astronomical Satellite, 151, **152**, 400
Infrared radiation, 124
 observation of, 147
Inner satellites of Jupiter, 248
Interacting binaries, 383–386
Interacting galaxies, 486–487
Intercloud medium, 397, 398
Intercrater plains, 204
Interference of light, 122
Interferometers, 150–151
 table of, 142
Intermediate main sequence, 426
 evolution, 423–433
International Astronomical Union, 63
International Cometary Explorer spacecraft, 300
International date line, 40
International Ultraviolet Explorer, 151, **152**
Interstellar absorption, 400, 401
Interstellar absorption lines, **396**, 397
Interstellar bubbles, **401**, 414, **415**, **438**
 role in star formation, 414
Interstellar cirrus, 400
Interstellar dust, *see* Dust
Interstellar grains, *see* Dust
Interstellar medium, 393–405
 absorption lines, **396**, 397
 Bok globules, **56**, **392**, 398–399, **399**
 coronal gas, 398

diffuse nebulae, 393–396, 398
 dust, 398–402
 giant molecular clouds, 404–405
 intercloud medium, 397, 398
 interstellar cirrus, 400
 neutral hydrogen clouds, 397, 398
 reflection nebulae, 399
 21-cm line, 397
Interstellar molecules, 402–405
 formation, 405
 spectrum, **403**
 table, 403
Interstellar polarization, 402
Interstellar reddening, 400
Inverse Compton effect, 514
Inverse P Cygni lines, 408
Inverse relationships, 98
Inverse square laws, 98
Io, 241, **242**, 245–246, **246**, **247**
Io plasma torus, 241, 242
Ionosphere, 169
Ions, 121
IRAS 16293-2422, **413**
IRC+10 216, **381**, 404
Iron meteorites, 286, **287**
Iron spectrum, **128**
Irregular clusters of galaxies, 482
Irregular galaxies, 11, 478
Irregular variables, 381
Irwin, James, **188**
Ishtar Terra, 207, **213**
Island arcs, 166
Island universes, 475
Isotopes, 119
Isotropic relation, 494

Janus, 261
Japan, 166
Jason, 55
Jets from active galaxies, 507–510
Jewel Box, **10**
Jewish calendar, 39, 69
Jovian planets, 7
Julian calendar, 39
Juno, 289
Jupiter, **8**, **233**, 234–248, **235**, **239**, **242**
 atmosphere, 238–241
 chemical composition, 239–240
 cloud belts and zones, 234
 data table, 234
 data table, rings, 248
 data table, satellites, 243
 density, 235, 237
 Great Red Spot, 241, **241**
 interior, 237–238
 internal heat, 235, 238
 magnetic field, 241, 272

magnetosphere, 241–243
map, 235
mass, 235
movement of, **81**
ovals, 240–241
radio radiation, 235–236
radius, 234, 270
rings, 248, **248**
rotation, 235–236
satellites, 243–248
spectrum, **240**
temperature, 235, 238
visit to, 247
winds, 236, 238–239

Kamioka neutrino detector, 329–330
Kant, Emmanuel, 475
Kappa Crucis, 365
Kapteyn's Star, 337
Keck telescopes, 142
Keeler, James, 256
Keenan, Philip C., 349
Kellman, Edith, 349
Kelvin-Helmholtz contraction, 327
Kelvin temperature scale, 100
Kepler, Johannes, 90–92, **91**, 521
    generalized laws of planetary
        motion, 102–103
    laws of planetary motion, 91–92
Kepler's star (supernova), 386
Kerr, Roy, 451
Kerr black hole, 451
Keyser, Pieter, 57
Kilogram, 98
Kiloparsec, 337
Kinetic energy, 99
Kinetic temperature, 169
Kirchhoff, Gustav, 127
Kirchhoff's laws, 128–129
Kirkwood, Daniel, 290
Kirkwood gaps, 290
Kitt Peak National Observatory, **136**
Kneeler, The, 55
Krakatau, 161
Krüger 60, **357**, 383, **383**
KT boundary, 192–193
Kuiper Airborne Observatory, 148,
    149
Kuiper belt, 303, 306

Lacaille, Abbé Nicolas de, 57
    southern sky map, 59
Lacerta, 57
Lagrange, Joseph, 261
Lagrangian points, 261
Lake Erie neutrino detector, 443
Lakshmi Planum, 207, 212, **213**
Lambda Cephei, 368

Lambda Orionis, **401**
Lammas Day, 40
Large Magellanic Cloud, 375, **376**
Lasers, **117**, 132
    lunar distance, 178
Late heavy bombardment, 187
Latitude, 17–18
    celestial definition, 20, 21
    length of degree, 18
    rising and setting of stars, 22–25
Lava, 158
Lava plateau, 161
Lavinia Planitia, **211**
Law of ellipses, 91
Law of gravity, 101–102
Laws of motion (Newton), 97–99
Laws of planetary motion, 91–92,
    102–103
Layered deposits (Mars), 229, **229**
Leap second, 180
Leap year, 39
Leavitt, Henrietta, 376
Lee, Tsung Dao, 532
Leibniz, 97
Lemaître, Abbé Georges, 524
Lenses, 138–140
Leo, 52, **52**, 53
    sickle of, 59
Leo I, **480**
Leo Minor, 57
Leonids, 303, **303**
Lepus, 54, **54**
Leverrier, Urbain, 105, 109
Libra, 52
Life, 172
    intelligent, 418–419
    Mars, 229–230
    other stars, 417–418
Lifetimes of stars, 424
Light, 122–125
    speed, 9, 110, 122
Light curves (eclipsing binary), 361,
    362
    Cepheid, 374
    Mira, 380
    nova, 384
    RR Lyrae, 378
    supernova, 387
Light echoes, 444, **444**
Light gathering power, 139
Light-year, 9, 334
Limb darkening, 313
LINERS, 507
Liquid state, 122
Lithosphere, 164
Little Dipper, **46**, 53, **53**
LMC X-3, 453
Local apparent solar time, 37

Local Group, 11, 482
    table, 480
Local mean solar time, 37
Local sidereal time, 41
Local standard of rest, 338–339
    orbit around Galaxy, 461
Local supercluster of galaxies, 498,
    499
Logarithms, 343
Long-period comets, 296
Long-period variables, see Mira
    variables
Longitude, 17–18
    determination of, 39
Look-back time, 527
Lord Kelvin, 326
Lorentz contraction, 110
Low tide, 178
Lowell, Percival, 106, 218
Lower main sequence, 425–426
Lucas, George, 263
Luminosity classes (stars), 348–349
    table of, 349
Luminous blue variables, 382–383
Lumpiness problem, 531
Lunae Planum, **223**, 223
Lunar module, 183, **183**
Lunar orbiters, 181
Lupus, 55
Lyman limit, 128
Lyman series, 128
Lynx, 57
Lyot, Bernard, 319
Lyra, 56, **56**

M 1, see Crab Nebula
M 3, **10**
M 5, **369**, 378
M 13, 368
M 15, 368, **369**
M 17, **402**
    in infrared, **402**
M 20, see Trifid Nebula
M 31, 10, **11**, **350**, 475, **475**, 479,
    483, 484, 492
    nucleus, 485
M 32, **11**, **350**, 476, 492
    nucleus, 486, 511
M 33, **377**, 475, 479, 483, 492
    nucleus, 485, **486**
M 42, see Orion Nebula
M 44, see Praesepe
M 45, see Pleiades
M 51, 475, **475**, 478, 479, 487
    nucleus, 486, **486**
M 63 (with supernova), **386**
M 67, 366
M 77, see NGC 1068

M 81, **477**, 478, 479
M 82, 478, 479, **479**
M 84, 482, **485**
M 86, 482, **485**
M 87, **480**, 482, 484, 511
  jet, **507**
M 101, **477**, 478
M 104, **465**
  nucleus, 486
Maat Mons, 208, **209**
Magellan, Ferdinand, 16
Magellan, 201, **201**
Magellanic Clouds, 375, **376**
Magma, 158
Magnetic activity cycle (solar), 321–323
  origin of, 322–323
Magnetic axis, 163
Magnetic braking, 352
Magnetic field, 117
  Earth, 163, 165, 173–174
  Galaxy, 402, 410–411, 464
  Jupiter, 241, 272
  Mars, 220, 229
  Mercury, 204
  Moon, 192
  Neptune, 271–272
  Saturn, 255, 272
  Sun, 298–299, 319–323
  Uranus, 271–272
  Venus, 207
Magnetic monopoles, 535
Magnetodisk, 241
Magnetosphere (Earth), 173–174
  Jupiter, 241–243
  Neptune, 274
  Saturn, 255
  Uranus, 273
Magnetotail, 173
Magnifying power, 139
Magnitude equation, 342
Magnitudes, 60, 339–342
  absolute, 341–342
  apparent, 339–341
  blue, 342
  bolometric, 341–342
  photographic, 339–341
  table of, 339
  visual, 339
Main sequence, 346, 423–426
  evolution, 425–426
  realms of, 425–426
  zero-age, 413
Man in the Moon, 181
Mantle of Earth, 163–164
Mantle of the Moon, 192
Mare Frigoris, 189
Mare Humorum, **188**
Mare Imbrium, **187**, 189
Mare Nectaris, 189

Mare Orientale, 189, **191**
Maria, lunar, 181, 187–191
Mariner 2 (Venus), 201
Mariner 10 (Mercury), 201
Mariners 4, 6, 7, and *9* (Mars), 218
Marginally open universe, 524–527
Mars, 216–231, **216**, **217**, **221**
  atmosphere, 217
  canals, 217–218
  chronology, 224
  core, 216
  day on, 228
  density, 216
  dunes, 228, **228**
  favorable opposition, 84
  interior, 229
  life, 229–230, 417
  magnetic field, 219, 229
  map, 221
  mass, 216
  oceans, 226, 227
  plains, 223
  poles, 228–229, **229**
  radius, 216
  regolith, 227–230, **227**, **229**
  rotation, 217
  satellites, 230–231, **230**
  seasons, 216
  sunset, **215**
  surface features, 216, 220–230
  temperature, 219
  volcanoes, 220–222
  water, 224–226
  weather, 220
Masers, 281
Mass, 98
  center of, 104
  Earth, 104
  elliptical galaxies, 484
  energy equivalence, 110
  galaxies, 484–485
  Galaxy, the, 465–468
  Jupiter, 235
  loss, *see* Stellar winds, Solar wind
  Mars, 216
  measurement of, 104
  Mercury, 198
  Moon, 178
  Neptune, 269
  Pluto, 281
  Saturn, 254
  spiral galaxies, 484
  stars, 359, 363–364
  stellar luminosity relation, 363–364
  Uranus, 269
  Venus, 198
  Sun, 311
Mass-energy, 110

Mass loss from stars, *see* Stellar winds
Mass loss from the Sun, *see* Solar wind
Mass-luminosity relation, 363–364
Matter-antimatter problem, 531
  solution, 532–533
Mauna Loa, **160**, 161
Maunder, E. W., 322
Maunder minimum of sunspots, 322
Maury, Antonia, 342, 348
Maxwell, James Clerk, 123, 256, 530
Maxwell Montes, 207, 212
Mayans, 34
May Day, 40
McNaught, Robert, 443
Mean sun, 37
Mechanics, 97
Mechanical energy, 99
Medusa, 55
Megachannel Extraterrestrial Assay, 419–420
Megaparsecs, 479
Mercury, 198–200, **198**, **203**
  atmosphere, 202–203
  data table, 199
  day on, 204
  density, 198
  distance (Copernicus), 198
  hot poles, 203
  ice, 203
  impact craters, 204
  interior, 204–205
  mass, 198
  origin, 205
  perihelion, 109, 111
  regolith, 203
  rotation, 201–202
  surface features, 203–204
  temperature, 200, 203
Meridian transit circle, 42, **42**
Meridians of longitude, 18
Merging galaxies, 487–489
Mesosphere, 169
Messier, Charles, 365
Messier Catalogue, 365
Metamorphic rocks, 159
Meteor crater, 192, **193**, 287
Meteor showers, 301–303
  table of, 302
Meteor storms, 303
Meteorites, 8, **9**, 157–158, 286–288
  ages, 158, 288
  chemical composition, 288
  chondrites, 287, **287**
  classes (table of), 287
  irons, 286, **287**
  lunar, 184, **185**
  Martian, 225, **225**

orbits, 288
  relation to asteroids, 291
  stones, 287
  stony-irons, 287–288
Meteoroids, 286
Meteors, 8, 158, 286, 301–304
Metis, 248
Metonic cycle, 69
Metric system, 4
Michelson, A. A., 347
Michelson interferometer, 348, **348**
Microscopium, 57
Microwave radiation, 124
Mid-Atlantic Ridge, 159, 161, 165
Midnight, 37
Midsummer night, 40
Milky Way, 9, **10**, 61–63, **63**, **333**, **457**, **460**, **466**
Milky Way Galaxy, *see* The Galaxy
Miller, Stanley, 417
Millet, Jean-François, 67
Millisecond pulsars, 448
Mimas, 261, **262**, 263
Minerals, 158
Minimum energy orbit, 107–109
Minkowski, Rudolph, 387
Mira, 373, 374, **378**, 379, 431
  light curve, 379
  spectrum, **379**
Mira variables, 379–381
  cause of variability, 380
  OH/IR stars, 381
  table of, 380
Miranda, 278, **278**, **279**
Mizar, 357
Mk 205 (quasar), **518**
Mk 1243 (Seyfert), **506**
Mk 1157 (Seyfert), **506**
MKK classes, *see* Luminosity classes
MKK atlas, **349**
Mock suns, 170, **171**
Modern constellations, 57–59
  table of, 58
Molecules, 121–122
  in Mira variable winds, 381
  spectra of, 131, **131**
Monochromatic light, 122
Monoceros OB 2, 367
Month, 40
Moon, 6, **177**, 178–194, **182**
  acceleration of gravity on, 178
  age of surface, 186–190
  basins, 187, 189, 190
  chronology, 189–190
  craters, 181–185
  data table, 178
  day on, 190
  density, 191
  diameter, 69

distance, 68, 74, 178, 180
eclipses, 73–75
escape velocity, 178
farside and nearside, 71
folklore, 68, 181
highlands, 181–186, **186**, 187
interior, 191–192
magnetic field, 192
map of, 182
maria, 181, 187–191
mass, 178
occultations of binary stars, 358
orbit, 68
origin, 193–194
phases, 69–72, **70**
radius, 178
relation to the zodiac, 52
rotation, 71
structure, 192
surface features, 181–185
synodic and sidereal periods, 72
temperature, 181
Moondogs, 171
Moonquakes, 181, 183
Moonrise and Moonset, 69–72
Morgan, William W., 349
Moslem calendar, 39, 69
Mount Everest, 159, 169
Mount St. Helens, **159**, 161
Mount Wilson 100-inch, 141
Mountain ranges, map, 159
Mountains, 159
Moving cluster method, 365–366
Mu Cephei, 368, 429, 442
Multiringed impact basins (Moon), 190
  Mars, 224
Murchison meteorite, 286
Musca, **56**
MXB 1636-53, 447

N stars, 344
  creation of, 431
Nadir, 20
Nasmyth focus, 141
Native American timekeeping, 69
Navigation, *see* Celestial navigation
Neap tide, 180
Nearside of the Moon, 71
Nebulium, 394–395
Negative hydrogen ion, 313
Negatively curved spacetime, 525
Neptune, **105**, **267**, **268**, 269–271, 273, **273**, 276, 278–280
  atmosphere, 269, 273–274
  clouds, 273–274, **274**
  data table, 269
  density, 269
  discovery, 105–106

  Great Dark Spot, 273–274, **273**
  interior, 270–271
  magnetic field, 271, 272
  magnetosphere, 274
  mass, 269
  radius, 269, 270
  rings, 270, 276, **276**
  rotation, 270
  satellites, 269, 270, 278–280
  temperature, 271, 274
  winds, 273
Nereid, 279
Neutral hydrogen clouds, 397, 398
Neutrino detectors, 328–329
Neutrinos, 327
  Big Bang, 533
  Sun, 329–330
  supernovae, 441, 443, **443**
Neutron degeneracy, 447
Neutron stars, 445–449
  binaries, 447–449
  limit to, 447
Neutrons, 117
New Astronomy of Kepler, 91
New General Catalogue, 365
New Madrid earthquake, 161
New Moon, 69
Newton, 99
Newton, Isaac, 97, **97**
  law of gravity, 100–102
  laws of motion, 97–100
  reflecting telescope, 140
Newtonian focus, 141, **144**
19 Cephei, 368
1992 QB1, 291, **291**, 304
1993 FW, 291
NGC 175, **477**
NGC 205, **11**, **350**, 476
NGC 253, **474**
NGC 524, **478**
NGC 891, **460**
NGC 1068, **506**
  nucleus, **506**
NGC 1073, **518**
NGC 1300, **477**
NGC 1302, **477**
NGC 2264, **407**
NGC 2297, **462**
NGC 2359, **439**
NGC 2859, **478**
NGC 3115, 486, **486**
NGC 3201, 368, **369**
NGC 3293, **355**
NGC 3367, **477**
NGC 3377, 476, **476**
NGC 4038, 487, **487**
NGC 4039, 487, **487**
NGC 4139, **205**
NGC 4261 nucleus and jets, **508**

NGC 4565, **9**, **401**
NGC 4603, **9**
NGC 4636, 476, **476**
NGC 4762, **478**
NGC 5128, 509, **509**, 511
NGC 6164-5, **438**
NGC 6240, 487, **488**
NGC 6397, 368
NGC 6826, **432**
Noctis Labyrinthus (Mars), 223, **223**
Nodes of orbit, 73
Non-Euclidean geometries, 525
Nonradial oscillation, 379
Nonthermal radiation, 235
Noon, 37
Normal spirals, 477
North, 20
North celestial pole, 20
North ecliptic pole, 31
North pole, 17
    view of sky from, 37
Northern Cross, 59
Northern lights, 173–174
Nova Aquilae 1918, 385
Nova Cygni 1975, 383, **384**
Nova Herculis 1934, 383
Nova Muscae 1991, 453
Nova remnants, 384
Novae, 383–385
    light curve, 384
    spectra, 384, **384**
    table of, 384
Nuclear burning, *see* Thermonuclear
    fusion
Nuclear force, *see* Strong force
Nuclear starburst galaxies, 478
Nuclei of galaxies, 485–486
Nucleons, 119
Nucleus of an atom, 117
Nucleus of a comet, 299–200, **299**
Nucleus of the Galaxy, 465, **466**, **467**
Numbers, 3

OB associations, 367–368
Oberon, 278, **278**
Objective of a telescope, 138
Oblate spheroid, 16
Obliquity of the ecliptic, 31
Observational selection, 287
Observatories, 143–145
Occultations, 244
Ocean stream, 56
Oceans, 158
    on Mars, 226, 227
Oceanus Borealis (Mars), 226, 227
Oceanus Procellarum, 189
OH/IR stars, 381, 431
OH masers (interstellar), 398
OH masers (OH/IR stars), 381

Olbers, Heinrich Wilhelm, 289, 521
Olbers' paradox, 521
Olympus Mons (Mars), 220, **222**
Omega, 528
Omega Centauri, 368, **368**
Omicron Persei, 368
On the Phaenomena of Aratos, 50
Oort, Jan, 304, 467
Oort comet cloud, 304, 306
Open clusters, 10, 364–366
    ages, 366, 424
    distances, 365–366
    table of, 364
Open universe, 524–527
Ophelia, 275
Ophiuchus, 55
Ophiuchus dark cloud, **413**
Opposition, 82
Optical double stars, 357
Optical radiation, 124
Optics, 137–142
Orbiting Solar Observatories, 312
Orbits, 100–109
    asteroids, 290–291
    geosynchronous, 106
    inclination, 68
    meaning of, 100
    meteorites, 288
    minimum energy, 107–109
    Moon, 68
    nodes, 73
    perturbations, 105
    several bodies, 105
Organic molecules, 121, 122
Orion, **49**, 54, **54**, **341**
    in infrared, **401**
Orion Nebula, 393, **393**, 395
    spectrum, **393**
Orion OB 1, 368
Orpheus, 56
Outer satellites of Jupiter, 248
Outflow channels (Mars), **224**, 225,
    225
Oxygen/neon/magnesium fusion, 440
Ozone, 125
    hole, 168

P Cygni, 383, 442
    spectrum, **383**
P Cygni lines, 383, **383**
Pacific Ocean, 159, 161, 166
PAHs, 405
Pallas, 289
Palomar Observatory, 5-meter
    telescope, 140
Palomar Observatory Sky Survey, 63,
    65
Pangaea, 167
Parabola, 102

Paraboloid, 140
Parallax, 43
    of stars, 43, 334–337
Parallel lines, 525
Parallel postulate of Euclid, 525
Parallels of latitude, 18
Parsec, 9, 334
Parsons, William, *see* Earl of Rosse
Partial eclipse of the Moon, 73
Pateras, 222, **223**
Pauli, Wolfgang, 427
Pauli exclusion principle, 427
Pavo, 57
Payne-Gaposchkin, Cecilia, 327, 344
Peculiar galaxies, 486–487
Pegasus, 55
    Great Square of, 59
Pele (Io), **247**
Pencil beam map of Universe, **502**
Penumbras of shadows, 73
    of sunspots, 319
Perfect Cosmological Principle, 522
Perfect gas law, 328
Perigee of Moon, 68
Perihelion, 30
Period, 25
Period-luminosity relation, 376–377
Periodic table, 318
Permanent polar caps (Mars), 228,
    229
Peroxides on Mars, 230
Perseids, 301, **301**
Perseus, 54–55, **55**
Perseus OB 2, 368
Perturbations of an orbit, 105
Phaeton, 302
Phases of the Moon, 69–72, **70**
    Earth, 71
    inferior planets, 85
Phobos, 230–231, **230**
    data, 231
Phoebe, 261
Pholus, 291, 294
Photino, 531
Photochemical reactions, 207
Photoelectric cell, 145
Photoelectric photometer, 145
Photography, 145
Photoionization, 394–395
Photons, 124–125
Photosphere (solar), 313–317
    currents in, 321–322, **321**
Photosynthesis, 172
Pi, 44
Piazzi, Giuseppe, 289
Pickering, Edward C., 342
Pioneer 10 (Jupiter), 236
Pioneer 11 (Jupiter, Saturn), 236, 253
Pioneer Venus Orbiter, **197**, 201

Pioneer Venus Probe Carrier, 201
Pisces, 53
Planck, Max, 125
Planck curves, 126
Planck time, 531
Planck's constant, 125
Planet X, 106
Planetary nebulae, 431–432
Planetesimals, 283
Planets, 7
   ancient, 81
   apparent motions, 83–86
   brightnesses, 82
   detection near other stars, 416
   discovery of outer planets, 104–106
   distances (Copernicus), 89
   formation, 305–307, 415
   inferior planets, 82
   Jovian, 7
   mythology, 81
   orbital eccentricities, 81–82
   orbital inclinations, 81–83
   orbits, 81–93
   other stars, 415–416
   pulsars, 448–449
   retrograde motion, 83
   sidereal periods, 81–83
   superior planets, 82
   synodic periods, 81, 83
   table of orbital characteristics, 81
   terrestrial, 7
Plasma, 173
Plasma cosmology, 536
Plasma tails, 295, 299
Plate tectonics (Earth), 166–167
   Mars, 223–224
   Venus, 211–212
Plates, 165–167
Plato, 15
Pleiades, 59, 365–367, **365**
Pleiades reflection nebula, **365**, 399
Plume tracks, 167
Plumes, 167
   Venus, 208
Pluto, **106**, **268**, 280–283, **280**
   albedo, 282
   atmosphere, 282
   chaotic orbit, 113
   data table, 281
   density, 282
   discovery, 106
   mass, 281
   radius, 282
   rotation, 281
   surface features, 282
   temperature, 283
   transits and occultations with
     Charon, 281
   visit to, 283

Polar axis, 143
Polar flattening (Earth), 16
   Saturn, 253, 254
Polar ring galaxies, 487
Polaris, 20, 45, **46**
Polarization, 401–402
Polycyclic aromatic hydrocarbons, 405
Pope Gregory XIII, 39
Population I, 460–464
   creation of, 470
   dust, 399–400
   structure of, 460
Population II, 464–465
   creation of, 470
Population III, 469–470
Populations I and II, 338–339, 458
   discovery of, 351
   location in the Galaxy, 350–352
   M 31, 350
   metal content, 361
   variable stars, 376, 379
Positively curved spacetime, 525
Positron, 327
Post meridiem, 37
Potential energy, 99, 102
Power-law spectra, 444
Praesepe, 365
Pre-main-sequence stars, 408–414
Precession, 44–46
   astrological signs, 52
   variations in, 46
Pressure, 163
Primary earthquake wave, 162
Prime focus, 140, 141
Prime meridian, 17, 18
Principia, 97
Procellarum basin, 189
Procyon, 54, **54**, 61
Procyon B, 347, 363
Project Ozma, 419
Prominences, 323, **324**, **326**
Proper motion, 337
Proper names of stars, 60
Prograde motion, 83
Proteus, 278–279
Proton-proton chain, 327–328
   in protostars, 413
Protons, 117
Protostars, 408–414
Proxima Centauri, 334, 357, 363
PSR 1257+12, **448**
Ptolemaic theory of Solar System,
   87–88
Ptolemy, 50, **50**, 87
   distance to Moon, 74
   theory of Solar System, 87–88
Pulsars, 446–449
   braking, 446–447
   Crab Nebula, 446

   distances, 446
   magnetic fields, 446
   millisecond, 448
   planets, 448–449
   supernova remnants, 447
Pulsations, 375
Puppis, 55
Purcell, Edward, 397
Pyroclastics, 161
Pythagoras, 15

Q0051-279 (quasar), 512, 513
Q0351+026 (quasar), **513**
Q0957+561 (quasar), 515–516, **515**
QSOs, *see* Quasars
Quantum efficiency, 145
Quantum foam, 532
Quantum mechanics, 112, 129–132
Quarks, 117
   creation in Big Bang, 532
Quartz, 158
Quasars, 511–518
   absorption lines, 512, 513
   evolution, 514–515
   gravitational lenses, 515–517
   interpretation, 512–513
   local hypothesis, 513, 517–518
   Lyman-alpha forest, 513, **514**
   number of, 514–515
   properties, 511–512
   radio quiet, 511
   redshift controversy, 517
   relation to active galaxies, 514
   spectra, 511–512
   superluminal velocities in, 517
   variability, 512, 513
   X-rays, 514, **515**
Quasi-stellar radio sources, *see*
   Quasars
Quiet sun, 312, 313–319

R Aquarii, 386
R Coronae Borealis, **382**
R Coronae Borealis stars, 382
R Leporis, 380
R Monocerotis, 409, **409**
r-process, 442
R stars, 344
   creation of, 431
Ra Patera (Io), **247**
Radar, 200
Radial velocity, 133
   of stars, 337
Radian, 44
Radiant (meteor shower), 301, 302
Radiant energy, 99
Radicals, 403
Radio galaxies, 507–510
Radio radiation, 124

Radio telescopes, 149–151
    table of largest, 142
Radioactive dating, 157
Radioactivity, 119–121
Radium, 119
Rainbows, 170
Rampart craters, 226, **226**
Ranger program, 181
Ravi Vallis (Mars), **224**
Rays, lunar, 184, **185**
Recombination, 394–395
Recombination lines, 394–395
Recurrent novae, 385–386
Red giant branch, *see* Giant branch
Redshifts of galaxies, 492–497, **492**
    controversy, 517
    corrections to, 492–493
    discovery, 475
    distances from, 498
    meaning, 494–496
    quasars, 511–512
Reflecting telescope, **136**, 140–142,
    140, 142, 143, 144
Reflection, 137
Reflection nebulae, 399
Refracting telescope, 138–140, **139**
Refraction, 137
    dispersion by, 138
Refractory elements, 189
Regolith (Moon), 191, **191**
    Mars, 227–230, **227, 229**
    Mercury, 203
    Venus, 210
Regression of lunar nodes, 73
Regular clusters of galaxies, 482
Relativistic beaming, 508–509
Relativistic effects, 110–112
Relativity, 109–112
    general theory, 111–112
    special theory, 109–112
    tests, 111–112
Research, 148
Resolving power, 140, 146
    of radio telescopes, 150
Resonances, orbital, 246, 259
Retrograde motion, 83–84
    explanation of, 84
    inferior planets, 85
Revolution
    Earth, 30–33
Rhea, **262**, 263
Rhea Mons, **211**
Rho Cassiopeiae, 382
Riccioli, Jean Baptiste, 357
Rigel, 54, **54**
Right ascension, 42–43
Rilles, 187, **188**
Ring of fire, 161
Ring of Fire eclipse, **76**

Ring Nebulae, 439
Ringlets, 257, **257**
Rings of Jupiter, 248, **248**
Rings of Neptune, 270, 276, **276**
    data table, 276
    dust, 278
Rings of Saturn, 253, **253, 255,**
    256–260
    Cassini division, 253, 258, 259
    computer simulation, **258**
    data table, 257
    dust, 258
    Encke division, 256, 258, **258**, 259
    F ring, 259, **259**, 263
    maps, 252, 257, 258, 259
    opacities, 258
    ringlets, 257, **257**
    structure, 258–259
    thickness, 258
Rings of Uranus, 270, 275–276, **275,**
    276
    data table, 275
    dust, 278
Robur Carolinium, 57
Roche limit, 259
Roche lobe, 385
Rock cycle, 158–159
Rockets, 108
Rocks, 158–159
Rocky mountains, 159
Rome and constellations, 50
Römer, Olaus, 244
Rosette Nebula, 395
Rosh Hashanah, 40, 69
Rotation (Earth), 17, 25–26
    black holes, 451
    galaxies, 484
    Galaxy, 460–462
    Jupiter, 235–236
    Mars, 217
    Mercury, 201–202
    Moon, 71
    Neptune, 270
    Pluto, 281
    Saturn, 253
    stars, 352–353
    Sun, 312
    Uranus, 270
    Venus, 206
Rotation curve of Galaxy, 460–462
Royal Observatory at Greenwich, 18
RR Lyrae, 378
RR Lyrae gap, 378, 428
RR Lyrae stars, 378–379
RS Ophiuchi, 385–386
Runaway greenhouse effect, 206
Runoff channels (Mars), **224**, 225,
    226
Russell, Henry Norris, 346

Rydberg, J. R., 128
Rydberg formula, 129

S5 0014+81 (quasar) spectrum, **514**
S Andromedae, 386, 475
s-process, 430–431
S stars, 344, **344**
    creation of, 431
Sabatelli, L., 80
Sacajewea (Venus), **213**
SAGE neutrino detector, 330
Sagitta, **56**
Sagittarius, 53, **105**
Sagittarius A, 465, **467**
Sagittarius A*, 465, **467**
Sakigake, 300
San Andreas fault, 159, **160**, 166
San Francisco Earthquake (1989),
    160
Sand dunes on Mars, 228, **228**
Saros, 77
Saros cycle, 77
Satellite(s), 6, 7
Satellite of Earth, *see* Moon
Satellites of Jupiter, 243–248
    data table, 243
    Galilean, 243–246
    inner satellites, 248
    map, 244
    outer satellites, 246–248
Satellites of Mars, 230, **230**, 231
    data table, 231
Satellites of Neptune, 269, 270,
    278–280
    data table, 279
    Triton, 279–280
Satellite of Pluto, 281–282
Satellites of Saturn, 253, 260–265
    data table, 260
    map, 261
    Titan, 253, 261, 263–265, **264**
Satellites of Uranus, 269, 277–278
    data table, 277
    map, 277
Saturn, **251**, 252–265, **252, 255, 256**
    atmosphere, 255–256
    chemical composition, 254
    clouds, 253, 253
    data table, 252
    density, 254
    Great White Spot, 256
    interior, 254–255
    internal heat, 254
    magnetic field, 255, 272
    magnetosphere, 255
    map, 252
    mass, 254
    polar flattening, 253, 254
    radius, 253, 270

rings, 253, **253**, **255**, 256–260, **257**
rotation, 253
satellites, 253, 260–265
temperature, 254
winds, 256
Saturn 1981 S13 satellite, **258**
Saturn V rocket, 183, **183**
Scaling factor of the Universe, 495–496
Scarps (Mercury), 204
Scattering of light, 168, 275
Schiaparellii, Giovanni, 217, 303
Schmidt telescope, 142, **143**
Schmidt, Maarten, 511
Schwabe, Heinrich, 320
Schwarzschild, Karl, 451
Schwarzschild black hole, 451
Science, 14
    interactions within, 534
Scientific method, 14–15, 218
Scooter (Neptune), 274
Scorpius, 52, **52**, 53, 54
Scorpius OB 2, 368
Scutum, 57
Search for extraterrestrial intelligence, 419–420
Seasonal polar caps (Mars), 217, **217**, 228
Seasons, 32–33
    on Mars, 216
    on Uranus, 273
Secondary craters, 184
Secondary mirrors, 141
Secondary earthquake waves, 162
Sector boundaries, 298–299
Sedimentary rocks, 158
Seeds (for galaxy formation), 535–536
Seeing disk, 140
Seismograph, 162, **162**
Semiregular variables, 381
Serpens, 55
70 Ophiuchi B, 359
Sextant, **42**, 43
Seyfert galaxies, 506–507
    type 1, 506
    type 2, 507
Shadow bands, 78
Shapley, Harlow, 370, 459
    Great Debate, 475
Shear wave, 162
Shearing (winds), 238
Shelton, Ian, 443
Shepherd satellites (Saturn), 259
    Neptune, 276
    Uranus, 275
Shield volcanoes, 161
    Mars, 220, 222
    Venus, 208–209

Shock wave, 173
Short–period comets, 296
Sickle of Leo, 59
Sidereal clock drive, 143
Sidereal day, 41
Sidereal hour, 41
Sidereal Messenger of Galileo, 93
Sidereal period (revolution), 72
    of Moon, 72
    of planets, 81–83
Sidereal time, 40–43
Sif Mons, 208, **209**
Sigma Orionis, 364
Signal-to-noise ratio, 336
Silicates, 158
Silicon fusion, 440
Single-lined spectroscopic binaries, 363
Singularity (black hole), 451
    in Big Bang, 530
Sinus Medii, **186**
Sirius, 54, **54**, 61, **347**
Sirius B, 347, **347**, 363, 364, 432
61 Cygni, 334
61 Cygni B, 352
Skylab, 312
Slipher, V. M., 475, 492, 501
Slitless spectrogram, **343**
Slitless spectrograph, 342
Small Magellanic cloud, 375, **376**
Smithsonian Astrophysical Observatory Atlas, 63, **65**
SNC meteorites, 225, **225**
Sobieski, John, 57
Socrates, 15
Sodium spectrum, **128**
Solar calendar, 39
Solar constant, 312
Solar cycle, 319–323
Solar dynamo, 322
Solar heating of Earth, 33
Solar Maximum Mission, 312
Solar model, 328–329
Solar nebula, 303
Solar oscillations, 314, 329
Solar seismology, 329
Solar System, 7–8
    creation, 304–307
    geocentric theories of, 86–88
Solar telescopes, 312–313
Solar terrestrial relations, 322
Solar time, 37–39
Solar wind, 173, 326
Solid-body tide, 180
Solid state, 122
Solstices, 32–33
Sosigenes, 39
South, 20
South celestial pole, 20

South ecliptic pole, 31
South pole, 17
Southern Cross, *see* Crux
Southern Sky Survey, 63
Space observatories, 151–152
    table of, 151
Space shuttle, 106, **107**
Space velocity, 337
Spacecraft, 107–108
Spaceflight, 106–108
Spacetime, 111
Special theory of relativity, 109–111
    Doppler effect, 496
Speckle interferometry, 281
Spectral classes, 342–346
    table of characteristics, 344
Spectral sequence, 342–346, **343**
    origin of, 345
    as temperature sequence, 344, 345
Spectrogram, 147, **147**
Spectrograph, 146, **146**
Spectroheliograms, 317
Spectroscope, 147
Spectroscopic binaries, 360–361
Spectroscopic distances, 349–350
Spectrum, *see* Electromagnetic spectrum
    of different elements, **128**
    of molecules, **131**
Spectrum lines, 127–133
    formation, 129–132
    formation in the Sun, 315, 317
Speed, 97
Speed of light, 9, 110
    first measurement of, 244
Spherical aberration, 140
Spherical spacetime, 525
Spicules, 317–318, **318**
Spin (electron), 397
Spin–flip, 397
Spiral arms, 9, 462–464
    density waves, 462
    in galaxies, 475
    star formation, 463
Spiral galaxies, **9**, 11
    classification, 477–478
    masses, 484
    rotation, 484
Splotches, 211, **211**
Spokes, 259, **259**
Spring, 32
Spring tide, 180
Sputnik 1, 108
Sputtering, 202
SS 433, 449, **449**
SS Cygni, 385
Standard candles, *see* Distance indicators for galaxies

Standard meridians, 39
Standard time, 39
Star formation, 408–414
    binary and multiple stars, 412
    bipolar flows, 409–410, **410**, **411**, 412
    birthline, 413
    brown dwarfs, 414–415
    dense cores, 410–411
    Herbig-Haro objects, 409, **409**, **410**, 412
    higher-mass stars, 414
    magnetic fields, 411–412
    solar-type stars, 411–414
    T Tauri stars, 408–409
Starburst galaxies, 478
Stars, 8–9
    activity and activity cycles, 352
    ages, 366, 424, 469
    binary, 357–363
    brightest (table), 340
    carbon stars, 344, 431
    closest (table), 335
    clusters, 364–379
    colors, 60, 127, 341
    distances, 334–337, 349–350, 365–366, 379, 399–401
    evolution, 348, 413, 423–455
    formation, 408–415
    galactic distribution, 351–352
    globular clusters, 368–370
    Hertzsprung-Russell diagram, 346–351
    locations in the sky, 21–25
    luminosity classes, 348–349
    magnitudes, 60, 339–342
    maps, 63–65
    mass loss, 427, 431
    mass-luminosity relation, 363–364
    masses, 359, 363–364
    motions, 337–339
    names, 59–62
    open clusters, 364–366
    populations of, 339–339
    radii, 346–348, 362
    rotation, 352–353
    spectra, 342–346
    table of star names, 62
    travel to, 418
    variable, 374–388
    winds, 382, 427, 431, 439
States of matter, 122
Static universe, 495
Steady State Theory, 522, 524
Stefan, Josef, 126
Stefan-Boltzmann law, 126
Stellar activity and activity cycles, 352

Stellar evolution, 348, 423–455
    asymptotic giant branch, 429–431
    binary stars, 433, 447–449
    black holes, 450–454
    clump, 428
    cycle of, 454–455
    evolutionary tracks, 413
    giants, 426–428
    globular clusters, 428
    helium flash, 428
    horizontal branch, 428
    lifetimes, 424
    main sequence, 425
    mass loss, 427, 431
    neutron stars, 446–449
    planetary nebulae, 431–432
    pulsars, 446–449
    supergiants, 429, 438–441
    supernovae, 441–443
    white dwarfs, 432–433
    zero-age main sequence, 413
Stellar lifetimes, 424
Stellar winds, 427, 431
    luminous blue variables, 382
    Sun, 123, 326
    Wolf Rayet stars, 439
Stephan's quintet, **482**
Stonehenge, 34, **34**
Stony meteorites, 287
Stony-iron meteorites, 287–288
Stratigraphy, 186
Stratosphere, 169
Strömgren, Bengt, 395
Strömgren sphere, 395
Strong force, 119
    particles of, 531
Subatomic particles, 117
Subduction, 166, 167
Subdwarfs, 351
Subgiants, 349
Sublimation, 203
Suisei, 300
Summer, 33
Summer triangle, **56**, 59
Summer solstice, 32
Sun, 6, **7**, 311–330, **311**, **313**
    active, 319–326
    chromosphere, 317–318
    circumpolarity of, 35
    composition, 315–317
    core of, 328–329
    data table, 311
    declination of, 33
    density, 311
    density of core, 328
    distance from Earth, 198, 200
    eclipses, 75–78
    effective temperature, 312

    envelope of, 328–329
    evolution, 425–426
    interior, 326–330
    luminosity, 312
    magnetic field, 298–299, 319–323
    mass, 311
    model of, 328–329
    neutrinos from, 329–330
    orbit around Galaxy, 460
    oscillations, 314–329
    passage through zodiac, 31
    photosphere, 313–317
    quiet, 312, 313–318
    radius, 311
    rotation, 312
    spectra, **314**, 315–317, **315**
    spectral class, 349
    spectral sequence, 342–346, **343**
Sundial, 37, 39
Sundogs, 170, **171**
Sunlike stars, 417
Sunrise and sunset, **29**, 33–37
    color, 169–170
Sunspot cycle, 320–323
Sunspots, 319, **319**, **320**
    magnetic fields, 320, **320**
Superclusters of galaxies, 497
    local, 498, 499
Supercooling of Universe, 532
Superfluid state, 447
Supergiant diffuse galaxies, 482
Supergiants, 346–347
    advanced evolution, 440–441
    evolution of, 438–441
    evolution to, 429
    iron core, 440–441
    spectra, **348**, **438**
Supergranules, 314
Superior conjunctions, 85
Superior planets, 82
    distances (Copernicus), 198
    motions of, 82–85
Superluminal velocities, 517
Supersymmetry, 531
Supernova of 1054, 386
Supernova 1987A, 386, 443–444, **443**
Supernova remnants, 387, **436**, 444–445, **445**, **446**
Supernovae, 386–387, 441–444
    creation of types, 442
    light curves, 387
    role in star formation, 445
    types, 387
Superstrings, 531
Surface brightness, 139
Surveying, 43
Swift, Jonathan, 231

Symbiotic stars, 386
Symmetry of forces, 531
    broken symmetry, 530, 532
Symmetry of matter-antimatter,
    532–533
Synchronous rotation, 180
Synchrotron radiation, 235, 236
Synodic period, 72
    of Moon, 72
    of planets, 81, 83
Syntaxis of Ptolemy, 50, 87
Synthesis of Elements in Stars,
    430
Systems I, II, and III, 235–236

T associations, 408
T Coronae Borealis, 385
T Tauri, **408**
    spectrum of, **408**
    structure, 408
T Tauri stars, 408–409
    spectra, 408
Tangential velocity, 133
    of stars, 337
Tarantula Nebula, 394
Tau Scorpii, 368
Taurus, **49**, 52, 54
Technetium in stars, 431
Tectonics (Earth), 166–167
    on Mars, 229
    on Venus, 211
Telescopes, 136–152
    apertures, 138
    eyepieces, 138, 139
    mountings, 143–144
    table of largest, 142
Telesto, 261
Temperature, 100
    cosmic microwave background
        radiation, 523
    determination, 127
    effective, 127
    Jupiter, 235, 238
    kinetic, 169
    Mars, 219
    Mercury, 200, 203
    Moon, 181
    Neptune, 271, 274
    Pluto, 283
    Saturn, 254
    scales, 100
    solar chromosphere, 317–318
    solar core, 328
    solar corona, 318
    stars, 344–345
    Sun, 312
    Uranus, 271, 273
    Venus, 200

Temperature gradient
    (photosphere), 313
    chromosphere, 317
Terrestrial planets, 7
Tesserae (Venus), **211**, 212, **212**
Tethys, 261, **262**
Tharsis bulge (Mars), 222, **223**
Theia Mons (Venus), **211**
Theory, 14
Theory of Everything, 531
Thermal radiation, 235
Thermonuclear fusion, 327–328
    carbon cycle, 423
    carbon fusion, 440
    oxygen/neon/magnesium fusion,
        440
    proton-proton chain, 327–328
    silicon fusion, 440
    triple-alpha process, 427–428
Thermosphere, 169
Theta^1 Orionis, 364
Theta Virginis (spectrum), **348**
Thin triangle, 44
Third quarter Moon, 71
30 Doradus, *see* Tarantula Nebula
30 Herculis, spectrum, **341**
Thomson, William, *see* Lord Kelvin
3° background radiation, *see* Cosmic
    microwave background
    radiation
3C 48 (quasar), 511, **511**, 512
3C 175, **514**
3C 273 (quasar), 511, **512**
    distances, 512–514
    jet, 511, **512**
    spectrum, **512**
    variable nature, 513
3C 279 (quasar), 513
3C 345 (quasar), **517**
3C 348 (galaxy), **488**
Tidal bulge, 179
Tidal locking of Mercury, 202
Tides, 178–180, **179**
    galaxies, 487, **487**
    length of day, 180
Time, 37–43
    Coordinated Universal Time, 180
    relativistic corrections, 112
    sidereal, 40–43
    solar, 37–39
    standard, 39
    Universal, 39
Time dilation, 110
Time zones, 38–39
Titan, 253, 261, 263–265, **264**
Titania, 278, **278**
Titius, Johann, 288
Tombaugh, Clyde, 106, 280

Torus, 241
Total eclipse of the Moon, 74, **75**
Total eclipse of the Sun, 76, **78**
Toutatis, 291, **292**
Transit of meridian, 22
Transits of satellites, 233
Transverse wave, 162
Triangle, sum of angles, 525
Trifid Nebula, 399, **400**
Triple-alpha process, 427–428
Tritium, 119
Triton, **267**, 270, 278–280, **280**
    similarity to Pluto, 282
Trojan asteroids, 291
Tropic of Cancer, 35, 45
Tropic of Capricorn, 35, 45
Tropics, 33–36
Troposphere, 169
Trumpler, Robert, 299
Tsiolkovsky, Konstantin, 108
Tully-Fisher relation, 485
Tunguska, 192
Tuning fork diagram, 478
21-cm line, 397
    establishment of galactic rotation,
        461–462
Twilight, 32
Twinkling of stars, 140
Tycho Brahe, 89–90, **89**, 374
Tycho (crater), **185**, 189
Tycho's star (supernova), 374, 386
    remnant of, **445**

Ultraviolet radiation, 124
Ulysses spacecraft, 313
Umbras of shadows, 73
    of sunspots, 319
Umbriel, 278, **278**
Uncompressed density, 162
Unidentified flying objects, 418, 453
Unification of forces, 530–531
Unit circle, 44
Universal time, 39
Universe, 3, 11, 521–539
    age, 11, 525–527
    center, 525
    creation of galaxies, 534–536
    critical density, 528
    density, 528, 533
    evolution, 532–537
    expanding, 495–496
    expansion of (observed), 11,
        494–497
    expansion models, 524–527
    formation of galaxies in, 534–537
    future of, 536–537
    horizon, 527
    inflation, 532, 533

local maps, 500, 502
look-back time, 527
Omega, 528
origin, 529, 532
scaling factor, 495–496
static, 494–495
tests of models, 528–529
Upper main sequence, 426
Uraniborg, 89, **89**
Uranium, 157, 158
Uranometria, 57, 63
Uranus, **105**, 268–273, **268**, **272**,
    275–278
  clouds, 271, **272**
  data table, 268
  density, 269
  discovery, 105
  interior, 270–271
  magnetic field, 271, 272
  mass, 269
  radius, 269, 270
  rings, 270, 275–276, **275**, **276**
  rotation, 270
  satellites, 269, 277–278
  temperature, 271, 273
  winds, 273
Urey, Harold, 417
Ursa Major, 53, **53**
Ursa Major Cluster of galaxies, **493**
Ursa Minor, **46**, 53, **53**
Utopia Planitia, 227, **227**

V2 rocket, 108
V404 Cygni, 453
V603 Aquilae, 385
Valles Marineris (Mars), **223**, 224
Van Allen radiation belts, 173
Van de Hulst, Hendrik, 397
Variable stars, 374–388
  Cepheids, 374–377
  dwarf Cepheids, 379
  dwarf novae, 385
  eruptive, 383–387
  flare stars, 383
  irregular, 381
  luminous blue variables, 382–383
  Miras, 379–381
  naming system, 378
  novae, 383–386
  R Coronae Borealis stars, 382
  recurrent novae, 385
  RR Lyrae stars, 378–379
  semiregular, 381
  symbiotic stars, 386
  supernovae, 386–387
  table of, 388
  T Tauri stars, 408–409
  ZZ Ceti stars, 379

Vega, 56, **56**, 363
  disk around, 415
  spectrum, **342**
Vega spacecraft, 201, 300, 352
  spectrum, **352**
Vela, 55
  Vela remnant, **436**, 444
Velocity, 97–98
  radial, 133
  tangential, 133
Velocity curves, 350
Velocity-distance relation, 492–494
Velocity equation, 337
Venera craft, 201, **210**
Venus, **86**, 198–200, **198**, 205–213,
    206
  atmosphere, 206–207
  core, 213
  data table, 199
  day on, 212
  density, 198
  distance (Copernicus), 198
  impact craters, 210–211
  interior, 213
  lava channels and flows, 208,
    209
  magnetic field, 207, 213
  map, 208
  mass, 198
  motions of in sky, 86
  rotation, 206
  surface, 210, **210**
  surface features, 207–213
  temperature, 200
  volcanism, 208–210
  weather, 207
Verne, Jules, 108
Vernal equinox, 32
Very Large Array, 150, **150**
Very Large Telescope, 142
Very long baseline interferometer
    (VLBI), 151
Vesta, 289
Viking (Mars), 218–219, **219**,
    226–230
Virgo, 52
Virgo Cluster of galaxies, 482, **485**,
    493
  distance, 497
Virtual photons, 530
Viscosity, 447
Visual binary stars, 357–358
Voids of galaxies, 499, 501
Volatile elements, 189
Volcanoes (Earth), 161
  classes, 161
  Io, 246, **247**
  Mars, 220, 222

Moon, 189
  Venus, 208–210, **209**
von Braun, Wernher, 108
Voyager 1, 263
Voyager 2, 109
Voyagers 1 and 2, 236–237, **236**
Vulpecula, **56**
VV Cephei, 362, 363, 368, 381

W particles, 531
W Virginis stars, 376
Waning phases of the Moon, 71
War of the Worlds, 218
Water cycle, 171
Wavelength of radiation, 122
Waxing phases of the Moon, 71
Weak force, 119
  particles of, 531
Weakly interacting massive
    particles, 531
Weather (Earth), 170–172
  Mars, 220
  Venus, 207
Week, 40
Weight, 99
Weightlessness, 106
Welles, Orson, 218
Wells, H. G., 218
West, 20
Wet quarter, 52
Whipple, Fred, 298
White dwarfs, 347
  densities, 364
  evolution to, 432–433
  spectra, 347, **347**, **348**
  supernovae, 442
  variables, 379
Wien, Wilhelm, 126
Wien's law, 126
Wind(s) (Earth), 170–171
  Jupiter, 238–239
  Mars, 220
  Neptune, 273
  Saturn, 256
  stellar, 382, 427, 431, 439
  Uranus, 273
  Venus, 207
Winter, 33
Winter triangle, **54**
Winter solstice, 33
Wolf-Rayet stars, 439–440
  ring nebulae, 439
  spectra, **438**
Wollaston, William, 127
Work, 99
Wright, Thomas, 475
Wrinkle ridges, 187, **188**
WWV, 39

X-ray binaries, 447
X-ray bursters, 447
X rays, 124
    quasars, 514, **515**
    radio galaxies, 509
Xi Persei, 368

Yang, Chen-Ning, 532
Year, 30
Yerkes 40-inch telescope, **139**
Yom Kippur, 40

Young, Thomas, 122
Young's experiment, 122
Yucatán peninsula crater, 193
Yuty crater (Mars), **226**

Z particles, 531
Zanstra, Hermann, 394
Zeeman effect, 319–320, **320**
Zenith, 20
Zenith distance, 20
Zero-age main sequence, 413

Zeta Aquilae (spectrum), **352**, 353
Zeta Aurigae, 362
Zeta Cancri, 357, 359
Zeta Persei, 368
Zodiac, 31, 52–53
Zodiacal light, 303
Zone of avoidance, 399, 400
Zones (Jupiter), 234
Zwicky, Fritz, 386
ZZ Ceti stars, 379